Therapeutic Potential of Plant Secondary Metabolites in the Treatment of Diseases and Drug Development

Therapeutic Potential of Plant Secondary Metabolites in the Treatment of Diseases and Drug Development

Editor

Pavel B. Drašar

MDPI • Basel • Beijing • Wuhan • Barcelona • Belgrade • Manchester • Tokyo • Cluj • Tianjin

Editor
Pavel B. Drašar
University of Chemistry and
Technology
Czech Republic

Editorial Office
MDPI
St. Alban-Anlage 66
4052 Basel, Switzerland

This is a reprint of articles from the Special Issue published online in the open access journal *Biomedicines* (ISSN 2227-9059) (available at: https://www.mdpi.com/journal/biomedicines/special_issues/Plant_Secondary_Metabolites_Drug).

For citation purposes, cite each article independently as indicated on the article page online and as indicated below:

LastName, A.A.; LastName, B.B.; LastName, C.C. Article Title. *Journal Name* **Year**, *Volume Number*, Page Range.

ISBN 978-3-0365-3881-5 (Hbk)
ISBN 978-3-0365-3882-2 (PDF)

Contents

About the Editor ix

Pavel B. Drašar
Plant Secondary Metabolites Used for the Treatment of Diseases and Drug Development
Reprinted from: *Biomedicines* **2022**, *10*, 576, doi:10.3390/biomedicines10030576 1

Thomas Linder, Eleni Papaplioura, Diyana Ogurlu, Sophie Geyrhofer, Scarlet Hummelbrunner, Daniel Schachner, Atanas G. Atanasov, Marko D. Mihovilovic, Verena M. Dirsch and Michael Schnürch
Investigation of Leoligin Derivatives as NF-κB Inhibitory Agents
Reprinted from: *Biomedicines* **2022**, *10*, 62, doi:10.3390/biomedicines10010062 3

Egor V. Shekunov, Svetlana S. Efimova, Natalia M. Yudintceva, Anna A. Muryleva, Vladimir V. Zarubaev, Alexander V. Slita and Olga S. Ostroumova
Plant Alkaloids Inhibit Membrane Fusion Mediated by Calcium and Fragments of MERS-CoV and SARS-CoV/SARS-CoV-2 Fusion Peptides
Reprinted from: *Biomedicines* **2021**, *9*, 1434, doi:10.3390/biomedicines9101434 19

Claudia Manca, Sébastien Lacroix, Francine Pérusse, Nicolas Flamand, Yvon Chagnon, Vicky Drapeau, Angelo Tremblay, Vincenzo Di Marzo and Cristoforo Silvestri
Oral Capsaicinoid Administration Alters the Plasma Endocannabinoidome and Fecal Microbiota of Reproductive- Aged Women Living with Overweight and Obesity
Reprinted from: *Biomedicines* **2021**, *9*, 1246, doi:10.3390/biomedicines9091246 39

David Kodr, Jarmila Stanková, Michaela Rumlová, Petr Džubák, Jiří Řehulka, Tomáš Zimmermann, Ivana Křížová, Soňa Gurská, Marián Hajdúch, Pavel B. Drašar and Michal Jurášek
Betulinic Acid Decorated with Polar Groups and Blue Emitting BODIPY Dye: Synthesis, Cytotoxicity, Cell-Cycle Analysis and Anti-HIV Profiling
Reprinted from: *Biomedicines* **2021**, *9*, 1104, doi:10.3390/biomedicines9091104 61

Daiana Mattoteia, Aniello Schiano Moriello, Orazio Taglialatela-Scafati, Pietro Amodeo, Luciano De Petrocellis, Giovanni Appendino, Rosa Maria Vitale and Diego Caprioglio
The Combined Effect of Branching and Elongation on the Bioactivity Profile of Phytocannabinoids. Part I: Thermo-TRPs
Reprinted from: *Biomedicines* **2021**, *9*, 1070, doi:10.3390/ biomedicines9081070 85

Zulal Özdemir, Uladzimir Bildziukevich, Martina Čapková, Petra Lovecká, Lucie Rárová, David Šaman, Michala Zgarbová, Barbora Lapuníková, Jan Weber, Oxana Kazakova and Zdeněk Wimmer
Triterpenoid–PEG Ribbons Targeting Selectivity in Pharmacological Effects
Reprinted from: *Biomedicines* **2021**, *9*, 951, doi:10.3390/biomedicines9080951 97

Wan-Yun Gao, Pei-Yi Chen, Hao-Jen Hsu, Ching-Yen Lin, Ming-Jiuan Wu and Jui-Hung Yen
Tanshinone IIA Downregulates Lipogenic Gene Expression and Attenuates Lipid Accumulation through the Modulation of LXRα/SREBP1 Pathway in HepG2 Cells
Reprinted from: *Biomedicines* **2021**, *9*, 326, doi:10.3390/biomedicines9030326 111

Dong Young Kang, Nipin Sp, Jin-Moo Lee and Kyoung-Jin Jang
Antitumor Effects of Ursolic Acid through Mediating the Inhibition of STAT3/PD-L1 Signaling in Non Small Cell Lung Cancer Cells
Reprinted from: *Biomedicines* **2021**, *9*, 297, doi:10.3390/biomedicines9030297 131

Marta Claudia Nocito, Arianna De Luca, Francesca Prestia, Paola Avena, Davide La Padula, Lucia Zavaglia, Rosa Sirianni, Ivan Casaburi, Francesco Puoci, Adele Chimento and Vincenzo Pezzi
Antitumoral Activities of Curcumin and Recent Advances to ImProve Its Oral Bioavailability
Reprinted from: *Biomedicines* **2021**, *9*, 1476, doi:10.3390/biomedicines9101476 . . . . **149**

Liyan Yang and Zhonglei Wang
Natural Products, Alone or in Combination with FDA-Approved Drugs, to Treat COVID-19 and Lung Cancer
Reprinted from: *Biomedicines* **2021**, *9*, 689, doi:10.3390/biomedicines9060689 . . . . **187**

Laurence Dinan, Waly Dioh, Stanislas Veillet and Rene Lafont
20-Hydroxyecdysone, from Plant Extracts to Clinical Use: Therapeutic Potential for the Treatment of Neuromuscular, Cardio-Metabolic and Respiratory Diseases
Reprinted from: *Biomedicines* **2021**, *9*, 492, doi:10.3390/biomedicines9050492 . . . . **215**

Søren Brøgger Christensen
Natural Products That Changed Society
Reprinted from: *Biomedicines* **2021**, *9*, 472, doi:10.3390/biomedicines9050472 . . . . **249**

Jackson M. J. Oultram, Joseph L. Pegler, Timothy A. Bowser, Luke J. Ney, Andrew L. Eamens and Christopher P. L. Grof
Cannabis sativa: Interdisciplinary Strategies and Avenues for Medical and Commercial Progression Outside of CBD and THC
Reprinted from: *Biomedicines* **2021**, *9*, 234, doi:10.3390/biomedicines9030234 . . . . **281**

About the Editor

Pavel B. Drašar

Education and recognition of professional experience: 2008—chartered scientist; 2004—full professor of organic chemistry; 2004—DSc in organic chemistry; 1997—EurChem; and 1993—CChem, FRSC. Scientific activity and professional positions: 2002—UCT Praha, educator and research worker; 1972–2002—Institute of Organic Chemistry and Biochemistry (IOCB), CAS. Board and committee membership: 2019—Chairman of the ECTN Label Committee; 2015–2018—ECTN President; 2004–2019—EuChemS ExComm Member; 1996—Vice President of the Czech Chemical Society; 1997—Chemicke Listy, Editor; 2015—Steroids (Elsevier) Editorial Board Member, Managing Guest Editor; 2018—MDPI Molecules, Guest Editor, and MDPI Biomedicines, Guest Editor; 1994—Alfred Bader Prize Committees, Member; and 2014—Isoprenoid Society General Secretary. Main areas of scientific interest: synthesis and evaluation of steroids and their conjugates, terpenes, alkaloids, brassinosteroids, carbohydrates, fluorescent labels for bioimaging, and the targeting of biologically active compounds by peptide vectors. Publication activity: 283 documents and 1669/1243 citations in WoS, h-index of 18, over 160 conferences, 18 books, and 39 patents.

Editorial

Plant Secondary Metabolites Used for the Treatment of Diseases and Drug Development

Pavel B. Drašar

Department of Chemistry of Natural Compounds, University of Chemistry and Technology, Technicka 5, 166 28 Prague, Czech Republic; pavel.drasar@vscht.cz; Tel.: +420-220-443-698

Citation: Drašar, P.B. Plant Secondary Metabolites Used for the Treatment of Diseases and Drug Development. *Biomedicines* **2022**, *10*, 576. https://doi.org/10.3390/biomedicines10030576

Received: 8 February 2022
Accepted: 21 February 2022
Published: 1 March 2022

Publisher's Note: MDPI stays neutral with regard to jurisdictional claims in published maps and institutional affiliations.

The importance of natural products in medicine, and in particular, plant secondary metabolites used for the treatment of diseases and drug development, has been obvious for several thousands of years. Thus, this Special Issue of MDPI's *Biomedicines* has collected the eight top issues from the field as regular full papers, namely, e.g., the investigation of leoligin derivatives as the transcription factor NF-κB, an essential mediator of inflammation NF-κB, inhibitory agents. A broad study was made possible using the modular total synthesis method of leoligin, which enabled modifications at two positions, yielding the investigation of the influence of these modifications on the biological activity [1].

Another study showed plant alkaloids inhibiting membrane fusion mediated by calcium and fragments of MERS-CoV and SARS-CoV/SARS-CoV-2 fusion peptides in search of the rationalization of the antiviral actions of plant alkaloids [2].

A further study showed that the oral use of the capsaicinoids (the pungent principles of chilli peppers and prototypical activators of the transient receptor potential of the vanilloid type-1 channel administration) alters the plasma endocannabinoidome and faecal microbiota of reproductive-aged women living with overweight and obesity [3].

An interesting paper described betulinic acid decorated with polar groups and blue-emitting BODIPY dye. This paper described the synthesis of betulinic acid derivatives, their cytotoxicity, cell-cycle analysis, and anti-HIV profiling. The results of this study suggest that betulinic acid has a similar mechanism of inhibition as described in bevirimat [4].

Other topics that were studied include the combined effects of branching and elongation of phytocannabinoids on their bioactivity profile [5], triterpenoid–PEG ribbons targeting selectivity in pharmacological effects [6] showing antimicrobial activity, especially on *Staphylococcus aureus*, *Pseudomonas aeruginosa*, and *Enterococcus faecalis*, and the downregulation of lipogenic gene expression and the attenuation of lipid accumulation through the modulation of the LXRα/SREBP1 pathway in HepG2 cells via tanshinone IIA [7].

A paper in this Special Issue also described the antitumor effects of ursolic acid, a pentacyclic triterpenoid derived from medicinal herbs, through the mediation of the inhibition of STAT3/PD-L1 signalling in non-small cell lung cancer cells [8].

Finally, this Special Issue contains five reviews. One concerns the antitumor activities of curcumin, a main bioactive component of the *Curcuma longa* L. rhizome, and the recent advancements to improve its oral bioavailability that mainly limited its use [9]. A second review shows natural products, alone or in combination with US Food and Drug Administration-approved drugs, used to treat COVID-19, which is a public health emergency of international concern, and lung cancer, a malignant tumour with the highest mortality rate, which has presented significant challenges to both human health and economic development [10]. The third review describes 20-hydroxyecdysone, a polyhydroxylated steroid, and its path from plant extracts to clinical use, mainly showing its therapeutic potential for the treatment of neuromuscular, cardio-metabolic, and respiratory diseases [11]. This review is connected to the fourth one, which names the natural products that "changed society", as, despite the impressive results achieved within the art of synthetic chemistry, natural products or modified natural products still constitute almost half of the drugs used for the treatment of cancer and diseases such as malaria, onchocerciasis (river blindness), and lymphatic filariasis (caused by parasites) [12]. The fifth review discusses *Cannabis sativa* in terms of interdisciplinary strategies and avenues for medical and commercial progression outside of CBD and THC use [13].

Conflicts of Interest: The author declares no conflict of interest.

References

1. Linder, T.; Papaplioura, E.; Ogurlu, D.; Geyrhofer, S.; Hummelbrunner, S.; Schachner, D.; Atanasov, A.G.; Mihovilovic, M.D.; Dirsch, V.M.; Schnürch, M. Investigation of Leoligin Derivatives as NF-κB Inhibitory Agents. *Biomedicines* **2022**, *10*, 62. [CrossRef] [PubMed]
2. Shekunov, E.V.; Efimova, S.S.; Yudintceva, N.M.; Muryleva, A.A.; Zarubaev, V.V.; Slita, A.V.; Ostroumova, O.S. SARS-CoV/SARS-CoV-2 Fusion Peptides. *Biomedicines* **2021**, *9*, 1434. [CrossRef] [PubMed]
3. Manca, C.; Lacroix, S.; Pérusse, F.; Flamand, N.; Chagnon, Y.; Drapeau, V.; Tremblay, A.; Di Marzo, V.; Silvestri, C. Oral Capsaicinoid Administration Alters the Plasma Endocannabinoidome and Fecal Microbiota of Reproductive-Aged Women Living with Overweight and Obesity. *Biomedicines* **2021**, *9*, 1246. [CrossRef] [PubMed]
4. Kodr, D.; Stanková, J.; Rumlová, M.; Džubák, P.; Řehulka, J.; Zimmermann, T.; Křížová, I.; Gurská, S.; Hajdúch, M.; Drašar, P.B.; et al. Betulinic Acid Decorated with Polar Groups and Blue Emitting BODIPY Dye: Synthesis, Cytotoxicity, Cell-Cycle Analysis and Anti-HIV Profiling. *Biomedicines* **2021**, *9*, 1104. [CrossRef] [PubMed]
5. Mattoteia, D.; Schiano Moriello, A.; Taglialatela-Scafati, O.; Amodeo, P.; De Petrocellis, L.; Appendino, G.; Vitale, R.M.; Caprioglio, D. The Combined Effect of Branching and Elongation on the Bioactivity Profile of Phytocannabinoids. Part I: Thermo-TRPs. *Biomedicines* **2021**, *9*, 1070. [CrossRef] [PubMed]
6. Özdemir, Z.; Bildziukevich, U.; Čapková, M.; Lovecká, P.; Rárová, L.; Šaman, D.; Zgarbová, M.; Lapuníková, B.; Weber, J.; Kazakova, O.; et al. Triterpenoid–PEG Ribbons Targeting Selectivity in Pharmacological Effects. *Biomedicines* **2021**, *9*, 951. [CrossRef] [PubMed]
7. Gao, W.-Y.; Chen, P.-Y.; Hsu, H.-J.; Lin, C.-Y.; Wu, M.-J.; Yen, J.-H. Tanshinone IIA Downregulates Lipogenic Gene Expression and Attenuates Lipid Accumulation through the Modulation of LXRα/SREBP1 Pathway in HepG2 Cells. *Biomedicines* **2021**, *9*, 326. [CrossRef] [PubMed]
8. Young Kang, D.; Sp, N.; Lee, J.-M.; Jang, K.-J. Antitumor Effects of Ursolic Acid through Mediating the Inhibition of STAT3/PD-L1 Signaling in Non-Small Cell Lung Cancer Cells. *Biomedicines* **2021**, *9*, 297. [CrossRef] [PubMed]
9. Nocito, M.C.; De Luca, A.; Prestia, F.; Avena, P.; La Padula, D.; Zavaglia, L.; Sirianni, R.; Casaburi, I.; Puoci, F.; Chimento, A.; et al. Antitumoral Activities of Curcumin and Recent Advances to Improve Its Oral Bioavailability. *Biomedicines* **2021**, *9*, 1476. [CrossRef] [PubMed]
10. Yang, L.; Wang, Z. Natural Products, Alone or in Combination with FDA-Approved Drugs, to Treat COVID-19 and Lung Cancer. *Biomedicines* **2021**, *9*, 689. [CrossRef] [PubMed]
11. Dinan, L.; Dioh, W.; Veillet, S.; Lafont, R. 20-Hydroxyecdysone, from Plant Extracts to Clinical Use: Therapeutic Potential for the Treatment of Neuromuscular, Cardio-Metabolic and Respiratory Diseases. *Biomedicines* **2021**, *9*, 492. [CrossRef] [PubMed]
12. Christensen, S.B. Natural Products That Changed Society. *Biomedicines* **2021**, *9*, 472. [CrossRef] [PubMed]
13. Oultram, J.M.J.; Pegler, J.L.; Bowser, T.A.; Ney, L.J.; Eamens, A.L.; Grof, C.P.L. Cannabis sativa: Interdisciplinary Strategies and Avenues for Medical and Commercial Progression Outside of CBD and THC. *Biomedicines* **2021**, *9*, 234. [CrossRef] [PubMed]

Article

Investigation of Leoligin Derivatives as NF-κB Inhibitory Agents

Thomas Linder [1], Eleni Papaplioura [1], Diyana Ogurlu [2], Sophie Geyrhofer [1], Scarlet Hummelbrunner [2], Daniel Schachner [2], Atanas G. Atanasov [2,3,4], Marko D. Mihovilovic [1], Verena M. Dirsch [2,*] and Michael Schnürch [1,*]

[1] Institute for Applied Synthetic Chemistry, TU Wien, Getreidemarkt 9/163, 1060 Vienna, Austria; thomas.linder@thermofisher.com (T.L.); eleni.papaplioura@tuwien.ac.at (E.P.); sophie.geyrhofer@gmail.com (S.G.); marko.mihovilovic@tuwien.ac.at (M.D.M.)

[2] Department of Pharmaceutical Sciences, University of Vienna, Althanstraße 14, 1090 Vienna, Austria; diyanaogurlu@hotmail.com (D.O.); scarlet.hummelbrunner@univie.ac.at (S.H.); daniel.schachner@univie.ac.at (D.S.); atanas.atanasov@univie.ac.at (A.G.A.)

[3] Ludwig Boltzmann Institute for Digital Health and Patient Safety, Medical University of Vienna, Spitalgasse 23, 1090 Vienna, Austria

[4] Institute of Genetics and Animal Biotechnology of the Polish Academy of Sciences, Jastrzebiec, 05-552 Magdalenka, Poland

* Correspondence: verena.dirsch@univie.ac.at (V.M.D.); michael.schnuerch@tuwien.ac.at (M.S.)

Abstract: The transcription factor NF-κB is an essential mediator of inflammation; thus, the identification of compounds that interfere with the NF-κB signaling pathway is an important topic. The natural products leoligin and 5-methoxyleoligin have served as a starting point for the development of NF-κB inhibitors. Using our modular total synthesis method of leoligin, modifications at two positions were undertaken and the effects of these modifications on the biological activity were investigated. The first modification concerned the ester functionality, where it was found that variations in this position have a significant influence, with bulky esters lacking Michael-acceptor properties being favored. Additionally, the substituents on the aryl group in position 2 of the tetrahydrofuran scaffold can vary to some extent, where it was found that a 3,4-dimethoxy and a 4-fluoro substitution pattern show comparable inhibitory efficiency.

Keywords: natural product synthesis; lignans; inflammation; NF-κB inhibition

Citation: Linder, T.; Papaplioura, E.; Ogurlu, D.; Geyrhofer, S.; Hummelbrunner, S.; Schachner, D.; Atanasov, A.G.; Mihovilovic, M.D.; Dirsch, V.M.; Schnürch, M. Investigation of Leoligin Derivatives as NF-κB Inhibitory Agents. *Biomedicines* **2022**, *10*, 62. https://doi.org/10.3390/biomedicines10010062

Academic Editors: Jun Lu and Pavel B. Drašar

Received: 23 October 2021
Accepted: 23 December 2021
Published: 28 December 2021

Publisher's Note: MDPI stays neutral with regard to jurisdictional claims in published maps and institutional affiliations.

1. Introduction

Leoligin [1] (Figure 1), a naturally occurring furan-type lignan, found in the roots of Edelweiss (*Leontopodium nivale* ssp. *alpinum*), was shown to display a pharmacological profile that suggests an overall anti-inflammatory activity. Reisinger et al. demonstrated that leoligin is able to reduce intimal hyperplasia and to decrease TNF-α (tumor necrosis factor α)-induced vascular cell adhesion molecule (VCAM)-1 expression in primary human endothelial cells [2], which is highly regulated by the nuclear factor kappa-light-chain-enhancer of activated B-cells (NF-κB) [3]. Leoligin was also identified as an inducer of macrophage cholesterol efflux, an activity that renders it a promising candidate in atherosclerosis-related experimental models [4].

In the context of a multi-disciplinary project on the synthesis of furan-type lignans with anti-inflammatory activity, several leoligin analogs were synthesized in our group and subjected to cell-based assays, which led to analogs that selectively inhibited VSMC (vascular smooth muscle cell) versus EC (endothelial cell) proliferation, which is advantageous in the treatment of vascular neointima formation [5]. Based on this finding and taking into account that drug-eluting stents releasing immunosuppressant drugs are able to reduce local VSMC proliferation, leoligin and derivatives thereof were exploited in drug-releasing experiments using an inexpensive stent model, aiming to determine their relative releasing properties [6]. Additionally, particular emphasis has been put on 5-methoxyleoligin, an interesting lignan that induces CYP26B1-dependent angiogenesis in vitro, and arteriogenesis in vivo, which are important benefits for post-myocardial infarction therapy [7].

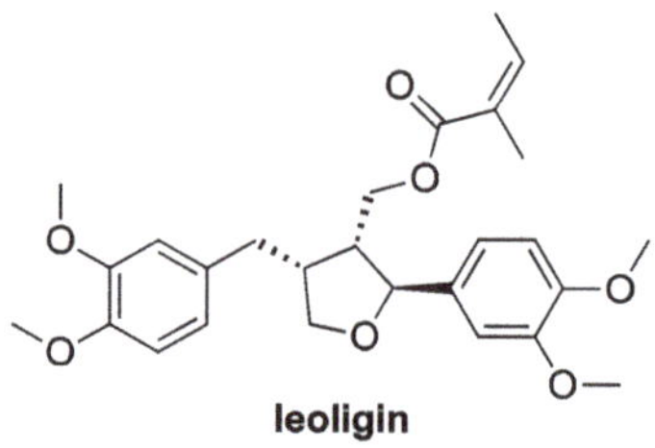

leoligin

Figure 1. Structure of leoligin.

The nuclear factor-κB (NF-κB) family of transcription factors consists of several protein members: NF-κB1 p50, NF-κB2 p52, p65 (also called RelA), RELB and c-REL [8]. NF-κB exists as an inactive complex bound to members of the inhibitor of κB (IκB) protein family. Proinflammatory stimuli, such as cytokines, activate the NF-κB signaling cascade, which leads to proteasomal degradation of the IκB protein, release of the transcription factor and its translocation into the nucleus, where it binds to respective response elements in the DNA as a dimeric complex [8]. Target genes regulated by NF-κB include a wide variety of proinflammatory genes. Thus, NF-κB plays a fundamental role in mediating all classical attributes of inflammation [9]. Hence, suppression of NF-κB is a topic of significant interest in medicinal chemistry. Over the years, a plethora of inhibitors has been identified with different mechanisms interfering with the NF-κB signaling pathway. They range from proteasome inhibitors such as bortezomib to IKK inhibitors such as parthenolide. Additionally, established drugs such as acetylsalicylic acid are reported to act as NF-κB inhibitors [10–12].

Leoligin itself is a weak inhibitor of the NF-κB pathway. It is therefore interesting to investigate how modifications to the parent leoligin motif would impact this activity. Within a previous study, we studied leoligin and a small set of leoligin analogs to identify structural features [4–6,13], that can improve either NF-κB inhibition or inhibition of the proliferation of VSMCs [5]. The present contribution is dedicated to reveal the structure–activity relationships of leoligin-like lignans affecting NF-κB. Herein, we report our investigations towards modifications of the leoligin motif that might impact such an activity.

2. Materials and Methods

2.1. General Information

Unless noted otherwise, reactants and reagents were purchased from commercial sources and used without further purification. Dry CH_2Cl_2, Et_2O, THF and MeOH were obtained from a dispensing system by passing commercial material through a cartridge containing activated alumina (PURESOLV, Innovative Technology, Oldham, UK), stored under dry nitrogen, and then used as such without further drying unless specified. **DMSO** was dried by treating commercial material with CaH_2 mesh at 150 °C under argon, followed by distillation under reduced pressure [14]. Deoxygenated and dry THF was obtained by refluxing and distilling pre-dried material (as described above) from sodium and benzophenone under argon.

Molecular sieves were activated by heating them to 200 °C for approximately 6 h in a high vacuum and were then stored under argon [15].

Melting ranges were determined using a Kofler-type Leica Galen III micro hot stage microscope or an SRS OptiMelt Automated Melting Point System and are uncorrected. Temperatures are reported in intervals of 0.5 °C.

Aluminum-backed Merck silica gel 60 with the fluorescence indicator F254 was used for thin layer chromatography (TLC). Spots were visualized under UV light (254 nm) and by staining with cerium ammonium molybdate (CAM) solution (20 g of ammonium pentamolybdate, 0.8 g of cerium(IV) ammonium sulfate, 400 mL of 10 v/v % sulfuric acid) as a general purpose reagent. Alcohols were also visualized with p-anisaldehyde solution (3.5 g p-anisaldehyde, 1.5 mL acetic acid, 5 mL sulfuric acid, 120 mL ethanol), and compounds pertaining double bonds were visualized with potassium permanganate solution (1.5 g potassium permanganate, 10 g potassium carbonate, 1 mL 10 w/w % NaOH, 200 mL water).

Specific rotation was measured using an Anton Parr MCP500 polarimeter (AntonPaar GmbH, Graz, Austria) and HPLC-grade solvents under conditions as specified individually. Values are reported in the form + or—specific rotation (concentration in terms of g/100 mL, solvent).

Analytical chromatography–spectroscopy gas chromatography–mass spectroscopy (GC-MS) was used to analyze samples of reaction products with sufficient volatility. The following instruments and columns were used: Thermo Scientific (Fisher Scientific GmbH, Schwerte, Germany) Finnigan Focus GC/Quadrupole DSQ II device using a helium flow of 2.0 mL/min, analyzing an m/z range from 50 to 650; BGB 5 (0.25 µm film; 30 m × 0.25 mm ID). Temperature gradients are as follows: Method A: 100 °C (2 min), to 280 °C in 10 min, 11 min hold-time at 280 °C (23 min); Method B: 80 °C (2 min), to 280 °C in 10 min (20 °C/min), 12 min hold-time at 280 °C (24 min); Method C: 100 °C (2 min), to 280 °C in 4.5 min (40 °C/min), 16.5 min hold-time at 280 °C (23 min); Method D: 100 °C (2 min), to 280 °C in 4.5 min (40 °C/min), 31.5 min hold-time at 280 °C (38 min); Method E: 100 °C (2 min), to 280 °C in 4.5 min (40 °C/min), 41.5 min hold-time at 280 °C (48 min). Data are reported in the form retention time; $m/z1$ (relative intensity in %), $m/z2$ (relative intensity in %), Only signals with $m/z \geq 90$ and relative intensity $\geq 15\%$ are given, except for the signal at 100% relative intensity, which is always given. Additionally, the molecular ion signal M+ is given regardless of its intensity or m/z; in cases where M+ was not visible due to excessive fragmentation, a characteristic fragment signal is identified instead.

High-pressure liquid chromatography (HPLC) was used to determine the enantiomeric excess of reaction products, using a Dionex UltiMate 3000 device (RS Diode Array Detector Fisher Scientific GmbH, Schwerte, Germany). Chiral separation columns and analysis conditions are specified individually. In all cases, retention times include appropriate guard cartridges containing the same stationary phase as the separation column.

Liquid chromatography–high-resolution mass spectroscopy (LC-HRMS) was used to confirm the exact molecular mass of reaction products by their quasi-molecular ions (M + H$^+$ or M + Na$^+$). The following two instruments were used: (1) Shimadzu Prominence HPLC device (DGU-20 A3 degassing unit, 2 × LC20AD binary gradient pump, SIL-20 A auto injector, CTO-20AC column oven, CBM-20A control module, and SPD-M20A diode array detector). Samples were eluted through a Phenomenex Kinetex precolumn (5 µm core shell ODS(3) phase; 4 mm × 2 mm ID) at 40 °C under conditions comprising gradients of H_2O/MeOH containing formic acid (0.1 v/v %), and then detected using a Shimadzu IT-TOF-MS by electrospray ionization (ESI) or atmospheric pressure chemical ionization (APCI), as indicated individually. Analyses were performed by E. Rosenberg (CTA, VUT) and L. Czollner (IAS, VUT); (2) Agilent 1100/1200 HPLC device (degassing unit, 1200SL binary gradient pump, column thermostat, and CTC Analytics HTC PAL autosampler). Samples were eluted through a silica-based Phenomenex C-18 Security guard cartridge (1.7 µm PD; 2.1 mm ID) at 40 °C under isocratic conditions comprising H_2O containing formic acid (0.1 v/v %)/MeOH containing formic acid (0.1 v/v %) in a ratio of 30:70 at a flow rate of 0.5 mL/min, and then detected using an Agilent 6230 LC-TOF-MS equipped with an Agilent Dual AJS ESI source by electrospray ionization (ESI). Analyses were performed by L. Czollner (IAS, VUT). Flash column chromatography was carried out on Merck silica gel 60 (40–63 µm), and separations were performed using a Büchi Sepacore system (dual Pump Module C-605, Pump Manager C-615, Fraction Collector C-660, and UV Monitor C-630 or UV Photometer C635).

Preparative high-performance liquid chromatography (preparative HPLC) was carried out on a Phenomenex Luna reverse-phase column (10 µm C18(2) phase, 100 A; 250 mm × 21.20 mm ID), and separations were performed using a Shimadzu LC-8A device (SIL-10AP autosampler, SPD-20 detector, and FRC-10A fraction collector (Shimadzu Österreich, Korneuburg, Austria). Reaction temperatures were measured externally (electronic thermometer connected to a heater–stirrer or low-temperature thermometer in the case of cryogenic reactions), unless otherwise noted.

Nuclear magnetic resonance spectroscopy (NMR) spectra were recorded from $CDCl_3$ or DMSO-d6 solutions on a Bruker AC 200 (200 MHz proton resonance frequency) or a Bruker Advanced UltraShield (400 MHz) spectrometer (as indicated individually) from Bruker Daltonik GmbH, Bremen, Germany, and chemical shifts are reported in ascending order in ppm relative to the nominal residual solvent signals, i.e., 1H: δ = 2.50 ppm (DMSO-d6); 13C: δ = 77.16 ppm ($CDCl_3$), δ = 39.52 ppm (DMSO-d6) [5,6]. For all 1H spectra in $CDCl_3$, however, shifts are reported relative to tetramethyl silane (TMS) as an internal standard (δ = 0 ppm) due to the interference of aromatic signals of many samples with the residual solvent signal of $CDCl_3$. For the 13C spectra, the J-modulated attached proton test (APT) or distortionless enhancement by polarization transfer (DEPT-135) pulse sequences were used to aid in the assignment.

NF-κB transactivation activity was determined by a luciferase reporter gene assay in HEK293 cells stably transfected with the NF-κB-driven luciferase reporter gene NF-κB-luc (293/NF-κB-luc cells, Panomics, Fremont, CA, USA, RC0014) [16]. Cells were stained with 2 µM cell tracker green (CTG, Thermo Scientific). After one hour, 4×10^4 cells per well were seeded in a 96-well plate in serum-free DMEM (4.5 g/L glucose) obtained from Lonza and supplemented with 2 mM glutamine, 100 U/mL benzylpenicillin and 100 µg/mL streptomycin. After incubation at 37 °C, 5% CO_2 overnight, cells were pre-treated on the next day with test samples for 1 h. Thereafter, cells were stimulated with 2 ng/mL human recombinant TNF-α (Sigma-Aldrich Handels Gmbh, Vienna, Austria) for 4 h to activate the NF-κB signaling pathway. Then, the medium was removed, and cells were lysed with luciferase reporter lysis buffer (E3971, Promega, Madison, WI, USA). Leoligin and its derivatives were tested at a concentration of 20 µM (screening) in at least three independent experiments. The sesquiterpene lactone parthenolide, an effective inhibitor of the NF-κB pathway [12], was used as a positive control at a concentration of 5 µM and 0.1% DMSO served as vehicle control. The luminescence of the firefly luciferase product and the CTG-derived fluorescence were quantified on a Tecan GeniosPro plate reader (Tecan, Grödig, Austria). The ratio of luminescence units to fluorescence units was calculated to account for differences in cell number. Results were expressed as fold changed relative to the vehicle control with TNFα, which was set to 1 [16]. CTG-fluorescence values used to estimate cell viability were also normalized to the vehicle control with TNFα. Compared to the vehicle control, treatments with fluorescence values below 0.75 were considered as toxic.

IC_{50} values were determined by the luciferase reporter assay, as described above. Dose–response curves were established by measuring compounds at concentrations of 20 µM, 10 µM, 5 µM and 1 µM, each concentration in three independent experiments. For statistical analyses, the IC_{50} values were calculated using nonlinear regression, with sigmoidal dose responses (GraphPad Software 4.03., Inc., San Diego, CA, USA).

2.2. Syntheses of Target Compounds via Steglich Esterification

2.2.1. ((2S,3R,4R)-4-(3,4-Dimethoxybenzyl)-2-(3,4-dimethoxyphenyl)tetrahydrofuran-3-yl)methyl cyclohex-1-enecarboxylate (3)

A reaction vessel was charged with a stirring bar, cyclohex-1-enecarboxylic acid (45.4 mg, 0.360 mmol, 4.0 equiv.) and 4-DMAP (1.1 mg, 9.0 µmol, 0.1 equiv.), and then evacuated and back-filled with argon using standard Schlenk technique. Dry CH_2Cl_2 (1.0 mL) was then added via syringe and the solution was cooled to 0 °C in an ice bath. The vessel was briefly opened, EDCI.HCl (63.8 mg, 0.333 mmol, 3.7 equiv.) was added in one go, and the mixture was stirred for 3 h at 0 °C. Meanwhile, a second vessel was charged with a stirring bar and starting material **18** (35.0 mg, 0.090 mmol, 1.00 equiv.), evacuated and back-filled with argon (3×), and DIPEA (78 μL, 0.45 mmol, 5.0 equiv.) was added via syringe. After 3 h, the solution containing the activated carboxylic acid was transferred to the second vial via syringe and stirred for 24 h at room temperature. The reaction solution was used directly for flash column chromatography (9 g silica, flow rate 20 mL/min, EtOAc/LP, 10:90 to 22:78 in 9 min, then 22:78 isocratically for 6 min, then to 62:38 in 30 min) to afford the title compound **3** with an overall yield of 60% as a colorless oil (Scheme 1). $[\alpha]_D^{20}$: +20.5 (c 2.64, MeOH) LC-HRMS (ESI): calculated for M + Na$^+$: 519.2353, found: 519.2378, Δ: 4.81 ppm. (log P)$_{calc}$: 5.95 ± 0.45. GC-MS (EI, 70 eV, Method E): 48.95 min; 496.3 (M$^+$, 1), 370.2 (15), 219.1 (32), 207.0 (24), 206.1 (16), 189.1 (17), 177.1 (17), 166.1 (15), 165.1 (89), 152.1 (16), 151.1 (100), 109.1 (39), 107.1 (19). ^{1}H NMR (200 MHz, CDCl$_3$): δ 1.51–1.71 (m, 4H), 2.10–2.27 (m, 4H), 2.50–2.95 (m, 4H), 3.76 (dd, ^2J = 8.5 Hz, ^3J = 6.2 Hz, 1H), 3.86 (s, 3H), 3.87 (s, 6H), 3.88 (s, 3H), 4.08 (dd, ^2J = 8.5 Hz, ^3J = 6.2 Hz, 1H), 4.26 (dd, ^2J = 11.3 Hz, ^3J = 7.1 Hz, 1), 4.42 (dd, ^2J = 11.3 Hz, ^3J = 6.5 Hz, 1H), 4.83 (d, ^3J = 6.3 Hz, 1H), 6.67–6.92 (m, 7H). ^{13}C NMR (50 MHz, CDCl$_3$): δ 21.5 (d), 22.1 (d), 24.2 (d), 25.9 (d), 33.4 (t), 42.8 (d), 49.2 (d), 56.00 (q), 56.02 (q), 56.03 (q), 56.04 (q), 62.6 (t), 72.9 (t), 83.3 (d), 109.1 (d), 111.1 (d), 111.4 (d), 112.0 (d), 118.3 (d), 120.6 (d), 130.1 (s), 132.8 (s), 135.1 (s), 140.5 (d), 147.6 (s), 148.5 (s), 149.1 (s), 149.1 (s), 167.5 (s).

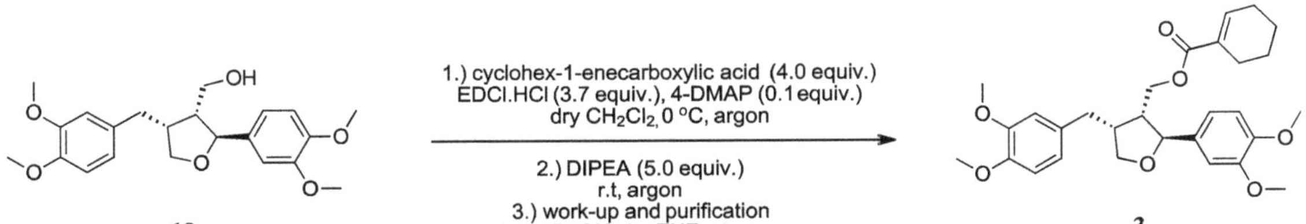

Scheme 1. Synthesis of cyclohexenyl derivative **3**.

2.2.2. ((2S,3R,4R)-4-(3,4-Dimethoxybenzyl)-2-(3,4-dimethoxyphenyl)tetrahydrofuran-3-yl)methyl cyclobutanecarboxylate (7)

Analogous to **3**, using starting material **18** (21.5 mg, 0.055 mmol, 1.00 equiv.), cyclobutanecarboxylic acid (12.7 mg, 0.127 mmol, 2.3 equiv.), EDCI.HCl (21.2 mg, 0.111 mmol, 2.0 equiv.), 4-DMAP (0.7 mg, 5.5 μmol, 0.1 equiv.) and DIPEA (24 μL, 0.14 mmol, 2.5 equiv.), and stirring the reaction for 24 h. The reaction solution was used directly for flash column chromatography (9 g silica, flow rate 20 mL/min, EtOAc/LP, 10:90 to 15:85 in 10 min, then to 25:75 in 5 min, then to 50:50 in 7 min) to afford the title compound **7** with an overall yield of 83% as a colorless oil. $[\alpha]_D^{23}$: +22.8 (c 1.27, MeOH). LC-HRMS (APCI): calculated for M + Na$^+$: 493.2197, found: 493.2214, Δ: 3.45 ppm. (log P)$_{calc}$: 4.62 ± 0.44. GC-MS (EI, 70 eV, Method D): 22.60 min; 470.1 (M$^+$, 4), 207.0 (52), 177.1 (16), 166.1 (17), 165.1 (50), 151.0 (100), 107.1 (15). ^{1}H NMR (200 MHz, CDCl$_3$): δ 1.79–2.06 (m, 2H), 2.07–2.37 (m, 4H), 2.47–2.93 (m, 4H), 3.11 (quint, ^3J = 8.6 Hz, 1H), 3.75 (dd, ^2J = 8.5 Hz, ^3J = 6.3 Hz, 1H), 3.86 (s, 3H), 3.87 (s, 6H), 3.89 (s, 3H), 4.07 (dd, ^2J = 8.5 Hz, ^3J = 6.3 Hz, 1H), 4.19 (dd, ^2J = 11.2 Hz, ^3J = 7.1 Hz, 1H), 4.38 (dd, ^2J = 11.2 Hz, ^3J = 6.9 Hz, 1H), 4.80 (d, ^3J = 6.4 Hz, 1H), 6.66–6.90 (m, 6H). ^{13}C NMR (50 MHz, CDCl$_3$): δ 18.6 (t), 25.4 (t), 25.4 (t), 33.3 (t), 38.2 (d), 42.7 (d), 49.2 (d), 56.0 (q), 62.6 (t), 72.9 (t), 83.1 (d), 109.0 (d), 111.1 (d), 111.5 (d), 112.0 (d), 118.2 (d), 120.6 (d), 132.7 (s), 135.1 (s), 147.6 (s), 148.6 (s), 149.1 (s), 149.2 (s), 175.4 (s).

2.2.3. ((2S,3R,4R)-4-(3,4-Dimethoxybenzyl)-2-(3,4-dimethoxyphenyl)tetrahydrofuran-3-yl)methylbutyrate (8)

Analogous to **3**, using starting material **18** (35.0 mg, 0.090 mmol, 1.00 equiv.), butyric acid (18.2 mg, 0.207 mmol, 2.3 equiv.), EDCI.HCl (34.5 mg, 0.180 mmol, 2.0 equiv.), 4-DMAP (1.1 mg, 9.0 μmol, 0.1 equiv.) and DIPEA (39 μL, 0.23 mmol, 2.5 equiv.). The reaction solution was used directly for flash column chromatography (9 g silica, flow rate

20 mL/min, EtOAc/LP, 10:90 to 22:78 in 9 min, then 22:78 isocratically for 6 min, then to 62:38 in 30 min) to afford the title compound **8** with an overall yield of 85% as a colorless oil. $[\alpha]_D^{20}$: +20.3 (c 1.77, MeOH). LC-HRMS (ESI): calculated for M + Na$^+$: 481.2197, found: 481.2207, Δ: 2.08 ppm. (log P)$_{calc}$:4.75 $\pm$ 0.44. GC-MS (EI, 70 eV, Method D): 20.50 min; 458.2 (M$^+$, 6), 219.1 (21), 165.1 (52), 152.1 (15), 151.1 (100). ^{1}H NMR (200 MHz, CDCl$_3$): δ 0.94 (t, ^3J = 7.4 Hz, 3H), 1.64 (sext, ^3J = 7.4 Hz, 2H), 2.26 (t, ^3J = 7.4 Hz, 2H), 2.47–2.93 (m, 4H), 3.75 (dd, ^2J = 8.6 Hz, ^3J = 6.2 Hz, 1H), 3.86 (s, 3H), 3.87 (s, 6H), 3.89 (s, 3H), 4.07 (dd, ^2J = 8.6 Hz, ^3J = 6.3 Hz, 1H), 4.19 (dd, ^2J = 11.2 Hz, ^3J = 7.1 Hz, 1H), 4.38 (dd, ^2J = 11.2 Hz, ^3J = 6.9 Hz, 1H), 4.79 (d, ^3J = 6.4 Hz, 1H), 6.64–6.92 (m, 6H). ^{13}C NMR (50 MHz, CDCl$_3$): δ 13.8 (q), 18.5 (t), 33.3 (t), 36.3 (t), 42.6 (d), 49.1 (d), 56.0 (q), 62.5 (t), 72.8 (t), 83.0 (d), 109.0 (d), 111.1 (d), 111.4 (d), 112.0 (d), 118.2 (d), 120.5 (d), 132.7 (s), 135.0 (s), 147.6 (s), 148.6 (s), 149.0 (s), 149.1 (s), 173.6 (s).

2.2.4. ((2S,3R,4R)-4-(3,4-Dimethoxybenzyl)-2-(3,4-dimethoxyphenyl)tetrahydrofuran-3-yl)methyl 3,3-dimethylbutanoate (10)

Analogous to **3**, using starting material **18** (30.0 mg, 0.077 mmol, 1.00 equiv.) and 3,3-dimethylbutanoic acid (20.6 mg, 0.178 mmol, 2.3 equiv.), EDCI.HCl (29.6 mg, 0.154 mmol, 2.0 equiv.), 4-DMAP (0.9 mg, 7.7 µmol, 0.1 equiv.) and DIPEA (34 µL, 0.19 mmol, 2.5 equiv.). The reaction solution was used directly for flash column chromatography (9 g silica, flow rate 20 mL/min, EtOAc/LP, 10:90 to 30:70 in 30 min) to afford the title compound **10** with an overall yield of 51% as a colorless oil. $[\alpha]_D^{23}$: +24.7 (c 1.01, MeOH). LC-HRMS (APCI): calculated for M+Na$^+$: 509.2510, found: 509.2512, Δ: 0.39 ppm. (log P)$_{calc}$: 5.45 $\pm$ 0.45. GC-MS (EI, 70 eV, Method D): 18.75 min; 486.1 (M$^+$, 4), 219.0 (17), 166.0 (17), 165.0 (45), 152.1 (15), 151.0 (100). ^{1}H NMR (200 MHz, CDCl$_3$): δ 1.03 (s, 9H), 2.20 (s, 2H), 2.47–2.63 (m, 2H), 2.64–2.82 (m, 1H), 2.88 (dd, ^2J = 12.5 Hz, ^3J = 4.2 Hz, 1H), 3.76 (dd, ^2J = 8.6 Hz, ^3J = 6.1 Hz, 1H), 3.86 (s, 3H), 3.87 (s, 3H), 3.87 (s, 3H), 3.89 (s, 3H), 4.06 (dd, ^2J = 8.6 Hz, ^3J = 6.2 Hz, 1H), 4.16 (dd, ^2J = 11.3 Hz, ^3J = 6.9 Hz, 1H), 4.36 (dd, ^2J = 11.3 Hz, ^3J = 7.2 Hz, 1H), 4.80 (d, ^3J = 6.4 Hz, 1H), 6.66–6.90 (m, 6H). ^{13}C NMR (50 MHz, CDCl$_3$): δ 29.8 (q), 30.9 (s), 33.2 (t), 42.6 (d), 48.1 (t), 49.3 (d), 56.0 (q), 62.3 (t), 72.8 (t), 82.9 (d), 109.0 (d), 111.2 (d), 111.5 (d), 112.0 (d), 118.2 (d), 120.6 (d), 132.7 (s), 135.1 (s), 147.7 (s), 148.6 (s), 149.1 (s), 149.2 (s), 172.4 (s).

2.2.5. ((2S,3R,4R)-4-(3,4-Dimethoxybenzyl)-2-(3,4-dimethoxyphenyl)tetrahydrofuran-3-yl)methyl2-ethylbutanoate (11)

Analogous to **3**, using starting material **18** (20.3 mg, 0.052 mmol, 1.00 equiv.) and 2-ethylbutanoic acid (14.0 mg, 0.120 mmol, 2.3 equiv.), EDCI.HCl (20.0 mg, 0.105 mmol, 2.0 equiv.), 4-DMAP (0.6 mg, 5.2 µmol, 0.1 equiv.) and DIPEA (23 µL, 0.13 mmol, 2.5 equiv.). The reaction solution was used directly for flash column chromatography (9 g silica, flow rate 20 mL/min, EtOAc/LP, 10:90 to 30:70 in 30 min) to afford the title compound **11** with an overall yield of 87% as a colorless oil. $[\alpha]_D^{23}$: +23.3 (c 0.88, MeOH).LC-HRMS (APCI): calculated for M + Na$^+$: 509.2510, found: 509.2533, Δ: 4.52 ppm. (log P)$_{calc}$: 5.63 $\pm$ 0.45. GC-MS (EI, 70 eV, Method E): 23.47 min; 486.2 (M$^+$, 7), 219.1 (32), 189.1 (15), 165.1 (56), 151.0 (100). ^{1}H NMR (200 MHz, CDCl$_3$): δ 0.90 (t, ^3J = 7.4 Hz, 6H), 1.42–1.75 (m, 4H), 2.15–2.30 (m, 1H), 2.47–2.82 (m, 3H), 2.88 (dd, ^2J = 12.6 Hz, ^3J = 4.0 Hz, 1H), 3.76 (dd, ^2J = 8.6 Hz, ^3J = 6.1 Hz, 1H), 3.87 (s, 6H), 3.87 (s, 3H), 3.89 (s, 3H), 4.06 (dd, ^2J = 8.6 Hz, ^3J = 6.2 Hz, 1H), 4.18 (dd, ^2J = 11.3 Hz, ^3J = 6.8 Hz, 1H), 4.41 (dd, ^2J = 11.3 Hz, ^3J = 7.1 Hz, 1H), 4.81 (d, ^3J = 6.4 Hz, 1H), 6.64–6.91 (m, 6H).

2.2.6. ((2S,3R,4R)-4-(3,4-Dimethoxybenzyl)-2-(4-fluorophenyl)tetrahydrofuran-3-yl)methyl 3,3-dimethylbutanoate (14)

Analogous to **3**, using starting material **19** (22.0 mg, 0.064 mmol, 1.00 equiv.), 3,3-dimethylbutanoic acid (16.9 mg, 0.146 mmol, 2.3 equiv.), EDCI.HCl (24.3 mg, 0.127 mmol, 2.0 equiv.), 4-DMAP (0.8 mg, 6.4 µmol, 0.1 equiv.) and DIPEA (28 µL, 0.16 mmol, 2.5 equiv.), and stirring for 13.5 in place of 24 h. The reaction solution was used directly for flash column chromatography (9 g silica, flow rate 20 mL/min, EtOAc/LP, 9:91 to 25:75 in 60 min) to afford the title compound **14** with an overall yield of 92% as a colorless oil (Scheme 2). $[\alpha]_D^{23}$: +15.7 (c 2.07, MeOH). LC-HRMS (APCI): calculated for M + H$^+$: 445.2385, found: 445.2398, Δ: 2.92 ppm. (log P)$_{calc}$: 5.76 ± 0.50. GC-MS (EI, 70 eV, Method D): 11.19 min; 444.1 (M$^+$, 7), 194.0 (15), 190.1 (17), 189.1 (19), 177.0 (39), 164.1 (19), 152.0 (28), 151.0 (100), 123.0 (42), 109.0 (24), 107.0 (16). ^{1}H NMR (200 MHz, CDCl$_3$): δ 1.03 (s, 9H), 2.19 (s, 2H), 2.43–2.62 (m, 2H), 2.62–2.82 (m, 1H), 2.86 (dd, ^2J = 12.5 Hz, ^3J = 4.2 Hz, 1H), 3.77 (dd, ^2J = 8.6 Hz, ^3J = 6.3 Hz, 1H), 3.86 (s, 3H), 3.86 (s, 3H), 4.06 (dd, ^2J = 8.6 Hz, ^3J = 6.2 Hz, 1H), 4.16 (dd, ^2J = 11.1 Hz, ^3J = 7.2 Hz, 1H), 4.37 (dd, ^2J = 11.3 Hz, ^3J = 6.9 Hz, 1H), 4.85 (d, ^3J = 6.1 Hz, 1H), 6.64–6.84 (m, 3H), 6.96–7.09 (m, 2H), 7.23–7.34 (m, 2H). ^{13}C NMR (50 MHz, CDCl$_3$): δ 29.8 (q), 30.9 (s), 33.1 (t), 42.6 (d), 48.1 (t), 49.5 (d), 56.00 (q), 56.04 (q), 62.2 (t), 72.9 (t), 82.7 (d), 111.5 (d), 112.0 (d), 115.4 (dd), 120.6 (d), 127.5 (dd), 132.6 (s), 138.5 (d), 147.7 (s), 149.1 (s), 162.4 (d), 172.4 (s).

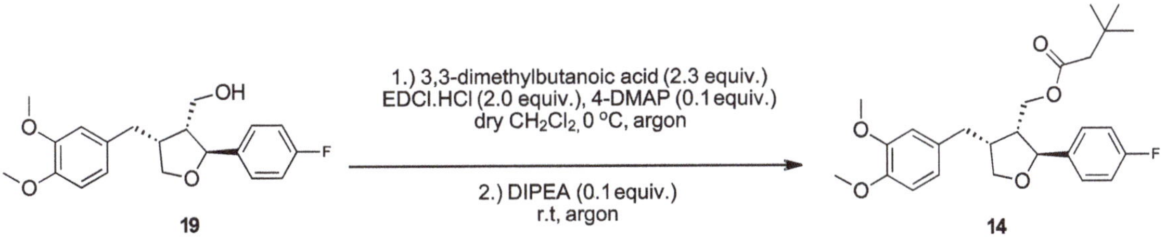

1.) 3,3-dimethylbutanoic acid (2.3 equiv.)
EDCI.HCl (2.0 equiv.), 4-DMAP (0.1 equiv.)
dry CH₂Cl₂, 0 °C, argon

2.) DIPEA (0.1 equiv.)
r.t, argon

19 → **14**

Scheme 2. Synthesis of para-fluoro derivative **14**.

2.2.7. ((2S,3R,4R)-4-(3,4-Dimethoxybenzyl)-2-(4-fluorophenyl)tetrahydrofuran-3-yl)methyl cyclohexanecarboxylate (17)

Analogous to **3**, using starting material **19** (31.2 mg, 0.090 mmol, 1.00 equiv.), cyclohexanecarboxylic acid (26.5 mg, 0.207 mmol, 2.3 equiv.), EDCI.HCl (34.5 mg, 0.180 mmol, 2.0 equiv.), 4-DMAP (1.1 mg, 9.0 μmol, 0.1 equiv.) and DIPEA (39 μL, 0.23 mmol, 2.5 equiv.). The reaction solution was used directly for flash column chromatography (9 g silica, flow rate 20 mL/min, EtOAc/LP, 0:100 to 15:85 in 16 min, then to 22:78 in 4 min) to afford the title compound **17** with an overall yield of 91% as a colorless oil. $[\alpha]_D^{20}$: +15.1 (c 2.10, MeOH). LC-HRMS (ESI): calculated for M+Na$^+$: 479.2204, found: 479.2185, Δ: −3.96 ppm. (log P)$_{calc}$: 6.06 ± 0.49. GC-MS (EI, 70 eV, Method C): 19.53 min; 456.3 (M$^+$, 4), 194.1 (24), 190.1 (24), 189.1 (23), 177.1 (44), 164.1 (26), 163.1 (16), 152.1 (30), 151.1 (100), 123.0 (47), 109.1 (33), 107.1 (18). ^{1}H NMR (200 MHz, CDCl$_3$): δ 1.11–1.95 (m, 10H), 2.17–2.34 (m, 1H), 2.42–2.62 (m, 2H), 2.62–2.79 (m, 1H), 2.85 (dd, ^2J = 12.4 Hz, ^3J = 4.3 Hz, 1H), 3.77 (dd, ^2J = 8.6 Hz, ^3J = 6.3 Hz, 1H), 3.86 (s, 3H), 3.87 (s, 3H), 4.07 (dd, ^2J = 8.6 Hz, ^3J = 6.3 Hz, 1H), 4.17 (dd, ^2J = 11.3 Hz, ^3J = 7.3 Hz, 1H), 4.38 (dd, ^2J = 11.3 Hz, ^3J = 6.7 Hz, 1H), 4.84 (d, ^3J = 6.2 Hz, 1H), 6.65–6.75 (m, 2H), 6.81 (d, ^3J = 7.9 Hz, 1H), 7.02 (dd, ^3J = 8.7 Hz, $^3J_{H\text{-}F}$ = 8.7 Hz, 2H), 7.29 (dd, ^3J = 8.6 Hz, $^4J_{H\text{-}F}$ = 5.4 Hz, 2H). ^{13}C NMR (50 MHz, CDCl$_3$): δ 25.5 (t), 25.8 (t), 29.1 (t), 29.1 (t), 33.2 (t), 42.7 (d), 43.3 (d), 49.5 (d), 55.97 (q), 56.00 (q), 62.3 (t), 72.9 (t), 82.7 (d), 111.4 (d), 111.9 (d), 115.4 (dd), 120.5 (d), 127.5 (dd), 132.6 (s), 138.5 (d), 147.7 (s), 149.1 (s), 162.3 (d), 176.0 (s).

3. Results

Molecules that could be used as NF-κB-inhibiting tool compounds for further investigation should possess an inhibitory potency comparable to that of parthenolide [17–20], a sesquiterpene lactone from feverfew *Tanacetum parthenium*, which modulates the NF-κB-mediated inflammatory response, e.g., in experimental atherosclerosis [21], and induces apoptosis in acute myelogenous leukemia (AML) cells [22,23]. Therefore, parthenolide was used as a positive control for comparing NF-κB IC$_{50}$ values of the synthetic lignan library. The IC$_{50}$ value of parthenolide in NF-κB inhibition was determined to be 1.7 μM in our assay system. The IC$_{50}$ of leoligin itself is around 20 μM [5]. For the purpose of this study, compounds which did not exert appreciable activity (no more than 50% inhibition at a single-dose concentration of 20 µM) were not considered for more detailed investigation. Thus, we screened a subset of synthetic analogs for NF-κB inhibition, modified at two positions, either at the ester functionality or at the aryl group in position 2.

The complete total synthesis of leoligin has previously been described by our group [5]. In this paper, it was hypothesized that ester functionalities which possess Michael-acceptor properties are detrimental for high NF-κB inhibition properties. However, this hypothesis was based only on the assessment of leoligin itself and four additional ester derivatives. Hence, the evaluation of more ester derivatives has become mandatory, in order to prove or disprove this theory. Consequently, a set of compounds with different ester groups was synthesized (experimental and analytical details of synthesized compounds can be found in the supporting information, Supplementary Materials). Various saturated, unsaturated and aromatic carboxylic acids were employed to carry out the esterification following either a Steglich or Mitsunobu protocol [5]. Several α,β-unsaturated carboxylic acids underwent successful Steglich esterification under mild conditions. However, leoligin had to be prepared by following a Mitsunobu protocol to avoid the double-bond isomerization in the ester moiety. This was otherwise likely under typical Steglich conditions due to EDCl-mediated (*N*-(3-dimethylaminopropyl)-*N'*-ethylcarbodiimide hydrochloride) acid activation and reversible Michael addition by the 4-dimethylaminophenol (4-DMAP) acylation catalyst, resulting in bond rotation to the thermodynamically more stable (*E*) configuration (Scheme 3).

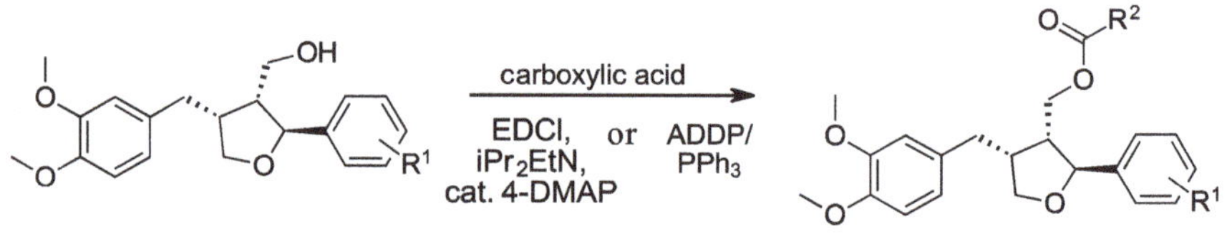

Scheme 3. General scheme for Steglich and Mitsunobu esterification.

Thus, to preserve the integrity of the Z-configuration of the α,β-unsaturated system during esterification, a Mitsunobu protocol was applied for the generation of leoligin (and generally, of angelic acid-bearing lignans), using diethyl azodicarboxylate (DEAD) as the reagent. As a side note, DEAD can be replaced by 1,1'-(azodicarbonyl)dipiperidine (ADDP) in order to simplify purification by column chromatography in some cases, because its reduction byproduct possesses significantly different polarity as compared with the target products [24,25]. Overall, 17 different leoligin derivatives were prepared, whereas our group has previously described the synthesis of some compounds reported here in the literature, but in a different context. For instance, leoligin and compounds **1**, **9** and **12** have already been investigated regarding NF-κB and VSMC proliferation inhibition [5], and the IC$_{50}$ values for those compounds are reproduced in the present manuscript. 5-Methoxyleoligin, derivatives **4** and **5**, have been explored as potential cholesterol efflux promoters [4], whereas compounds **2** and **4** have been characterized as farnesoid X receptor agonists and modulators of cholesterol transport [26]. Last but not least, compounds **6**, **13**, **15** and **16** have been included in a patent application [27].

The synthesized ester derivatives were screened in a first round at a single dose of 20 µM. IC$_{50}$ values of promising analogs were determined by dose–response measurements at 1 to 20 µM. Higher concentrations were not tested, because only compounds with IC$_{50}$ values lower than that of the natural occurring leoligin were of interest. The parent compounds leoligin and 5-methoxyleoligin showed only a very moderate inhibitory effect on NF-κB, with 19.7 µM and >20 µM, respectively (Figure 2).

According to our previous report [5], the initial hypothesis was that leoligin and 5-methoxyleoligin might very well inhibit the NF-κB pathway more effectively, if it were not for the angelic acid ester providing a point of attack in the metabolism of the cell because it is a good Michael acceptor in which addition to the double bond system can easily occur with a range of nucleophiles. For this purpose, esters **1**, **2** and **3** were evaluated (Table 1). Indeed, we observed significantly higher NF-κB inhibition for all three derivatives, with IC$_{50}$ values of 5.3 μM for **1** (data already reported in [5]), 6.5 μM for **2** and 4.9 μM for **3**, respectively. Compound **1** is likely least prone to Michael addition due to steric shielding of the β-carbon in the α,β-unsaturated system, but this position should also be less reactive in compounds **2** and **3**, because embedding the double bond in a cyclic system increases its stability. To further test this hypothesis, compound **4** was prepared, which carries a phenyl ring in the ester moiety and is not a Michael acceptor at all. Interestingly, no inhibitory activity at 20 μM concentration was observed in this case (Table 1).

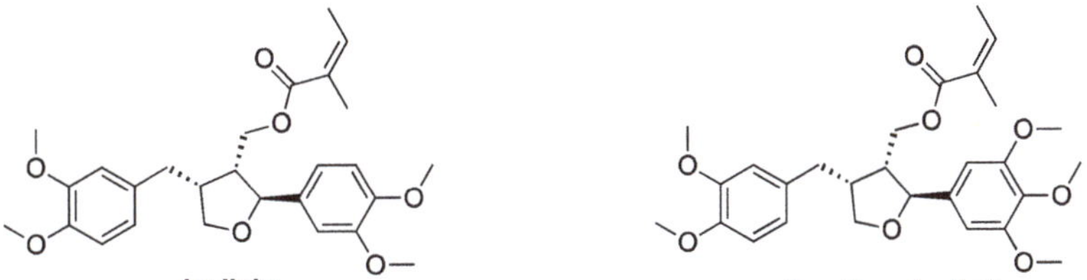

leoligin
NF-κB: IC$_{50}$= 19.7 μM

5-methoxyleoligin
NF-κB: IC$_{50}$ >20 μM

Figure 2. Structures of leoligin and 5-methoxyleoligin.

Table 1. Structures and IC$_{50}$ values of tested leologin derivatives.

Entry	Structure/No	IC$_{50}$/μM
1	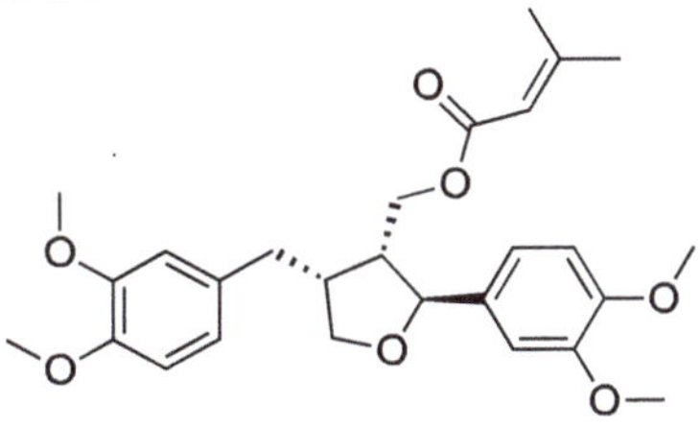 **1**	5.3
2	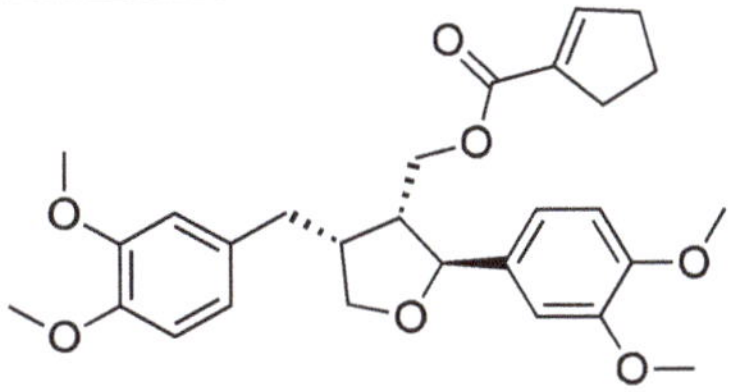 **2**	6.5
3	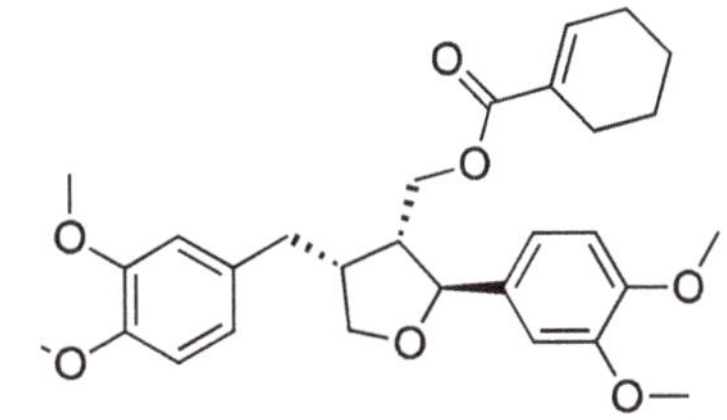 **3**	4.9

Table 1. *Cont.*

| 4 | 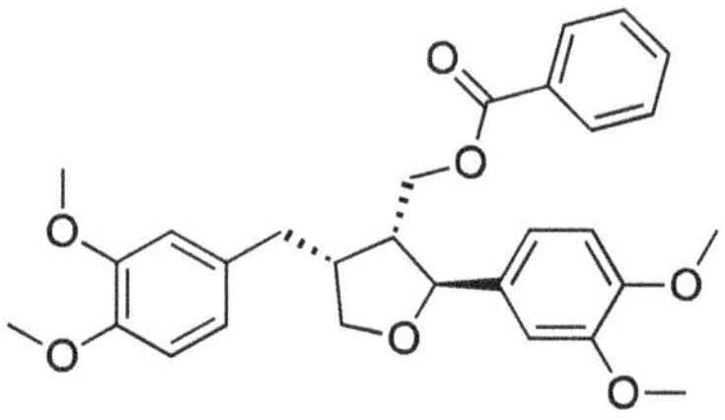 | >20 |

4

| 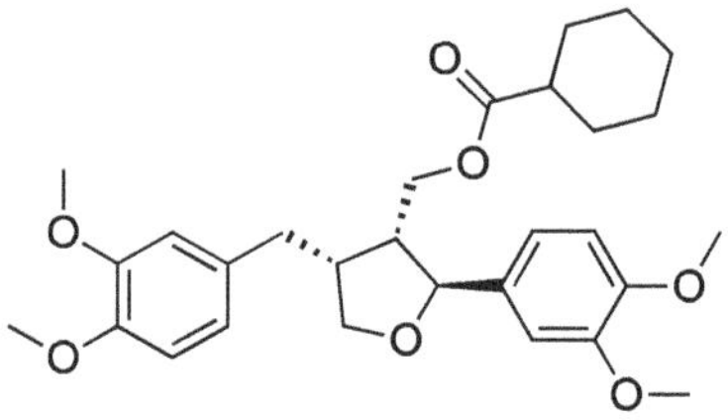 5 | | 3.2 |

5

| 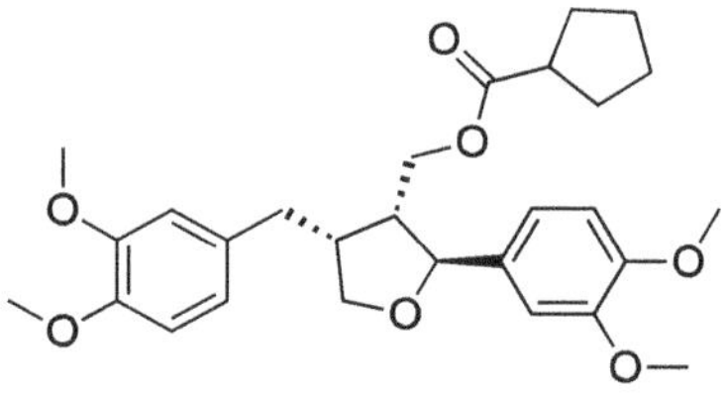 6 | | 8.0 |

6

| 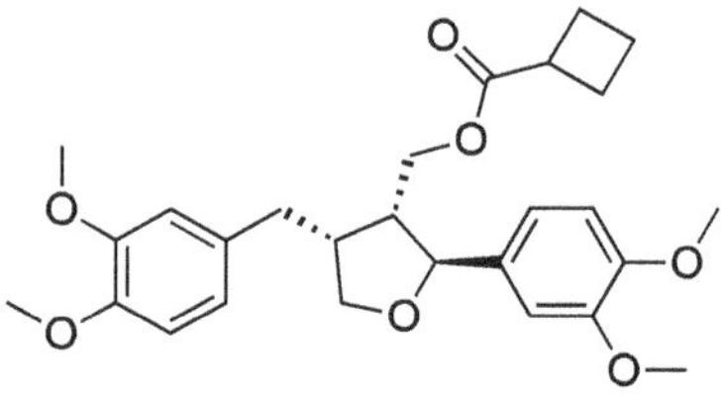 7 | | 6.4 |

7

| 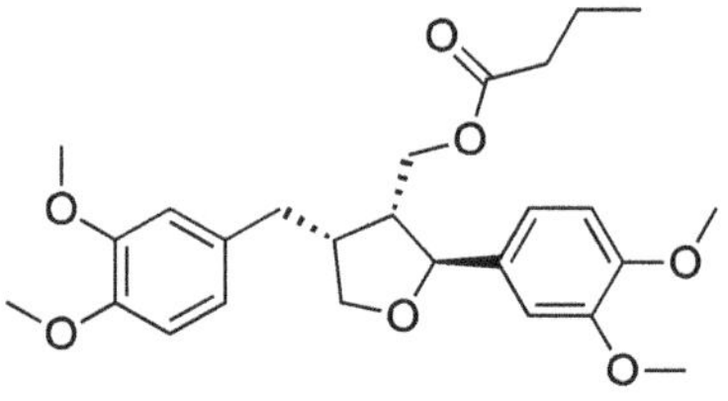 8 | | 6.3 |

8

Table 1. *Cont.*

9	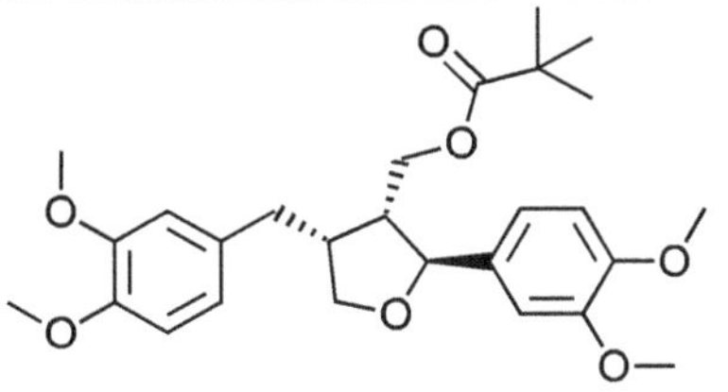 **9**	2.2
10	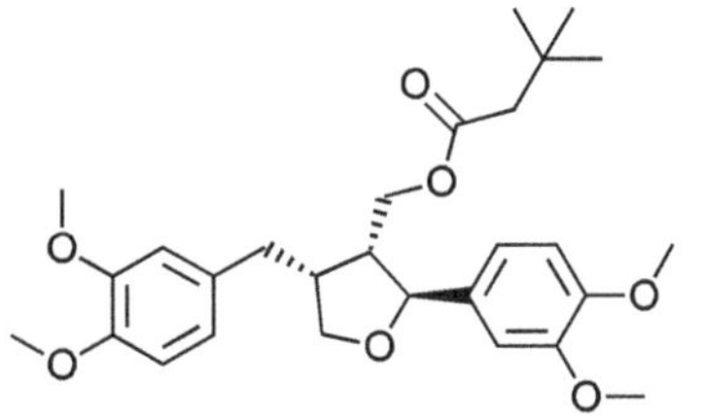 **10**	4.7
11	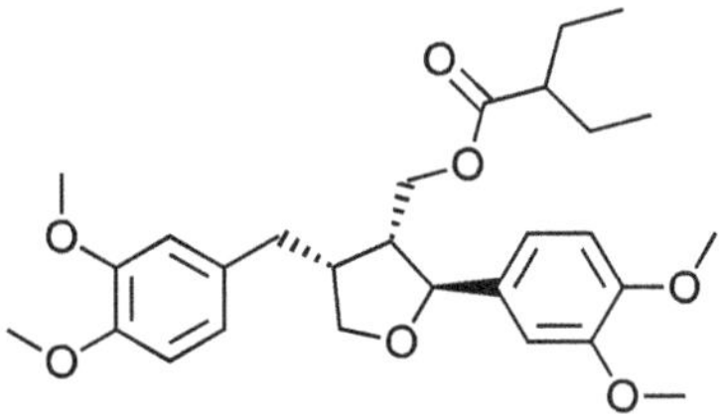 **11**	4.9
12	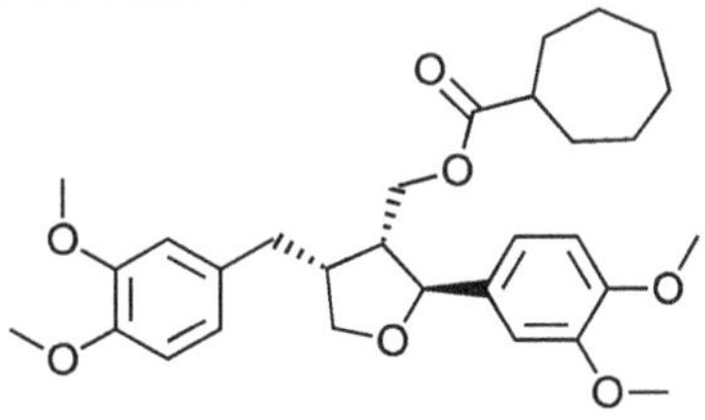 **12**	1.6
13	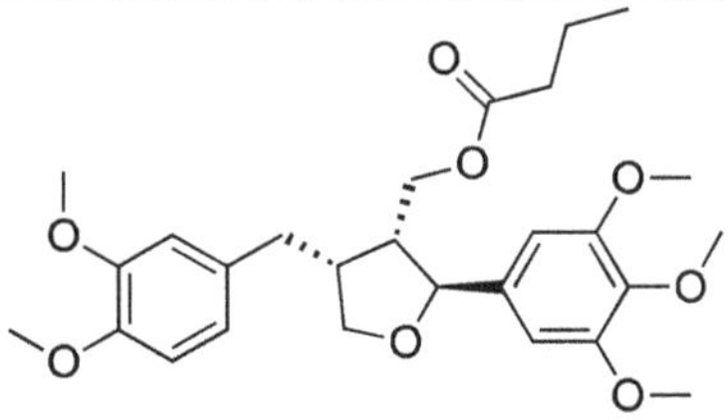 **13**	>20

Table 1. *Cont.*

14	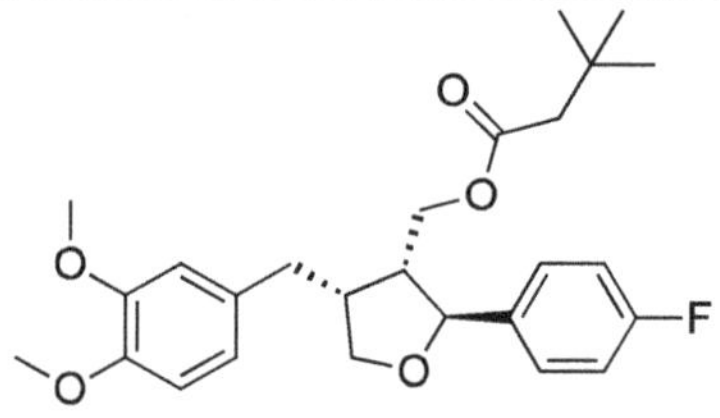	3.7
	14	
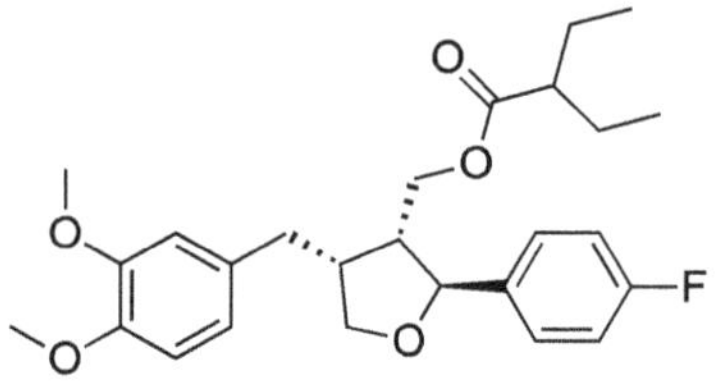 15		5.3
	15	
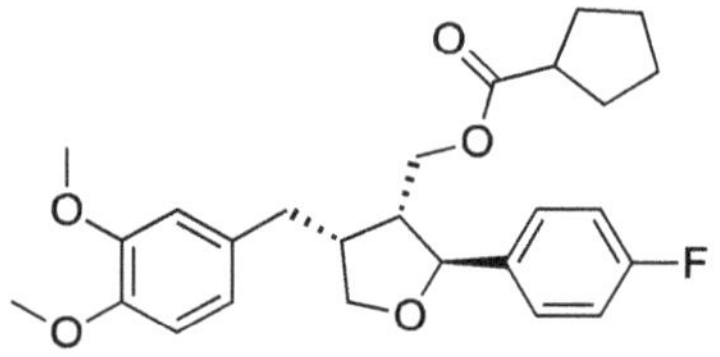 16		3.9
	16	
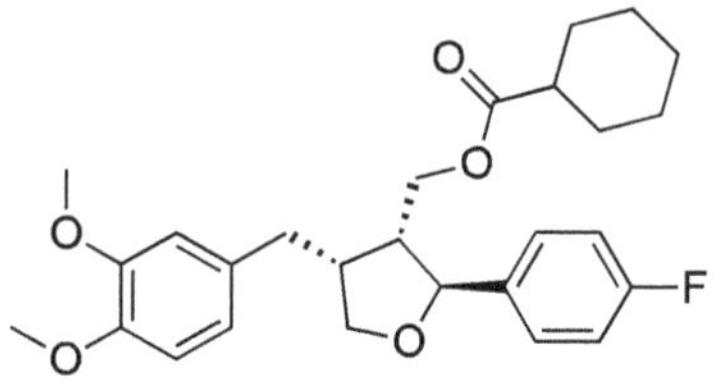 17		6.0
	17	

The question arose as to whether the extended π-system in **4** is detrimental for the desired activity and whether a π-system is required at all. Therefore, we tested compound **5** as the saturated analog of **4** where the phenyl ring is replaced by a cyclohexyl ring. The result was remarkable, because **5** gave an IC_{50} value as low as 3.2 µM, indicating that the π-system in **4** is indeed detrimental (Table 1). Decreasing the ring size to cyclopentyl (compound **6**, IC_{50} 8.0 µM) and cyclobutyl (compound **7**, IC_{50} 6.4 µM) still gave active compounds with somewhat higher IC_{50} values (Table 1). Additionally, an open chain derivative, *n*-propyl derivative **8**, was tested as well and showed inhibitory activity, with an IC_{50} value of 6.3 µM, demonstrating that neither a π-system nor a cyclic system is necessarily required for NF-κB inhibition.

Out of the derivatives investigated thus far, cyclohexylester **5** is the most sterically demanding. Hence, in the next step we investigated how much steric bulk is tolerated in the ester position and whether even larger groups could have a beneficial effect. The *t*-butyl group was a natural choice to test for steric effects, and in fact, compound **9** exhibited a very low IC_{50} of 2.2 µM (Table 1). Similarly, 3,3-dimethylbutanoate **10** also showed significant NF-κB inhibition with an IC_{50} of 4.7 µM as well as 3-pentyl derivative **11** (IC_{50}: 4.9 µM). However, the lowest IC_{50} of 1.6 µM we observed in this study was found when we installed the cycloheptyl ring in compound **12** (Table 1). At this point, we could conclude that aliphatic esters of substantial size are required to obtain a strong inhibitory effect on NF-κB activation.

Next, we investigated whether modifications to the substitution pattern on the C2-aryl ring influence NF-κB inhibition. First, 5-methoxyleoligin analog **13** (akin to compound **8** from the leoligin series, Table 1) was prepared in order to see whether aliphatic esters have a beneficial effect on this scaffold as well. However, in this case, no inhibitory activity was observed at 20 μM. Consequently, no further derivatives of this series were investigated.

In a previous study, we had already investigated modifications to the aryl ring at position 2. Therein, an activity profile in favor of VSMC proliferation inhibition without cytotoxicity was observed when substituting the 3,4-dimethoxy groups with *p*-fluorine [5]. NF-κB inhibition was rather weak, but for these derivatives, we still relied on the angelic acid ester which was shown to be detrimental for NF-κB inhibitory activity in the present study. Hence, we prepared *p*-fluorine derivatives with different ester functionalities and subjected them to our test assay. Table 1 compares the observed activities of the corresponding 3,4-dimethoxy ester variations with their 4-fluoro counterparts. In this comparison, similar potencies were observed for cyclic and acyclic aliphatic esters. For instance, compounds **14** and **10** (bearing the 3,3-dimethylbutanoate) exhibited very similar NF-κB inhibition (IC$_{50}$ of 3.7 μM and 4.7 μM, respectively). The same was true for compounds **15** and **11** (IC$_{50}$ of 5.3 μM and 4.9 μM, respectively) and all other pairs of compounds (IC$_{50}$ of 3.9 μM for **16** vs. 8.0 μM for **6**, and 6.0 μM for **17** vs. 3.2 μM for **5**).

4. Conclusions

In conclusion, we could show that modifications to the ester functionality have a tremendous impact on the inhibition of NF-κB activation. Metabolic stability of the ester group plays an important role and steric bulk is well tolerated and even required in this position. On the other hand, a π-system, either in the form of a phenyl ring or an isolated double bond, is not required to obtain the desired activity. Additionally, the substituents on the aryl ring in C2-position have an effect because an additional methoxy group was detrimental, but a single fluoro substituent in *para*-position led to similar activity, again with the proper ester in place. Overall, we succeeded in revealing the structure–activity relationship of leoligin-derived derivatives in the context of NF-κB inhibition, paving the way for further structural refinements, e.g., via investigating further changes to the C2-aryl moiety as well as to the C4-aryl ring.

Supplementary Materials: The following supporting information can be downloaded at: https://www.mdpi.com/article/10.3390/biomedicines10010062/s1, Experimental details for all prepared compounds including analytical data.

Author Contributions: T.L., investigation, writing—review and editing; E.P., writing—original draft preparation; D.O., investigation; S.G., investigation; S.H., formal analysis, visualization, investigation; D.S., formal analysis, visualization, investigation; A.G.A., methodology, writing—review and editing, investigation; M.D.M., conceptualization, methodology, supervision, project administration, funding acquisition; V.M.D., conceptualization, methodology, writing—original draft preparation, supervision, writing—review and editing, project administration, funding acquisition; and M.S., conceptualization, methodology, writing—original draft preparation, writing—review and editing, supervision. All authors have read and agreed to the published version of the manuscript.

Funding: This research was funded by the Austrian Science Fund (FWF) within research grants S10710 and S10704 (NFN 'Drugs from Nature Targeting Inflammation). Open Access Funding by the University of Vienna.

Institutional Review Board Statement: Not applicable.

Informed Consent Statement: Not applicable.

Data Availability Statement: The data that support the findings of this study are available from the corresponding author upon reasonable request.

Conflicts of Interest: The authors declare no conflict of interest.

References

1. Schwaiger, S.; Adams, M.; Seger, C.; Ellmerer, E.P.; Bauer, R.; Stuppner, H. New Constituents of *Leontopodium alpinum* and their in vitro Leukotriene Biosynthesis Inhibitory Activity. *Planta Med.* **2004**, *70*, 978–985. [CrossRef]
2. Reisinger, U.; Schwaiger, S.; Zeller, I.; Messner, B.; Stigler, R.; Wiedemann, D.; Mayr, T.; Seger, C.; Schachner, T.; Dirsch, V.M.; et al. Leoligin, the major lignan from Edelweiss, inhibits intimal hyperplasia of venous bypass grafts. *Cardiovasc. Res.* **2009**, *82*, 542–549. [CrossRef]
3. Weber, C.; Erl, W.; Pietsch, A.; Ströbel, M.; Ziegler-Heitbrock, H.W.; Weber, P.C. Antioxidants inhibit monocyte adhesion by suppressing nuclear factor-kappa B mobilization and induction of vascular cell adhesion molecule-1 in endothelial cells stimulated to generate radicals. *Arter. Thromb. J. Vasc. Biol.* **1994**, *14*, 1665–1673. [CrossRef]
4. Linder, T.; Geyrhofer, S.; Papaplioura, E.; Wang, L.; Atanasov, A.G.; Stuppner, H.; Dirsch, V.M.; Schnürch, M.; Mihovilovic, M.D. Design and Synthesis of a Compound Library Exploiting 5-Methoxyleoligin as Potential Cholesterol Efflux Promoter. *Molecules* **2020**, *25*, 662. [CrossRef]
5. Linder, T.; Liu, R.; Atanasov, A.G.; Li, Y.; Geyrhofer, S.; Schwaiger, S.; Stuppner, H.; Schnürch, M.; Dirsch, V.M.; Mihovilovic, M.D. Leoligin-inspired synthetic lignans with selectivity for cell-type and bioactivity relevant for cardiovascular disease. *Chem. Sci.* **2019**, *10*, 5815–5820. [CrossRef] [PubMed]
6. Czollner, L.; Papaplioura, E.; Linder, T.; Liu, R.; Li, Y.; Atanasov, A.G.; Dirsch, V.M.; Schnürch, M.; Mihovilovic, M.D. A silver-coated copper wire as inexpensive drug eluting stent model: Determination of the relative releasing properties of leoligin and derivatives. *Mon. Chem.-Chem. Mon.* **2020**. [CrossRef]
7. Messner, B.; Kern, J.; Wiedemann, D.; Schwaiger, S.; Türkcan, A.; Ploner, C.; Trockenbacher, A.; Aumayr, K.; Bonaros, N.; Laufer, G.; et al. 5-Methoxyleoligin, a Lignan from Edelweiss, Stimulates CYP26B1-Dependent Angiogenesis In Vitro and Induces Arteriogenesis in Infarcted Rat Hearts In Vivo. *PLoS ONE* **2013**, *8*, e58342. [CrossRef]
8. Sun, S.-C. The non-canonical NF-κB pathway in immunity and inflammation. *Nat. Rev. Immunol.* **2017**, *17*, 545–558. [CrossRef]
9. Ghosh, S.; Hayden, M.S. New regulators of NF-kappaB in inflammation. *Nat. Rev. Immunol.* **2008**, *8*, 837–848. [CrossRef] [PubMed]
10. Gilmore, T.D.; Herscovitch, M. Inhibitors of NF-κB signaling: 785 and counting. *Oncogene* **2006**, *25*, 6887–6899. [CrossRef]
11. Freitas, R.H.C.N.; Fraga, C.A.M. NF-κB-IKKβ Pathway as a Target for Drug Development: Realities, Challenges and Perspectives. *Curr. Drug Targets* **2018**, *19*, 1933–1942. [CrossRef] [PubMed]
12. Hehner, S.P.; Hofmann, T.G.; Dröge, W.; Schmitz, M.L. The antiinflammatory sesquiterpene lactone parthenolide inhibits NF-kappa B by targeting the I kappa B kinase complex. *J. Immunol.* **1999**, *163*, 5617–5623.
13. Wang, L.; Ladurner, A.; Latkolik, S.; Schwaiger, S.; Linder, T.; Hošek, J.; Palme, V.; Schilcher, N.; Polanský, O.; Heiss, E.H.; et al. Leoligin, the Major Lignan from Edelweiss (*Leontopodium nivale* subsp. *alpinum*), Promotes Cholesterol Efflux from THP-1 Macrophages. *J. Nat. Prod.* **2016**, *79*, 1651–1657. [CrossRef]
14. Schwetlick, K. *Organikum*; Wiley-VCH: Weinheim, Germany, 2009.
15. Gao, Y.; Klunder, J.M.; Hanson, R.M.; Masamune, H.; Ko, S.Y.; Sharpless, K.B. Catalytic asymmetric epoxidation and kinetic resolution: Modified procedures including in situ derivatization. *J. Am. Chem. Soc.* **1987**, *109*, 5765–5780. [CrossRef]
16. Fakhrudin, N.; Waltenberger, B.; Cabaravdic, M.; Atanasov, A.G.; Malainer, C.; Schachner, D.; Heiss, E.H.; Liu, R.; Noha, S.M.; Grzywacz, A.M.; et al. Identification of plumericin as a potent new inhibitor of the NF-κB pathway with anti-inflammatory activity in vitro and in vivo. *Br. J. Pharmacol.* **2014**, *171*, 1676–1686. [CrossRef] [PubMed]
17. Wang, M.; Li, Q. Parthenolide could become a promising and stable drug with anti-inflammatory effects. *Nat. Prod. Res.* **2015**, *29*, 1092–1101. [CrossRef]
18. Seca, A.M.L.; Silva, A.M.S.; Pinto, D.C.G.A. Parthenolide and Parthenolide-Like Sesquiterpene Lactones as Multiple Targets Drugs. *Stud. Nat. Prod. Chem.* **2017**, *52*, 337–372.
19. Agatonovic-Kustrin, S.; Morton, D.W. The Current and Potential Therapeutic Uses of Parthenolide. *Stud. Nat. Prod. Chem.* **2018**, *58*, 61–91.
20. Freund, R.R.A.; Gobrecht, P.; Fischer, D.; Arndt, H.-D. Advances in chemistry and bioactivity of parthenolide. *Nat. Prod. Rep.* **2020**, *37*, 541–565. [CrossRef] [PubMed]
21. Lopez-Franco, O.; Hernández-Vargas, P.; Ortiz-Muñoz, G.; Sanjuán, G.; Suzuki, Y.; Ortega, L.; Blanco, J.; Egido, J.; Gómez-Guerrero, C. Parthenolide Modulates the NF-κB–Mediated Inflammatory Responses in Experimental Atherosclerosis. *Arter. Thromb. Vasc. Biol.* **2006**, *26*, 1864–1870. [CrossRef] [PubMed]
22. Guzman, M.L.; Rossi, R.M.; Karnischky, L.; Li, X.; Peterson, D.R.; Howard, D.S.; Jordan, C.T. The sesquiterpene lactone parthenolide induces apoptosis of human acute myelogenous leukemia stem and progenitor cells. *Blood* **2005**, *105*, 4163–4169. [CrossRef]
23. Ghantous, A.; Sinjab, A.; Herceg, Z.; Darwiche, N. Parthenolide: From plant shoots to cancer roots. *Drug Discov. Today* **2013**, *18*, 894–905. [CrossRef]
24. Mitsunobu, O. The Use of Diethyl Azodicarboxylate and Triphenylphosphine in Synthesis and Transformation of Natural Products. *Synthesis* **1981**, *1981*, 1–28. [CrossRef]

25. Swamy, K.C.K.; Kumar, N.N.B.; Balaraman, E.; Kumar, K.V.P.P. Mitsunobu and Related Reactions: Advances and Applications. *Chem. Rev.* **2009**, *109*, 2551–2651. [CrossRef] [PubMed]
26. Ladurner, A.; Linder, T.; Wang, L.; Hiebl, V.; Schuster, D.; Schnürch, M.; Mihovilovic, M.D.; Atanasov, A.G.; Dirsch, V.M. Characterization of a Structural Leoligin Analog as Farnesoid X Receptor Agonist and Modulator of Cholesterol Transport. *Planta Med.* **2020**, *86*, 1097–1107. [CrossRef] [PubMed]
27. Mihovilovic, M.; Linder, T.; Dirsch, V.M.; Atanasov, A.; Bernhard, D.; Stuppner, H.; Schwaiger, S. Leoligin Derivatives as Smooth Muscle Cell Proliferation Inhibitors. U.S. Patent Application WO2015196228; CAN:164:148815, 30 December 2015.

Article

Plant Alkaloids Inhibit Membrane Fusion Mediated by Calcium and Fragments of MERS-CoV and SARS-CoV/SARS-CoV-2 Fusion Peptides

Egor V. Shekunov [1], Svetlana S. Efimova [1,*], Natalia M. Yudintceva [1], Anna A. Muryleva [2], Vladimir V. Zarubaev [2], Alexander V. Slita [2] and Olga S. Ostroumova [1]

[1] Institute of Cytology of Russian Academy of Sciences, Tikhoretsky 4, 194064 Saint Petersburg, Russia; e.shekunov@alumni.nsu.ru (E.V.S.); yudintceva@incras.ru (N.M.Y.); ostroumova@incras.ru (O.S.O.)

[2] Saint-Petersburg Pasteur Institute of Epidemiology and Microbiology, Mira 14, 197101 Saint Petersburg, Russia; anna.murka09099@mail.ru (A.A.M.); zarubaev@gmail.com (V.V.Z.); a_slita@yahoo.com (A.V.S.)

*   Correspondence: efimova@incras.ru

Abstract: To rationalize the antiviral actions of plant alkaloids, the ability of 20 compounds to inhibit calcium-mediated fusion of lipid vesicles composed of phosphatidylglycerol and cholesterol was investigated using the calcein release assay and dynamic light scattering. Piperine, tabersonine, hordenine, lupinine, quinine, and 3-isobutyl-1-methylxanthine demonstrated the most potent effects (inhibition index greater than 50%). The introduction of phosphatidylcholine into the phosphatidylglycerol/cholesterol mixture led to significant changes in quinine, hordenine, and 3-isobutyl-1-methylxanthine efficiency. Comparison of the fusion inhibitory ability of the tested alkaloids, and the results of the measurements of alkaloid-induced alterations in the physical properties of model membranes indicated a potent relationship between a decrease in the cooperativity of the phase transition of lipids and the ability of alkaloids to prevent calcium-mediated vesicle fusion. In order to use this knowledge to combat the novel coronavirus pandemic, the ability of the most effective compounds to suppress membrane fusion induced by fragments of MERS-CoV and SARS-CoV/SARS-CoV-2 fusion peptides was studied using the calcein release assay and confocal fluorescence microscopy. Piperine was shown to inhibit vesicle fusion mediated by both coronavirus peptides. Moreover, piperine was shown to significantly reduce the titer of SARS-CoV2 progeny in vitro in *Vero* cells when used in non-toxic concentrations.

Keywords: alkaloids; membrane fusion; viral fusion inhibitor; antiviral therapy; COVID-19

Citation: Shekunov, E.V.; Efimova, S.S.; Yudintceva, N.M.; Muryleva, A.A.; Zarubaev, V.V.; Slita, A.V.; Ostroumova, O.S. Plant Alkaloids Inhibit Membrane Fusion Mediated by Calcium and Fragments of MERS-CoV and SARS-CoV/ SARS-CoV-2 Fusion Peptides. *Biomedicines* **2021**, *9*, 1434. https:// doi.org/10.3390/biomedicines9101434

Academic Editor: Pavel B. Drašar

Received: 26 August 2021
Accepted: 7 October 2021
Published: 10 October 2021

Publisher's Note: MDPI stays neutral with regard to jurisdictional claims in published maps and institutional affiliations.

1. Introduction

More than a third of FDA-approved compounds are natural substances of plant origins or their derivatives [1]. Plant compounds play a major role in the pharmaceutical industry due to their safety and relative cheapness [2–4]. Alkaloids are a wide class of nitrogen-containing compounds with basic properties numbering more than 8000 representatives [5]. Alkaloids demonstrate multiple biological activities, including antiviral [6], antibacterial [7], and antifungal actions [8]. Alkaloids were shown to be effective against *Retroviridae* [9–12], *Togaviridae* [13], *Herpesviridae* [14,15], *Flaviviridae* [16–19], *Filoviridae* [20], and *Orthomyxoviridae* [21,22]. Due to the global epidemiological situation, the ability of alkaloids to effectively fight viruses of the *Coronaviridae* family, including SARS-CoV-2, is of particular interest [23–27]. The use of alkaloids against COVID-19 were recently explored in clinical practices [28]. The literature indicates that the mechanisms of the antiviral activities of alkaloids are pleiotropic. For example, suppression of HIV by alkaloids, fagaronine, nitidine, columbamine, berberine, palmatine, and coraline is mediated by the inhibition of reverse transcriptase [29], while antiviral activity against the hepatitis B virus is related to the reduction of the activity of the p38 MAPK protein [30]. The extensive antiviral activity of alkaloids might suggest a more general and fundamental mechanism of their activity. Many viruses (influenza virus, HIV, SARS-CoV-2, Ebola virus, Zika virus, etc.), against which alkaloids are effective, are enveloped, i.e., they contain a lipid membrane [31]. The fusion of viral and cell membranes is a necessary stage for the viral development cycle, and blocking this stage is an effective therapeutic strategy [32]. Currently, there are several works devoted to the inhibition of viral fusion through alkaloid–peptide interactions [33,34]. It was revealed that alkaloids can inhibit viral fusion through interactions with cellular factors [35] and viral proteins [17,18,25,26,36]. However, the contribution of the lipid matrix and its biophysical properties to the fusion remains underestimated [37,38]. Research devoted to the study of fusion inhibitors, targeting the physical properties of viral and cell membranes, is still not sufficient [39].

Model lipid bilayers are usually used to rationalize and better understand the membrane fusion process [40–42]. Namely, the use of systems with specified physical parameters provides an accurate interpretation of the results and make it possible to reveal the molecular basis of the studied processes [42–44].

The aim of this work was to determine whether the antiviral activity of alkaloids is related to the ability to inhibit the fusion of the viral and cell membranes. The ability of 20 structurally different alkaloids to inhibit the calcium-induced fusion of lipid vesicles was evaluated, and the dependence of the alkaloid inhibitory activity on the lipid composition of fusing liposomes was studied. In order to demonstrate the relationship between changes in the physical properties of the lipid matrix and the suppression of membrane fusion under the action of alkaloids, the alkaloid-induced changes in the transmembrane distribution of lateral pressure and electrostatic potential were also assessed. To show that the data obtained could have direct implications for clinical practices, in terms of using plant alkaloids as fusion inhibitors, the alkaloid ability to inhibit lipid vesicle fusion triggered by crucial fragments of MERS-CoV and SARS-CoV/SARS-CoV-2 fusion peptides was additionally tested. The most effective compound, piperine, was also evaluated for in vitro antiviral activity against SARS-CoV-2 for *Vero* cells.

2. Materials and Methods

2.1. Materials

All chemicals used were of reagent grade. Synthetic 1,2-dioleoyl-sn-glycero-3-phospho-(1′-rac-glycerol) (DOPG), 1,2-dioleoyl-sn-glycero-3-phosphocholine (DOPC), 1-palmitoyl-2-oleoyl-sn-glycero-3-phosphocholine (POPC), sphingomyelin (Brain, Porcine) (SM), 1,2-dipalmitoyl-sn-glycero-3-phospho-(1′-rac-glycerol) (DPPG), cholesterol (CHOL), and 1,2-dipalmitoyl-sn-glycero-3-phosphoethanolamine-N-(lissamine rhodamine B sulfonyl) (Rh-DPPE) were obtained from Avanti® Polar Lipids. Nonactin, KCl, CaCl$_2$, PEG-8000, calcein, Triton X-100, Sephadex G-50, HEPES, DMSO, phosphate buffer solution (PBS), sorbitol, and alkaloids, atropine, (−)-lupinine, (−)-cotinine, berberine chloride, quinine, melatonin, caffeine, 1,7-dimethylxanthine, 3,9-dimethylxanthine, theophylline, 3-isobutyl-1-methylxanthine, 7-(β-hydroxyethyl) theophylline, pentoxifylline, hordenine, (±)-synephrine, colchicine, capsaicin, dihydrocapsaicin, and tabersonine, were purchased from Sigma-Aldrich Company Ltd. (Gillingham, UK). The chemical structures of the tested alkaloids are presented in Figure S1 (Supplementary Materials).

DMEM and DMEM/F12 culture mediums, fetal bovine serum (FBS), trypsin solution, and Dulbecco's phosphate buffer solution (DPBS) were purchased from Gibco™ (Life Technologies, Paisley, Scotland).

The modeling fusion peptides, FP-SARS-CoV-2 (RSFIEDLLFNKVT) and FP-MERS-CoV (RSAIEDLLFDKVT), were synthesized by the IQ Chemical LLC (Saint-Petersburg, Russia). The purity of peptides was ≥98%.

2.2. Calcein Leakage Assay

The choice of membrane-forming lipids is related to the enrichment of the viral lipid envelope with negatively charged and neutral glycerophospholipids, CHOL, and SM [45–47]. Small unilamellar vesicles were prepared from the mixtures of DOPG/CHOL (80/20 mol%), DOPC/DOPG/CHOL (40/40/20 mol%), and POPC/SM/CHOL (60/20/20 mol%) by extrusion. The resulting liposome suspension contained 3 mM lipid. Lipid stock in chloroform was dried under a gentle stream of nitrogen. Dry lipid film was hydrated by a buffer (35 mM calcein, 10 mM HEPES, pH 7.4). The suspension was subjected to five freeze–thaw cycles and then passed through a 100 nm nuclepore polycarbonate membrane for 13 times. The calcein that was not entrapped in the vesicles was removed by gel filtration with a Sephadex G-50 column to replace the buffer outside the liposomes with calcein-free solution (0.15 M NaCl, 10 mM HEPES, pH 7.4). Calcein inside the vesicles fluorescence very poorly, because of strong self-quenching at millimolar concentrations, while the fluorescence of disengaged calcein in the surrounding media is due to liposome fusion [41].

Here, $CaCl_2$ and PEG-8000 were used as widely recognized fusion inducers [48–51]. The results of adding the different concentrations of $CaCl_2$ into aqueous solutions, bathing calcein-loaded DOPG/CHOL (80/20 mol%) and DOPC/DOPG/CHOL (40/40/20 mol%) liposomes, are shown in Figure S2a,b (Supplementary Materials), respectively. Moreover, 20 and 40 mM $CaCl_2$ was used to induce the fusion of DOPG/CHOL and DOPC/DOPG/CHOL liposomes, respectively.

The structure/function study performed by [49] showed that the fragment of MERS-CoV spike protein S2 subunit (RSARSAIEDLLFDKV), which is highly conserved throughout the coronavirus family, induces the dipalmitoylphosphatidylcholine/SM/CHOL liposome syncytium formation, and the IEDLLF core sequence has a key role in membrane perturbation. A shorter α-helical fragment of MERS-CoV fusion peptide (RSAIEDLLFD-KVT) and the homologous highly conserved fragment of SARS-CoV/SARS-CoV-2 fusion peptides (RSFIEDLLFNKVT) [52] were chosen to mimic viral associated fusion. The chosen peptides were further referred to FP-MERS-CoV and FP-SARS-CoV-2, respectively. FP-SARS-CoV-2 and FP-MERS-CoV were added to POPC/SM/CHOL vesicles (60/20/20 mol%) to induce membrane fusion. The results of adding the different concentrations of peptides into aqueous solutions, bathing calcein-loaded POPC/SM/CHOL vesicles, are shown in Figure S2c,d (Supplementary Materials). Moreover, 50 µM FP-SARS-CoV-2 and 200 µM FP-MERS-CoV were used to induce liposome fusion to test the inhibitory action of alkaloids.

To test the fusion inhibitory action of alkaloids, a fusion inducer (calcium or modeling fusion peptides at the appropriate concentrations) was added to liposomes, pre-incubated with 400 µM of alkaloids (except for 40 µM of tabersonine). Since tabersonine at a concentration of 400 µM completely inhibited the calcium-mediated fusion of both DOPG/CHOL and DOPC/DOPG/CHOL vesicles, in order to understand whether its inhibitory effect depended on the lipid composition of liposomes, we reduced its concentration by 10-fold. Both concentrations of tabersonine (40 and 400 µM) were used to test its inhibitory action on the FP-SARS-CoV-2 and FP-MERS-CoV mediated fusion.

We found that the osmotic pressure gradient resulting from the addition of a fusion inducer or an inhibitor did not affect the leakage of calcein from liposomes (data not shown).

The degree of calcein release was determined using a Fluorat-02-Panorama spectrofluorometer (Lumex, Saint-Petersburg, Russia). The excitation wavelength was 490 nm and the emission wavelength was 520 nm. Triton X-100 was added to a final concentration of 1% to each sample in order to completely disrupt liposomes, and the intensity after releasing the total amount of calcein from liposomes was measured.

To describe the fusion of the liposomal membranes, the relative fluorescence of leaked calcein (RF, %) was calculated using the following formula:

$$RF = \frac{I - I_0}{I_{\max}/0.9 - I_0} \cdot 100\%, \tag{1}$$

where I and I_0 were the calcein fluorescence intensities in the sample, in the presence and absence of the fusion inducer/inhibitor, respectively, and I_{max} was the maximum fluorescence of the sample after lysis of liposomes by Triton X-100. Factor of 0.9 was introduced to calculate the dilution of the sample by Triton X-100.

The inhibition index (II) was calculated to characterize the relative efficiency of tested alkaloids to inhibit the fusion of liposomes compared to the action of the fusion inducer alone on the same liposome preparation:

$$II = \frac{RF_U - RF_S}{RF_U} \cdot 100\%, \tag{2}$$

where RF_U and RF_S were the maximum relative fluorescence of the leaked calcein at the vesicle fusion induced by the calcium or modeling fusion peptides in the absence (RF_U, $U = Ca^{2+}$/FP-MERS-CoV/FP-SARS-CoV-2) and presence of alkaloids (RF_S), respectively. The estimation of RF_S was made, taking into account the intrinsic effect of alkaloids (the calcein leakage due to the potent lipid disordering action of alkaloid alone, RF_A). Figure S3 and Table S1 (Supplementary Materials) present the kinetics and maximum values of RF_A at the addition of different alkaloids into aqueous solutions, bathing calcein-loaded DOPG/CHOL (80/20 mol%) (Figure S3a), DOPC/DOPG/CHOL (40/40/20 mol%) (Figure S3b), and POPC/SM/CHOL (60/20/20 mol%) vesicles (Figure S3c). Time dependences of calcein leakage were fitted with two-exponential functions with characteristic times, t_1 and t_2, related to the fast and slow components of the release process.

2.3. Confocal Fluorescence Microscopy

The visualization of changes in the morphological features and behavior of vesicles in the suspension under the action of fusion inducers was performed by marking the liposome membranes by the fluorescent labeled lipid, Rh-DPPE. Giant unilamellar vesicles were prepared from the mixture of POPC/SM/CHOL (60/20/20 mol%) and 1 mol% of fluorescent lipid probe Rh-DPPE by the electroformation method (standard protocol, 3 V, 10 Hz, 1 h, 25 °C) using Nanion Vesicle Prep Pro (Munich, Germany). The resulting liposome suspension contained 1 mM lipid in 0.5 M sorbitol solution. The measurements of the liposome fusion were added; 10% of PEG-8000, 50 µM of FP-SARS-CoV-2, and 200 µM of MERS-CoV were used to induce membrane fusion and were incubated for 1–2 min at room temperature (25 ± 1 °C). Vesicles were imaged through an oil immersion objective (65 ×/1.4HCX PL) using an Olympus (Hamburg, Germany). A helium-neon laser with a wavelength of 561 nm was used to excite Rh-DPPE. The temperature during observation was controlled by air heating/cooling in a thermally insulated camera.

2.4. Dynamic Light Scattering

Small unilamellar vesicles composed of DOPG/CHOL (80/20 mol%) and POPC/SM/CHOL (60/20/20 mol%) were prepared by the extrusion and were treated with 400 µM of cotinine, melatonin, 3-isobutyl-1-methylxanthine, lupinine, piperine, hordenine, and 40 or 400 µM of tabersonine for 30 min before the addition of 20 mM $CaCl_2$, 50 µM FP-SARS-CoV-2, and 200 µM MERS-CoV solution. Three different control samples were used: unmodified liposomes and vesicles incubated with fusion inducer, and the inhibitor alone. We found that the addition of alkaloid alone into the liposome suspension did not result in a noticeable alteration in vesicle size (data not shown). The hydrodynamic diameter (d, nm) and zeta-potential (ζ, mV) were determined on a Malvern Zetasizer Nano ZS 90 (Malvern Instruments Ltd., Malvern, United Kingdom) by gradual titration of liposome suspension in PBS at 25 °C.

2.5. Membrane Boundary Potential Measurements

Virtually solvent-free planar lipid bilayers were prepared using a monolayer-opposition technique [53] on a 50-µm diameter aperture in a 10-µm thick Teflon film separating the two compartments of the Teflon chamber. The steady-state conductance of K^+-nonactin was modulated via the two-sided addition of the colchicine, cotinine, 1,7-dimethylxanthine, capsaicin, synephrine, 3-isobutyl-1-methylxanthine, lupinine, piperine, tabersonine, and hordenine from different mM stock solutions in ethanol or methanol to the membrane-bathing solution (0.1 M KCl, pH 7.4) to obtain a final concentration ranging from 5 µM to 1 mM. The aperture was pretreated with hexadecane. Lipid bilayers were made from DOPG/CHOL (80/20 mol%). The conductance of the lipid bilayers was determined by measuring membrane conductance (G) at a constant transmembrane voltage (V = 50 mV). In the subsequent calculations, the membrane conductance was assumed to be related to the membrane boundary potential (φ_b), the potential drop between the aqueous solution and the membrane hydrophobic core, by the Boltzmann distribution [54]:

$$G \sim \xi \cdot C \exp\left(-\frac{ze\phi_b}{kT}\right) \tag{3}$$

where ξ is the ion mobility, ze is the ion charge, k is the Boltzmann constant, and T is the absolute temperature.

Ag/AgCl electrodes with 1.5% agarose/2 M KCl bridges were used to apply V and measure G.

The current was measured using an Axopatch 200B amplifier (Molecular Devices, LLC, Orleans Drive, Sunnyvale, CA, USA) in the voltage clamp mode. Data were digitized using a Digidata 1440A and analyzed using pClamp 10.0 (Molecular Devices, LLC, Orleans Drive, Sunnyvale, CA, USA) and Origin 8.0 (OriginLab Corporation, Northampton, MA, USA). Data were acquired at a sampling frequency of 5 kHz using low-pass filtering at 200 Hz.

All experiments were performed at room temperature (25 °C).

2.6. Differential Scanning Microcalorimetry

Differential scanning microcalorimetry experiments were performed by a µDSC 7EVO (Setaram, Caluire-et-Cuire, France). Giant unilamellar vesicles were prepared from a mixture of DPPG/CHOL (90/10 mol%) by the electroformation method (standard protocol, 3 V, 10 Hz, 1 h, 55 °C). The liposome suspension contained 3 mM lipid and was buffered by 5 mM HEPES at pH 7.4. The tested alkaloids (colchicine, cotinine, 1,7-dimethylxanthine, capsaicin, synephrine, 3-isobutyl-1-methylxanthine, lupinine, piperine, and hordenine) were added to aliquots to obtain a final concentration up to 400 µM of alkaloids (except for 40 µM of tabersonine). The liposomal suspension was heated at a constant rate of 0.2 C·min^{-1}. The reversibility of the thermal transitions was assessed by reheating the sample immediately after the cooling step from the previous scan. The temperature dependence of the excess heat capacity was analyzed using Calisto Processing (Setaram, Caluire-et-Cuire, France). The peaks on the thermograms were characterized by the maximum temperatures of the main phase transition (T_m) of DPPG/CHOL mixture and the half-width of the main peak ($T_{1/2}$) indicating the size of the lipid cooperative unit.

2.7. In Vitro Antiviral Activity Assessment

To analyze the antiviral activity of piperine, monkey kidney epithelial cells, *Vero* (ATCC CCL81) were grown in culture vials (Nunc, Roskilde, Denmark) in DMEM culture medium supplemented with 10% FBS. The cell concentration was adjusted to 5×10^5 cells/mL, and cells were seeded into 96-well culture plates (0.1 mL per well). The plates were incubated for 24 h at 37 °C in an atmosphere of 5% CO_2. Piperine was dissolved in DMSO, and then two-fold serial dilutions of 200 to 1.56 µg/mL in serum-free DMEM/F12 nutrient medium were prepared. Afterwards, 100 µL/well of each dilution of piperine was added to *Vero* cells and incubated for 1 h at 37 °C in a 5% CO_2 atmosphere. Serum-free

DMEM/F12 medium was added to control wells instead of piperine. Aliquots of SARS-CoV-2 (10^3 TCID50/mL) (isolate 17612) were mixed 1:1 with dilutions of piperine and incubated for 1 h at 37 °C. After 1 h incubation, piperine was removed from the plates, 100 μL/well of the SARS-CoV2/piperine mixture was added to the corresponding wells and incubated for 1 h at 37 °C in a 5% CO_2 atmosphere, the virus-containing medium was added to the wells of the viral control, and serum-free DMEM/F12 was added to the cell control. The virus was then removed, the cells were washed three times with DPBS, 100 μL/well of dilutions of piperine (200–1.56 μg/mL) were added to the wells, serum-free DMEM/F12 medium was added to the control wells, and the plates were incubated for 24 h at 37 °C in a 5% CO_2 atmosphere. After 24 h, supernatants from the experimental and control wells with the virus were collected and titrated by $TCID_{50}$ assay in *Vero* cells after incubation for 72 h. The visual assessment of the cytopathogenic effect (CPE) of the virus was performed using an Olympus CKX41 light inverted microscope (Olympus, Tokyo, Japan).

2.8. Statistical Analysis

The control RF_{Ca2+}-, $RF_{FP\text{-}MERS\text{-}CoV}$-, $RF_{FP\text{-}SARS\text{-}CoV\text{-}2}$-, t_1-, and t_2-values, characterizing the vesicle fusion induced by $CaCl_2$, FP-SARS-CoV-2 or FP-MERS-CoV in the absence of alkaloids, were averaged from 5 to 9 independent experiments. The experimental RF_S-, t_1-, and t_2-values, characterizing the $CaCl_2$-, FP-SARS-CoV-2- and FP-MERS-CoV-mediated fusion of liposomes pretreated with different alkaloids, were averaged from 2 to 4 independent experiments. All RF-, t_1-, and t_2-values were presented as *mean ± standard error* of the mean ($p \leq 0.05$). The magnitudes of d, ζ, T_m, $T_{1/2}$, and $\Delta\varphi_b$ were averaged from 3 to 5 independent experiments and presented as *mean ± standard deviation* ($p \leq 0.05$).

To prove the statistical significance of difference between RF-, t_1-, t_2-, d-, ζ-, T_m-, and $T_{1/2}$-values at the addition of alkaloids and the control values of related parameters (in the absence of alkaloids) the nonparametric signed-rank U-test (Mann–Whitney–Wilcoxon's U-test) was employed (*—$p \leq 0.01$, **—$p \leq 0.05$).

Pearson's correlation coefficient was applied to estimate the potent relationship between the inhibitory action of alkaloids on the vesicle fusion and the alkaloid-induced alterations in physical properties of lipid bilayers (ΔT_m, $\Delta T_{1/2}$, $\Delta\varphi_b$). The coefficients were calculated using Microsoft Excel (Microsoft Corp., Redmond, WA, USA).

3. Results and Discussion

3.1. Effect of Alkaloids on $CaCl_2$-Mediated Fusion of Negatively Charged Liposomes

Figure S4a–c (Supplementary Materials) shows the calcein leakage resulted from 20 mM $CaCl_2$-mediated fusion of DOPG/CHOL-liposomes in the absence and presence of different alkaloids at 400 μM concentration (40 μM of tabersonine). Some alkaloids were able to inhibit calcium-mediated fusion, and this ability strictly depended on the alkaloid type (Figure S4b,c). Table 1 summarizes the mean values of maximum marker leakage caused by calcium-induced fusion of DOPG/CHOL vesicles pretreated by different alkaloids (RF_S). The mean maximum leakage at the addition of fusion inducer alone is about 90%. A significant and reliable decrease in RF_S (indicating the inhibition of calcium-mediated fusion) is caused by pretreatment of DOPG/CHOL vesicles with dihydrocapsaicin (RF_S is about 70%), capsaicin, synephrine (RF_S is about 50%), 3-isobutyl-1-methylxanthine, quinine, piperine (RF_S is about 30%), tabersonine, hordenine (RF_S is about 20%), and lupinine (RF_S is about 10%).

Table 1. The parameters characterized the effects of alkaloids on the calcein release caused by the calcium-mediated fusion of liposomes composed of DOPG/CHOL (80/20 mol%) and DOPC/DOPG/CHOL (40/40/20 mol%).

Alkaloid	DOPG/CHOL			DOPC/DOPG/CHOL		
	RF_S, %	t_1, min	t_2, min	RF_S, %	t_1, min	t_2, min
$CaCl_2$	92 ± 7	9 ± 2	68 ± 6	82 ± 1	5 ± 3	51 ± 12
3,9-dimethylxanthine	96 ± 2	2 ± 1 *	20 ± 1 *	69 ± 7	2 ± 1	82 ± 21
caffeine	94 ± 5	3 ± 2	38 ± 20	65 ± 8*	1 ± 1	94 ± 22
7-(β-hydroxyethyl)theophylline	94 ± 6	3 ± 1	36 ± 1 *	75 ± 1	1 ± 1	75 ± 31
colchicine	94 ± 4	1 ± 1 *	7 ± 1 *	87 ± 3	2 ± 1	14 ± 1 *
pentoxifylline	92 ± 6	2 ± 1 *	36 ± 13	64 ± 8 *	1 ± 1	60 ± 14
cotinine	87 ± 12	7 ± 2	46 ± 6 **	68 ± 2 *	3 ± 1	20 ± 9 *
1,7-dimethylxanthine	85 ± 3	8 ± 4	33 ± 5 *	60 ± 15	1 ± 1	64 ± 30
atropine	81 ± 6	7 ± 1	50 ± 9 **	52 ± 6 *	8 ± 1	119 ± 9 *
theophylline	79 ± 8	7 ± 3	61 ± 19	83 ± 2	11 ± 1	97 ± 7 *
melatonin	76 ± 9	10 ± 1	119 ± 24 *	84 ± 8	5 ± 3	51 ± 28
berberine	74 ± 11	2 ± 1 *	27 ± 6 *	92 ± 6	5 ± 2	33 ± 13
dihydrocapsaicin	67 ± 2 *	7 ± 1	48 ± 21	56 ± 4 *	2 ± 1	24 ± 11
capsaicin	53 ± 9 *	2 ± 1 *	14 ± 2 *	56 ± 2 *	3 ± 1	14 ± 5 *
synephrine	46 ± 20 *	5 ± 1	85 ± 12	77 ± 5	12 ± 4	64 ± 30
3-isobutyl-1-methylxanthine	32 ± 12 *	3 ± 1 *	68 ± 12	84 ± 4	9 ± 1	80 ± 24
quinine	27 ± 4 *	3 ± 1 **	118 ± 23 **	66 ± 3 *	3 ± 1	27 ± 10
piperine	27 ± 10 *	2 ± 1 *	27 ± 11 **	24 ± 1 *	2 ± 1	8 ± 2 *
tabersonine	21 ± 7 *	2 ± 1 *	11 ± 2 *	1 ± 1 *	– &	– &
hordenine	20 ± 16 *	1 ± 1 *	17 ± 6 *	82 ± 4	4 ± 3	50 ± 17
lupinine	13 ± 10 *	1 ± 1 *	90 ± 12	38 ± 4 *	6 ± 3	42 ± 3

RF_S—the maximum leakage of fluorescent marker caused by the fusion of liposomes composed of DOPG/CHOL (80/20 mol%) and DOPC/DOPG/CHOL (40/40/20 mol%) mediated by 20 and 40 mM $CaCl_2$ respectively. Liposomes have been incubated with 400 μM of alkaloids (except for 40 μM of tabersonine) for 30 min before the addition of $CaCl_2$. t_1 and t_2—the time constants characterizing the fast (1) and slow (2) components of marker release kinetics (the time dependences of marker leakage were fitted by two-exponential functions). *—$p \leq 0.01$, **—$p \leq 0.05$ (Mann–Whitney–Wilcoxon's test, $CaCl_2$ alone vs. $CaCl_2$ + alkaloid). &—cannot be determined.

Table 1 also demonstrates the kinetic parameters of liposome fusion in the presence of alkaloids. It should be noted that the time dependences of the marker release due to vesicle fusion are well-described by two-exponential dependences with characteristic times related to fast and slow components (t_1 and t_2, respectively). Analysis of Table 1 indicates that t_1 and t_2 values vary in the ranges of about 1–10 and 10–120 min, respectively. The fast component might characterize the calcium sorption on the negatively charged DOPG-enriched membranes and the subsequent aggregation of the lipid vesicles, while the slow one might be related to the topological rearrangements of membrane lipids and further dye leakage at the liposome fusion. Colchicine, capsaicin, and tabersonine significantly accelerated the marker release kinetics at calcium-mediated fusion of DOPG/CHOL vesicles (about 5–10-fold decrease in both characteristic times is observed) (Table 1). Melatonin and quinine drastically slowed down the fusion process increasing t_2 from about 70 to more than 100 min (Table 1).

To characterize the relative efficiency of different alkaloids to inhibit calcium-mediated DOPG/CHOL liposome fusion the inhibition index (*II*) was calculated using Formula (2) for single preparation of liposome suspension, and then was averaged between the various preparations (Figure 1a). Taking into account the *II* value, the tested alkaloids were divided into three groups: ineffective (*II* does not exceed 15%); of low and medium efficiency

(*II* is in the range of 16–50%), and of high efficiency (*II* is more than 50%). The first group includes 3,9-dimethylxanthine, caffeine, 7-(β-hydroxyethyl)theophylline, colchicine, pentoxifylline, cotinine, 1,7-dimethylxanthine, atropine, theophylline, melatonin, and berberine. The efficiency in the second group increases in the series: dihydrocapsaicin (*II* is about 25%) ≤ capsaicin ≈ synephrine (*II* is about 45%). The latter (the most effective) group includes 3-isobutyl-1-methylxanthine (*II* is about 65%), piperine ≈ quinine (*II* is about 70%), tabersonine (*II* is about 75%), hordenine (*II* is about 80%), and lupinine (*II* is about 85%).

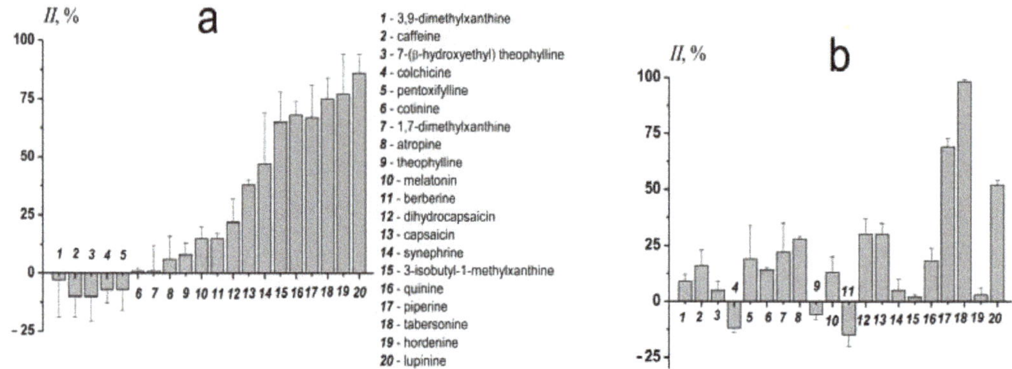

Figure 1. The inhibition index (*II*) characterizing the ability of tested alkaloids to suppress the fusion of vesicles made from DOPG/CHOL (80/20 mol%) (**a**) and DOPC/DOPG/CHOL (40/40/20 mol%) (**b**). Liposomes were incubated with 400 μM of alkaloids (except for 40 μM of tabersonine) for 30 min before the addition of CaCl₂.

Figure 2a shows the results of size measurements after calcium addition to unmodified DOPG/CHOL liposomes and vesicles pretreated by alkaloids of various anti-fusogenic activity, cotinine, melatonin, 3-isobutyl-1-methylxanthine, lupinine, piperine, tabersonine, and hordenine. The diameter of DOPG/CHOL liposomes in the absence of any modifiers is equal to 80 ± 20 nm. CaCl₂ induces liposome fusion, and the vesicle size increased up to 145 ± 40 nm (Figure 2a). Cotinine and melatonin practically did not affect the size-increasing effect of calcium, as the liposome diameter was about 140 nm. The other tested compounds (lupinine, 3-isobutyl-1-methylxanthine, tabersonine, piperine, and hordenine) significantly inhibited calcium-mediated liposome fusion: the diameter of the vesicles was equal to 77 ± 13 nm (Figure 2a). The data obtained are in good agreement with the *II* values assessed by the calcein release assay (Figure 1a).

Figure 2b presents the results of estimation of the ζ-potential of DOPG/CHOL liposomes in the presence of 20 mM CaCl₂ without and with different alkaloids. The ζ-potential of unmodified DOPG/CHOL liposomes was equal to about −70 mV. The addition of calcium led to an increase in ζ-potential up to about −50 mV. The ζ-potential value of DOPG/CHOL vesicles pretreated with alkaloids was 10–20 mV less than that of untreated ones (Figure 2b). The observed incomplete compensation of negative surface charge of DOPG-enriched liposomes was expected at the introduction of positively charged calcium ions and alkaloid molecules (lupinine, tabersonine, and hordenine). The disagreement of alkaloid abilities to inhibit calcium-mediated vesicle fusion and to increase ζ-potential (Figure 2a,b) clearly demonstrates that the anti-fusogenic activity of alkaloids is not due to partial compensation of the negative membrane surface charge by alkaloids or a competition between alkaloids and calcium ions for the interaction with negatively charged lipid groups.

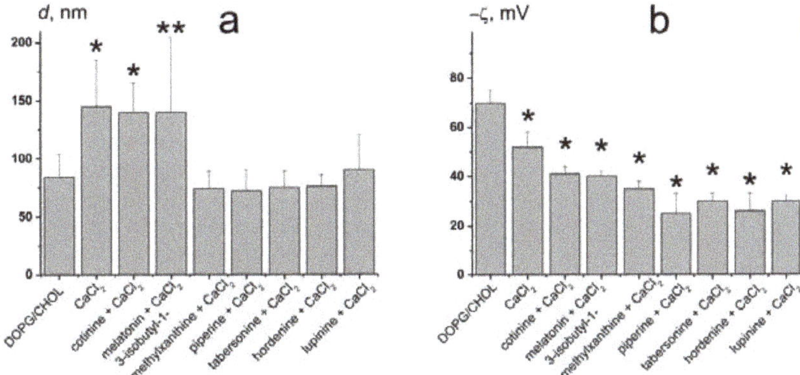

Figure 2. The diameter (d, nm) (**a**) and the ζ-potential (**b**) of the DOPG/CHOL (80/20 mol%) liposomes before and after addition of 20 mM CaCl$_2$ to unmodified vesicles and liposomes pretreated by 400 μM of cotinine, melatonin, 3-isobutyl-1-methylxanthine, lupinine, piperine, hordenine, or 40 μM of tabersonine. *—$p \leq 0.01$, **—$p \leq 0.05$ (Mann–Whitney–Wilcoxon's test, untreated liposomes vs. vesicles in the presence of CaCl$_2$ or/and alkaloids).

To study the dependence of the inhibition effect of alkaloids on the lipid composition of fusing vesicles, DOPC was introduced into membrane composition. It should be noted that the cell membranes are enriched of phosphatidylcholine species compared to virions [47,55,56]. The decrease in the proportion of negatively charged DOPG in membrane composition from 80 to 40 mol% was accompanied by an increase in CaCl$_2$ concentration from 20 to 40 mM to reach the RF_S value of about 80%. Figure S4d–f (Supplementary Materials) demonstrates the calcein leakage caused by the calcium-mediated fusion of DOPC/DOPG/CHOL liposomes. RF_S resulted from calcium-mediated fusion of DOPC/DOPG/CHOL vesicles pretreated by different alkaloids reliably decreases in the series: caffeine ≈ pentoxifylline ≈ cotinine ≈ quinine (RF_S is about 65%) ≥ atropine ≈ dihydrocapsaicin ≈ capsaicin (RF_S is about 55%) > lupinine (RF_S is about 40%) > piperine (RF_S is about 25%) > tabersonine (RF_S is equal to 1%) (Table 1).

The kinetic parameters of the time dependence of the calcein leakage caused by calcium-mediated fusion of DOPC/DOPG/CHOL liposomes are shown in Table 1. Colchicine and capsaicin accelerate the release kinetics due to calcium-mediated fusion of DOPC/DOPG/CHOL vesicles as well as of DOPG/CHOL liposomes. The huge inhibition of calcium-mediated fusion of DOPC/DOPG/CHOL liposomes in the presence of tabersonine did not allow for determining the characteristic times of dye release kinetics. Similar to DOPG/Chol vesicles, the pretreatment of DOPC/DOPG/CHOL liposomes with colchicine and capsaicin led to about five-fold decrease in t_2-value. Unlike the case of DOPG/CHOL vesicles, the kinetics of calcein leakage due to fusion of DOPC/DOPG/CHOL liposomes was also accelerated by piperine (by six times), while theophylline and atropine slightly slowed down the fusion process, increasing t_2 from 50 to about 100 and 120 min, respectively (Table 1).

Figure 1b shows the II values, characterizing the relative ability of different alkaloids to suppress the fusion of DOPC/DOPG/CHOL vesicles. Moreover, 3,9-dimethylxanthine, 7-(β-hydroxyethyl)theophylline, colchicine, 1,7-dimethylxanthine, theophylline, melatonin, berberine, synephrine, 3-isobutyl-1-methylxanthine, and hordenine did not demonstrate a significant ability to inhibit the fusion of DOPC/DOPG/CHOL liposomes. Caffeine, pentoxifylline, cotinine, atropine, dihydrocapsaicin, capsaicin, and quinine were characterized by low or medium efficiency to suppress the fusion of vesicles made of ternary lipid mixture, while the II values in the presence of lupinine, piperine, and tabersonine were about 50, 70, and 100% (Figure 1b). Comparison of Figure 1a,b clearly demonstrates that the anti-fusogenic activity of alkaloids strictly depends on the lipid composition of the vesicles.

3.2. The Influence of Alkaloids on the Physical Properties of the Model Lipid Membrane

It has been repeatedly shown that the elastic characteristics of the lipid bilayer play a huge role in the membrane fusion. The ordering of the lipid head groups and acyl chains, transbilayer lateral pressure profile, phase state of lipids, spontaneous curvature of the membrane, area per lipid molecule, and other characteristics can influence fusion [57]. Many of these parameters influence each other, but all of them are directly dependent on the lipid composition of the membrane. The key role of lipids is also supported by the fact that viral infections can alter cell lipid synthesis, regulating it according to their needs [58]. The ability of alkaloids and their derivatives to change the physical properties of a bilayer upon intercalation was demonstrated in a number of studies [59–62]. In particular, alkaloids affect the lipid packing [63,64], change lipid phase transition temperatures [65,66], etc. Recently, we performed a detailed study of alkaloid effects on the physical properties of the POPC membrane [67]. To understand the relationship between the alkaloid effects on the fusion and modulation of lipid matrix under their action, the influence of tested alkaloids on the properties of PG and CHOL-enriched bilayers were performed.

Figure 3 presents the typical heating thermograms of DPPG/CHOL-liposomes in the absence (control) and in the presence of tested alkaloids. The value of the phase transition temperature (T_m) of DPPG/CHOL mixture, and the half-width of the main peak ($T_{1/2}$) are summarized in Table 2. T_m is the point at which thermally induced lipid melting occurs [68], while $T_{1/2}$ describes the sharpness of the phase transition or the inverse cooperativity of this process [69]. The temperature of the main transition (T_m) of untreated DPPG/CHOL vesicles is equal to 40.4 °C, with a half-width of the peak ($T_{1/2}$) of about 0.8 °C (Figure 3, control).

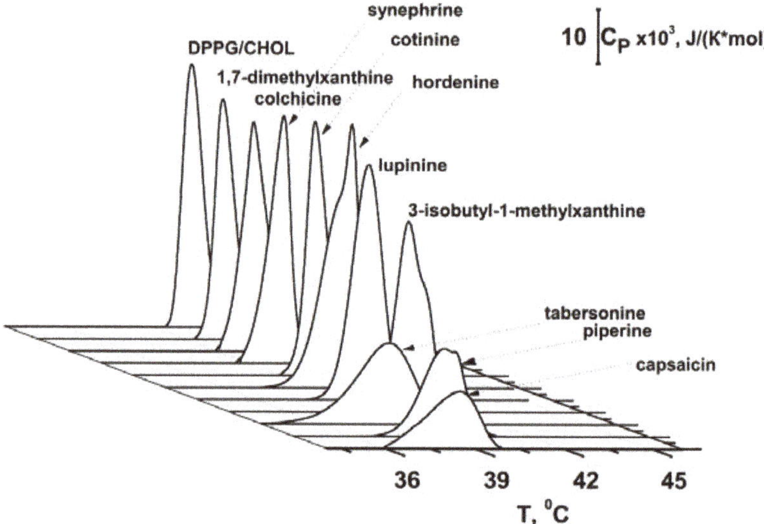

Figure 3. Heating thermograms of DPPG/CHOL (90/10 mol%) liposomes in the absence and presence of 400 µM of 1,7-dimethylxanthine, piperine, hordenine, cotinine, 3-isobutyl-1-methylxanthine, lupinine, synephrine, capsaicin, colchicine, and 40 µM of tabersonine.

Table 2. The parameters characterized the effects of alkaloids on the physical properties of lipid bilayers: T_m—the main transition temperature of DPPG/CHOL (90/10 mol%); $T_{1/2}$—the half-width of the main peak in the presence of 400 µM of alkaloids (40 µM of tabersonine); $\Delta\varphi_b$(max)—the maximum changes in the boundary potential of DOPG/CHOL (80/20 mol%) membranes.

Alkaloid	T_m, °C	$T_{1/2}$, °C	$\Delta\varphi_b$(max), mV
control	40.4 ± 0.0	0.8 ± 0.0	–
colchicine	40.4 ± 0.0	0.8 ± 0.0	-24 ± 6
cotinine	40.4 ± 0.0	0.8 ± 0.0	-7 ± 3
1,7-dimethylxanthine	40.4 ± 0.0	0.8 ± 0.0	-5 ± 2
capsaicin	38.3 ± 0.2 *	2.0 ± 0.1 *	-70 ± 7
synephrine	40.1 ± 0.1 *	1.2 ± 0.1 *	-25 ± 3
3-isobutyl-1-methylxanthine	39.9 ± 0.2 *	1.2 ± 0.1 *	-18 ± 4
piperine	39.5 ± 0.3 *	1.5 ± 0.2 *	-39 ± 8
tabersonine	38.5 ± 0.2 *	1.9 ± 0.2 *	58 ± 9
hordenine	39.7 ± 0.2 *	1.2 ± 0.1 *	8 ± 3
lupinine	39.7 ± 0.1 *	1.5 ± 0.2 *	-30 ± 3

*—$p \leq 0.01$ (Mann–Whitney–Wilcoxon's test, control vs. alkaloid).

Colchicine, cotinine, and 1,7-dimethylxanthine did not affect the thermotropic phase behavior of DPPG/CHOL mixture (Figure 3, Table 2). Synephrine, 3-isobutyl-1-methylxanthine, lupinine, and hordenine decreased the T_m by approximately 0.3 to 0.7 °C and increased the $T_{1/2}$ by 0.4 to 0.6 °C (Figure 3, Table 2). The data are in agreement with the study of hordenine influence on the phase behavior of dimyristoylphosphatidylglycerol [61]. The introduction of capsaicin, piperine, and tabersonine was accompanied by a sharp decrease in T_m (0.9–2.1 °C) and increase in $T_{1/2}$ (0.7–1.2 °C) (Figure 3, Table 2). The correlation coefficients between $LogD_{o/w}$ (Table S1 Supplementary Materials) and $-\Delta T_m$- and $\Delta T_{1/2}$-values (Table 2) are in the range of 0.74–0.77, demonstrating a good correlation between the lipophilicity of the alkaloid molecules and their ability to alter lipid packing.

The correlation coefficient between the alkaloid-induced changes in T_m of DPPG/CHOL (Table 2) and their *II* values in DOPG/CHOL vesicles (Figure 1a) was equal to 0.52, the corresponding coefficient characterizing the interdependence of alkaloid-induced changes in $T_{1/2}$ (Table 2) and the *II* values (Table 1) was equal to 0.63. The correlation coefficient between the alkaloid-induced changes in T_m (Table 2) and their *II* values in DOPC/DOPG/CHOL vesicles (Table 1) was equal to 0.64, while the interdependence of compound-induced changes in $T_{1/2}$ (Table 2) and alkaloid *II* values of DOPC/DOPG/CHOL liposomes (Figure 1b) was characterized by a coefficient of 0.68. The observed correlation between the parameters characterizing the lipid melting, especially the cooperativity of lipid phase transition, and the fusion inhibition index indicates a relationship between the ability of alkaloids to disorder membrane lipids and to inhibit the liposome fusion. The incorporation of alkaloids into the polar region of the membrane, leading to an increase in lateral pressure, positive curvature stress, and an area per lipid molecule in this region, contributes to both a reduction in the temperature/cooperativity of the phase transition and an increase in the energy of fusion intermediates characterized by lipid surfaces of a negative curvature. Moreover, a comparison of the data obtained (Figure 3, Table 2) and our recently published results [65] demonstrate that the tested alkaloids have different effects on the phase behavior of DPPG/CHOL and DPPC. This fact might be related to the observed difference in their ability to inhibit the fusion of DOPG/CHOL and DOPC/DOPG/CHOL vesicles (Figure 1, Table 1). Alkaloid-induced curvature stress should depend on the depth of molecule insertion into lipid bilayer and its orientation in the membrane. The significant inhibitory action of 3-isobutyl-1-methylxanthine on the fusion of DOPG/CHOL liposomes might be related to the induction of the high positive curvature stress due to both the electrostatic interaction between xanthine residue and DOPG on the membrane surface and the tendency of the isobutyl side chain to be embedded into the membrane hydrocarbon core. Xanthine derivatives with lower hydrophobicity, i.e., caffeine, pentoxifylline, 1,7-dimethylxanthine, 3,9-dimethylxanthine, and 7-(β-hydroxyethyl)theophylline (Table S1, Supplementary Materials) did not inhibit calcium-mediated fusion of DOPG/CHOL liposomes. Introduction of DOPC into the membrane lipid composition was accompanied by a two-fold decrease in the DOPG content, which led to significant changes in the membrane orientation and embedment of xanthines. Thus, caffeine and pentoxifylline demonstrated a weak inhibitory effect on a fusion of DOPC/DOPG/CHOL vesicles, while 3-isobutyl-1-methylxanthine did not affect this process. The complete loss of the inhibiting ability of β-phenylethylamine derivatives, synephrine and hordenine, in membranes with lower content of DOPG compared to DOPG-enriched bilayers, could be explained in a similar way. An increase in the inhibitory effect of tabersonine on the fusion of DOPC/DOPG/CHOL vesicles compared to DOPG/CHOL liposomes might indicate its deeper immersion into the DOPC/DOPG/CHOL membranes than into DOPG/CHOL bilayers due to a decrease in the electrostatic interactions between positively charged tabersonine and negatively charged DOPG molecules.

Considering that liposome fusion is triggered by calcium ions, we also considered the possibility of alkaloids influencing fusion by changing the boundary potential of the membranes. Table 2 shows the maximum changes in φ_b at the adsorption of different alkaloids ($-\Delta\varphi_b(\max)$). The dependences of $\Delta\varphi_b$ on the concentrations of tested alkaloids are presented in Figure S5 (Supplementary Materials). Tabersonine significantly increased the φ_b of DOPG/CHOL membranes (by about 60 mV), while capsaicin dramatically decreased this value (by about -70 mV) (Figure S5 in the Supplementary Materials, Table 2). Colchicine, synephrine, 3-isobutyl-1-methylxanthine, lupinine, and piperine led to reduction in φ_b by about 20–40 mV (Figure S5 in the Supplementary Materials, Table 2). Cotinine, 1,7-dimethylxanthine, and hordenine practically did not alter the φ_b-value (Figure S5 Supplementary Materials, Table 2). There was no correlation between $\Delta\varphi_b$ of DOPG/CHOL membranes after the addition of alkaloids and their *II* values. This fact demonstrates the minor role of changes in the bilayer electrical properties by alkaloids in their ability to inhibit the calcium-mediated membrane fusion.

3.3. Liposome Fusion Mediated by Fragments of Coronavirus Fusion Peptides

The COVID-19 pandemic has caused major challenges to healthcare systems across the globe. The lack of specific treatment, virus mutations, the emergence of new strains that have become resistant to vaccines, and many other factors, have led to the search for new drugs to treat COVID-19. To demonstrate the pharmacological applications of alkaloids as coronavirus fusion inhibitors—the abilities of the most effective compounds to suppress the membrane fusion induced by FP-MERS-CoV and FP-SARS-CoV-2 were studied. These peptides are able to induce calcein release due to fusion of POPC/SM/CHOL liposomes (Figure S2c,d Supplementary Materials), and are not effective to produce the fusion of DOPG/CHOL vesicles (data not shown).

To further validate the fusogenic properties of FP-MERS-CoV and FP-SARS-CoV-2, the confocal fluorescence microscopy of giant unilamellar vesicles composed of POPC/SM/CHOL was performed. Figure 4 presents the fluorescence micrographs of POPC/SM/CHOL liposomes in the absence (Figure 4a) and presence of 10% of the well-established fusion inducer PEG-8000 (Figure 4b), 50 μM FP-SARS-CoV-2 (Figure 4c), and 200 μM FP-MERS-CoV (Figure 4d). Both peptides at the indicated concentrations caused deformation and aggregation of lipid vesicles, and increased in size. FP-SARS-CoV-2, unlike FP-MERS-CoV, also induced the formation of multilamellar and multivesicular liposomes (Figure 4c). In general, the morphological picture under the action of both peptides is similar to that which occurred when PEG-8000 was added (Figure 4b). Thus, the results obtained confirm the fusogenic activity of FP-SARS-CoV-2 and FP-MERS-CoV.

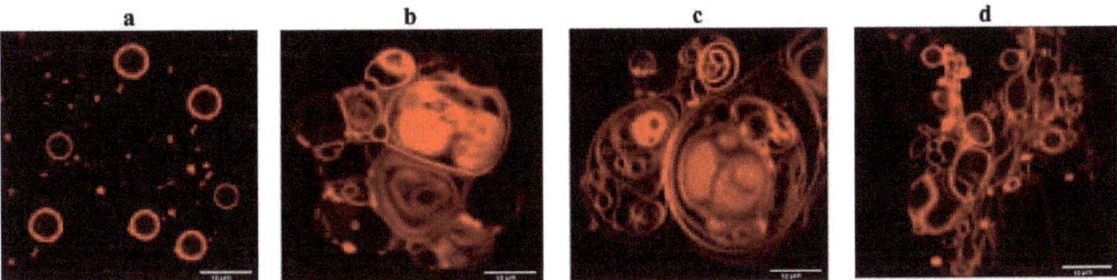

Figure 4. Fluorescence micrographs of giant unilamellar vesicles made from POPC/SM/CHOL (60/20/20 mol%) and 1 mol% of fluorescent lipid probe Rh-DPPE in the absence of any modifiers (**a**), and in the presence of 10% of PEG-8000 (**b**), 50 µM of FP-SARS-CoV-2 (**c**), and 200 µM of FP-MERS-CoV (**d**).

Figure 5a,d present the effects of alkaloids that effectively inhibit calcium-mediated fusion of DOPG/CHOL vesicles, 3-isobutyl-1-methylxanthine, piperine, tabersonine, hordenine, and lupinine, on the calcein leakage caused by FP-SARS-CoV-2- and FP-MERS-CoV-mediated fusion of POPC/SM/CHOL liposomes, respectively. In the absence of alkaloids RF_s-value produced by 50 µM of FP-SARS-CoV-2 and 200 µM of FP-MERS-CoV was about 70 (Figure 5a) and 80% (Figure 5d), respectively. Table 3 summarizes the mean values of maximum marker leakage caused by peptide-induced fusion of POPC/SM/CHOL vesicles pretreated by different alkaloids and the kinetic parameters of liposome fusion in the presence of alkaloids. RF_S caused by FP-SARS-CoV-2-mediated fusion of POPC/SM/CHOL vesicles significantly decreased in the presence of piperine (RF_S is about 30%) and tabersonine (RF_S is about 50%). Moreover, piperine was also effective against FP-MERS-CoV-mediated fusion (RF_S is about 30%), while tabersonine was not characterized by an ability to inhibit FP-MERS-CoV-mediated fusion of POPC/SM/CHOL liposomes. The fusion-induced leakage in the presence of 3-isobutyl-1-methylxanthine, tabersonine, hordenine, and lupinine was characterized by faster kinetics compared to the absence of alkaloids, while highly active piperine slightly slowed down kinetics of FP-MERS-CoV-induced fusion by increasing t_2-value.

Figure 5b,e show the alkaloid *II*-values clearly demonstrating the ability of piperine to significantly inhibit both FP-SARS-CoV-2- and FP-MERS-CoV-mediated fusion of POPC/SM/CHOL liposomes, and the selectivity of tabersonine anti-fusogenic action. The observed selectivity of tabersonine action against liposome fusion induced by FP-SARS-CoV-2 and FP-MERS-CoV should be researched further.

To further demonstrate the anti-fusogenic activity of piperine, the vesicle size before and after the addition of FP-SARS-CoV-2 and FP-MERS-CoV to the unmodified and alkaloid pretreated POPC/SM/CHOL-liposomes was measured by dynamic light scattering (Figure 5c,f). In the absence of the any modifiers, the diameters of POPC/SM/CHOL-vesicles were equal to 110 ± 15 nm. Addition of the peptide FP-SARS-CoV-2 and FP-MERS-CoV led to an increased in size, up to 290 ± 40 nm and 190 ± 15 nm, respectively (Figure 5c,f). In the presence of piperine, the modeling fusion peptides were not able to increase the diameter of POPC/SM/CHOL-liposomes due to strong inhibition of vesicle fusion by the alkaloid. Figure 5c also demonstrates the ability of tabersonine to prevent the increase of liposome size under FP-SARS-CoV-2 action. The invariability of the liposome ζ-potential upon the addition of FP-SARS-CoV-2- and FP-MERS-CoV to both unmodified and alkaloid pretreated POPC/SM/CHOL liposomes is shown in Figure S6 (Supplementary Materials).

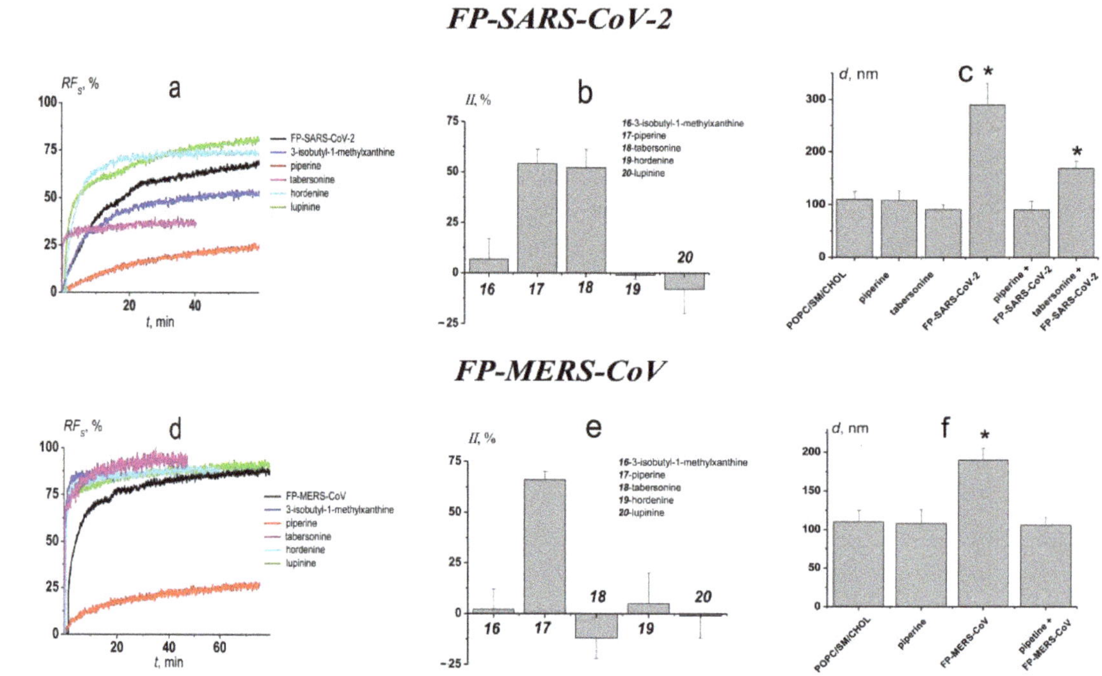

Figure 5. The effects of the alkaloids on FP-SARS-CoV-2 (upper panel) and FP-MERS-CoV (lower panel) mediated fusion of POPC/SM/CHOL (60/20/20 mol%) vesicles. (**a,d**) The time dependence of relative fluorescence of calcein (RF_S, %) leaked due to the fusion induced by 50 μM of FP-SARS-CoV-2; (**a**) and 200 μM of FP-MERS-CoV; (**d**) in the absence and presence of alkaloids. Liposomes were incubated with 400 μM of alkaloids for 30 min before addition of the peptides. The relationship between the color line and the alkaloid is given in the figure. (**b,e**) The inhibition index (*II*) characterizing the ability of tested alkaloids to suppress the fusion induced by 50 μM of FP-SARS-CoV-2 (**b**); and 200 μM of FP-MERS-CoV (**e**). (**c,f**) The diameter (*d*, nm) of the POPC/SM/CHOL liposomes before and after addition of 50 μM of FP-SARS-CoV-2 (**c**) and 200 μM of FP-MERS-CoV (**f**) to vesicles pretreated with 400 μM of piperine or tabersonine. *—$p \leq 0.01$ (Mann–Whitney–Wilcoxon's test, untreated liposomes vs. vesicles in the presence of fusion peptides or/and alkaloids).

Table 3. The parameters characterized the effects of alkaloids on the calcein release caused by the FP-SARS-CoV-2 and FP-MERS-CoV-mediated fusion of liposomes composed of POPC/SM/CHOL.

Alkaloid	50 μM FP-SARS-CoV-2			200 μM FP-MERS-CoV		
	RF_S, %	t_1, min	t_2, min	RF_S, %	t_1, min	t_2, min
inducer	77 ± 6	11 ± 1	70 ± 4	80 ± 5	3 ± 2	22 ± 14
3-isobutyl-1-methylxanthine	66 ± 5	3 ± 1 *	15 ± 2 *	78 ± 6	2 ± 1	15 ± 5
piperine	33 ± 3 *	9 ± 3	73 ± 6	29 ± 2*	4 ± 1	48 ± 1 *
tabersonine	53 ± 9 **	1 ± 1*	27 ± 13 *	79 ± 7	2 ± 1	12 ± 2
hordenine	74 ± 8	3 ± 1*	18 ± 4 *	75 ± 10	2 ± 1	21 ± 1
lupinine	76 ± 4	1 ± 1*	31 ± 9 *	80 ± 6	1 ± 1	26 ± 1

RF_S—the maximum leakage of fluorescent marker caused by the fusion of liposomes composed of POPC/SM/CHOL (60/20/20 mol%) mediated by 50 μM FP-SARS-CoV-2 or 200 μM FP-MERS-CoV. Liposomes have been incubated with 400 μM of alkaloids for 30 min before the addition of $CaCl_2$. t_1 and t_2—the time constants characterizing the fast (1) and slow (2) components of marker release kinetics (the time dependences of marker leakage were fitted by two-exponential functions). *—$p \leq 0.01$, **—$p \leq 0.05$ (Mann–Whitney–Wilcoxon's test, fusion peptide alone vs. fusion peptide + alkaloid).

3.4. Antiviral Evaluation

Piperine is an alkaloid found in black pepper, one of the most widely used spices. The in vitro and in vivo antiviral activity of the alkaloid against the MERS-CoV in Vero cells, and in a mice model, was already demonstrated [70]. Here, we evaluated in vitro antiviral activity of piperine against the SARS-CoV-2 virus.

According to results of the cytotoxicity analysis performed by Hegeto et al. [71], the value of IC_{50} of piperine against *Vero* cells was equal to 183.33 µg/mL. Based on this finding, the maximum piperine concentration of 200 µg/mL was chosen. The incubation with 200 µg/mL of piperine for 72 h led to death of all cells in wells; all lower piperine concentrations used were nontoxic to *Vero* cells (data not shown).

To determine the infectious activity of viral progeny, ten-fold dilutions of the supernatants collected from the experimental wells and viral controls in serum-free DMEM/F12 medium were prepared. The resulting dilutions were added to a 96-well culture plate with 80–90% *Vero* cell monolayer and incubated for 72 h. After that, visual assessment of the cytopathogenic effect (CPE) of the virus was performed. The control samples without any additives demonstrated that the virus titer was 10^4 TCID$_{50}$/mL. Samples with a piperine concentration of 1.56 µg/mL resulted in 10^3 TCID$_{50}$/mL; alkaloid concentrations of 3.12–25 µg/mL had 10^2 TCID$_{50}$/mL, and in samples with piperine, concentrations of 50 and 100 µg/mL CPE were not observed at all. Thus, the significant reduction of the titer of SARS-CoV2 progeny in *Vero* cells at 1.56–100 µg/mL of piperine was clearly demonstrated.

4. Conclusions

New broad-spectrum antiviral drugs are needed due to the increase in the number of viral infections. Virus fusion inhibitors could be effective antiviral agents because fusion is a necessary step for the lifecycle of the virus. We used plant secondary metabolites, in particular, alkaloids, as a new class of fusion inhibitors. It was shown that the ability of alkaloids to inhibit calcium-mediated fusion of liposomes depends on their lipid composition. A correlation between the anti-fusogenic activity of alkaloids and their disordering effects on membrane lipids was found. Additionally, the ability of piperine to suppress the fusion induced by fragments of the coronavirus fusion peptides (MERS-CoV and SARS-CoV/SARS-CoV-2) was demonstrated. We also showed that piperine dramatically reduced the titer of SARS-CoV2 progeny in vitro in *Vero* cells when used in non-toxic concentrations. Thus, we hypothesize that the antiviral activity of piperine is related to its lipid-associated action. Moreover, to our knowledge, except for one bioassay with SARS-CoV-2-S pseudotyped particles [72], this is the first study demonstrating the in vitro activity of piperine against SARS-CoV-2 virus. It should be noted that piperine with curcumin is being studied in several clinical trials for the treatment of COVID-19 [73–75]. According to [75], administration of oral curcumin with piperine substantially reduced the morbidity and mortality due to COVID-19. We hope that, over time, the number of effective alkaloids in the treatment of COVID-19 will increase. The use of plant metabolites can successfully complete the existing therapeutic strategies, which will make it possible to combat the SARS-CoV-2 virus more effectively.

Supplementary Materials: The following are available online at https://www.mdpi.com/article/10.3390/biomedicines9101434/s1, Figure S1: The chemical structures of the tested alkaloids, Figure S2: The time dependence of relative fluorescence of calcein leaked from DOPG/CHOL (80/20 mol%) (a), DOPC/DOPG/CHOL (40/40/20 mol%) (b), and POPC/SM/CHOL (60/20/20 mol%) (c,d) vesicles induced by different concentrations of CaCl$_2$ (a,b), FP-SARS-CoV-2 (c) and FP-MERS-CoV (d). The relationship between the color line and the concentration of the fusion inducers is given in the figure; Figure S3: The time dependence of relative fluorescence of calcein (RF_A, %) leaked from DOPG/CHOL (80/20 mol%) (a), DOPC/DOPG/CHOL (40/40/20 mol%) (b), and POPC/SM/CHOL (60/20/20 mol%) (c) vesicles induced by alkaloids alone. Alkaloids were added into the liposomal suspension up to a concentration of 400 µM (except for 40 µM of tabersonine in DOPG/CHOL and DOPC/DOPG/CHOL vesicles) at the initial moment. The relationship between the color line and the compound is given in the figure; Figure S4. The time dependence of relative fluorescence of calcein (RF_S, %) leaked by the fusion of DOPG/CHOL (80/20 mol%) (a–c) and DOPC/DOPG/CHOL (40/40/20 mol%) (d–f) vesicles induced by 20 mM of $CaCl_2$ in the absence (*black line*) and presence of alkaloids. Liposomes were incubated with 400 µM of alkaloids (except for 40 µM of tabersonine) for 30 min before the addition of $CaCl_2$. The relationship between the color line and the alkaloid is given in the figure; Figure S5: The dependence of the changes in the membrane boundary potential ($\Delta\varphi_b$) on the concentration of colchicine (*), cotinine (▲), 1,7-dimethylxanthine (■), capsaicin (○), synephrine (◇), 3-isobutyl-1-methylxanthine (•), lupinine (□), piperine (▼), tabersonine (♦), and hordenine (◄). The membranes were composed of DOPG/CHOL (80/20 mol%) and bathed in 0.1 M KCl at pH 7.4. V = 50 mV, Figure S6: The ζ-potential of the POPC/SM/CHOL (60/20/20 mol%) liposomes before and after addition of 50 µM of FP-SARS-CoV-2 or 200 µM of FP-MERS-CoV to unmodified and pretreated by 400 µM of piperine vesicles; Table S1: The parameters characterized the alkaloid molecules and their influence on the calcein leakage from liposomes of different composition.

Author Contributions: E.V.S.—investigation, analysis, validation, and writing (original draft); S.S.E.—investigation, analysis, and validation; N.M.Y.—investigation, A.A.M.—investigation, V.V.Z.—investigation, A.V.S.—investigation, analysis, and writing (review draft), O.S.O.—conceptualization, project administration, supervision, analysis, and writing (original draft, review and editing). All authors have read and agreed to the published version of the manuscript.

Funding: This research was financially supported by the joint-stock company Evalar.

Conflicts of Interest: The authors declare that the research was conducted in the absence of any commercial or financial relationships that could be construed as a potential conflict of interest.

References

1. Patridge, E.; Gareiss, P.; Kinch, M.S.; Hoyer, D. An analysis of FDA-approved drugs: Natural products and their derivatives. *Drug Discov. Today* **2016**, *21*, 204–207. [CrossRef] [PubMed]
2. Zhao, J.; Shan, T.; Mou, Y.; Zhou, L. Plant-derived bioactive compounds produced by endophytic fungi. *Mini Rev. Med. Chem.* **2011**, *11*, 159–168. [CrossRef] [PubMed]
3. Mani, J.S.; Johnson, J.B.; Steel, J.C.; Broszczak, D.A.; Neilsen, P.M.; Walsh, K.B.; Naiker, M. Natural product-derived phytochemicals as potential agents against coronaviruses: A review. *Virus Res.* **2020**, *284*, 197989. [CrossRef]
4. Ngwa, W.; Kumar, R.; Thompson, D.; Lyerly, W.; Moore, R.; Reid, T.E.; Lowe, H.; Toyang, N. Potential of flavonoid-inspired phytomedicines against COVID-19. *Molecules* **2020**, *25*, 2707. [CrossRef]
5. Kishimoto, S.; Sato, M.; Tsunematsu, Y.; Watanabe, K. Evaluation of biosynthetic pathway and engineered biosynthesis of alkaloids. *Molecules* **2016**, *21*, 1078. [CrossRef]
6. Xu, W.; Zhang, M.; Liu, H.; Wei, K.; He, M.; Li, X.; Hu, D.; Yang, S.; Zheng, Y. Antiviral activity of aconite alkaloids from *Aconitum carmichaelii* Debx. *Nat. Rrod. Res.* **2019**, *33*, 1486–1490. [CrossRef]
7. Cushnie, T.P.; Cushnie, B.; Lamb, A.J. Alkaloids: An overview of their antibacterial, antibiotic-enhancing and antivirulence activities. *Int. J. Antimicrob. Agents* **2014**, *44*, 377–386. [CrossRef]
8. Khan, H.; Mubarak, M.S.; Amin, S. Antifungal potential of alkaloids as an emerging therapeutic target. *Curr. Drug Targets* **2017**, *18*, 1825–1835. [CrossRef]
9. Boyd, M.R.; Hallock, Y.F.; Cardellina, J.H., 2nd; Manfredi, K.P.; Blunt, J.W.; McMahon, J.B.; Buckheit, R.W., Jr.; Bringmann, G.; Schäffer, M.; Cragg, G.M. Anti-HIV michellamines from *Ancistrocladus korupensis*. *J. Med. Chem.* **1994**, *37*, 1740–1745. [CrossRef]
10. Houghton, P.J.; Woldemariam, T.Z.; Khan, A.I.; Burke, A.; Mahmood, N. Antiviral activity of natural and semi-synthetic chromone alkaloids. *Antivir. Res.* **1994**, *25*, 235–244. [CrossRef]
11. Bodiwala, H.S.; Sabde, S.; Mitra, D.; Bhutani, K.K.; Singh, I.P. Synthesis of 9-substituted derivatives of berberine as anti-HIV agents. *Eur. J. Med. Chem.* **2011**, *46*, 1045–1049. [CrossRef]
12. Wan, Z.; Lu, Y.; Liao, Q.; Wu, Y.; Chen, X. Fangchinoline inhibits human immunodeficiency virus type 1 replication by interfering with gp160 proteolytic processing. *PLoS ONE* **2012**, *7*, e39225. [CrossRef]
13. Varghese, F.S.; Kaukinen, P.; Gläsker, S.; Bespalov, M.; Hanski, L.; Wennerberg, K.; Kümmerer, B.M.; Ahola, T. Discovery of berberine, abamectin and ivermectin as antivirals against chikungunya and other alphaviruses. *Antivir. Res.* **2016**, *126*, 117–124. [CrossRef]
14. Liu, X.; Wang, Y.; Zhang, M.; Li, G.; Cen, Y. Study on the inhibitory effect of cepharanthine on herpes simplex type-1 virus (HSV-1) in vitro. *Zhong Yao Cai* **2004**, *27*, 107–110. [PubMed]
15. Chin, L.W.; Cheng, Y.W.; Lin, S.S.; Lai, Y.Y.; Lin, L.Y.; Chou, M.Y.; Chou, M.C.; Yang, C.C. Anti-herpes simplex virus effects of berberine from Coptidis rhizoma, a major component of a Chinese herbal medicine, Ching-Wei-San. *Arch. Virol.* **2010**, *155*, 1933–1941. [CrossRef] [PubMed]
16. Liou, J.T.; Chen, Z.Y.; Ho, L.J.; Yang, S.P.; Chang, D.M.; Liang, C.C.; Lai, J.H. Differential effects of triptolide and tetrandrine on activation of COX-2, NF-kappaB, and AP-1 and virus production in dengue virus-infected human lung cells. *Eur. J. Pharmacol.* **2008**, *589*, 288–298. [CrossRef]

17. Hung, T.C.; Jassey, A.; Liu, C.H.; Lin, C.J.; Lin, C.C.; Wong, S.H.; Wang, J.Y.; Yen, M.H.; Lin, L.T. Berberine inhibits hepatitis C virus entry by targeting the viral E2 glycoprotein. *Phytomedicine* **2019**, *53*, 62–69. [CrossRef]

18. Gao, Y.; Tai, W.; Wang, N.; Li, X.; Jiang, S.; Debnath, A.K.; Du, L.; Chen, S. Identification of novel natural products as effective and broad-spectrum anti-zika virus inhibitors. *Viruses* **2019**, *11*, 1019. [CrossRef]

19. Warowicka, A.; Nawrot, R.; Goździcka-Józefiak, A. Antiviral activity of berberine. *Arch. Virol.* **2020**, *165*, 1935–1945. [CrossRef]

20. Sakurai, Y.; Kolokoltsov, A.A.; Chen, C.C.; Tidwell, M.W.; Bauta, W.E.; Klugbauer, N.; Grimm, C.; Wahl-Schott, C.; Biel, M.; Davey, R.A. Ebola virus. Two-pore channels control Ebola virus host cell entry and are drug targets for disease treatment. *Science* **2015**, *347*, 995–998. [CrossRef]

21. Dang, Z.; Jung, K.; Zhu, L.; Lai, W.; Xie, H.; Lee, K.H.; Huang, L.; Chen, C.H. Identification and synthesis of quinolizidines with anti-influenza a virus activity. *ACS Med. Chem. Lett.* **2014**, *5*, 942–946. [CrossRef]

22. Dai, J.P.; Wang, Q.W.; Su, Y.; Gu, L.M.; Deng, H.X.; Chen, X.X.; Li, W.Z.; Li, K.S. Oxymatrine inhibits influenza a virus replication and inflammation via TLR4, p38 MAPK and NF-κB pathways. *Int. J. Mol. Sci.* **2018**, *19*, 965. [CrossRef]

23. Li, S.Y.; Chen, C.; Zhang, H.Q.; Guo, H.Y.; Wang, H.; Wang, L.; Zhang, X.; Hua, S.N.; Yu, J.; Xiao, P.G.; et al. Identification of natural compounds with antiviral activities against SARS-associated coronavirus. *Antivir. Res.* **2005**, *67*, 18–23. [CrossRef] [PubMed]

24. Kim, H.Y.; Shin, H.S.; Park, H.; Kim, Y.C.; Yun, Y.G.; Park, S.; Shin, H.J.; Kim, K. *In vitro* inhibition of coronavirus replications by the traditionally used medicinal herbal extracts, *Cimicifuga rhizoma*, *Meliae cortex*, *Coptidis rhizoma*, and *Phellodendron cortex*. *J. Clin. Virol.* **2008**, *41*, 122–128. [CrossRef] [PubMed]

25. Kim, D.E.; Min, J.S.; Jang, M.S.; Lee, J.Y.; Shin, Y.S.; Song, J.H.; Kim, H.R.; Kim, S.; Jin, Y.H.; Kwon, S. Natural bis-benzylisoquinoline alkaloids-tetrandrine, fangchinoline, and cepharanthine, inhibit human coronavirus OC43 infection of MRC-5 human lung cells. *Biomolecules* **2019**, *9*, 696. [CrossRef]

26. Fielding, B.C.; da Silva Maia Bezerra Filho, C.; Ismail, N.; Sousa, D.P. Alkaloids: Therapeutic potential against human coronaviruses. *Molecules* **2020**, *25*, 5496. [CrossRef] [PubMed]

27. Wu, R.; Wang, L.; Kuo, H.D.; Shannar, A.; Peter, R.; Chou, P.J.; Li, S.; Hudlikar, R.; Liu, X.; Liu, Z.; et al. An update on current therapeutic drugs treating COVID-19. *Curr. Pharmacol. Rep.* **2020**, *11*, 1–15. [CrossRef] [PubMed]

28. Della-Torre, E.; Della-Torre, F.; Kusanovic, M.; Scotti, R.; Ramirez, G.A.; Dagna, L.; Tresoldi, M. Treating COVID-19 with colchicine in community healthcare setting. *Clin. Immunol.* **2020**, *217*, 108490. [CrossRef]

29. Tan, G.T.; Pezzuto, J.M.; Kinghorn, A.D.; Hughes, S.H. Evaluation of natural products as inhibitors of human immunodeficiency virus type 1 (HIV-1) reverse transcriptase. *J. Nat. Prod.* **1991**, *54*, 143–154. [CrossRef]

30. Kim, S.Y.; Kim, H.; Kim, S.W.; Lee, N.R.; Yi, C.M.; Heo, J.; Kim, B.J.; Kim, N.J.; Inn, K.S. An effective antiviral approach targeting hepatitis B virus with NJK14047, a novel and selective biphenyl amide p38 mitogen-activated protein kinase inhibitor. *Antimicrob. Agents Chemother.* **2017**, *61*, e00214–e00217. [CrossRef]

31. Rey, F.A.; Lok, S.M. Common features of enveloped viruses and implications for immunogen design for next-generation vaccines. *Cell* **2018**, *172*, 1319–1334. [CrossRef]

32. Vigant, F.; Santos, N.C.; Lee, B. Broad-spectrum antivirals against viral fusion. *Nat. Rev. Microbiol.* **2015**, *13*, 426–437. [CrossRef]

33. Rout, J.; Swain, B.C.; Tripathy, U. *In silico* investigation of spice molecules as potent inhibitor of SARS-CoV-2. *J. Biomol. Struct. Dyn.* **2020**, 1–15. [CrossRef]

34. Junior, A.G.; Tolouei, S.; Dos Reis Lívero, F.A.; Gasparotto, F.; Boeing, T.; de Souza, P. Natural agents modulating ACE-2: A review of compounds with potential against SARS-CoV-2 infections. *Curr. Pharm. Des.* **2021**, *27*, 1588–1596. [CrossRef]

35. Lv, X.Q.; Zou, L.L.; Tan, J.L.; Li, H.; Li, J.R.; Liu, N.N.; Dong, B.; Song, D.Q.; Peng, Z.G. Aloperine inhibits hepatitis C virus entry into cells by disturbing internalisation from endocytosis to the membrane fusion process. *Eur. J. Pharmacol.* **2020**, *883*, 173323. [CrossRef]

36. Enkhtaivan, G.; Muthuraman, P.; Kim, D.H.; Mistry, B. Discovery of berberine based derivatives as anti-influenza agent through blocking of neuraminidase. *Bioorg. Med. Chem.* **2017**, *25*, 5185–5193. [CrossRef]

37. Buzón, V.; Cladera, J. Effect of cholesterol on the interaction of the HIV GP41 fusion peptide with model membranes. Importance of the membrane dipole potential. *Biochemistry* **2006**, *45*, 15768–15775. [CrossRef] [PubMed]

38. Barz, B.; Wong, T.C.; Kosztin, I. Membrane curvature and surface area per lipid affect the conformation and oligomeric state of HIV-1 fusion peptide: A combined FTIR and MD simulation study. *Biochim. Biophys. Acta* **2008**, *1778*, 945–953. [CrossRef]

39. Matsuda, K.; Hattori, S.; Komizu, Y.; Kariya, R.; Ueoka, R.; Okada, S. Cepharanthine inhibited HIV-1 cell-cell transmission and cell-free infection via modification of cell membrane fluidity. *Bioorg. Med. Chem. Lett.* **2014**, *24*, 2115–2117. [CrossRef] [PubMed]

40. Ashkenazi, A.; Viard, M.; Unger, L.; Blumenthal, R.; Shai, Y. Sphingopeptides: Dihydrosphingosine-based fusion inhibitors against wild-type and enfuvirtide-resistant HIV-1. *FASEB J.* **2012**, *26*, 4628–4636. [CrossRef] [PubMed]

41. Pattnaik, G.P.; Chakraborty, H. Coronin 1 derived tryptophan-aspartic acid containing peptides inhibit membrane fusion. *Chem. Phys. Lipids* **2018**, *217*, 35–42. [CrossRef]

42. Sardar, A.; Lahiri, A.; Kamble, M.; Mallick, A.I.; Tarafdar, P.K. Translation of mycobacterium survival strategy to develop a lipopeptide based fusion inhibitor. *Angew. Chem. Int. Ed. Engl.* **2021**, *60*, 6101–6106. [CrossRef]

43. Moreno, M.R.; Guillén, J.; Pérez-Berna, A.J.; Amorós, D.; Gómez, A.I.; Bernabeu, A.; Villalaín, J. Characterization of the interaction of two peptides from the N terminus of the NHR domain of HIV-1 gp41 with phospholipid membranes. *Biochemistry* **2007**, *46*, 10572–10584. [CrossRef]

44. Franquelim, H.G.; Veiga, A.S.; Weissmüller, G.; Santos, N.C.; Castanho, M.A. Unravelling the molecular basis of the selectivity of the HIV-1 fusion inhibitor sifuvirtide towards phosphatidylcholine-rich rigid membranes. *Biochim. Biophys. Acta* **2010**, *1798*, 1234–1243. [CrossRef] [PubMed]

45. Kalvodova, L.; Sampaio, J.L.; Cordo, S.; Ejsing, C.S.; Shevchenko, A.; Simons, K. The lipidomes of vesicular stomatitis virus, semliki forest virus, and the host plasma membrane analyzed by quantitative shotgun mass spectrometry. *J. Virol.* **2009**, *83*, 7996–8003. [CrossRef]

46. Merz, A.; Long, G.; Hiet, M.S.; Brügger, B.; Chlanda, P.; Andre, P.; Wieland, F.; Krijnse-Locker, J.; Bartenschlager, R. Biochemical and morphological properties of hepatitis C virus particles and determination of their lipidome. *J. Biol. Chem.* **2011**, *286*, 3018–3032. [CrossRef]

47. Gerl, M.J.; Sampaio, J.L.; Urban, S.; Kalvodova, L.; Verbavatz, J.M.; Binnington, B.; Lindemann, D.; Lingwood, C.A.; Shevchenko, A.; Schroeder, C.; et al. Quantitative analysis of the lipidomes of the influenza virus envelope and MDCK cell apical membrane. *J. Cell Biol.* **2012**, *196*, 213–221. [CrossRef]

48. Chicka, M.C.; Hui, E.; Liu, H.; Chapman, E.R. Synaptotagmin arrests the SNARE complex before triggering fast, efficient membrane fusion in response to Ca^{2+}. *Nat. Struct. Mol. Biol.* **2008**, *15*, 827–835. [CrossRef]

49. Lai, A.L.; Millet, J.K.; Daniel, S.; Freed, J.H.; Whittaker, G.R. The SARS-CoV fusion peptide forms an extended bipartite fusion platform that perturbs membrane order in a calcium-dependent manner. *J. Mol. Biol.* **2017**, *429*, 3875–3892. [CrossRef]

50. Yang, Q.; Guo, Y.; Li, L.; Hui, S.W. Effects of lipid headgroup and packing stress on poly(ethylene glycol)-induced phospholipid vesicle aggregation and fusion. *Biophys. J.* **1997**, *73*, 277–282. [CrossRef]

51. Lentz, B.R. PEG as a tool to gain insight into membrane fusion. *Eur. Biophys. J.* **2007**, *36*, 315–326. [CrossRef]

52. Khelashvili, G.; Plante, A.; Doktorova, M.; Weinstein, H. Ca^{2+}-dependent mechanism of membrane insertion and destabilization by the SARS-CoV-2 fusion peptide. *Biophys. J.* **2021**, *120*, 1105–1119. [CrossRef]

53. Montal, M.; Muller, P. Formation of bimolecular membranes from lipid monolayers and study of their electrical properties. *Proc. Nat. Acad. Sci. USA* **1972**, *65*, 3561–3566. [CrossRef] [PubMed]

54. Andersen, O.S.; Finkelstein, A.; Katz, I.; Cass, A. Effect of phloretin on the permeability of thin lipid membranes. *J. Gen. Physiol.* **1976**, *67*, 749–771. [CrossRef] [PubMed]

55. Folli, C.; Calderone, V.; Ottonello, S.; Bolchi, A.; Zanotti, G.; Stoppini, M.; Berni, R. Identification, retinoid binding, and x-ray analysis of a human retinol-binding protein. *Proc. Natl. Acad. Sci. USA* **2001**, *98*, 3710–3715. [CrossRef] [PubMed]

56. Martín-Acebes, M.A.; Merino-Ramos, T.; Blázquez, A.B.; Casas, J.; Escribano-Romero, E.; Sobrino, F.; Saiz, J.C. The composition of West Nile virus lipid envelope unveils a role of sphingolipid metabolism in flavivirus biogenesis. *J. Virol.* **2014**, *88*, 12041–12054. [CrossRef] [PubMed]

57. Akimov, S.A.; Molotkovsky, R.J.; Kuzmin, P.I.; Galimzyanov, T.R.; Batishchev, O.V. Continuum models of membrane fusion: Evolution of the theory. *Int. J. Mol. Sci.* **2020**, *21*, 3875. [CrossRef]

58. Ketter, E.; Randall, G. Virus impact on lipids and membranes. *Annu. Rew. Virol.* **2019**, *6*, 319–340. [CrossRef]

59. de Lima, V.R.; Caro, M.S.; Munford, M.L.; Desbat, B.; Dufourc, E.; Pasa, A.A.; Creczynski-Pasa, T.B. Influence of melatonin on the order of phosphatidylcholine-based membranes. *J. Pineal Res.* **2010**, *49*, 169–175. [CrossRef]

60. Torrecillas, A.; Schneider, M.; Fernández-Martínez, A.M.; Ausili, A.; de Godos, A.M.; Corbalán-García, S.; Gómez-Fernández, J.C. Capsaicin fluidifies the membrane and localizes itself near the lipid-water interface. *ACS Chem. Neurosci.* **2015**, *6*, 1741–1750. [CrossRef]

61. Lebecque, S.; Crowet, J.M.; Lins, L.; Delory, B.M.; du Jardin, P.; Fauconnier, M.L.; Deleu, M. Interaction between the barley allelochemical compounds gramine and hordenine and artificial lipid bilayers mimicking the plant plasma membrane. *Sci. Rep.* **2018**, *8*, 9784. [CrossRef]

62. Ashrafuzzaman, M. Amphiphiles capsaicin and triton X-100 regulate the chemotherapy drug colchicine's membrane adsorption and ion pore formation potency. *Saudi J. Biol. Sci.* **2021**, *28*, 3100–3109. [CrossRef] [PubMed]

63. Ashrafuzzaman, M.; Tseng, C.Y.; Duszyk, M.; Tuszynski, J.A. Chemotherapy drugs form ion pores in membranes due to physical interactions with lipids. *Chem. Biol. Drug Des.* **2012**, *80*, 992–1002. [CrossRef] [PubMed]

64. Pentak, D. *In vitro* spectroscopic study of piperine-encapsulated nanosize liposomes. *Eur. Biophys. J.* **2016**, *45*, 175–186. [CrossRef]

65. Zidovetzki, R.; Sherman, I.W.; Atiya, A.; De Boeck, H. A nuclear magnetic resonance study of the interactions of the antimalarials chloroquine, quinacrine, quinine and mefloquine with dipalmitoylphosphatidylcholine bilayers. *Mol. Biochem. Parasitol.* **1989**, *35*, 199–207. [CrossRef]

66. Swain, J.; Kumar Mishra, A. Location, partitioning behavior, and interaction of capsaicin with lipid bilayer membrane: Study using its intrinsic fluorescence. *J. Phys. Chem. B* **2015**, *119*, 12086–12093. [CrossRef]

67. Efimova, S.S.; Zakharova, A.A.; Ostroumova, O.S. Alkaloids modulate the functioning of ion channels produced by antimicrobial agents via an influence on the lipid host. *Front. Cell Dev. Biol.* **2020**, *8*, 537. [CrossRef]

68. Taylor, K.M.; Morris, R.M. Thermal analysis of phase transition behaviour in liposomes. *Thermochim. Acta* **1995**, *248*, 289–301. [CrossRef]

69. McElhaney, R.N. The use of differential scanning calorimetry and differential thermal analysis in studies of model and biological membranes. *Chem. Phys. Lipids* **1982**, *30*, 229–259. [CrossRef]

70. Zakaria, M.Y.; Fayad, E.; Althobaiti, F.; Zaki, I.; Abu Almaaty, A.H. Statistical optimization of bile salt deployed nanovesicles as a potential platform for oral delivery of piperine: Accentuated antiviral and anti-inflammatory activity in MERS-CoV challenged mice. *Drug Deliv.* **2021**, *28*, 1150–1165. [CrossRef]

71. Hegeto, L.A.; Caleffi-Ferracioli, K.R.; Nakamura-Vasconcelos, S.S.; Almeida, A.L.; Baldin, V.P.; Nakamura, C.V.; Siqueira, V.L.D.; Scodro, R.B.L.; Cardoso, R.F. In vitro combinatory activity of piperine and anti-tuberculosis drugs in *Mycobacterium tuberculosis*. *Tuberculosis (Edinb.)* **2018**, *111*, 35–40. [CrossRef] [PubMed]

72. Primary qHTS to Identify Inhibitors of SARS-CoV-2 Cell Entry. Available online: https://pubchem.ncbi.nlm.nih.gov/bioassay/1645846 (accessed on 14 July 2021).

73. Miryan, M.; Bagherniya, M.; Sahebkar, A.; Soleimani, D.; Rouhani, M.H.; Iraj, B.; Askari, G. Effects of curcumin-piperine co-supplementation on clinical signs, duration, severity, and inflammatory factors in patients with COVID-19: A structured summary of a study protocol for a randomised controlled trial. *Trials* **2020**, *21*, 1027. [CrossRef]

74. Askari, G.; Alikiaii, B.; Soleimani, D.; Sahebkar, A.; Mirjalili, M.; Feizi, A.; Iraj, B.; Bagherniya, M. Effect of curcumin-pipeine supplementation on clinical status, mortality rate, oxidative stress, and inflammatory markers in critically ill ICU patients with COVID-19: A structured summary of a study protocol for a randomized controlled trial. *Trials* **2021**, *22*, 434. [CrossRef] [PubMed]

75. Pawar, K.S.; Mastud, R.N.; Pawar, S.K.; Pawar, S.S.; Bhoite, R.R.; Bhoite, R.R.; Kulkarni, M.V.; Deshpande, A.R. Oral curcumin with piperine as adjuvant therapy for the treatment of COVID-19: A randomized clinical trial. *Front. Pharmacol.* **2021**, *12*, 669362. [CrossRef] [PubMed]

biomedicines

Article

Oral Capsaicinoid Administration Alters the Plasma Endocannabinoidome and Fecal Microbiota of Reproductive-Aged Women Living with Overweight and Obesity

Claudia Manca [1,2,3,†,‡], Sébastien Lacroix [2,4,†], Francine Pérusse [4,5], Nicolas Flamand [1,2,3], Yvon Chagnon [1,2,3], Vicky Drapeau [3,4,6], Angelo Tremblay [3,4,5], Vincenzo Di Marzo [1,2,3,4,7,8,*] and Cristoforo Silvestri [1,2,3,4,*]

[1] Département de Médecine, Faculté de Médecine, Université Laval, Québec, QC G1V 0A6, Canada; claumanca@unica.it (C.M.); nicolas.flamand@criucpq.ulaval.ca (N.F.); yvon-c.chagnon.1@ulaval.ca (Y.C.)
[2] Canada Excellence Research Chair on the Microbiome-Endocannabinoidome Axis in Metabolic Health, Québec, QC G1V 4G5, Canada; sebastien.lacroix.8@ulaval.ca
[3] Institut Universitaire de Cardiologie et de Pneumologie de Québec, Université Laval, Québec, QC G1V 4G5, Canada; vicky.drapeau@fse.ulaval.ca (V.D.); angelo.tremblay@kin.ulaval.ca (A.T.)
[4] Nutrition, Santé et Société (NUTRISS), Institut sur la Nutrition et les Aliments Fonctionnels (INAF), Université Laval, Québec, QC G1V 0A6, Canada; francine.perusse@kin.ulaval.ca
[5] Département de Kinésiologie, Faculté de Médecine, Université Laval, Québec, QC G1V 0A6, Canada
[6] Département d'éducation Physique, Faculté des Sciences de l'éducation, Université Laval, Québec, QC G1V 0A6, Canada
[7] Faculté des Sciences de l'Agriculture et de l'Alimentation, Université Laval, Québec, QC G1V 0A6, Canada
[8] Institute of Biomolecular Chemistry, National Research Council, 80078 Pozzuoli, Italy
*   Correspondence: vincenzo.dimarzo@criucpq.ulaval.ca (V.D.M.); cristoforo.silvestri@criucpq.ulaval.ca (C.S.); Tel.: +418-656-8711 (ext. 7263) (V.D.M.); +418-656-8711 (ext. 7229) (C.S.)
†   These authors contributed equally.
‡   Current address: Department of Biomedical Science, University of Cagliari, 09042 Monserrato, Italy.

Citation: Manca, C.; Lacroix, S.; Pérusse, F.; Flamand, N.; Chagnon, Y.; Drapeau, V.; Tremblay, A.; Di Marzo, V.; Silvestri, C. Oral Capsaicinoid Administration Alters the Plasma Endocannabinoidome and Fecal Microbiota of Reproductive-Aged Women Living with Overweight and Obesity. *Biomedicines* **2021**, *9*, 1246. https://doi.org/10.3390/biomedicines 9091246

Academic Editor: Pavel B. Drašar

Received: 30 July 2021
Accepted: 13 September 2021
Published: 17 September 2021

Publisher's Note: MDPI stays neutral with regard to jurisdictional claims in published maps and institutional affiliations.

Abstract: Capsaicinoids, the pungent principles of chili peppers and prototypical activators of the transient receptor potential of the vanilloid type-1 (TRPV1) channel, which is a member of the expanded endocannabinoid system known as the endocannabinoidome (eCBome), counteract food intake and obesity. In this exploratory study, we examined the blood and stools from a subset of the participants in a cohort of reproductive-aged women with overweight/obesity who underwent a 12-week caloric restriction of 500 kcal/day with the administration of capsaicinoids (two capsules containing 100 mg of a capsicum annuum extract (CAE) each for a daily dose of 4 mg of capsaicinoids) or a placebo. Samples were collected immediately before and after the intervention, and plasma eCBome mediator levels (from 23 participants in total, 13 placebo and 10 CAE) and fecal microbiota taxa (from 15 participants in total, 9 placebo and 6 CAE) were profiled using LC–MS/MS and 16S metagenomic sequencing, respectively. CAE prevented the reduced caloric-intake-induced decrease in beneficial eCBome mediators, i.e., the TRPV1, GPR119 and/or PPARα agonists, *N*-oleoyl-ethanolamine, *N*-linoleoyl-ethanolamine and 2-oleoyl-glycerol, as well as the anti-inflammatory *N*-acyl-ethanolamines *N*-docosapentaenyl-ethanolamine and *N*-docosahexaenoyl-ethanolamine. CAE produced few but important alterations in the fecal microbiota, such as an increased relative abundance of the genus *Flavonifractor*, which is known to be inversely associated with obesity. Correlations between eCBome mediators and other potentially beneficial taxa were also observed, thus reinforcing the hypothesis of the existence of a link between the eCBome and the gut microbiome in obesity.

Keywords: capsaicinoids; endocannabinoidome; microbiota; overweight; obesity; food intake; lipidomics; metabolism

39

1. Introduction

Non-communicable diseases, such as heart disease, stroke, cancer, chronic respiratory diseases and diabetes, are the leading causes of mortality in the world [1]. Among the common, modifiable risk factors, such as alcohol consumption, unhealthy diets, insufficient physical activity, increased blood pressure, blood sugar and cholesterol, obesity represents a growing major public health problem in most countries, with an incidence that has nearly tripled since 1975 [2]. As such, obesity (particularly visceral obesity in subjects living with overweight) and the related comorbidities impose an enormous burden at the individual, public health and economic levels [3].

Notwithstanding the idea that some nutrients are more obesogenic than others, overweight and obesity occur when energy intake exceeds energy expenditure, causing a state of positive energy balance and a consequent increase in body mass, of which 60–80 percent is usually body fat [4]. Since a positive energy balance is a problem for many people in Western countries, a healthy lifestyle, including a balanced diet, is necessary to prevent or reverse an excess energy intake over expenditure. However, due to the multifactorial nature of obesity—encompassing hereditary factors, ethnic differences, environment and individual behavior—efficient strategies to regulate energy intake that lead to significant and sustained weight loss are not well defined, and, to date, no clear non-surgical solution has been developed [4,5]. Indeed, pharmaceutical drugs, such as orlistat, which is a potent and specific inhibitor of intestinal lipases that can reduce body weight with variable efficacy, can lead to gastrointestinal, hepatic and kidney injuries [6,7]. Recently, the glucagon-like peptide-1 analog semaglutide, in conjunction with calorie restriction, induced significantly more weight loss than a placebo [8]. Other anti-obesity drugs were withdrawn from the market due to severe adverse effects, including increased cardiovascular risks, mood disorders and even suicidal susceptibility [9]. Increasing efforts are therefore aimed at developing novel strategies, including active natural ingredients and functional foods, which could facilitate appetite control, to promote favorable changes in body energy stores and exert long-term benefits on metabolic health [6].

Among a plethora of identified natural anti-obesity compounds, such as celastrol (from the roots of the thunder god vine) [10], the stilbenoid resveratrol (from grapes and red wine), genistein (an isoflavone from soy), glycyrrhizin (from licorice), quercetin, ethanolic extracts from ginseng roots and green tea extract [11], chili pepper, with its more than 200 active constituents, was suggested to display energy-expenditure-stimulating, anorexigenic, anti-hypertensive, vasodilatory, cardioprotective and, finally, anti-obesity actions [12]. In particular, capsaicin, a non-nutritional food constituent and the main pungent principle in hot red peppers and its major active compound, seems to have multiple beneficial effects in the treatment of pain, inflammation, cancer and oxidative stress, among others [12–17], and was suggested as an anti-obesity therapeutic [18]. Capsaicin was observed to decrease appetite and subsequent protein and fat intake in Japanese females and to lower energy intake in Caucasian males [19]. These effects were potentially related to an increase in sympathetic nervous system (SNS) activity. This was in agreement with the observation that pre-prandial intake of capsaicin, be it in capsules or diluted in tomato juice, increases satiety, thus reducing energy intake [20]. Capsaicin was also found to decrease energy intake when tested in combination with caffeine [21]. Epidemiological evidence suggested that the incidence of obesity is lower in individuals who consume capsaicinoid-containing foods [22]. Furthermore, post-weight loss fat oxidation following a 4-week very low energy dietary intervention was better maintained over 3 months by individuals that were supplemented with capsaicin [23]. Finally, a recent meta-analysis of controlled human trials evidenced how dietary capsaicinoids produce beneficial effects on metabolic parameters, such as serum total cholesterol levels, also in non-obese subjects and independently from effects on BMI and body fat [24].

Capsaicin's positive metabolic effects appear to be due, at least in part, to the modulation of the transient receptor potential vanilloid subtype 1 (TRPV1) channel, which plays a critical role in the regulation of metabolic health, including body weight, glucose and lipid metabolism, white and brown adipose tissue biology and the cardiovascular system [25,26]. It was demonstrated that the activation of TRPV1 by capsaicin can attenuate abnormal glucose homeostasis by stimulating insulin secretion and increase glucagon-like peptide 1 (GLP-1) levels [27,28]. Although it is well accepted that much of the effect is caused by stimulation of the TRPV1 receptor, other mechanisms of action were described. Capsaicin can indeed play its role in a receptor-independent manner through the inactivation of nuclear factor κB (NF-κB) and activation of peroxisome proliferator-activated receptor γ (PPARγ) [29]. Another mechanism is represented by the regulation of the production of the endocannabinoids (eCBs), which are lipid mediators that act as major regulators of energy homeostasis [30–32] and, together with a plethora of eCB-related molecules, receptors and metabolic enzymes, constitute the endocannabinoidome (eCBome), a complex lipid-signaling system [33]. Some of these molecules, such as long-chain unsaturated mono-acyl-glycerols (MAGs) and, particularly, *N*-acyl-ethanolamines (NAEs), are well-established endogenous activators of TRPV1 channels [34,35]. Interestingly, the eCBome is increasingly understood to communicate with the gut microbiome in the context of metabolic disorders [36–38]. Similarly, capsaicin modulates the gut microbiota population, which represents a critical factor for the anti-obesity effects that are exerted by this natural compound, contributing to improving glucose homeostasis through increasing short-chain fatty acids, regulating gastrointestinal hormones and inhibiting pro-inflammatory cytokines [39,40]. A very recent study suggested that such effects might be at least in part mediated by the activation of TRPV1 since the ablation of this ligand-activated channel from the rat intestine was found to cause gut dysbiosis, as evidenced by the decreased abundance of beneficial taxa and production of short-chain fatty acids, and by the impairment of colonic mucus secretion [41].

These elements prompted us to investigate the impact of the intake of capsaicinoids (i.e., capsaicin and its TRPV1-active analogs that are present in *Capsicum annuum*) on the plasma eCBome and gut microbiome in individuals with overweight/obesity. For this purpose, we took advantage of the availability of a limited number of blood and fecal samples from a clinical trial where the effect of a combination of dietary capsaicin (provided in the form of *Capsicum annuum* extract (CAE) capsules) with a low-caloric intervention was tested on appetite sensations, energy intake and expenditure, body weight and fat distribution in free-living, physically non-trained reproductive-aged women with overweight and obesity.

2. Materials and Methods

2.1. Ethical Approval

This study was conducted according to the guidelines laid down in the Declaration of Helsinki and all procedures involving human subjects were approved by the Research Ethics Committee of Health Sciences of Laval University (protocol 2015-041 A-4 R-2/01-05-2018). Written informed consent was obtained from all the participants before entering the study. The clinical trial was registered in the public trials registry ClinicalTrials.gov (accessed on 30 July 2021) (registration identification number NCT04874701).

2.2. Subjects and Treatment Protocol

Sixty-one reproductive-aged women with overweight or obesity ($25 \, \text{kg/m}^2 \leq \text{BMI} \leq 35 \, \text{kg/m}^2$) aged between 18 and 50 years had their health status assessed by a physician to confirm their eligibility against inclusion/exclusion criteria. Eligible volunteers could not be pregnant, had to have stable body weight, needed to be free from chronic diseases, could not take any medication or have restrictive dietary habits or food allergies. Sixty-one volunteers were enrolled and underwent a first session (visit 1 (V1)), during which overnight-fasted participants were assessed for their resting metabolic rate and substrate oxidation. Blood and stool samples were collected, preferably before breakfast. Volunteers were randomly distributed into two groups: the CAE group (30 participants that took two CAE capsules per day (100 mg *Capsicum annuum*/capsule containing 2% capsaicinoids; 1.2% capsaicin,

0.7% dihydrocapsaicin and 0.1% nordihydrocapsaicin; OmniActive Health Technologies)) and the placebo group (31 participants). The capsules were taken at breakfast. The composition of CAE and placebo capsules are detailed in Table 1.

Table 1. Composition of the OmniActive Health Technologies CAE and placebo capsules.

	OmniActive Health Technologies CAE Capsules	Placebo Capsules
Medicinal ingredient	*Capsicum annuum* extract (100 mg)	Corn starch (230 mg)
Extract (per capsule)	15:1 DHE 1.5 mg (dry herb)	
Potency (per capsule)	0.7% Dihydrocapsaicin, 0.1% nordihydrocapsaicin and 1.2% capsaicin (2% capsaicinoids)	
Non-medicinal ingredients	Cellulose gum, hypromellose	Cellulose gum, hypromellose
Pharmaceutical glaze	Sucrose	Sucrose

Each participant received a 12-week personalized dietary plan with a 500 kcal/day energy restriction (RMR × physical activity level − 500 kcal). Specifically, in order to respect the restriction, the volunteers were provided with a precise meal plan. Every two weeks, the participants met the dietitian in order to ensure adequate compliance with the prescribed diet, to measure their heart rate and blood pressure, to verify the adequacy of capsule consumption and to provide motivational support. After the 12 weeks of supplementation (visit 7 (V7)), the subjects participated in a second testing session that was the same as the first one. Body composition (body fat and fat-free mass) was measured using dual X-ray absorptiometry at the beginning and the end of the program. Seventeen participants did not complete the study, resulting in a dropout rate of 27%. Furthermore, one subject displayed a body weight gain of 8 kg, which was considered as an unacceptable outlier from a clinical and a statistical standpoint and was removed from the analysis. Twenty-two placebo- and 21 CAE-treated women completed the study. However, only a limited subset of participants could be subjected to an assessment of fecal microbiota composition (15 participants in total, 9 placebo and 6 CAE) and plasma eCBome mediator levels (23 participants in total, 13 placebo and 10 CAE) at V1 and V7 due to the fact that such measurements were introduced in the protocol only after the start of the study (see Table S1 for an overview of CAE and placebo samples).

2.3. Assessment of Anthropometric Parameters

Height, weight and waist circumference were measured at baseline and every two weeks over the course of the intervention. Height measurements were performed using a Stadiometer Holtain. Body weight was measured with a Tanita Body Composition Monitor scale and waist circumference (located between the bottom of the ribs and the iliac crest) was measured using a tape graduated in centimeters.

2.4. Measurement of Appetite Sensations and Eating Behaviors

Appetite sensations (hunger, desire to eat, fullness, prospective food consumption) were assessed using a visual analog scale (VAS) before a standardized breakfast containing 599 kcal was offered using the procedures of Drapeau et al. [42]. Participants then received two CAE capsules or a placebo and were then monitored for appetite sensations using the VAS, calorimetry measurements and blood pressure every 30 min during a 180-min period. After 3.5 h, participants were offered a lunch buffet that was used to measure ad libitum energy and macronutrient intake [43]. The VAS was used to measure appetite every 30 min during a 240 min period following consumption of the meal. At the end of the afternoon, the participants were invited to consume an ad libitum standardized dinner. Appetite and blood pressure were measured before and immediately after consumption of the meal. These procedures were repeated on the last visit following the 12-week intervention period.

As for the first testing session of the protocol, which assessed the effects of a standardized breakfast on post-prandial appetite sensations, satiety was measured as a score of post-prandial appetite feelings. As for the second testing session, which was aimed at evaluating the buffet-type meal served at lunchtime, as well as variations in appetite sensations in the afternoon and energy/macronutrient intake at dinner time, satiety was mostly measured as the ad libitum energy intake at lunchtime. Since the ad libitum energy intake was not the same for each subject, we calculated the satiety quotient, which was determined as the change in afternoon appetite sensations relative to the ad libitum energy intake at lunchtime [42,44]. These observations were completed using the measurement of energy intake at dinner time, which provided a more complete outcome on cumulative energy intake at a time that CAE was entirely made available.

2.5. Measurement of Eating Behavior

Eating behaviors were assessed at baseline and at 12 weeks with the 51-item Three-Factor Eating Questionnaire (TFEQ), which measures habitual cognitive restraint, disinhibition and susceptibility for hunger [45], and with the Food Cravings Questionnaire [46].

2.6. Lipid Extraction and HPLC–MS/MS for the Analysis of eCBome Mediators

eCBome mediators were extracted from V1 and V7 plasma samples of a total of 23 participants (13 placebo and 10 CAE). Sample processing and LC–MS/MS parameters were exactly as described by Turcotte and collaborators [47]. The quantification of eCBome-related mediators (Table S2) was carried out by an HPLC system interfaced with the electrospray source of a Shimadzu 8050 triple quadrupole mass spectrometer and using multiple reaction monitoring in positive ion mode for the compounds and their deuterated homologs.

In the case of unsaturated monoacylglycerols, the data are presented as 1/2-monoacylglycerols (2-MAGs) but represent the combined signals from the 2- and 1(3)-isomers since the latter were most likely generated from the former via acyl migration from the *sn*-2 to the *sn*-1 or *sn*-3 position.

2.7. DNA Extraction and 16S rRNA Gene Sequencing

DNA was extracted from the V1 and V7 fecal samples of a total of 18 participants (12 placebo and 6 CAE) using the QIAmp PowerFecal DNA kit (Qiagen, Hilden, Germany) according to the manufacturers' instructions. The DNA concentrations of the extracts were measured fluorometrically with the Quant-iT PicoGreen dsDNA Kit (Thermo Fisher Scientific, Waltham, MA, USA) and the DNA samples were stored at $-20\,^{\circ}$C until the 16S rDNA library preparation. Briefly, 1 ng of DNA was used as a template and the V3–V4 region of the 16S rRNA gene was amplified by polymerase chain reaction (PCR) using the QIAseq 16S Region Panel protocol in conjunction with the QIAseq 16S/ITS 384-Index I (Sets A, B, C, D) kit (Qiagen, Hilden, Germany). The 16S metagenomic libraries were eluted in 30 µL of nuclease-free water and 1 µL was qualified with a Bioanalyser DNA 1000 Chip (Agilent, Santa Clara, CA, USA) to verify the amplicon size (expected size ~600 bp) and quantified with a Qubit (Thermo Fisher Scientific, Waltham, MA, USA). Libraries were then normalized and pooled to 4 nM, denatured and diluted to a final concentration of 10 pM and supplemented with 5% PhiX control (Illumina, San Diego, CA, USA). Sequencing (2×300 bp paired-end) was performed using the MiSeq Reagent Kit V3 (600 cycles) on an Illumina MiSeq System. Data were processed using the DADA2 pipeline [48] and assignation to bacterial taxa was obtained using the Silva v132 reference database. Sequences that were present in fewer than 3 samples were filtered out and bacterial abundances were normalized using the Cumulative Sum Scaling (CSS, *MetagenomeSeq* R package) [49]. Alpha-diversity indices were obtained using the *Phyloseq* R package.

2.8. Statistical Analyses

Values are presented as mean ± standard error. The effects of interventions on eCBome mediators were assessed using Student's *t*-test comparing the pre- and post-intervention (i.e., visit 1 (V1) and visit 7 (V7), respectively) levels, for placebo and CAE individually. Post-intervention levels were also compared between placebo and CAE groups. Variations in fecal microbiota compositions were evaluated using two-way analysis of variance (ANOVA) and Tukey's HSD post hoc tests to assess the treatment (placebo vs. CAE), time (before vs. after dietary intervention, i.e., visit 1 (V1) and visit 7 (V7), respectively) and treatment × time interactions. Correlations between the eCBome mediators, microbial taxa, anthropometric parameters and appetite sensations were obtained using repeated measures correlation tests (*rmcorr* package), which accounted for the within-individual relationships between paired measures [50]. A *p*-value of 0.05 was selected as reflecting a statistically significant effect.

3. Results

3.1. Trial Completion and Side Effects

As shown in Table S1, 43 participants completed the trial and 17 dropped out of the study. Out of the 61 participants, 39 (14 placebo and 25 CAE) reported symptoms. Table S1 presents the list of symptoms reported by the two groups of participants. Most side-effects were minor and related to gastrointestinal disturbance and were more frequent in the CAE group. One subject experienced an adverse event that necessitated hospitalization. Of the 17 participants that dropped out of the study (6 placebo and 11 CAE), only three did so because of side effects, whilst the other dropouts were due to personal reasons.

3.2. Effect of CAE and Caloric Restriction on Body Weight, BMI, Waist Circumference and Fat Mass

Full details of the effects of CAE on these endpoints, which were assessed in the whole cohort, will be reported elsewhere. For the purpose of this article, we believe that it will suffice to state that caloric restriction, alone or supplemented with CAE, induced significant reductions in body weight, BMI, waist circumference and fat mass percentage from baseline to V7. Reductions of anthropometric variables associated with CAE were non-significantly different than for the placebo treatment: body weight (↓ 3.81% vs. ↓ 3.08%), BMI (↓ 3.97% vs. ↓ 2.34%), waist circumference (↓ 2.83% vs. ↓ 2.18%) and fat mass percentage (↓ 4.03% vs. ↓ 3.20%), though trends toward significance were observed for decreased body weight (*p* = 0.1) and fat-free mass (*p* = 0.08) (Table 2).

Table 2. Response to weight loss and CAE interventions. The *t*-test (paired where appropriate) *p*-values were computed for selected anthropometric parameters (BMI, body weight, waist circumference, fat mass percentage, fat-free mass); ns: not statistically significant.

Variable	Treatment	V1		V7		*p*-Value	*p*-Value	*p*-Value
		Mean	**SE**	**Mean**	**SE**	**Treatment**	**Time**	**Interaction**
BMI	Placebo	29.47	0.7	28.78	0.7	ns	0.0001	ns
	CAE	28.99	0.7	27.84	0.8			
Body weight (kg)	Placebo	81.2	2.1	78.7	2.1	0.1	<0.0001	0.1
	CAE	76.1	2.4	73.2	2.4			
Waist Circumference (cm)	Placebo	91.8	1.8	89.8	1.8	ns	<0.0001	ns
	CAE	88.4	2.1	85.9	2.1			
Fat mass (%)	Placebo	40.6	0.9	39.3	0.9	ns	<0.0001	ns
	CAE	39.7	1.1	38.1	1.1			
Fat-free mass (kg)	Placebo	45.5	1.0	45.2	1.0	0.08	0.02	ns (0.08)
	CAE	43.1	1.1	42.5	1.1			

More importantly for the purpose of this article, in the subset of patients whose plasma eCBome mediator levels and fecal microbiota composition were determined, waist circumferences were reduced in both the placebo and CAE groups ($\downarrow$ 1.98 $\pm$ 0.65 and $\downarrow$ 1.45 $\pm$ 0.77, respectively; p = 0.009 vs. V1 for both), while body weight and BMI were non-significantly different from baseline. Interestingly, only within the CAE group was the fat mass percentage was reduced ($\downarrow$ 3.73 $\pm$ 1.25, p = 0.01) and fat-free mass increased ($\uparrow$ 2.47 $\pm$ 0.79, p = 0.01).

3.3. Effect of CAE and Caloric Restriction on Measures of Appetite Sensation and Eating Behavior

Full details of the effects of CAE on these endpoints, which were assessed in the whole cohort, will be reported elsewhere. For the purpose of this article, we believe that it will suffice to state that caloric-restriction-induced weight loss caused an increase in self-administered VAS-derived desire to eat, hunger and prospective food consumption. However, no additional effect could be observed with CAE (Table 3). No significant effect of either caloric restriction alone or supplementation with CAE was observed using the TFEQ hunger subscale (TFEQ-HUN) (Table 4).

Table 3. Effects of weight loss and CAE on pre- and postprandial appetite sensation, as assessed using the visual analogous scale. Desire to eat (DE), hunger (Hun), satiety (Sat) and prospective food consumption (PFC) were evaluated before (V1) and after the interventions (V7). The area under the curve (AUC) was measured for the first hour (60 min) and the subsequent two hours (60–180 min) following the breakfast test; ns: not statistically significant.

Variable	Treatment	V1		V7		*p*-Value	*p*-Value	*p*-Value
		Mean	SE	Mean	SE	Treatment	Time	Interaction
DE_60 (mm)	Placebo	1576	277	1792	281	ns	0.1	ns
	CAE	1265	311	1678	311			
DE_180 (mm)	Placebo	5308	551	6253	560	0.07	0.02	ns
	CAE	3842	619	5017	619			
Hun_AUC_60	Placebo	1246	262	1775	267	ns	0.08	ns
	CAE	1122	294	1401	294			
Hun_AUC60_180	Placebo	4800	573	5832	583	ns	0.04	ns
	CAE	4225	644	5023	644			
Sat_AUC_60	Placebo	6536	299	6353	304	ns	ns	ns
	CAE	6596	336	6762	336			
Sat_AUC60_180	Placebo	10,698	651	10,028	659	ns	ns	ns
	CAE	10,846	732	11,116	732			
PFC_AUC_60	Placebo	2032	293	2190	297	ns	ns	ns
	CAE	1786	329	1941	329			
PFC_AUC60_180	Placebo	6049	539	6818	548	ns	0.04	ns
	CAE	5102	606	6026	606			

Table 4. Effects of weight loss and CAE supplementation on feelings of hunger, as measured with the Three-Factor Eating Questionnaire (TFEQ) before (V1) and after the interventions (V7). Note that this table includes the TFEQ values from all individuals that completed the trial; ns: not statistically significant.

Variable	Treatment	V1		V7		*p*-Value	*p*-Value	*p*-Value
		Mean	SE	Mean	SE	Treatment	Time	Interaction
TFEQ_hunger	Placebo	5.96	0.64	4.17	0.64	ns	ns	0.09
	CAE	5.27	0.72	5.75	0.73			

We then sought to investigate whether these observations could also be reported in the subset of participants that had both their plasma eCBome and fecal microbiota composition characterized (Table 5). Pre-meal evaluation of the amount of food that could be eaten (DQuaN) was drastically increased by weight loss that was either induced by caloric restriction alone or in conjunction with CAE supplementation. The pre-meal feeling of hunger (BSenF) increased during caloric restriction-induced weight loss (placebo V1 vs. V7). The CAE supplementation did not seem to influence any subjective measures of hunger or satiety. Conversely, susceptibility to hunger, as derived from the TFEQ hunger subscale (TFEQ-HUN) was reduced by caloric restriction only (placebo V1 vs. V7), which was an effect that seemed to be due to reduced feelings of hunger arising from internal sources (TFEQ-HIL).

Table 5. Effects of weight loss and CAE supplementation on appetite sensations in self-administered VAS-derived and Three-Factor Eating Questionnaire (TFEQ) measured variables before (V1) and after the interventions (V7). DQuaN, pre-meal evaluation of the amount of food that could be eaten; BEnvM, pre-meal desire to eat; BSenF, feeling of hunger; TFEQ-HUN, habitual feelings of hunger; TFEQ-HEL, feelings of hunger arising from external sources; TFEQ-HIL, feelings of hunger arising from internal sources. This table represents the measurements made in the subset of participants that had both their plasma eCBome and fecal microbiota composition characterized. Effects of the Capsimax supplementation on appetite sensations were measured before (0) and 30 or 180 min following breakfast; ns: not statistically significant.

Variable	Treatment	V1		V7		*p*-Value	*p*-Value	*p*-Value
		Mean	SE	Mean	SE	Treatment	Time	Interaction
DQuaN0	Placebo	28.1	5.0	104.6	4.9	ns	<0.001	ns
	CAE	26.6	5.9	99.8	4.7	ns	<0.001	ns
DQuaN180	Placebo	78.3	6.8	80.5	4.8	ns	ns	ns
	CAE	70.6	6.8	76.7	5.2	ns	ns	ns
BEnvM_30	Placebo	85.9	6.9	99.9	3.6	ns	ns	ns
	CAE	82.6	7.2	84.5	6.7	ns	ns	ns
BEnvM180	Placebo	39.2	5.9	40.5	4.7	ns	ns	ns
	CAE	43.5	6.5	42.2	6.8	ns	ns	ns
BSenF_30	Placebo	81.3	7.1	94.4	5.6	ns	ns	ns
	CAE	86.7	6.9	80.8	6.0	ns	ns	ns
BSenF180	Placebo	34.2	5.1	42.7	4.5	ns	ns	ns
	CAE	47.2	6.6	39.7	5.4	ns	ns	ns
TFEQ-HUN	Placebo	5.5	0.7	4.2	0.5	ns	0.04	ns
	CAE	5.7	0.7	5.9	0.9	ns	ns	ns
TFEQ-HIL	Placebo	1.8	0.4	1.3	0.3	ns	0.05	ns
	CAE	1.8	0.4	1.9	0.5	ns	ns	ns
TFEQ-HEL	Placebo	2.4	0.3	1.8	0.3	ns	ns	ns
	CAE	2.7	0.3	2.7	0.4	ns	ns	ns

3.4. Effect of CAE and Caloric Restriction on Plasma ECBome Mediator Levels

As mentioned above, this endpoint could be measured only in a relatively small subset of the participants from the larger study (i.e., 13 placebo and 10 CAE); nevertheless, its assessment was one of the two major objectives of this article. Caloric restriction alone (placebo V1 vs. V7) induced important reductions in the plasma levels of *N*-oleoyl-ethanolamine (OEA), *N*-oleoyl-ethanolamine (LEA), *N*-docosapentaenoyl-ethanolamine (DPEA) and 1,2-oleoyl-glycerol (1/2-OG), as well as in the levels of the omega-3 fatty acid docosahexaenoic acid and its eCBome derivative *N*-docosahexaenoyl-ethanolamine (DHEA) (Figure 1). However, the addition of CAE prevented all these decreases.

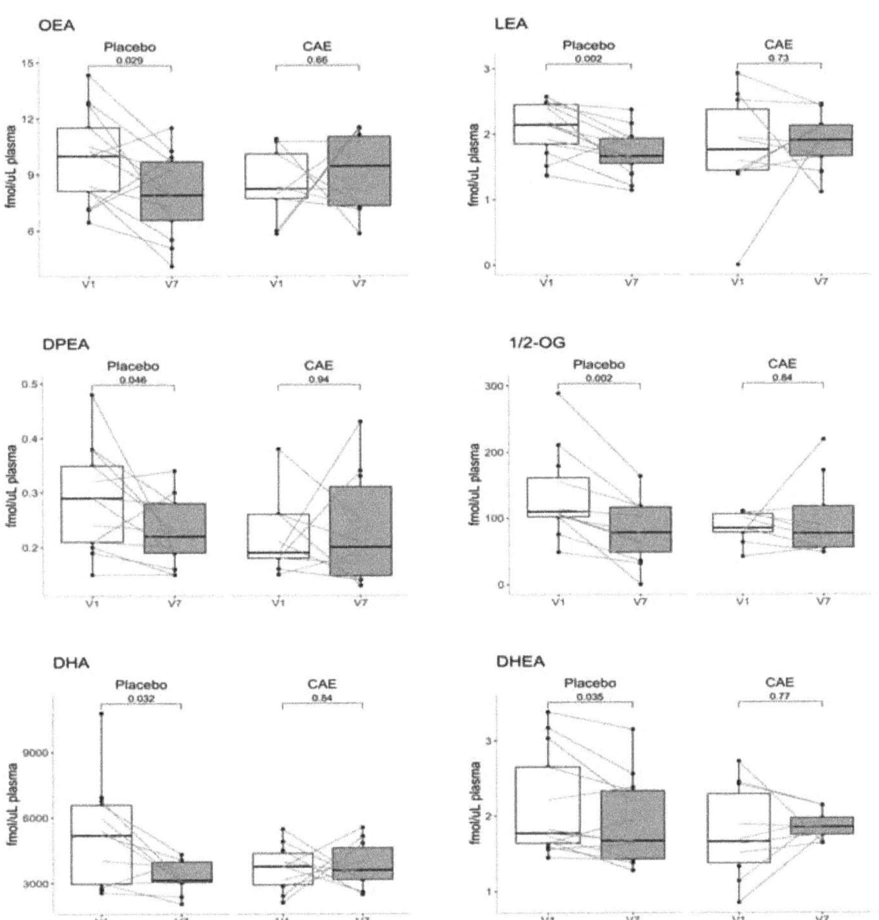

Figure 1. Effect of caloric restriction without or with CAE on plasma eCBome mediator levels. eCBome mediator levels in the plasma of a subset of 23 participants (13 placebo and 10 CAE) before (visit 1 (V1)) and after (visit 7 (V7)) the interventions. The levels of the eCBome-related mediators measured using HPLC-MS/MS are expressed in femtomoles per microlitre of plasma. Each median is represented by the middle line of each boxplot, Q1 and Q3 are represented by the bottom and top of the boxes, respectively, and whiskers denote 1.5 × IQR. OEA, *N*-oleoyl-ethanolamine; LEA, *N*-linoleoyl-ethanolamine; DPEA, *N*-docosapentaenoyl-ethanolamine; 1/2-OG, 1/2-oleoyl-glycerol; DHA, docosahexaenoic acid; DHEA, *N*-docosahexaenoyl-ethanolamine.

3.5. Effect of CAE and Caloric Restriction on Fecal Microbiota Composition

As mentioned above, this endpoint could be measured only in a relatively small subset of the participants to the larger study (i.e., 9 placebo and 6 CAE), where its assessment was the other major objective of this article. Caloric restriction caused an increase in the alpha diversity of the fecal microbiota at V7, which was, however, counteracted by CAE administration, possibly also due to a trending difference between the V1 placebo and V1 CAE groups (Figure 2A). The Firmicutes/Bacteriodetes ratio was not significantly reduced following caloric restriction, either in the presence or absence of CAE (Figure 2B). Accordingly, no changes in the relative abundance of fecal microbiota phyla were observed following caloric restriction. However, a trend ($p = 0.07$) toward increased Bacteroidetes was observed following CAE administration (Figure 2C). At the class level, Clostridia ($p = 0.043$) and Actinobacteria ($p = 0.021$) were increased at visit 7 with CAE as compared to the placebo controls (Figure 2D). Finally, and perhaps more importantly, at the genus level, Flavonifractor was increased by CAE over the course of the treatment but not by caloric restriction alone (Figure 2E). However, Faecalibacterium, Ruminococcaceae_UCG-004 and Ruminococcaceae_UCG-010 showed significant differences in their relative abundances at baseline (V1) between placebo and CAE groups, which may have obfuscated CAE-induced effects (Figure S1). Nevertheless, Faecalibacterium showed a trend toward significance ($p = 0.14$) for increased abundance in response to CAE treatment (V1 vs. V7).

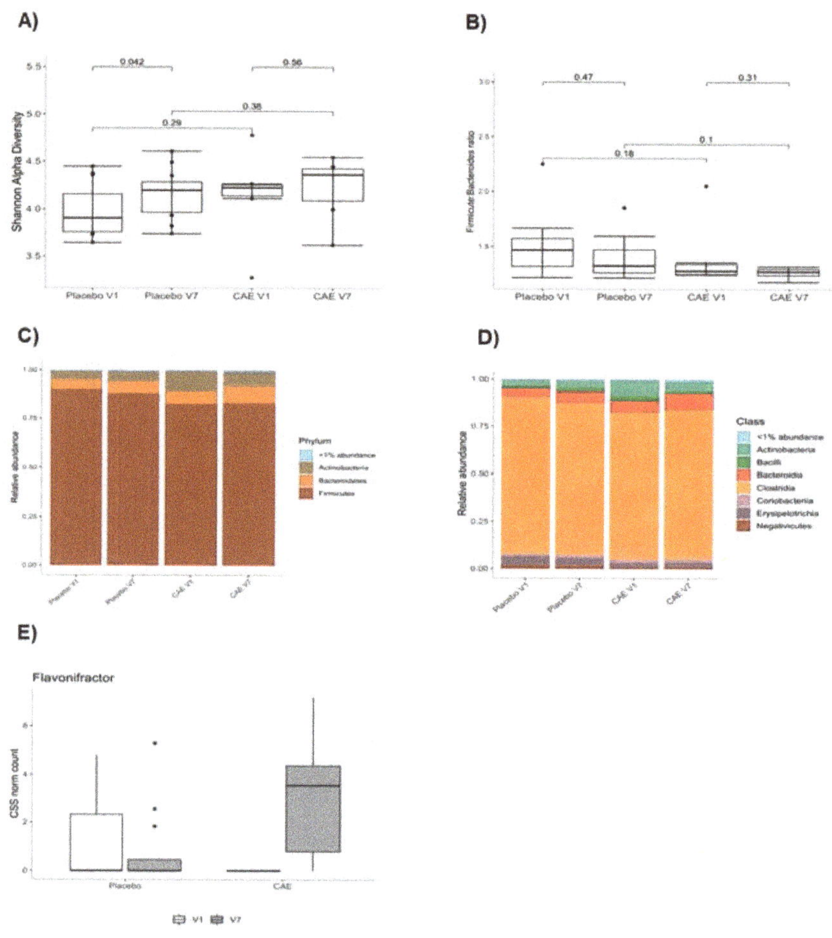

Figure 2. Effect of caloric restriction without or with CAE on the fecal microbiota composition. Alpha diversity of the fecal microbiota (**A**), Firmicutes:Bacteriodetes ratio (**B**) and the relative abundances of fecal microbiota phyla (**C**), classes (**D**) and the Flavonifractor genera (**E**). The data were obtained from stool samples of a subset of 25 (19 placebo and 6 CAE) reproductive-aged women with overweight or obesity before (visit 1 (V1)) and after (visit 7 (V7)) the interventions.

3.6. Correlations between Plasma ECBome Mediator Levels, Anthropometric Measures and Fecal Microbiota Composition

Next, we assessed whether any correlation existed between the plasma levels of eCBome mediators and other outcome measures within individuals for which the appropriate data were collected. Several eCBome mediators correlated with the relative abundances of the phyla, families and genera of the fecal microbiota (Figure 3). In particular, among the pyla, MAGs appeared to correlate negatively with Actinobacteria and Firmicutes (in the case of 1/2-OG and 1/2-DPG the correlations were statistically significant), whereas NAEs correlated positively with Bacteroidetes (significantly for the endocannabinoid anandamide (AEA) and for DHEA), Synergistetes (significantly for AEA, OEA, LEA and DHEA) and Verrucomicrobia (significantly for DHEA, *N*-palmitoyl-ethanolamine (PEA) and *N*-stearoyl-ethanolamine (SEA)). Significant lipid-class-independent correlations were also found with some phyla, i.e., for Cyanobacteria (negative), which was only significantly correlated with the LEA levels, and Verrucomicrobia (positive), which was significantly correlated with the endocannabinoid 2-arachidonoyl-glycerol (2-AG) (calculated together with its 1-isomer (1/2-AG)) and several NAEs (Figure 3A). Stronger correlations were found at the family and genus levels (Figure 3B,C). Most relevant correlations, namely, those that occurred between the eCBome mediators that were found here to be differentially affected by caloric restriction and CAE treatment (Figure 1), include: (1) At the family level, Akkermansiaceae positively correlated with DHEA; Desulfovibrionaceae with DPEA and DHEA; Synergistaceae with OEA, LEA, DPEA and DHEA; and Tannerellaceae with OEA, DHEA and 2-OG. Meanwhile, Atopobiaceae and Prevotellaceae correlated negatively with 2-OG. (2) At the genus level, Akkermansia, Coprococcus 2, Escherichia/Shigetta, Lachnospiraceae UCG-008 and Merdibacter positively correlated with PEA, Bacteroides, Erysipelotrichaceae UCG003, Dialister, Flavonifractor, Parabacteroides and 2-OG; Marvinbryantia with OEA and DHEA; Marvinbryantia and Ruminococcus 2 with LEA; Parabacteroides and the Ruminococcaceae NK4A214 group with DHEA; and Turicibacter with 2-OG.

We also found significant correlations between several fecal microbiota taxa and body weight, BMI and fat mass (FM) (Figure S2). Interestingly, families that negatively correlated with BMI, weight and FM were positively correlated with fat-free mass (FFM) (even when only non-statistically significant trends were observed and vice versa). Therefore, some of the above-mentioned correlations might, at least in part, be mediated by variations of these anthropometric measures. However, of the taxa mentioned above, only the Akkermansia genus (from the Verrucomicrobia phylum and Akkermansiaceae family) was correlated (negatively) with body weight, BMI and fat mass (Figure S2C), and not with the eCBome mediators that correlated with these parameters (except for 1/2-AG). This suggests that these correlations were independent of the variations in these anthropometric measures. Conversely, the positive correlations between Akkermansia (and associated phylogenic levels) and 1/2-AG levels (Figure 3B) suggest that this mediator correlated negatively with BMI and body weight (Figure S3).

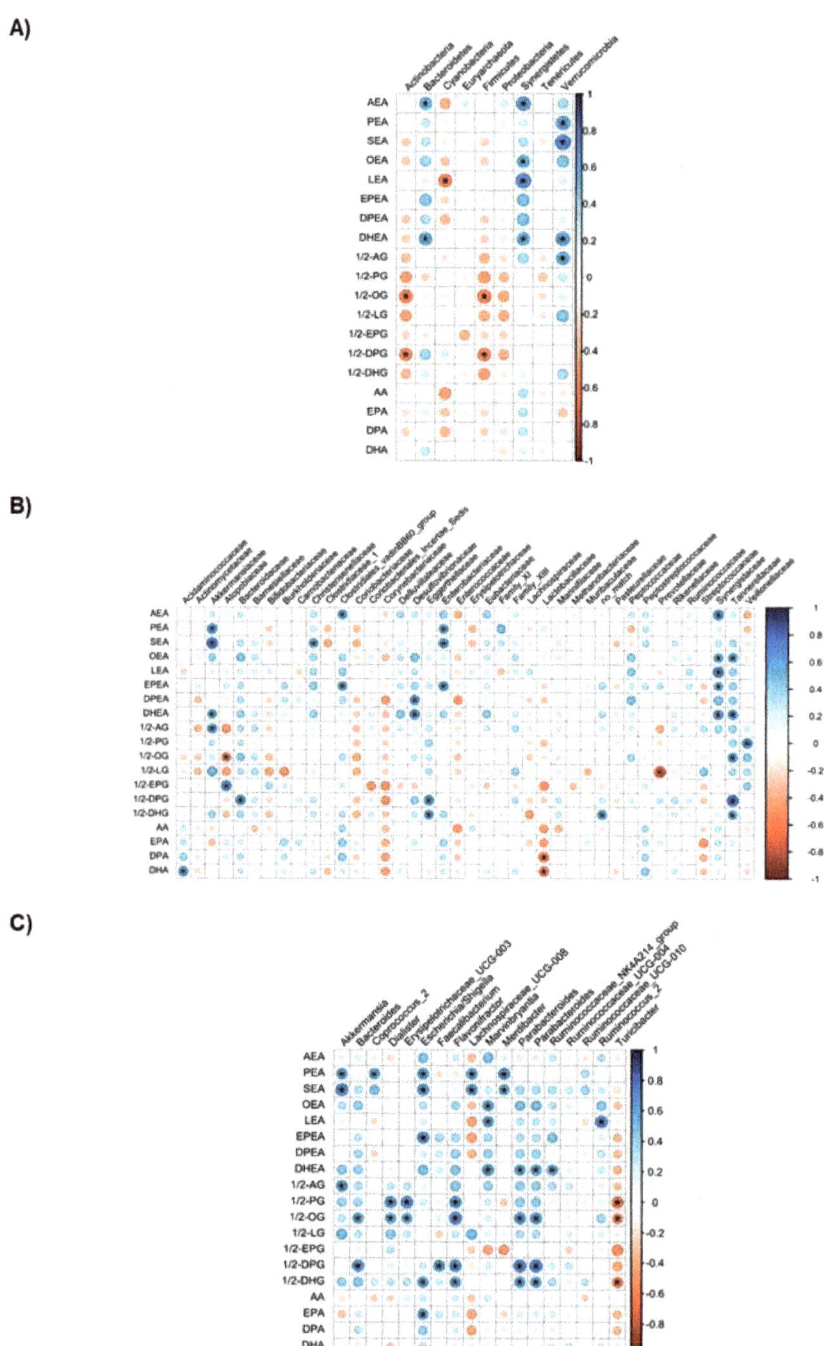

Figure 3. Correlations between the plasma levels of eCBome mediators and the relative abundance of phyla (**A**), families (**B**) and genera (**C**) of the fecal microbiota of 15 (9 placebo and 6 CAE) reproductive-aged women with overweight or obesity. The correlation analysis was performed using repeated measures correlation tests (rmcorr package). * indicates a correlation coefficient with *p*-value < 0.05.

3.7. Correlations between Plasma ECBome Mediator Levels and Appetite Sensations and Eating Behaviors

Our investigation of the associations between plasma eCBome mediator levels and appetite and satiety measures were motivated by previously reported roles of some eCBome mediators and receptors on regulating food intake [51] (Figure 4). The amount of food that could be eaten (DQuaN) was inversely related to levels of LEA, while pre-meal desire to eat (BEnvM) and feeling of hunger (BSenF) were inversely related to DPA and both DPA and DHA, respectively.

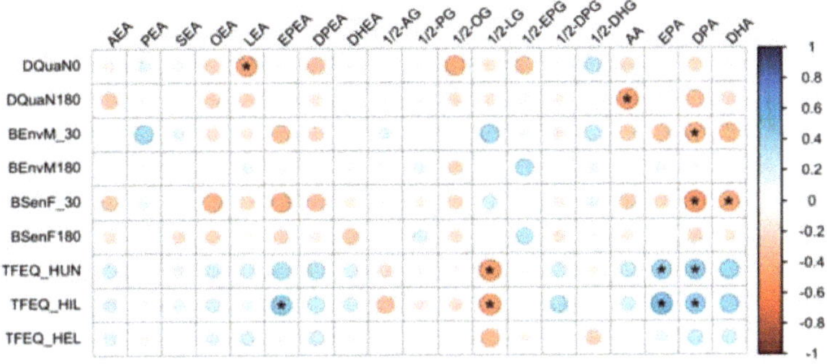

Figure 4. Correlations between the plasma levels of eCBome mediators and the appetite and satiety measures of 15 (9 placebo and 6 CAE) reproductive-aged women with overweight or obesity. The correlation analysis was performed using repeated measures correlation tests (rmcorr package). Only correlation coefficients with *p*-value < 0.05 are displayed. DQuaN, pre-meal evaluation of the amount of food that could be eaten; BEnvM, pre-meal desire to eat; BSenF, feeling of hunger; TFEQ-HUN, habitual feelings of hunger; TFEQ-HIL, feelings of hunger arising from internal sources. * indicates a correlation coefficient with *p*-value < 0.05.

A habitual feeling of hunger (TFEQ-HUN) was inversely related to circulating levels of 1/2-LG and subscale of hunger and hunger from internal (e.g., affect, arousal or stress) cues (TFEQ-HUN and HIL) were positively correlated to the levels of long-chain omega-3 precursors EPA and DPA. Conversely, hunger from external food cues, which are mostly related to situations or objects that have previously been associated with food intake, did not show any correlation with the mediators analyzed.

4. Discussion

The main novel result of this exploratory study is that a capsaicinoid-based supplement produced effects on the eCBome and the gut microbiota, potentially facilitating an amelioration of the metabolic status in women with overweight and obesity. These effects may represent early adaptations that could ultimately result in beneficial effects on food intake, body weight and fat accumulation, as possibly suggested by the present finding showing that, in the subset of patients that underwent plasma eCBome and gut microbiota assessment, CAE performed better than the placebo at enhancing caloric-restriction-induced fat mass reduction. However, in the whole cohort (whose plasma eCBome and fecal microbiome could not be entirely assessed for logistical reasons), CAE enhanced the effects of caloric restriction on body weight, waist circumference and fat mass only in a non-statistically significant manner. It is possible that either a more or a less pronounced caloric-restriction-induced weight/fat loss must be achieved in order to allow for the beneficial effects on the eCBome and gut microbiota that are induced by CAE to produce a stronger impact also on these two anthropometric measures. Indeed, previous studies showed that the same CAE used here, under the same dosing protocol, does produce reductions in body weight and fat mass in a larger cohort of both male and female individuals with lower starting BMI and in the absence of concomitant caloric restriction [52,53]. Future clinical studies, as well as a more thorough examination of the data from the whole cohort than that performed in this article, which was focused on the effects of CAE on the plasma eCBome and fecal microbiota composition, need to be awaited in order to draw any definitive conclusion as to what extent these natural products can be used as a therapy against obesity and hyperphagia.

Indeed, capsaicin is one of the most studied natural products and food (spice) components in the context of metabolic disorders and related cardiovascular risk [3,54,55]. Epidemiological studies suggested that hot chili pepper consumption may positively affect some parameters of the metabolic syndrome, particularly hypertension and body fat [12,56]. A meta-analysis of clinical studies supported the notion that, especially at high doses, capsaicin positively affects energy expenditure and appetite regulation, though contradictory results were often observed in studies with subjects living with overweight and obesity [57]. Several preclinical studies with oral capsaicin supported the possible use of this compound to treat metabolic disorders, also due to effects on food intake, satiety and energy expenditure [28,58,59]. In some of these animal studies, capsaicin effects were ascribed, at least in part, to its capability of modifying the composition of the gut microbiota, for example, in a manner that enhances short-chain fatty acid-producing bacteria [39,40,60]. In one case, these beneficial and gut-microbiota-mediated effects of capsaicin were shown to be due to counteracting the CB1 overactivation by endocannabinoids in obesity [39]. However, in only a few of these experimental studies, the role of TRPV1, which is the major target for capsaicin, was investigated and demonstrated [41], despite the fact that also this ligand-activated channel is known to be strongly implicated in several aspects of energy metabolism control, ranging from food intake and satiety mechanisms to adipogenesis, white adipocyte browning and brown adipocyte activation [25,61–64]. In the present study, we assessed the effects of a 12 weeks dietary supplementation of CAE (equivalent to 4 mg capsaicinoids/day, of which 2.4 mg were capsaicin, 1.4 mg were dihydrocapsaicin and 0.2 mg were nordihydrocapsaicin) in reproductive-aged women with overweight and obesity undergoing a calorie-reduced diet (−500 kcal/day) on the gut microbiota and the expanded endocannabinoid system known as the eCBome. These are two major players that control energy homeostasis, independent from their effects on body weight and food intake, and that are disrupted in obesity and other metabolic disorders [32,37]. We report for the first time that CAE inhibited changes in circulating eCBome lipids induced by the caloric restriction in control participants and induced alterations in the gut microbiota in directions that may favor beneficial actions on the metabolic consequences of obesity. Indeed, some eCBome mediators, such as OEA and other non-CB1-activating NAEs, can be considered as biomarkers of metabolic health since their endogenous concentrations in blood or tissues correlate with, among others, increased lipolysis and insulin sensitivity and reduced body fat, dyslipidemia and systemic inflammation, with or without body-weight-reducing interventions [65–67], as well as with the dietary intake of the corresponding fatty acids [68].

As mentioned above, capsaicin was shown in experimental models of obesity to affect both endocannabinoid signaling by counteracting CB1 activity, and gut microbiota composition [39,60]. TRPV1 activation by this compound in isolated cells expressing the human recombinant protein was found to lead to the biosynthesis of NAEs, including AEA, OEA and PEA [30]. More recently, TRPV1 activation was also shown to increase MAG levels [69]. Here we observed that CAE counteracted a time-dependent reduction in the circulating levels of NAEs observed in placebo-treated women. In particular, the treatment reversed the reduction in the levels of the anorectic and lipolysis-inducing and energy-expenditure-inducing mediator OEA, which acts via the activation of TRPV1, as well as of the pro-lipolytic peroxisome proliferator-activated receptor α (PPARα) and, possibly, the GLP-1-release stimulating GPCR, GPR119 [70–72], with no effect on AEA nor on the other CB1-activating mediator, 2-AG. CAE also counteracted the reduction of the levels of LEA and 2-OG, two other endogenous TRPV1 and GPR119 agonists [73,74], and of the anti-inflammatory NAEs, namely, DHEA and DPEA [74–76]. The decrease in non-CB1 activating NAEs by caloric restriction is likely to represent an adaptive, negative feedback mechanism, and was observed for OEA also in the small intestine of rats following food deprivation [77]. By being partly in agreement with the aforementioned in vitro studies showing that TRPV1 activation by capsaicin stimulates NAE biosynthesis [30,69], our present findings may suggest that, in the subset of participants of our cohort in whom the plasma eCBome was measured, capsaicinoids, administered as CAE, were acting via TRPV1 channels. More importantly, these findings indicate that CAE positively affected the peripheral levels of metabolically beneficial NAEs without changing those of CB1-activating, and hence potentially obesity-exacerbating, endocannabinoids, i.e., AEA and 2-AG.

Several studies have linked the anti-obesity effects of capsaicin to modulation of the gut microbiome, including increasing the abundance of the "anti-obesity" species *Akkermansia muciniphila* [40,60,78]. Results from faecal microbiota transplantation, as well as antibiotics treatments experiments, suggest that capsaicin-induced anti-metabolic endotoxemia effects were at least partly dependent on modulation of the microbiome in mice [39]. CAE also produced some effects on gut microbiota composition in our study by both counteracting the increased diversity observed with caloric restriction only and affecting the relative abundance of some taxa. In particular, it tended to increase the *Bacteroidetes* phylum, increase the *Clostridia* class and increase the *Flavonifractor* genus. While the former two might exert beneficial metabolic effects, the increase in *Flavonifractor* might relate to the capability of this genus to metabolize flavonoids and, in particular, curcumin, with which capsaicin has structural homology. Accordingly, dietary curcumin or curcumin-containing turmeric were also shown to increase the abundance of *Flavonifractor sp.* in human fecal microbiota in vitro [79] or in vivo [80]. Species belonging to this genus, such as *F. plautii*, were suggested to be decreased in obesity/increased in leanness [81] and play anti-inflammatory actions [82]. Interestingly, it appears that an individual's gut microbiota composition can modulate the response to capsaicin. In a trial involving healthy adults, it was observed that a broad separation of participants into *Bacteroides* and *Prevotella* enterotypes similarly stratified the participants into either strong or weak capsaicin responders, respectively, with respect to capsaicin-induced increases in *Faecalibacterium* abundance and serum incretin (GIP and GLP-1) levels [83]. While we did observe a trend toward an increase in *Faecalibacterium* abundance in response to CAE (data not shown; $p = 0.14$, V1 vs. V7 CAE), the results may have been confounded by the fact that baseline *Faecalibacterium* abundance was significantly lower in the CAE V1 group as compared to the placebo V1 group ($p = 0.014$). We previously proposed that this difference between enterotypes may be due to microbiome-mediated modulation of TRPV1 responsiveness to capsaicin [37], since it was shown that the probiotic *Lactobacillus reuteri* can inhibit TRPV1 activity in mesenteric neurons [84]. Thus, the relatively small changes in the gut microbiome that we observed here may in part be due to the participants harboring capsaicin-resistant microbiomes, which may be generalizable and explain why capsaicin effects on energy expenditure were reported to be blunted in obese as compared to lean individuals [57]. Nevertheless, we were expecting to see stronger effects of CAE on fecal microbiota composition. It is possible that some effects, as mentioned above, were masked by the fact that the relative abundance of some taxa were different at the baseline (V1) in the two control groups, which was a limitation of this study. However, in agreement with previous studies showing the existence of a bilateral communication between the host eCBome and the gut microbiome [85], we did find several correlations between the eCBome mediators whose concentrations in the plasma were modified by CAE and numerous gut bacterial taxa, also at the genus level. Of these, some seemed to be particularly relevant to the potential beneficial metabolic effects of capsaicin. The most obvious of these was undoubtedly *Akkermansia*, which positively correlated with PEA levels (as did the *Akkermansiaceae* family). *Akkermansia* abundance is decreased in obesity and a recent clinical trial showed that supplementation, especially with heat-inactivated *Akkermansia*, improved metabolic parameters in obese participants, including decreasing hepatic inflammation markers [86]. Interestingly, PEA is a potent anti-inflammatory compound with reported metabolic benefits in models of obesity, including a reduction in high-fat diet-induced liver inflammation [87,88]. In contrast, *Enterobacteriaceae*, a lipopolysaccharide-producing family that is associated with obesity and metabolic endotoxemia [89], was also positively correlated with PEA. *Flavonifractor*, the only genus that was found to be specifically upregulated by CAE in our study, was positively associated with levels of the GPR119 agonist 2-OG (see above). *Flavonifractor* was found to be associated with non-obese as compared to obese individuals and, when provided to mice on a high-fat diet, it attenuated adipose tissue inflammation [81,82]. Similarly, *Parabacteroides* and *Dialister* were also correlated with 2-OG, and different species of the former were tied to the beneficial metabolic effects of fiber and to ameliorate diet-induced metabolic dysfunction [90,91], while the latter was associated with improved insulin sensitivity [92], though paradoxically also with resistance to weight loss in response to a volumetric diet [93]. Taken together, it is clear that the assessment of the significance of these correlations will require further investigation in order to understand whether they have any causal effects on obesity and associated complications.

The CAE-encapsulating technology used here allowed for a controlled release that delivered effective levels of capsaicinoids in the small intestine without the stomach upset that may result from unprotected extracts of red chili peppers. Although we did not aim to compare the difference between such technology and capsaicinoids ingested as supplements with some specific foods, we believe that it is possible that such difference might exist. In particular, capsule release in the small intestine may bypass mouth- and stomach-specific activation of TRPV1 and modify the sensory detection of taste differences in the mouth.

5. Conclusions

In summary, in the present exploratory clinical study, the oral administration of CAE to reproductive-aged women living with overweight and obesity produced important changes in the plasma eCBome, which may be indicative of potential beneficial effects at the level of hyperphagia (through increased levels of an anorectic factor, such as OEA), insulin sensitivity (through increases of endogenous agonists of GPR119) and inflammation (through the increase of anti-inflammatory mediators like DPEA and DHEA). However, while we did not see any significant correlation between eCBome mediators and the appetite and satiety measures, here we also did not assess measures of insulin sensitivity or chronic inflammation. This, together with the relatively small number of participants from whom we were allowed to sample blood and stools for eCBome and gut microbiome measurements, represent limitations of this study. Nevertheless, we suggest that some of the observed effects of CAE on eCBome mediators and the gut microbiome, the latter of which probably also represents an attempt of the gut microbial ecosystem to deal with capsaicin and its metabolism, may contribute to potential metabolically beneficial effects of these natural products. In agreement with this hypothesis, in the subset of patients that underwent plasma eCBome and fecal microbiota assessments, CAE performed significantly better than the placebo at reducing fat mass. The finding of strong correlations between several eCBome mediators and fecal microbiota taxa should also be considered in this context and reinforces the hypothesis of the existence of a connection between these two metabolically relevant signaling systems in the framework of overweight and obesity.

Supplementary Materials: The following are available online at https://www.mdpi.com/article/10.3390/biomedicines9091246/s1. Table S1: Reported symptoms and inventory of collected samples, Table S2: List of the lipid mediators used as deuterated internal standards for LC/MS-MS analyses, Figure S1: Effect of Capsimax and caloric restriction on relative abundance of fecal bacterial genera, Figure S2: Correlation between fecal microbiota taxa at phylum (A), classed (B) and genera (C) levels and BMI, waist circumference, fat mass (FM) and fat-free mass (FFM), Figure S3: Correlations between the plasma levels of eCBome mediators and BMI, waist circumference, fat mass (FM) and fat-free mass (FFM).

Author Contributions: Conceptualization, V.D., A.T., V.D.M. and C.S.; Data curation, C.M., S.L., N.F. and Y.C.; Formal analysis, C.M., S.L., N.F. and Y.C.; Methodology, F.P. and V.D.; Writing—original draft, C.M., S.L., A.T., V.D.M. and C.S.; Writing—review and editing, C.M., S.L., F.P., N.F., Y.C., V.D., A.T., V.D.M. and C.S. All authors have read and agreed to the published version of the manuscript.

Funding: This work was supported by the Canada Research Excellence Chair in the Microbiome-Endocannabinoidome Axis in Metabolic Health (CERC-MEND), which is funded by the Tri-Agency of the Canadian Federal Government (The Canadian Institutes of Health Research (CIHR), the Natural Sciences and Engineering Research Council of Canada (NSERC) and the Social Sciences and Humanities Research Council of Canada (SSHRC)) to V.D., as well as by the Canadian Foundation of Innovation (to V.D.) and the Sentinelle Nord-Apogée program (to Université Laval). C.M. received a post-doctoral grant from the Joint International Research Unit for Chemical and Biomolecular Research on the Microbiome and its impact on Metabolic Health and Nutrition (UMI-MicroMeNu), which is partly supported by the Sentinelle Nord/Apogée program.

Institutional Review Board Statement: This study was conducted according to the guidelines laid down in the Declaration of Helsinki and all procedures involving human subjects in the study protocol were approved by the Research Ethics Committee of Health Sciences of Laval University (protocol 2015-041 A-4 R-2/01-05-2018). The clinical trial was registered in the public trials registry ClinicalTrials.gov (accessed on 30 July 2021) (registration identification number NCT04874701).

Informed Consent Statement: Written informed consent was obtained from all the participants before entering the study.

Data Availability Statement: Data is contained within the article or Supplementary Materials.

Acknowledgments: The authors wish to thank Cyril Martin for his help with the LC–MS/MS measurements.

Conflicts of Interest: The authors declare no conflict of interest. The funders had no role in the design of the study; in the collection, analyses, or interpretation of data; in the writing of the manuscript; or in the decision to publish the results.

References

1. Non Communicable Diseases. Available online: https://www.who.int/news-room/fact-sheets/detail/noncommunicable-diseases (accessed on 26 January 2021).
2. Obesity and Overweight. Available online: https://www.who.int/news-room/fact-sheets/detail/obesity-and-overweight (accessed on 26 January 2021).
3. Zheng, J.; Zheng, S.; Feng, Q.; Zhang, Q.; Xiao, X. Dietary Capsaicin and Its Anti-Obesity Potency: From Mechanism to Clinical Implications. *Biosci. Rep.* **2017**, *37*. [CrossRef]
4. Hill, J.O.; Wyatt, H.R.; Peters, J.C. Energy Balance and Obesity. *Circulation* **2012**, *126*, 126–132. [CrossRef]
5. Hruby, A.; Hu, F.B. The Epidemiology of Obesity: A Big Picture. *Pharmacoeconomics* **2015**, *33*, 673–689. [CrossRef]
6. Yanovski, S.Z.; Yanovski, J.A. Long-Term Drug Treatment for Obesity: A Systematic and Clinical Review. *JAMA* **2014**, *311*, 74. [CrossRef]
7. Filippatos, T.D.; Derdemezis, C.S.; Gazi, I.F.; Nakou, E.S.; Mikhailidis, D.P.; Elisaf, M.S. Orlistat-Associated Adverse Effects and Drug Interactions: A Critical Review. *Drug Saf.* **2008**, *31*, 53–65. [CrossRef] [PubMed]
8. Wilding, J.P.H.; Batterham, R.L.; Calanna, S.; Davies, M.; van Gaal, L.F.; Lingvay, I.; McGowan, B.M.; Rosenstock, J.; Tran, M.T.D.; Wadden, T.A.; et al. Once-Weekly Semaglutide in Adults with Overweight or Obesity. *N. Engl. J. Med.* **2021**. [CrossRef] [PubMed]
9. Dietrich, M.O.; Horvath, T.L. Limitations in Anti-Obesity Drug Development: The Critical Role of Hunger-Promoting Neurons. *Nat. Rev. Drug Discov.* **2012**, *11*, 675–691. [CrossRef] [PubMed]
10. Liu, J.; Lee, J.; Salazar Hernandez, M.A.; Mazitschek, R.; Ozcan, U. Treatment of Obesity with Celastrol. *Cell* **2015**, *161*, 999–1011. [CrossRef] [PubMed]
11. Martel, J.; Ojcius, D.M.; Chang, C.-J.; Lin, C.-S.; Lu, C.-C.; Ko, Y.-F.; Tseng, S.-F.; Lai, H.-C.; Young, J.D. Anti-Obesogenic and Antidiabetic Effects of Plants and Mushrooms. *Nat. Rev. Endocrinol.* **2017**, *13*, 149–160. [CrossRef]
12. Varghese, S.; Kubatka, P.; Rodrigo, L.; Gazdikova, K.; Caprnda, M.; Fedotova, J.; Zulli, A.; Kruzliak, P.; Büsselberg, D. Chili Pepper as a Body Weight-Loss Food. *Int. J. Food Sci. Nutr.* **2017**, *68*, 392–401. [CrossRef]
13. Richards, B.L.; Whittle, S.L.; Buchbinder, R. Neuromodulators for Pain Management in Rheumatoid Arthritis. *Cochrane Database Syst. Rev.* **2012**, *1*. [CrossRef]
14. Zheng, L.; Chen, J.; Ma, Z.; Liu, W.; Yang, F.; Yang, Z.; Wang, K.; Wang, X.; He, D.; Li, L.; et al. Capsaicin Enhances Anti-Proliferation Efficacy of Pirarubicin via Activating TRPV1 and Inhibiting PCNA Nuclear Translocation in 5637 Cells. *Mol. Med. Rep.* **2016**, *13*, 881–887. [CrossRef]

15. Liu, Y.-P.; Dong, F.-X.; Chai, X.; Zhu, S.; Zhang, B.-L.; Gao, D.-S. Role of Autophagy in Capsaicin-Induced Apoptosis in U251 Glioma Cells. *Cell. Mol. Neurobiol.* **2016**, *36*, 737–743. [CrossRef]
16. Dairam, A.; Fogel, R.; Daya, S.; Limson, J.L. Antioxidant and Iron-Binding Properties of Curcumin, Capsaicin, and *S*-Allylcysteine Reduce Oxidative Stress in Rat Brain Homogenate. *J. Agric. Food Chem.* **2008**, *56*, 3350–3356. [CrossRef]
17. Singh, U.; Bernstein, J.A. Intranasal Capsaicin in Management of Nonallergic (Vasomotor) Rhinitis. *Prog Drug Res.* **2014**, *68*, 147–170. [CrossRef] [PubMed]
18. Tremblay, A.; Arguin, H.; Panahi, S. Capsaicinoids: A Spicy Solution to the Management of Obesity? *Int J. Obes* **2016**, *40*, 1198–1204. [CrossRef] [PubMed]
19. Yoshioka, M.; St-Pierre, S.; Drapeau, V.; Dionne, I.; Doucet, E.; Suzuki, M.; Tremblay, A. Effects of Red Pepper on Appetite and Energy Intake. *Br. J. Nutr.* **1999**, *82*, 115–123. [CrossRef] [PubMed]
20. Westerterp-Plantenga, M.S.; Smeets, A.; Lejeune, M.P.G. Sensory and Gastrointestinal Satiety Effects of Capsaicin on Food Intake. *Int. J. Obes.* **2005**, *29*, 682–688. [CrossRef]
21. Yoshioka, M.; Doucet, E.; Drapeau, V.; Dionne, I.; Tremblay, A. Combined Effects of Red Pepper and Caffeine Consumption on 24 h Energy Balance in Subjects given Free Access to Foods. *Br. J. Nutr.* **2001**, *85*, 203–211. [CrossRef]
22. Wahlqvist, M.L.; Wattanapenpaiboon, N. Hot Foods—Unexpected Help with Energy Balance? *Lancet* **2001**, *358*, 348–349. [CrossRef]
23. Lejeune, M.P.G.M.; Kovacs, E.M.R.; Westerterp-Plantenga, M.S. Effect of Capsaicin on Substrate Oxidation and Weight Maintenance after Modest Body-Weight Loss in Human Subjects. *Br. J. Nutr.* **2003**, *90*, 651–659. [CrossRef] [PubMed]
24. Kelava, L.; Nemeth, D.; Hegyi, P.; Keringer, P.; Kovacs, D.K.; Balasko, M.; Solymar, M.; Pakai, E.; Rumbus, Z.; Garami, A. Dietary Supplementation of Transient Receptor Potential Vanilloid-1 Channel Agonists Reduces Serum Total Cholesterol Level: A Meta-Analysis of Controlled Human Trials. *Crit. Rev. Food Sci. Nutr.* **2021**, 1–11. [CrossRef]
25. Chen, J.; Li, L.; Li, Y.; Liang, X.; Sun, Q.; Yu, H.; Zhong, J.; Ni, Y.; Chen, J.; Zhao, Z.; et al. Activation of TRPV1 Channel by Dietary Capsaicin Improves Visceral Fat Remodeling through Connexin43-Mediated Ca2+ Influx. *Cardiovasc. Diabetol.* **2015**, *14*, 22. [CrossRef]
26. Sun, F.; Xiong, S.; Zhu, Z. Dietary Capsaicin Protects Cardiometabolic Organs from Dysfunction. *Nutrients* **2016**, *8*, 174. [CrossRef]
27. Gram, D.X.; Ahrén, B.; Nagy, I.; Olsen, U.B.; Brand, C.L.; Sundler, F.; Tabanera, R.; Svendsen, O.; Carr, R.D.; Santha, P.; et al. Capsaicin-Sensitive Sensory Fibers in the Islets of Langerhans Contribute to Defective Insulin Secretion in Zucker Diabetic Rat, an Animal Model for Some Aspects of Human Type 2 Diabetes: Systemic Capsaicin in ZDF Rats. *Eur. J. Neurosci.* **2007**, *25*, 213–223. [CrossRef] [PubMed]
28. Wang, P.; Yan, Z.; Zhong, J.; Chen, J.; Ni, Y.; Li, L.; Ma, L.; Zhao, Z.; Liu, D.; Zhu, Z. Transient Receptor Potential Vanilloid 1 Activation Enhances Gut Glucagon-Like Peptide-1 Secretion and Improves Glucose Homeostasis. *Diabetes* **2012**, *61*, 2155–2165. [CrossRef] [PubMed]
29. Kang, J.-H.; Kim, C.-S.; Han, I.-S.; Kawada, T.; Yu, R. Capsaicin, a Spicy Component of Hot Peppers, Modulates Adipokine Gene Expression and Protein Release from Obese-Mouse Adipose Tissues and Isolated Adipocytes, and Suppresses the Inflammatory Responses of Adipose Tissue Macrophages. *FEBS Lett.* **2007**, *581*, 4389–4396. [CrossRef]
30. Di Marzo, V.; Lastres-Becker, I.; Bisogno, T.; De Petrocellis, L.; Milone, A.; Davis, J.B.; Fernandez-Ruiz, J.J. Hypolocomotor Effects in Rats of Capsaicin and Two Long Chain Capsaicin Homologues. *Eur. J. Pharmacol.* **2001**, *420*, 123–131. [CrossRef]
31. Acharya, N.; Penukonda, S.; Shcheglova, T.; Hagymasi, A.T.; Basu, S.; Srivastava, P.K. Endocannabinoid System Acts as a Regulator of Immune Homeostasis in the Gut. *Proc. Natl. Acad. Sci. USA* **2017**, *114*, 5005–5010. [CrossRef]
32. Silvestri, C.; di Marzo, V. The Endocannabinoid System in Energy Homeostasis and the Etiopathology of Metabolic Disorders. *Cell Metab.* **2013**, *17*, 475–490. [CrossRef]
33. Di Marzo, V.; Wang, J. (Eds.) *The Endocannabinoidome the World of Endocannabinoids and Related Mediators*; Academic Press: London, UK, 2014.
34. Zygmunt, P.M.; Ermund, A.; Movahed, P.; Andersson, D.A.; Simonsen, C.; Jönsson, B.A.G.; Blomgren, A.; Birnir, B.; Bevan, S.; Eschalier, A.; et al. Monoacylglycerols Activate TRPV1—A Link between Phospholipase C and TRPV. *PLoS ONE* **2013**, *8*, e81618. [CrossRef]
35. Movahed, P.; Jönsson, B.A.G.; Birnir, B.; Wingstrand, J.A.; Jørgensen, T.D.; Ermund, A.; Sterner, O.; Zygmunt, P.M.; Högestätt, E.D. Endogenous Unsaturated C18 N-Acylethanolamines Are Vanilloid Receptor (TRPV1) Agonists. *J. Biol. Chem.* **2005**, *280*, 38496–38504. [CrossRef] [PubMed]
36. Cani, P.D.; Plovier, H.; Van Hul, M.; Geurts, L.; Delzenne, N.M.; Druart, C.; Everard, A. Endocannabinoids—At the Crossroads between the Gut Microbiota and Host Metabolism. *Nat. Rev. Endocrinol.* **2016**, *12*, 133–143. [CrossRef] [PubMed]
37. Di Marzo, V.; Silvestri, C. Lifestyle and Metabolic Syndrome: Contribution of the Endocannabinoidome. *Nutrients* **2019**, *11*, 1956. [CrossRef] [PubMed]
38. Dione, N.; Lacroix, S.; Taschler, U.; Deschênes, T.; Abolghasemi, A.; Leblanc, N.; Di Marzo, V.; Silvestri, C. Mgll Knockout Mouse Resistance to Diet-Induced Dysmetabolism Is Associated with Altered Gut Microbiota. *Cells* **2020**, *9*, 2705. [CrossRef] [PubMed]
39. Kang, C.; Wang, B.; Kaliannan, K.; Wang, X.; Lang, H.; Hui, S.; Huang, L.; Zhang, Y.; Zhou, M.; Chen, M.; et al. Gut Microbiota Mediates the Protective Effects of Dietary Capsaicin against Chronic Low-Grade Inflammation and Associated Obesity Induced by High-Fat Diet. *mBio* **2017**, *8*, e00470-17. [CrossRef]

40. Song, J.-X.; Ren, H.; Gao, Y.-F.; Lee, C.-Y.; Li, S.-F.; Zhang, F.; Li, L.; Chen, H. Dietary Capsaicin Improves Glucose Homeostasis and Alters the Gut Microbiota in Obese Diabetic Ob/Ob Mice. *Front. Physiol.* **2017**, *8*, 602. [CrossRef]

41. Kumar, V.; Mahajan, N.; Khare, P.; Kondepudi, K.K.; Bishnoi, M. Role of TRPV1 in Colonic Mucin Production and Gut Microbiota Profile. *Eur. J. Pharmacol.* **2020**, *888*, 173567. [CrossRef]

42. Drapeau, V.; King, N.; Hetherington, M.; Doucet, E.; Blundell, J.; Tremblay, A. Appetite Sensations and Satiety Quotient: Predictors of Energy Intake and Weight Loss. *Appetite* **2007**, *48*, 159–166. [CrossRef]

43. Arvaniti, K.; Richard, D.; Tremblay, A. Reproducibility of Energy and Macronutrient Intake and Related Substrate Oxidation Rates in a Buffet-Type Meal. *Br. J. Nutr* **2000**, *83*, 489–495. [CrossRef]

44. Green, S.M.; Delargy, H.J.; Joanes, D.; Blundell, J.E. A Satiety Quotient: A Formulation to Assess the Satiating Effect of Food. *Appetite* **1997**, *29*, 291–304. [CrossRef]

45. Stunkard, A.J.; Messick, S. The Three-Factor Eating Questionnaire to Measure Dietary Restraint, Disinhibition and Hunger. *J. Psychosom. Res.* **1985**, *29*, 71–83. [CrossRef]

46. Cepeda-Benito, A.; Gleaves, D.H.; Fernández, M.C.; Vila, J.; Williams, T.L.; Reynoso, J. The Development and Validation of Spanish Versions of the State and Trait Food Cravings Questionnaires. *Behav. Res. Ther.* **2000**, *38*, 1125–1138. [CrossRef]

47. Turcotte, C.; Archambault, A.; Dumais, É.; Martin, C.; Blanchet, M.; Bissonnette, E.; Ohashi, N.; Yamamoto, K.; Itoh, T.; Laviolette, M.; et al. Endocannabinoid Hydrolysis Inhibition Unmasks That Unsaturated Fatty Acids Induce a Robust Biosynthesis of 2-arachidonoyl-glycerol and Its Congeners in Human Myeloid Leukocytes. *FASEB J.* **2020**, *34*, 4253–4265. [CrossRef] [PubMed]

48. Callahan, B.J.; McMurdie, P.J.; Rosen, M.J.; Han, A.W.; Johnson, A.J.A.; Holmes, S.P. DADA2: High-Resolution Sample Inference from Illumina Amplicon Data. *Nat. Methods* **2016**, *13*, 581–583. [CrossRef] [PubMed]

49. Paulson, J.N.; Stine, O.C.; Bravo, H.C.; Pop, M. Differential Abundance Analysis for Microbial Marker-Gene Surveys. *Nat. Methods* **2013**, *10*, 1200–1202. [CrossRef] [PubMed]

50. Bakdash, J.Z.; Marusich, L.R. Repeated Measures Correlation. *Front. Psychol.* **2017**, *8*, 456. [CrossRef]

51. Rochefort, G.; Provencher, V.; Castonguay-Paradis, S.; Perron, J.; Lacroix, S.; Martin, C.; Flamand, N.; Di Marzo, V.; Veilleux, A. Intuitive Eating Is Associated with Elevated Levels of Circulating Omega-3-Polyunsaturated Fatty Acid-Derived Endocannabinoidome Mediators. *Appetite* **2021**, *156*, 104973. [CrossRef] [PubMed]

52. Urbina, S.L.; Roberts, M.D.; Kephart, W.C.; Villa, K.B.; Santos, E.N.; Olivencia, A.M.; Bennett, H.M.; Lara, M.D.; Foster, C.A.; Purpura, M.; et al. Effects of Twelve Weeks of Capsaicinoid Supplementation on Body Composition, Appetite and Self-Reported Caloric Intake in Overweight Individuals. *Appetite* **2017**, *113*, 264–273. [CrossRef] [PubMed]

53. Rogers, J.; Urbina, S.L.; Taylor, L.W.; Wilborn, C.D.; Purpura, M.; Jäger, R.; Juturu, V. Capsaicinoids Supplementation Decreases Percent Body Fat and Fat Mass: Adjustment Using Covariates in a Post Hoc Analysis. *BMC Obes.* **2018**, *5*, 22. [CrossRef]

54. McCarty, M.F.; DiNicolantonio, J.J.; O'Keefe, J.H. Capsaicin May Have Important Potential for Promoting Vascular and Metabolic Health. *Open Heart* **2015**, *2*, e000262. [CrossRef]

55. Sanati, S.; Razavi, B.M.; Hosseinzadeh, H. A Review of the Effects of Capsicum Annuum L. and Its Constituent, Capsaicin, in Metabolic Syndrome. *Iran. J. Basic Med. Sci.* **2018**, *21*, 439. [CrossRef]

56. Shi, Z.; Riley, M.; Brown, A.; Page, A. Chilli Intake Is Inversely Associated with Hypertension among Adults. *Clin. Nutr. ESPEN* **2018**, *23*, 67–72. [CrossRef] [PubMed]

57. Ludy, M.-J.; Moore, G.E.; Mattes, R.D. The Effects of Capsaicin and Capsiate on Energy Balance: Critical Review and Meta-Analyses of Studies in Humans. *Chem. Senses* **2012**, *37*, 103–121. [CrossRef] [PubMed]

58. Kang, J.-H.; Tsuyoshi, G.; Han, I.-S.; Kawada, T.; Kim, Y.M.; Yu, R. Dietary Capsaicin Reduces Obesity-Induced Insulin Resistance and Hepatic Steatosis in Obese Mice Fed a High-Fat Diet. *Obesity* **2010**, *18*, 780–787. [CrossRef] [PubMed]

59. Kwon, D.Y.; Kim, Y.S.; Ryu, S.Y.; Cha, M.-R.; Yon, G.H.; Yang, H.J.; Kim, M.J.; Kang, S.; Park, S. Capsiate Improves Glucose Metabolism by Improving Insulin Sensitivity Better than Capsaicin in Diabetic Rats. *J. Nutr. Biochem.* **2013**, *24*, 1078–1085. [CrossRef]

60. Shen, W.; Shen, M.; Zhao, X.; Zhu, H.; Yang, Y.; Lu, S.; Tan, Y.; Li, G.; Li, M.; Wang, J.; et al. Anti-Obesity Effect of Capsaicin in Mice Fed with High-Fat Diet Is Associated with an Increase in Population of the Gut Bacterium Akkermansia Muciniphila. *Front. Microbiol.* **2017**, *8*, 272. [CrossRef] [PubMed]

61. Lee, E.; Jung, D.Y.; Kim, J.H.; Patel, P.R.; Hu, X.; Lee, Y.; Azuma, Y.; Wang, H.-F.; Tsitsilianos, N.; Shafiq, U.; et al. Transient Receptor Potential Vanilloid Type-1 Channel Regulates Diet-Induced Obesity, Insulin Resistance, and Leptin Resistance. *FASEB J.* **2015**, *29*, 3182–3192. [CrossRef] [PubMed]

62. Zhang, L.L.; Yan Liu, D.; Ma, L.Q.; Luo, Z.D.; Cao, T.B.; Zhong, J.; Yan, Z.C.; Wang, L.J.; Zhao, Z.G.; Zhu, S.J.; et al. Activation of Transient Receptor Potential Vanilloid Type-1 Channel Prevents Adipogenesis and Obesity. *Circ. Res.* **2007**, *100*, 1063–1070. [CrossRef]

63. Baboota, R.K.; Singh, D.P.; Sarma, S.M.; Kaur, J.; Sandhir, R.; Boparai, R.K.; Kondepudi, K.K.; Bishnoi, M. Capsaicin Induces "Brite" Phenotype in Differentiating 3T3-L1 Preadipocytes. *PLoS ONE* **2014**, *9*, e103093. [CrossRef]

64. Baskaran, P.; Krishnan, V.; Ren, J.; Thyagarajan, B. Capsaicin Induces Browning of White Adipose Tissue and Counters Obesity by Activating TRPV1 Channel-Dependent Mechanisms: TRPV1 Activates Browning of WAT to Counter Obesity. *Br. J. Pharmacol.* **2016**, *173*, 2369–2389. [CrossRef]

65. Montecucco, F.; Lenglet, S.; Quercioli, A.; Burger, F.; Thomas, A.; Lauer, E.; Silva, D.A.R.; Mach, F.; Vuilleumier, N.; Bobbioni-Harsch, E.; et al. Gastric Bypass in Morbid Obese Patients Is Associated with Reduction in Adipose Tissue Inflammation via N-Oleoylethanolamide (OEA)-Mediated Pathways. *Thromb. Haemost.* **2015**, *113*, 838–850. [CrossRef]

66. Fanelli, F.; Mezzullo, M.; Repaci, A.; Belluomo, I.; Ibarra Gasparini, D.; Di Dalmazi, G.; Mastroroberto, M.; Vicennati, V.; Gambineri, A.; Morselli-Labate, A.M.; et al. Profiling Plasma N-Acylethanolamine Levels and Their Ratios as a Biomarker of Obesity and Dysmetabolism. *Mol. Metab.* **2018**, *14*, 82–94. [CrossRef]

67. Azar, S.; Sherf-Dagan, S.; Nemirovski, A.; Webb, M.; Raziel, A.; Keidar, A.; Goitein, D.; Sakran, N.; Shibolet, O.; Tam, J.; et al. Circulating Endocannabinoids Are Reduced Following Bariatric Surgery and Associated with Improved Metabolic Homeostasis in Humans. *Obes. Surg.* **2019**, *29*, 268–276. [CrossRef] [PubMed]

68. Castonguay-Paradis, S.; Lacroix, S.; Rochefort, G.; Parent, L.; Perron, J.; Martin, C.; Lamarche, B.; Raymond, F.; Flamand, N.; di Marzo, V.; et al. Dietary Fatty Acid Intake and Gut Microbiota Determine Circulating Endocannabinoidome Signaling beyond the Effect of Body Fat. *Sci. Rep.* **2020**, *10*, 15975. [CrossRef] [PubMed]

69. Manchanda, M.; Leishman, E.; Sangani, K.; Alamri, A.; Bradshaw, H.B. Activation of TRPV1 by Capsaicin or Heat Drives Changes in 2-Acyl Glycerols and N-Acyl Ethanolamines in a Time, Dose, and Temperature Dependent Manner. *Front. Cell Dev. Biol.* **2021**, *9*, 611952. [CrossRef] [PubMed]

70. Ahern, G.P. Activation of TRPV1 by the Satiety Factor Oleoylethanolamide. *J. Biol. Chem.* **2003**, *278*, 30429–30434. [CrossRef] [PubMed]

71. Lauffer, L.M.; Iakoubov, R.; Brubaker, P.L. GPR119 Is Essential for Oleoylethanolamide-Induced Glucagon-like Peptide-1 Secretion from the Intestinal Enteroendocrine L-Cell. *Diabetes* **2009**, *58*, 1058–1066. [CrossRef]

72. Im, D.-S. GPR119 and GPR55 as Receptors for Fatty Acid Ethanolamides, Oleoylethanolamide and Palmitoylethanolamide. *Int. J. Mol. Sci.* **2021**, *22*, 1034. [CrossRef] [PubMed]

73. Syed, S.K.; Bui, H.H.; Beavers, L.S.; Farb, T.B.; Ficorilli, J.; Chesterfield, A.K.; Kuo, M.-S.; Bokvist, K.; Barrett, D.G.; Efanov, A.M. Regulation of GPR119 Receptor Activity with Endocannabinoid-like Lipids. *Am. J. Physiol. Endocrinol. Metab.* **2012**, *303*, E1469–E1478. [CrossRef]

74. Di Marzo, V. New Approaches and Challenges to Targeting the Endocannabinoid System. *Nat. Rev. Drug Discov* **2018**, *17*, 623–639. [CrossRef] [PubMed]

75. Figueroa, J.D.; Cordero, K.; Serrano-Illan, M.; Almeyda, A.; Baldeosingh, K.; Almaguel, F.G.; De Leon, M. Metabolomics Uncovers Dietary Omega-3 Fatty Acid-Derived Metabolites Implicated in Anti-Nociceptive Responses after Experimental Spinal Cord Injury. *Neuroscience* **2013**, *255*, 1–18. [CrossRef] [PubMed]

76. Wojdasiewicz, P.; Poniatowski, Ł.A.; Turczyn, P.; Frasuńska, J.; Paradowska-Gorycka, A.; Tarnacka, B. Significance of Omega-3 Fatty Acids in the Prophylaxis and Treatment after Spinal Cord Injury in Rodent Models. *Mediat. Inflamm.* **2020**, *2020*, 3164260. [CrossRef] [PubMed]

77. Izzo, A.A.; Piscitelli, F.; Capasso, R.; Marini, P.; Cristino, L.; Petrosino, S.; di Marzo, V. Basal and Fasting/Refeeding-Regulated Tissue Levels of Endogenous PPAR-α Ligands in Zucker Rats. *Obesity* **2010**, *18*, 55–62. [CrossRef] [PubMed]

78. Baboota, R.K.; Murtaza, N.; Jagtap, S.; Singh, D.P.; Karmase, A.; Kaur, J.; Bhutani, K.K.; Boparai, R.K.; Premkumar, L.S.; Kondepudi, K.K.; et al. Capsaicin-Induced Transcriptional Changes in Hypothalamus and Alterations in Gut Microbial Count in High Fat Diet Fed Mice. *J. Nutr. Biochem.* **2014**, *25*, 893–902. [CrossRef] [PubMed]

79. Peterson, C.T.; Rodionov, D.A.; Iablokov, S.N.; Pung, M.A.; Chopra, D.; Mills, P.J.; Peterson, S.N. Prebiotic Potential of Culinary Spices Used to Support Digestion and Bioabsorption. *Evid. Based Complement. Altern. Med.* **2019**, *2019*, 8973704. [CrossRef] [PubMed]

80. Leyva-Diaz, A.A.; Hernandez-Patlan, D.; Solis-Cruz, B.; Adhikari, B.; Kwon, Y.M.; Latorre, J.D.; Hernandez-Velasco, X.; Fuente-Martinez, B.; Hargis, B.M.; Lopez-Arellano, R.; et al. Evaluation of Curcumin and Copper Acetate against Salmonella Typhimurium Infection, Intestinal Permeability, and Cecal Microbiota Composition in Broiler Chickens. *J. Anim. Sci. Biotechnol.* **2021**, *12*, 1–12. [CrossRef]

81. Kasai, C.; Sugimoto, K.; Moritani, I.; Tanaka, J.; Oya, Y.; Inoue, H.; Tameda, M.; Shiraki, K.; Ito, M.; Takei, Y.; et al. Comparison of the Gut Microbiota Composition between Obese and Non-Obese Individuals in a Japanese Population, as Analyzed by Terminal Restriction Fragment Length Polymorphism and next-Generation Sequencing. *BMC Gastroenterol.* **2015**, *15*, 100. [CrossRef]

82. Mikami, A.; Ogita, T.; Namai, F.; Shigemori, S.; Sato, T.; Shimosato, T. Oral Administration of Flavonifractor Plautii Attenuates Inflammatory Responses in Obese Adipose Tissue. *Mol. Biol. Rep.* **2020**, *47*, 6717–6725. [CrossRef]

83. Kang, C.; Zhang, Y.; Zhu, X.; Liu, K.; Wang, X.; Chen, M.; Wang, J.; Chen, H.; Hui, S.; Huang, L.; et al. Healthy Subjects Differentially Respond to Dietary Capsaicin Correlating with Specific Gut Enterotypes. *J. Clin. Endocrinol. Metab.* **2016**, *101*, 4681–4689. [CrossRef]

84. Perez-Burgos, A.; Wang, L.; McVey Neufeld, K.-A.; Mao, Y.-K.; Ahmadzai, M.; Janssen, L.J.; Stanisz, A.M.; Bienenstock, J.; Kunze, W.A. The TRPV1 Channel in Rodents Is a Major Target for Antinociceptive Effect of the Probiotic *Lactobacillus Reuteri* DSM 17938: Visceral Analgesic Effect of a *Lactobacillus* Depends on TRPV. *J. Physiol.* **2015**, *593*, 3943–3957. [CrossRef] [PubMed]

85. Iannotti, F.A.; Di Marzo, V. The Gut Microbiome, Endocannabinoids and Metabolic Disorders. *J. Endocrinol.* **2021**, *248*, R83–R97. [CrossRef]

86. Depommier, C.; Everard, A.; Druart, C.; Plovier, H.; Van Hul, M.; Vieira-Silva, S.; Falony, G.; Raes, J.; Maiter, D.; Delzenne, N.M.; et al. Supplementation with Akkermansia Muciniphila in Overweight and Obese Human Volunteers: A Proof-of-Concept Exploratory Study. *Nat. Med.* **2019**, *25*, 1096–1103. [CrossRef]

87. Hoareau, L.; Buyse, M.; Festy, F.; Ravanan, P.; Gonthier, M.-P.; Matias, I.; Petrosino, S.; Tallet, F.; D'Hellencourt, C.L.; Cesari, M.; et al. Anti-Inflammatory Effect of Palmitoylethanolamide on Human Adipocytes. *Obesity* **2009**, *17*, 431–438. [CrossRef]

88. Annunziata, C.; Lama, A.; Pirozzi, C.; Cavaliere, G.; Trinchese, G.; Di Guida, F.; Nitrato Izzo, A.; Cimmino, F.; Paciello, O.; De Biase, D.; et al. Palmitoylethanolamide Counteracts Hepatic Metabolic Inflexibility Modulating Mitochondrial Function and Efficiency in Diet-induced Obese Mice. *FASEB J.* **2020**, *34*, 350–364. [CrossRef]

89. Peters, B.A.; Shapiro, J.A.; Church, T.R.; Miller, G.; Trinh-Shevrin, C.; Yuen, E.; Friedlander, C.; Hayes, R.B.; Ahn, J. A Taxonomic Signature of Obesity in a Large Study of American Adults. *Sci. Rep.* **2018**, *8*, 9749. [CrossRef] [PubMed]

90. Wu, T.-R.; Lin, C.-S.; Chang, C.-J.; Lin, T.-L.; Martel, J.; Ko, Y.-F.; Ojcius, D.M.; Lu, C.-C.; Young, J.D.; Lai, H.-C. Gut Commensal *Parabacteroides Goldsteinii* Plays a Predominant Role in the Anti-Obesity Effects of Polysaccharides Isolated from *Hirsutella Sinensis*. *Gut* **2019**, *68*, 248–262. [CrossRef]

91. Wang, K.; Liao, M.; Zhou, N.; Bao, L.; Ma, K.; Zheng, Z.; Wang, Y.; Liu, C.; Wang, W.; Wang, J.; et al. Parabacteroides Distasonis Alleviates Obesity and Metabolic Dysfunctions via Production of Succinate and Secondary Bile Acids. *Cell Rep.* **2019**, *26*, 222–235.e5. [CrossRef]

92. Naderpoor, N.; Mousa, A.; Gomez-Arango, L.; Barrett, H.; Dekker Nitert, M.; de Courten, B. Faecal Microbiota Are Related to Insulin Sensitivity and Secretion in Overweight or Obese Adults. *JCM* **2019**, *8*, 452. [CrossRef] [PubMed]

93. Muñiz Pedrogo, D.A.; Jensen, M.D.; Van Dyke, C.T.; Murray, J.A.; Woods, J.A.; Chen, J.; Kashyap, P.C.; Nehra, V. Gut Microbial Carbohydrate Metabolism Hinders Weight Loss in Overweight Adults Undergoing Lifestyle Intervention With a Volumetric Diet. *Mayo Clin. Proc.* **2018**, *93*, 1104–1110. [CrossRef]

 *biomedicines* 

Article

Betulinic Acid Decorated with Polar Groups and Blue Emitting BODIPY Dye: Synthesis, Cytotoxicity, Cell-Cycle Analysis and Anti-HIV Profiling

David Kodr [1,†], Jarmila Stanková [2,†], Michaela Rumlová [3], Petr Džubák [2], Jiří Řehulka [2], Tomáš Zimmermann [1], Ivana Křížová [3], Soňa Gurská [2], Marián Hajdúch [2], Pavel B. Drašar [1] and Michal Jurášek [1,*]

[1] Department of Chemistry of Natural Compounds, University of Chemistry and Technology Prague, 16628 Prague, Czech Republic; david.kodr@vscht.cz (D.K.); tomas.zimmermann@vscht.cz (T.Z.); pavel.drasar@vscht.cz (P.B.D.)
[2] Institute of Molecular and Translational Medicine, Faculty of Medicine and Dentistry, Palacký University and University Hospital in Olomouc, 77900 Olomouc, Czech Republic; jarmila.stankova@upol.cz (J.S.); petr.dzubak@upol.cz (P.D.); jiri.rehulka@upol.cz (J.Ř.); sona.gurska@upol.cz (S.G.); marian.hajduch@upol.cz (M.H.)
[3] Department of Biotechnology, University of Chemistry and Technology Prague, 16628 Prague, Czech Republic; michaela.rumlova@vscht.cz (M.R.); ivana.krizova@vscht.cz (I.K.)
* Correspondence: michal.jurasek@vscht.cz
† These authors have contributed equally to this work.

Citation: Kodr, D.; Stanková, J.; Rumlová, M.; Džubák, P.; Řehulka, J.; Zimmermann, T.; Křížová, I.; Gurská, S.; Hajdúch, M.; Drašar, P.B.; et al. Betulinic Acid Decorated with Polar Groups and Blue Emitting BODIPY Dye: Synthesis, Cytotoxicity, Cell-Cycle Analysis and Anti-HIV Profiling. *Biomedicines* **2021**, *9*, 1104. https://doi.org/10.3390/biomedicines9091104

Academic Editor: Jun Lu

Received: 3 August 2021
Accepted: 21 August 2021
Published: 28 August 2021

Publisher's Note: MDPI stays neutral with regard to jurisdictional claims in published maps and institutional affiliations.

Abstract: Betulinic acid (BA) is a potent triterpene, which has shown promising potential in cancer and HIV-1 treatment. Here, we report a synthesis and biological evaluation of 17 new compounds, including BODIPY labelled analogues derived from BA. The analogues terminated by amino moiety showed increased cytotoxicity (e.g., BA had on CCRF-CEM $IC_{50} > 50$ µM, amine **3** IC_{50} 0.21 and amine **14** IC_{50} 0.29). The cell-cycle arrest was evaluated and did not show general features for all the tested compounds. A fluorescence microscopy study of six derivatives revealed that only **4** and **6** were detected in living cells. These compounds were colocalized with the endoplasmic reticulum and mitochondria, indicating possible targets in these organelles. The study of anti-HIV-1 activity showed that **8**, **10**, **16**, **17** and **18** have had $IC_{50i} > 10$ µM. Only completely processed p24 CA was identified in the viruses formed in the presence of compounds **4** and **12**. In the cases of **2**, **8**, **9**, **10**, **16**, **17** and **18**, we identified not fully processed p24 CA and p25 CA-SP1 protein. This observation suggests a similar mechanism of inhibition as described for bevirimat.

Keywords: betulinic acid; BODIPY; bevirimat; cytotoxicity; cancer; cell-cycle; fluorescent microscopy; maturation inhibitor

1. Introduction

Betulinic acid (BA) is a natural pentacyclic triterpene of the lupane type (Figure 1). Despite its low solubility in aqueous solutions, this substance is gaining attention with its wide range of interesting biological activity. BA is often derivatized to increase solubility, enhance the therapeutic effect, and target the drug to the specific site of action [1]. BA shows a significant degree of selectivity for cytotoxicity against a variety of tumour cells mboxciteB2-biomedicines-1332342,B3-biomedicines-1332342,B4-biomedicines-1332342 and activity against HIV-1 [5]. There are several possible mechanisms of action of BA (reviewed in [6]), which provide an advantage in the development of resistance to one of the mechanisms and may thus find application in the treatment of tumours resistant to current chemotherapeutics [6]. One is the direct action of BA on the mitochondrial membrane, leading to an increase of outer membrane permeability, its depolarization and release of cytochrome *c* into the cytosol. It is then responsible for triggering apoptosis [7]. Among other effects of BA, reactive oxygen species can be formed causing non-specific damage to mitochondria [8,9], followed by the induction of caspase activity [10]. BA exhibits topoisomerase I28 inhibitory activity, and through the proteasome-dependent independent regulatory pathway, is responsible for the function of the transcription factors Sp1, Sp3 and Sp4 inhibition [11]. It is also able to inhibit the activation of the stress transcription factor NF-κB [12]. A slightly different way in which tumour growth is inhibited is a complete or partial slowing of angiogenesis [13]. Later studies have shown that the antiangiogenic effect is achieved via modulation of mitochondria [14].

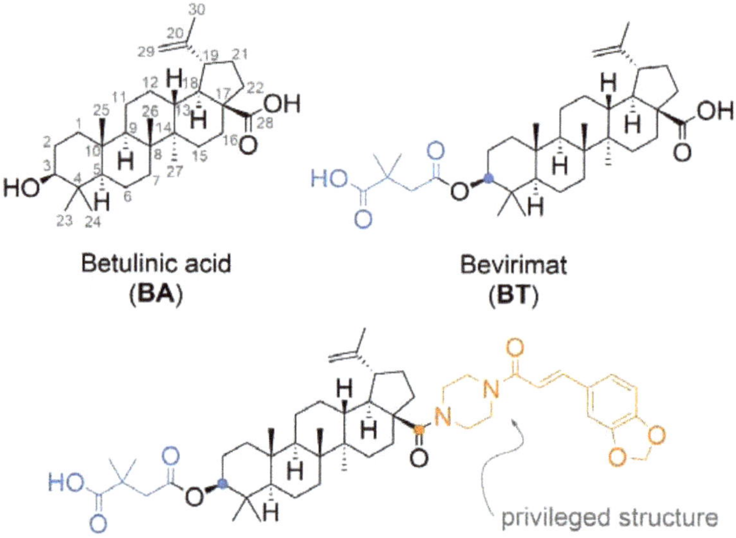

Figure 1. Chemical structure of betulinic acid and its derivatives.

BA has been shown to have anti-HIV-1 activity in the past. Although the test results were not groundbreaking, and the effect was observed only at relatively high concentrations [5] This discovery inevitably led to the synthesis of several other analogues. One of the derivatives with strong anti-HIV-1 activity was 3-*O*-(3,3-dimethylsuccinyl) betulinic acid, known as bevirimat (Figure 1, BT) [15]. BT acts as an inhibitor of HIV-1 particle maturation. Inhibition of viral particle maturation appears to be a critical point of therapeutic intervention. During the maturation phase, the viral protease cleaves the Gag polyprotein while releasing the individual structural proteins. The final step is the cleavage of p25 CA-SP1 to a functional p24 CA protein. Inhibition of the last step of maturation results in virus particles with aberrantly formed mature cores that are incapable of further infection [16]. BT advanced to the second phase of clinical testing [17–19], during which virus reduction was observed in only 40–50% of patients. The remainder of the patients developed resistance due to natural polymorphic variation in the Gag polyprotein [20]. With this result, the clinical studies were terminated.

Given the important features of BA mentioned above, it is no surprise that many research groups addressed it. Hundreds of derivatives have been prepared over the last few decades. However, with derivatization, for example, the expected effect disappeared, resistance developed rapidly, or toxicity to normal cells increased dramatically. For anti-HIV derivatives, several so-called "privileged structures" (Figure 1), structural motifs that can be the basis for the design of an effective drug, were found [21,22]. BA is most often chemically modified at C-3 and C-28 positions. Addition to the double bond between carbon atoms C-20 and C-30 usually does not significantly enhance activity, on the contrary, the activity often disappears. This finding generally applies to both anti-cancer and anti-HIV effects [23–25]. Recent works have confirmed that the presence of an extra amine group introduced by conjugation into a BA molecule can significantly increase antitumour potency [26,27].

This work presents the preparation and biological evaluation of new analogues of BA and BT containing an amino group. In the past, fluorescent analogues of BA labelled with green-emitting BODIPY (4,4-difluoro-4-bora-3a,4a-diaza-*s*-indacene) [28,29] and red-emitting Rhodamine B [30] were synthesized to study its localization and trafficking in living cells. In this work, we synthesized and studied new derivatives of BA and BT labelled at C-3 and C-28 positions using a small blue-emitting BODIPY dye.

2. Materials and Methods

2.1. Chemical Synthesis

Aluminium silica gel sheets for detection in UV light (TLC Silica gel 60 F254, Merck, Darmstadt, Denmark) were used for thin-layer chromatography (TLC), subsequent visualization was proceeded by a diluted solution of sulfuric acid in methanol and plates were heated. Silica gel (30–60 μm, SiliTech, MP Biomedicals, Costa Mesa, CA, USA) was used for column chromatography. NMR Spectra were recorded by Agilent-MR DDR2 (Santa Clara, CA, USA). HRMS were measured by LTQ ORBITRAP VELOS with HESI+/HESI-ionization (Thermo Scientific, Waltham, MA, USA). For microwave synthesis, an Initiator Classic 355,301 (Biotage, Uppsala, Sweden) was used.

The following chemicals were purchased from TCI Europe (Zwijndrecht, Belgium): *N,N,N*-triethylamine—Et$_3$N (>99%), 4-dimethylaminopyridine—4-DMAP (>99%), 1-(3-dimethylaminopropyl)-3-ethylcarbodiimide hydrochloride—EDCI (>98%), *N,N'*-dicyclohexylcarbodiimide—DCC (>98%), 1-hydroxybenzotriazole monohydrate—HOBt (>97%), triphenylphosphine—PPh$_3$ (>95%), *p*-toluenesulfonic acid monohydrate—*p*-TsOH (>98%), and palladium on carbon—Pd/C (10%). The following chemicals were purchased from Sigma-Aldrich (Prague, Czech Republic): 3-azidopropylamine (≥95%), 1-(2-*N*-boc-aminoethyl)piperazine (≥95%), β-alanine (99%). Betulinic acid (BA) was purchased from Betulinines (Stříbrná skalice, Czech Republic).

The solvents for column chromatography and reactions were purchased from PENTA (Praha, Czech Republic) and were used without further distillation.

Compound Synthesis and Characterization

8-*N*-(3-Azidopropyl)amino-4,4-difluoro-4-bora-3a,4a-diaza-*s*-indacene (**BODIPY-N$_3$**)

To a solution of **BODIPY-SMe** (205 mg, 0.86 mmol) in DCM (10 mL), 3-azidopropylamine (95 mg, 0.95 mmol) was added and the mixture was stirred for 30 min at RT. The solvents were evaporated under reduced pressure and the residue was taken up with AcOEt and the product was precipitated by the addition of hexanes. **BODIPY-N$_3$** (243 mg, 0.83 mmol) was obtained as a yellowish solid in 97% yield. R$_F$ = 0.55 in hexanes-AcOEt 1:1. ^{1}H NMR (400 MHz, CD3OD) δ ppm: 2.11 (quin, J = 6.7 Hz, 2 H), 3.56 (t, J = 6.5 Hz, 2 H), 3.87 (t, J = 7.4 Hz, 2 H), 6.39 (br. s., 1 H), 6.55 (br. s, 1 H), 7.32 (br. s, 1 H), 7.36 (br. s., 2 H), 7.57 (s, 1 H). ^{13}C NMR (101 MHz, CD$_3$OD) δ ppm: 27.00, 44.08, 48.76, 112.67, 113.99, 115.78, 123.16, 130.78, 133.76, 148.87. HRMS-ESI: calculated 290.12628 Da, found *m/z* 291.13312 [M+H]$^+$.

8-*N*-(3-Aminopropyl)amino-4,4-difluoro-4-bora-3a,4a-diaza-*s*-indacene (**BODIPY-NH$_2$**)

To a solution of **BODIPY-N$_3$** (150 mg, 0.52 mmol) in AcOEt (8 mL), was added Pd/C (80 mg) and the mixture was stirred under hydrogen atmosphere for 2 h. The catalyst was filtered off and the solvents were evaporated under reduced pressure. The residue was taken up with AcOEt and the product was obtained after precipitation with hexane. **BODIPY-NH$_2$** (96 mg, 0.36 mmol) was obtained as yellow solid in 70% yield. R$_F$ = 0.15 in DCM-MeOH 20:1 (*v/v*). ^{1}H NMR (400 MHz, CD$_3$OD) δ ppm: 1.98 (quin, J = 6.7 Hz, 2 H), 2.87 (t, J = 6.7 Hz, 2 H), 3.83 (t, J = 7.0 Hz, 2 H), 6.37 (br. s, 1 H), 6.52 (br. s, 1 H), 7.27 (br. s, 1 H), 7.30–7.36 (m, 2 H), 7.55 (br. s, 1 H). ^{13}C NMR (101 MHz, CD$_3$OD) δ ppm: 29.67, 39.12, 45.50, 112.53, 113.78, 115.51, 123.15, 130.40, 133.46, 148.73. HRMS-ESI: calculated 264.13578 Da, found *m/z* 263.12810 [M-H]$^-$.

8-*N*-(β-Alanyl)amino-4,4-difluoro-4-bora-3a,4a-diaza-*s*-indacene (**BODIPY-CO$_2$H**)

To a solution of **BODIPY-SMe** (220 mg, 0.92 mmol) in DMSO (5 mL), was added a solution of β-Ala (91 mg, 1.02 mmol) in H_2O (2 mL). The mixture was stirred at 30 °C for 16 h. The solvents were removed under reduced pressure and the residue was diluted with $CHCl_3$ (100 mL). The product was precipitated by the addition of cyclohexane. **BODIPY-CO$_2$H** (175 mg, 0.63 mmol) was obtained as yellow solids in 68% yield. R_F = 0.38 in hexanes-AcOEt 1:1. ^{1}H NMR (400 MHz, CD$_3$OD) δ ppm: 2.87 (t, J = 6.9 Hz, 2 H), 4.00 (t, J = 6.9 Hz, 2 H), 6.35 (br. s., 1 H), 6.53 (br. s., 1 H), 7.27–7.31 (m, 1 H), 7.33 (br. s., 2 H), 7.55 (br. s., 1 H). ^{13}C NMR (101 MHz, CD$_3$OD) δ ppm: 31.71, 42.48, 112.67, 114.02, 115.89, 123.26, 130.78, 133.83, 148.75, 173.03. HRMS-ESI: calculated 279.09906 Da, found m/z 278.09139 [M-H]$^-$.

(3β)-*N*-(3-Azidopropyl)-3-hydroxylup-20(29)-ene-28-amide (**1**)

To a solution of BA (200 mg, 0.44 mmol) and 4-DMAP (59 mg, 0.48 mmol) in DMF (3 mL), 3-azidopropylamine (53 mg, 0.53 mmol), HOBt (65 mg, 0.48 mmol) and EDCI (93 mg, 0.48 mmol) were sequentially added. The mixture was stirred at RT for 36 h. Solvents were removed under reduced pressure and the residue was chromatographed twice (i. DCM-MeOH 100:1, *v/v* ii. DCM→DCM-MeOH 70:1, *v/v*). Azide **1** (200 mg, 0.37 mmol) was obtained as white solids in 84% yield. R_F = 0.48 in DCM-MeOH 40:1 (*v/v*). ^{1}H NMR (400 MHz, CDCl$_3$) δ ppm: 0.63–0.70 (m, 1 H), 0.75 (s, 3 H), 0.81 (s, 3 H), 0.83–0.91 (m, 1 H), 0.93 (s, 3 H), 0.96 (s, 3 H), 0.96 (s, 3 H), 0.97–1.05 (m, 1 H), 1.12–1.65 (m, 19 H), 1.67 (s, 3 H), 1.69–1.72 (m, 1 H), 1.75–1.82 (m, 2 H), 1.90–1.95 (m, 1 H), 2.38–2.48 (m, 1 H), 3.08 3.20 (m, 2 H), 3.32 (s, 4 H), 4.58 (s, 1 H), 4.73 (s, 1 H), 5.85 (t, J = 5.9 Hz, 1 H); Figure S1. ^{13}C NMR (101 MHz, CDCl$_3$) δ ppm: 14.29, 15.01, 15.79, 15.80, 17.93, 19.13, 20.57, 25.26, 27.05, 27.63, 28.68, 29.10, 30.51, 33.40, 34.04, 36.61, 36.84, 37.40, 38.05, 38.36, 38.49, 40.40, 42.11, 46.38, 49.31, 49.74, 50.27, 55.02, 55.30, 78.58, 109.01, 150.50, 175.96; Figure S3. HRMS-ESI: calculated 538.42468 Da, found m/z 561.41394 [M+Na]$^+$; Figure S2.

4-{[(3β)-28-[(3-Azidopropyl)amino]-28-oxolup-20(29)-ene-3-yl]oxy}-2,2-dimethyl-4-oxobutanoic acid (**2**)

To a solution of **1** (200 mg, 0.37 mmol) and 4-DMAP (73 mg, 0.59 mmol) in THF (2 mL), 2,2-dimethylsucccinic anhydride (238 mg, 1.86 mmol) and *p*-TsOH were added and the mixture was stirred for 2 h at 130 °C in microwave reactor (MW). The mixture was diluted with H_2O (20 mL) and extracted with DCM (4 × 15 mL). Combined organic layers were dried over Na$_2$SO$_4$, filtered and the solvents were evaporated under reduced pressure. The residue was chromatographed two times (i. DCM-MeOH 100:1, *v/v*; ii. hexanes-AcOEt 3:1, *v/v*). Compound **2** (94 mg, 0.14 mmol) was obtained as white solids in 38% yield. R_F = 0.60 in hexanes-AcOEt, 1:1. ^{1}H NMR (400 MHz, CDCl$_3$) δ ppm: 0.73–0.76 (m, 1 H), 0.79 (s, 3 H), 0.82 (s, 6 H), 0.92 (s, 3 H), 0.95 (s, 3 H), 0.97–1.01 (m, 1 H), 1.12 1.17 (m, 1 H), 1.27 (s, 3 H), 1.29 (s, 3 H), 1.30–1.67 (m, 17 H), 1.67 (s, 3 H), 1.71 (d, J = 7.0 Hz, 1 H), 1.75–1.81 (m, 2 H), 1.90–1.93 (m, 1 H), 2.39–2.46 (m, 1 H), 2.52–2.68 (m, 2 H), 3.08 3.14 (m, 1 H), 3.22–3.33 (m, 2 H), 3.33–3.43 (m, 4 H), 4.45–4.50 (m, 1 H), 4.58 (s, 1 H), 4.73 (s, 1 H), 5.86 (t, J = 5.9 Hz, 1 H); Figure S4. ^{13}C NMR (101 MHz, CDCl3) δ ppm: 14.60, 16.14, 16.46, 18.13, 19.44, 20.92, 23.59, 24.99, 25.27, 25.55, 25.58, 27.88, 29.02, 29.42, 30.83, 33.73, 34.30, 36.97, 37.08, 37.70, 37.72, 38.40, 40.44, 40.76, 42.45, 44.69, 46.74, 49.65, 50.06, 50.50, 55.47, 55.65, 81.51, 109.42, 109.99, 150.81, 170.95, 176.38, 182.48; Figure S5. HRMS-ESI: calculated 666.47202 Da, found m/z 667.47921 [M+H]$^+$, 689.46125 [M+Na]$^+$ and 705.43463 [M+K]$^+$; Figure S6.

(3β)-*N*-(3-Aminopropyl)-3-hydroxylup-20(29)-ene-28-amide (**3**) [31]

A solution of **2** (339 mg, 0.63 mmol) and PPh$_3$ (248 mg, 0.95 mmol) in THF (10 mL) was stirred for 3h at RT. Water (1 mL) was added and the mixture was stirred for additional 20 h at RT. Solvents were evaporated under reduced pressure and the residue was chromatographed (CHCl$_3$-MeOH 20:1, *v/v* + 0.5% Et$_3$N → 10:1, *v/v* + 0.5% Et$_3$N). Compound **3** (278 mg, 0.54 mmol) was obtained as white solids in 86% yield. R_F = 0.15 in DCM-MeOH 10:1 (*v/v*) + 0.5% Et$_3$N. ^{1}H NMR (400 MHz, CD$_3$OD) δ ppm: 0.70–0.75 (m, 1 H), 0.77 (s, 3 H), 0.88 (s, 3 H), 0.91–0.96 (m, 1 H), 0.97 (s, 3 H), 0.99 (s, 3 H), 1.02 (s, 3 H), 1.04–1.10 (m, 1 H), 1.15–1.69 (m, 20 H), 1.71 (s, 3 H), 1.80–1.95 (m, 2 H), 2.10–2.19 (m, 1 H), 2.55–2.64 (m, 1 H), 2.69 (t, J = 6.9 Hz, 2 H), 3.07–3.18 (m, 2 H), 3.20–3.32 (m, 2 H), 4.60 (s, 1 H), 4.72 (s, 1 H);

Figure S7. ^{13}C NMR (101 MHz, CD$_3$OD) δ ppm: 13.80, 14.78, 15.46, 15.49, 18.07, 18.37, 20.78, 25.58, 26.64, 27.28, 29.22, 30.57, 32.01, 32.75, 34.24, 35.77, 36.95, 37.53, 38.11, 38.25, 38.57, 38.72, 40.61, 42.14, 46.67, 50.00, 50.68, 55.49, 55.57, 78.26, 108.63, 150.86, 177.88; Figure S8. HRMS-ESI: calculated 512.43418 Da, found *m/z* 513.44206 [M+H]$^+$; Figure S9.

(3β)-*N*-[*N'*-(4,4-Difluoro-4-bora-3a,4a-diaza-*s*-indacene-8-yl)-3-aminopropyl]-3-hydroxylup-20(29)-ene-28-amide (**4**)

To a solution of BA (50 mg, 0.11 mmol) and **BODIPY-NH$_2$** (32 mg, 0.12 mmol) in DMF (3 mL), 4-DMAP (15 mg, 0.12 mmol), HOBt (16 mg, 0.12 mmol) and EDCI (23 mg, 0.12 mmol) were added. The mixture was stirred for 20h at RT. The solvents were evaporated under reduced pressure and residue was chromatographed (hexanes-AcOEt 1:1). Compound **4** (56 mg, 0.08 mmol) was obtained as yellow-green solid in 73% yield. R$_F$ = 0.27 in hexanes-AcOEt 1:1. ^{1}H NMR (400 MHz, CDCl$_3$) δ ppm: 0.66–0.70 (m, 1 H), 0.74 (s, 3 H), 0.80 (s, 3 H), 0.84–0.89 (m, 1 H), 0.90 (s, 3 H), 0.96 (s, 3 H), 1.00 (s, 3 H), 1.01–1.12 (m, 1 H), 1.17–1.71 (m, 20 H), 1.72 (s, 3 H), 1.73–1.80 (m, 2 H), 1.87–2.04 (m, 4 H), 2.48 2.55 (m, 1 H), 3.14–3.21 (m, 2 H), 3.28–3.36 (m, 2 H), 3.73 (q, J = 5.5 Hz, 2 H), 4.65 (s, 1 H), 4.77 (s, 1 H), 6.27 (t, J = 5.9 Hz, 1 H), 6.45 (br. s., 1 H), 6.50 (br. s., 1 H), 7.12 (br. s, 1 H), 7.51 (br. s., 1 H), 7.67 (br. s., 2 H), 9.73 (t, J = 5.3 Hz, 1 H); Figure S10. ^{13}C NMR (101 MHz, CDCl$_3$) δ ppm: 14.67, 15.35, 16.12, 16.15, 18.25, 19.46, 20.96, 25.61, 27.38, 27.96, 29.15, 29.53, 30.88, 33.53, 34.34, 36.75, 37.17, 37.94, 38.48, 38.72, 38.84, 40.81, 42.48, 44.48, 46.90, 50.10, 50.58, 55.35, 55.72, 78.95, 109.78, 113.37, 114.28, 116.74, 122.42, 125.66, 131.71, 134.28, 134.30, 149.07, 150.40, 179.08; Figure S11. HRMS-ESI: calculated 702.48556 Da, found *m/z* 725.47504 [M+Na]+ and 741.44867 [M+K]+; Figure S12.

(3β)-*N*-(3-Azidopropyl)-3-[*N '*-(4,4-difluoro-4-bora-3a,4a-diaza-*s*-indacene-8-yl)-β-alanyl]-oxy-lup-20(29)-ene-28-amide (**5**)

To a solution of **1** (130 mg, 0.24 mmol) and 4-DMAP (59 mg, 0.48 mmol) in dry DCM (5 mL), **BODIPY-CO$_2$H** (101 mg, 0.36 mmol) and DCC (100 mg, 0.48 mmol) were added. The mixture was stirred at RT for 16 h. DCU was filtered off and the solvents were removed under reduced pressure. The crude was chromatographed (DCM-MeOH 100:1, *v/v*), and the material thus obtained was dissolved in AcOEt and precipitated by the addition of hexanes and chromatographed once again (DCM-MeOH 100:1, *v/v*) to obtain pure **5** (160 mg, 0.20 mmol) as yellow solid in 83% yield. R$_F$ = 0.62 in DCM-MeOH 40:1 (*v/v*). ^{1}H NMR (400 MHz, CDCl$_3$) δ ppm: 0.78–0.82 (m, 1 H), 0.85 (s, 3 H), 0.85 (s, 6 H), 0.94 (s, 3 H), 0.97 (s, 3 H), 0.99–1.05 (m, 2 H), 1.14–1.18 (m, 1 H), 1.22–1.66 (m, 17 H), 1.69 (s, 3 H), 1.72 (m, 1 H), 1.75–1.81 (m, 2 H), 1.89–1.97 (m, 1 H), 2.42–2.50 (m, 1 H), 2.80 (t, J = 6.3 Hz, 2 H), 3.09–3.16 (m, 1 H), 3.23–3.39 (m, 4 H), 3.94 (q, J = 6.1 Hz, 2 H), 4.56–4.60 (m, 1 H), 4.60 (s, 1 H), 4.74 (s, 1 H), 5.83 (t, J = 5.9 Hz, 1 H), 6.44 (br. s., 2 H), 7.01 (br. s., 2 H), 7.45–7.68 (m, 2 H), 7.73 (t, J = 5.3 Hz, 1 H); Figure S13. ^{13}C NMR (101 MHz, CDCl$_3$) δ ppm: 14.59, 16.17, 16.19, 16.54, 18.14, 19.47, 20.96, 23.73, 25.54, 28.06, 29.04, 29.43, 30.85, 32.63, 33.74, 34.27, 36.98, 37.11, 37.70, 37.87, 38.33, 38.39, 40.77, 42.48, 42.53, 46.74, 55.42, 55.65, 76.70, 77.02, 77.34, 82.86, 109.44, 114.00, 114.79, 115.12, 123.08, 132.60, 135.52, 147.98, 150.82, 171.54, 176.32; Figure S14. HRMS-ESI: calculated 799.51318 Da, found *m/z* 822.50287 [M+Na]$^+$ and 838.47620 [M+K]$^+$; Figure S15.

(3β)-*N*-(3-Aminopropyl)-3-[*N '*-(4,4-difluoro-4-bora-3a,4a-diaza-s-indacene-8-yl)-β-alanyl]-oxy-lup-20(29)-ene-28-amide (**6**)

To a solution of **5** (60 mg, 0.08 mmol) in dry THF (3 mL), PPh$_3$ (26 mg, 0.10 mmol) was added. After stirring for 3h, H$_2$O was added and the mixture was stirred for 20 h. Solvents were removed under reduced pressure and the product was chromatographed (DCM-MeOH 9:1, *v/v* → MeOH-H$_2$O 100:1, *v/v*) to obtain **6** (25 mg, 0.03 mmol) as yellow solids in 43% yield. R$_F$ = 0.12 in DCM-MeOH 10:1 (*v/v*) + 0.5% Et$_3$N. ^{1}H NMR (400 MHz, CDCl$_3$) δ ppm: 0.79–0.82 (m, 1 H), 0.85 (s, 3 H), 0.85 (s, 6 H), 0.95 (s, 3 H), 0.97 (s, 3 H), 0.98–1.03 (m, 2 H), 1.12–1.16 (m, 1 H), 1.26–1.65 (m, 22 H), 1.69 (s, 3 H), 1.72–1.77 (m, 2 H), 1.92–1.98 (m, 1 H), 2.45–2.51 (m, 1 H), 2.77–2.85 (m, 4 H), 3.11–3.18 (m, 1 H), 3.30–3.40 (m, 2 H), 3.98 (t, J = 6.3 Hz, 2 H), 4.57–4.61 (m, 2 H), 4.74 (s, 1 H), 6.40 (t, J = 5.5 Hz, 1 H), 6.46 (br. s., 2 H), 6.99–7.09 (m, 2 H), 7.49–7.72 (m, 3 H); Figure S16. ^{13}C NMR (101 MHz, CDCl$_3$)

δ ppm: 14.59, 16.17, 16.19, 16.54, 18.14, 19.48, 20.97, 23.74, 25.56, 28.07, 29.43, 30.89, 32.44, 32.67, 33.70, 34.28, 37.11, 37.52, 37.66, 37.87, 38.33, 38.49, 40.09, 40.76, 42.48, 42.52, 46.74, 50.07, 50.55, 55.42, 55.61, 82.90, 109.34, 114.15, 114.82, 115.40, 122.45, 132.64, 135.57, 148.06, 150.99, 171.58, 176.32; Figure S17. HRMS-ESI: monoisotopic mass 773.52268 Da, found m/z 774.53046 $[M+H]^+$ and 772.51587 $[M-H]^-$; Figure S18.

Tert-butyl-(3-{[(3β)-3-hydroxy-28-oxolup-20(29)-ene-28-yl]amino}propyl)-carbamate (**7**) [23]

To a solution of BA (1.00 g, 2.19 mmol) and 4-DMAP (295 mg, 2.41 mmol) in DMF (12 mL), *N*-boc-1,3-diaminopropane (459 mg, 2.63 mmol), HOBt (326 mg, 2.41 mmol) and EDCI (462 mg, 2.41 mmol) were added. The mixture was stirred for 48 h at RT. The solvents were evaporated under reduced pressure and the residue was chromatographed (DCM-MeOH 100:1, v/v). Compound **7** (900 mg, 1.47 mmol) was obtained as white solid in 67% yield. R_F = 0.34 in DCM-MeOH 40:1 (v/v). ^{1}H NMR (400 MHz, CDCl$_3$) δ ppm: 0.65 0.70 (m, 1 H), 0.75 (s, 3 H), 0.81 (s, 3 H), 0.85–0.90 (m, 1 H), 0.93 (s, 3 H), 0.96 (s, 3 H), 0.97 (s, 3 H), 0.98–1.05 (m, 2 H), 1.13–1.20 (m, 1 H), 1.22–1.44 (m, 8 H), 1.45 (s, 9 H), 1.46 1.68 (m, 11 H), 1.69 (s, 3 H), 1.69–1.81 (m, 2 H), 1.88–2.00 (m, 1 H), 2.00–2.07 (m, 1 H), 2.42–2.50 (m, 1 H), 3.10–3.28 (m, 5 H), 3.30–3.39 (m, 1 H), 4.59 (s, 1 H), 4.74 (s, 1 H), 4.91 (br. s., 1 H), 6.33 (br. s., 1 H); Figure S19. ^{13}C NMR (101 MHz, CDCl$_3$) δ ppm: 14.64, 15.35, 16.12, 16.14, 18.28, 19.48, 20.94, 25.64, 27.41, 27.97, 28.40 (s, 3 C) 29.48, 30.54, 30.92, 33.65, 34.39, 35.28, 37.20, 37.75, 38.51, 38.72, 38.84, 40.76, 42.47, 46.66, 50.04, 50.64, 55.38, 55.76, 67.07, 78.96, 79.32, 109.27, 151.05, 156.65, 176.50; Figure S20. HRMS-ESI: calculated 612.48661 Da, found m/z 613.49394 $[M+H]^+$, 635.47609 $[M+Na]^+$ and 651.44943 $[M+K]^+$; Figure S21.

4-{[(3β)-28-({3-[(*Tert*-butoxycarbonyl)amino]propyl}amino)-28-oxolup-20(29)-ene-3–yl]oxy}-2,2-dimethyl-4-oxobutanoic acid (**8**)

To a solution of **7** (400 mg, 0.65 mmol) and 4-DMAP (128 mg, 1.04 mmol) in THF (4 mL), 2,2-dimethylsuccinic anhydride (418 mg, 3.26 mmol) and *p*-TsOH were added. The mixture was stirred for 2 h at 130 °C in a microwave reactor. The mixture was poured into water and extracted with DCM (4 × 20 mL). Combined organic layers were washed with KHSO$_4$ (3 × 5 mL) and brine. The organic layer was dried over Na$_2$SO$_4$ and the solvents were evaporated under reduced pressure. The residue was chromatographed (DCM-MeOH 40:1, v/v + 1% Et$_3$N) to obtain **8** (260 mg, 0.35 mmol) as white solids in 54% yield. R_F = 0.18 in DCM-MeOH 40:1 (v/v) + 1% Et$_3$N. ^{1}H NMR (400 MHz, CDCl$_3$) δ ppm: 0.74–0.77 (m, 1 H), 0.80 (s, 3 H), 0.82 (s, 6 H), 0.92 (s, 3 H), 0.96 (s, 3 H), 0.97–1.02 (m, 1 H), 1.13–1.17 (m, 1 H), 1.25–1.27 (m, 2 H), 1.28 (s, 3 H), 1.30 (s, 3 H), 1.31–1.44 (m, 7 H), 1.45 (s, 9 H), 1.46–1.68 (m, 10 H), 1.69 (s, 3 H), 1.70–1.82 (m, 2 H), 1.90–1.97 (m, 1 H), 2.03 2.07 (m, 1 H), 2.41–2.48 (m, 1 H), 2.53–2.70 (m, 2 H), 3.10–3.40 (m, 7 H), 4.46–4.51 (m, 1 H), 4.59 (s, 1 H), 4.74 (s, 1 H), 4.94 (br. s., 1 H), 6.43 (br. s., 1 H); Figure S22. ^{13}C NMR (101 MHz, CDCl$_3$) δ ppm: 14.61, 16.09, 16.15, 16.47, 18.14, 19.45, 20.95, 23.60, 25.02, 25.58, 25.61, 27.90, 28.39, 29.46, 30.49, 30.88, 33.61, 34.30, 35.35, 37.08, 37.70, 37.92, 38.40, 38.52, 40.44, 40.72, 40.76, 42.46, 44.71, 46.64, 49.99, 50.52, 55.47, 55.80, 79.40, 81.55, 109.37, 150.97, 156.80, 171.00, 176.72, 182.24; Figure S23. HRMS-ESI: calculated 740.53395 Da, found m/z 741.54119 $[M+H]^+$ and 763.52302 $[M+Na]^+$; Figure S24.

4-{[(3β)-28-[(3-Aminopropyl)amino]-28-oxolup-20(29)-ene-3–yl]oxy}-2,2-dimethyl-4-oxo-butanooic acid hydrochloride (**9**)

Compound **6** (130 mg, 0.18 mmol) was dissolved in CHCl$_3$ (1.5 mL) and 2 M HCl solution in Et$_2$O (3 mL) was added slowly. The mixture was stirred for 1 h at RT under argon atmosphere. Solvents were evaporated under reduced pressure and the residue was sonicated for 20 min in Et$_2$O (5 mL). The product was collected by filtration and dried in vacuo. Compound **9** (102 mg, 0.16 mmol) was obtained as white solids in 91% yield. R_F = 0.1 in DCM-MeOH 9:1 (v/v). ^{1}H NMR (400 MHz, CD$_3$OD) δ ppm: 0.81–0.83 (m, 1 H), 0.86 (s, 6 H), 0.88 (s, 1 H), 0.89 (s, 3 H), 0.98 (s, 3 H), 1.02 (s, 3 H), 1.03–1.11 (m, 1 H), 1.19 (m, 1 H), 1.25 (s, 3 H), 1.26 (s, 3 H), 1.27–1.32 (m, 1 H), 1.36–1.65 (m, 15 H), 1.69 (s, 3 H), 1.70–1.76 (m, 2 H), 1.80–1.90 (m, 5 H), 2.10–2.15 (m, 1 H), 2.52–2.65 (m, 3 H), 2.90 2.96 (m, 2 H), 3.06–3.13 (m, 1 H), 3.23–3.30 (m, 2 H), 4.43–4.48 (m, 1 H), 4.59 (s, 1 H), 4.70 (s, 1 H), 7.89 (t, J = 5.9 Hz, 1 H); Figure S25. ^{13}C NMR (101 MHz, CD$_3$OD) δ ppm: 13.64, 15.36, 15.43,

15.66, 17.85, 18.13, 20.76, 23.25, 24.43, 24.79, 25.49, 27.06, 27.71, 29.22, 30.51, 32.59, 34.06, 35.22, 36.85, 36.86, 37.42, 37.58, 38.02, 38.03, 38.18, 39.94, 40.60, 42.13, 44.23, 49.90, 50.50, 55.46, 55.65, 81.22, 108.66, 150.77, 171.53, 179.14, 179.17; Figure S26. HRMS-ESI: calculated 640.48152 Da, found m/z 641.48955 $[M+H]^+$; Figure S27.

4-{[(3β)-28-{[*N*-(4,4-Difluoro-4-bora-3a,4a-diaza-s-indacene-8-yl)-3-aminopropyl]amino}-28-oxolup-20(29)-ene-3-yl]oxy}-2,2-dimethyl-4-oxobutanoic acid (**10**)

Compound **9** (15 mg, 0.02 mmol) and **BODIPY-SMe** (6 mg, 0.03 mmol) were dissolved in the mixture of $CHCl_3$ (2 mL) and THF (1 mL). To this solution, one drop of Et_3N was added. The mixture was stirred 30 min at RT. Solvents were evaporated under reduced pressure and the residue was chromatographed (DCM-MeOH 19:1, v/v + 1% Et_3N). The mixture was dissolved in AcOEt and washed with $KHSO_4$ (10% solution, 3×5 mL) and brine (1 × 5 mL). Organic layer was dried over Na_2SO_4 and the solvents were evaporated. Compound **10** (14 mg, 0.02 mmol) as yellow-green solid in 74% yield. R_F = 0.73 in DCM-MeOH 9:1 (v/v). ^{1}H NMR (400 MHz, $CDCl_3$) δ ppm: 0.73–0.77 (m, 1 H), 0.78 (s, 3 H), 0.81 (s, 3 H), 0.82 (s, 3 H), 0.83–0.85 (m, 1 H), 0.89 (s, 3 H), 0.99 (s, 3 H), 1.00–1.10 (m, 1 H), 1.16–1.21 (m, 1 H), 1.29 (s, 3 H), 1.31 (s, 3 H), 1.34–1.67 (m, 15 H), 1.72 (s, 3 H), 1.74 1.79 (m, 2 H), 1.87–2.03 (m, 5 H), 2.49–2.71 (m, 3 H), 3.14–3.20 (m, 1 H), 3.33 (q, J = 5.4 Hz, 2 H), 3.74 (q, J = 5.5 Hz, 2 H), 4.46–4.51 (m, 1 H), 4.65 (s, 1 H), 4.77 (s, 1 H), 6.24 (t, J = 6.1 Hz, 1 H), 6.44 (br. s., 1 H), 6.51 (br. s., 1 H), 7.12 (br. s., 1 H), 7.51 (br. s., 1 H), 7.68 (br. s., 2 H), 9.71 (t, J = 5.5 Hz, 1 H); Figure S28. ^{13}C NMR (101 MHz, $CDCl_3$) δ ppm: 14.63, 16.11, 16.15, 16.47, 18.09, 19.43, 20.99, 23.59, 24.97, 25.62, 27.88, 29.16, 29.51, 30.86, 33.56, 34.26, 36.73, 37.06, 37.70, 37.89, 38.40, 38.49, 40.45, 40.81, 42.47, 44.40, 44.70, 46.88, 50.08, 50.47, 55.46, 55.71, 81.51, 109.83, 113.40, 114.28, 116.67, 122.36, 125.65, 131.78, 134.36, 149.06, 150.35, 171.00, 179.13, 182.28; Figure S29. HRMS-APCI: calculated 830.53291 Da, found m/z 829.52699 $[M-H]^-$; Figure S30.

(3β)-28-(4-{2-[(*Tert*-butoxycarbonyl)amino]ethyl}piperazine-1-yl)-28-oxolup-20(29)-ene-3-yl acetate (**12**)

To a solution of compound **11** (1.27 g, 2.55 mmol) in DCM (20 mL), oxalyl chloride (1.2 mL) in DCM (10 mL) and 3 drops of DMF were added. After stirring for 2 h at RT, the solvents were co-evaporated with toluene (3 × 20 mL). Chloride thus obtained was dissolved in DCM (35 mL), and 1-(2-*N*-boc-aminoethyl)piperazine (876 mg, 3.82 mmol) followed by Et_3N (0.42 mL) were added. After stirring for 16 h at RT, the mixture was diluted with DCM (20 mL) and washed with brine (3 × 30 mL). The organic layer was dried over Na_2SO_4 and the solvents were evaporated under reduced pressure. The residue was chromatographed ($CHCl_3$-MeOH 100:1 → 50:1, v/v) to obtain product **12** (720 mg, 1.01 mmol) as white solids in 40% yield. R_F = 0.16 in DCM-MeOH 100:1 (v/v). ^{1}H NMR (400 MHz, $CDCl_3$) δ ppm: 0.77–0.80 (m, 1 H), 0.83 (s, 3 H), 0.84 (s, 3 H), 0.85 (s, 3 H), 0.94 (s, 3 H), 0.95 (s, 3 H), 0.97–1.00 (m, 1 H), 1.12–1.17 (m, 1 H), 1.28–1.42 (m, 9 H), 1.46 (s, 9 H), 1.48–1.66 (m, 7 H), 1.68 (s, 3 H), 1.70–1.74 (m, 1 H), 1.81–1.87 (m, 1 H), 1.93–1.98 (m, 1 H), 2.04 (s, 3 H), 2.07–2.11 (m, 1 H), 2.42 (br. s., 4 H), 2.49 (br. s., 2 H), 2.83–2.90 (m, 1 H), 2.94–3.01 (m, 1 H), 3.25 (br. s., 2 H), 3.61 (br. s., 4 H), 4.44–4.49 (m, 1 H), 4.58 (s, 1 H), 4.72 (s, 1 H), 4.98 (br. s., 1 H); Figure S31. ^{13}C NMR (101 MHz, $CDCl_3$) δ ppm: 14.62, 16.11, 16.24, 16.46, 18.18, 19.63, 21.15, 21.30, 23.70, 25.61, 27.93, 28.41, 29.79, 31.30, 32.46, 32.49, 34.31, 35.91, 36.86, 36.95, 37.14, 37.80, 38.41, 40.68, 41.85, 45.65, 50.76, 52.65, 53.12, 54.52, 55.52, 57.15, 79.30, 80.97, 109.16, 151.30, 155.90, 170.99, 173.46; Figure S32. HRMS-ESI: calculated 709.53937 Da, found m/z 710.54562 $[M+H]^+$ and 732.52546 $[M+Na]^+$; Figure S33.

Tert-butyl-(2-{4-[(3β)-3-hydroxy-28-oxolup-20(29)-ene-28-yl]piperazine-1-yl}ethyl)-carbamate (**13**)

To compound **12** (700 mg, 0.99 mmol) in MeOH (18 mL) and THF (9 mL), 4 M NaOH solution (9 mL) was added. The mixture was stirred for 2 h at RT. The mixture was neutralized by 1 M HCl solution and extracted with DCM (4 × 40 mL). Combined organic layers were washed with saturated brine (2 × 50 mL) and dried over Na_2SO_4. Solvents were evaporated under reduced pressure and the residue was chromatographed (DCM-MeOH 40:1, v/v) to obtain **13** (310 mg, 0.46 mmol) as white solids in 47% yield. R_F = 0.33 in

DCM-MeOH 20:1 (v/v). ^{1}H NMR (400 MHz, CDCl$_3$) δ ppm: 0.66–0.69 (m, 1 H), 0.75 (s, 3 H), 0.82 (s, 3 H), 0.86–0.90 (m, 1 H), 0.93 (s, 3 H), 0.96 (br. s, 6 H), 1.12–1.17 (m, 1 H), 1.23–1.29 (m, 2 H), 1.29–1.42 (m, 8 H), 1.45 (s, 9 H), 1.48–1.66 (m, 7 H), 1.68 (s, 3 H), 1.69–1.73 (m, 1 H), 1.81–1.87 (m, 1 H), 1.93–1.98 (m, 1 H), 2.06–2.11 (m, 1 H), 2.42 (br. s., 4 H), 2.48 (br. s., 2 H), 2.84–2.90 (m, 1 H), 2.94–3.00 (m, 1 H), 3.15–3.20 (m, 1 H), 3.25 (br. s, 2 H), 3.61 (br. s., 4 H), 4.57 (s, 1 H), 4.72 (s, 1 H), 4.99 (br. s., 1 H); Figure S34. ^{13}C NMR (101 MHz, CDCl$_3$) δ ppm: 14.67, 15.34, 16.11, 16.18, 18.30, 19.66, 21.14, 25.66, 27.43, 27.98, 28.41, 29.80, 31.32, 32.48, 32.50, 34.39, 35.91, 36.88, 36.97, 37.23, 38.74, 38.85, 40.67, 41.87, 45.62, 50.86, 52.67, 53.11, 54.53, 55.47, 57.15, 78.97, 79.28, 109.10, 151.33, 155.91, 173.45; Figure S35. HRMS-ESI: calculated 667.52881 Da, found m/z 668.53693 [M+H]$^+$; Figure S36.

(3β)-28-[4-(2-Aminoethyl)piperazine-1-yl]-3-hydroxylup-20(29)-ene-28-one hydrochloride (**14**)

To compound **13** (200 mg, 0.30 mmol) in CHCl$_3$ (2.8 mL), 2 M HCl solution in Et$_2$O (5.1 mL) was slowly added. The mixture was stirred for 16 h at RT under argon. Solvents were removed under reduced pressure and the crude was sonicated for 20min in Et$_2$O (5 mL), collected by filtration and dried. Compound **14** (142 mg, 0.25 mmol) was isolated as white solid in 84% yield. ^{1}H NMR (400 MHz, CD$_3$OD) δ ppm: 0.69–0.73 (m, 1 H), 0.75 (s, 3 H), 0.86 (s, 3 H), 0.88–0.93 (m, 1 H), 0.95 (s, 3 H), 0.96 (s, 3 H), 1.01 (s, 3 H), 1.02–1.07 (m, 1 H), 1.22–1.67 (m, 18 H), 1.70 (s, 3 H), 1.73–1.77 (m, 1 H), 1.82–1.88 (m, 1 H), 1.97–2.02 (m, 1 H), 2.11–2.15 (m, 1 H), 2.79–2.86 (m, 1 H), 2.89–2.95 (m, 1 H), 3.11–3.15 (m, 1 H), 3.35–3.81 (m, 11 H), 4.59 (s, 1 H), 4.70 (s, 1 H); Figure S37. ^{13}C NMR (101 MHz, CD$_3$OD) δ ppm: 13.65, 14.68, 15.27, 15.38, 18.04, 18.29, 20.86, 25.52, 26.63, 26.67, 27.18, 29.66, 30.92, 31.88, 33.62, 33.63, 34.19, 36.94, 36.96, 38.53, 38.68, 40.54, 41.64, 45.78, 50.78, 52.19, 52.37, 53.15, 54.58, 55.52, 78.24, 108.66, 150.85, 174.45; Figure S38. HRMS-ESI: calculated 567.47638 Da, found m/z 568.4832 [M+H]$^+$; Figure S39.

(3β)-28-{4-[N-(4,4-Difluoro-4-bora-3a,4a-diaza-s-indacene-8-yl)-2-aminoethyl]piperazine-1-yl}-3-hydroxylup-20(29)-ene-28-one (**15**)

To compound **14** (45 mg, 0.08 mmol) and **BODIPY-SMe** (21 mg, 0.09 mmol) in the mixture of DCM (5 mL) and THF (2.5 mL), was added one drop of Et$_3$N and the mixture was stirred 30 min at RT. Solvents were evaporated under reduced pressure and the residue was chromatographed (DCM-MeOH 100:1, v/v) to obtain **15** (42 mg, 0.06 mmol) as yellow solids in 70% yield. R$_F$ = 0.19 in DCM-MeOH 100:1 (v/v). ^{1}H NMR (400 MHz, CDCl$_3$) δ ppm: 0.67–0.71 (m, 1 H), 0.76 (s, 3 H), 0.83 (s, 3 H), 0.87–0.91 (m, 1 H), 0.95 (s, 3 H), 0.97 (s, 6 H), 1.16–1.21 (m, 1 H), 1.21–1.32 (m, 3 H), 1.38–1.66 (m, 15 H), 1.69 (s, 3 H), 1.72–1.76 (m, 1 H), 1.82–1.89 (m, 1 H), 1.93–1.98 (m, 1 H), 2.06–2.12 (m, 1 H), 2.55 (br. s., 4 H), 2.82–2.90 (m, 3 H), 2.95–3.01 (m, 1 H), 3.17–3.21 (m, 1 H), 3.70 (br. s., 4 H), 3.75 (br. s., 2 H), 4.60 (s, 1 H), 4.74 (s, 1 H), 6.41 (br. s., 1 H), 6.53 (br. s., 1 H), 6.91 (br. s., 1 H), 7.13 (br. s., 1 H), 7.51 (br. s., 1 H), 7.72 (br. s., 1 H), 7.92 (br. s., 1 H); Figure S40. ^{13}C NMR (101 MHz, CDCl$_3$) δ ppm: 14.68, 15.34, 16.18, 18.32, 19.64, 21.14, 25.45, 25.64, 27.43, 27.97, 29.86, 31.31, 32.52, 34.42, 35.93, 36.92, 37.24, 38.74, 38.86, 40.69, 41.80, 41.90, 45.62, 50.84, 52.53, 52.65, 54.26, 54.60, 55.46, 78.98, 109.29, 113.76, 114.93, 122.67, 123.28, 125.04, 132.47, 135.78, 147.77, 151.12, 173.64; Figure S41. HRMS-ESI: calculated 757.52776 Da, found m/z 780.5171 [M+Na]$^+$ and 796.4904 [M+K]$^+$; Figure S42.

4-{[(3β)-28-(4-{2-[(*Tert*-butoxycarbonyl)amino]ethyl}piperazine-1-yl)-28-oxolup-20(29)-ene-3-yl]oxy}-2,2-dimethyl-4-oxobutanoic acid (**16**)

To a solution of **13** (500 mg, 0.75 mmol) and 4-DMAP (146 mg, 1.20 mmol) in THF (5 mL), 2,2-dimethylsuccinic anhydride (480 mg, 3.74 mmol) and a catalytic amount of *p*-TsOH were added. The reaction was stirred for 2 h at 130 °C in a microwave reactor. The mixture was diluted with H$_2$O and extracted with DCM (4 × 20 mL). Combined organic layers were dried over Na$_2$SO$_4$ and the solvents were removed under reduced pressure. The crude was chromatographed twice (i. toluene-AcOEt 1:1 → AcOEt; ii. toluene-AcOEt 1:1 + 1% Et$_3$N → AcOEt → DCM-MeOH 9:1, v/v). Compound **16** (328 mg, 0.41 mmol) was obtained as white solids in 53% yield. R$_F$ = 0.43 in DCM-MeOH 9:1 (v/v). ^{1}H NMR (400 MHz, CDCl$_3$) δ ppm: 0.74–0.77 (m, 1 H), 0.81 (s, 6 H), 0.83 (s, 3 H), 0.93 (s, 3 H), 0.95 (s,

3 H), 1.12–1.16 (m, 1 H), 1.27 (br. s., 6 H), 1.32–1.41 (m, 8 H), 1.44 (s, 9 H), 1.47–1.65 (m, 7 H), 1.68 (s, 3 H), 1.70–1.74 (m, 1 H), 1.80–1.86 (m, 1 H), 1.90–1.95 (m, 1 H), 2.05–2.09 (m, 1 H), 2.49–2.69 (m, 6 H), 2.82–2.89 (m, 1 H), 2.93–3.00 (m, 1 H), 3.29 (br. s., 2 H), 3.61 (br. s., 4 H), 3.71 (br. s., 2 H), 4.45–4.50 (m, 1 H), 4.58 (s, 1 H), 4.72 (s, 1 H), 5.15 (br. s., 1 H); Figure S43. ^{13}C NMR (101 MHz, CDCl$_3$) δ ppm: 14.67, 16.06, 16.21, 16.57, 18.23, 19.64, 21.14, 23.66, 23.68, 25.57, 25.58, 25.92, 25.94, 27.91, 28.40, 29.82, 31.30, 32.36, 32.39, 34.26, 35.96, 36.69, 36.82, 37.12, 37.76, 38.39, 38.41, 40.63, 41.87, 45.60, 50.72, 52.60, 52.94, 54.49, 55.53, 57.19, 57.20, 109.24, 128.19, 129.00, 151.20, 155.99, 173.43; Figure S44. HRMS-ESI: calculated 795.57615 Da, found m/z 796.58292 [M+H]$^+$; Figure S45.

4-{[(3β)-28-[4-(2-Aminoethyl)piperazine-1-yl]-28-oxolup-20(29)-ene-3-yl]oxy}-2,2-dimethyl-4-oxobutanoic acid hydrochloride (**17**)

To compound **16** (150 mg, 0.19 mmol) in CHCl$_3$ (2 mL), 2 M HCl solution in Et$_2$O (3.2 mL) was added. The mixture was stirred for 2 h at RT under argon atmosphere. Solvents were evaporated and the crude was sonicated for 20 min in Et$_2$O (5 mL). The precipitate was collected and dried in vacuo. Compound **17** (102 mg, 0.16 mmol) was obtained as white solids in 91% yield. ^{1}H NMR (400 MHz, CD$_3$OD) δ ppm: 0.82–0.85 (m, 1 H), 0.87 (br. s, 6 H), 0.91 (s, 3 H), 0.98 (s, 3 H), 0.99–1.02 (m, 1 H), 1.04 (s, 3 H), 1.05–1.10 (m, 1 H), 1.27 (s, 3 H), 1.28 (s, 3 H), 1.29–1.69 (m, 8 H), 1.71 (s, 3 H), 1.73–1.78 (m, 2 H), 1.83–1.89 (m, 1 H), 1.98–2.04 (m, 1 H), 2.13–2.18 (m, 1 H), 2.54–2.67 (m, 2 H), 2.81–2.87 (m, 1 H), 2.91–2.97 (m, 1 H), 3.05–3.77 (m, 11 H), 4.45–4.49 (m, 1 H), 4.61 (s, 1 H), 4.72 (s, 1 H); Figure S46. ^{13}C NMR (101 MHz, CD$_3$OD) δ ppm: 13.72, 15.25, 15.40, 15.67, 17.87, 18.29, 20.87, 20.90, 23.27, 24.44, 24.79, 25.43, 25.45, 27.09, 29.66, 30.91, 31.85, 33.65, 34.06, 36.89, 36.92, 37.44, 38.21, 39.94, 40.54, 41.67, 44.24, 45.78, 50.65, 52.19, 52.32, 53.16, 54.56, 55.52, 81.25, 108.69, 150.83, 171.53, 174.40, 179.15; Figure S47. HRMS-ESI: calculated 695.52372 Da, found m/z 696.53153 [M+H]$^+$; Figure S48.

4-{[(3β)-28-{4-[*N*-(4,4-Difluoro-4-bora-3a,4a-diaza-*s*-indacene-8-yl)-2-aminoethyl]piperazine-1-yl}-28-oxolup-20(29)-ene-3-yl]oxy}-2,2-dimethyl-4-oxobutanoic acid (**18**)

To compound **17** (50 mg, 0.07 mmol) and **BODIPY-SMe** (19 mg, 0.08 mmol) in the mixture of CHCl$_3$ (5 mL) and THF (3 mL), a drop of Et$_3$N was added. The mixture was stirred 30 min at RT. Solvents were evaporated and the residuum was chromatographed (DCM-MeOH 20:1, v/v + 0.5% Et$_3$N). Compound **18** (39 mg, 0.04 mmol) was obtained as yellowish solid in 62% yield. R$_F$ = 0.22 in DCM-MeOH 20:1 (v/v) + 0.5% Et$_3$N. ^{1}H NMR (400 MHz, CDCl$_3$) δ ppm: 0.75–0.78 (m, 1 H), 0.81 (s, 3 H), 0.83 (s, 3 H), 0.84 (s, 3 H), 0.88 0.92 (m, 1 H), 0.94 (s, 3 H), 0.96 (s, 3 H), 1.15–1.19 (m, 1 H), 1.28 (s, 3 H), 1.30 (s, 3 H), 1.35 1.42 (m, 7 H), 1.45–1.51 (m, 2 H), 1.51–1.64 (m, 6 H), 1.69 (s, 3 H), 1.72–1.76 (m, 1 H), 1.83 1.89 (m, 1 H), 1.93–1.98 (m, 1 H), 2.06–2.12 (m, 1 H), 2.51–2.70 (m, 6 H), 2.82–2.89 (m, 3 H), 2.94–3.00 (m, 1 H), 3.68 (br. s., 4 H), 3.75 (br. s., 2 H), 4.46–4.51 (m, 1 H), 4.59 (s, 1 H), 4.73 (s, 1 H), 6.39 (br. s., 1 H), 6.52 (br. s., 1 H), 6.91 (br. s., 1 H), 7.13 (br. s., 1 H), 7.50 (br. s., 1 H), 7.71 (br. s., 1 H), 7.96 (br. s., 1 H); Figure S49. ^{13}C NMR (101 MHz, CDCl$_3$) δ ppm: 14.67, 16.13, 16.21, 16.50, 18.18, 19.60, 21.17, 23.65, 25.05, 25.60, 27.92, 29.68, 29.84, 31.29, 32.49, 34.32, 35.94, 36.88, 37.12, 37.74, 38.42, 40.48, 40.68, 41.88, 41.96, 44.75, 45.65, 50.71, 52.57, 52.61, 54.28, 54.57, 55.55, 81.47, 109.33, 113.68, 114.87, 122.61, 123.38, 125.04, 132.41, 135.68, 147.79, 151.11, 171.11, 173.64, 182.29; Figure S50. HRMS-ESI: calculated 885.57511 Da, found m/z 884.56855 [M-H]$^-$; Figure S51.

2.2. Biochemistry

2.2.1. Cell Lines

All cells (if not indicated otherwise) were purchased from the American Tissue Culture Collection (ATCC; Manassas, VA, USA). The highly chemosensitive CCRF-CEM line is derived from T lymphoblastic leukaemia, K562 represent cells of chronic myelogenous leukaemia. Colorectal adenocarcinoma HCT116 cell line and its p53 gene knockout counterpart (HCT116p53−/−, Horizon Discovery Ltd., Cambridge, UK) were used as models to assess the impact of p53 deficiency on cell line sensitivity. A549 cells are derived from lung adenocarcinoma and U2OS from human osteosarcoma. CEM-DNR and K562-Tax are well-characterized daunorubicin and paclitaxel-resistant sublines of CCRF-CEM and K562. The CEM-DNR resistant cells overexpress the P-glycoprotein and LRP protein, the K562-Tax overexpress P-glycoprotein but is losing the expression of LRP, which is present at parental K562 cell line. P-glycoprotein belongs to the ABC transporters' family and is involved in the primary and acquired multidrug resistance phenomenon by the efflux of toxic compounds, LRP protein is involved in the lysosomal degradation. MRC-5 and BJ cell lines were used as a non-tumour control and represent human fibroblasts. The cells were maintained in Nunc/Corning 80 cm^2 plastic tissue culture flasks and cultured in cell culture medium according to ATCC or Horizon recommendations (DMEM/RPMI 1640 with 5 g/L-glucose, 2 mM glutamine, 100 U/mL penicillin, 100 mg/mL streptomycin, 10% fetal calf serum, and $NaHCO_3$).

2.2.2. MTS Assay

To perform the cytotoxicity MTS assay, cell suspensions were prepared and diluted according to the cell type and the expected target cell density (25,000–35,000 cells/mL) based on cell growth characteristics. Cells were added by an automatic pipettor (30 µL) into 384 well microtiter plates. All tested compounds were dissolved in 100% DMSO and four-fold dilutions of the intended test concentration were added in 0.15 µL aliquots at time zero to the microtiter plate wells by the echo-acoustic liquid handler Echo550 (Labcyte, San Jose, CA, USA). The experiments were performed in technical duplicates and at least three biological replicates. The cells were incubated with the tested compounds for 72 h at 37 °C, in a 5% CO_2 atmosphere at 99% humidity. At the end of the incubation period, the cells were assayed by using the MTS test. Aliquots (5 µL) of the MTS stock solution were pipetted into each well and incubated for an additional 1–4 h. After this incubation period, the optical density (OD) was measured at 490 nm with an Envision microplate reader (Perkin Elmer, Waltham, Massachusetts, USA). Tumour cell survival (TCS) was calculated using the following equation: TCS = ($OD_{\text{drug-exposed well}}$/mean $OD_{\text{control wells}}$) × 100%. The IC_{50} value, the drug concentration that is lethal to 50% of the tumour cells, was calculated from the appropriate dose-response curves in Dotmatics software (The Old Monastery, Windhill, Bishop´s Stortford, Herts, UK).

2.2.3. Cell Cycle and Apoptosis Analysis

CCRF-CEM cells were seeded in 6-well plates at a density of 1 × 106/well. After 24 h, compounds at concentrations corresponding to 1× or 5 × IC_{50} were added to the wells and incubated for 24 h. Cells were then harvested, washed with cold 1 × PBS and fixed in ice-cold 70% ethanol. Fixed cells were incubated overnight at −20 °C, washed in hypotonic citrate buffer, treated with RNase (50 µg mL^{-1}) and incubated with propidium iodide for 15 min. DNA content was analysed using Becton Dickinson flow cytometer and cell cycle data were analysed in the program ModFitLT (Verity, Carrollton, TX, USA). Apoptosis was measured in a logarithmic model expressing the percentage of the particles with propidium content lower than cells in G0/G1 phase (<G1) of the cell cycle. The mitotic marker pH3Ser10 antibody (Sigma) and secondary anti-mouse-FITC antibody (Sigma) were used for labelling and subsequent flow cytometry analysis of ethanol-fixed CCRF-CEM cells.

2.2.4. BrDU Incorporation Analysis

Cells were cultivated as in the method above and pulse-labelled with 10 µM 5-bromo-2-deoxyuridine (BrDU) for 30 min before collection to the test tubes. The cells were washed with cold 1 × PBS and fixed in ice-cold 70% ethanol. Before analysis, they were washed with 1 × PBS and incubated in 2M HCl for 30 min at room temperature. Following neutralization with 0.1M $Na_2B_4O_7$ (borax), the cells were washed with 0.5% Tween-20 and 1% BSA in 1 × PBS. The cell pellets were stained using a primary anti-BrdU antibody (Exbio, Vestec, Czech Republic) for 30 min at room temperature and a secondary anti mouse FITC antibody (Sigma). The samples were then incubated with propidium iodide (0.1 mg mL^{-1}), treated with RNase A (0.5 mg mL^{-1}) for 1 h at room temperature in the dark and analysed as above.

2.2.5. BrU Incorporation Analysis

Cells were cultured, treated as above, pulse-labelled with 1 mM 5-bromouridine (BrU) for 30 min and fixed in 1% buffered paraformaldehyde with 0.05% NP-40 at room temperature for 15 min. Following overnight incubation at 4 °C, they were washed with 1% glycine in $1 \times$ PBS, washed with $1 \times$ PBS again and stained with primary anti-BrdU antibody cross-reacting to BrU (Exbio) for 30 min and secondary anti-mouse-FITC antibody (Sigma). The analysis was performed similarly to the BrDU analysis.

2.2.6. Fluorescent Microscopy

U2OS cell line (ATCC, USA) was transduced with premade lentiviral particles (Vectalis-TaKaRa, Japan) with sequences that express fluorescent protein tag mCherry targeted to specific subcellular locations. All cell lines were prepared according to the vendor's instructions. The U2OS-Nuc cell line was prepared by using rLV.EF1.mCherry-Nuc-9 (cat. n. 0023VCT), containing a NLS sequence that imports protein into the nucleus. The U2OS-ER cell line was transduced by rLV.EF1.mCherry-ER-9 (cat. n. 0025VCT), which contains a calreticulin signal sequence and a KDEL sequence that associates protein with the endoplasmic reticulum. The U2OS-GA cell line was transduced by rLV.EF1.mCherry-Golgi-9 (cat. n. 0022VCT), containing a human GT precursor, a protein localized in Golgi Apparatus. The U2OS-Mito cell line was prepared by using rLV.EF1.mCherry-Mito-9 (cat. n. 0024VCT), containing a mitochondrial targeting sequence.

U2OS cells with fluorescent fusion proteins (density 1.0×103 per well) were seeded into 384 CellCarrier plates (Perkin Elmer, Waltham, MA, USA) and pre-incubated for 24 h at 37 °C and 5% CO_2. The attached cells were treated with tested compounds in concentration 10 µM for 1 h and subsequently rinsed with fresh media. The live-cell imaging was performed by Cell Voyager CV7000 (Yokogawa, Tokyo, Japan) spinning disc confocal microscopy system at 37 °C in a 5% CO_2 atmosphere. Live cells were monitored by a $60 \times$ water immersion objective. The fluorescent signal was excited by lasers (405 nm and 561 nm) and the emission was filtered by bandpass filters (BP 445/45 and BP 595/20). All images were post-processed, and Pearson's and Mander's coefficients were calculated using the JACoP plugin in Image-J software.

2.2.7. VSV-G Pseudotyped HIV-1 Particles Production

HIV-1 particles were obtained from HEK 293 cells, cotransfected by a combination of three vectors: packaging psPAX2 vector encoding HIV Gag, Pol, Tat and Rev, reporter/transfer pWPXLd-GFP vector encoding LTR, RRE and GFP as a reporter, and envelope pHEF-VSV-G vector, encoding vesicular stomatitis virus Env, VSV-G. The psPAX2 vector [32] was kindly provided by Dr. Luban, the pWPXLd-GFP and pHEF-VSV-G vectors were purchased from Addgene (Watertown, MA, USA).

HEK-293 cells were grown in Dulbecco's Modified Eagle Medium (DMEM, Sigma) supplemented with 10% fetal bovine serum (Sigma) and 1% L-glutamine (Sigma) at 37 °C under 5% CO_2. A day before transfection, cells were plated at 3×105 cells per well. The following day, cells were transfected with the appropriate vectors using polyethylenimine (PEI, 1 mg/mL) at a 2:1 PEI:DNA ratio. Four hours post-transfection, the culture medium was replaced with fresh DMEM, containing various concentrations of tested compounds, solubilized in DMSO. At 48 h post-transfection, the culture media containing released virions were harvested, filtered through 0.45-µm pores membrane and used for immunochemical quantification and characterization by ELISA and Western blot using rabbit anti-HIV-1 CA antibody.

2.2.8. Single-Round Infectivity Assay

The infectivity was determined similarly as described earlier [33–35]. Briefly, 48 h post-transfection, the culture media from HEK 293 cells transfected with psPAX2, pWPXLd-GFP and pHEF-VSV-G vectors at a ratio 1:1:1 in the presence of tested compounds were collected and filtered through a 0.45-μm filter. HIV-1 CA content was determined by ELISA [33]. The freshly seeded HEK 293 cells were infected with ELISA-normalized amounts of VSV-G pseudotyped HIV-1 particles and incubated for 48 h. The cells were fixed with 2% paraformaldehyde and transferred to a FACS tube. Quantification of GFP-positive cells was performed using a BD FACS Aria III flow cytometer (BD Life Sciences, San Jose, CA, USA).

2.2.9. Western Blot

At 48 h post-transfection, 100 μL aliquots of virus-containing culture media were combined with 20 μL of PLB (6×) and the samples were analysed by Western blot using rabbit anti-HIV-1 CA (in house production). Proteins were resolved by reducing SDS-PAGE (12%) and blotted onto a nitrocellulose membrane. The antigen-antibody complexes were detected by Clarity™ Western ECL Substrate (Biorad, Hercules, CA, USA) and visualized using the FUSION 7S system (Vilber Lourmat, Marne-la-Vallée, France).

3. Results and Discussion

3.1. Chemistry

The synthesis of fluorescent labels was based on 8-thiomethyl BODIPY (**BODIPY-SMe**; Figure 2), which was prepared in our laboratory, according to the procedure previously described in the literature [36]. The thiomethyl group is reactive towards amines. After this reaction, secondary amines are formed with significant fluorescence characterized by emission in the blue region of the spectrum. For the preparation of betulinic acid conjugates, a carboxy-terminated derivative (**BODIPY-CO$_2$H**, Figure 2) was prepared by reaction of **BODIPY-SMe** and β-alanine [37] and an amino-terminated derivative by reaction with 3-azidopropan-1-amine [38] and reduction of azide (**BODIPY-N$_3$**, Figure 2) to amine (**BODIPY-NH$_2$**, Figure 2) by catalytic hydrogenation.

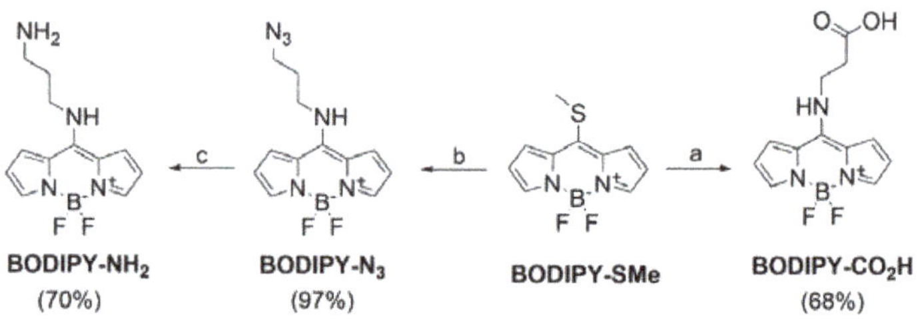

Figure 2. Synthesis of functionalized BODIPY dyes. Reagents and conditions: (**a**) β-Ala, DMSO-H$_2$O, 30 °C, 12 h; (**b**) 3-azidopropylamine, DCM, 30 min, RT; (**c**) H$_2$, Pd/C, AcOEt, 2h, RT.

Betulinoyl azidopropylamide (*N*-(3-azidopropyl)-3β-hydroxylup-20(29)-en-28-amide) **1** was prepared by reacting BA with 3-azidopropan-1-amine using carbodiimide chemistry (Figure 3A). The reaction was catalysed by EDCI (1-ethyl-3-(3-dimethylaminopropyl)carbodiimide) in the presence of 4-DMAP (4-dimethylaminopyridine) and HOBt (1-hydroxybenzotriazole). The bevirimat derivative **2** was prepared from compound **1** by reaction with 2,2-dimethylsuccinic anhydride according to a protocol reported in the literature [39]. By Staudinger reduction [40] catalysed by triphenylphosphine in aqueous THF, the azido group of compound **1** was reduced to amino derivative **3**. In an effort to reduce derivative **2** by the same method, a non-separable mixture of products was obtained. By the reaction of BA with **BODIPY-NH$_2$** catalysed by DCC (*N*,*N'*-dicyclohexylcarbodiimide) in the presence of 4-DMAP, derivative **4** was obtained. This reaction proceeded without difficulty in good yield (Figure 3A). Azide 1 was further conjugated at the C-3 position with **BODIPY-CO$_2$H** by Steglich esterification [41] to produce derivative **5**. The azide group of derivative **5** was reduced by Staudinger reduction to amine **6**. When attempting to modify compound **4** with dimethylsuccinic anhydride under the conditions used to prepare derivative **2**, degradation of the fluorescent label occurred, probably due to too high a temperature. Therefore, another synthetic procedure using a *tert*-butoxycarbonyl protecting group (Boc) on the terminal amino group was chosen for the synthesis of other "aminopropyl" derivatives (Figure 3B). The *N*-Boc-1,3-diaminopropane linker was conjugated to BA to give compound **7**, which could already be used to prepare the bevirimat derivative **8**. The protecting group was removed in an acidic environment to give amine **9**. From compound **9**, a fluorescently labelled derivative of bevirimat was prepared by the reaction with **BODIPY-SMe**.

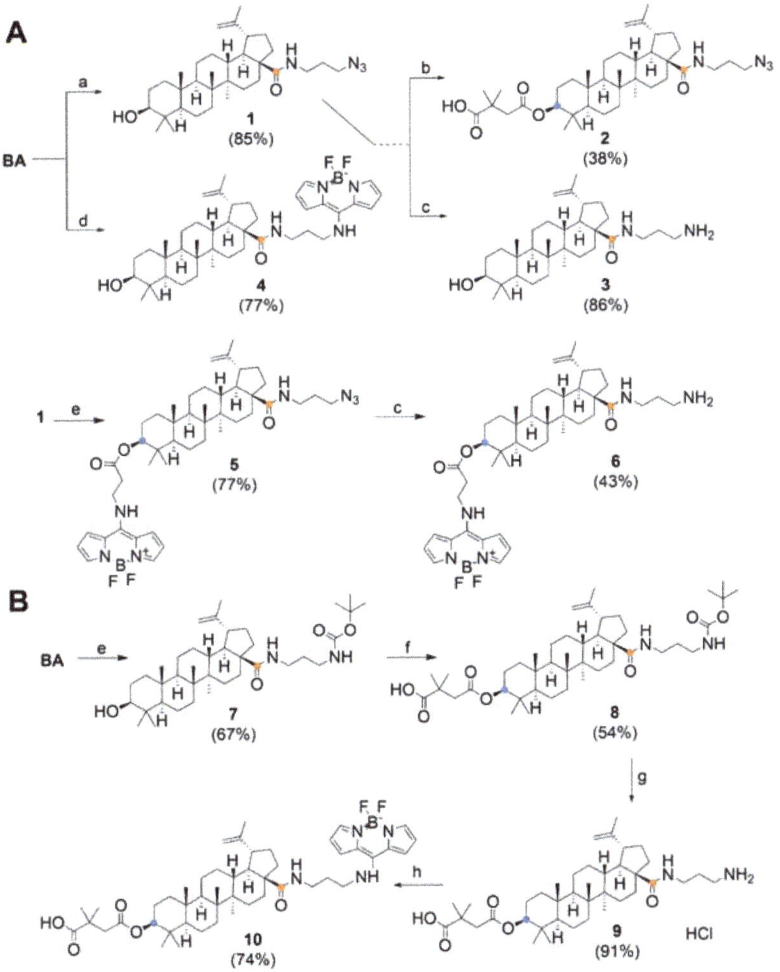

Figure 3. Synthesis of "aminopropyl" derivatives by azide reduction (panel **A**) and Boc chemistry (panel **B**). Reagents and conditions: (a) 3-azidopropylamine, 4-DMAP, HOBt, EDC, DMF, 36 h, RT; (b) 2,2-dimethylsuccinic anhydride, 4-DMAP, *p*-TsOH, THF, 2 h, MW-130 °C; (c) PPh$_3$, THF/H$_2$O, 23 h; (d) BODIPY-NH$_2$, 4-DMAP, DCC, DCM, 12 h, RT; (e) BODIPY-CO$_2$H, DCC, 4-DMAP, DCM, 12 h, RT; (f) *N*-Boc-1,3-diaminopropane, EDC, HOBt, 4-DMAP, DMF, 48 h, RT; (g) 2M HCl/Et$_2$O, 12 h, RT (under argon); (h) BODIPY-SMe, CHCl$_3$-THF, Et$_3$N, 30 min, RT.

Analogous to the synthetic procedure shown in Figure 3B, a series of substances with a piperazine linker at position C-28 was prepared (Figure 4). The introduction of the piperazine motif was chosen on the basis of promising results for the so-called privileged structures published previously [21]. The exception was that the C-3 hydroxyl was first acetylated to produce compound **11**, and in the next step, the carboxyl group was activated to reactive chloride. After condensation with 1-(2-*N*-Boc-aminoethyl) piperazine, tertiary amide **12** was obtained. Deacetylation of **12** occurred in a relatively low yield; however, part of the starting material was recovered during the separation of the reaction mixture.

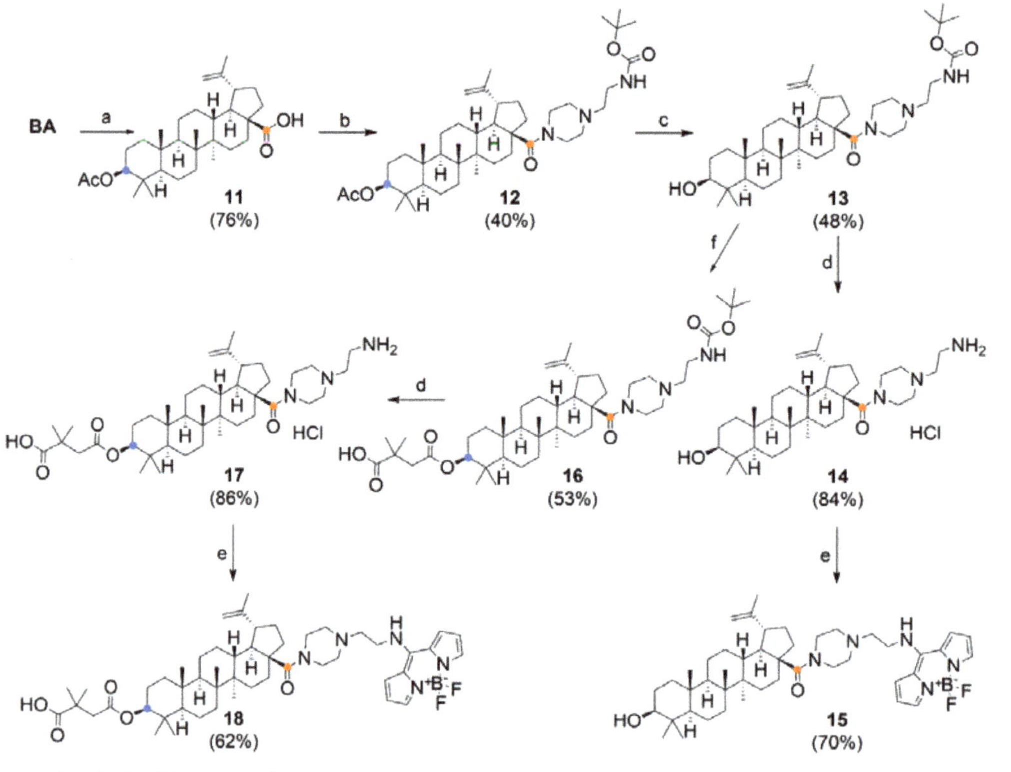

Figure 4. Synthesis of "piperazinyl" derivatives. Reagents and conditions: (a) Ac$_2$O, pyridine, 12 h, RT; (b) i. (COCl)$_2$, DCM, DMF, 2 h, RT, ii. 1-(2-*N*-boc-aminoethyl)piperazine, Et$_3$N, DCM, 12 h, RT; (c) 4M NaOH, THF-MeOH, 3 h, RT; (d) 2M HCl/Et$_2$O, 12 h, RT (under argon); (e) BODIPY-SMe, CHCl$_3$-THF, Et$_3$N, 30 min, RT; (f) 2,2-dimethyl succinic anhydride, 4-DMAP, *p*-TsOH, THF, 2 h, MW-130 °C.

Experimental details of the preparation of substances are described in Section 2.1. and the NMR, HRMS spectra (Figures S1–S51) and photochemical properties (Figure S52 and Table S1) of the substances are shown in the Supplementary Material.

3.2. Cytotoxicity on a Panel of Cell Lines

The in vitro cytotoxicity of derivatives was assessed using MTS assay on the normal human foreskin and lung fibroblasts BJ and MRC-5 and cancer cell lines of a different histogenetic type (Table 1). Under the experimental conditions, BA and BT showed a weak or medium cytotoxic effect directed against cancer cell lines. Structures **5**, **15** and **16** did not induce any cytotoxic effect in the entire cell line panel at the maximal tested concentration. Derivatives **4**, **9**, **12**, **17** and **18** were inactive against the entire cell line panel except for the CCRF-CEM lymphoblastic leukaemia cell line. IC$_{50}$ values obtained for these compounds in the sensitive cell line CCRF-CEM were between 5.76 and 23.65 µM. Derivatives **3**, **6**, **13** and **14** exerted high cytotoxicity against the entire cell line panel, including normal fibroblasts. The most potent compounds in the study were structures **3** and **14** with IC_{50} values 0.21 and 0.29 µM in CCRF-CEM. Derivatives **2**, **8** and **10** displayed medium cytotoxicity across the cell line panel. Derivatives **1** and **7** showed activity only against selected cell lines in the panel. Betulinic acid intermediate **11** was not tested. The MTS assay did not reveal any effect directed specifically against cancer cell lines, IC_{50} values calculated for normal fibroblast and cancer cell lines were highly comparable. Resistant sublines CEM-DNR and K562-Tax displayed for some compounds different sensitivity compared to their parental cell lines. As expected, a lower sensitivity was observed in the CEM-DNR resistant subline. The biggest difference in favour of CEM-DNR was observed for derivatives **6** and **3**. **BA** and **13** showed an opposite profile in CEM-DNR and **1**, **2**, **3**, **8** and **10** in the K562-Tax resistant subline, proposing better activity in resistant cell lines. Based on this data, we can speculate that there is a different mechanism in the elimination of cytotoxic derivatives. Several tested compounds are probably substrates of the P-glycoprotein as **4**, **6**, **13**, **16**, **17** and **18**. However, not all data are in conclusion with P-glycoprotein transport, and we think that several tested derivatives could be substrates for LRP protein. Higher LRP expression in CEM-DNR and lower in K562-Tax correlates with cytotoxicity of the derivatives **1**, **2**, **3**, **8**, **10**. Derivative **13** is not active in the highly chemosensitive CCRF-CEM cell line, but comparable activity was observed in all tested cell lines, including non-tumour lines.

3.3. Cell Cycle, Apoptosis and DNA/RNA Synthesis

To reveal cytostatic effects, we examined proliferation markers and cell cycle profile of the sensitive CCRF-CEM cell line following a 24 h incubation with the derivatives (Table 2).

Exposure to $1 \times IC_{50}$ and $5 \times IC_{50}$ concentrations of derivatives did not induce DNA fragmentation with the exception of high doses of **2** and **8**. Treatment with $1 \times IC_{50}$ concentrations did not modulate cell cycle profile while $5 \times IC_{50}$ concentration led in all samples to a more pronounced effect. The treatment with $5 \times IC_{50}$ derivatives **2**, **12** and **14** increased the percentage of cells in the S-phase by about 50% compared to untreated control. Nevertheless, there was not any other prominent effect on the cell cycle profile or cell cycle arrest. To assess the impact of structures on CCRF-CEM proliferation potential, we monitored mitotic marker pH3Ser10 and proliferation marker BrDU after 24 h incubation with the compounds. Analysis of mitotic marker showed a low rate of cell division in cells treated with $5 \times IC_{50}$ concentration of derivatives **2**, **3**, **6**, **12** and **14**. Derivatives **3**, **6**, **8**, **10**, and **14** reduced the fraction of proliferating BrDU positive CCRF-CEM cells. In contrast, structures **2**, **12** and **18** increased the percentage of cells incorporating BrDU into the DNA during pulse labelling. The complementary BrU based method of monitoring newly synthesized RNA in cells pre-incubated for 24 with the selected derivatives revealed stalled RNA synthesis induced by $5 \times IC_{50}$ concentration of derivatives **2**, **3**, **8**, **10**, **14** and **17**. Compound **18** at a high concentration increased the percentage of BrU positive cells. Such an increase indicates the high transcription activity as a mark of replication stress leading to DNA damage and cell death [42]. Although there was observed a slight modulation of cell cycle profile induced by compound derivatives **2**, **12** and **14**, the overall cell cycle data indicates that there is no general cytostatic effect of tested betulinic acid derivatives.

Table 1. Summary of cytotoxic activities (IC$_{50}$, μM).

Compound / Cell Line [a]	BA	BT	1	2	3	4	6	7	8	9	10	12	13	14	16	17	18
CCRF-CEM	>50	12.82	8.98	8.14	0.21	23.65	1.55	>50	8.18	22.48	2.92	9.62	>50	0.29	>50	5.76	8.61
CEM-DNR	23.05	22.17	16.84	10.23	1.22	>50	11.53	>50	9.24	>50	7.00	>50	4.76	0.35	>50	>50	>50
K562	>50	23.60	>50	25.99	0.90	>50	5.25	>50	19.18	>50	23.19	>50	5.09	0.40	47.90	>50	>50
K562-Tax	>50	22.03	10.58	15.94	0.37	>50	31.80	>50	13.30	>50	12.21	>50	8.77	0.52	>50	>50	>50
A549	22.68	23.06	>50	22.87	1.80	>50	6.65	21.29	18.84	>50	13.55	>50	5.15	1.26	44.93	>50	47.80
HCT116	>50	14.17	>50	19.40	0.82	>50	3.85	>50	13.21	>50	7.92	>50	6.02	0.39	46.63	>50	30.84
HCT116p53−/−	>50	18.20	>50	29.24	0.44	>50	3.39	>50	21.56	>50	8.80	>50	4.99	0.44	44.76	>50	45.50
U2OS	29.69	27.63	29.04	22.22	0.89	>50	5.00	18.39	17.16	>50	12.38	>50	4.16	0.42	44.62	44.58	>50
MRC-5	>50	>50	>50	24.19	2.59	>50	8.07	17.58	23.14	>50	14.12	>50	5.18	1.58	44.61	>50	>50
BJ	>50	>50	>50	25.33	1.91	>50	8.37	20.69	21.54	>50	15.49	>50	5.36	1.59	47.63	>50	>50

[a] Cytotoxic activity was determined by MTS assay following 3-day incubation. Values represent means of IC$_{50}$ from three independent experiments with SD ranging from 10–25% of the average values. Tested cell lines: CCRF-CEM (childhood T acute lymphoblastic leukaemia), CEM-DNR (CCRF-CEM daunorubicin resistant), K562 (chronic myelogenous leukaemia), K562-Tax (K562 paclitaxel-resistant), A549 (lung adenocarcinoma), HCT116 (colorectal cancer), HCT116p53−/− (null p53 gene), and U2OS (osteosarcoma). Normal human cell lines: MRC-5 and BJ (normal cycling fibroblasts). BA, betulinic acid; BT, bevirimat.

Table 2. Effect of cytotoxic compounds on cell cycle, apoptosis and DNA/RNA synthesis in CCRF-CEM lymphoblasts (% of positive cells).

Compound	<G1	G0/G1	S	G2/M	pH3$^{\text{Ser10}}$ [a]	BrDU [b]	BrU [c]
control	2.20	40.77	37.64	21.60	1.77	37.22	40.77
1 ●	1.36	37.03	44.13	18.84	1.53	36.54	52.69
1 ●	1.93	42.33	35.87	21.80	1.71	37.99	44.47
2 ●	4.33	39.88	46.65	13.47	1.33	52.07	28.60
2 ●	41.17	23.94	56.21	19.85	0.58	22.75	4.05
3 ●	1.30	35.07	36.54	28.40	2.04	41.88	50.56
3 ●	3.53	31.73	41.18	27.09	0.16	2.06	0.40
6 ●	2.23	33.39	49.74	16.87	1.50	44.97	42.25
6 ●	5.24	37.16	39.52	23.32	0.50	7.10	26.05
8 ●	2.64	31.93	47.72	20.35	1.70	46.44	33.61
8 ●	39.65	40.78	37.66	21.56	1.25	8.51	1.85
10 ●	3.80	37.98	44.59	17.43	1.44	42.25	35.04
10 ●	10.69	44.58	42.32	13.10	1.25	10.18	13.02
12 ●	2.45	33.64	43.16	23.20	1.40	59.73	41.98
12 ●	3.01	29.75	56.69	13.55	0.68	64.07	34.00
14 ●	8.63	21.46	56.29	22.25	0.15	24.55	1.56
14 ●	8.16	27.28	51.82	20.90	0.18	0.52	1.53
17 ●	2.89	34.83	46.96	18.21	1.56	48.07	26.70
17 ●	4.04	43.42	38.47	18.10	1.49	29.41	3.39
18 ●	2.06	36.06	44.31	19.63	1.25	55.81	46.98
18 ●	2.97	38.01	44.12	17.87	1.55	49.64	70.23

Flow cytometry analysis was used for quantification of cell cycle distribution and apoptotic cells with a concentration of compounds equal to $1 \times \text{IC}_{50}$ (●) and $5 \times \text{IC}_{50}$ (●) values. [a] phospho-Histone (Ser10); [b] 5-bromo-2-deoxyuridine; [c] BrU, 5-bromouridine.

3.4. Live Cells Imaging

The group of six derivatives of BA and BODIPY was studied on the U2OS-Nuc cell line with the nucleus labelled by fluorescein protein mCherry. The functionalized BODIPY dyes (**BODIPY-CO$_2$H** and **BODIPY-NH$_2$**), as well as precursor **BODIPY-SMe**, were used as a control. All fluorescent microscopic images of this pilot experiment are shown in Figure S53. To achieve a better specificity of the staining, we have focused on the short incubation with the fluorescent conjugates. After short incubation (1 h), conjugates **4** and **6** out of this group of derivatives were localized in living cells, but only with the weak signal in the nucleus of the studied cell line (Figure 5B—Pearson's and Mander's coefficients). The functionalized BODIPY dyes were not detected in the U2OS-Nuc cell line and thus it is highly possible that cellular uptake of conjugates **4** and **6** is due to their groups on BA residue. Other studied derivatives of BA and BODIPY were not detected in living cells under our experimental conditions; however, it is possible that the signal can be observed at later intervals. **BODIPY-SMe** is reactive due to the 8-thiomethyl group and it was predicted to penetrate cell compartments; this was confirmed by fluorescent microscopy.

To further study the cellular localization of conjugates **4** and **6**, we decided to continue with fluorescent microscopy on cell lines with fluorescently labelled structures of mitochondria, endoplasmic reticulum, and Golgi apparatus, which are the most published targets of BA [29,30]. The results of these colocalization experiments are shown in Figure 5. Both conjugates demonstrated presence in multiple cellular structures. Pearson's coefficient (Figure 5B) showed the highest correlation of signal in U2OS-ER cell line, and then in U2OS-Mito cell line and the lowest correlation was measured in U2OS-GA cell line. When we expressed colocalization by Mander's coefficient (overlap of red channel compared to the blue channel), which is more specific for colocalization calculation of signal presented in multiple cellular structures, the obtained data showed both conjugates **4** and **6** almost perfectly label mitochondria and endoplasmic reticulum. The lowest colocalization signal was again detected in the U2OS-GA cell line. **BODIPY-SMe** was used based on the data from the pilot experiment as a positive control with perfect colocalization in all studied cell lines. Images with entire microscopic fields are shown in Figures S54–S56.

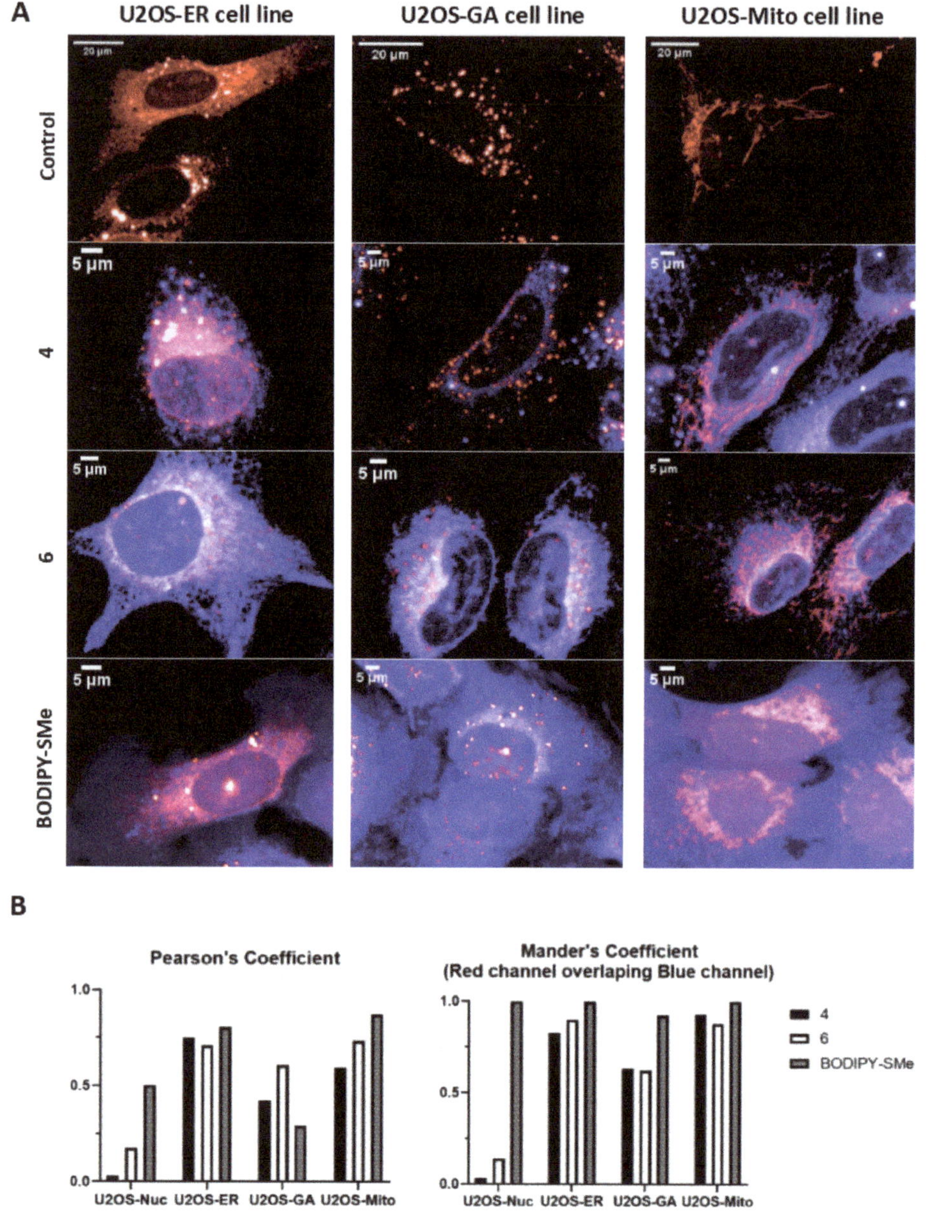

Figure 5. Live cell imaging and colocalization experiments of active compounds (**4**, **6**) and **BODIPY-SMe** (panel **A**) and visualization of Pearson´s and Mander´s coefficient (panel **B**).

To conclude our results from fluorescent microscopy study of six derivatives of BA and BODIPY, only conjugates **4** and **6** are detected in living cells under our experimental conditions (1 h following the treatment). Furthermore, we were able to almost perfectly colocalize both conjugates with cellular structures as endoplasmic reticulum and mitochondria, which is in agreement with data published in the past [21]. Compound **4** has, in its structure, fluorophore attached to the carboxyl group at the C-28 position, thus it is more similar to the pristine structure of BA and has low cytotoxicity (close to free BA). Conversely, compound **6** contains a conjugated amine at the C-28 position and the fluorophore is attached to the hydroxyl group at the C-3 position of BA. Its cytotoxicity is markedly more pronounced than in the case of substance **4**. It is clear that the "polar head" of the molecule is responsible for cytotoxicity. Moreover, this moiety can be used for the intracellular targeted delivery, or organelle/mitochondrion targeting, as is described by the previous research works [43,44]. As the localization of both compounds is similar, it is likely that the direct target remained unchanged, but the effect of compound **6** was potentiated by the presence of free amine moiety in the molecule. The localization of the compounds in lipid rich compartments (mitochondria, endoplasmic reticulum) can also be explained by the lipid character of the BA and its analogues. The calculated values of lipophilicity (logP) of the substances are close to BA (Table S2). The acidity constants (pKa) are indicative and their reproducibility is difficult because, in comparison with BA, the compounds described here are mostly in the form of amide or ester derivatives.

3.5. Anti-HIV Activity

Bevirimat (3-O-(3′,3′-dimethylsuccinyl) betulinic acid) and its derivatives were shown to be maturation inhibitors of HIV-1 [45–47]. By binding to the CA-SP1 region of HIV-1 Gag polyprotein, bevirimat prevents HIV-1 protease-mediated release of C-terminal part of CA from a spacer peptide 1 (SP1) [48]. This results in a block of the final step of virus maturation and subsequently abolishes HIV-1 infectivity. An atomic model of HIV-1 CA-SP1 suggested that this inhibitor stabilizes the CA-SP1 structure, thus preventing the proteolytic cleavage [49]. Although bevirimat is a potent inhibitor of HIV-1 maturation, its clinical development was discontinued in 2010 due to the bevirimat resistance caused by Gag SP1 natural polymorphism (Q6, V7 and T8) [50–52]. However, bevirimat derivatives with modification at the C-28 position seem to overcome the problem with HIV-1 resistance [53,54]. Here, using VSV-G pseudotyped HIV-1 particles, we tested the effect of 17 BA derivatives on HIV-1 maturation and infectivity. The 50% cytotoxic concentration (IC$_{50}$) of the compounds was first evaluated by Resazurin assay. Two of the tested compounds, **3** and **14**, were highly toxic to HEK 293 cells at a concentration lower than 5 μM and significant cytotoxicity was also found for compound **6** (IC$_{50}$ 12 μM) (Table 3).

Table 3. Cytotoxicity and anti-HIV-1 activity of the tested compounds [a].

Compd.	1	2	4	5	6	7	8	9	10	12	13	15	16	17	18
● IC$_{50}$ [μM]	>40	36.4	>40	>40	12.0	>40	37.8	>40	>40	>40	>40	>40	>40	>40	>40
● IC$_{50i}$ [μM]	>50	11.7	44.1	>50	n.d.	>50	1.4	14.0	8.4	31.9	>50	>50	9.1	7.6	7.1

[a] HEK 293 cells were grown in the presence or absence of tested compounds at a concentration ranging from 5 to 40 μM. The viability of the cells was determined by Resazurin assay 48 h later (●). To determine the effect of the compounds on HIV-1 infectivity (●), HEK 293 cells were transfected with the lentiviral vectors and treated with the tested compounds. The cells producing HIV-1 particles in the presence or absence of DMSO (at a final concentration of 1%) were used as controls. At 48 h post-transfection, the content of HIV-1 capsid (CA) protein from the culture media was quantified by ELISA and normalized amounts of VSV-G pseudotyped HIV-1 viruses were used to infect fresh HEK 293 cells. HIV-1 infectivity was determined 48 h later by quantification of GFP-producing cells by flow cytometry. The 50% infection inhibition (IC$_{50i}$) was defined as the concentration of the compound that reduced the HIV-1 infectivity by 50% compared to the untreated controls.

Apart from these three cytotoxic compounds, 14 fewer toxic compounds were used in the HIV-1 single-round infectivity assay. HIV-1 particles pseudotyped with VSV glycoproteins were produced in HEK 293 cells in the presence of tested compounds. At 48 h post-transfection, the content of HIV-1 capsid (CA) protein from the culture media was quantified by ELISA and normalized amounts of VSV-G pseudotyped HIV-1 viruses were used to infect fresh HEK 293 cells. At 48 h post-infection, the HIV-1 infectivity was determined by quantification of GFP-producing cells by flow cytometry. The 50% infection inhibition (IC$_{50i}$) was defined as the concentration of the compound that reduced the HIV-1 infectivity by 50% compared to the untreated controls (Table 3). The compounds **1**, **7**, **13**, **15** and **5** did not exhibit any potent anti-HIV-1 activity (data not shown). Conversely, compounds **2**, **4**, **9** and **12** inhibited anti-HIV-1 activity with IC$_{50i}$ from 11.7 to 44.1 μM. The compounds **8**, **10**, **16**, **17** and **18** inhibited HIV-1 with IC$_{50i}$ below 10 μM (Table 3). To analyse whether these bevirimat derivatives also act as maturation inhibitors of CA-SP1 cleavage, the HIV-1 virions released from the HEK 293 cells treated with the selected compounds (**2**, **4**, **8**, **9**, **10**, **12**, **16**, **17** and **18**) were analysed by Western blot using anti-HIV-1 CA antibody (Figure 6).

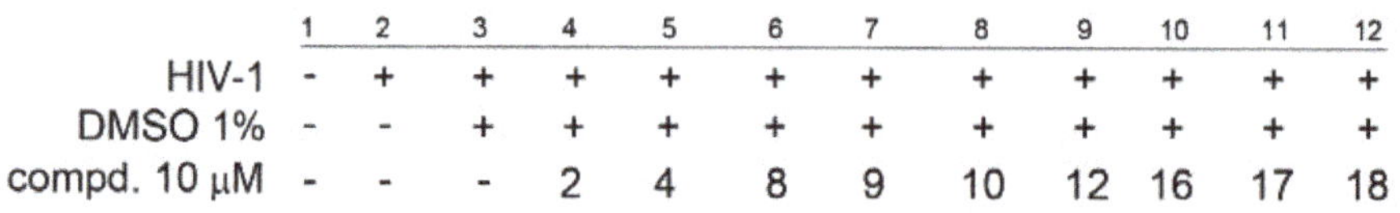

Figure 6. Effect of selected tested compounds on CA-SP1 processing of HIV-1 Gag polyprotein. HEK 293 cells produced HIV-1 particles pseudotyped with VSV-glycoproteins in the absence (lanes 2 and 3) or presence of selected tested compounds (lanes 4–12). At 48 h post-transfection, VSV-G pseudotyped HIV-1 viruses released from the HEK 293 cells were analysed by Western blot using an anti-HIV-1 CA antibody (duplicate of blot shown in Figure S57).

Only completely processed p24 CA of molecular weight of 24 kDa was identified in the viruses formed in the presence of compounds **4** and **12**. However, in the samples treated with compounds **2**, **8**, **9**, **10**, **16**, **17** and **18**, we identified not only fully processed p24 CA, but also p25 CA-SP1 protein. This observation suggests a similar mechanism of inhibition as described for bevirimat, i.e., the block of the final step of HIV-1 maturation.

4. Conclusions

This study describes the synthesis and biological evaluation of 17 betulinic acid derivatives. The biological profiling revealed that BA derivatives **3** and **14** with modification at C-28 show increased cytotoxicity. However, the cytotoxicity was not specifically directed against cancer cell lines and was not associated with cell cycle arrest. The most effective compounds with sub-micromolar IC$_{50}$ values **3** and **14** possess a hydroxyl group at C-3, whereas structures with a succinyl hemiester group displayed medium cytotoxicity or were inactive. The study introduced six original structures with BODIPY moiety linked to the lupane skeleton. BODIPY conjugates **4**, **5**, **15** and **18** showed low or no cytotoxic activity. In contrast, BODIPY derivative **6** induced strong and derivative **10** medium cytotoxicity in the entire cell line panel, although they do not share any similar substituents at positions C-3 and C-28. The cellular localization of BODIPY conjugates was further studied in U2OS cells using fluorescent microscopy. Fluorescent derivatives **4** and **6** colocalized with endoplasmic reticulum and mitochondria, which is in agreement with previous studies showing interaction with the processes and proteins localized in these organelles [55,56]. Uncoupling of the mitochondrial respiration, followed by radical burst and mitochondrial membrane disruption, is one of the well-described effects of betulin and betulinic acid [57–60]. Thus, we believe that reliable tools to study the derivatives of BA on living cells were established. The anti-HIV-1 activity showed that compounds **2**, **8**, **9**, **10**, **16** and **18** with IC$_{50i}$ lower than 10 μM did not fully process the p24 CA and p25 CA-SP1 proteins, suggesting a similar mechanism of inhibition as described for bevirimat.

Supplementary Materials: The following are available online at https://www.mdpi.com/article/10.3390/biomedicines9091104/s1. Supplementary Figures S1–S52 and Table S1 document the analytical identification (NMR, HRMS, UV-Vis and fluorescence). Table S2: Calculated physical properties (pKa and logP) of the derivatives. Figures S53–S56 show supplementary pictures from fluorescent microscopy. Figure S57: Effect of selected tested compounds on CA-SP1 processing of HIV-1 Gag polyprotein (a duplicate of western blot showed in Figure 5. in the article).

Author Contributions: M.J., P.D., P.B.D., M.R. and M.H. conceived and designed the experiments; D.K., T.Z., M.J., P.D., I.K., J.S., S.G., J.Ř. and P.D. performed the experiments; M.J., P.D. and M.R. analysed the data; M.J., P.D. and M.R. prepared the manuscript; P.B.D. and M.H. edited the article. All authors have read and agreed to the published version of the manuscript.

Funding: This work was supported by GA CR (CZ) GA20-19906S, by the Czech Ministry of Education, Youth and Sports (CZ-OPENSCREEN-LM2018130, and EATRIS-CZ-LM2018133, Czech-BioImaging-LM2018129) and internal grant of Palacky University (IGA_LF_2021_038).

Conflicts of Interest: The authors declare that they have no known competing financial interests or personal relationships that could have appeared to influence the work reported in this paper.

References

1. Sousa, J.L.C.; Freire, C.S.R.; Silvestre, A.J.D.; Silva, A.M.S. Recent developments in the functionalization of betulinic acid and its natural analogues: A route to new bioactive compounds. *Molecules* **2019**, *24*, 355. [CrossRef]
2. Zuco, V.; Supino, R.; Righetti, S.C.; Cleris, L.; Marchesi, E.; Gambacorti-Passerini, C.; Formelli, F. Selective cytotoxicity of betulinic acid on tumor cell lines, but not on normal cells. *Cancer Lett.* **2002**, *175*, 17–25. [CrossRef]
3. Pisha, E.; Chai, H.; Lee, I.S.; Chagwedera, T.E.; Farnsworth, N.R.; Cordell, G.A.; Beecher, C.W.W.; Fong, H.H.S.; Kinghorn, A.D.; Brown, D.M.; et al. Discovery of betulinic acid as a selective inhibitor of human melanoma that functions by induction of apoptosis. *Nat. Med.* **1995**, *1*, 1046–1051. [CrossRef] [PubMed]
4. Noda, Y.; Kaiya, T.; Kohda, K.; Kawazoe, Y. Enhanced cytotoxicity of some triterpenes toward leukemia L1210 cells cultured in low pH media: Possibility of a New Mode of Cell Killing. *Chem. Pharm. Bull.* **1997**, *45*, 1665–1670. [CrossRef] [PubMed]
5. Fujioka, T.; Kashiwada, Y.; Kilkuskie, R.E.; Cosentino, L.M.; Ballas, L.M.; Jiang, J.B.; Janzen, W.P.; Chen, I.-S.; Lee, K.-H. Anti-AIDS agents, 11. Betulinic acid and platanic acid as anti-HIV principles from Syzigium claviflorum, and the anti-HIV activity of structurally related triterpenoids. *J. Nat. Prod.* **1994**, *57*, 243–247. [CrossRef]
6. Kodr, D.; Rumlová, M.; Zimmermann, T.; Džubák, P.; Drašar, P.; Jurášek, M. Antitumor and anti-HIV derivatives of betulinic acid. *Chem. Listy* **2020**, *114*, 658–667.
7. Fulda, S.; Friesen, C.; Los, M.; Scaffidi, C.; Mier, W.; Benedict, M.; Nunez, G.; Krammer, P.H.; Peter, M.E.; Debatin, K.M. Betulinic acid triggers CD95 (APO-1/Fas)- and p53-independent apoptosis via activation of caspases in neuroectodermal tumors. *Cancer Res.* **1997**, *57*, 4956–4964. [PubMed]
8. Fulda, S.; Scaffidi, C.; Susin, S.A.; Krammer, P.H.; Kroemer, G.; Peter, M.E.; Debatin, K.M. Activation of mitochondria and release of mitochondrial apoptogenic factors by betulinic acid. *J. Biol. Chem.* **1998**, *273*, 33942–33948. [CrossRef]
9. Liu, W.K.; Ho, J.C.K.; Cheung, F.W.K.; Liu, B.P.L.; Ye, W.C.; Che, C.T. Apoptotic activity of betulinic acid derivatives on murine melanoma B16 cell line. *Eur. J. Pharmacol.* **2004**, *498*, 71–78. [CrossRef]
10. Fulda, S.; Jeremias, I.; Steiner, H.H.; Pietsch, T.; Debatin, K.M. Betulinic acid: A new cytotoxic agent against malignant brain-tumor cells. *Int. J. Cancer* **1999**, *82*, 435–441. [CrossRef]
11. Mullauer, F.B.; van Bloois, L.; Daalhuisen, J.B.; Ten Brink, M.S.; Storm, G.; Medema, J.P.; Schiffelers, R.M.; Kessler, J.H. Betulinic acid delivered in liposomes reduces growth of human lung and colon cancers in mice without causing systemic toxicity. *Anti-Cancer Drug* **2011**, *22*, 223–233. [CrossRef]
12. Takada, Y.; Aggarwal, B.B. Betulinic acid suppresses carcinogen-induced NF-kappa B activation through inhibition of I kappa B alpha kinase and p65 phosphorylation: Abrogation of cyclooxygenase-2 and matrix metalloprotease-9. *J. Immunol.* **2003**, *171*, 3278–3286. [CrossRef]
13. Melzig, M.F.; Bormann, H. Betulinic acid inhibits aminopeptidase N activity. *Planta Med.* **1998**, *64*, 655–657. [CrossRef]
14. Kwon, H.J.; Shim, J.S.; Kim, J.H.; Cho, H.Y.; Yum, Y.N.; Kim, S.H.; Yu, J. Betulinic acid inhibits growth factor-induced in vitro angiogenesis via the modulation of mitochondrial function in endothelial cells. *Jpn. J. Cancer Res.* **2002**, *93*, 417–425. [CrossRef]
15. Kashiwada, Y.; Nagao, T.; Hashimoto, A.; Ikeshiro, Y.; Okabe, H.; Cosentino, L.M.; Lee, K.-H. Anti-AIDS agents 38. Anti-HIV activity of 3-O-acyl ursolic acid derivatives. *J. Nat. Prod.* **2000**, *63*, 1619–1622. [CrossRef] [PubMed]
16. Sundquist, W.I.; Krausslich, H.G. HIV-1 Assembly, budding, and maturation. *Cold Spring Harb. Perspect. Med.* **2012**, *2*, a006924. [CrossRef] [PubMed]
17. Smith, P.F.; Ogundele, A.; Forrest, A.; Wilton, J.; Salzwedel, K.; Doto, J.; Allaway, G.P.; Martin, D.E. Phase I and II study of the safety, virologic effect, and pharmacokinetics/pharmacodynamics of single-dose 3-O-(3′,3′-dimethylsuccinyl)betulinic acid (bevirimat) against human immunodeficiency virus infection. *Antimicrob. Agents Chemother.* **2007**, *51*, 3574–3581. [CrossRef] [PubMed]

18. Martin, D.E.; Blum, R.; Wilton, J.; Doto, J.; Galbraith, H.; Burgess, G.L.; Smith, P.C.; Ballow, C. Safety and pharmacokinetics of bevirimat (PA-457), a novel inhibitor of human immunodeficiency virus maturation, in healthy volunteers. *Antimicrob. Agents Chemother.* **2007**, *51*, 3063. [CrossRef]

19. Martin, D.E.; Blum, R.; Doto, J.; Galbraith, H.; Ballow, C. Multiple-Dose Pharmacokinetics and safety of bevirimat, a novel inhibitor of HIV maturation, in healthy volunteers. *Clin. Pharmacokinet.* **2007**, *46*, 589–598. [CrossRef] [PubMed]

20. Margot, N.A.; Gibbs, C.S.; Miller, M.D. Phenotypic susceptibility to bevirimat in isolates from HIV-1-infected patients without prior exposure to bevirimat. *Antimicrob. Agents Chemother.* **2010**, *54*, 2345–2353. [CrossRef] [PubMed]

21. Zhao, Y.; Gu, Q.; Morris-Natschke, S.L.; Chen, C.-H.; Lee, K.-H. Incorporation of privileged structures into bevirimat can improve activity against wild-type and bevirimat-resistant HIV-1. *J. Med. Chem.* **2016**, *59*, 9262–9268. [CrossRef]

22. Zhao, Y.; Chen, C.-H.; Morris-Natschke, S.L.; Lee, K.-H. Design, synthesis, and structure activity relationship analysis of new betulinic acid derivatives as potent HIV inhibitors. *Eur. J. Med. Chem.* **2021**, *215*, 113287. [CrossRef]

23. Mukherjee, R.; Jaggi, M.; Rajendran, P.; Siddiqui, M.J.A.; Srivastava, S.K.; Vardhan, A.; Burman, A.C. Betulinic acid and its derivatives as anti-angiogenic agents. *Bioorg. Med. Chem. Lett.* **2004**, *14*, 2181–2184. [CrossRef]

24. Kim, J.Y.; Koo, H.M.; Kim, D.S.H.L. Development of C-20 modified betulinic acid derivatives as antitumor agents. *Bioorg. Med. Chem. Lett.* **2001**, *11*, 2405–2408. [CrossRef]

25. Chowdhury, A.R.; Mandal, S.; Mittra, B.; Sharma, S.; Mukhopadhyay, S.; Majumder, H.K. Betulinic acid, a potent inhibitor of eukaryotic topoisomerase I: Identification of the inhibitory step, the major functional group responsible and development of more potent derivatives. *Med. Sci. Monit.* **2002**, *8*, BR254–BR265. [PubMed]

26. Bildziukevich, U.; Rarova, L.; Janovska, L.; Saman, D.; Wimmer, Z. Enhancing effect of cystamine in its amides with betulinic acid as antimicrobial and antitumor agent in vitro. *Steroids* **2019**, *148*, 91–98. [CrossRef] [PubMed]

27. Bildziukevich, U.; Vida, N.; Rárová, L.; Kolář, M.; Šaman, D.; Havlíček, L.; Drašar, P.; Wimmer, Z. Polyamine derivatives of betulinic acid and beta-sitosterol: A comparative investigation. *Steroids* **2015**, *100*, 27–35. [CrossRef] [PubMed]

28. Brandes, B.; Hoenke, S.; Fischer, L.; Csuk, R. Design, synthesis and cytotoxicity of BODIPY-FL labelled triterpenoids. *Eur. J. Med. Chem.* **2020**, *185*, 111858. [CrossRef] [PubMed]

29. Krajčovičová, S.; Staňková, J.; Džubák, P.; Hajdúch, M.; Soural, M.; Urban, M. A synthetic approach for the rapid preparation of BODIPY conjugates and their use in imaging of cellular drug uptake and distribution. *Chem. Eur. J.* **2018**, *24*, 4957–4966. [CrossRef] [PubMed]

30. Sommerwerk, S.; Heller, L.; Kerzig, C.; Kramell, A.E.; Csuk, R. Rhodamine B conjugates of triterpenoic acids are cytotoxic mitocans even at nanomolar concentrations. *Eur. J. Med. Chem.* **2017**, *127*, 1–9. [CrossRef]

31. Pal, A.; Ganguly, A.; Chowdhuri, S.; Yousuf, M.; Ghosh, A.; Barui, A.K.; Kotcherlakota, R.; Adhikari, S.; Banerjee, R. Bis-arylidene oxindole-betulinic acid conjugate: A fluorescent cancer cell detector with potent anticancer activity. *ACS Med. Chem. Lett.* **2015**, *6*, 612–616. [CrossRef] [PubMed]

32. Rumlová, M.; Křížová, I.; Keprová, A.; Hadravová, R.; Doležal, M.; Strohalmová, K.; Pichová, I.; Hájek, M.; Ruml, T. HIV-1 protease-induced apoptosis. *Retrovirology* **2014**, *11*, 37. [CrossRef]

33. Dostálková, A.; Kaufman, F.; Křížová, I.; Kultová, A.; Strohalmová, K.; Hadravová, R.; Ruml, T.; Rumlová, M. Mutations in the basic region of the Mason-Pfizer monkey virus nucleocapsid protein affect reverse transcription, genomic RNA packaging, and the virus assembly site. *J. Virol.* **2018**, *92*, e00106-18. [CrossRef]

34. Křížová, I.; Hadravová, R.; Štokrová, J.; Günterová, J.; Doležal, M.; Ruml, T.; Rumlová, M.; Pichová, I. The G-patch domain of Mason-Pfizer monkey virus is a part of reverse transcriptase. *J. Virol.* **2012**, *86*, 1988. [CrossRef] [PubMed]

35. Strohalmová-Bohmová, K.; Spiwok, V.; Lepšík, M.; Hadravová, R.; Křížová, I.; Ulbrich, P.; Pichová, I.; Bednárová, L.; Ruml, T.; Rumlová, M. Role of Mason-Pfizer monkey virus CA-NC spacer peptide-like domain in assembly of immature particles. *J. Virol.* **2014**, *88*, 14148. [CrossRef]

36. Goud, T.V.; Tutar, A.; Biellmann, J.-F. Synthesis of 8-heteroatom-substituted 4,4-difluoro-4-bora-3a,4a-diaza-s-indacene dyes (BODIPY). *Tetrahedron* **2006**, *62*, 5084–5091. [CrossRef]

37. Kim, D.; Ma, D.; Kim, M.; Jung, Y.; Kim, N.H.; Lee, C.; Cho, S.W.; Park, S.; Huh, Y.; Jung, J.; et al. Fluorescent labeling of protein using blue-emitting 8-amino-BODIPY derivatives. *J. Fluoresc.* **2017**, *27*, 2231–2238. [CrossRef]

38. Chang, Y.-T.; Alamudi, S.H.; Satapathy, R.; Su, D. Background-free fluorescent probes for live cell imaging. US Patent WO2017078623A1, 2017.

39. Qian, K.; Bori, I.D.; Chen, C.-H.; Huang, L.; Lee, K.-H. Anti-AIDS agents 90. Novel C-28 modified bevirimat analogues as potent HIV maturation inhibitors. *J. Med. Chem.* **2012**, *55*, 8128–8136. [CrossRef] [PubMed]

40. Staudinger, H.; Meyer, J. Über neue organische Phosphorverbindungen III. Phosphinmethylenderivate und Phosphinimine. *Helv. Chim. Acta* **1919**, *2*, 635–646. [CrossRef]

41. Neises, B.; Steglich, W. Simple method for the esterification of carboxylic acids. *Angew. Chem. Int. Ed.* **1978**, *17*, 522–524. [CrossRef]

42. Kotsantis, P.; Silva, L.M.; Irmscher, S.; Jones, R.M.; Folkes, L.; Gromak, N.; Petermann, E. Increased global transcription activity as a mechanism of replication stress in cancer. *Nat. Commun.* **2016**, *7*, 13087. [CrossRef] [PubMed]

43. Fantin, V.R.; St-Pierre, J.; Leder, P. Attenuation of LDH-A expression uncovers a link between glycolysis, mitochondrial physiology, and tumor maintenance. *Cancer Cell* **2006**, *9*, 425–434. [CrossRef] [PubMed]

44. Zielonka, J.; Joseph, J.; Sikora, A.; Hardy, M.; Ouari, O.; Vasquez-Vivar, J.; Cheng, G.; Lopez, M.; Kalyanaraman, B. Mitochondria-targeted triphenylphosphonium-based compounds: Syntheses, mechanisms of action, and therapeutic and diagnostic applications. *Chem. Rev.* **2017**, *117*, 10043–10120. [CrossRef] [PubMed]
45. Evers, M.; Poujade, C.; Soler, F.; Ribeill, Y.; James, C.; Lelievre, Y.; Gueguen, J.C.; Reisdorf, D.; Morize, I.; Pauwels, R.; et al. Betulinic acid derivatives: A new class of human immunodeficiency virus type 1 specific inhibitors with a new mode of action. *J. Med. Chem.* **1996**, *39*, 1056–1068. [CrossRef]
46. Kashiwada, Y.; Hashimoto, F.; Cosentino, L.M.; Chen, C.H.; Garrett, P.E.; Lee, K.H. Betulinic acid and dihydrobetulinic acid derivatives as potent anti-HIV agents. *J. Med. Chem.* **1996**, *39*, 1016–1017. [CrossRef]
47. Soler, F.; Poujade, C.; Evers, M.; Carry, J.C.; Henin, Y.; Bousseau, A.; Huet, T.; Pauwels, R.; DeClercq, E.; Mayaux, J.F.; et al. Betulinic acid derivatives: A new class of specific inhibitors of human immunodeficiency virus type 1 entry. *J. Med. Chem.* **1996**, *39*, 1069–1083. [CrossRef]
48. Li, F.; Goila-Gaur, R.; Salzwedel, K.; Kilgore, N.R.; Reddick, M.; Matallana, C.; Castillo, A.; Zoumplis, D.; Martin, D.E.; Orenstein, J.M.; et al. PA-457: A potent HIV inhibitor that disrupts core condensation by targeting a late step in Gag processing. *Proc. Natl. Acad. Sci. USA* **2003**, *100*, 13555–13560. [CrossRef]
49. Schur, F.K.M.; Obr, M.; Hagen, W.J.H.; Wan, W.; Jakobi, A.J.; Kirkpatrick, J.M.; Sachse, C.; Krausslich, H.G.; Briggs, J.A.G. An atomic model of HIV-1 capsid-SP1 reveals structures regulating assembly and maturation. *Science* **2016**, *353*, 506–508. [CrossRef]
50. Adamson, C.S.; Sakalian, M.; Salzwedel, K.; Freed, E.O. Polymorphisms in Gag spacer peptide 1 confer varying levels of resistance to the HIV-1 maturation inhibitor bevirimat. *Retrovirology* **2010**, *7*, 1–8. [CrossRef]
51. Lu, W.X.; Salzwedel, K.; Wang, D.; Chakravarty, S.; Freed, E.O.; Wild, C.T.; Li, F. A single polymorphism in HIV-1 subtype C SP1 is sufficient to confer natural resistance to the maturation inhibitor bevirimat. *Antimicrob. Agents Chemother.* **2011**, *55*, 3324–3329. [CrossRef]
52. Van Baelen, K.; Salzwedel, K.; Rondelez, E.; Van Eygen, V.; De Vos, S.; Verheyen, A.; Steegen, K.; Verlinden, Y.; Allaway, G.P.; Stuyver, L.J. Susceptibility of human immunodeficiency virus type 1 to the maturation inhibitor bevirimat is modulated by baseline polymorphisms in Gag spacer peptide 1. *Antimicrob. Agents Chemother.* **2009**, *53*, 2185–2188. [CrossRef]
53. Coric, P.; Turcaud, S.; Souquet, F.; Briant, L.; Gay, B.; Royer, J.; Chazal, N.; Bouaziz, S. Synthesis and biological evaluation of a new derivative of bevirimat that targets the Gag CA-SP1 cleavage site. *Eur. J. Med. Chem.* **2013**, *62*, 453–465. [CrossRef]
54. Wang, D.; Lu, W.X.; Li, F. Pharmacological intervention of HIV-1 maturation. *Acta Pharm. Sin. B* **2015**, *5*, 493–499. [CrossRef]
55. Gu, M.; Zhao, P.; Zhang, S.Y.; Fan, S.J.; Yang, L.; Tong, Q.C.; Ji, G.; Huan, C. Betulinic acid alleviates endoplasmic reticulum stress-mediated nonalcoholic fatty liver disease through activation of farnesoid X receptors in mice. *Brit. J. Pharmacol.* **2019**, *176*, 847–863. [CrossRef]
56. Ye, Y.Q.; Zhang, T.; Yuan, H.Q.; Li, D.F.; Lou, H.X.; Fan, P.H. Mitochondria-targeted lupane triterpenoid derivatives and their selective apoptosis-inducing anticancer mechanisms. *J. Med. Chem.* **2017**, *60*, 6353–6363. [CrossRef] [PubMed]
57. Dubinin, M.V.; Semenova, A.A.; Ilzorkina, A.I.; Mikheeva, I.B.; Yashin, V.A.; Penkov, N.V.; Vydrina, V.A.; Ishmuratov, G.Y.; Sharapov, V.A.; Khoroshavina, E.I.; et al. Effect of betulin and betulonic acid on isolated rat liver mitochondria and liposomes. *Biochim. Biophys. Acta-Biomembr.* **2020**, *1862*. [CrossRef] [PubMed]
58. Dubinin, M.V.; Semenova, A.A.; Nedopekina, D.A.; Davletshin, E.V.; Spivak, A.Y.; Belosludtsev, K.N. Effect of F16-betulin conjugate on mitochondrial membranes and its role in cell death initiation. *Membranes* **2021**, *11*, 352. [CrossRef]
59. Dubinin, M.V.; Semenova, A.A.; Ilzorkina, A.I.; Penkov, N.V.; Nedopekina, D.A.; Sharapov, V.A.; Khoroshavina, E.I.; Davletshin, E.V.; Belosludtseva, N.V.; Spivak, A.Y.; et al. Mitochondria-targeted prooxidant effects of betulinic acid conjugated with delocalized lipophilic cation F16. *Free Radic. Bio. Med.* **2021**, *168*, 55–69. [CrossRef] [PubMed]
60. Wang, X.; Lu, X.C.; Zhu, R.L.; Zhang, K.X.; Li, S.; Chen, Z.J.; Li, L.X. Betulinic acid induces apoptosis in differentiated PC12 cells via ROS-mediated mitochondrial pathway. *Neurochem. Res.* **2017**, *42*, 1130–1140. [CrossRef] [PubMed]

biomedicines

Article

The Combined Effect of Branching and Elongation on the Bioactivity Profile of Phytocannabinoids. Part I: Thermo-TRPs

Daiana Mattoteia [1,†], Aniello Schiano Moriello [2,3,†], Orazio Taglialatela-Scafati [4], Pietro Amodeo [2], Luciano De Petrocellis [2], Giovanni Appendino [1], Rosa Maria Vitale [2,*] and Diego Caprioglio [1,*]

1   Dipartimento di Scienze del Farmaco, Università del Piemonte Orientale, Largo Donegani 2, 28100 Novara, Italy; daiana.mattoteia@uniupo.it (D.M.); giovanni.appendino@uniupo.it (G.A.)
2   Institute of Biomolecular Chemistry, National Research Council (ICB-CNR), Via Campi Flegrei 34, 80078 Pozzuoli, Italy; aniello.schianomoriello@icb.cnr.it (A.S.M.); pamodeo@icb.cnr.it (P.A.); luciano.depetrocellis@icb.cnr.it (L.D.P.)
3   Epitech Group SpA, Saccolongo, 35100 Padova, Italy
4   Dipartimento di Farmacia, Università di Napoli Federico II, Via Montesano 49, 80131 Napoli, Italy; scatagli@unina.it
*   Correspondence: rmvitale@icb.cnr.it (R.M.V.); diego.caprioglio@uniupo.it (D.C.); Tel.: +39-081-8675316 (R.M.V.); +39-0321-375843 (D.C.)
†   These authors contributed equally to this work.

Citation: Mattoteia, D.; Schiano Moriello, A.; Taglialatela-Scafati, O.; Amodeo, P.; De Petrocellis, L.; Appendino, G.; Vitale, R.M.; Caprioglio, D. The Combined Effect of Branching and Elongation on the Bioactivity Profile of Phytocannabinoids. Part I: Thermo-TRPs. *Biomedicines* **2021**, *9*, 1070. https://doi.org/10.3390/biomedicines9081070

Academic Editor: Pavel B. Drašar

Received: 28 July 2021
Accepted: 18 August 2021
Published: 23 August 2021

Publisher's Note: MDPI stays neutral with regard to jurisdictional claims in published maps and institutional affiliations.

Abstract: The affinity of cannabinoids for their CB_1 and CB_2 metabotropic receptors is dramatically affected by a combination of α-branching and elongation of their alkyl substituent, a maneuver exemplified by the *n*-pentyl -> α,α-dimethylheptyl (DMH) swap. The effect of this change on other cannabinoid end-points is still unknown, an observation surprising since thermo-TRPs are targeted by phytocannabinoids with often sub-micromolar affinity. To fill this gap, the α,α-dimethylheptyl analogues of the five major phytocannabinoids [CBD (**1a**), Δ^8-THC (**6a**), CBG (**7a**), CBC (**8a**) and CBN (**9a**)] were prepared by total synthesis, and their activity on thermo-TRPs (TRPV1-4, TRPM8, and TRPA1) was compared with that of one of their natural analogues. Surprisingly, the DMH chain promoted a shift in the selectivity toward TRPA1, a target involved in pain and inflammatory diseases, in all investigated compounds. A comparative study of the putative binding modes at TRPA1 between DMH-CBC (**8b**), the most active compound within the series, and CBC (**8a**) was carried out by molecular docking, allowing the rationalization of their activity in terms of structure–activity relationships. Taken together, these observations qualify DMH-CBC (**8b**) as a non-covalent TRPA1-selective cannabinoid lead that is worthy of additional investigation as an analgesic and anti-inflammatory agent.

Keywords: phytocannabinoids; cannabichromene; thermos-TRPs; TRPA1; α,α-dimethylheptyl effect

1. Introduction

Phytocannabinoids from cannabis (*Cannabis sativa* L.) are characterized by a linear alkyl substituent spanning one to six carbons bound to their resorcinol core, with the pentyl substitution being by far the most common one [1]. The existence of branched chain cannabinoids of the *iso-* and *ante-iso* series is, in principle, plausible from a biogenetic standpoint, since the alkylresorcinol moiety of phytocannabinoids is of ketide derivation, and branching would therefore simply require the replacement of the acetate-derived starter with one derived from a branched amino acid. However, branched phytocannabinoids of this type have only been tentatively detected as trace constituents of cannabis, largely remaining unconfirmed curiosities in its inventory of constituents [1]. Conversely, non-biogenetic branching at the benzyl carbon of the alkyl residue has played a critical role in research on cannabis and cannabinoids ever since the early synthesis of these compounds by Adams in the early forties of the past century [2]. Adams was unsuccessful in the isolation of the narcotic principle of cannabis, but the generation of a mixture of narcotic tetrahydrocannabinols from the acidic treatment of cannabidiol (CBD, Figure 1, **1a**) made him correctly postulate a structure of this type for this elusive principle, whose structure was eventually established as **2** by Mechoulam two decades later [3].

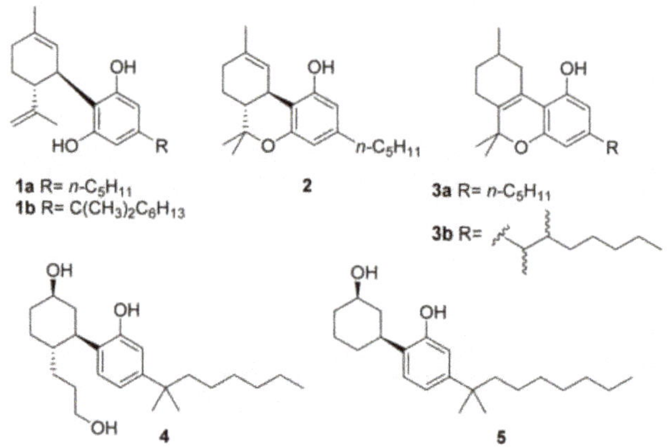

1a R= *n*-C$_5$H$_{11}$
1b R= C(CH$_3$)$_2$C$_6$H$_{13}$

2

3a R= *n*-C$_5$H$_{11}$

3b R=

4

5

Figure 1. Structures of CBD (**1a**), its α,α-dimethylheptyl analogue (**1b**), Δ^9-THC (**2**), $\Delta^{6a,10a}$ THC (**3a**), pyrahexyl (**3b**), CP-55.940 (**4**), and cannabicyclohexanol (**5**).

In the course of these studies, Adams discovered that $\Delta^{6a,10a}$-tetrahydrocannabinol (**3a**), an unnatural compound relatively easily available by total synthesis, could replicate most of the activity of the elusive narcotic principle of cannabis, and investigated the activity of analogues where the *n*-pentyl substituent of the model compound was replaced by longer, shorter, and branched alkyl groups [2]. During these studies, the remarkable potency of compounds bearing substituents on the benzyl and the homobenzyl positions of the alkyl group was discovered. One of these compounds, named pyrahexyl (**3b**) [4], was next developed as an incapacitating non-lethal war weapon and as anti-riot agent, but these studies terminated when the US signed the convention on the ban of chemical weapons in the early sixties [2]. A decade later, capitalizing on these observations, Pfizer chemists developed a series of ultra-potent and less lipophilic analogues of Δ^9-tetrahydrocannabinol (**2**) that proved instrumental in identifying the two cannabinoid receptors, and that are still used today as a reference for cannabinoid activity, as exemplified by CP-55,940 (**4**). Unfortunately, these compounds are also popular designer drugs for herbal incenses (spices), as shown by cannabicyclohexanol (**5**), the first synthetic cannabinoid discovered in this type of product [4].

Phytocannabinoids are multi-target agents, capable of modulating a variety of endpoints that includes not only metabotropic receptors (CB$_1$, CB$_2$), but also enzymes, nuclear receptors, and ionotropic receptors belonging to the family of thermo-TRPs (TRPV1-V4, TRPM8, TRPA1) [5]. In particular, thermo-TRPs represent relevant pharmacological targets in a wide array of pathological conditions ranging from pain to inflammatory and respiratory diseases [6], including those linked to COVID-19 [7]. Surprisingly, no systematic attempt has been made so far to investigate the effect of branching and elongation on the overall biological profile of phytocannabinoids. To fill this gap, we have synthesized the α,α-dimethylheptyl analogues of the five major phytocannabinoids (cannabidiol (CBD, **1a**), Δ^8-tetrahydrocannabinol (Δ^8-THC, Figure 2, **6a**) (This biologically similar Δ^8-analogue of **3** was used because of its major stability and ease of synthesis.), cannabigerol (CBG, **7a**), cannabichromene (CBC, **8a**), and cannabinol (CBN, **9a**)) and compared their profile of activity against thermo-TRPs with the one of their natural analogues.

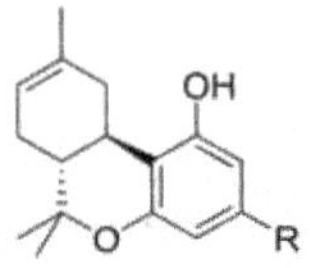

6a R= *n*-C$_5$H$_{11}$
6b R= C(CH$_3$)$_2$C$_6$H$_{13}$

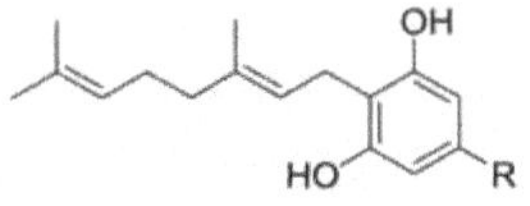

7a R= *n*-C$_5$H$_{11}$
7b R= C(CH$_3$)$_2$C$_6$H$_{13}$

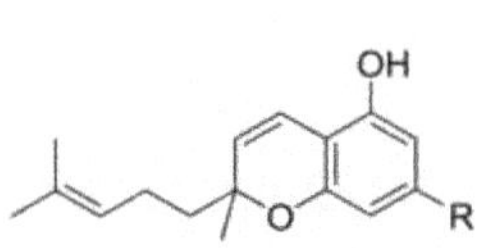

8a R= *n*-C$_5$H$_{11}$
8b R= C(CH$_3$)$_2$C$_6$H$_{13}$
8c R= *n*-C$_7$H$_{15}$
8d R= C(CH$_3$)$_2$C$_4$H$_9$

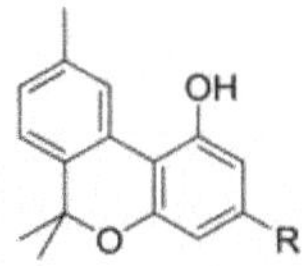

9a R= *n*-C$_5$H$_{11}$
9b R= C(CH$_3$)$_2$C$_6$H$_{13}$

Figure 2. Structure of major phytocannabinoids Δ^8-THC (**6a**), CBG (**7a**), CBC (**8a**), and CBN (**9a**) and their α,α-dimethylheptyl analogues (**6b–9b**).

2. Materials and Methods

2.1. Synthesis

2.1.1. General Experimental Procedures

IR spectra were recorded on an Avatar 370 FT-IR Techno-Nicolet apparatus. ^{1}H (400 MHz) and ^{13}C (100 MHz) NMR spectra were measured on a Bruker Avance 400 MHz spectrometer. Chemical shifts were referenced to the residual solvent signal (CDCl$_3$: δ_H = 7.21, δ_C = 77.0). Homonuclear ^{1}H connectivities were determined by the correlation spectroscopy (COSY) experiment. One-bond heteronuclear ^{1}H-^{13}C connectivities were determined with the heteronuclear single quantum coherence (HSQC) spectroscopy experiment. Two-and three-bond ^{1}H-^{13}C connectivities were determined by gradient two-dimensional (2D) heteronuclear multiple bond correlation (HMBC) experiments optimized for a $^{2,3}J$ = 9 Hz. Low- and high-resolution electrospray ionization mass spectrometry (ESI-MS) data were determined on an LTQ OrbitrapXL (Thermo Scientific) mass spectrometer. Reactions were monitored by thin-layer chromatography (TLC) on Merck 60 F254 (0.25 mm) plates, visualized by staining with 5% H$_2$SO$_4$ in EtOH and heating. Organic phases were dried with Na$_2$SO$_4$ before evaporation. Chemical reagents and solvents were purchased from Sigma-Aldrich, TCI Europe, or Fluorchem, and were used without further purification unless stated otherwise. Petroleum ether with a boiling point of 40–60 °C was used. Silica gel 60 (70–230 mesh) was used for gravity column chromatography (GCC). All work-up solutions were dried with Na$_2$SO$_4$ before evaporation.

2.1.2. Depentyl-α,α-dimethylheptyl-cannabidiol (DMH-CBD, **1b**)

Under a nitrogen atmosphere, BF$_3$OEt$_2$ (1.5 mL/g substrate, 150 μL) was added to a stirred suspension of Al$_2$O$_3$ (10 g/g of substrate, 1 g) in dry DCM (10 mL). After stirring for 15 min at room temperature, the suspension was heated to 40 °C for 1 min, and then the resorcinol **10** (100 mg, 0.423 mmol) and (1*S*,4*R*)-*p*-mentha-2,8-dien-1-ol (11.52 mg, 0.34 mmol, 0.8 molar equiv.) were sequentially added. The reaction was stirred at 40 °C for 10 s. and then quenched with 5 mL of sat. Na$_2$CO$_3$. The resulting biphasic system was extracted with EtOAc, and the organic phase was washed with brine, dried, and concentrated under reduced pressure. Purification by flash column chromatography on silica gel (PE 100% to PE-EtOAc 95:5 as eluent) gave **1b** (94 mg, 60%, Rf: 0.75 (PE-EtOAc 9:1)) as a colorless oil. ^{1}H NMR (400 MHz, CDCl$_3$): δ 6.25–6.23 (br s, 2H), 5.90–6.05 (br s, 1H), 5.56 (s, 1H), 4.65 (s, 1H), 4.54 (s, 1H), 4.55 (s, 1H), 3.85 (br s, 1H), 2.30–2.05 (m, 2H), 1.79 (s, 3H), 1.63 (s, 3H), 1.45–1.50 (m, 2H), 1.21 (br s, 12 H), 0.95–1.05 (br s, 2H), 0.83 (t, J = 7.5 Hz, 3H); ^{13}C NMR (100 MHz, CDCl$_3$): δ 150.22, 149.5, 140.0, 124.1, 113.4, 110.7, 46.0,

44.6, 37.5, 37.3, 31.8, 30.4, 29.9, 28.7, 28.6, 28.4, 24.6, 23.6, 22.6, 20.7, 14.0; HRESIMS *m/z* [M + H]$^+$ 371.2934 (calcd for $C_{25}H_{39}O_2$, 371.2950).

2.1.3. Depentyl-α,α-dimethylheptyl-Δ^8-tetrahydrocannabinol (DMH-Δ^8-THC, **6b**)

To a stirred suspension of 10 (450 mg, 1.69 mmol) in toluene (2.5 mL), p-toluensulfonic acid (PTSA, 60 mg, 0.34 mmol, 0.2 molar equiv.), and (1*S*,4*R*)-*p*-mentha-2,8-dien-1-ol (11,290 mg, 1.86 mmol, 1.1 molar equiv.) were sequentially added. The solution was heated at 120 °C for two hours and then cooled to room temperature and quenched with brine. The biphasic system was extracted with EtOAc, and the organic phase was dried and evaporated. Purification by flash column chromatography on silica gel (PE 100% to PE/DCM 8:2 as eluent) gave **6b** (532 mg, 85%, Rf: 0.90 (PE-EtOAc 9:1)) as a brown oil. ^{1}H NMR (400 MHz, CDCl$_3$): δ 6.42 (d, *J* = 1.8 Hz, 1H), 6.25 (d, *J* = 1.8 Hz, 1H), 5.49–5.41 (m, 1H), 4.79–4.71 (m, 1H), 3.22 (dd, *J* = 16.4, 4.4 Hz, 1H), 2.72 (td, *J* = 10.9, 4.7 Hz, 1H), 2.19–2.13 (m, 1H), 1.96–1.78 (m, 4H), 1.73 (s, 3H), 1.54–1.50 (m, 2H), 1.41 (s, 3H), 1.33–1.16 (m, 10H), 1.14 (s, 3H), 1.11–1.06 (m, 2H), 0.87 (t, *J* = 6.9 Hz, 3H); ^{13}C NMR (100 MHz, CDCl$_3$): δ 154.5, 154.45, 150.0, 134.7, 119.3, 110.1, 108.0, 105.4, 44.8, 44.4, 37.3, 36.0, 31.8, 31.5, 30.0, 28.7, 28.6, 27.9, 27.6, 24.6, 23.5, 22.6, 18.5, 14.1; HRESIMS *m/z* [M + H]$^+$ 371.2934 (calcd for $C_{25}H_{39}O_2$, 371.2950).

2.1.4. Depentyl-α,α-dimethylheptyl-cannabigerol (DMH-CBG, **7b**)

Under a nitrogen atmosphere, BF$_3$Et$_2$O (1.5 mL/g substrate, 150 μL) was added to a stirred suspension of Al$_2$O$_3$ (10 g/g of substrate, 1 g) in dry DCM (10 mL). After stirring 15 min at room temperature, the suspension was heated to 40 °C for 1 min, and then the resorcinol 10 (100 mg, 0.42 mmol) and geraniol (12, 150 μL, 130 mg, 0.87 mmol, 2 mol. equivalents) were sequentially added. The reaction was stirred at 40 °C for 48 h, and then quenched with 20 mL of 2 *M* H$_2$SO$_4$. The reaction mixture was extracted with EtOAc, and the organic phase was washed with brine, dried, and concentrated under reduced pressure. Purification by flash column chromatography on silica gel (PE 100% to PE-EtOAc 95:5 as eluent) gave **1b** [94 mg, 60%, Rf: 0.75 (PE-EtOAc 9:1)] as a colorless oil. Purification by flash column chromatography on silica gel (PE 100% to PE-EtOAc 95:5 as eluent) gave **7b** (101 mg, 64%, Rf: 0.75 (PE-EtOAc 9:1)) as a colorless oil. ^{1}H NMR (400 MHz, CDCl$_3$): δ 6.40 (s, 2H), 5.32 (t, *J* = 6.6 Hz, 1H), 5.11 (br s, 2H), 5.08 (t, *J* = 6.7 Hz, 1H), 3.42 (d, *J* = 7.1 Hz, 2H), 2.21–2.07 (m, 4H), 1.84 (s, 3H), 1.70 (s, 3H), 1.62 (s, 3H), 1.55–1.51 (m, 2H), 1.29–1.16 (m, 12H) 1.12–1.05 (m, 2H), 0.87 (t, *J* = 6.9 Hz, 3H); ^{13}C NMR (100 MHz, CDCl$_3$): δ 154.5, 150.0, 139.0, 132.0, 123.7, 121.7, 110.2, 106.1, 44.4, 39.7, 37.4, 31.8, 30.0, 28.8, 26.4, 25.6, 24.6, 22.7, 22.3, 17.7, 16.2, 14.0; HRESIMS *m/z* [M + H]$_+$ 373.3084 (calcd for $C_{25}H_{41}O_2$, 373.3106).

2.1.5. Depentyl-α,α-dimethylheptyl-cannabichromene (DMH-CBC, **8b**)

To a stirred suspension of 10 (350 mg, 1.48 mmol) in toluene (10 mL), *n*-butylamine (141 μL, 104 mg, 1.42 mmol, 1 molar equiv.) was added. The solution was heated to 60 °C for 10 min, then citral (13, 241 μL, 217 mg, 1.42 mmol, 1 molar equiv.) was added and the solution was refluxed overnight. The reaction was cooled to room temperature and quenched with 20 mL 2*M* H$_2$SO$_4$, and then extracted with EtOAc. The organic phase was washed with brine, dried, and concentrated under reduced pressure. Purification by flash column chromatography on silica gel (PE 100% to PE-EtOAc 95:5 as eluent) gave 210 mg **8b** (210 mg, 40%, Rf: 0.85 (PE-EtOAc 9:1)) as a brown oil. ^{1}H NMR (400 MHz, CDCl$_3$): δ 6.61 (d, *J* = 10.0 Hz, 1H), 6.39 (s, 1H), 6.25 (s, 1H), 5.51 (d, *J* = 10.0 Hz, 1H), 5.09 (t, *J* = 7.4 Hz, 1H), 2.32–2.02 (m, 2H), 1.76–1.62 (m, 2H), 1.65 (s, 3H), 1.57 (s, 3H), 1.55–1.44 (m, 2H), 1.39 (s, 3H), 1.31–1.12 (m, 6H), 1.19 (s, 6H), 1.11–0.99 (m, 2H), 0.94–0.60 (m, 3H); ^{13}C NMR (100 MHz, CDCl$_3$): δ 153.7, 152.0, 150.7, 131.6, 127.4, 124.1, 116.7, 107.0, 106.7, 105.5, 78.2, 44.4, 41.0, 37.7, 31.7, 30.0, 28.7, 26.2, 25.6, 24.6, 22.6, 17.6, 14.0; HRESIMS *m/z* [M + H]$^+$ 371.2935 (calcd for $C_{25}H_{39}O_2$, 371.2950). The same protocol was used for the preparation of **8c** and **8d**, starting from 5-*n*-heptylresorcinol and 5-α,α-dimethylpentylresorcinol, respectively. Depentyl-5-*n*-heptylcannabichromene (**8c**): brown oil (42%, Rf: 0.87 (PE-EtOAc 9:1)). ^{1}H

NMR (400 MHz, CDCl$_3$): δ 6.66 (d, J = 10.0 Hz, 1H), 6.26 (s, 1H), 6.16 (s, 1H), 5.50 (d, J = 10.0 Hz, 1H), 5.12 (t, J = 7.7 Hz, 1H), 2.45 (t, J = 7.9 Hz, 2H), 2.18–2.08 (m, 2H), 1.81–1.67 (m, 2H), 1.69 (s, 3H), 1.60 (s, 3H), 1. 59–1.50 (m, 2H), 1.40 (s, 3H), 1.37–1.16 (m, 8H), 0.90 (t, J = 6.9 Hz 3H); ^{13}C NMR (100 MHz, CDCl$_3$): δ 154.0, 151.3, 144.7, 131.6, 127.0, 124.2, 117.0, 108.9, 107.8, 107.1, 78.1, 41.0, 36.0, 31.8, 31.0, 29.3, 29.2, 26.2, 25.7, 22.7, 22.6, 17.6, 14.1; HRESIMS m/z [M + H]$^+$ 346.2624 (calcd for C$_{23}$H$_{35}$O$_2$, 343.2637). Depentyl-5-α,α-dimethylpentylcannabichromene (**8d**): brown oil [45%, Rf: 0.88 (PE-EtOAc 9:1)]. ^{1}H NMR (400 MHz, CDCl$_3$): δ 6.67 (d, J = 10.0 Hz, 1H), 6.42 (s, 1H), 6.31 (s, 1H), 5.52 (d, J = 10.0 Hz, 1H), 5.13 (t, J = 7.2 Hz, 1H), 2.18–2.12 (m, 2H), 1.80–1.65 (m, 2H), 1.69 (s, 3H), 1.55 (s, 3H), 1.56–1.49 (m, 2H), 1.43 (s, 3H), 1.27–1.19 (m, 2H), 1.24 (s, 6H), 1.13–1.03 (m, 2H), 0.85 (t, J = 7.3 Hz 3H); ^{13}C NMR (100 MHz, CDCl$_3$): δ 153.7, 151.9, 151.1, 131.6, 127.3, 124.2, 117.0, 106.8, 106.7, 105.7, 78.2, 46.8, 44.2, 41.0, 37.7, 35.4, 28.8, 26.9, 26.3, 25.7, 23.4, 22.7, 14.1; HRESIMS m/z [M + H]$^+$ 343.2625 (calcd for C$_{23}$H$_{35}$O$_2$, 343.2637).

2.1.6. Depentyl-α,α-dimethylheptyl-cannabinol (DMH-CBN, **9b**)

To a stirred suspension of **9b** (210 mg, 0.57 mmol) in toluene (40 mL), iodine (291 mg, 1.15 mmol, 2 eq) was added. The solution was refluxed for three hours then cooled to room temperature and quenched with sat. Na$_2$SO$_3$. The mixture was extracted with EtOAc, dried, and concentrated under reduced pressure. Purification by flash column chromatography on silica gel (PE 100% to PE-CH$_2$Cl$_2$ 9:1 as eluent) gave **9b** (173 mg, 83%, Rf: 0.85 (PE-EtOAc 9:1)) as a brown oil. ^{1}H NMR (400 MHz, CDCl$_3$): δ 8.20 (s, 1H), 7.18 (d, J = 7.9 Hz, 1H), 7.10 (d, J = 8.9 Hz, 1H), 6.59 (d, J = 1.8 Hz, 1H), 6.44 (d, J = 1.8 Hz, 1H), 2.41 (s, 3H), 1.63 (s, 6H), 1.58–1.54 (m, 2H), 1.27 (s, 6H), 1.26–1.13 (m, 6H), 0.86 (t, J = 6.9 Hz, 3H); ^{13}C NMR (100 MHz, CDCl$_3$): δ 154.3, 152.7, 151.8, 136.9, 136.9, 127.6, 127.5, 126.3, 122.6, 108.6, 108.3, 107.6, 44.4, 37.6, 31.7, 29.9, 28.6, 27.1, 24.6, 22.6, 21.5, 14.0. HRESIMS m/z [M + H]+ 367.2623 (calcd for C$_{25}$H$_{35}$O$_2$, 367.2637).

2.2. TRP Modulatory Activity

Effects of compounds on intracellular Ca^{2+} concentration ([Ca^{2+}]$_i$) were determined using Fluo-4, a selective intracellular fluorescent probe for Ca^{2+}. Assays of rat TRPA1, TRPV2, TRPV3, TRPV4, and TRPM8 or human TRPV1 mediated elevation of intracellular Ca^{2+} in transfected HEK-293 cells were performed with a continuous monitoring of [Ca^{2+}]$_i$ during the experiments [8]. Briefly, human embryonic kidney (HEK-293) cells, stably transfected with rat TRPA1, TRPV2, TRPV3, TRPV4, and TRPM8 or human TRPV1 (selected by geneticin 600 µg mL^{-1}) or not transfected, were cultured in EMEM+ 2 mM glutamine + 1% nonessential amino acids +10% FBS and maintained at 37 °C with 5% CO$_2$. TRP-HEK-293 cells express stably high levels of TRP transcripts, while these transcripts were virtually absent in wild-type HEK-293 cells as checked by real-time PCR. On the day of the experiment, the cells were loaded for 1 h in the dark at room temperature with Fluo-4 AM (4 µM in DMSO containin 0.02% Pluronic F-127). The cells were rinsed, resuspended in Tyrode's solution (145 mM NaCl, 2.5 mM KCl, 1.5 mM CaCl$_2$, 1.2 mM MgCl$_2$, 10 mM D-glucose, and 10 mM HEPES, pH 7.4), and transferred to a quartz cuvette of a spectrofluorimeter (PerkinElmer LS50B; λEX = 488 nm, λEM = 516 nm) equipped with a PTP-1 fluorescence Peltier system (PerkinElmer Life and Analytical Sciences, Waltham, MA, USA) under continuous stirring. Cell fluorescence before and after the addition of various concentrations of test compounds was measured, normalizing the effects against the response to ionomycin (IM, 4 µM). The values of the effect on [Ca^{2+}]$_i$ in HEK-293 cells not transfected were used as baseline and subtracted from the values obtained from transfected cells. For TRPA1, agonist efficacy was expressed as a percentage of the effect on [Ca^{2+}]$_i$ observed with 100 µM allylisothiocyanate. The potency of the compounds (EC$_{50}$ values) is determined as the concentration required to produce half-maximal increases in [Ca^{2+}]$_i$. In the TRPV3 assay, TRPV3-expressing HEK293 cells were first sensitized with the structurally unrelated agonist 2-aminoethoxydiphenyl borate (100 µM). Antagonist/desensitizing behavior is evaluated against the agonist of the TRP analyzed by adding the compounds directly in the quartz cuvette 5 min before stimulation of cells with the agonist. Allylisothiocyanate (100 µM) was used for TRPA1, capsaicin (0.1 µM) for TRPV1, cannabidiol (2 µM) for TRPV2, thymol (100 µM) for TRPV3, GSK1016790A (10 nM) for TRPV4, and icilin (0.25 µM) for TRPM8. The IC_{50} value was expressed as the concentration exerting a half-maximal inhibition of agonist effect, taking as 100% the effect on $[Ca^{2+}]_i$ exerted by the agonist alone. Dose-response curve fitting (sigmoidal dose-response variable slope) and parameter estimation were performed with Graph-Pad Prism8 (GraphPad Software Inc., San Diego, CA, USA). All determinations were performed at least in triplicate.

2.3. Molecular Docking

Docking studies were performed with AutoDock 4.21 [9]. The rTRPA1 model, along with the ligands, were processed with AutoDock Tools (ADT) package version 1.5.6rc1 to merge non polar hydrogens, calculate Gasteiger charges, and select rotatable sidechain bonds. Grid dimensions of $60 \times 60 \times 50$, centered in the putative binding pocket, were generated with the program AutoGrid 4.2 included in Auto-dock 4.2 distribution, with a spacing of 0.375 Å. Docking runs were carried out by either keeping the whole protein fixed or allowing the rotation of selected residues (Leu873, Leu884, Phe912, Met915 and Met949). A total of 100 molecular AutoDock docking runs for each docking calculation were performed adopting a Lamarckian genetic algorithm (LGA) and the following associated parameters: 100 individuals in a population with a maximum of 15 million energy evaluations and a maximum of 37,000 generations, followed by 300 iterations of Solis and Wets local search. Flexibility was used for all rotatable bonds of both docked ligands. The representative poses for each receptor were selected for the subsequent energy minimization with Amber16 package [10] using ff14SB force field for the protein, and gaff parameters for the ligand [11,12].

3. Results

3.1. Synthesis

All α,α-DMH analogs were prepared from 5-(1,1-dimethylhepty)-resorcinol (**10**), in turn prepared from 3,5-dimethoxybenzoic acid [13]. Terpenylation was then carried out mutuating chemistry developed for the synthesis of the corresponding natural n-pentyl phytocannabinoids. C-Menthylation with (4R)-2,9-p-menthadien-1-ol (**11**) afforded either DMH-CBD (**1b**) or DMH-Δ^8-THC (**6b**) depending on the reaction conditions [14], while geranylation with BF_3 on alumina gave DMH-CBG (**7b**) [15], and chromenyation with citral (**13**) under basic conditions generated DMH-CBC (**8b**) [16], next aromatized to DMH-CBN (**9b**) by treatment with iodine (Scheme 1) [17].

3.2. Biological Evaluation

All DMH-derivatives retained the full agonistic profile on TRPA1 typical of their naturally occurring *n*-pentyl analogues, along with a comparable efficacy and ability to desensitize this channel (Table 1). Conversely, the replacement of the pentyl chain with the branched DMH group was detrimental for the activity toward TRPV1–4 and TRPM8 channels (Tables 1 and 2). Thus, while CBD (**1a**) and CBG (**7a**) are TRPV1 antagonists in a sub/low micromolar range, their correspondent DMH derivatives (**1b** and **7b**, respectively) were inactive against this receptor. Similarly, while the natural derivatives are antagonists in a low micromolar range at TRPV4, the activity was lost for all DMH-derivatives, that were only weak inhibitors of TRPV2, with IC_{50} ranging from 9 to >50 µM. As for TRPV3, DMH-Δ^8-THC (**6b**) was the only compound active as an inhibitor at a low micromolar range (IC_{50} of ~2 µM), with a slightly better potency than Δ^8-THC (**6a**). At TRPM8, al DMH analogues acted as pure antagonists albeit with a lower potency than the natural derivatives, except DMH-CBC (**8b**), which was totally inactive, and with only DMH-CBN (**9b**) retaining a certain affinity (IC_{50}~1 µM).

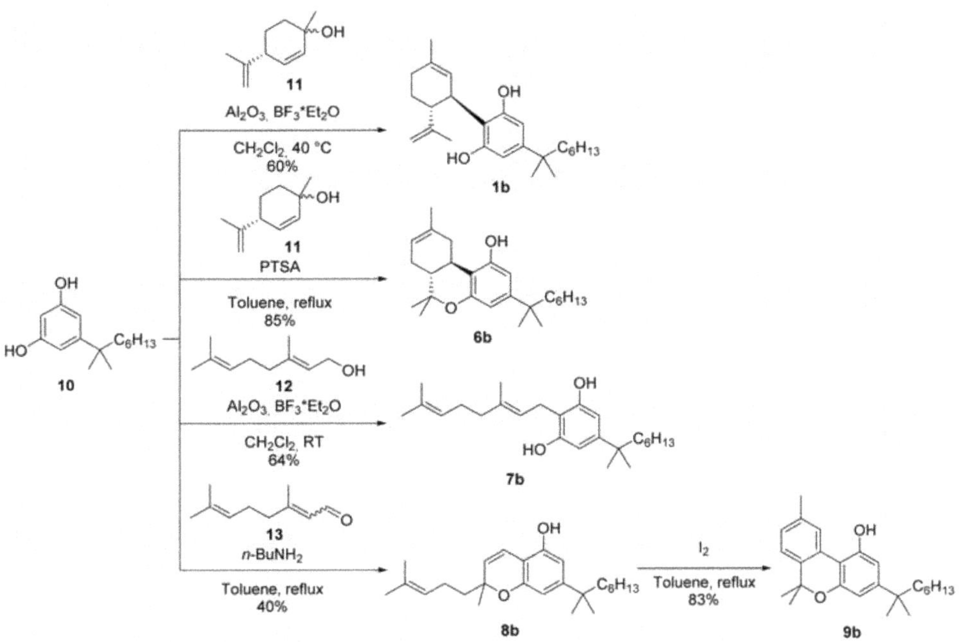

Scheme 1. Synthesis of α,α-dimethylheptyl analogues of major phytocannabinoids DMH-CBD (**1b**), DMH-Δ^8-THC (**6b**), DMH-CBG (**7b**), DMH-CBC (**8b**) and DMH-CBN (**9b**).

Table 1. Efficacy, potency, and inhibitory effect on TRPA1, TRPV1, and TRPM8 channels.

	TRPA1			TRPV1			TRPM8
	Efficacy (% ITC 100 μM)	Potency EC_{50} μM	IC_{50} [a] TRPA1 μM	Efficacy (% IM 4 μM)	Potency EC_{50}	IC_{50} [b] TRPV1 μM	IC_{50} [c] TRPM8 μM
1b	133.4 ± 5.3	0.40 ± 0.15	0.51 ± 0.03	<10	NA	>100	13.4 ± 2.1
1a [18]	115.9 ± 4.6	0.11 ± 0.05	0.16 ± 0.05	44.7 ± 0.02	1.0 ± 0.1	0.6 ± 0.05	0.06 ± 0.01
6b	124.2 ± 4.0	4.8 ± 1.1	6.0 ± 0.9	15.3	>100	>100	39.9 ± 6.9
6a [18]	117.0 ± 12.0	0.23 ± 0.03		<10	NA		0.16 ± 0.01
9b	132.3 ± 7.9	2.1 ± 0.9	3.2 ± 0.6	<10	NA	>100	0.98 ± 0.12
9a [18]	83.3 ± 12.0	0.18 ± 0.02	0.40 ± 0.04	<10		>50	0.21 ± 0.05
7b	99.2 ± 4.5	9.1 ± 2.0	1.7 ± 0.15	<10	NA	>100	9.2 ± 0.8
7a [18]	99.9 ± 1.1	0.70 ± 0.03	13.0 ± 4.8	33.8 ± 2.3	1.3 ± 0.5	2.6 ± 0.5	0.16 ± 0.02
8b	120.4 ± 2.8	0.76 ± 0.12	0.32 ± 0.01	18.9 ± 0.2	12.8 ± 0.9	>100	>100
8a [18]	119.4 ± 3.1	0.09 ± 0.01	0.37 ± 0.05	<10		>50	40.7 ± 0.6
8c	97.8 ± 3.3	1.2 ± 0.2	3.8 ± 0.35	<10	NA	>100	>100
8d	98.7 ± 1.9	0.15 ± 0.03	0.40 ± 0.01	<10	NA	>100	23.6 ± 7.4

Reference agonists: [a] AITC 100 μM; [b] Capsaicin 0.1 μM; [c] Icilin 0.25 μM.

Table 2. Efficacy, potency, and inhibitory effect on TRPA1, TRPV1, and TRPM8 channels.

	TRPV2			TRPV3			TRPV4		
	Efficacy (% IM 4 μM)	Potency EC$_{50}$ μM	IC$_{50}$ [a] TRPV2 μM	Efficacy (% IM 4 μM)	Potency EC$_{50}$ μM	IC$_{50}$ [c] TRPV3 μM	Efficacy (% IM 4 μM)	Potency EC$_{50}$ μM	IC$_{50}$ [e] TRPV4 μM
1b	<10	NA	16.8 ± 0.2	15.8 ± 0.4	11.0 ± 1.1	32.6 ± 5.1	34.0 ± 1.3	NA	>100
1a [18,19]	40.5 ± 1.6	1.25 ± 0.23	4.5 ± 0.7 [b]	50.1 ± 4.8	3.7 ± 1.6	0.9 ± 0.3 [d]	16.7 ± 1.0	0.8 ± 0.3	1.1 ± 0.1 [f]
6b	<10	NA	45.8 ± 3.9	53.8 ± 1.6	0.14 ± 0.03	2.1 ± 0.6	<10	NA	>100
6a [18,19]	53.0 ± 1.4	0.65 ± 0.05	0.8 ± 0.01	18.2 ± 1.0	9.5 ± 1.9	32.3 ± 2.1	<10	NA	15.2 ± 2.7
6b	10.5 ± 0.02	0.25 ± 0.03	>50	20.9 ± 3.0	41.4 ± 21.6	>100	12.1 ± 0.2	16.1 ± 21.6	>100
6a [18,19]	39.9 ± 2.1	19.0 ± 3.7	15.7 ± 2.1	13.1 ± 2.4	5.3 ± 2.7	9.4 ± 0.1	15.3 ± 1.5	16.1 ± 4.5	5.4 ± 0.8
7b	65.3 ± 0.7	9.8 ±0.6	39.4 ± 4.5	64.6 ± 4.8	39.7 ± 0.03	35.4 ± 3.6	34.4 ± 5.2	39.0 ± 21.6	35.4 ± 3.2
7a [18,19]	73.6 ± 1.2	1.7 ± 0.08	1.5 ± 0.2	18.5 ± 0.7	1.0 ± 0.2	>50	23.7 ± 1.8	5.1 ± 1.6	1.3 ± 0.1
8b	76.1 ± 0.6	>50	22.6 ± 0.2	41.6 ± 1.8	>50	>50	<10	NA	>100
8a [18,19]	<10	NA	6.5 ± 1.6	20.2 ± 0.4	1.9 ± 0.2	>100	22.9 ± 1.2	0.6 ± 0.2	9.9 ± 1.2
8c	<10	NA	9.0± 0.8	<10	NA	>100	<10	NA	>100
8d	<10	NA	13.3± 1.1	<10	NA	>100	<10	NA	>100

Reference agonists: [a] CBD 2 μM; [b] LPC 3 μM; [c] Thymol 100 μM; [d] Carvacrol 1 mM; [e] GSK1016790A 10 nM; [f] 4αPDD 1 μM.

To better rationalize the selectivity of the DMH-derivatives toward TRPA1 and dissect the relative contribution of chain elongation and branching, two point-mutated analogues of the most potent compound (DMH-CBC (**8b**)) were prepared according to the general synthetic sequence summarized in Scheme 1 starting from a different 5-alkylresorcinol. In the *n*-heptyl chromene **8c**, the alkyl chain is linear and expanded by two carbon atoms compared to CBC (**8a**), while in the α,α-dimethylpentyl analogue **8d** the benzyl branching is implanted on a pentyl chain. As shown in Table 1, the *n*-heptyl derivative **8c** was less potent than the dimethyl-pentyl analogue **8d** (IC$_{50}$ = 3.8 and 0.40 μM, respectively). Hence, the α,α-dimethyl substitution is the major contributor to the activity of the DHM derivative (IC$_{50}$ = 0.32 μM).

3.3. Molecular Docking Studies on CBC and CBC-DMH at rTRPA1

To shed light on the putative binding modes of DMH-CBC (**8a**) on TRPA1 channels, a molecular docking study was carried out using our previous homology model of rat-TRPA1 [20]. Since the compound is a racemate, both enantiomers were considered, using CBC (**8a**) as a reference compound. Different docking runs were carried out by either keeping the protein side chains rigid or allowing flexibility for selected residues, as described in detail in the Materials and Methods section. The best docking poses for both CBC and DMH-CBC, selected on the basis of binding energy value and visual inspection, were subjected to energy minimization (Figure 3). Both enantiomers of CBC and DMH-CBC engage H-bonding with their phenolic hydroxyl and Thr877 (monB). (*R*)-CBC points its pentyl chain in a hydrophobic cleft formed by Ile953 (monA), Met956 (monB), and Phe880 (monB), while the methyl group of the pentyl chain forms a CH-π interaction with Phe880 and the chromene methyl is surrounded by the hydrophobic residues Ile881 (monB), Leu885 (monB), Phe912 (monB), and Ile945 (monA). Conversely, (*S*)-CBC adopts a flipped orientation with the pentyl chain now pointing toward Phe912 (monB), engaging a CH-π interaction with the aromatic ring of this amino acid. The same arrangement of (*S*)-CBC is adopted by both enantiomers of CBC-DMH (**8b**), with the gem-dimethyls sandwiched between Phe912 (monB) and Ile881 (mon B). In this arrangement, the alkyl chain stabilizes the complex by hydrophobic interactions, counteracting the higher flexibility of the heptyl chain in comparison with the pentyl one and in accordance with the lower activity of the *n*-heptyl chain analogue, with the length of the CBC pentyl chain being consistent with alternative CH-π interaction with Phe880/Phe912 within the binding pocket. The two DMH-CBC enantiomers share the same orientation of the DMH-chain but show a distinct orientation of the chromene methyl group: in the *R* enantiomer, the methyl group points toward Phe880, and the terpenoid chain is hosted in a hydrophobic cleft at the interfaces between two monomers. Conversely, in the *S* enantiomer the orientation of these group is just the opposite.

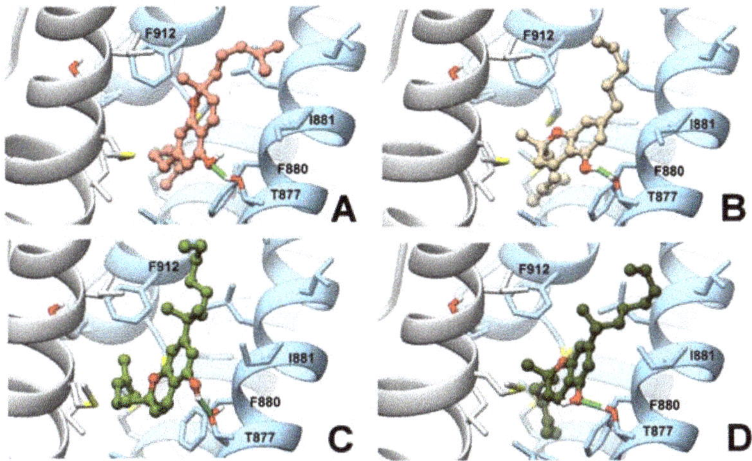

Figure 3. Representative complexes of the rTRPA1 model with (R)-CBC (colored in salmon, panel (**A**)), (S)-CBC (colored in tan, panel (**B**)), (R)-CBC-DMH (colored olive drab, panel (**C**)) and (S)-CBC-DMH (colored in dark green, panel (**D**)). A ribbon representation is used for the protein backbone and sticks for protein side chains of residues within 5 Å from the ligand, in ball & stick representation. H-bonds are shown as green sticks. Carbon atoms are painted according to receptor subunits. Sulfur, oxygen, and polar hydrogen atoms are painted yellow, red, and white, respectively.

4. Discussion

The serendipitous discovery by Adams that a combination of branching and elongation of the alkyl chain boosts the narcotic potency of tetrahydrocannabinols provided the basis for the synthesis of highly potent molecular probes, eventually leading to the characterization of the two cannabinoid receptors (CB1 and CB2) and to the synthesis of nabilone (Cesamet), a clinically efficacious synthetic analogue of Δ^9-THC [2]. Despite the relevance of this maneuver for CB_1 and CB_2 affinity, and its potential to bias activity toward specific targets, its effect has not yet been investigated on other high-affinity targets of phytocannabinoids like the thermo-TRPs (TRPV1-V4, TRPM8 and TRPA1) [5]. This is surprising, since one major issue plaguing the bench-to-bed translation of the biological activity of cannabinoids is their promiscuous end-point profile, which spans metabotropic and ionotropic receptors as well as enzymes and transcription factors. For Δ^9-THC, the interaction with the two cannabinoid receptors dominates, albeit not fully recapping, the clinical profile, while the other phytocannabinoids show a constellation of end-points characterized by a generally modest affinity for macromolecular end-points, with sub-micromolar potency basically limited to some thermo-TRPs [5].

To systematically address this issue, we have synthesized the α,α-dimethylheptyl analogues of the five major phytocannabinoids and have comparatively evaluated their activity on thermo-TRPs. The results showed that, by quenching affinity for the other thermo-TRPs, the replacement of the linear pentyl substituent with an α,α-dimethylheptyl residue induces a selectivity switch toward TRPA1 in all major phytocannabinoid chemotypes. Molecular docking studies rationalized these results in terms of binding, capitalizing our previously developed homology model of ratTRPA1 [20]. Structural and mutagenic experiments have identified two binding sites for non-electrophilic agonists, centered, respectively, on hThr874 (rThr877, uniprot ID:Q6RI86, Site 1) and hTyr840 (Site 2). However, only Site1 was considered, since, unlike phytocannabinoids and their DMH analogues, the selective Site 2 agonist GNE551 lacks channel desensitizing properties. Moreover, Tyr840, critical for GNE551 binding, is not conserved between human and rat hortologues, being replaced by a Phe residue in the latter. The results of the docking studies suggest the *gem*-dimethyls on the benzyl carbon provide an additional anchoring site to engage the binding pocket into non-covalent interactions, rationalizing the retention of affinity for TRPA1. This maneuver is detrimental for the other thermos-TRPs presumably because of the incapacity to accommodate the steric and lipophilicity changes associated with benzyl branching into less broadly tuned binding sites [21].

Non-covalent modulators of TRPA1 are relatively few compared to the covalent ones [22], and their biological profile is critically dependent on their binding mode. Site 2 ligands like GNE551 do not elicit channel desensitization and tachyphylaxis, thus inducing persistent pain insensitive to TRPA1 antagonists [23]. Conversely, Site 1 ligands induce only transient pain, followed by desensitization and analgesia [24]. The desensitization of TRPA1 has the potential to translate into beneficial effects for a wide range of inflammatory conditions [25], traditionally focused in the realm of inflammation and pain [20]. However, growing evidence of involvement in the SARS-CoV-2 infection through the modulation of inflammation, pain, and fever, has provided an additional therapeutic area of investigation, with documented proof-of-concept clinical cases of the beneficial effects of sulforafane, a TRPA1 and Nrf2-interacting nutrient, on the course of the infection [7].

5. Conclusions

Replacement of the *n*-pentyl side chain with an α,α-dimethylheptyl group quenches the exuberant interaction of phytocannabinoids with thermo-TRPs, making them selective non-covalent modulators of TRPA1. Within the DMH-analogues of the most common phytocannabinoids, DMH-CBC (**8b**) emerged as the most potent compound in the series, and the molecular details of its interaction with this ion channel were investigated and compared to those of the parent compound CBC (**8a**). Both compounds are racemic, but only the enantiomers of the natural product CBC showed a distinct orientation at TRPA1 binding sites. These results provide a basis for a broader comparison that will also include additional end points of phytocannabinoids and their endogenous analogues, like anandamide and palmitoylethanolamide.

Author Contributions: Conceptualization, G.A., D.C., R.M.V., O.T.-S. and L.D.P.; methodology, R.M.V., P.A. and L.D.P.; formal analysis, A.S.M., R.M.V. and L.D.P.; investigation, P.A., R.M.V., A.S.M., L.D.P. and D.M.; visualization, P.A., writing—original draft preparation, D.C. and R.M.V.; writing—review and editing, G.A., L.D.P. and O.T.-S.; funding acquisition, G.A., O.T.-S. and L.D.P. All authors have read and agreed to the published version of the manuscript.

Funding: This research was funded by MIUR Italy (PRIN2017, Project 2017WN73PL, bioactivity-directed exploration of the phytocannabinoid chemical space).

Institutional Review Board Statement: Not applicable.

Informed Consent Statement: Not applicable.

Data Availability Statement: Not applicable.

Conflicts of Interest: The authors declare no conflict of interest. The funder had no role in the design of the study; in the collection, analyses, or interpretation of data; in the writing of the manuscript; or in the decision to publish the results.

References

1. Hanuš, L.O.; Meyer, S.M.; Munoz, E.; Taglialatela-Scafati, O.; Appendino, G. Phytocannabinoids: A unified critical inventory. *Nat. Prod. Rep.* **2016**, *33*, 1357–1392. [CrossRef] [PubMed]
2. Appendino, G. The early history of cannabinoid research. *Rend. Fis. Acc. Lincei* **2020**, *31*, 919–929. [CrossRef]
3. Gaoni, Y.; Mechoulam, R. Isolation, Structure, and Partial Synthesis of an Active Constituent of Hashish. *J. Am. Chem. Soc.* **1964**, *86*, 1646–1647. [CrossRef]
4. Appendino, G.; Minassi, A.; Taglialatela-Scafati, O. Recreational drug discovery: Natural products as lead structures for the synthesis of smart drugs. *Nat. Prod. Rep.* **2014**, *31*, 880–904. [CrossRef]

5. Ligresti, A.; De Petrocellis, L.; Di Marzo, V. From phytocannabinoids to cannabinoid receptors and endocannabinoids: Pleiotropic physiological and pathological roles through complex pharmacology. *Physiol. Rev.* **2016**, *96*, 1593–1659. [CrossRef]
6. Vitale, R.M.; Schiano Moriello, A.; De Petrocellis, L. Natural compounds and synthetic drugs targeting the ion-otropic cannabinoid members of Transient Receptor Potential (TRP) channels. In *New Tools to Interrogate En-docannabinoid Signalling: From Natural Compounds to Synthetic Drugs*; Maccarrone, M., Ed.; Royal Society of Chemistry: London, UK, 2020; pp. 201–300.
7. Bousquet, J.; Czarlewski, W.; Zuberbier, T.; Mullol, J.; Blain, H.; Cristol, J.-P.; De La Torre, R.; Pizarro Lozano, N.; Le Moing, V.; Bedbrook, A.; et al. Potential Interplay between Nrf2, TRPA1, and TRPV1. *Int. Arch. Allergy Immunol.* **2021**, *182*, 324–338. [CrossRef]
8. Moriello, A.S.; De Petrocellis, L. Assay of TRPV1 Receptor Signaling. *Methods Mol. Biol.* **2016**, *1412*, 65–76. [CrossRef] [PubMed]
9. Morris, G.; Huey, R.; Lindstrom, W.; Sanner, M.F.; Belew, R.K.; Goodsell, D.S.; Olson, A.J. AutoDock4 and AutoDockTools4: Automated docking with selective receptor flexibility. *J. Comput. Chem.* **2009**, *30*, 2785–2791. [CrossRef]
10. Case, D.A.; Betz, R.M.; Cerutti, D.S.; Cheatham, T.E.; Darden, T.A.; Duke, R.E.; Giese, T.J.; Gohlke, H.; Goetz, A.W.; Homeyer, N.; et al. *AMBER 2016*; University of California: San Francisco, CA, USA, 2016.
11. James, A.; Maier, J.A.; Martinez, C.; Kasavajhala, K.; Wickstrom, L.; Hauser, K.E.; Simmerling, C. ff14SB: Im-proving the accuracy of protein side chain and backbone parameters from ff99SB. *J. Chem. Theory Comput.* **2015**, *11*, 3696–3713. [CrossRef]
12. Wang, J.; Wolf, R.M.; Caldwell, J.W.; Kollman, P.A.; Case, D.A. Development and testing of a general amber force field. *J. Comput. Chem.* **2004**, *25*, 1157–1174. [CrossRef] [PubMed]
13. Harrington, P.E.; Stergiades, I.A.; Erickson, J.; Makriyannis, A.; Tius, M. Synthesis of functionalized canna-binoids. *J. Org. Chem.* **2000**, *65*, 6576–6582. [CrossRef]
14. Bloemendal, V.R.L.J.; Van Hest, J.C.M.; Rutjes, F.P.J.T. Synthetic pathways to tetrahydrocannabinol (THC): An overview. *Org. Biomol. Chem.* **2020**, *18*, 3203–3215. [CrossRef]
15. Baek, S.-H.; Du Han, S.; Yook, C.N.; Kim, Y.C.; Kwak, J.S. Synthesis and antitumor activity of cannabigerol. *Arch. Pharmacal Res.* **1996**, *19*, 228–230. [CrossRef]
16. Pollastro, F.; Caprioglio, D.; Del Prete, D.; Rogati, F.; Minassi, A.; Taglialatela-Scafati, O.; Munoz, E.; Appendino, G. Cannabichromene. *Nat. Prod. Commun.* **2018**, *13*, 1189–1194. [CrossRef]
17. Pollastro, F.; Caprioglio, D.; Marotta, P.; Schiano Moriello, A.; De Petrocellis, L.; Taglialatela-Scafati, O.; Appendino, G. Iodine-promoted aromatization of p-menthane-type phytocannabinoid. *J. Nat. Prod.* **2018**, *81*, 630–633. [CrossRef] [PubMed]
18. De Petrocellis, L.; Ligresti, A.; Moriello, A.S.; Allarà, M.; Bisogno, T.; Petrosino, S.; Stott, C.G.; Di Marzo, V. Effects of canna-binoids and cannabinoid-enriched cannabis extracts on TRP channels and endocannabinoid metabolic enzymes. *Br. J. Phar-macol.* **2011**, *162*, 1479–1494. [CrossRef]
19. De Petrocellis, L.; Orlando, P.; Moriello, A.S.; Aviello, G.; Stott, C.; Izzo, A.; Di Marzo, V. Cannabinoid actions at TRPV channels: Effects on TRPV3 and TRPV4 and their potential relevance to gastrointestinal inflammation. *Acta Physiol.* **2011**, *204*, 255–266. [CrossRef] [PubMed]
20. Chianese, G.; Lopatriello, A.; Schiano-Moriello, A.; Caprioglio, D.; Mattoteia, D.; Benetti, E.; Ciceri, D.; Arnoldi, L.; De Combarieu, E.; Vitale, R.M.; et al. Cannabitwinol, a Dimeric Phytocannabinoid from Hemp, Cannabis sativa L., Is a Selective Thermo-TRP Modulator. *J. Nat. Prod.* **2020**, *83*. [CrossRef] [PubMed]
21. Zhao, Y.; McVeigh, B.M.; Moiseenkova-Bell, V.Y. Structural Pharmacology of TRP Channels. *J. Mol. Biol.* **2021**, *433*, 166914. [CrossRef]
22. Nilius, B.; Appendino, G.; Owsianik, G. The transient receptor potential channel TRPA1: From gene to patho-physiology. *Pflugers Arch.* **2012**, *464*, 425–458. [CrossRef] [PubMed]
23. Liu, C.; Reese, R.; Vu, S.; Rougé, L.; Shields, S.D.; Kakiuchi-Kiyota, S.; Chen, H.; Johnson, K.; Shi, Y.P.; Chernov-Rogan, T.; et al. A Non-covalent Ligand Reveals Biased Agonism of the TRPA1 Ion Channel. *Neuron* **2020**, *109*, 273–284.e4. [CrossRef] [PubMed]
24. Chandrabalan, A.; McPhillie, M.J.; Morice, A.H.; Boa, A.N.; Sadofsky, L.R. N-Cinnamoylanthranilates as hu-man TRPA1 modulators: Structure–activity relationships and channel binding sites. *Eur. J. Med. Chem.* **2019**, *170*, 141–156. [CrossRef] [PubMed]
25. Nilius, B.; Appendino, G. Spices: The Savory and Beneficial Science of Pungency. *Rev. Physiol. Biochem. Pharmacol.* **2013**, *164*, 1–76. [CrossRef] [PubMed]

 *biomedicines*

Article

Triterpenoid–PEG Ribbons Targeting Selectivity in Pharmacological Effects

Zulal Özdemir [1,2,†], Uladzimir Bildziukevich [1,2,†], Martina Čapková [1,†], Petra Lovecká [3], Lucie Rárová [4,‡], David Šaman [5], Michala Zgarbová [5,‡], Barbora Lapuníková [5,‡], Jan Weber [5], Oxana Kazakova [6] and Zdeněk Wimmer [1,2,*]

[1] Department of Chemistry of Natural Compounds, University of Chemistry and Technology in Prague, Technická 5, 16628 Prague 6, Czech Republic; zulalozdemr@gmail.com (Z.Ö.); vmagius@gmail.com (U.B.); m.capkova.bigy@seznam.cz (M.Č.)

[2] Isotope Laboratory, Institute of Experimental Botany of the Czech Academy of Sciences, Vídeňská 1083, 14220 Prague 4, Czech Republic

[3] Department of Biochemistry and Microbiology, University of Chemistry and Technology in Prague, Technická 5, 16628 Prague 6, Czech Republic; loveckap@vscht.cz

[4] Department of Experimental Biology, Faculty of Science, Palacký University, Šlechtitelů 27, 78371 Olomouc, Czech Republic; lucie.rarova@upol.cz

[5] Institute of Organic Chemistry and Biochemistry of the Czech Academy of Sciences, Flemingovo náměstí 2, 16610 Prague 6, Czech Republic; nmrsaman@gmail.com (D.Š.); michala.zgarbova@uochb.cas.cz (M.Z.); barbora.lapunikova@uochb.cas.cz (B.L.); jan.weber@uochb.cas.cz (J.W.)

[6] Ufa Institute of Chemistry of the Ufa Federal Research Centre of the Russian Academy of Sciences, 71, pr. Oktyabrya, 450054 Ufa, Russia; obf@anrb.ru

[*] Correspondence: wimmerz@vscht.cz or wimmer@biomed.cas.cz

[†] Z.Ö., U.B. and M.Č. contributed equally.

[‡] L.R., cytotoxicity screening tests; M.Z. and B.L., antiviral tests.

Citation: Özdemir, Z.; Bildziukevich, U.; Čapková, M.; Lovecká, P.; Rárová, L.; Šaman, D.; Zgarbová, M.; Lapuníková, B.; Weber, J.; Kazakova, O.; et al. Triterpenoid–PEG Ribbons Targeting Selectivity in Pharmacological Effects. *Biomedicines* **2021**, *9*, 951. https://doi.org/10.3390/biomedicines9080951

Academic Editor: Jun Lu

Received: 18 June 2021
Accepted: 30 July 2021
Published: 3 August 2021

Publisher's Note: MDPI stays neutral with regard to jurisdictional claims in published maps and institutional affiliations.

Abstract: (1) Background: To compare the effect of selected triterpenoids with their structurally resembling derivatives, designing of the molecular ribbons was targeted to develop compounds with selectivity in their pharmacological effects. (2) Methods: In the synthetic procedures, Huisgen 1,3-dipolar cycloaddition was applied as a key synthetic step for introducing a 1,2,3-triazole ring as a part of a junction unit in the molecular ribbons. (3) Results: The antimicrobial activity, antiviral activity, and cytotoxicity of the prepared compounds were studied. Most of the molecular ribbons showed antimicrobial activity, especially on *Staphylococcus aureus*, *Pseudomonas aeruginosa*, and *Enterococcus faecalis*, with a 50–90% inhibition effect ($c = 25\ \mu g\cdot mL^{-1}$). No target compound was effective against HSV-1, but **8a** displayed activity against HIV-1 ($EC_{50} = 50.6 \pm 7.8\ \mu M$). Cytotoxicity was tested on several cancer cell lines, and **6d** showed cytotoxicity in the malignant melanoma cancer cell line (G-361; $IC_{50} = 20.0 \pm 0.6\ \mu M$). Physicochemical characteristics of the prepared compounds were investigated, namely a formation of supramolecular gels and a self-assembly potential in general, with positive results achieved with several target compounds. (4) Conclusions: Several compounds of a series of triterpenoid molecular ribbons showed better pharmacological profiles than the parent compounds and displayed certain selectivity in their effects.

Keywords: triterpenoid; molecular ribbon; Huisgen 1,3-dipolar cycloaddition; amide bond; multifunctional PEG3 derivative; antimicrobial activity; anti-HIV activity; cytotoxicity; supramolecular self-assembly

1. Introduction

Triterpenoids represent plant secondary metabolites with increasing importance due to the spectra of their pharmacological activity and ability to self-assemble into the supramolecular systems, often forming gels in different solvents, including aqueous media [1]. Oleanolic acid (**1a**), morolic acid (**1b**), and moronic acid (**1c**) are triterpenoids of the oleanane series, bearing a 6-6-6-6-6 pentacyclic skeleton (Figure 1). While oleanolic acid has been one of the most frequently investigated triterpenoids, morolic and moronic acid have been so far much less studied.

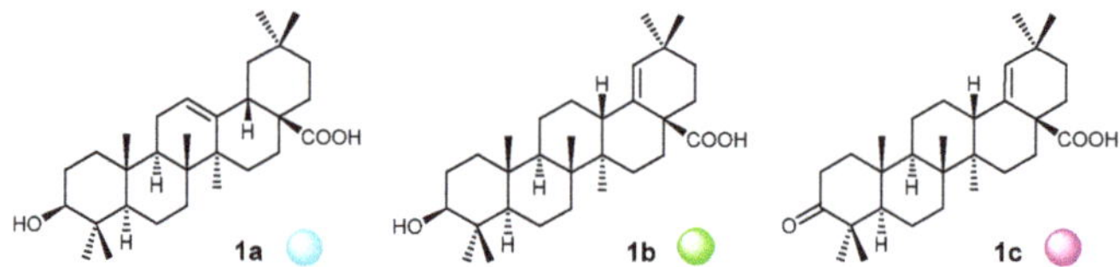

Figure 1. Structures of oleanolic acid (**1a**), morolic acid (**1b**), and moronic acid (**1c**).

Oleanolic acid (**1a**) has been isolated from several hundred of plant species, and it is frequently used in the pharmaceutical and food industry, as reviewed recently [2]. The most important source of **1a** is plants of the genus *Oleaceae* [2]. Due to its lipophilic nature, it appears mainly in olive oil, but it is also present in the olive leaves [3]. Oleanolic acid (**1a**) has been proven to display inhibition activity against HIV and HSV [4–6]. It shows important antimicrobial activity, namely against G+ bacteria (*Staphylococcus aureus, S. epidermidis, Enterococcus faecalis, E. faecium, Streptococcus pneumoniae*, or *Bacillus subtilis*) [7–9]. It is also active against *Mycobacterium tuberculosis*, a Gram staining-type of pathogenic bacterium [10]. Its activity against G– bacteria is low or nil [11]. Oleanolic acid (**1a**) also displays an antidiabetic effect [12]. One of the possible mechanisms consists in the inhibition of glycogen phosphorylase, which is an enzyme capable of liberating glucose from glycogen, and the glucose level in blood is affected [13]. The other possible mechanism is connected with a possible inhibition of proteintyrosine phosphatase (PTP-1B), which is a part of a de-phosphorylation process of the insulin receptor [14]. The last but not least possible mechanism consists of a direct support of insulin secretion [15]. Oleanolic acid (**1a**) has been proven to display an anti-inflammatory effect, either due to the inhibition of transaminases or due to the inhibition of histamine secretion from mastocytes [16], and it also shows antitumor activity, as proven in several tumor cell lines [17]. It is also active against hepatitis of the C type through depressing of the gene types HCV-1b and 2a-JFH1 virus replication [18–20], and it inhibits HCV proteases [18]. However, the mechanism has not yet been clarified [19].

Morolic acid (**1b**) was first isolated from the wood of *Mora excelsa* in 1950 [21], and moronic acid (**1c**) was found in the roots of *Ozoroa mucronata* in 1979 [22]. Later on, both triterpenoids were found in other plants as well, often together [23]. Morolic acid (**1b**) and moronic acid (**1c**) are active against HIV and HSV [24]. Their antimicrobial activity is different: **1c** is more potential than **1b** in the number of affected microorganisms [25]. They also display anti-hypertensive and hypoglycemic effects [26]. Antidiabetic effect is explained either by a mechanism of inhibition of tyrosin phosphatase (PTP-1B; cf. above) [27] or by inhibiting 11β-hydroxysteroid dehydrogenase 1 (11β-HSD 1) that is a mediator of intracellular production of cortisol [28]. Both triterpenoids also display anti-inflammatory effect. Moronic acid (**1c**) was proven to show inhibition of tumor cells, in contrast to **1b** [29]. Moronic acid (**1c**) can also be synthesized in several steps from betulin, which is a lupane-type triterpenoid [30,31].

Pentacyclic triterpenoids have been used to control biofilm-related infections caused by *S. aureus* and other pathogens [32]. Patients, which were intubated due to the SARS-CoV-2 infection nowadays, often suffer from infections caused by microorganisms forming bacterial films that occur in the intubation equipment [33,34]. Therefore, an investigation of novel antimicrobial agents potentially capable of controlling bacterial films may contribute to augmenting the health of the patients seriously infected with SARS-CoV-2. These findings make this investigation even more important, because the designed molecular ribbons are based solely on natural products from sustainable sources, and their pharmacological effect is targeted to enhance the effect of their parent triterpenoids.

Since plant triterpenoids can also be found in vegetables, medicinal plants, spices, etc., they become natural constituents of the human diet. The literature data show that the average daily intake varies between 250 and 400 mg per day in humans, depending on the specific part of the world [35]. This is an important item of information in consequence with the different types of pharmacological activity resulting from this investigation.

The objectives of the presented investigation consisted of several partial tasks. (a) The first is to design procedures for the synthesis of selected molecular ribbons based on the selected triterpenoids connected through the multifunctional PEG3 junction units, and by the newly formed amide bonds and the 1,4-disubstituted 1,2,3-triazole rings. Generally, PEG-mediated motifs have been used in pharmacologically active compounds, such as steroids [36], alkaloids [37], or triterpenoids [38,39]. A modification of biomolecules by the PEG motifs resulted in compounds having a longer blood-circulation time than the corresponding parent biomolecules [40,41]. A formation of a covalent bond between the PEG spacer and the biologically active compounds is one of the promising techniques to design novel drug molecules with improving therapeutic effect [40,41]. Due to the ester-based protecting groups to be employed during the synthetic procedures described here, using any other ester group in the target molecules is excluded, while a combination of amide bonds and 1,4-disubstituted 1,2,3-triazole rings does not interfere with introducing and removing any protecting group. (b) The second objective is to target the investigation on antimicrobial and antiviral activity, and cytotoxicity screening tests of the triterpenoid molecular ribbons to show their advantages and/or disadvantages in comparison with their parent single triterpenoid molecules. (c) The third objective is to investigate supramolecular characteristics of the target molecular ribbons and to evaluate the importance of the formed supramolecular systems for potential pharmacological applications in comparison with their parent triterpenoids.

2. Materials and Methods

2.1. General

The NMR measurements (Bruker, Berlin, Germany) were performed on a Bruker AVANCE II 600 MHz spectrometer equipped with a 5 mm TCI cryoprobe in a 5 mm tube in different solvents. The ^{1}H NMR and the ^{13}C NMR spectra were recorded at 600.13 MHz and 150.90 MHz in $CDCl_3$ or CD_3OD using tetramethylsilane (δ = 0.0–$CDCl_3$) or signal of solvent (δ = 3.31 or 49.50 for ^{1}H/^{13}C–CD_3OD) as internal references. ^{1}H NMR data are presented in the following order: chemical shift (δ) expressed in ppm, multiplicity (s, singlet; d, doublet; t, triplet; q, quartet; m, multiplet), number of protons, coupling constants in Hertz. For unambiguous assignment of both ^{1}H and ^{13}C signals, 2D NMR, ^{1}H,^{13}C gHSQC, and gHMBC spectra were measured using standard parameters sets and pulse programs delivered by the producer of the spectrometer. Infrared spectra were measured with a Nicolet iS-5 FT-IR spectrometer (Thermofisher Scientific, Waltham, MA, USA). UV-VIS spectra were measured on a Specord 210 spectrometer (Jena Analytical, Germany). TLC was carried out on silica gel plates (Merck 60F$_{254}$), and the visualization was performed by both the UV detection and spraying with the methanolic solution of phosphomolybdic acid (5%), followed by heating. For column chromatography, silica gel 60 (0.063–0.200 mm) from Merck was used. Analytical HPLC was carried out on a TSP (Thermoseparation Products, Boston, MA, USA) instrument equipped with a ConstaMetric 4100 Bio pump and a SpectroMonitor 5000 UV DAD. The analyses of the products were performed on a reverse phase Nucleosil 120-5 C18 column (250 × 4 mm; Watrex, Prague, Czech Republic) using a methanol/water mixture (9:1, *v/v*) as the mobile phase at 0.5 to 1.0 mL.min^{-1}. The eluate was monitored at 220, 254, and 275 nm, and the UV spectra were run from 200 to 300 nm. All compounds were purified until their purity was >99%. Their analytical data are summarized in the Supplementary material, part 1, Synthetic procedures and analytical data. All chemicals and solvents were purchased from regular commercial sources in analytical grade, and the solvents were purified by general methods before use. Triterpenoids were purchased from Dr. Jan Šarek–Betulinines (www.betulinines.com, (accessed on 14 June 2021).

2.2. Synthetic Procedures

2.2.1. Procedure A

Benzyl bromide (2.5 eq) and potassium carbonate (3 eq) were added to a solution of the starting compound (1 eq) in DMF (15 mL), and the reaction mixture was stirred at r.t. for 24 h. The mixture was washed with brine, the organic layer was extracted with benzene, and then, it was dried over sodium sulfate. The solid phase was filtered off, and the solvent was evaporated under reduced pressure. The crude residue was purified by column chromatography on silica gel, using a chloroform/ethanol mixture (200:1) as a mobile phase.

2.2.2. Procedure B

Sodium hydride (8 eq) was added to a solution of the starting compound (1 eq) in THF (20 mL), and the reaction mixture was stirred at r.t. for 1 h. Then, an additional amount of sodium hydride (8 eq) was added, and stirring continued for an additional 2 h. Finally, propargyl bromide (3 eq) was added, and the reaction mixture was stirred at r.t. for 7 days. The solvent was evaporated, and the residue was washed with brine and extracted with chloroform. After drying over sodium sulfate, the solvent was evaporated, and the crude residue was purified by column chromatography on silica gel, using chloroform as a mobile phase.

2.2.3. Procedure C

Acetic anhydride (1.44 eq), EDIPA (1.8 eq), and DMAP (0.13 eq) were added to a solution of the starting compound (1 eq) in THF (20 mL). The reaction mixture was stirred at 100 °C for 3 h. Then, an additional amount of acetic anhydride (0.77 eq) and EDIPA (0.9 eq) were added, and heating continued for an additional 1 h. The reaction was stopped by adding water. The organic layer was extracted with chloroform, and the extract was washed with saturated solution of sodium bicarbonate and dried over sodium sulfate. After filtering off the solid phase, solvents were evaporated, and the crude residue was purified by column chromatography on silica gel, using a chloroform/ethanol mixture (100:1) as a mobile phase.

2.2.4. Procedure D

Propargyl bromide (1.5 eq) and potassium carbonate (3 eq) were added to a solution of the starting compound in DMF (15 mL). The reaction mixture was stirred at r.t. for 24 h. It was quenched with brine and extracted with benzene. After drying the extract over sodium sulfate, the solid phase was filtered off, and the solvent was evaporated. The crude residue was purified by column chromatography on silica gel, using a chloroform/petroleum ether mixture (1:3) as a mobile phase.

2.2.5. Procedure E

Oxalyl chloride (7 eq) was added to a solution of the starting compound (1 eq) in DCM (30 mL), and the reaction mixture was stirred at r.t. for 3 h. The solvent was evaporated under reduced pressure, and the residue was re-dissolved in freshly distilled DCM (30 mL). Azido-PEG3-amide (1.1 eq) and EDIPA (2.5 eq) were added, and the reaction mixture was stirred at r.t. for 24 h. The reaction mixture was quenched with brine, extracted with chloroform, and the extract was dried over sodium sulfate. The solid phase was filtered off, and the solvents were evaporated under reduced pressure. The residue was purified by column chromatography on silica gel using a chloroform/ethanol mixture (200:1) as a mobile phase.

2.2.6. Procedure F

The starting compounds bearing either a propargyl group (1 eq) or azide group (1.05 eq) were dissolved in DCM (30 mL). Then, a mixture of aqueous solution of copper sulfate pentahydrate and TBTA (0.05 M) was added, which was followed by the addition of sodium ascorbate (1 eq). The reaction mixture was stirred at r.t. for 24 h, and then, it was quenched with water. The organic layer was extracted with chloroform, and the extract was dried over sodium sulfate. The solid phase was filtered off, and the solvent was evaporated under reduced pressure. The crude residue was purified by column chromatography on silica gel, using a chloroform/ethanol mixture (100:1) as a mobile phase.

2.2.7. Procedure G

Catalyst (10% Pd/C, 0.67 eq) was added to a solution of the starting compound (1 eq) in a mixture of ethanol/THF (1:1, 30 mL) in a bottle equipped with a septum. Air was removed from the bottle, it was filled with hydrogen, and the reaction mixture was stirred at r.t. for 24 h. To stop the hydrogenation reaction, hydrogen was removed, and the catalyst was filtered off. Solvents were evaporated under reduced pressure, and the crude residue was purified by column chromatography on silica gel, using a chloroform/ethanol mixture (100:1) as a mobile phase.

2.2.8. Procedure H

Lithium hydroxide monohydrate (5 eq) was added to a solution of the starting compound (1 eq) in methanol (30 mL). The reaction mixture was heated to 80 °C for 5 h. Then, the solvent was evaporated under reduced pressure, and the crude residue was purified by column chromatography on silica gel, using a chloroform/ethanol mixture (100:1) as a mobile phase.

2.3. Antimicrobial Screening Tests

The antimicrobial activity of **1a–1c, 8a, 12a, 8b, 12b, 4c, 6c, 6d**, and **8d** was studied in vitro, using *Bacillus cereus*, *Escherichia coli*, *Pseudomonas aeruginosa*, *Staphylococcus aureus*, and *Enterococcus faecalis*. The resazurine test was used to evaluate this activity.

The cell inoculum was made in Mueller–Hinton broth (10 mL). Colonies of cells were inoculated from the solid medium that was cultivated for 24 h under optimum conditions. A part of the inoculum was diluted to get suspension with an optical density 0.6 to 0.9 M_cF. An aliquot amount (1 mL) was taken and diluted with additional volume of the medium (19 mL). The stored solutions of the tested compounds were made at a concentration $c = 200$ $\mu g.mL^{-1}$. The stored suspensions of the cells, MHB medium, solutions of the antibiotics specific for the given bacteria (vancomycin for *S. aureus* and *E. faecalis*, kanamycin for *E. coli* and *P. aeruginosa*, and tetracycline for *B. cereus*), and the solutions of the compounds were pipetted into the microtitration plates. Subsequently, a concentration sequence of the tested compounds was made ($c = 100, 50$, and 25 $\mu g.mL^{-1}$). The content of the microtitration plates was cultivated for 24 h under optimum conditions for the given bacteria. A solution of resazurine was made, based on a dilution of its stored solution ($c = 0.15$ $mg.mL^{-1}$; 2 mL) into the MHB medium (18 mL). Then, it was pipetted into the microtitration plates that were cultivated for an additional 1 h. Finally, fluorescence measurement (excitation at 560 nm, emission at 590 nm) was made, and the inhibition of the bacterium was determined.

2.4. Cytotoxicity Screening Tests and Cell Cultures Used

Description of the experimental procedure used in the cytotoxicity screening test was already published [42,43]. The following cancer cell lines were used in the investigation: T-lymphoblastic leukemia, CEM; breast carcinoma, MCF7; cervical carcinoma, HeLa; and malignant melanoma, G-361. Human foreskin fibroblasts, BJ, were used as reference cells. All of the above cell lines were obtained from the American Type Culture Collection (Manassas, VA, USA). African green monkey kidney Vero cells were obtained from the

European Collection of Cell Cultures (Salisbury, UK) and human T-cell leukemia MT-4 cells were obtained through the NIH AIDS Reagent Program, Division of AIDS, NIAID, NIH, from Dr. Douglas Richman. All cells, but MT-4 cells, were cultured in DMEM (Dulbecco's Modified Eagle Medium, Sigma, St. Louis, MO, USA). MT-4 cells were maintained in RPMI1640 medium. Media used were supplemented with 10% fetal bovine serum, 2 mM L-glutamine, and 1% penicillin–streptomycin (all Sigma, St. Louis, MO, USA). The cell lines were maintained under standard cell culture conditions at 37 °C and 5% CO_2 in a humid environment. Cells were subcultured twice or three times a week using the standard trypsinization procedure.

2.5. Antiviral Tests

The anti-HIV-1 and anti-HSV-1 activity and cytotoxicity in MT-4 cells and Vero cells, respectively, were determined as previously described [44,45]. Three-fold dilution of the compounds were tested in triplicate, using HIV-1 (laboratory strain NL4-3) and HSV-1 (strain HF), respectively. Tenofovir alafenamide and acyclovir were used as controls. Compounds were not toxic up to 50 µM against MT-4 and Vero cells.

2.6. Supramolecular Self-Assembly Studies

Ability of the target compounds **8a**, **12a**, **8b**, **12b**, **4c**, **6c**, **6d**, and **8d** was investigated by (a) measuring UV-VIS absorption in methanol/water mixtures with a constant concentration of the studied compound, and (b) investigating their ability to form gel in different solvents.

2.6.1. Ability to Form Supramolecular Gel

The solutions of the studied compounds **8a**, **12a**, **8b**, **12b**, **4c**, **6c**, **6d**, and **8d** (1%, w/v) in the appropriate solvent were mixed with ultrasound for 30 min, then heated to the boiling points of the used solvents, and then allowed to cool down to the r.t. for 24 h. An example of the resulting gels that were produced from **8a**, **4c**, **6c**, and **6d** is shown in the Supplementary Material, Figure S1.

2.6.2. UV-VIS Spectrometry

The self-assembly of the selected compounds **8a**, **12a**, **8b**, **12b**, **4c**, **6c**, **6d**, and **8d** was measured using UV absorption in methanol/water mixtures with a constant concentration of the studied compound for 7 days in one-hour intervals on the first day, and in 24 h intervals all following days. The stock solutions of the studied compounds were prepared at a concentration of 1 mg·mL^{-1} in water. Then, a series of methanol/water mixtures were prepared starting from a methanol/water ratio 100/0 down to 0/100 in 20% steps. The stock solutions of the studied compounds (0.15 mL) were added separately to each vial containing methanol/water mixtures (3 mL), and the UV spectra were recorded in the wavelength range of 200–400 nm. The most remarkable changes in these spectra were observed for methanol/water ratios 40/60 and 60/40 in their mixtures. For the compounds **8a**, **12a**, **12b** and **6d**, the details are presented in the Supplementary Material; see Figure S2.

3. Results and Discussion

3.1. Synthetic Procedures

3.1.1. Procedure A

Esterification of the carboxyl group of the starting triterpenoid (**1a** or **1b**) was made by a substitution reaction using benzyl bromide in DMF under the presence of potassium carbonate [46]. Compounds **2a**/**2b** were prepared using the procedure A (Scheme 1).

3.1.2. Procedure B

Synthesis of propargyl ether of the starting triterpenoid (**1a** or **1b**) consisted in a formation of the corresponding alcoholate using sodium hydride in THF, followed by its reaction with propargylic bromide [46]. Compounds **3a**/**3b** were prepared using the procedure B (Scheme 1).

3.1.3. Procedure C

Esterification of the free hydroxyl group of the starting triterpenoid (**1a** or **1b**) was made with acetic anhydride in THF, using EDIPA as a base and DMAP as a reaction enhancer [43]. Compounds **4a**/**4b** were prepared using the procedure C (Scheme 1).

3.1.4. Procedure D

Synthesis of propargyl ester of the starting triterpenoid (**1c**) was made by a substitution reaction of the carboxyl group of **1c** with propargylic bromide in DMF in the presence of potassium carbonate [47]. Compound **2c** was prepared using the procedure D (Scheme 2).

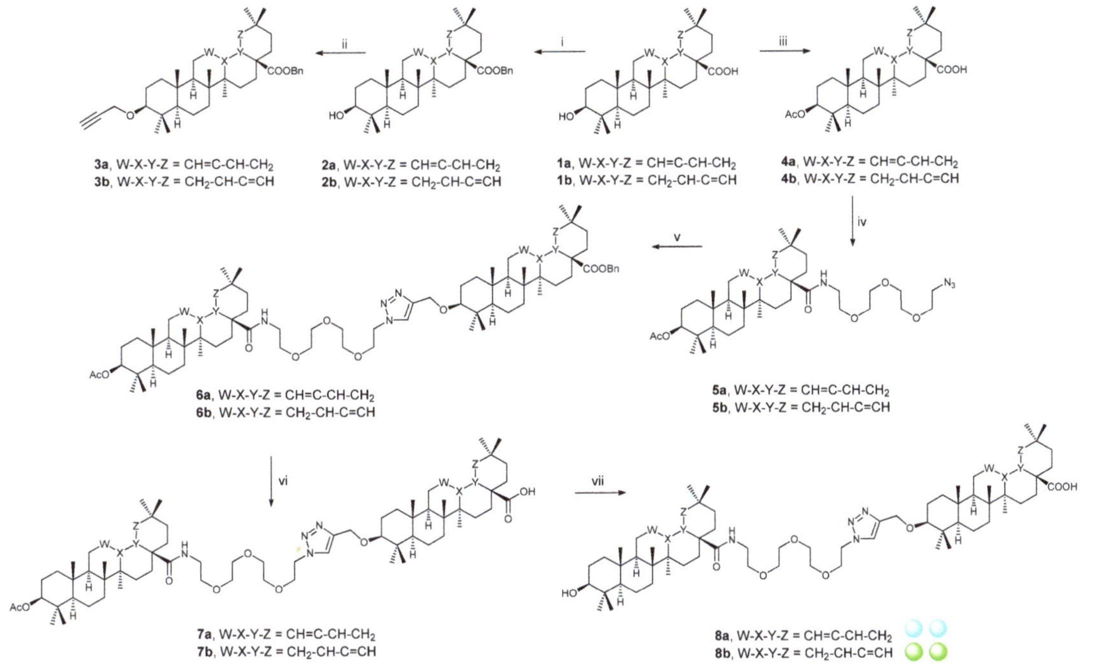

3a, W-X-Y-Z = CH=C-CH-CH$_2$
3b, W-X-Y-Z = CH$_2$-CH-C=CH

2a, W-X-Y-Z = CH=C-CH-CH$_2$
2b, W-X-Y-Z = CH$_2$-CH-C=CH

1a, W-X-Y-Z = CH=C-CH-CH$_2$
1b, W-X-Y-Z = CH$_2$-CH-C=CH

4a, W-X-Y-Z = CH=C-CH-CH$_2$
4b, W-X-Y-Z = CH$_2$-CH-C=CH

6a, W-X-Y-Z = CH=C-CH-CH$_2$
6b, W-X-Y-Z = CH$_2$-CH-C=CH

5a, W-X-Y-Z = CH=C-CH-CH$_2$
5b, W-X-Y-Z = CH$_2$-CH-C=CH

7a, W-X-Y-Z = CH=C-CH-CH$_2$
7b, W-X-Y-Z = CH$_2$-CH-C=CH

8a, W-X-Y-Z = CH=C-CH-CH$_2$
8b, W-X-Y-Z = CH$_2$-CH-C=CH

Scheme 1. Synthetic procedure, part 1. Reagents and conditions. (i) BnBr, K$_2$CO$_3$, DMF, r.t., 24 h; (ii) NaH, propargyl bromide, THF, r.t., 7 days; (iii) Ac$_2$O, EDIPA, DMAP, THF, 100 °C, 4 h; (iv) (a) oxalyl chloride, DCM, r.t., 3 h, (b) N$_3$-PEG3-NH$_2$, EDIPA, r.t., 24 h; (v) **3a/3b**, CuSO$_4$·5H$_2$O, TBTA, sodium ascorbate, DCM/H$_2$O, r.t., 24 h; (vi) 10% Pd/C, H$_2$, ethanol/THF (1:1), r.t., 24 h; (vii) LiOH·H$_2$O, methanol, 80 °C, 5 h.

3.1.5. Procedure E

An amide bond formation was made by converting free carboxyl group of the starting triterpenoid (**1a–1c**) into the corresponding acyl chloride using oxalyl chloride in DCM, followed by its reaction with the corresponding amine in DCM using EDIPA as a base [43]. Compounds **5a/5b**, **9a/9b**, **3c**, and **7d** were prepared using the procedure E (Schemes 1–3).

3.1.6. Procedure F

Cu(I)-Catalyzed Huisgen 1,3-dipolar cycloaddition was made using components bearing either a propargylic group (**3a/3b**, **2c**) or azide group (**5a/5b**, **9a/9b**, **3c**, and **7d**). Cu(I) ions were generated in situ by reducing copper(II) sulfate by sodium ascorbate. TBTA was used as reaction initiator [46]. Compounds **6a/6b, 10a/10b, 4c, 5c/5d,** and **8d** were prepared using the procedure F (Schemes 1–3).

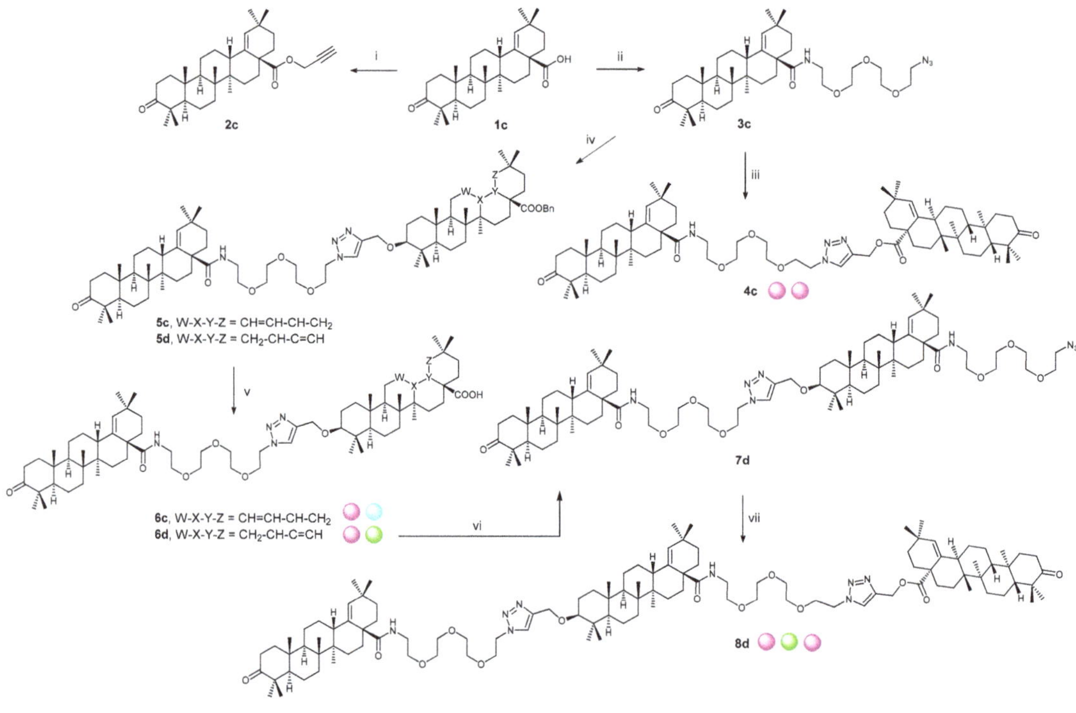

Scheme 2. Synthetic procedure, part 2. Reagents and conditions. (i) propargyl bromide, K_2CO_3, DMF, r.t., 24 h; (ii) (a) oxalyl chloride, DCM, r.t., 3 h, (b) N_3-PEG3-NH_2, EDIPA, r.t., 24 h; (iii) **2c**, $CuSO_4 \cdot 5H_2O$, TBTA, sodium ascorbate, DCM/H_2O, r.t., 24 h; (iv) **3a/3b**, $CuSO_4 \cdot 5H_2O$, TBTA, sodium ascorbate, DCM/H_2O, r.t., 24 h; (v) 10% Pd/C, H_2, ethanol/THF (1:1), r.t., 24 h; (vi) (a) oxalyl chloride, DCM, r.t., 3 h, (b) N_3-PEG3-NH_2, EDIPA, r.t., 24 h; (vii) **2c**, $CuSO_4 \cdot 5H_2O$, TBTA, sodium ascorbate, DCM/H_2O, r.t., 24 h.

3.1.7. Procedure G

Removing of the protecting benzyl group was made by palladium-catalyzed hydrogenation of the selected starting compound (**6a/6b, 10a/10b,** and **5c/5d**) [46]. Compounds **7a/7b, 11a/11b,** and **6c/6d** were prepared using the procedure G (Schemes 1–3).

3.1.8. Procedure H

Removing of the protecting acetyl group was made by alkaline hydrolysis of the esters (**7a/7b** and **11a/11b**) by lithium hydroxide in methanol [43]. Compounds **8a/8b** and **12a/12b** were prepared using the procedure H (Schemes 1 and 3).

3.2. Cytotoxicity

Cytotoxicity in the malignant melanoma cancer cell line (G-361) was only found for **6d** (IC_{50} = 20.0 ± 0.6 μM), even if all target compounds were subjected to the cytotoxicity screening tests. The compound **6d**, a hybrid molecular ribbon based on **1b** and **1c**, was inactive in CEM, MCF7, and HeLa cancer cell lines (IC_{50} > 50 μM), and was not toxic to the human fibroblasts (BJ; IC_{50} > 50 μM). Therefore, **6d** is selectively cytotoxic in the G-361 cancer cell line.

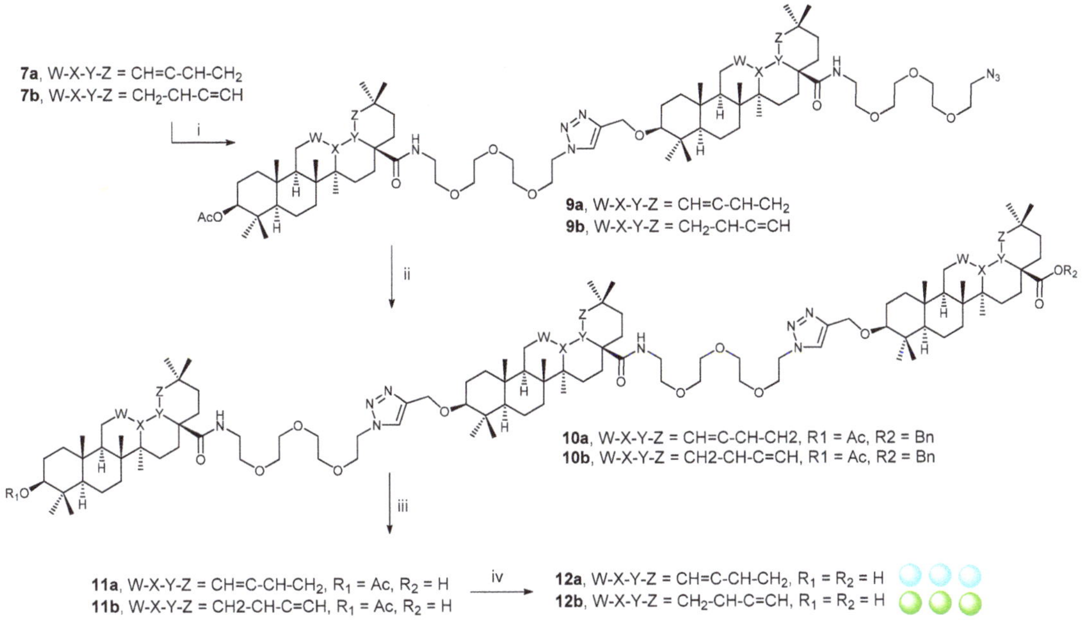

Scheme 3. Synthetic procedure, part 3. Reagents and conditions. (i) (a) oxalyl chloride, DCM, r.t., 3 h, (b) N_3-PEG3-NH_2, EDIPA, r.t., 24 h; (ii) **3a/3b**, $CuSO_4 \cdot 5H_2O$, TBTA, sodium ascorbate, DCM/H_2O, r.t., 24 h; (iii) 10% Pd/C, H_2, ethanol/THF (1:1), r.t., 24 h; (iv) $LiOH \cdot H_2O$, methanol, 80 °C, 5 h.

3.3. Antimicrobial Activity

Antimicrobial activity of the target compounds was tested on *Staphylococcus aureus*, *Escherichia coli*, *Bacillus cereus*, *Pseudomonas aeruginosa*, and *Enterococcus faecalis* using three different concentrations ($c = 25, 50, 100$ µg·mL^{-1}). The results achieved with the lowest concentration of the tested compound ($c = 25$ µg·mL^{-1}) are presented in Table 1. When higher concentrations of the tested compounds were used, inhibition of microorganism growth increased as well, even though for some compounds, certain variations were observed (cf. Supplementary Material, Table S1). The irregularity in the antimicrobial activity values in a dependence on the concentrations of the tested compounds may be connected with the supramolecular characteristics of the molecular ribbons (cf. Supplementary Material, Table S1). We had already observed such behavior in the investigation of cytotoxicity of another series of triterpenoid-based compounds [43,47]. The highest inhibition activity for *S. aureus* was observed when it is treated with **8a**, which is a dimer-like molecular ribbon based on **1a**. While **1a** was not active, dimer-like molecular ribbon **8a** showed increasing inhibition activity. However, trimer-like molecular ribbon **12a**, also based on **1a**, showed lower activity than **8a**, which can be caused either by a decrease in solubility or by size of the compound for possible penetration into the microorganism or by the effect of supramolecular self-assembly (cf. Supplementary Material). The same trend of inhibition activity was observed when *E. coli* was treated with **1a** or its ribbons. Interestingly, in the tests with *S. aureus*, when **1a** was replaced by **1b**, which only differs from **1a** by the position of the double bond in the skeleton, the formed trimer-like molecular ribbon **12b** exhibited the highest activity in comparison with **1b** and **8b**, presenting the importance of small structural variations on the inhibition activity. While **1c** was not active on *S. aureus*, its dimer-like molecular ribbon **4c** was slightly active. Hybrid dimer-like molecular ribbons **6c** and **6d** were synthesized as well, by combining **1c** with **1a** or **1b**. While the hybrid **6c** was less active than **8a**, hybrid **6d** was more active than **8d**. It can be assumed that the hybridization of **1c** (moronic acid) with **1b** (morolic acid) was more favorable than its hybridization with **1a** (oleanolic acid) when the inhibition activity results were taken into consideration.

Table 1. Selected data of the antimicrobial activity [a,b].

Microorganism	Compound	Inhibition [%] (c = 25 μg·mL^{-1}) [c]
Staphylococcus aureus (+)	**1a**	–
Staphylococcus aureus	**8a**	52 ± 7.8 [e]
Staphylococcus aureus	**6c**	13 ± 2.0
Staphylococcus aureus	**12a**	28 ± 4.2
Staphylococcus aureus	**1b**	21 ± 3.2
Staphylococcus aureus	**8b**	–
Staphylococcus aureus	**6d**	25 ± 3.8
Staphylococcus aureus	**12b**	40 ± 6.0 [e]
Staphylococcus aureus	**1c**	–
Staphylococcus aureus	**4c**	11 ± 1.7
Escherichia coli (+)	**1a**	–
Escherichia coli	**8a**	51 ± 7.7 [e]
Escherichia coli	**12a**	–
Pseudomonas aeruginosa (−)	**1c**	–
Pseudomonas aeruginosa	**4c**	50 ± 7.5 [e]
Enterococcus faecalis (+)	**1a** [d]	87 ± 13.0
Enterococcus faecalis	**1b**	38 ± 5.7
Enterococcus faecalis	**1c**	–
Enterococcus faecalis	**4c**	62 ± 9.3 [e]
Enterococcus faecalis	**8d**	38 ± 5.7
Bacillus cereus (+)	**1a** [d]	76 ± 11.4
Bacillus cereus	**1b** [d]	78 ± 11.7
Bacillus cereus	**1c**	–

[a] A comparison of the antimicrobial activity of a target compound is always followed by the antimicrobial activity of its parent triterpenoid; [b] antibiotics (c = 25 μg·mL^{-1}, causing 99% inhibition) were used as positive control: vancomycin for *S. aureus* and *E. faecalis*, kanamycin for *E. coli* and *P. aeruginosa*, and tetracycline for *B. cereus*); [c] hyphen represents no inhibition; [d] the highest inhibition of a microorganism found for **1a** and **1b** at the given concentration; [e] These molecular ribbons displayed higher inhibition of the microorganism than its parent triterpenoid(s).

In turn, for *B. cereus*, the highest inhibition activity was achieved when it is treated with **1a** or **1b** rather than by the trimer-like or dimer-like molecular ribbons, although **1c** (moronic acid) did not show any activity. Similar inhibition activities were observed with **1a** (oleanolic acid) and **1b** (morolic acid), showing the importance of the presence of the (C3)-OH group of triterpenoid for possible activity on *B. cereus*. This group is replaced with a ketone group in case of **1c** (moronic acid). In contrast to these results, dimer-like molecular ribbon **4c** showed the highest activity against *P. aeruginosa*, which is the only G− bacterial strain used in this investigation, while the other dimer-like molecular ribbons exhibited lower inhibition activity values than **4c**. Parent triterpenoids **1a**–**1c** were not active at all, which is in accordance with the literature data [11], demonstrating no or nil antimicrobial activity of this types of triterpenoids on the G− bacteria. The dimer-like molecular ribbon **4c** displayed much higher antimicrobial activity on *E. faecalis* than **1c** that showed no inhibition of the tested microorganisms. However, the highest activity was still observed when *E. faecalis* was treated with **1a**.

The target molecular ribbons are the new compounds, and no data on their pharmacological activity have been so far available in the literature. Therefore, the found results cannot be discussed in the context with the literature data, with the only exception presented above. The data have proven that most of the target molecular ribbons display higher inhibition of the studied microorganisms than their parent triterpenoids **1a**–**1c**. Table S1 (cf. Supplementary material) shows data on the dose–response dependence of antimicrobial activity of the target molecular ribbons **8a**, **12a**, **8b**, **12b**, **4c**, **6c**, **6d**, and **8d** versus their parent compounds **1a–1c**. All doses mentioned there at the maximal concentration used in the screening tests ($c = 100 \ \mu g \cdot mL^{-1}$) are much lower doses than humans may consume with plant-based food (vegetables, medicinal plants, spices, etc.) through which they may receive 250–400 mg per day in average [35].

3.4. Antiviral Activity

The only compound of this series (**8a**) showed low activity against HIV-1 in MT-4 ($EC_{50} \pm SE = 50.6 \pm 7.8 \ \mu M$). However, the found EC_{50} value is several orders of magnitude higher than those for reference compound, tenofovir alafenamide ($EC_{50} \pm SE = 0.0097 \pm 0.0023 \ \mu M$). No target compound displayed activity against HSV-1.

3.5. Gel formation and Supramolecular Self-Assembly

The target compounds were tested to show their ability to form supramolecular gels. This investigation was made because we have found a relationship between cytotoxicity and the ability of compounds to form supramolecular aggregates capable of affecting cytotoxicity [43,47], and similar results have recently been published by other authors [48,49]. Among the studied series of compounds, the only **4c**, **6c**, and **6d** produced a gel in glycerol; **8a** produced a gel in ethylene glycol (Supplementary material, Figure S1). In turn, self-assembly can be achieved with other target compounds based on spectral measurements. UV spectra measured in a mixture of water/methanol with changing ratio of both solvents in 20% steps, but containing a constant concentration of the studied compound, show irregularities in the absorbance maxima sequence or even in wavelengths of the signal maximum, if self-assembly takes place. This effect of irregularity in the absorbance maxima sequence was observed with several target compounds **8a** and **12a**. For more details, see the Supplementary Material (Supplementary material, Figure S2).

4. Conclusions

During this investigation, we have designed several molecular ribbons based on the combinations of oleanolic, morolic, and moronic acid (**1a–1c**) in the ribbons. The multifunctional PEG3 unit was used as a junction chain, together with the 1,2,3-triazole ring and the amide bond, between the triterpenoid molecules to form molecular ribbons. The synthetic procedures recently developed in the team were used, and some of them were optimized. The target structural ribbons were subjected to the antimicrobial and antiviral activity as well as cytotoxicity screening tests. The only target compound **6d** displayed mild and selective cytotoxicity against melanoma cells G-361 ($IC_{50} = 20.0 \pm 0.6 \ \mu M$). The only compound **8a** showed low but again selective activity against HIV-1 in MT-4 cells ($EC_{50} \pm SE = 50.6 \pm 7.8 \ \mu M$). However, more target compounds displayed antimicrobial activity. Compounds were tested in three concentrations ($c = 100, 50,$ and $25 \ \mu g \cdot mL^{-1}$), and the complete results are presented in the Supplementary Material, Table S1, while selected results are shown in Table 1. Among the highest observed inhibition, **8a** inhibited *S. aureus* and *E. coli* in >50% ($c = 25 \ \mu g \cdot mL^{-1}$), **12b** inhibited *S. aureus* in >40% ($c = 25 \ \mu g \cdot mL^{-1}$), and **4c** inhibited *E. faecalis* in >62% ($c = 25 \ \mu g \cdot mL^{-1}$). In turn, **1a** and **1b** inhibited *B. cereus* in >77% ($c = 25 \ \mu g \cdot mL^{-1}$), and **1a** inhibited *E. faecalis* in >87% ($c = 25 \ \mu g \cdot mL^{-1}$).

Since the inhibition of several types of microorganisms, including *S. aureus*, has been found, the molecular ribbons, namely **8a** and **12b**, displaying higher antimicrobial activity on *S. aureus* than their parent triterpenoids **1a** and **1b**, can also be considered for use in controlling bacterial films of *S. aureus* in the intubation equipment of patients suffering from complicated SARS-CoV-2 infections, as reported recently [33,34], to prevent additional microbial infection from the intubation equipment, using molecular ribbons based on the natural products (cf. Table 1 and Table S1 in the Supplementary Material).

Four target compounds (**4c**, **6c**, **6d**, and **8a**) formed gels either in glycerol or in ethylene glycol. Data from the UV spectrometry achieved under the described conditions gave proof that compounds **8a** and **12a** formed self-aggregated systems detected by the instrument in the methanol/water mixtures.

Several of the synthesized triterpenoid–PEG molecular ribbons gave evidence that the investigation of more complex structures based on plant triterpenoids can be important, because several of those structures either displayed enhancing pharmacological effects different from those shown by their parent monomeric triterpenoids or displayed existing supramolecular characteristics, including the ability to form supramolecular gels in contrast to their parent triterpenoids. However, a clear relation between the antimicrobial activity and supramolecular characteristics cannot be presented because we do not have enough indisputable data at present, and, therefore, this topic has been currently under more detailed investigation.

Supplementary Materials: The following material is available online at https://www.mdpi.com/article/10.3390/biomedicines9080951/s1: Details on the synthetic procedures and analytical data of the prepared compounds, including the scanned NMR spectra of the target molecular ribbons, additional details on antimicrobial screening tests, and details on gel formation and other supramolecular characteristics.

Author Contributions: Conceptualization, Z.W., P.L. and O.K.; methodology, Z.Ö. and U.B.; validation, L.R., D.Š. and J.W.; formal analysis, Z.W.; investigation, Z.Ö., U.B., M.Č., M.Z. and B.L.; resources, Z.W.; data curation, Z.Ö., U.B. and M.Č.; writing—original draft preparation, Z.W., Z.Ö., U.B., P.L. and J.W.; writing—review and editing, Z.W. and O.K.; supervision, Z.W.; project administration, Z.W.; funding acquisition, Z.W. All authors have read and agreed to the published version of the manuscript.

Funding: FV30300 (MPO: Z.Ö., U.B., P.L. and Z.W.), and CZ.02.1.01/0.0/0.0/16_019/0000738 (European Regional Developmental Fund Project "Centre for Experimental Plant Biology"; L.R.). U.B. thanks the Academy of Sciences of the Czech Republic for funding his post-doctoral stay (grant No. L200381952). Antiviral testing was funded by the Institute of Organic Chemistry and Biochemistry (RVO 61388963). This work was also supported from the grant (No. A1_FPBT_2021_002) of the Specific University Research.

Institutional Review Board Statement: Not applicable.

Informed Consent Statement: Not applicable.

Data Availability Statement: The data presented in this study are available on request from the corresponding author.

Conflicts of Interest: The authors declare no conflict of interest.

References

1. Bag, B.G.; Majumdar, R. Self-assembly of renewable nano-sized triterpenoids. *Chem. Rec.* **2017**, *17*, 841–873. [CrossRef]
2. Pollier, J.; Goossens, A. Oleanolic acid. *Phytochemistry* **2012**, *77*, 10–15. [CrossRef]
3. Jaeger, S.; Trojan, H.; Kopp, T.; Laszczyk, M.N.; Scheffler, A. Pentacyclic triterpene distribution in various plants—Rich sources for a new group of multi-potent plant extracts. *Molecules* **2009**, *14*, 2016–2031. [CrossRef] [PubMed]
4. Kashiwada, Y.; Wang, H.-K.; Nagao, T.; Kitanaka, S.; Yasuda, I.; Fujioka, T.; Yamagishi, T.; Cosentino, L.M.; Kozuka, M.; Okabe, H.; et al. Anti-AIDS agents. 30. Anti-HIV activity of oleanolic acid, pomolic acid, and structurally related triterpenoids. *J. Nat. Prod.* **1998**, *61*, 1090–1095. [CrossRef]
5. Mengoni, F.; Lichtner, M.; Battinelli, L.; Marzi, M.; Mastroianni, C.M.; Vullo, V.; Mazzanti, G. In vitro anti-HIV activity of oleanolic acid on infected human mononuclear cells. *Planta Med.* **2002**, *68*, 111–114. [CrossRef] [PubMed]
6. Mukherjee, H.; Ojha, D.; Bag, P.; Chandel, H.S.; Bhattacharyya, S.; Chatterjee, T.K.; Mukherjee, P.K.; Chakraborti, S.; Chattopadhyay, D. Anti-herpes virus activities of Achyranthes aspera: An Indian ethnomedicine, and its triterpene acid. *Microbiol. Res.* **2013**, *168*, 238–244. [CrossRef] [PubMed]
7. Woldemichael, G.M.; Singh, M.P.; Maiese, W.M.; Timmermann, B.N. Constituents of antibacterial extract of Caesalpinia paraguariensis Burk. *Z. Naturforsch. C J. Biosci.* **2003**, *58*, 70–75. [CrossRef]
8. Horiuchi, K.; Shiota, S.; Hatano, T.; Yoshida, T.; Kuroda, T.; Tsuchiya, T. Antimicrobial activity of oleanolic acid from Salvia officinalis and related compounds on vancomycin-resistant enterococci (VRE). *Biol. Pharm. Bull.* **2007**, *30*, 1147–1149. [CrossRef] [PubMed]

9.   Szakiel, A.; Ruszkowski, D.; Grudniak, A.; Kurek, A.; Wolska, K.I.; Doligalska, M.; Janiszowska, W. Antibacterial and antiparasitic activity of oleanolic acid and its glycosides isolated from marigold (Calendula officinalis). *Planta Med.* **2008**, *74*, 1709–1715. [CrossRef]

10.   Bamuamba, K.; Gammon, D.W.; Meyers, P.; Dijoux-Franca, M.-G.; Scott, G. Anti-mycobacterial activity of five plant species used as traditional medicines in the Western Cape Province (South Africa). *J. Ethnopharmacol.* **2008**, *117*, 385–390. [CrossRef]

11.   Kuete, V.; Wabo, G.F.; Ngameni, B.; Mbaveng, A.T.; Metuno, R.; Etoa, F.-X.; Ngadjui, B.T.; Beng, V.P.; Meyer, J.J.M.; Lall, N. Antimicrobial activity of the methanolic extract, fractions and compounds from the stem bark of Irvingia gabonensis (Ixonanthaceae). *J. Ethnopharmacol.* **2007**, *114*, 54–60. [CrossRef]

12.   Wu, D.; Zhang, Q.; Yu, Y.; Zhang, Y.; Zhang, M.; Liu, Q.; Zhang, E.; Li, S.; Song, G. Oleanolic acid, a novel endothelin A receptor antagonist, alleviated high glucose-induced cardiomyocytes injury. *Am. J. Chin. Med.* **2018**, *46*, 1187–1201. [CrossRef]

13.   Kurukulasuriya, R.; Link, J.T.; Madar, D.J.; Pei, Z.; Richards, S.J.; Rohde, J.J.; Souers, A.J.; Szczepankiewicz, B.G. Potential drug targets and progress towards pharmacologic inhibition of hepatic glucose production. *Curr. Med. Chem.* **2003**, *10*, 123–153. [CrossRef] [PubMed]

14.   Klaman, L.D.; Boss, O.; Peroni, O.D.; Kim, J.K.; Martino, J.L.; Zabolotny, J.M.; Moghal, N.; Lubkin, M.; Kim, Y.-B.; Sharpe, A.H.; et al. Increased energy expenditure, decreased adiposity, and tissue-specific insulin sensitivity in protein-tyrosine phosphatase 1B-deficient mice. *Mol. Cell. Biol.* **2000**, *20*, 5479–5489. [CrossRef]

15.   Teodoro, T.; Zhang, L.; Alexander, T.; Yue, J.; Vranic, M.; Volchuk, A. Oleanolic acid enhances insulin secretion in pancreatic β-cells. *FEBS Lett.* **2008**, *582*, 1375–1380. [CrossRef] [PubMed]

16.   Singh, G.B.; Singh, S.; Bani, S.; Gupta, B.D.; Banerjee, S.K. Anti-inflammatory activity of oleanolic acid in rats and mice. *J. Pharm. Pharmacol.* **1992**, *44*, 456–458. [CrossRef]

17.   Akihisa, T.; Kamo, S.; Uchiyama, T.; Akazawa, H.; Banno, N.; Taguchi, Y.; Yasukawa, K. Cytotoxic activity of Perilla frutescens var. japonica leaf extract is due to high concentrations of oleanolic and ursolic acids. *J. Nat. Med.* **2006**, *60*, 331–333. [CrossRef]

18.   Ma, C.-M.; Wu, X.-H.; Masao, H.; Wang, X.-J.; Kano, Y. HCV protease inhibitory, cytotoxic and apoptosis-inducing effects of oleanolic acid derivatives. *J. Pharm. Pharm. Sci.* **2009**, *12*, 243–248. [CrossRef] [PubMed]

19.   Wang, H.; Wang, Q.; Xiao, S.-L.; Yu, F.; Ye, M.; Zheng, Y.-X.; Zhao, C.-K.; Sun, D.-A.; Zhang, L.-H.; Zhou, D.-M. Elucidation of the pharmacophore of echinocystic acid, a new lead for blocking HCV entry. *Eur. J. Med. Chem.* **2013**, *64*, 160–168. [CrossRef]

20.   Kong, L.; Li, S.; Liao, Q.; Zhang, Y.; Sun, R.; Zhu, X.; Zhang, Q.; Wang, J.; Wu, X.; Fang, X.; et al. Oleanolic acid and ursolic acid: Novel hepatitis C virus antivirals that inhibit NS5B activity. *Antivir. Res.* **2013**, *98*, 44–53. [CrossRef]

21.   Barton, D.H.R.; Brooks, C.J.W. Morolic acid, a triterpenoid sapogenin. *J. Am. Chem. Soc.* **1950**, *72*, 3314. [CrossRef]

22.   Hostettmann-Kaldas, M.; Nakanishi, K. Moronic acid, a simple triterpenoid keto acid with antimicrobial activity isolated from Ozoroa mucronata. *Planta Med.* **1979**, *37*, 358–360. [CrossRef] [PubMed]

23.   Hamburger, M.; Riese, U.; Graf, H.; Melzig, M.F.; Ciesielski, S.; Baumann, D.; Dittmann, K.; Wegner, C. Constituents in evening primrose oil with radical scavenging, cyclooxygenase, and neutrophil elastase inhibitory activities. *J. Agric. Food Chem.* **2002**, *50*, 5533–5538. [CrossRef]

24.   Paduch, R.; Kandefer-Szerzen, M.; Trytek, M.; Fiedurek, J. Terpenes: Substances useful in human healthcare. *Arch. Immunol. Ther. Exp.* **2007**, *55*, 315–327. [CrossRef]

25.   Gehrke, I.T.S.; Neto, A.T.; Pedroso, M.; Mostardeiro, C.P.; Da Cruz, I.B.M.; Silva, U.F.; Ilha, V.; Dalcol, I.I.; Morel, A.F. Antimicrobial activity of Schinus lentiscifolius (Anacardiaceae). *J. Ethnopharmacol.* **2013**, *148*, 486–491. [CrossRef] [PubMed]

26.   Lopez-Martinez, S.; Navarrete-Vazquez, G.; Estrada-Soto, S.; Leon-Rivera, I.; Rios, M.Y. Chemical constituents of the hemiparasitic plant Phoradendron brachystachyum DC Nutt (Viscaceae). *Nat. Prod. Res.* **2013**, *27*, 130–136. [CrossRef]

27.   Ramirez-Espinosa, J.J.; Rios, M.Y.; Lopez-Martinez, S.; Lopez-Vallejo, F.; Medina-Franco, J.L.; Paoli, P.; Camici, G.; Navarrete-Vazquez, G.; Ortiz-Andrade, R.; Estrada-Soto, S. Antidiabetic activity of some pentacyclic acid triterpenoids, role of PTP-1B: In vitro, in silico, and in vivo approaches. *Eur. J. Med. Chem.* **2011**, *46*, 2243–2251. [CrossRef] [PubMed]

28.   Ramirez-Espinosa, J.J.; Garcia-Jimenez, S.; Rios, M.Y.; Medina-Franco, J.L.; Lopez-Vallejo, F.; Webster, S.P.; Binnie, M.; Ibarra-Barajas, M.; Ortiz-Andrade, R.; Estrada-Soto, S. Antihyperglycemic and sub-chronic antidiabetic actions of morolic and moronic acids, in vitro and in silico inhibition of 11β-HSD 1. *Phytomedicine* **2013**, *20*, 571–576. [CrossRef]

29.   Giner-Larza, E.M.; Manez, S.; Giner, R.M.; Recio, M.C.; Prieto, J.M.; Cerda-Nicolas, M.; Rios, J.L. Anti-inflammatory triterpenes from Pistacia terebinthus Galls. *Planta Med.* **2002**, *68*, 311–315. [CrossRef] [PubMed]

30.   Flekhter, O.B.; Medvedeva, N.I.; Tolstikov, G.A.; Galin, F.Z.; Yunusov, M.S.; Mai, H.N.T.; Tien, L.V.; Savinova, I.V.; Boreko, E.I.; Titov, L.P.; et al. Synthesis of olean-18(19)-ene derivatives from botulin. *Russ. J. Bioorg. Chem.* **2009**, *35*, 233–239. [CrossRef] [PubMed]

31.   Khusnutdinova, E.F.; Medvedeva, N.I.; Kazakov, D.V.; Kukovinets, O.S.; Lobov, A.N.; Suponitsky, K.Y.; Kazakova, O.B. An efficient synthesis of moronic and heterobetulonic acids from allobetulin. *Tetrahedron Lett.* **2016**, *57*, 148–151. [CrossRef]

32.   Chung, P.Y. Novel targets of pentacyclic triterpenoids in Staphylococcus aureus: A systematic review. *Phytomedicine* **2020**, *73*, 152933. [CrossRef]

33.   Knight, G.M.; Glover, R.E.; McQuaid, C.F.; Olaru, I.D.; Gallandat, K.; Leclerc, Q.J.; Fuller, N.M.; Willcocks, S.J.; Hasan, R.; van Kleef, E.; et al. Antimicrobial resistance and COVID-19: Intersections and implications. *eLife* **2021**, *10*, e64139. [CrossRef] [PubMed]

34. Ferrando, M.L.; Coghe, F.; Scano, A.; Carta, K.G.; Orru, G. Co-infection of Streptococcus pneumoniae in respiratory infections caused by SARS-CoV-2. *Biointerface Res. Appl. Chem.* **2021**, *11*, 12170–12177.
35. Furtado, N.A.J.C.; Pirson, L.; Edelberg, H.; Miranda, L.M.; Loira-Pastoriza, C.; Preat, V.; Larondelle, Y.; André, C.M. Pentacyclic triterpene bioavailability: An overview of in vitro and in vivo studies. *Molecules* **2017**, *22*, 400. [CrossRef]
36. Le Devedec, F.; Fuentealba, D.; Strandman, S.; Bohne, C.; Zhu, X.X. Aggregation behavior of pegylated bile acid derivatives. *Langmuir* **2012**, *28*, 13431–13440. [CrossRef]
37. Castillo, P.M.; de la Mata, M.; Casula, M.F.; Sanchez-Alcazar, J.A.; Zaderenko, A.P. PEGylated versus non-PEGylated magnetic nanoparticles as camptothecin delivery system. *Beilstein J. Nanotechnol.* **2014**, *5*, 1312–1319. [CrossRef]
38. Zacchigna, M.; Cateni, F.; Drioli, S.; Procida, G.; Altieri, T. PEG–ursolic acid conjugate: Synthesis and in vitro release studies. *Sci. Pharm.* **2014**, *82*, 411–422. [CrossRef]
39. Medina-O'Donnell, M.; Rivas, F.; Reyes-Zurita, F.J.; Martinez, A.; Martin-Fonseca, S.; Garcia-Granados, A.; Ferrer-Martin, R.M.; Lupianez, J.A.; Parra, A. Semi-synthesis and antiproliferative evaluation of PEGylated pentacyclic triterpenes. *Eur. J. Med. Chem.* **2016**, *118*, 64–78. [CrossRef] [PubMed]
40. Pasut, G.; Veronese, F.M. State of the art in PEGylation: The great versatility achieved after forty years of research. *J. Control. Release* **2012**, *161*, 461–472. [CrossRef] [PubMed]
41. Kolate, A.; Baradia, D.; Patil, S.; Vhora, I.; Kore, G.; Misra, A. PEG—A versatile conjugating ligand for drugs and drug delivery systems. *J. Control. Release* **2014**, *192*, 67–81. [CrossRef] [PubMed]
42. Bildziukevich, U.; Vida, N.; Rárová, L.; Kolář, M.; Šaman, D.; Havlíček, L.; Drašar, P.; Wimmer, Z. Polyamine derivatives of betulinic acid and β-sitosterol: A comparative investigation. *Steroids* **2015**, *100*, 27–35. [CrossRef] [PubMed]
43. Bildziukevich, U.; Malík, M.; Özdemir, Z.; Rárová, L.; Janovská, L.; Šlouf, M.; Šaman, D.; Šarek, J.; Nonappa, N.; Wimmer, Z. Spermine amides of selected triterpenoid acids: Dynamic supramolecular system formation influences the cytotoxicity of the drugs. *J. Mater. Chem. B* **2020**, *8*, 484–491. [CrossRef]
44. Humpolíčková, J.; Weber, J.; Stárková, J.; Mašínová, E.; Günterová, J.; Flaisigová, I.; Konvalinka, J.; Majerová, T. Inhibition of the precursor and mature forms of HIV-1 protease as a tool for drug evaluation. *Sci. Rep.* **2018**, *8*, 10438. [CrossRef] [PubMed]
45. Konč, J.; Tichý, M.; Pohl, R.; Hodek, J.; Džubák, P.; Hajdúch, M.; Hocek, M. Sugar modified pyrimido[4,5-b]indole nucleosides: Synthesis and antiviral activity. *Med. Chem. Commun.* **2017**, *8*, 1856–1862. [CrossRef]
46. Özdemir, Z.; Rybková, M.; Vlk, M.; Šaman, D.; Rárová, L.; Wimmer, Z. Synthesis and pharmacological effects of diosgenin-betulinic acid conjugates. *Molecules* **2020**, *25*, 3546. [CrossRef]
47. Özdemir, Z.; Šaman, D.; Bertula, K.; Lahtinen, M.; Bednárová, L.; Pazderková, M.; Rárová, L.; Wimmer, Z. Rapid self-healing and thixotropic organogelation of amphiphilic oleanolic acid–spermine conjugates. *Langmuir* **2021**, *37*, 2693–2706. [CrossRef]
48. Ha, W.; Zhao, X.-B.; Zhao, W.-H.; Tang, J.-J.; Shi, Y.P. A colon-targeted podophyllotoxin nanoprodrug: Synthesis, characterization, and supramolecular hydrogel formation for the drug combination. *J. Mater. Chem. B* **2021**, *9*, 3200–3209. [CrossRef]
49. Keum, C.; Hong, J.; Kim, D.; Lee, S.-Y.; Kim, H. Lysozyme-instructed self-assembly of amino-acid-functionalized perylene diimide for multidrug-resistant cancer cells. *ACS Appl. Mater. Interfaces* **2021**, *13*, 14866–14874. [CrossRef]

biomedicines

Article

Tanshinone IIA Downregulates Lipogenic Gene Expression and Attenuates Lipid Accumulation through the Modulation of LXRα/SREBP1 Pathway in HepG2 Cells

Wan-Yun Gao [1,†], Pei-Yi Chen [2,3,†], Hao-Jen Hsu [4], Ching-Yen Lin [3], Ming-Jiuan Wu [5] and Jui-Hung Yen [1,3,*]

[1] Institute of Medical Sciences, Tzu Chi University, Hualien 970, Taiwan; 102712131@gms.tcu.edu.tw
[2] Center of Medical Genetics, Hualien Tzu Chi Hospital, Buddhist Tzu Chi Medical Foundation, Hualien 970, Taiwan; pyc571@gmail.com
[3] Department of Molecular Biology and Human Genetics, Tzu Chi University, Hualien 970, Taiwan; jouyuan22@gmail.com
[4] Department of Life Science, Tzu Chi University, Hualien 970, Taiwan; hjhsu32@mail.tcu.edu.tw
[5] Department of Biotechnology, Chia Nan University of Pharmacy and Science, Tainan 717, Taiwan; mingjiuanwu@gmail.com
*   Correspondence: imyenjh@mail.tcu.edu.tw or imyenjh@hotmail.com; Tel.: +886-3-856-5301 (ext. 2683)
†   These authors contributed equally.

Citation: Gao, W.-Y.; Chen, P.-Y.; Hsu, H.-J.; Lin, C.-Y.; Wu, M.-J.; Yen, J.-H. Tanshinone IIA Downregulates Lipogenic Gene Expression and Attenuates Lipid Accumulation through the Modulation of LXRα/SREBP1 Pathway in HepG2 Cells. *Biomedicines* **2021**, 9, 326. https://doi.org/10.3390/biomedicines9030326

Academic Editor: Pavel B. Drašar

Received: 21 February 2021
Accepted: 21 March 2021
Published: 23 March 2021

Publisher's Note: MDPI stays neutral with regard to jurisdictional claims in published maps and institutional affiliations.

Abstract: Abnormal and excessive accumulation of lipid droplets within hepatic cells is the main feature of steatosis and nonalcoholic fatty liver disease (NAFLD) or metabolic-associated fatty liver disease (MAFLD). Dysregulation of lipogenesis contributes to hepatic steatosis and plays an essential role in the pathological progress of MAFLD. Tanshinone IIA is a bioactive phytochemical isolated from *Salvia miltiorrhiza* Bunge and exhibits anti-inflammatory, antiatherosclerotic and antihyperlipidemic effects. In this study, we aimed to investigate the lipid-lowering effects of tanshinone IIA on the regulation of lipogenesis, lipid accumulation, and the underlying mechanisms in hepatic cells. We demonstrated that tanshinone IIA can significantly inhibit the gene expression involved in de novo lipogenesis including FASN, ACC1, and SCD1, in HepG2 and Huh 7 cells. Tanshinone IIA could increase phosphorylation of ACC1 protein in HepG2 cells. We further demonstrated that tanshinone IIA also could suppress the fatty-acid-induced lipogenesis and TG accumulation in HepG2 cells. Furthermore, tanshinone IIA markedly downregulated the mRNA and protein expression of SREBP1, an essential transcription factor regulating lipogenesis in hepatic cells. Moreover, we found that tanshinone IIA attenuated liver X receptor α (LXRα)-mediated lipogenic gene expression and lipid droplet accumulation, but did not change the levels of LXRα mRNA or protein in HepG2 cells. The molecular docking data predicted tanshinone IIA binding to the ligand-binding domain of LXRα, which may result in the attenuation of LXRα-induced transcriptional activation. Our findings support the supposition that tanshinone IIA possesses a lipid-modulating effect that suppresses lipogenesis and attenuates lipid accumulation by modulating the LXRα/SREBP1 pathway in hepatic cells. Tanshinone IIA can be potentially used as a supplement or drug for the prevention or treatment of MAFLD.

Keywords: MAFLD; tanshinone IIA; phytochemical; lipogenesis; lipid accumulation; LXRα

1. Introduction

The liver is a major organ in the regulation of lipid metabolism and is critical for lipid acquisition, storage, export, and utilization as energy upon the oxidation of fatty acids [1]. The disruption of hepatic lipid homeostasis, such as that realized when balance between lipid acquisition through fatty-acid uptake and de novo lipogenesis is disrupted, and lipid deposition may lead to fat retention within the liver and the consequential development of nonalcoholic fatty liver disease (NAFLD), renamed as metabolic-associated fatty liver disease (MAFLD) [2–5]. MAFLD is recognized as the most common form of chronic liver disease worldwide, and it can cause deleterious clinical problems, including end-stage liver diseases and cancer, in some cases. MAFLD is frequently associated with insulin resistance, type 2 diabetes mellitus, hyperlipidemia, and obesity [6,7] The main feature of MAFLD pathogenesis is the accumulation of lipids in the liver. MAFLD manifests in liver conditions ranging from simple triglyceride (TG) and cholesterol accumulation to steatosis. Abnormal and excessive lipid accumulation in the liver has the potential to induce inflammatory injury in hepatocytes and lead to nonalcoholic steatohepatitis (NASH). NASH can progress to advanced fibrosis, cirrhosis, and hepatocellular carcinoma, and result in increases in liver-related mortality [8,9]. To date, no approved agents are available for the prevention and treatment of MAFLD. Thus, the development of therapeutic compounds for MAFLD is currently being actively pursued [10,11].

The lipid biosynthesis process plays the most important role in the regulation of fat accumulation in the liver. In hepatic cells, the first step of de novo fatty-acid synthesis is the conversion of acetyl-CoA to malonyl-CoA by acetyl-CoA carboxylase 1 (ACC1), then malonyl-CoA is converted to palmitate through a reaction catalyzed by a multifunctional fatty-acid synthase (FASN). Palmitate may further undergo elongation catalyzed by long-chain elongase, desaturation catalyzed by stearoyl-CoA desaturase 1 (SCD1), and esterification for triglyceride production or exportation by very low-density lipoprotein (VLDL) particles [12–15]. It has been demonstrated that increases in de novo lipogenesis in the liver may cause hypertriglyceridemia and play an essential role in MAFLD pathogenesis [16]. MAFLD patients were reported to have increased abnormal de novo lipogenesis even in fasting condition. The clinical data also showed that failure to regulate the process of de novo lipogenesis may cause hepatic lipid accumulation in obese patients with MAFLD [17]. These findings indicate that modulation of de novo lipogenesis in the liver can provide useful strategies to prevent hepatic steatosis and the development of MAFLD.

The transcriptional regulation of de novo lipogenesis is predominantly regulated by sterol regulatory element binding protein 1 (SREBP1), which binds to the sterol regulatory element (SRE) and upregulates the gene expression of FASN, ACC1, and SCD1, thereby increasing TG production and accumulation in hepatic cells [18]. It is known that SREBP-1c expression is stimulated by the transcription factor liver X receptor α (LXRα), a ligand-activated nuclear hormone receptor [19]. LXRα and retinoid X receptor (RXR) can form a heterodimer that binds to LXR-responsive elements (LXREs) in the promoter of target genes. LXRα binding to ligands causes the release of repressors and the recruitment of coactivators that interact with the LXRα/RXR heterodimer for the induction of target gene transcription [20,21]. It has been demonstrated that distinct LXRα ligands, such as synthetic nonsteroidal ligands or endogenous oxysterols, induce LXRα activation and promote de novo lipogenesis in the liver, which may lead to elevated levels of plasma and hepatic lipid [22–25]. Modulation of the LXRα/SREBP1-dependent lipogenesis pathway may be beneficial for reducing hepatic lipid accumulation and treating MAFLD.

Tanshinone IIA is a main lipophilic diterpenoids isolated from *Salvia miltiorrhiza* Bunge, also known as Danshen, with antioxidant, anti-inflammation, antiatherosclerosis, and anticancer properties that can be used in the prevention or treatment of hepatic fibrosis, neurodegeneration, and cardiovascular diseases [26,27]. Recently, several in vitro and in vivo studies have suggested that tanshinone IIA can modulate lipid homeostasis. Tanshinone IIA can attenuate oxidized-LDL (oxLDL) uptake by macrophage and inhibit foam-cell formation [28,29]. Tanshinone IIA can also increase LDLR levels and LDL uptake by cells through the downregulation of PCSK9 expression in HepG2 cells [30] Tanshinone IIA was reported to attenuate atherosclerotic lesions in hyperlipidemic and hypercholesterolemic animal models [31,32] Tanshinone IIA regulated the SREBP2/PCSK9 pathway, increased the levels of HDL, and decreased lipid deposition in the livers of hyperlipidemic rats [33,34]. In rats fed a high-fat diet, tanshinone IIA treatment reduced fat accumulation and partly ameliorated oxidative stress, inflammation, and apoptosis in the livers [35].

These recent findings suggest that tanshinone IIA can modulate hepatic lipid metabolism and homeostasis; however, the underlying mechanisms by which tanshinone IIA regulates hepatic lipid homeostasis remain unclear. In this study, we aimed to investigate the lipid-modulating effect and underlying mechanisms of tanshinone IIA, particularly focusing on its effects on lipogenic gene expression and lipid accumulation in hepatic cells.

2. Materials and Methods

2.1. Chemicals

Tanshinone IIA, 3-(4,5-dimethylthiazol-2-yl)-2,5-diphenyltetrazolium bromide (MTT), dimethyl sulfoxide (DMSO), palmitic acid, oleic acid, insulin–transferrin–selenium (ITS), T0901317, 24(S),25-epoxycholesterol, and other chemicals were purchased from Sigma-Aldrich Co. (St. Louis, MO, USA) unless otherwise indicated. DMEM, fetal bovine serum (FBS) and nonessential amino acids (NEAAs) were purchased from Thermo Fisher Scientific, Inc. (Rockford, IL, USA).

2.2. Cell Culture and Compound Treatment

HepG2 and Huh7 cells were cultured in growth medium composed of DMEM containing 10% FBS and 1x nonessential amino acid (NEAA) solution in a 5% CO_2 incubator at 37 °C. The high-fat medium formula was described in previous literature and slightly modified [36,37]. For the high-fat medium preparation, stock solutions of palmitic acid (0.25 mM) and oleic acid (1 mM) were prepared and precomplexed with 10% FBS in DMEM medium, respectively, for dilution of the desired final concentration. The high-fat medium contained DMEM, 10% FBS, 1xNEAA, palmitic acid (0.125 mM), oleic acid (0.5 mM), and insulin–transferrin–selenium (ITS). For compound treatment for 24 h in the high-fat medium condition, cells were treated with vehicle (0.1% DMSO) or tanshinone IIA (1–10 µM) in growth medium, and after 20 h, the medium was replaced with high-fat medium containing the indicated compounds (vehicle or tanshinone IIA) and incubated for an additional 4 h. For LXRα ligand treatment, cells were pretreated with vehicle or tanshinone IIA (10 µM) for 1 h, followed by incubation with T0901317 (1 µM) or 24(S),25-epoxycholesterol (40 µM) for 24 h and harvested for further studies.

2.3. Analysis of Cell Viability by MTT Assay

The viability of cells was determined by MTT assay as described previously [38]. Cells were treated with vehicle or tanshinone IIA (1–20 µM) for 24 h. These cells were incubated with the MTT reagent (1 mg/mL) at 37 °C for 3 h. The medium was removed, and the cells were washed with PBS. Purple crystals were dissolved in DMSO, and cell viability was determined by the absorbance measured at 550 nm using a microplate reader.

2.4. Analysis of Cell Cytotoxicity by LDH Release Assay

The cytotoxicity of cells was determined by LDH assay as described previously [39] Cells were treated with vehicle or tanshinone IIA (1–20 µM) for 24 h. Supernatants were harvested and LDH activity was measured using an LDH Cytotoxicity Assay Kit II (Abcam, Cambridge, MA, USA) according to the manufacturer's protocol. The absorbance of samples was determined at 450 nm using a microplate reader.

2.5. Reverse-Transcription Quantitative Polymerase Chain Reaction (RT-qPCR) Analysis

The RT-qPCR analysis was carried out as described previously [40] Cells were treated with vehicle or tanshinone IIA (5 and 10 µM) for 24 h, and cellular RNA was extracted using a FavorPrep™ blood/cultured cell total RNA purification mini kit (FAVORGEN Biotech, Ping-Tung, Taiwan). Reverse transcription was carried out with a high-capacity cDNA reverse transcription kit (Thermo Fisher Scientific, MA, USA). Quantitative real-time PCR was performed in a mixture containing cDNA, human-specific primers (Table S1), and Maxima SYBR Green/ROX qPCR Master Mix (Thermo Fisher Scientific) on a Roche LightCycler®-480 real-time PCR system. The $\Delta\Delta C_t$ method was used to measure the relative differences in mRNA expression between experimental groups and normalize with the mRNA expression of GAPDH in the same samples.

2.6. Western Blot Analysis

Western blot analysis was carried out as described previously [41]. Cells were treated with vehicle or tanshinone IIA (5 and 10 μM) for 24 h. For preparation of total cellular proteins, the cells were harvested with RIPA buffer (Thermo Fisher Scientific). For nuclear-extract preparation, the cells were harvested using NE-PER nuclear and cytoplasmic extraction reagent (Thermo Fisher Scientific). The samples were separated by 10% SDS-PAGE and transferred to PVDF membranes (PerkinElmer, Boston, MA, USA). The blots were incubated with the following primary antibodies at 4 °C for 24 h: anti-FASN (A6273) (1:2000) (ABclonal, Woburn, MA, USA), anti-ACC1 (#3676) (1:1000) and anti-phospho-ACC1 (Ser79) (#3661) (1:1000) (Cell Signaling Technology, Danvers, MA), anti-SCD1 (ab39969) (1:1000) and anti-LXRα (ab41902) (1:1000) (Abcam), anti-HDAC2 (GTX112957) (1:3000) (GeneTex, Irvine, CA, USA), anti-SREBP1 (sc-13551) (1:200) (Santa Cruz Biotechnology, Santa Cruz, CA, USA) and anti-actin (MAB1501) (1:30000) (Thermo Fisher Scientific). The blots were incubated with the appropriate HRP-conjugated secondary antibodies, and the amount of each protein was measured by Amersham ECL™ prime Western blotting detection reagent. The chemiluminescent signal was visualized using Amersham Hyperfilm™ ECL film (GE Healthcare, Buckinghamshire, UK).

2.7. Staining of Neutral Lipid Droplets

Lipid droplets were stained by a specifically lipophilic fluorescent dye as described previously [42,43]. HepG2 cells were seeded in poly-L-lysine-coated coverslips and treated with vehicle or tanshinone IIA in growth medium or high-fat medium as described above. The cells on coverslips were fixed in cold 4% paraformaldehyde at room temperature for 20 min followed by incubation with 0.5 μM BODIPY493/503 fluorescent dye (Thermo Fisher Scientific) for 15 min. The coverslips were washed three times with PBS and mounted using VECTASHIELD HardSet™ mounting medium with 4′,6-diamidino-2-phenylindole (DAPI) (Vector Laboratories, Burlingame, CA, USA). The imaged were observed and photographed with a ZEISS LSM 900 laser confocal microscope (ZEISS Microscopy, Jena, Germany). The mean fluorescent intensities of the lipid droplets per cell were analyzed and quantified in at least 200 cells in 25–30 randomly selected fields (63x) from independent replicates using ZEISS ZEN lite software (ZEN 2.3) (ZEISS Microscopy).

2.8. Analysis of Hepatic Triglyceride (TG)

The cells were treated with vehicle or tanshinone IIA, and cell lysates were prepared by using sonication. The triglyceride (TG) was extracted from cell lysates as described previously [44]. The amounts of TG were measured with a Triglyceride Colorimetric Assay Kit (Cayman Chemical, Ann Arbor, MI, USA) according to the manufacturer's protocol. The TG content was measured as the absorbance at 530 nm using a microplate reader.

2.9. Molecular Docking of Compounds to Binding Domain of LXRα Protein

The molecular docking of compounds to nuclear receptor LXRα was performed using Molecular Operating Environment software (MOE2019.01). The DOCK module with "induced fit" refinement to enhance the accuracy of the MOE2019.01 software program was used to predict the preferable binding sites between LXRα and the respective compounds (tanshinone IIA, T0901317, and 24 (S),25-epoxycholesterol). The compounds were manually built with the MOE software package for docking with the LXRα binding domain (PDB code: 2ACL, LXRα with compound SB313987). Water molecules in the crystal structure were removed, missing short loops were added with the MOE software, and the energy was minimized before molecular docking. The score was based on GBVI/WSA ΔG, a force-field-based scoring function that estimates the binding free energy of a ligand to a receptor. The preferable binding sites for each compound were determined based on the lowest binding free energy, which was the lowest S value in the scoring function.

2.10. Statistical Analysis

At least three independent experiments were performed, and each experiment was repeated three times. The data are expressed as the means $\pm$ SD. The data were analyzed using Student's *t*-test (for comparisons of data with two groups) and one-way ANOVA with Tukey's test for post hoc analysis (for comparisons of data with multiple groups), and a *p*-value of <0.05 was considered statistically significant.

3. Results

3.1. Effects of Tanshinone IIA on the Viability of Hepatic Cell Lines

To investigate the cytotoxic effect of tanshinone IIA (Figure 1a) on hepatic cells, HepG2 cells were treated with vehicle (0.1% DMSO) or tanshinone IIA (1, 2, 5, 10, and 20 μM) for 24 h. Cell cytotoxicity induced by tanshinone IIA was determined using an MTT assay and validated by an LDH release assay. The data showed that vehicle (0.1% DMSO) and tanshinone IIA (1–10 μM) induced no cytotoxicity in the HepG2 cells (Figure 1b,c). However, higher concentration of tanshinone IIA (20 μM) slightly reduced the cell viability as compared with vehicle-treated group. Similar results for cell viability also were found in the tanshinone IIA-treated Huh 7 hepatic cell line (Figure S1). Therefore, doses of 5 and 10 μM of tanshinone IIA were used in the subsequent experiments.

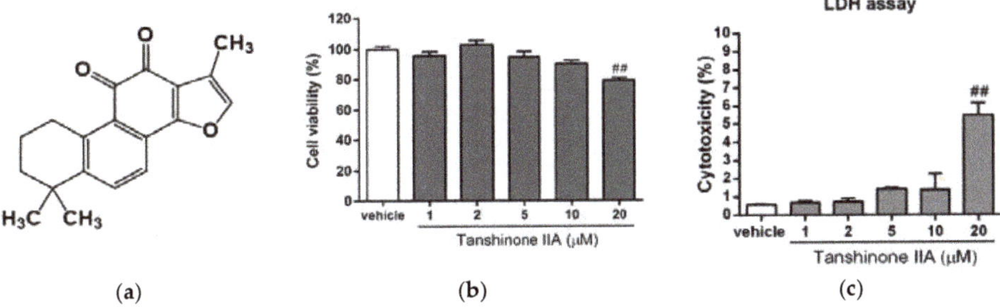

Figure 1. Effects of tanshinone IIA on HepG2 cell viability. (**a**) The chemical structure of tanshinone IIA. HepG2 cells were treated with vehicle (0.1% DMSO) or tanshinone IIA (1, 2, 5, 10, and 20 μM) for 24 h. (**b**) Cell viability was measured using an MTT assay. (**c**) Cytotoxicity was measured using LDH release assay. The data represent the mean $\pm$ SD of three independent experiments. ## *p* < 0.01 indicates significant differences compared to the vehicle-treated cells.

3.2. Effects of Tanshinone IIA on FASN, ACC1, and SCD1 Expression in Hepatic Cell Lines

De novo lipogenesis plays a critical role in regulating the amount of lipid accumulation in hepatic cells. To investigate whether tanshinone IIA can regulate the expression of endogenous de novo lipogenesis-related genes, we examined the effect of tanshinone IIA treatment on the mRNA expression of fatty-acid synthase (FASN), acetyl-CoA carboxylase 1 (ACC1), and stearoyl-CoA desaturase 1 (SCD1) in HepG2 cells. Tanshinone IIA (5 and 10 μM) significantly reduced the mRNA expression of FASN, ACC1, and SCD1 by 0.85 $\pm$ 0.09- and 0.55 $\pm$ 0.18-fold; 0.81 $\pm$ 0.08- and 0.67 $\pm$ 0.10-fold; and 0.64 $\pm$ 0.14- and 0.45 $\pm$ 0.13-fold in HepG2 cells, respectively, when compared to the vehicle-treated cells (1.00 $\pm$ 0.05, 1.00 $\pm$ 0.08, and 1.01 $\pm$ 0.14) (*p* < 0.05 and *p* < 0.01, respectively) (Figure 2a–c). Similar results showed that tanshinone IIA could significantly inhibit the mRNA expression of FASN, ACC1, and SCD1 genes in Huh 7 cells (Figure S2a–c).

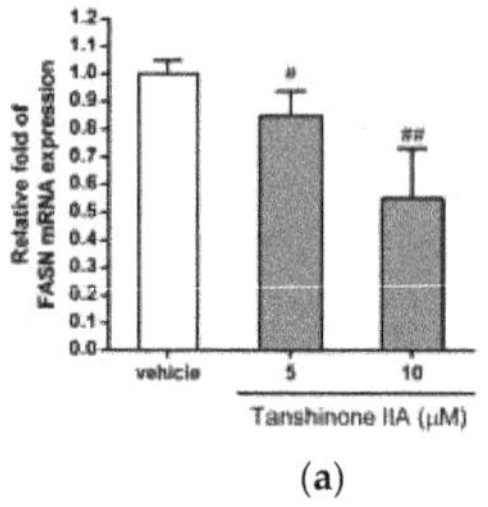 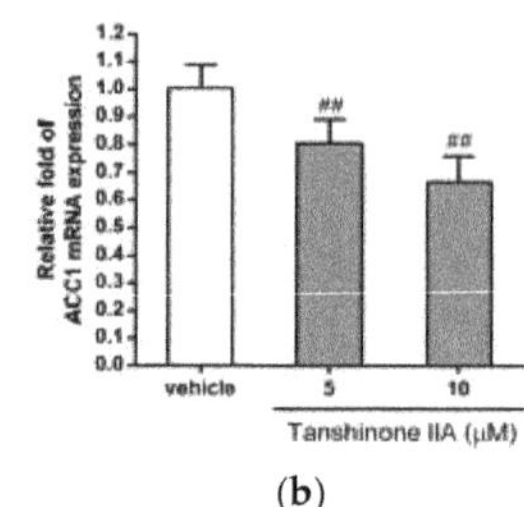 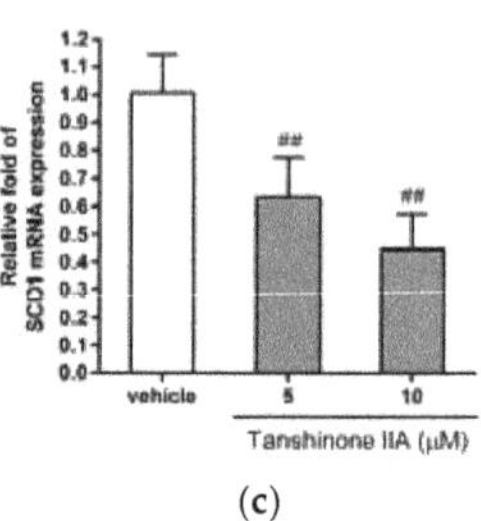

(a)   (b)   (c)

Figure 2. Effects of tanshinone IIA on FASN, ACC1, and SCD1 mRNA expression in HepG2 cells. HepG2 cells were treated with vehicle (0.1% DMSO) or tanshinone IIA (5 and 10 μM) for 24 h. The mRNA expression of (**a**) FASN, (**b**) ACC1, and (**c**) SCD1 was measured by RT-qPCR analysis. The data represent the mean ± SD of three independent experiments. # $p < 0.05$ and ## $p < 0.01$ indicate significant differences compared to the vehicle-treated cells.

Furthermore, we examined the levels of the FASN, ACC1, and SCD1 proteins in tanshinone IIA-treated HepG2 cells. The cells treated with tanshinone IIA (5 and 10 μM) had significantly decreased levels of FASN, ACC1, and SCD1 protein expression, by 0.80 ± 0.13- and 0.67 ± 0.18-fold; 0.57 ± 0.19- and 0.45 ± 0.17-fold; and 0.66 ± 0.19- and 0.43 ± 0.14-fold in HepG2 cells, respectively, when compared to the vehicle-treated cells (1.00 ± 0.03, 1.01 ± 0.08, and 1.00 ± 0.08) ($p < 0.05$ and $p < 0.01$, respectively) (Figure 3a–f). These results indicated that tanshinone IIA can significantly inhibit the mRNA and protein expression of these genes, which are essential for de novo lipogenesis in hepatic cells.

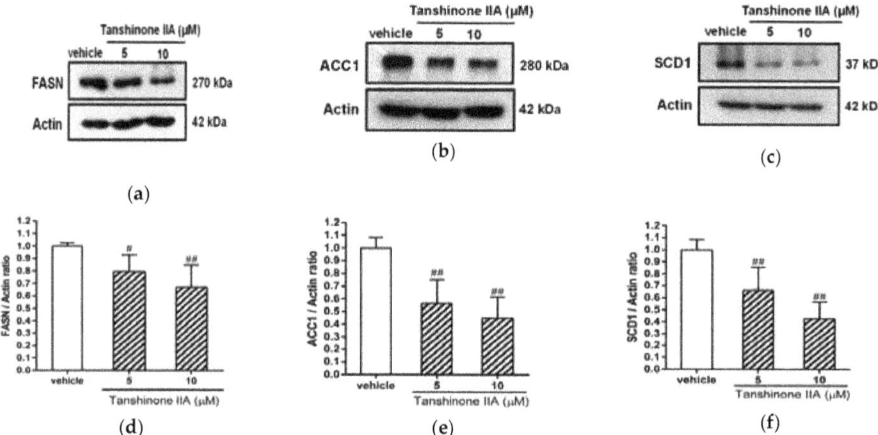

Figure 3. Effects of tanshinone IIA on FASN, ACC1, and SCD1 protein expression in HepG2 cells. HepG2 cells were treated with vehicle (0.1% DMSO) or tanshinone IIA (5 and 10 μM) for 24 h. The protein expression of (**a**) FASN, (**b**) ACC1, and (**c**) SCD1 was measured by Western blot analysis. A representative blot is shown. The normalized intensities of (**d**) FASN, (**e**) ACC1, and (**f**) SCD1 proteins versus actin are presented as the mean ± SD of three independent experiments. # $p < 0.05$ and ## $p < 0.01$ indicate significant differences compared to the vehicle-treated cells.

3.3. Effects of Tanshinone IIA on ACC1 Phosphorylation in HepG2 Cells

The enzyme ACC1 plays a critical role in fatty-acid synthesis and has been identified to be inactivated by phosphorylation of serine 79, leading to lipogenesis inhibition [45]. Therefore, we investigated the effect of tanshinone IIA on ACC1 phosphorylation in HepG2 cells. Tanshinone IIA (5 and 10 μM) significantly increased the ratio of phosphorylated ACC1 (p-ACC1)/ACC1 proteins by 1.41 ± 0.01- and 2.11 ± 0.05-fold in HepG2 cells, respectively, when compared to the vehicle-treated group (1.00 ± 0.08) ($p < 0.01$) (Figure 4a,b).

Moreover, we examined the effect of tanshinone IIA on ACC1 phosphorylation in HepG2 cells incubated with high-fat medium. The ratio of p-ACC1/ACC1 in the cells treated with high-fat medium was slightly reduced by 0.80 ± 0.04-fold as compared to cells treated with normal growth medium (1.00 ± 0.07) ($p < 0.01$). Tanshinone IIA significantly increased the p-ACC1/ACC1 ratio by 2.47 ± 0.08-fold, when compared to the vehicle-treated group (0.80 ± 0.04) ($p < 0.01$) (Figure 4c,d). This data indicated that tanshinone IIA could promote phosphorylation of ACC1 protein in hepatic cells.

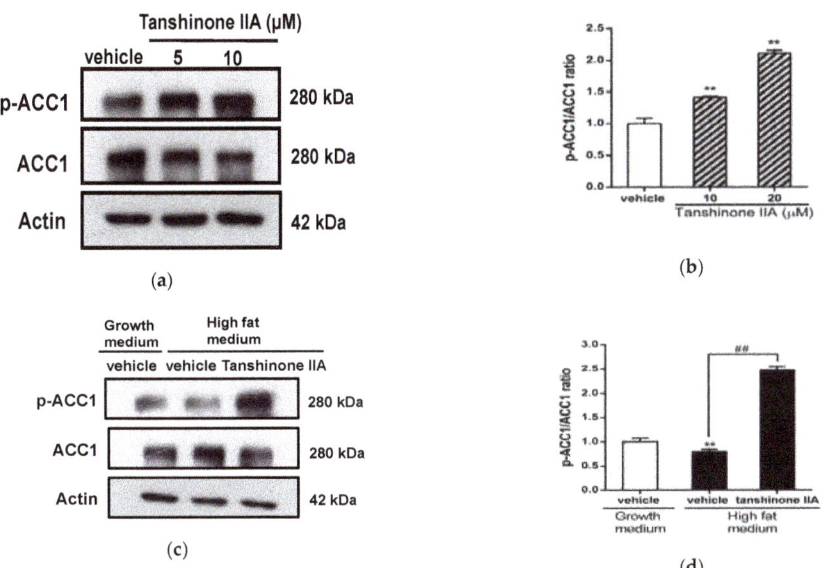

Figure 4. Effects of tanshinone IIA on ACC1 phosphorylation in HepG2 cells. (**a**) HepG2 cells were treated with vehicle (0.1% DMSO) or tanshinone IIA (5 and 10 µM) in normal growth medium. The protein levels of phosphorylated ACC1 at serine 79 (p-ACC1), ACC1, and actin was measured by Western blot analysis. A representative blot is shown. (**b**) The normalized intensity of p-ACC1 protein versus ACC1 is presented as the mean $\pm$ SD of three independent experiments. ** $p < 0.01$ indicates significant differences compared to the vehicle-treated cells. (**c**) HepG2 cells were treated with vehicle (0.1% DMSO) or tanshinone IIA (10 µM) in normal growth medium or high-fat medium. The phosphorylated ACC1 (p-ACC1), ACC1, and actin was measured by Western blot analysis. A representative blot is shown. (**d**) The normalized intensity of p-ACC1 versus ACC1 is presented as the mean $\pm$ SD of three independent experiments. ** $p < 0.01$ indicates significant differences compared to the vehicle-treated cells incubated with growth medium. ## $p < 0.01$ indicates significant differences compared to the vehicle-treated cells.

3.4. Effects of Tanshinone IIA on the Fatty-Acid-Induced Lipogenesis and TG Accumulation in HepG2 Cells

To investigate the effect of tanshinone IIA on the fatty-acid-induced lipogenesis and TG accumulation, first, we examined whether tanshinone IIA affected cell viability of HepG2 cells incubated in high-fat medium. The MTT assay data showed that the viability of cells treated with tanshinone IIA (1, 2, 5, and 10 µM) was $97.7 \pm 3.1\%$, $96.8 \pm 5.6\%$, $94.6 \pm 1.8\%$, and $92.4 \pm 4.5\%$ as compared to cells incubated in normal growth medium ($100.4 \pm 1.8\%$), respectively (Figure 5a). This data showed that tanshinone IIA did not induce significant cytotoxicity in cells incubated with high-fat medium. Next, the effect of tanshinone IIA on the palmitate/oleate-induced lipogenesis was investigated. The HepG2 cells incubated with high-fat medium showed increased levels of FASN, ACC1, and SCD1 mRNA expression as compared with that cultured in growth medium. Tanshinone IIA significantly decreased the amounts of FASN, ACC1, and SCD1 mRNA in cells incubated with normal growth or high-fat medium, when compared to their effects on vehicle-treated cells ($p < 0.01$) (Figure 5b–d). The effect of tanshinone IIA on intracellular triglyceride (TG) accumulation was examined, and the data showed that TG contents in cells incubated with high-fat medium were significantly increased as compared with cells treated with normal growth medium ($p < 0.01$). The TG contents were significantly reduced by tanshinone IIA (10 μM) treatment in HepG2 cells cultured in growth medium or high-fat medium (Figure 5e). These data indicated that tanshinone IIA can significantly ameliorate fatty-acid-induced lipogenesis and TG accumulation in hepatic cells.

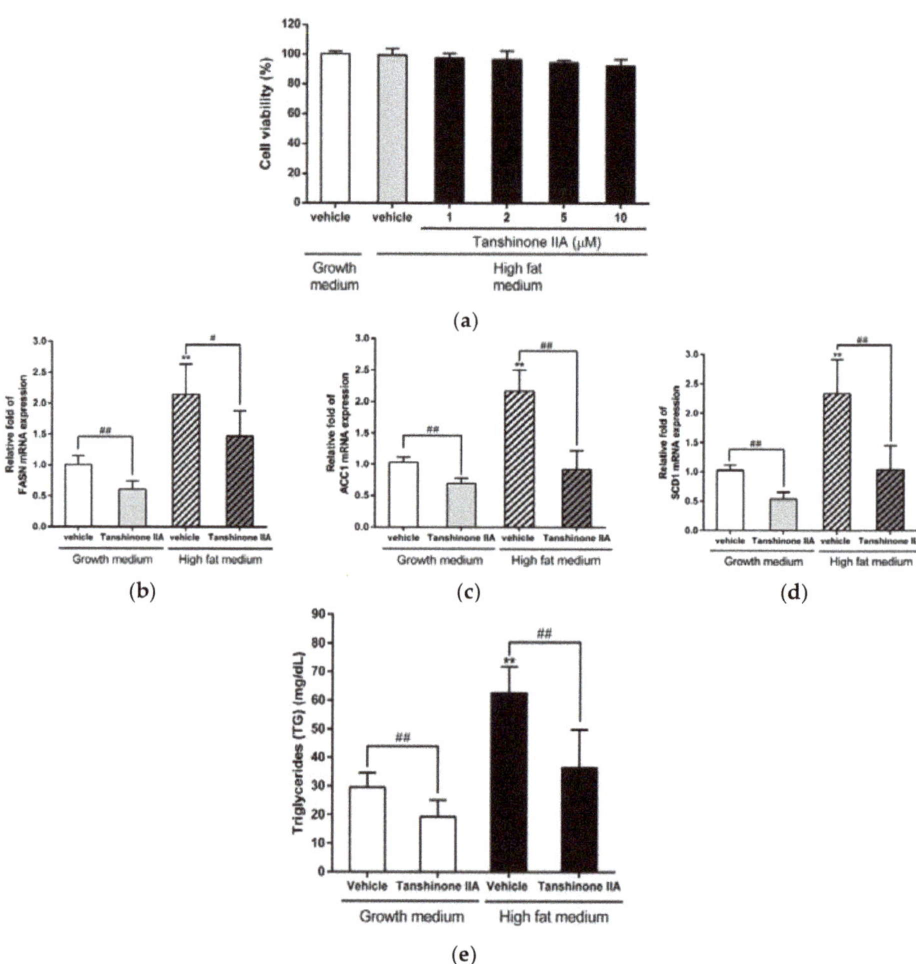

Figure 5. Effects of tanshinone IIA on the fatty-acid-induced lipogenesis and TG accumulation in HepG2 cells. HepG2 cells were treated with vehicle (0.1% DMSO) or tanshinone IIA (1–10 μM) in growth medium (DMEM with 10% FBS and 1xNEAA), and after 20 h, the medium was replaced with high-fat medium containing the indicated compounds (vehicle or tanshinone IIA). (**a**) Cell viability was measured using an MTT assay. The mRNA expression of (**b**) FASN, (**c**) ACC1, and (**d**) SCD1 was measured by RT-qPCR analysis. The data represent the mean ± SD of three independent experiments. # $p < 0.05$ and ## $p < 0.01$ indicate significant differences compared to the vehicle-treated cells. (**e**) Intracellular triglyceride (TG) levels of vehicle- or tanshinone-IIA-treated HepG2 cells were measured from three independent replicates, and the data represent the mean ± SD. ** $p < 0.01$ indicates significant differences compared to the vehicle-treated cells incubated with growth medium. ## $p < 0.01$ indicate significant differences compared to the tanshinone IIA-untreated cells.

3.5. Effects of Tanshinone IIA on SREBP1 Expression in HepG2 Cells

SREBP-1c is an important transcriptional regulator of FASN, ACC1, and SCD1 gene expression in hepatic cells. Therefore, we further investigated the effect of tanshinone IIA on SREBP-1c expression in HepG2 cells. Tanshinone IIA (5 and 10 μM) significantly reduced the level of SREBP-1c mRNA by 0.65 ± 0.15- and 0.37 ± 0.10-fold in HepG2 cells, respectively, when compared to the vehicle-treated group (1.01 ± 0.11) ($p < 0.01$) (Figure 6a). Because the SREBP1 antibody used was unable to distinguish between the SREBP-1a and -1c, we used the SREBP1 to mention in this data.. We found that tanshinone IIA (5 and 10 μM) decreased the amount of nuclear mature SREBP1 protein by 0.73 ± 0.19- and 0.35 ± 0.17-fold, respectively, when compared to the vehicle-treated group (1.01 ± 0.03) ($p < 0.05$ and $p < 0.01$, respectively) (Figure 6b,c).

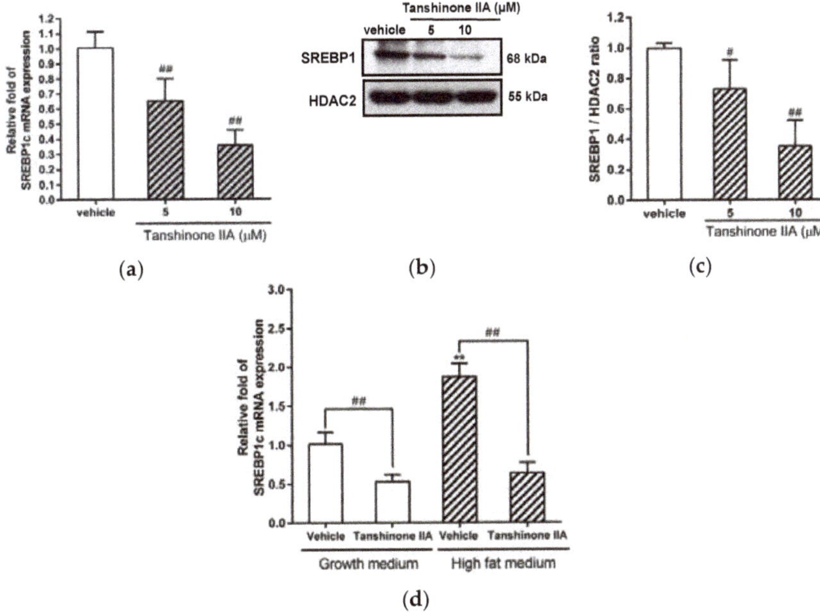

Figure 6. Effects of tanshinone IIA on SREBP1 gene expression. HepG2 cells were treated with vehicle (0.1% DMSO) or tanshinone IIA (5 and 10 μM) for 24 h. (**a**) The mRNA expression of SREBP-1c was measured by RT-qPCR analysis. The data represent the mean $\pm$ SD of three independent experiments. ## $p < 0.01$ indicates significant differences compared to the vehicle-treated cells. (**b**) Nuclear extracts were prepared from vehicle- or tanshinone-IIA-treated cells, and the level of nuclear SREBP1 protein was analyzed by Western blot analysis. A representative blot is shown. (**c**) The normalized intensity of SREBP1 protein versus HDAC2 is presented as the mean $\pm$ SD of three independent experiments. # $p < 0.05$ and ## $p < 0.01$ indicate significant differences compared to the vehicle-treated cells. (**d**) HepG2 cells were treated with vehicle (0.1% DMSO) or tanshinone IIA (10 μM) in normal growth medium or high-fat medium. The mRNA expression of SREBP-1c was measured by RT-qPCR analysis. The data represent the mean $\pm$ SD of three independent experiments. ** $p < 0.01$ indicates significant differences compared to the vehicle-treated cells incubated with growth medium. ## $p < 0.01$ indicates significant differences compared to the vehicle-treated cells.

Furthermore, we investigated the effect of tanshinone IIA on SREBP1c expression in HepG2 cells incubated with high-fat medium. The cells incubated with high-fat medium showed an elevated level of SREBP1c mRNA expression by 1.87 ± 0.16-fold as compared with the cells cultured in growth medium (1.01 ± 0.15) ($p < 0.01$). Tanshinone IIA significantly decreased the amount of SREBP1c mRNA by 0.64 ± 0.13-fold in cells incubated with high-fat medium, when compared to its effect on vehicle-treated cells ($p < 0.01$) (Figure 6d). In addition to SREBP1c, recently, carbohydrate regulatory binding protein (ChREBP) also was reported to regulate the FASN and ACC1 gene expression for glucose-induced lipogenesis [2,46]. We further examined the effect of tanshinone IIA on ChREBP expression in HepG2 cells. We found that tanshinone IIA (5 and 10 μM) could not change the level of ChREBP mRNA expression (Figure S3). These above results indicated that tanshinone IIA downregulates SREBP1 gene expression, which may lead to reduce expression of lipogenic genes in hepatic cells.

3.6. Effects of Tanshinone IIA on LXRα-Mediated Transcriptional Activity in HepG2 Cells

SREBP-1c and some lipogenic enzymes such as FASN are the direct targets for transcriptional activation by LXRα in hepatic cells. To explore the effect of tanshinone IIA on LXRα-mediated SREBP1 expression, we first examined whether tanshinone IIA could change LXRα gene expression in HepG2 cells. We found that tanshinone IIA did not change the mRNA or nuclear protein levels of LXRα in the HepG2 cells (Figure 7a–c). We further examined the effect of tanshinone IIA on the ligand-induced LXR activity for SREBP-1c mRNA expression. HepG2 cells were pretreated with vehicle or tanshinone IIA (10 μM) for 1 h followed by treatment with LXRα agonists; namely, T0901317 (Figure 7d) and 24(S),25-epoxycholesterol (Figure 7e), for an additional 24 h. As shown in Figure 7f, the level of SREBP-1c mRNA was significantly increased by T0901317 (1 μM) and 24(S),25-epoxycholesterol (40 μM) treatment of the HepG2 cells. When cells were cotreated with tanshinone IIA and agonists, the T0901317- and 24(S),25-epoxycholesterol-induced SREBP-1c mRNA expression was significantly attenuated by approximately 60% and 67%, respectively ($p < 0.01$). Furthermore, we found that T0901317- and 24(S),25-epoxycholesterol-induced FASN mRNA expression was also significantly decreased by approximately 56% and 45%, respectively, in tanshinone IIA-treated cells ($p < 0.01$) (Figure 7g). Moreover, to validate the regulatory effect of tanshinone IIA on LXRα-mediated transcription, mRNA expression of other downstream target genes, such as ACC1 and SCD1 [24,47], and GLUT1 [48], a negatively regulated gene by LXRα ligands, were also examined. T0901317-induced ACC1 and SCD1 mRNA expression was significantly decreased by tanshinone IIA in HepG2 cells (Figure S4a,b). Tanshinone IIA also could reverse the T0901317-decreased GLUT1 expression to basal level in HepG2 cells (Figure S4c). These above results verified that tanshinone IIA was involved in the regulation of LXRα-mediated transcriptional activity in hepatic cells.

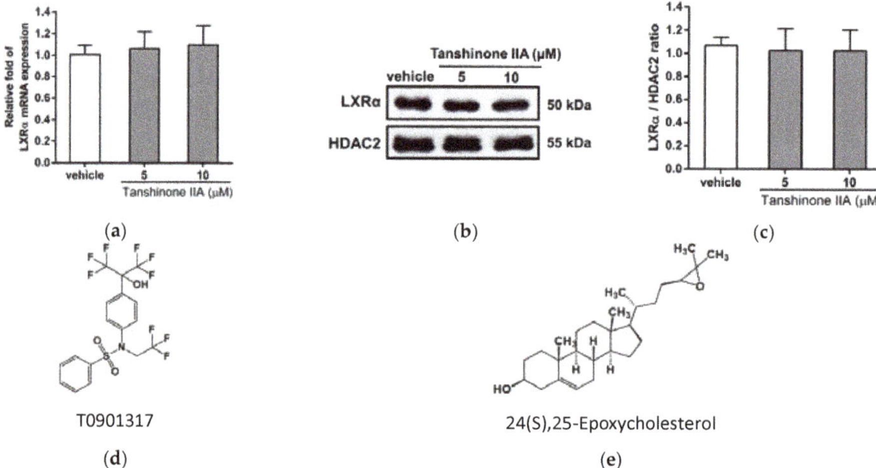

Figure 7. *Cont.*

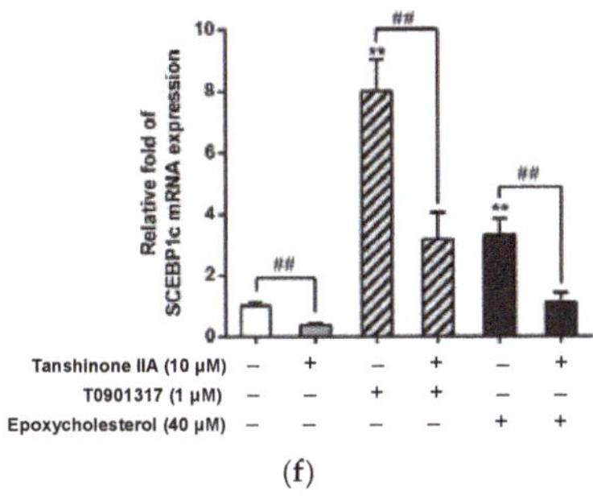

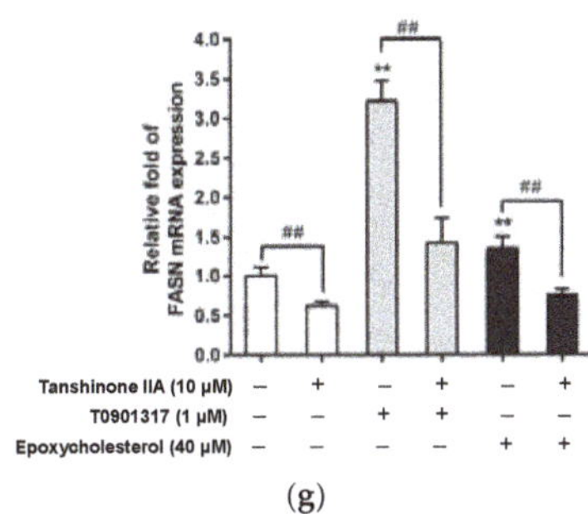

(f)　　　　　　　　　　　　　　　(g)

Figure 7. Effects of tanshinone IIA on LXRα expression and ligand-induced LXRα activation in HepG2 cells. HepG2 cells were treated with vehicle (0.1% DMSO) or tanshinone IIA (5 and 10 μM) for 24 h. (**a**) The mRNA expression of LXRα was measured by RT-qPCR analysis. The data are presented as the mean ± SD of three independent experiments. (**b**) The nuclear protein expression of LXRα and HDAC2 was measured by Western blot analysis. A represent blot is shown. (**c**) Normalized intensities of LXRα versus HDAC2 is presented as the mean ± SD of three independent experiments. The chemical structure of (**d**) T0901317 and (**e**) 24(S),25-epoxycholesterol HepG2 cells were pretreated with vehicle (0.1% DMSO) or tanshinone IIA (10 μM) for 1 h followed by treatment with agonist T0901317 (1 μM) or 24(S),25-epoxycholesterol (40 μM) for 24 h. The mRNA levels of (**f**) SREBP-1c and (**g**) FASN were measured by RT-qPCR analysis. ** $p < 0.01$ indicates a significant difference compared to compound-untreated group. ## $p < 0.01$ indicates significant differences compared to cells treated with the agonist alone.

3.7. Tanshinone IIA Attenuates LXRα-Mediated Lipid-Droplet Accumulation in HepG2 Cells

To investigate the effect of tanshinone IIA on lipid accumulation in HepG2 cells, the neutral lipid droplets in HepG2 cells were assessed using BODIPY493/503 fluorescent dye staining and analyzed under a confocal microscope. The cells incubated with high-fat medium showed an increased level of lipid droplets as compared with the cells cultured in normal growth medium (Figure 8a,b). Tanshinone IIA could significantly reduce the amount of lipid droplets in cells incubated with growth medium and high-fat medium by approximately 37% and 32%, respectively, when compared to its effect on vehicle-treated cells (Figure 8c). We further investigated the effect of tanshinone IIA on LXRα-mediated lipid-droplet accumulation in HepG2 cells. The data from Figure 8a–c showed that lipid accumulation induced by the T0901317 (1 μM) in cells with growth or high-fat medium was increased by approximately 25% and 33%, respectively, when compared to vehicle-treated groups. When cells were cotreated with T0901317 and tanshinone IIA (10 μM), the T0901317-induced lipid droplets were significantly attenuated by approximately 52% and 49% in cells incubated with growth medium and high-fat medium, respectively. These data indicated that tanshinone IIA can attenuate LXRα-mediated lipid droplets accumulation in HepG2 cells.

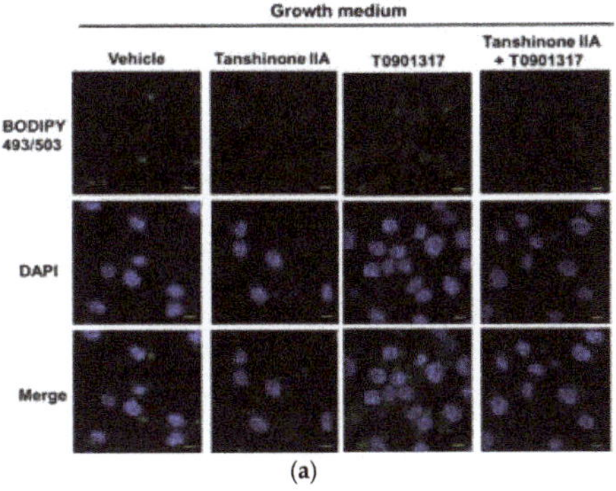

(a)

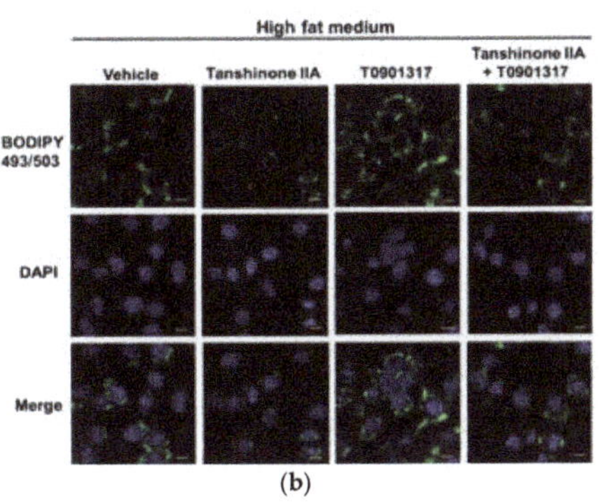

(b)

Figure 8. *Cont.*

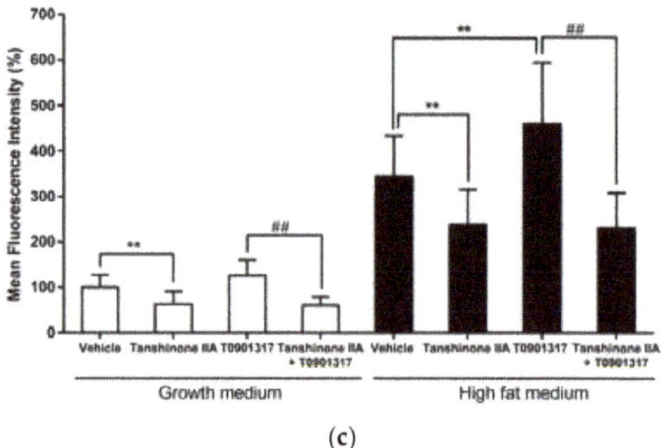

(c)

Figure 8. Effects of tanshinone IIA on lipid-droplet accumulation in HepG2 cells. HepG2 cells were pretreated with vehicle (0.1% DMSO) or tanshinone IIA (10 μM) for 1 h followed by treatment with T0901317 (1 μM) in (**a**) growth medium or (**b**) high-fat medium. The lipid droplets were stained by BODIPY493/503 fluorescent dye (green), and nuclei were stained by DAPI (blue). Stained cells were observed and photographed with a confocal microscope. The scale bar is 10 μm. (**c**) Quantification of lipid droplets accumulation. The mean fluorescent intensities of the lipid droplets per cell were quantified in 25–30 randomly selected fields (>200 cells) from three independent replicates, and the data represent the mean ± SD from three independent experiments. ** $p < 0.01$ indicates significant differences compared to the vehicle-treated cells. ## $p < 0.01$ indicates significant differences compared to the cells treated with T0901317 alone.

3.8. Tanshinone IIA Docks to the Ligand-Binding Domain of LXRα

To investigate whether tanshinone IIA can bind to LXRα, the molecular docking program was performed as described in the Materials and Methods section. The docking results showed that tanshinone IIA, T0901317, and 24(S),25-epoxycholesterol can dock to LXRα binding pocket with different docking scores (tanshinone IIA: −7.33; T0901317: −8.47; and 24(S),25-epoxycholesterol: −10.32) (Figure 9a–c). The superposition result showed that all these compounds docked into the ligand-binding domain of LXRα at similar poses (Figure 9d). The ligand-receptor interaction maps showed that the tanshinone IIA was surrounded by many hydrophobic amino-acid residues such as F255, L258, A259, M296, and F313 (Figure 9e). These data indicated that the binding pocket of LXRα is suitable for lipophilic molecule interactions and tanshinone IIA may serve as a ligand specifically bound to this ligand-binding domain for modulation of LXRα activity.

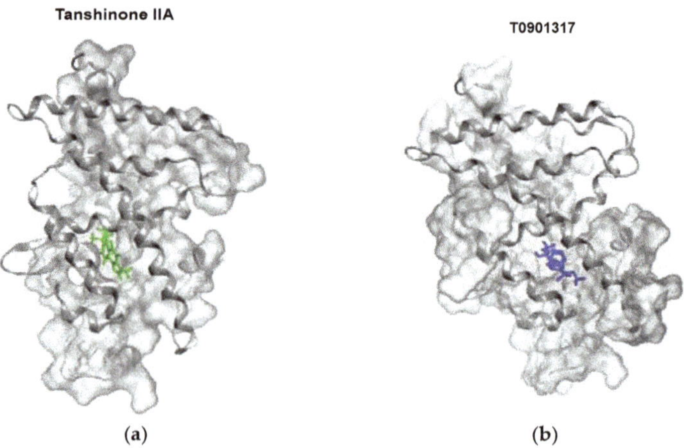

(**a**)　　　　　　　　　　　　　　　　　(**b**)

Figure 9. *Cont.*

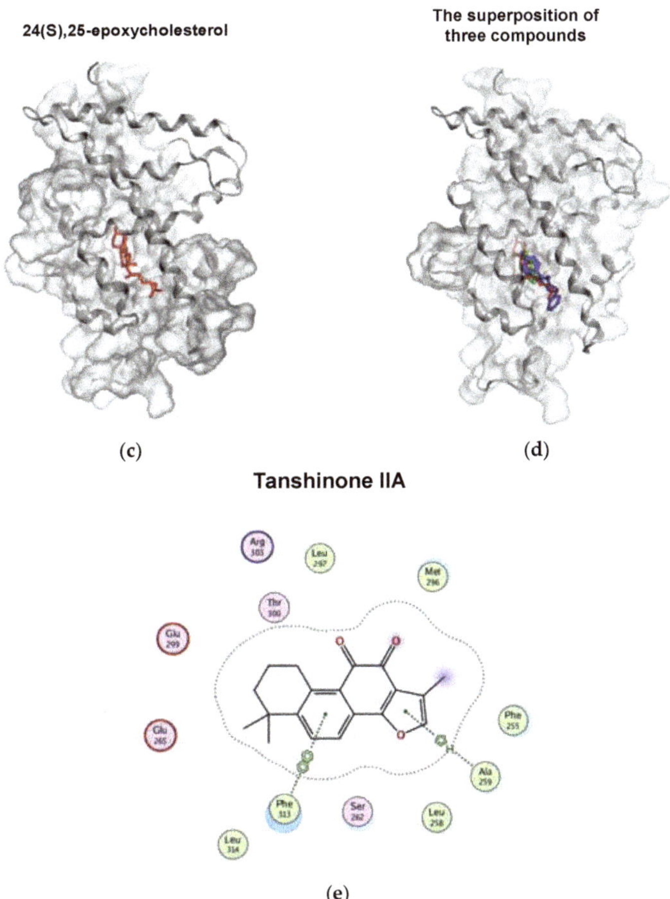

Figure 9. Tanshinone IIA docks to the ligand-binding domain of LXRα. The poses of compounds docked preferentially to the ligand-binding domain of LXRα. The receptor is shown as a gray ribbon, and the compounds are shown in (**a**) green (tanshinone IIA), (**b**) blue (T0901317), and (**c**) red (24(S),25-epoxycholesterol). (**d**) The superposition of these three compounds bound to the ligand-binding domain pocket of LXRα. Two-dimensional interaction map of LXRα with (**e**) tanshinone IIA. Many hydrophobic amino- acid residues (green color) surrounded the tanshinone IIA.

4. Discussion

Abnormal accumulation of lipid droplets within hepatic cells is the hallmark of MAFLD. MAFLD patients may benefit from early treatment to reduce hepatic or cardiovascular complications [49]. Tanshinone IIA, a pharmacologically bioactive phytochemical extracted from the dried root of *Salvia miltiorrhiza*, has been demonstrated to have putative effects on antiatherosclerosis, antihyperlipidemia and antiadipogenesis [50]. Recently, tanshinone IIA and its derived compound, sodium tanshinone IIA sulfonate, were reported to ameliorate hepatic steatosis or fatty liver [35,51,52]. However, studies of the molecular effects of tanshinone IIA on hepatic steatosis and its role in the prevention or treatment of MAFLD remain limited. In this study, we demonstrated that tanshinone IIA could significantly reduce the amounts of lipid-droplet deposition within HepG2 cells. Tanshinone IIA could increase the phosphorylation of ACC1 proteins, which may lead to inhibit hepatic lipogenesis. Tanshinone IIA downregulated the expression of lipogenic enzymes, including FASN, ACC1, and SCD1, through suppression of SREBP1 expression and LXRα-mediated transcriptional activation, resulting in reducing lipogenesis and lipid accumulation in the HepG2 cells. Moreover, we predicted that tanshinone IIA binds to the ligand-binding domain of LXRα to suppress its transcriptional activity (Figure 10). These findings support the hypothesis that tanshinone IIA possesses lipid-modulating activity and can potentially serve as a novel agent for the prevention or treatment of hepatic steatosis or MAFLD.

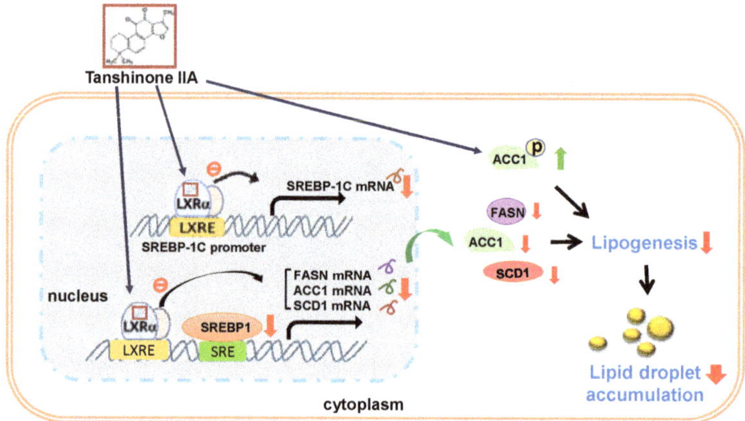

Figure 10. A hypothetical model for the tanshinone-IIA-mediated attenuation of lipid accumulation in hepatic cells. Tanshinone IIA promotes ACC1 phosphorylation to attenuate fatty-acid synthesis and lipogenesis. Tanshinone IIA down-regulates FASN, ACC1, and SCD1 expression by inhibiting the LXRα-mediated transcriptional activation, resulting in a reduction in endogenous and fatty-acid-induced lipogenesis as well as lipid-droplet accumulation in hepatic cells.

Hepatic steatosis results from an imbalance between lipid supplementation and clearance in the liver. Emerging studies have demonstrated that dysregulation of lipogenesis contributes to hepatic steatosis and plays an essential role in the pathologic progress of MAFLD. The increases in hepatic TG content measured in MAFLD patients showed that approximately 30% of the contribution came from de novo biosynthesis [17]. In this study, tanshinone IIA downregulated the mRNA and protein expression of lipogenic enzymes, including FASN, ACC1, and SCD1, in HepG2 cells. Our data also showed tanshinone IIA markedly reduced TG contents and lipid-droplet accumulation. It has been reported that modulation of de novo lipogenesis in hepatocytes is important for the attenuation of hepatic steatosis and treatment of MAFLD. Recent studies using animal models have shown that knocking out or knocking down the hepatic enzymes involved in lipogenesis or desaturation of long-chain fatty acids may reduce TG concentration and attenuate steatosis [16,53,54]. The present study revealed that tanshinone IIA can reduce the levels of lipogenic enzymes in hepatocytes, which may lead to decreased hepatic TG contents and lipid accumulation. Saturated-fat diets could increase lipogenic gene expression and hepatic steatosis in mice [55]. The TG-derived fatty-acid profile showed that higher saturated fatty-acid content and low levels of polyunsaturated fatty acid increased the proinflammatory cytokine to enhance inflammation [56]. In this study, inhibition of palmitate/oleate-induced lipogenesis by tanshinone IIA may be associated with reduction of TG accumulation. Whether tanshinone IIA could increase the unsaturated fatty-acid levels and reduce saturated fatty-acid content in cells treated with high-fat medium remains unclear. Analysis of the profile of TG-derived saturated and mono/polyunsaturated fatty acid in lipid droplets from tanshinone IIA-treated cells needs to be further investigated.

In this study, tanshinone IIA increased the ratio of p-ACC1/ACC1 in HepG2 cells. These data indicated that tanshinone IIA could promote ACC1 phosphorylation, which may lead to inhibit its enzyme activity for the decrease of de novo lipid biosynthesis. It is known that AMPK activation directly phosphorylated ACC1 at serine 79 and suppressed its activity, and reduced FASN expression to attenuate fatty-acid-induced hepatic steatosis. AMPK activation also decreased fatty-acid biosynthesis through reducing the SREBP1c levels or nuclear translocation to repress lipogenic gene expression and lipid accumulation [57,58]. Recently, AMPK activator was reported to modulate hepatic saturated/unsaturated fatty-acid composition, and suppress hepatic macrophage infiltration in the development of MAFLD [59]. Whether tanshinone IIA can activate AMPK signaling to regulate lipogenic gene expression and reduce hepatic steatosis remains unclear and needs to be further investigated.

Transcriptional modulation of the lipogenic enzymes in hepatocytes is an interesting idea for developing therapeutic strategies for hepatic steatosis. The transcription factor SREBP1 regulates the transcription rate of genes involved in the fatty-acid synthesis pathway. SREBP1 is expressed in two isoforms, SREBP-1a and SREBP-1c, can be translated as a precursor protein bound to the endoplasmic reticulum (ER) membrane, and must be transported to the Golgi and activated by a cleavage process to release a mature form to the nucleus [60]. SREBP1 mediates the regulation of hepatic genes involved in carbohydrate and lipid metabolism. SREBP-1c was reported to control the mRNA expression of FASN, ACC1, and SCD1 in hepatocytes [61,62]. Additionally, hepatic SREBP-1c was highly expressed in patients with MAFLD [63]. In this study, we found that tanshinone IIA significantly decreased the levels of SREBP-1c mRNA and the mature form of nuclear SREBP1 protein. These results revealed that tanshinone IIA can inhibit the transcription of FASN, ACC1, and SCD1 via the modulation of nuclear SREBP1 levels in hepatic cells.

The essential role of SREBP1 in lipogenesis has been demonstrated in gain- and loss-of-function studies in vitro and in vivo. Overexpressing a mature form of SREBP-1c in hepatocytes or mice can activate the lipogenic pathway and induce steatosis [64,65]. In contrast, mice lacking SREBP-1c fail to express of lipogenic enzymes or undergo TG synthesis in response to fasting and refeeding [23]. The data we presented here demonstrated that tanshinone IIA could cause a significant decrease in the level of the mature SREBP1 protein. These findings indicate that tanshinone-IIA-mediated nuclear SREBP1 reduction leads to the downregulated expression of lipogenic enzymes and attenuates lipid accumulation in hepatic cells. SREBP1 protein is also known to play a critical role in hepatic carbohydrate metabolism [66]. Whether tanshinone-IIA-mediated SREBP1 downregulation modulates glucose metabolism in hepatic cells needs to be further investigated.

The SREBP-1c gene is upregulated by the nuclear receptor LXRα. LXR is expressed in two isoforms, LXRα and LXRβ, and LXRs are ligand-dependent nuclear receptors that regulate cholesterol and lipid metabolism [67,68]. LXRα is mainly expressed in the liver, kidney, adipose tissue, and intestine, whereas LXRβ is ubiquitously expressed [69]. LXRα agonist treatment has been reported to induce lipogenesis, because they bind to LXREs within the promoter region of SREBP-1c, FASN, ACC1, and SCD1 to increase mRNA transcription in hepatic cells [24,47,70,71]. LXRα and its lipogenic target genes were also reported to be highly expressed in patients with MAFLD [72]. In this study, we found that the LXR agonist enhanced lipogenic genes and contributed to lipid accumulation as expected in HepG2 cells. In tanshinone-IIA-treated cells, LXRα-induced lipid-droplet accumulation can be significantly attenuated. Lipid droplets are dynamic lipid-rich organelles that have been shown to function in TG and energy storage, and as a protective strategy for fatty-acid-induced lipotoxicity in the liver. However, an imbalance of lipid-droplet formation and mobilization can lead to excess lipid droplets and TG accumulation in hepatic cells, which is a pathological condition for MAFLD [73,74]. In this study, tanshinone IIA significantly ameliorated both fatty-acid- and LXRα-induced lipid-droplet accumulation in HepG2 cells. This finding suggested that tanshinone IIA may modulate lipid-droplet formation and homeostasis in the liver.

The LXRα protein is composed of an N-terminal-activating domain, a DNA-binding domain, a ligand-binding domain (LBD), and a C-terminal activation domain. The LBD of LXRα has been reported to form a hydrophobic pocket and may interact with small-molecule compounds. The binding affinity of LXR to its ligand could be predicted by molecular-docking analysis [75]. In this study, molecular docking predicted that tanshinone IIA specifically binds to the LBD of LXRα. The lipophilic tanshinone IIA seems preferentially to interact with hydrophobic residues within the LBD pocket of LXRα. Thus, we suggested that the LXRα-ligand-induced lipid accumulation could be significantly ameliorated by tanshinone IIA. LXR agonists were reported to increase expression of their downstream target genes such as ACC1 and SCD1, and decrease GLUT1 expression. In this study, tanshinone IIA could counteract the effect of the LXRα ligand T0901317 on these target genes' expression; however, LXRα mRNA and protein levels were not changed. These results supported that tanshinone IIA could serve as a novel LXRα ligand to modulate the LXRα–SREBP1 axis in hepatic cells. The precise mechanisms underlying the suppression of LXRα activation by tanshinone IIA remain unclear. The determination of tanshinone IIA as an LXR antagonist, an inhibitor, or a ligand competitor that blocks the LXR-RXR heterodimer needs further functional studies, with subjects such as GLUT1 or other downstream targets involved in glucose or lipid metabolism.

LXRα plays essential roles in fatty-acid and cholesterol homeostasis, and in control of inflammation and immunity [76,77]. LXR activation may control cholesterol efflux from foam cells, induce cholesterol transporter expression, and inhibit inflammatory mediators in the artery wall and macrophage, which may lead to reduced vascular inflammation and atherosclerosis [49]. Therefore, LXR agonists are an option to be used for the treatment of atherosclerosis, diabetes, and other disorders [78]. However, LXR agonists activated hepatic lipogenesis and promoted TG accumulation in MAFLD, which limit the value of clinical application. In this study, tanshinone IIA may have served as a ligand and counteracted the LXRα–SREBP1 axis to reduce hepatic lipid accumulation. Recently, tanshinone IIA treatment was reported to reduced macrophage infiltration and nuclear factor (NF)-κB activation in macrophages for inhibiting inflammatory responses [79]. Tan et al. reported that tanshinone IIA increased cholesterol efflux and reduced lipid accumulation in macrophages, leading to a reduction in the development of aortic atherosclerosis [80]. Wen et al. reported that tanshinone IIA inhibited oxidized LDL-induced NLRP3 inflammasome activation in mouse macrophages and ameliorated atherogenesis [81]. These findings indicated that tanshinone IIA possessed the potential activities for anti-inflammation and antiatherosclerosis. In this study, we suggested that tanshinone IIA could attenuate ligand-induced LXRα activation and may be an LXRα antagonist or a ligand competitor for downregulation of lipogenic gene expression in hepatic cells. Whether tanshinone IIA affects the anti-inflammatory effect and cholesterol accumulation by LXRα-dependent pathways in extra-hepatocytes remains to be clarified, and needs to be considered with caution due to its adverse effect in therapeutic strategy for treatment of MAFLD. Recently, Griffett et al. reported that a hepato-selective LXR inverse agonist could specifically suppress LXR activity in hepatic cells to reduce lipogenesis and prevent MAFLD [82]. Whether tanshinone IIA could serve as an inverse agonist to reduce lipogenesis without promoting cholesterol accumulation in extra-hepatic tissue remains to be further investigated. In this study, we realized our results were found only on the in vitro lipid-lowering effects of tanshinone IIA on hepatic cancer cell lines. To verify the impact of treatment with tanshinone IIA in vivo, the primary human hepatocytes and animal models of MAFLD can be used in further studies of tanshinone IIA on modulation of lipogenesis and lipid accumulation.

5. Conclusions

In this study, for the first time, we demonstrated that tanshinone IIA exerts a promising lipid-modulating effect to attenuate lipid-droplet accumulation and downregulate the expression of genes involved in de novo lipogenesis by regulating the LXRα/SREBP1 pathway in hepatic cells. Our findings revealed that tanshinone IIA may serve as a novel supplement or drug to regulate LXRα activity and counter unwanted lipid overproduction in the liver for the prevention or treatment of MAFLD.

Supplementary Materials: The following are available online at https://www.mdpi.com/2227-9059/9/3/326/s1, Table S1: The primer pairs used in RT-qPCR, Figure S1: Effects of tanshinone IIA on Huh 7 cell viability, Figure S2: Effects of tanshinone IIA on FASN, ACC1, and SCD1 mRNA expression in Huh 7 cells, Figure S3: Effects of tanshinone IIA on ChREBP mRNA expression in HepG2 cells, Figure S4: Effects of tanshinone IIA on LXRα-mediated transcriptional activity in HepG2 cells.

Author Contributions: J.-H.Y. and P.-Y.C. conceived and designed the experiments; W.-Y.G., H.-J.H., and C.-Y.L. performed the experiments; W.-Y.G., H.-J.H., P.-Y.C., and J.-H.Y. analyzed the data; M.-J.W., J.-H.Y., and P.-Y.C. contributed reagents/materials; W.-Y.G., P.-Y.C., and J.-H.Y. wrote the paper. All authors have read and agreed to the published version of the manuscript.

Funding: This research was supported by grants MOST-105-2320-B-320-004-MY3 (funding approval date 1 August 2016) and MOST 108-2320-B-320-002-MY3 (to J.-H.Y.) (funding approval date 1 August 2019), and MOST 108-2635-B-303-001 (to P.-Y.C.) (funding approval date 1 August 2019) from the Ministry of Science and Technology, Taiwan.

Institutional Review Board Statement: Not applicable.

Informed Consent Statement: Not applicable.

Data Availability Statement: Not applicable.

Acknowledgments: We are grateful for the support from the Core Research Laboratory, Tzu Chi University for the assistance in the confocal image analysis.

Conflicts of Interest: The authors declare no conflict of interest.

References

1. Nguyen, P.; Leray, V.; Diez, M.; Serisier, S.; Le Bloc'H, J.; Siliart, B.; Dumon, H. Liver lipid metabolism. *J. Anim. Physiol. Anim. Nutr.* **2008**, *92*, 272–283. [CrossRef]
2. Ipsen, D.H.; Lykkesfeldt, J.; Tveden-Nyborg, P. Molecular mechanisms of hepatic lipid accumulation in non-alcoholic fatty liver disease. *Cell. Mol. Life Sci.* **2018**, *75*, 3313–3327. [CrossRef]
3. Byrne, C.D.; Targher, G. NAFLD: A multisystem disease. *J. Hepatol.* **2015**, *62*, S47–S64. [CrossRef] [PubMed]
4. Shiha, G.; Alswat, K.; Al Khatry, M.; I Sharara, A.; Örmeci, N.; Waked, I.; Benazzouz, M.; Al-Ali, F.; Hamed, A.E.; Hamoudi, W.; et al. Nomenclature and definition of metabolic-associated fatty liver disease: A consensus from the Middle East and north Africa. *Lancet Gastroenterol. Hepatol.* **2021**, *6*, 57–64. [CrossRef]
5. Eslam, M.; Newsome, P.N.; Sarin, S.K.; Anstee, Q.M.; Targher, G.; Romero-Gomez, M.; Zelber-Sagi, S.; Wong, V.W.-S.; Dufour, J.-F.; Schattenberg, J.M.; et al. A new definition for metabolic dysfunction-associated fatty liver disease: An international expert consensus statement. *J. Hepatol.* **2020**, *73*, 202–209. [CrossRef]
6. Younossi, Z.M.; Koenig, A.B.; Abdelatif, D.; Fazel, Y.; Henry, L.; Wymer, M. Global epidemiology of nonalcoholic fatty liver disease-Meta-analytic assessment of prevalence, incidence, and outcomes. *Hepatology* **2016**, *64*, 73–84. [CrossRef]
7. Adams, L.A.; Anstee, Q.M.; Tilg, H.; Targher, G. Non-alcoholic fatty liver disease and its relationship with cardiovascular disease and other extrahepatic diseases. *Gut* **2017**, *66*, 1138–1153. [CrossRef]
8. Fazel, Y.; Koenig, A.B.; Sayiner, M.; Goodman, Z.D.; Younossi, Z.M. Epidemiology and natural history of non-alcoholic fatty liver disease. *Metab. Clin. Exp.* **2016**, *65*, 1017–1025. [CrossRef]
9. Loomba, R.; Chalasani, N. The Hierarchical Model of NAFLD: Prognostic Significance of Histologic Features in NASH. *Gastroenterology* **2015**, *149*, 278–281. [CrossRef]
10. Sumida, Y.; Yoneda, M. Current and future pharmacological therapies for NAFLD/NASH. *J. Gastroenterol.* **2018**, *53*, 362–376. [CrossRef] [PubMed]
11. Adams, L.A.; Angulo, P. Treatment of non-alcoholic fatty liver disease. *Postgrad. Med. J.* **2006**, *82*, 315–322. [CrossRef]
12. Katsurada, A.; Iritani, N.; Fukuda, H.; Matsumura, Y.; Nishimoto, N.; Noguchi, T.; Tanaka, T. Effects of nutrients and hormones on transcriptional and post-transcriptional regulation of fatty acid synthase in rat liver. *JBIC J. Biol. Inorg. Chem.* **1990**, *190*, 427–433. [CrossRef] [PubMed]
13. Katsurada, A.; Iritani, N.; Fukuda, H.; Matsumura, Y.; Nishimoto, N.; Noguchi, T.; Tanaka, T. Effects of nutrients and hormones on transcriptional and post-transcriptional regulation of acetyl-CoA carboxylase in rat liver. *Eur. J. Biochem.* **1990**, *190*, 435–441. [CrossRef]
14. Jakobsson, A.; Westerberg, R.; Jacobsson, A. Fatty acid elongases in mammals: Their regulation and roles in metabolism. *Prog. Lipid Res.* **2006**, *45*, 237–249. [CrossRef] [PubMed]

15. Ntambi, J.M. Dietary regulation of stearoyl-CoA desaturase 1 gene expression in mouse liver. *J. Biol. Chem.* **1992**, *267*, 10925–10930. [CrossRef]

16. Postic, C.; Girard, J. Contribution of de novo fatty acid synthesis to hepatic steatosis and insulin resistance: Lessons from genetically engineered mice. *J. Clin. Investig.* **2008**, *118*, 829–838. [CrossRef] [PubMed]

17. Donnelly, K.L.; Smith, C.I.; Schwarzenberg, S.J.; Jessurun, J.; Boldt, M.D.; Parks, E.J. Sources of fatty acids stored in liver and secreted via lipoproteins in patients with nonalcoholic fatty liver disease. *J. Clin. Investig.* **2005**, *115*, 1343–1351. [CrossRef]

18. Eberlé, D.; Hegarty, B.; Bossard, P.; Ferré, P.; Foufelle, F. SREBP transcription factors: Master regulators of lipid homeostasis. *Biochimie* **2004**, *86*, 839–848. [CrossRef]

19. Higuchi, N.; Kato, M.; Shundo, Y.; Tajiri, H.; Tanaka, M.; Yamashita, N.; Kohjima, M.; Kotoh, K.; Nakamuta, M.; Takayanagi, R.; et al. Liver X receptor in cooperation with SREBP-1c is a major lipid synthesis regulator in nonalcoholic fatty liver disease. *Hepatol. Res.* **2008**, *38*, 1122–1129. [CrossRef]

20. Edwards, P.A.; Kennedy, M.A.; Mak, P.A. LXRs; oxysterol-activated nuclear receptors that regulate genes controlling lipid homeostasis. *Vasc. Pharm.* **2002**, *38*, 249–256. [CrossRef]

21. Perissi, V.; Aggarwal, A.; Glass, C.K.; Rose, D.W.; Rosenfeld, M.G. A Corepressor/Coactivator Exchange Complex Required for Transcriptional Activation by Nuclear Receptors and Other Regulated Transcription Factors. *Cell* **2004**, *116*, 511–526. [CrossRef]

22. Schultz, J.R.; Tu, H.; Luk, A.; Repa, J.J.; Medina, J.C.; Li, L.; Schwendner, S.; Wang, S.; Thoolen, M.; Mangelsdorf, D.J.; et al. Role of LXRs in control of lipogenesis. *Genes Dev.* **2000**, *14*, 2831–2838. [CrossRef]

23. Liang, G.; Yang, J.; Horton, J.D.; Hammer, R.E.; Goldstein, J.L.; Brown, M.S. Diminished Hepatic Response to Fasting/Refeeding and Liver X Receptor Agonists in Mice with Selective Deficiency of Sterol Regulatory Element-binding Protein-1c. *J. Biol. Chem.* **2002**, *277*, 9520–9528. [CrossRef] [PubMed]

24. Chu, K.; Miyazaki, M.; Man, W.C.; Ntambi, J.M. Stearoyl-Coenzyme A Desaturase 1 Deficiency Protects against Hypertriglyceridemia and Increases Plasma High-Density Lipoprotein Cholesterol Induced by Liver X Receptor Activation. *Mol. Cell. Biol.* **2006**, *26*, 6786–6798. [CrossRef] [PubMed]

25. Wagner, M.; Zollner, G.; Trauner, M. Nuclear receptors in liver disease. *Hepatology* **2011**, *53*, 1023–1034. [CrossRef] [PubMed]

26. Chen, Z.; Xu, H. Anti-Inflammatory and Immunomodulatory Mechanism of Tanshinone IIA for Atherosclerosis. *Evid. Based Complement. Altern. Med.* **2014**, *2014*, 1–6. [CrossRef] [PubMed]

27. Shang, Q.; Xu, H.; Huang, L. Tanshinone IIA: A Promising Natural Cardioprotective Agent. *Evid. Based Complement. Altern. Med.* **2012**, *2012*, 1–7. [CrossRef]

28. Xu, S.; Liu, Z.; Huang, Y.; Le, K.; Tang, F.; Huang, H.; Ogura, S.; Little, P.J.; Shen, X.; Liu, P. Tanshinone II-A inhibits oxidized LDL-induced LOX-1 expression in macrophages by reducing intracellular superoxide radical generation and NF-kappaB activation. *Transl. Res.* **2012**, *160*, 114–124. [CrossRef]

29. Liu, Z.; Wang, J.; Huang, E.; Gao, S.; Li, H.; Lu, J.; Tian, K.; Little, P.J.; Shen, X.; Xu, S.; et al. Tanshinone IIA suppresses cholesterol accumulation in human macrophages: Role of heme oxygenase-1. *J. Lipid Res.* **2014**, *55*, 201–213. [CrossRef]

30. Chen, H.-C.; Chen, P.-Y.; Wu, M.-J.; Tai, M.-H.; Yen, J.-H. Tanshinone IIA Modulates Low Density Lipoprotein Uptake via Down-Regulation of PCSK9 Gene Expression in HepG2 Cells. *PLoS ONE* **2016**, *11*, e0162414. [CrossRef]

31. Chen, W.; Tang, F.; Xie, B.; Chen, S.; Huang, H.; Liu, P. Amelioration of atherosclerosis by tanshinone IIA in hyperlipidemic rabbits through attenuation of oxidative stress. *Eur. J. Pharmacol.* **2012**, *674*, 359–364. [CrossRef] [PubMed]

32. Tang, F.; Wu, X.; Wang, T.; Wang, P.; Li, R.; Zhang, H.; Gao, J.; Chen, S.; Bao, L.; Huang, H.; et al. Tanshinone II A attenuates atherosclerotic calcification in rat model by inhibition of oxidative stress. *Vasc. Pharmacol.* **2007**, *46*, 427–438. [CrossRef] [PubMed]

33. Jia, L.; Song, N.; Yang, G.; Ma, Y.; Li, X.; Lu, R.; Cao, H.; Zhang, N.; Zhu, M.; Wang, J.; et al. Effects of Tanshinone IIA on the modulation of miR-33a and the SREBP-2/Pcsk9 signaling pathway in hyperlipidemic rats. *Mol. Med. Rep.* **2016**, *13*, 4627–4635. [CrossRef] [PubMed]

34. Jia, L.-Q.; Zhang, N.; Xu, Y.; Chen, W.-N.; Zhu, M.-L.; Song, N.; Ren, L.; Cao, H.-M.; Wang, J.-Y.; Yang, G.-L. Tanshinone IIA affects the HDL subfractions distribution not serum lipid levels: Involving in intake and efflux of cholesterol. *Arch. Biochem. Biophys.* **2016**, *592*, 50–59. [CrossRef] [PubMed]

35. Huang, L.; Ding, W.; Wang, M.-Q.; Wang, Z.-G.; Chen, H.-H.; Chen, W.; Yang, Q.; Lu, T.-N.; He, J.-M. Tanshinone IIA ameliorates non-alcoholic fatty liver disease through targeting peroxisome proliferator-activated receptor gamma and toll-like receptor 4. *J. Int. Med. Res.* **2019**, *47*, 5239–5255. [CrossRef]

36. Luo, W.-J.; Cheng, T.-Y.; Wong, K.-I.; Fang, W.-H.; Liao, K.-M.; Hsieh, Y.-T.; Su, K.-Y. Novel therapeutic drug identification and gene correlation for fatty liver disease using high-content screening: Proof of concept. *Eur. J. Pharm. Sci.* **2018**, *121*, 106–117. [CrossRef]

37. Oliveira, A.F.M.; Da Cunha, D.A.; Ladriere, L.; Igoillo-Esteve, M.; Bugliani, M.; Marchetti, P.; Cnop, M. In vitro use of free fatty acids bound to albumin: A comparison of protocols. *BioTechniques* **2015**, *58*, 228–233. [CrossRef]

38. Yen, J.-H.; Lin, C.-Y.; Chuang, C.-H.; Chin, H.-K.; Wu, M.-J.; Chen, P.-Y. Nobiletin Promotes Megakaryocytic Differentiation through the MAPK/ERK-Dependent EGR1 Expression and Exerts Anti-Leukemic Effects in Human Chronic Myeloid Leukemia (CML) K562 Cells. *Cells* **2020**, *9*, 877. [CrossRef]

39. McKenzie, B.A.; Mamik, M.K.; Saito, L.B.; Boghozian, R.; Monaco, M.C.; Major, E.O.; Lu, J.-Q.; Branton, W.G.; Power, C. Caspase-1 inhibition prevents glial inflammasome activation and pyroptosis in models of multiple sclerosis. *Proc. Natl. Acad. Sci. USA* **2018**, *115*, E6065–E6074. [CrossRef] [PubMed]

40. Chang, H.-Y.; Wu, J.-R.; Gao, W.-Y.; Lin, H.-R.; Chen, P.-Y.; Chen, C.-I.; Wu, M.-J.; Yen, J.-H. The Cholesterol-Modulating Effect of Methanol Extract of Pigeon Pea (*Cajanus cajan* (L.) Millsp.) Leaves on Regulating LDLR and PCSK9 Expression in HepG2 Cells. *Molecules* **2019**, *24*, 493. [CrossRef] [PubMed]

41. Chen, S.-F.; Chen, P.-Y.; Hsu, H.-J.; Wu, M.-J.; Yen, J.-H. Xanthohumol Suppresses Mylip/Idol Gene Expression and Modulates LDLR Abundance and Activity in HepG2 Cells. *J. Agric. Food Chem.* **2017**, *65*, 7908–7918. [CrossRef]

42. Spangenburg, E.E.; Pratt, S.J.P.; Wohlers, L.M.; Lovering, R.M. Use of BODIPY (493/503) to visualize intramuscular lipid droplets in skeletal muscle. *J. Biomed. Biotechnol.* **2011**, *2011*, 598358. [CrossRef]

43. Durandt, C.; van Vollenstee, F.A.; Dessels, C.; Kallmeyer, K.; de Villiers, D.; Murdoch, C.; Potgieter, M.; Pepper, M.S. Novel flow cytometric approach for the detection of adipocyte subpopulations during adipogenesis. *J. Lipid Res.* **2016**, *57*, 729–742. [CrossRef] [PubMed]

44. Zhu, X.; Bian, H.; Wang, L.; Sun, X.; Xu, X.; Yan, H.; Xia, M.; Chang, X.; Lu, Y.; Li, Y.; et al. Berberine attenuates nonalcoholic hepatic steatosis through the AMPK-SREBP-1c-SCD1 pathway. *Free Radic. Biol. Med.* **2019**, *141*, 192–204. [CrossRef] [PubMed]

45. Hardie, D.G. AMPK: A key regulator of energy balance in the single cell and the whole organism. *Int. J. Obes.* **2008**, *32*, S7–S12. [CrossRef] [PubMed]

46. Iizuka, K.; Horikawa, Y. ChREBP: A Glucose-activated Transcription Factor Involved in the Development of Metabolic Syndrome. *Endocr. J.* **2008**, *55*, 617–624. [CrossRef] [PubMed]

47. Talukdar, S.; Hillgartner, F.B. The mechanism mediating the activation of acetyl-coenzyme A carboxylase-α gene transcription by the liver X receptor agonist T0-901317. *J. Lipid Res.* **2006**, *47*, 2451–2461. [CrossRef] [PubMed]

48. Xiong, T.; Li, Z.; Huang, X.; Lu, K.; Xie, W.; Zhou, Z.; Tu, J. TO901317 inhibits the development of hepatocellular carcinoma by LXRα/Glut1 decreasing glycometabolism. *Am. J. Physiol. Liver Physiol.* **2019**, *316*, G598–G607. [CrossRef]

49. Ducheix, S.; Montagner, A.; Theodorou, V.; Ferrier, L.; Guillou, H. The liver X receptor: A master regulator of the gut–liver axis and a target for non alcoholic fatty liver disease. *Biochem. Pharmacol.* **2013**, *86*, 96–105. [CrossRef]

50. Ren, J.; Fu, L.; Nile, S.H.; Zhang, J.; Kai, G. Salvia miltiorrhiza in Treating Cardiovascular Diseases: A Review on Its Pharmacological and Clinical Applications. *Front. Pharmacol.* **2019**, *10*, 753. [CrossRef]

51. Li, X.-X.; Lu, X.-Y.; Zhang, S.-J.; Chiu, A.P.; Lo, L.H.; Largaespada, D.A.; Chen, Q.-B.; Keng, V.W. Sodium tanshinone IIA sulfonate ameliorates hepatic steatosis by inhibiting lipogenesis and inflammation. *Biomed. Pharmacother.* **2019**, *111*, 68–75. [CrossRef]

52. Yang, G.; Jia, L.; Wu, J.; Ma, Y.; Cao, H.; Song, N.; Zhang, N. Effect of tanshinone IIA on oxidative stress and apoptosis in a rat model of fatty liver. *Exp. Ther. Med.* **2017**, *14*, 4639–4646. [CrossRef]

53. Mao, J.; DeMayo, F.J.; Li, H.; Abu-Elheiga, L.; Gu, Z.; Shaikenov, T.E.; Kordari, P.; Chirala, S.S.; Heird, W.C.; Wakil, S.J. Liver-specific deletion of acetyl-CoA carboxylase 1 reduces hepatic triglyceride accumulation without affecting glucose homeostasis. *Proc. Natl. Acad. Sci. USA* **2006**, *103*, 8552–8557. [CrossRef] [PubMed]

54. Miyazaki, M.; Flowers, M.T.; Sampath, H.; Chu, K.; Otzelberger, C.; Liu, X.; Ntambi, J.M. Hepatic stearoyl-CoA desaturase-1 deficiency protects mice from carbohydrate-induced adiposity and hepatic steatosis. *Cell Metab.* **2007**, *6*, 484–496. [CrossRef] [PubMed]

55. Lin, J.; Yang, R.; Tarr, P.T.; Wu, P.H.; Handschin, C.; Li, S.; Yang, W.; Pei, L.; Uldry, M.; Tontonoz, P.; et al. Hyperlipidemic effects of dietary saturated fats mediated through PGC-1beta coactivation of SREBP. *Cell* **2005**, *120*, 261–273. [CrossRef]

56. Yamada, K.; Mizukoshi, E.; Sunagozaka, H.; Arai, K.; Yamashita, T.; Takeshita, Y.; Misu, H.; Takamura, T.; Kitamura, S.; Zen, Y.; et al. Response to Importance of confounding factors in assessing fatty acid compositions in patients with non-alcoholic steatohepatitis. *Liver Int.* **2014**, *35*, 1773. [CrossRef]

57. Ahmed, M.H.; Byrne, C.D. Modulation of sterol regulatory element binding proteins (SREBPs) as potential treatments for non-alcoholic fatty liver disease (NAFLD). *Drug Discov. Today* **2007**, *12*, 740–747. [CrossRef]

58. Li, Y.; Xu, S.; Mihaylova, M.M.; Zheng, B.; Hou, X.; Jiang, B.; Park, O.; Luo, Z.; Lefai, E.; Shyy, J.Y.-J.; et al. AMPK Phosphorylates and Inhibits SREBP Activity to Attenuate Hepatic Steatosis and Atherosclerosis in Diet-Induced Insulin-Resistant Mice. *Cell Metab.* **2011**, *13*, 376–388. [CrossRef] [PubMed]

59. Peng, X.; Li, J.; Wang, M.; Qu, K.; Zhu, H. A novel AMPK activator improves hepatic lipid metabolism and leukocyte trafficking in experimental hepatic steatosis. *J. Pharmacol. Sci.* **2019**, *140*, 153–161. [CrossRef]

60. Shimano, H.; Yahagi, N.; Amemiya-Kudo, M.; Hasty, A.H.; Osuga, J.-I.; Tamura, Y.; Shionoiri, F.; Iizuka, Y.; Ohashi, K.; Harada, K.; et al. Sterol Regulatory Element-binding Protein-1 as a Key Transcription Factor for Nutritional Induction of Lipogenic Enzyme Genes. *J. Biol. Chem.* **1999**, *274*, 35832–35839. [CrossRef]

61. Horton, J.D.; Shah, N.A.; Warrington, J.A.; Anderson, N.N.; Park, S.W.; Brown, M.S.; Goldstein, J.L. Combined analysis of oligonucleotide microarray data from transgenic and knockout mice identifies direct SREBP target genes. *Proc. Natl. Acad. Sci. USA* **2003**, *100*, 12027–12032. [CrossRef]

62. Ferré, P.; Foufelle, F. Hepatic steatosis: A role for de novo lipogenesis and the transcription factor SREBP-1c. *Diabetes Obes. Metab.* **2010**, *12*, 83–92. [CrossRef] [PubMed]

63. Kohjima, M.; Enjoji, M.; Higuchi, N.; Kato, M.; Kotoh, K.; Yoshimoto, T.; Fujino, T.; Yada, M.; Yada, R.; Harada, N.; et al. Re-evaluation of fatty acid metabolism-related gene expression in nonalcoholic fatty liver disease. *Int. J. Mol. Med.* **2007**, *20*, 351–358. [CrossRef]

64. Foretz, M.; Guichard, C.; Ferré, P.; Foufelle, F. Sterol regulatory element binding protein-1c is a major mediator of insulin action on the hepatic expression of glucokinase and lipogenesis-related genes. *Proc. Natl. Acad. Sci. USA* **1999**, *96*, 12737–12742. [CrossRef]

65. Bécard, D.; Hainault, I.; Azzout-Marniche, D.; Bertry-Coussot, L.; Ferré, P.; Foufelle, F. Adenovirus-mediated overexpression of sterol regulatory element binding protein-1c mimics insulin effects on hepatic gene expression and glucose homeostasis in diabetic mice. *Diabetes* **2001**, *50*, 2425–2430. [CrossRef] [PubMed]

66. Wong, R.H.F.; Sul, H.S. Insulin signaling in fatty acid and fat synthesis: A transcriptional perspective. *Curr. Opin. Pharmacol.* **2010**, *10*, 684–691. [CrossRef]

67. Zhao, C.; Dahlman-Wright, K. Liver X receptor in cholesterol metabolism. *J. Endocrinol.* **2009**, *204*, 233–240. [CrossRef]

68. Bełtowski, J. Liver X Receptors (LXR) as Therapeutic Targets in Dyslipidemia. *Cardiovasc. Ther.* **2008**, *26*, 297–316. [CrossRef]

69. Auboeuf, D.; Rieusset, J.; Fajas, L.; Vallier, P.; Frering, V.; Riou, J.P.; Staels, B.; Auwerx, J.; Laville, M.; Vidal, H. Tissue Distribution and Quantification of the Expression of mRNAs of Peroxisome Proliferator-Activated Receptors and Liver X Receptor- in Humans: No Alteration in Adipose Tissue of Obese and NIDDM Patients. *Diabetes* **1997**, *46*, 1319–1327. [CrossRef]

70. Repa, J.J.; Liang, G.; Ou, J.; Bashmakov, Y.; Lobaccaro, J.-M.A.; Shimomura, I.; Shan, B.; Brown, M.S.; Goldstein, J.L.; Mangelsdorf, D.J. Regulation of mouse sterol regulatory element-binding protein-1c gene (SREBP-1c) by oxysterol receptors, LXRalpha and LXRbeta. *Genes Dev.* **2000**, *14*, 2819–2830. [CrossRef] [PubMed]

71. Joseph, S.B.; Laffitte, B.A.; Patel, P.H.; Watson, M.A.; Matsukuma, K.E.; Walczak, R.; Collins, J.L.; Osborne, T.F.; Tontonoz, P. Direct and Indirect Mechanisms for Regulation of Fatty Acid Synthase Gene Expression by Liver X Receptors. *J. Biol. Chem.* **2002**, *277*, 11019–11025. [CrossRef] [PubMed]

72. Lima-Cabello, E.; García-Mediavilla, M.V.; Miquilena-Colina, M.E.; Vargas-Castrillón, J.; Lozano-Rodríguez, T.; Fernández-Bermejo, M.; Olcoz, J.L.; González-Gallego, J.; García-Monzón, C.; Sánchez-Campos, S. Enhanced expression of pro-inflammatory mediators and liver X-receptor-regulated lipogenic genes in non-alcoholic fatty liver disease and hepatitis C. *Clin. Sci.* **2010**, *120*, 239–250. [CrossRef]

73. Seebacher, F.; Zeigerer, A.; Kory, N.; Krahmer, N. Hepatic lipid droplet homeostasis and fatty liver disease. *Semin. Cell Dev. Biol.* **2020**, *108*, 72–81. [CrossRef]

74. Gluchowski, N.L.; Becuwe, M.; Walther, T.C.; Farese, R.V., Jr. Lipid droplets and liver disease: From basic biology to clinical implications. *Nat. Rev. Gastroenterol. Hepatol.* **2017**, *14*, 343–355. [CrossRef] [PubMed]

75. Wang, M.; Thomas, J.; Burris, T.P.; Schkeryantz, J.; Michael, L.F. Molecular determinants of LXRalpha agonism. *J. Mol. Graph. Model.* **2003**, *22*, 173–181. [CrossRef]

76. Lee, S.D.; Tontonoz, P. Liver X receptors at the intersection of lipid metabolism and atherogenesis. *Atheroscler.* **2015**, *242*, 29–36. [CrossRef] [PubMed]

77. Pascual-García, M.; Valledor, A.F. Biological Roles of Liver X Receptors in Immune Cells. *Arch. Immunol. Ther. Exp.* **2012**, *60*, 235–249. [CrossRef] [PubMed]

78. Viennois, E.; Mouzat, K.; Dufour, J.; Morel, L.; Lobaccaro, J.-M.; Baron, S. Selective liver X receptor modulators (SLiMs): What use in human health? *Mol. Cell. Endocrinol.* **2012**, *351*, 129–141. [CrossRef] [PubMed]

79. Ma, J.; Hou, D.; Wei, Z.; Zhu, J.; Lu, H.; Li, Z.; Wang, X.; Li, Y.; Qiao, G.; Liu, N. Tanshinone IIA attenuates cerebral aneurysm formation by inhibiting the NFkappaBmediated inflammatory response. *Mol. Med. Rep.* **2019**, *20*, 1621–1628. [PubMed]

80. Tan, Y.L.; Ou, H.X.; Zhang, M.; Gong, D.; Zhao, Z.W.; Chen, L.Y.; Xia, X.D.; Mo, Z.C.; Tang, C.K. Tanshinone IIA Promotes Macrophage Cholesterol Efflux and Attenuates Atherosclerosis of apoE-/- Mice by Omentin-1/ABCA1 Pathway. *Curr. Pharm. Biotechnol.* **2019**, *20*, 422–432. [CrossRef]

81. Wen, J.; Chang, Y.; Huo, S.; Li, W.; Huang, H.; Gao, Y.; Lin, H.; Zhang, J.; Zhang, Y.; Zuo, Y.; et al. Tanshinone IIA attenuates atherosclerosis via inhibiting NLRP3 inflammasome activation. *Aging* **2020**, *13*, 910–932. [CrossRef] [PubMed]

82. Griffett, K.; Solt, L.A.; El-Gendy, B.E.-D.M.; Kamenecka, T.M.; Burris, T.P. A Liver-Selective LXR Inverse Agonist That Suppresses Hepatic Steatosis. *ACS Chem. Biol.* **2012**, *8*, 559–567. [CrossRef] [PubMed]

Article

Antitumor Effects of Ursolic Acid through Mediating the Inhibition of STAT3/PD-L1 Signaling in Non-Small Cell Lung Cancer Cells

Dong Young Kang [1,†], **Nipin Sp** [1,†], **Jin-Moo Lee** [2] **and Kyoung-Jin Jang** [1,*]

[1] Department of Pathology, School of Medicine, Institute of Biomedical Science and Technology, Konkuk University, Chungju 27478, Korea; kdy6459@kku.ac.kr (D.Y.K.); nipinsp@konkuk.ac.kr (N.S.)

[2] Pharmacological Research Division, National Institute of Food and Drug Safety Evaluation, Osong Health Technology Administration Complex, Cheongju-si 28159, Korea; elzem@korea.kr

* Correspondence: jangkj@konkuk.ac.kr; Tel.: +82-2-2030-7839

† Authors contributed equally to this paper.

Abstract: Targeted therapy based on natural compounds is one of the best approaches against non-small cell lung cancer. Ursolic acid (UA), a pentacyclic triterpenoid derived from medicinal herbs, has anticancer activity. Studies on the molecular mechanism underlying UA's anticancer activity are ongoing. Here, we demonstrated UA's anticancer activity and the underlying signaling mechanisms. We used Western blotting and real-time quantitative polymerase chain reaction for molecular signaling analysis. We also used in vitro angiogenesis, wound healing, and invasion assays to study UA's anticancer activity. In addition, we used tumorsphere formation and chromatin immunoprecipitation assays for binding studies. The results showed that UA inhibited the proliferation of A549 and H460 cells in a concentration-dependent manner. UA exerted anticancer effects by inducing G0/G1 cell cycle arrest and apoptosis. It also inhibited tumor angiogenesis, migration, invasion, and tumorsphere formation. The molecular mechanism underlying UA activity involves UA's binding to epidermal growth factor receptor (EGFR), reducing the level of phospho-EGFR, and thus inhibiting the downstream JAK2/STAT3 pathway. Furthermore, UA reduced the expressions of vascular endothelial growth factor (VEGF), metalloproteinases (MMPs) and programmed death ligand-1 (PD-L1), as well as the formation of STAT3/MMP2 and STAT3/PD-L1 complexes. Altogether, UA exhibits anticancer activities by inhibiting MMP2 and PD-L1 expression through EGFR/JAK2/STAT3 signaling.

Keywords: ursolic acid; NSCLC; tumorsphere; EGFR; STAT3; MMP2; PD-L1

Citation: Kang, D.Y.; Sp, N.; Lee, J.-M.; Jang, K.-J. Antitumor Effects of Ursolic Acid through Mediating the Inhibition of STAT3/PD-L1 Signaling in Non-Small Cell Lung Cancer Cells. *Biomedicines* **2021**, *9*, 297. https://doi.org/10.3390/biomedicines9030297

Academic Editor: Pavel B. Drašar

Received: 12 February 2021
Accepted: 11 March 2021
Published: 13 March 2021

Publisher's Note: MDPI stays neutral with regard to jurisdictional claims in published maps and institutional affiliations.

1. Introduction

Non-small cell lung cancer (NSCLC), one of the common cancers, accounts for approximately 85% of all lung cancer incidence, 70% of which may lead to metastasis, resulting in death. Various factors can induce tumorigenesis in NSCLC; tobacco is considered the primary factor [1]. There are some standard treatment methods for cancer. Targeted therapy is an effective method, as the alteration of molecular signaling pathways is a well-known factor of NSCLC tumorigenesis. Aberrations in major signaling pathways, such as epidermal growth factor receptor (EGFR), tyrosine kinases, and molecular pathways that control cancer hallmarks such as angiogenesis, cell cycle, apoptosis, and proteosome regulation, finally result in metastasis. EGFR overexpression is associated with approximately 80% of NSCLC incidence with a poor prognosis rate [2]. Several commercially available inhibitors that target EGFR are used to treat lung cancer, but they cause some side effects and lack survival benefits [3]. Vascular endothelial growth factor (VEGF) is a ligand that binds to VEGF receptors, especially VEGF-R2, which is necessary for tumor progression processes such as vascular permeability and angiogenesis. VEGF is also considered an impactful target in lung cancer treatment [4].

The major checkpoints are uncontrolled in tumorigenesis, leading to cell cycle progression, the inability to induce apoptosis, and the suppression of antitumor molecules or tumor suppressor genes. Aberrations in cell division or cell cycle process could induce uncontrolled cell division, causing tumorigenesis. Cyclin-dependent kinases (CDKs) are critical regulators of the cell cycle; CDK inhibitors play a suppressive role and prevent normal cells from becoming cancerous. A CDK inhibitor, p21 (p21$^{WAF1/Cip1}$), arrests the cell cycle when DNA damages occur [5]. It acts as a tumor suppressor in response to stimuli. It can also inhibit tumor suppressor protein 53 (p53)-independent cell proliferation; as such, p21 is also known as a master effector of the tumor suppressor pathways [6]. The proteins p21 and p27 (KIP1) can inhibit CDK proteins as well as cyclin proteins such as cyclin D1; therefore, their anticancer activity depends on the enhancement of p21 or p27 and the down-regulation of cyclin and CDK proteins [7]. Cell cycle arrest leads to apoptosis and DNA damage response mechanisms. When a normal cell cannot induce apoptosis or DNA repair mechanisms by stimuli such as mutation, this condition promotes tumorigenesis. The loss of the ability to induce apoptosis is a critical factor in tumor development, and the method could also be used for anticancer treatments [8,9].

Promoting angiogenesis is considered one of the hallmarks of cancer [10,11]. Cancer cell growth, differentiation, and tumor metastasis mainly depend on angiogenesis triggered by stimuli [12]. Tumor cells need vascular support for the blood flow to supply enough oxygen for cell growth; the lack of vascular support may even lead to apoptosis [13]. During angiogenesis, multiple angiogenic activators such as VEGF, basic fibroblast growth factor (bFGF), angiogenin, transforming growth factor (TGF)-α, TGF-β, and tumor necrosis factor (TNF)-α become activated to promote the molecular signaling of tumor angiogenesis [14]. VEGF activation also promotes the activation of metalloproteinases (MMPs), which destroy the extracellular matrix and allow endothelial cells to migrate and invade, resulting in tumor metastasis [15]. Thus, VEGF and MMP proteins are considered critical targets in cancer treatment [16].

Cancer stem cells (CSCs) cause cancer recurrence and promote tumorigenesis. Gene mutations or stimuli can activate CSCs by activating angiogenesis, suppressing cell cycle arrest, inducing apoptosis, and finally resulting in metastasis [17]. Traditional cancer therapies reduce tumor size, but CSCs are resistant to conventional therapies. Targeted therapy against CSCs could eliminate the risk of cancer recurrence [18]. Tumorsphere formation has been widely used to evaluate the stemness properties of cancer cells, whereby differentiated cells undergo cell death, and CSCs survive and proliferate to form tumorspheres [19]. Tumorsphere formation is considered a treatment platform for CSCs in NSCLC cells. Lung tumorspheres are highly resistant to many chemotherapeutic agents. This idea could be useful for developing new drugs against NSCLC [20,21]. The tumorspheres from cancer cells give support to the independence of external stimuli. This can provide an experimental system for identifying the factors that play vital roles in the molecular signaling of CSCs during the screening of a new drug against NSCLC [22].

EGFR mutation or overexpression is often observed in NSCLC cells. It signals toward its downstream targets, which then translocate into the nucleus to promote transcription and tumor progression. Janus kinase 2 (JAK2) and signal transducer and activator of transcription 3 (STAT3) signaling is an essential pathway in human cancers, as well as CSCs, acting by regulating inflammatory cytokines such as interleukin (IL)-6 [23]. The JAK2/STAT3 pathway participates in cancer cell survival, proliferation and progression by regulating multiple processes, such as epithelial–mesenchymal transition (EMT), which is required for tumor metastasis, and molecular signals that control other cancer hallmarks [24]. The programmed death ligand-1 (PD-L1)/programmed cell death protein 1 (PD-1) pathway is a vital checkpoint for tumor-induced immune escape that is mediated through T-cell exhaustion. In NSCLC, PD-L1 (CD274) is found to be overexpressed and regulated through EGFR/JAK/STAT3 signaling [25,26]. Some studies showed that high PD-L1 expression was associated with tumor metastasis, cancer recurrence, and tumor invasion; PD-L1 could be considered an independent element in evaluating immunotherapy during metastasis [27,28]. As such, PD-L1 could play a crucial role in the immune microenvironment between the primary tumor and the secondary metastatic tumor; PD-L1 can help increase the understanding of cancer's response to immunotherapy and develop PD-L-targeted therapy [29].

Targeted anticancer therapy using natural compounds is an effective approach because the natural compounds are efficacious and have fewer adverse effects. Ursolic acid (UA) is a pentacyclic triterpenoid derived from fruits and medicinal herbs with pharmaceutical and biological effects [30]. It can act against various cancer-related processes, such as the induction of apoptosis, the suppression of inflammatory responses, tumor metastasis, angiogenesis, and antioxidation. On the other hand, UA derivatives are also found to have pharmacological applications related to disease prevention [31]. The molecular signaling of UA is primarily linked to pro-inflammatory cytokines such as IL-7, IL-17, IL-1β, TNF-α or cyclooxygenase-2, and nitric oxide synthase through nuclear factor-κB, the primary factor in inflammatory responses to external stimuli [32]. In breast cancer and gastric cancer cells, UA induces cell cycle arrest and inhibits cell proliferation by inducing intrinsic and extrinsic pathways of apoptosis in vitro as well as in vivo [33,34]. UA can also induce cancer cell death and reduced tumor growth by regulating the autophagy-related gene 5-dependent autophagy in cervical cancer cells [35]. In NSCLC, UA has been found to have anticancer effects through the inhibition of autophagy and the suppression of TGF-β1-induced EMT, via regulating integrin αVβ5/MMPs signaling [36,37]. However, the role of UA signaling in the inhibition of PD-L1 in NSCLC remains to be elucidated.

In this study, we aim to determine UA's anticancer effects on processes such as cell cycle arrest, apoptosis, angiogenesis, migration, invasion, and tumorsphere formation in NSCLC cells. We also aimed to investigate PD-L1's role in UA-mediated anticancer activities and the underlying molecular mechanisms.

2. Materials and Methods

2.1. Antibodies and Cell Culture Reagents

Roswell Park Memorial Institute-1640 (RPMI-1640) medium, penicillin–streptomycin solution, and trypsin-EDTA (0.05%) (Gibco, Thermo Fisher Scientific, Inc., Waltham, MA, USA) were purchased. UA (U6753) and fetal bovine serum (FBS) (Sigma-Aldrich, Merck KGaA, St. Louis, MO, USA) were obtained. The primary antibodies against CDK4 (sc-260), cyclin E (sc-481), VEGF (sc-507), MMP9 (sc-13520), and β-actin (sc-47778) with anti-mouse (sc-516102) and anti-rabbit (sc-2357) secondary antibodies (Santa Cruz Biotechnology, Dallas, TX, USA) were procured. The antibodies against p21 (#2974), p27 (#3686), pEGFR (#3777), EGFR (#4267), pJAK2 (#3776), JAK2 (#3230), pSTAT3 (#9145), and STAT3 (#9139) (Cell Signaling Technology, Beverly, MA, USA) were obtained. The antibodies against SOX2 (#MAB4423), OCT4 (#MABD76), NANOG (#MABD24), and MMP3 (#AB2963) were supplied by Merck Millipore (Burlington, MA, USA). The Cyclin D1 (ab6152) antibody (Abcam, Cambridge, MA, USA), MMP2 (E90317) antibody (EnoGene, New York, NY, USA), and the PD-L1 (R30949) antibody (NSJ Bioreagents, San Diego, CA, USA) were procured.

2.2. Cell Culture and Treatment

A549 (no. 10185) and H460 (no. 30177, Korean Cell Line Bank, Seoul, South Korea) cell lines were cultured in RPMI-1640 supplemented with 10% FBS and 1% penicillin–streptomycin at 37 °C and 5% CO_2. The cells were grown to 80% confluency, gently washed twice with phosphate-buffered saline (PBS), and treated with various concentrations of UA at 37 °C for different durations according to the experimental design.

2.3. Cell Proliferation Inhibition

Cell viability was measured using the 3-(4,5-Dimethylthiazol-2-yl)-2,5-Diphenyltetrazolium Bromide (MTT) assay. Briefly, the cells were resuspended in RPMI-1640 and seeded in 96-well culture plates at a density of 3×10^3 cells per well 1 day before drug treatment. During drug treatment, the cells were incubated with increasing concentrations of UA from

1 to 100 µM for 24 h. The vehicle control was incubated with a fresh medium containing dimethyl sulfoxide (DMSO). Following drug treatment, MTT at 5 mg/mL was added to the cells to incubate for 4 h at 37 °C. The resulting formazan crystals were dissolved in DMSO and read at the absorbance of 560 nm on an ultra-multifunctional microplate reader (Tecan, Durham, NC, USA). All measurements were performed in triplicates, and the experiments were repeated at least three times.

2.4. DAPI Staining and Morphological Analysis

Chromatin condensation during apoptosis was examined with DAPI staining solution (ab228549, Abcam, Cambridge, MA, USA). A549 and H460 cells were seeded in 6-well plates at a density of 1.5×10^5 cells/well, treated with 10 or 20 µM UA for 24 h, and washed twice with PBS. They were then fixed with 200 µL of 100% methanol for 10 min, and washed twice with PBS. Then, 500 µL of 300 nM DAPI staining solution was added. The cells were washed twice with PBS, mounted with the mounting solution on microscope slides, and observed with fluorescence microscopy (Olympus IX71/DP72, Tokyo, Japan).

2.5. Western Blotting

Whole-cell lysates were prepared by incubating the untreated or UA-treated A549 and H460 cells on ice with the RIPA lysis buffer (20–188; EMD Millipore) containing protease and phosphatase inhibitors. Protein concentrations were measured using the Bradford assay (Thermo Fisher Scientific, Inc., Waltham, MA, USA). Equal amounts of protein at 100 µg/well were resolved with 10–15% sodium dodecyl sulphate–polyacrylamide gel electrophoresis (SDS-PAGE). The separated proteins were then transferred onto nitrocellulose membranes. The blots were blocked for 1 h with 5% skim milk (BD Biosciences, San Jose, CA, USA) in TBS-T buffer (20 mM Tris–HCl) (Sigma-Aldrich; Merck KGaA, St. Louis, MO, USA), pH 7.6, 137 mM NaCl (Formedium, Norfolk, UK; NaCO₃), and 0.1 × Tween 20 (Scientific Sales, Inc., Oak Ridge, TN, USA). Next, the membranes were incubated with primary antibodies (0.5–1 µg/mL) diluted in 5% bovine serum albumin (EMD Millipore) overnight at 4 °C on a shaker, washed with TBS-T, incubated with HRP-conjugated secondary antibodies (1–2 µg/mL) for 1 h at room temperature, and detected using a Femto clean enhanced chemiluminescence solution kit (GenDEPOT; 77449; Katy, TX, USA) and an LAS-4000 imaging device (Fujifilm, Tokyo, Japan).

2.6. Reverse Transcriptase–Quantitative Polymerase Chain Reaction (RT-qPCR)

Total RNA was extracted from the cells using the RNeasy Mini Kit (Qiagen GmbH, Hilden, Germany). The extracted RNA was quantified with a spectrophotometer (NanoDrop™ 1000, Thermo Fisher Scientific, Inc., Waltham, MA, USA) at 230 nm and reverse transcribed into cDNA at 42 °C for 1 h and 95 °C for 5 min using a first-strand cDNA synthesis kit (Bioneer Corporation, Daejeon, South Korea) and oligo d(T) primers. The RT-PCR premix kit (Bioneer Corporation, Daejeon, Korea) was used to amplify *CCND1*, *CCNE1*, *CDK4*, *CDKN1A*, *CDKN1B*, *MMP2*, *MMP3*, *MMP9*, *VEGF*, *CD274*, *SOX2*, OCT4 (*POU5F1*), *NANOG* and *GAPDH* cDNA with the corresponding primers (Bioneer Corporation) (Table S1). The qPCR mix consisted of 2 µL of diluted cDNA, 1 µL each of the forward and reverse primers at 100 pM, and 10 µL of the TB Green Advantage Premix (Takara Bio, Shiga, Japan). The qPCR process included initial denaturation at 95 °C for 5 min, followed by 40 cycles of denaturation at 95 °C for 40 s, annealing at 58 °C for 40 s, extension at 72 °C for 40 s, and a final extension at 72 °C for 5 min in a thermal cycler (C1000 Thermal Cycler, Bio-Rad, Hercules, CA, USA). The relative expression of the target genes was normalized to the expression of *GAPDH* and calculated using the Cp values. All the measurements were performed in triplicates.

2.7. Cell Cycle Analysis

The DNA content of the UA-treated and non-treated cells was determined using a BD Cycletest Plus DNA Reagent Kit (BD Biosciences, CA, USA). Approximately 5×10^5

cells were treated with or without UA for 24 h, washed with PBS, and permeabilized with trypsin. The RNA interaction with propidium iodide (PI) was neutralized by treating the cells with RNase and trypsin inhibitor. The samples were then stained with PI for 30 min in the dark at room temperature and analyzed using a FACSCalibur flow cytometer (BD Biosciences, San Jose, CA, USA).

2.8. Apoptosis Analysis

Annexin V-FITC (BD556547, BD Pharmingen, CA, USA) was used to measure the level of apoptosis in A549 and H460 cells. UA-treated cells were washed with PBS, resuspended in the binding buffer to a concentration of 1×10^6 cells, and stained with annexin V-FITC and PI for 10 min in the dark at room temperature. The percentage of apoptotic cells was measured by flow cytometry using the FACSCalibur (BD) and analyzed using the FlowJo software (v10).

2.9. In Vitro Angiogenesis Assay

ECMatrix (ECM625, Merck KGaA, 64293 Darmstadt, Germany) was thawed at 4 °C overnight. Then, pre-chilled 96-well plates were incubated with 50 µL diluted ECMatrix at 37 °C for 1 h for ECMatrix to solidify. Next, 150 µL of human vascular endothelial cells (HUVECs) at 1×10^4 with or without UA were added to the solidified matrix and incubated at 37 °C for 12 h. Afterward, endothelial cell formation was assayed with the in vitro angiogenesis kit (Millipore, Billerica, MA, USA) using a microscope. The focus was placed on distinct areas, and the tubes that formed were counted according to the kit's instructions.

2.10. Matrigel Invasion Assay

The invasion assay was conducted with ready-to-use Transwell invasion chambers pre-coated with Matrigel (BD Biocoat, MA, USA). First, 5×10^4 cells were added to each insert. The drug-containing media were added to the receiver plate, and the inserts containing cells were placed onto it. After a 24 h incubation in a humidified chamber at 37 °C, the cells on the upper surface were removed using a cotton swab, and the cells that had invaded the inserts' apical surface were stained with crystal violet and observed under a microscope. The focus was placed on four distinct areas, and the cells were counted.

2.11. Wound Healing Assay

A549 and H460 cells were seeded in 6-well plates at 1×10^5 cells/well in RPMI-1640 media and incubated for 24 h. After the cells became confluent, the cell layers were scratched with a pipette tip, washed with PBS to remove the debris, and treated with UA for 24 h. The control cells were not treated. The wound edges were photographed at different time intervals under a microscope to measure the relative area of wound closure using ImageJ software (NIH Image, Bethesda, MD, USA).

2.12. Tumorsphere Formation Assay

The A549 and H460 cells were cultured with or without UA and STAT3 siRNA in DMEM/F12 media containing growth supplements, EGF, bFGF, and B27, in low-attachment 6-well plates. The treatment day was considered day 0, and the incubation with UA continued for 14 days. Photographs were taken on days 0, 7, and 14 under a microscope. Total RNA and protein were isolated from the tumorsphere and analyzed using real-time qPCR and western blotting, respectively.

2.13. siRNA Transfection

The A549 and H460 cells were seeded in 6-well plates at 1×10^6 cells per well, grown to 60% confluence, and transfected with STAT3 siRNA (sc-29493; Santa Cruz Biotechnology, Dallas, TX, USA) using the lipofectamine transfection reagent (Thermo Fisher Scientific,

Inc., Waltham, MA, USA). After 24 h, the transfected cells were then cultured with or without UA for another 24 h under the same cell culture conditions.

2.14. Molecular Docking

Molecular docking was used to identify UA's binding site in the EGF receptor (EGFR) using PyRx software (v0.99) in the AuotDock Vina platform (Scripps Research Institute, San Diego, CA, USA). UA's 3D structure was obtained from PubChem (ID: 64945) and EGFR's 3D structure from Protein Data Bank (PDB ID: 2GS2). The obtained docked products were visualized using PyMol software (0.8).

2.15. Chromatin Immunoprecipitation (ChIP) Assay

A ChIP assay was performed using an imprint chromatin immunoprecipitation kit according to the manufacturer's protocol. The A549 and H460 cells were fixed with 1% formaldehyde, quenched with 1.25 M glycine, resuspended in the nuclei preparation buffer, and sonicated in shearing buffer under optimized conditions. The sheared DNA was diluted in dilution buffer at the ratio of 1:1. The diluted supernatant was then incubated with the antibody against STAT3 in pre-coated wells for 90 min. Normal mouse IgG and anti-RNA polymerase II were used for negative and positive controls, respectively. The unbound DNA was washed off with the IP wash buffer, whereas the bound DNA was collected by cross-link reversal using a DNA release buffer containing proteinase K. The released DNAs and the DNA from the internal controls were purified using the GenElute Binding Column G and quantified using specific MMP2 and PD-L1 primers (Table S1) by real-time qPCR.

2.16. Statistical Analyses

All the experiments were performed at least three times. The results are expressed as the mean $\pm$ standard error of the mean. Statistical analyses were conducted using one-way analysis of variance (ANOVA) or Student's *t*-test. One-way ANOVA was performed with Tukey's test for post hoc analysis. The analyses were performed using the SAS 9.3 software program (SAS Institute, Inc., Cary, NC, USA). A difference with a *p*-value <0.05 (*) was considered statistically significant.

3. Results

3.1. UA Inhibits NSCLC Cell Proliferation in a Concentration-Dependent Manner

First, we hypothesized that UA (structure in Figure 1A) could induce the apoptosis of NSCLC cells, A549 and H460. We examined UA's inhibition of A549 and H460 cell proliferation at increasing concentrations from 1 to 100 µM for 24 h. We observed a concentration-dependent inhibition of cell proliferation starting at 20 µM UA on A549 and H460 cells (Figure 1B). Approximately 50% of the A549 and H460 cells died when treated with 20 and 30 µM UA, respectively. Thus, we used 10 and 20 µM UA for our further studies to show UA's concentration-dependent effects. We also evaluated the apoptosis of the A549 and H460 cells at 10 and 20 µM UA, respectively, using morphological analysis with DAPI staining (Figure 1C). We observed a reduction in cell number at increasing concentrations of UA, confirming the anticancer activity of the natural compound, UA.

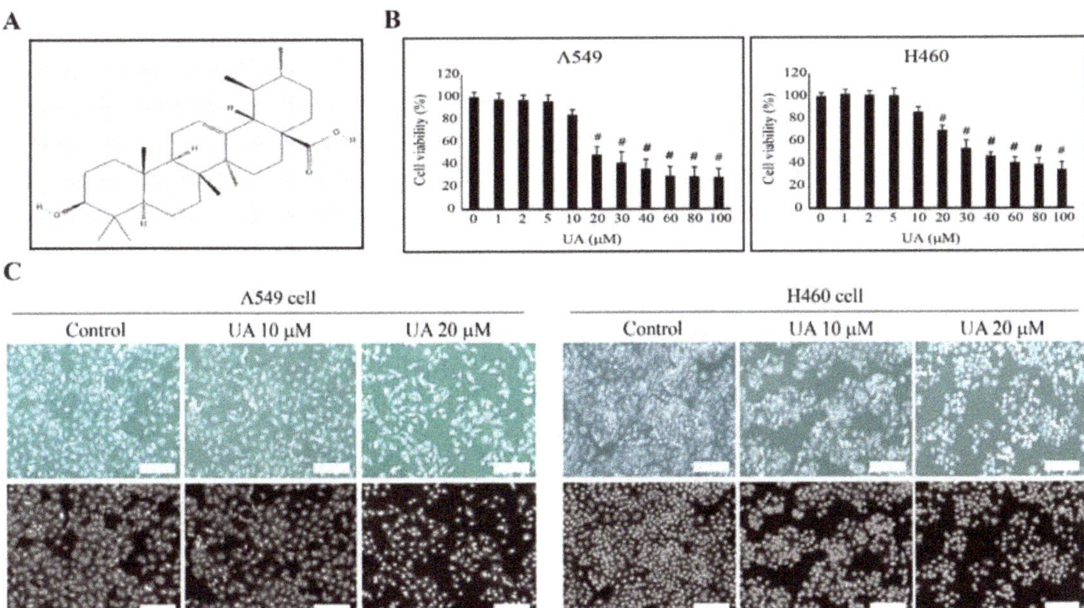

Figure 1. Ursolic acid (UA) inhibited non-small cell lung cancer (NSCLC) cell proliferation. (**A**) The structure of UA. (**B**) The MTT assay showed that the cell proliferation of A549 and H460 was inhibited by increasing UA concentrations for 24 h. (**C**) UA induced nuclear aberrations in NSCLC cells. Phase-contrast microscopy images showing abnormal nucleus formation induced by 10 and 20 μM UA in A549 and H460 cells. Representative photographs are presented (scale bar: 100 μm). The values of three independent experiments performed in triplicates (n = 3) were represented as mean ± SEM. The controls were set to 100. # $p < 0.001$ vs. control.

3.2. UA Induces Cell Cycle Arrest and Apoptosis in NSCLC Cells

We then examined the mechanisms underlying the inhibition of NSCLC cell proliferation by UA. We hypothesized that UA could induce cell cycle arrest and apoptosis. First, we analyzed the cell cycle checkpoint proteins in the UA-treated A549 and H460 cells using western blotting, and found reduced levels of CDK4, cyclin D1 and cyclin E, and increased levels of tumor suppressor proteins p21 and p27, with UA treatment (Figure 2A). We confirmed these patterns in the mRNA levels of the genes by observing that UA significantly inhibited the expressions of *CDK4*, *CCND1* and *CCNE1*, and upregulated the expressions of *CDKN1A* and *CDKN1B* in the NSCLC cells (Figure 2B). These results suggested UA's ability to induce cell cycle arrest in NSCLC cells. Then, we analyzed the cell cycle distribution of the UA-treated A549 and H460 cells using flow cytometry. We found an arrest in the G0/G1 phase in the cell cycle caused by UA treatment (Figure 2C), suggesting UA's ability to induce apoptosis. Therefore, we analyzed apoptosis induction by UA in NSCLC cells using flow cytometry, and observed an increased number of dead cells caused by 20 μM UA in both the A549 and H460 cells (Figure 2D). These results suggested that UA could be a candidate drug against NSCLC.

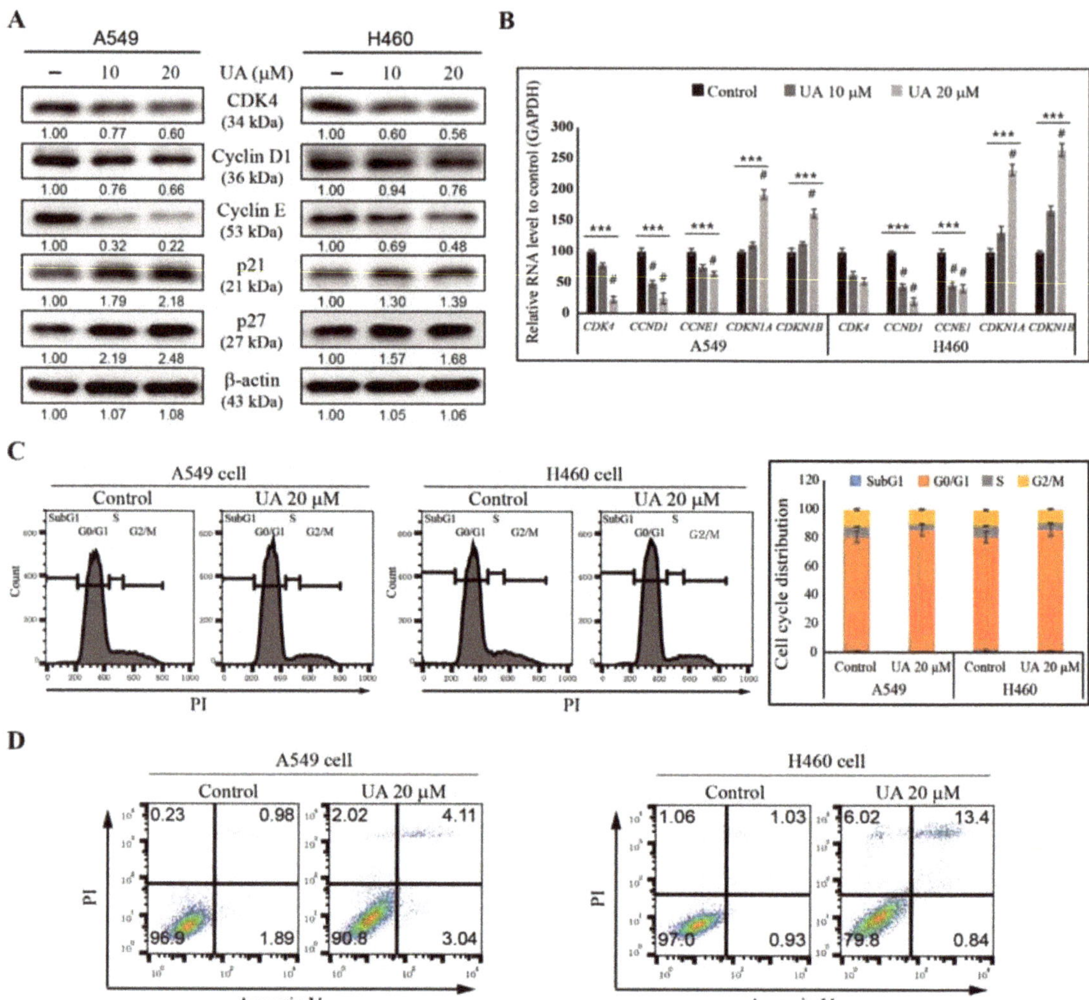

Figure 2. UA induced cell cycle arrest and apoptosis in NSCLC cells. (**A**) The western blotting for CDK4, cyclin D1, cyclin E1, p21 (CDKN1A), and p27 (CDKN1B) in A549 and H460 cells treated with 10 and 20 μM UA, respectively, for 24 h. β-actin was used as the housekeeping gene. (**B**) RT-qPCR analysis of the expression of cell cycle checkpoint genes *CCND1*, *CCNE1*, *CDK4*, *CDKN1A*, and *CDKN1B* in A549 and H460 cells after treatment with 10 and 20 μM UA, respectively, for 24 h. The Cp values were normalized to *GAPDH* mRNA. Controls are set to 100. (**C**) Flow cytometry analysis with PI staining of the cell cycle distribution in A549 and H460 cells after treatment with 20 μM UA for 24 h, with a graphical representation of G0/G1 arrest by UA in NSCLC cells. (**D**) Flow cytometry analysis using annexin V and PI staining in A549 and H460 cells after treatment with 20 μM UA for 24 h. *** $p < 0.001$ (ANOVA test), and # $p < 0.001$ vs. control.

3.3. UA Inhibits Angiogenesis, Invasion, and Migration of NSCLC Cells

We have observed the induction of cell cycle arrest and apoptosis by UA in A549 and H460 NSCLC cells. Because UA plays a role against other cancer processes, such as tumor angiogenesis, migration, and invasion, we evaluated UA's ability to suppress angiogenesis. HUVEC were used for in vitro angiogenesis assay with an extracellular matrix solution (Figure 3A). Angiogenesis was measured via tube formation with HUVECs in an extracellular matrix gel. The treatments of 10 and 20 μM of UA significantly reduced the tube formations in A549 and H460 cells, respectively (Figure 3B), indicating UA's inhibition of angiogenesis. We examined UA for the inhibition of angiogenesis, and analyzed the proteins derived from HUVECs treated with or without UA with western blotting. UA-treated cells showed decreased VEGF levels and phosphorylated STAT3, which played a crucial role in angiogenesis; the total amount of STAT3 remained the same for the control and UA treatments (Figure 3C). These results suggest that UA can inhibit angiogenesis. Then, we studied whether UA could suppress tumor metastasis using an invasion assay with Matrigel (Figure 3D) and a wound healing migration assay (Figure 3E). We found decreased cell invasion and wound closure with increasing concentrations of UA in both A549 and H460 cells. These results suggest the antimetastatic activity of UA against NSCLC cells.

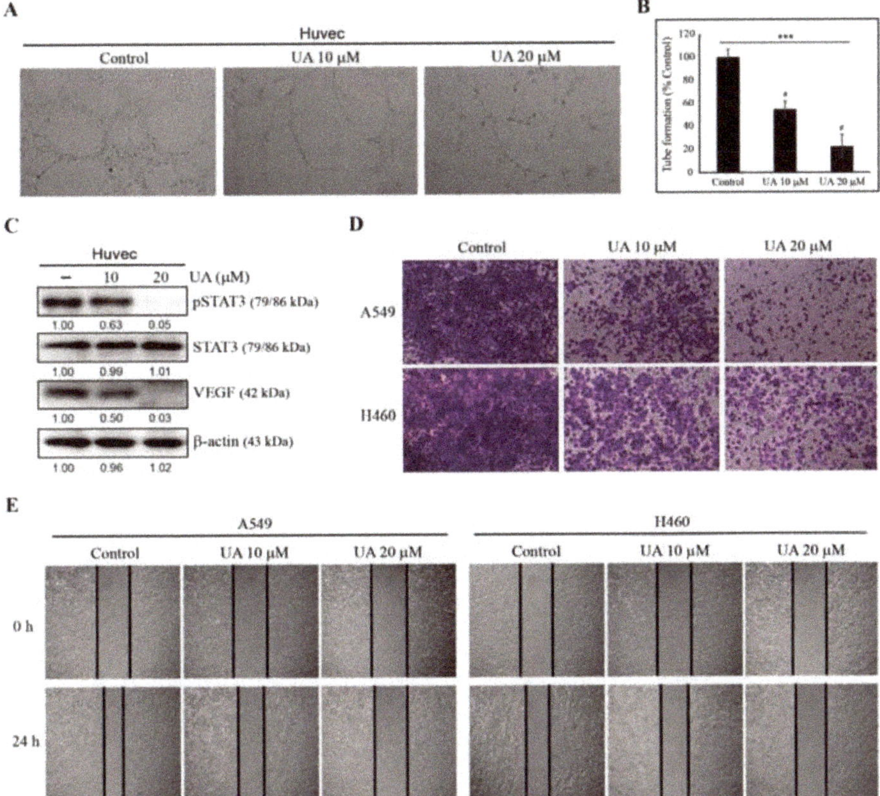

Figure 3. UA inhibited NSCLC cell angiogenesis, invasion, and migration. (**A**) The inhibition of angiogenesis by 10 and 20 μM UA during the in vitro angiogenesis assay of A549 and H460 cells, respectively. (**B**) Graphical representation of the in vitro angiogenesis assay showing the relative inhibition of tube formation. (**C**) Western blotting of A549 and H460 cells treated with 10 and 20 μM UA, respectively, for 24 h for phospho-STAT3, STAT3, and VEGF. The relative levels of the proteins were determined by densitometry and normalized to that of β-actin, which was set to 100. The data were confirmed by repeating the experiment three times. (**D**) The Matrigel invasion assay showing the inhibition of invasion in A549 and H460 cells treated with 10 and 20 μM UA, respectively, for 24 h. Scale bars: 100 μm. (**E**) The wound healing assay demonstrated the inhibition of migration in A549 and H460 cells treated with 10 and 20 μM UA, respectively, for 0 and 24 h. Scale bars: 50 μm. Statistical analysis was conducted using ANOVA test; *** $p < 0.001$; # $p < 0.001$ vs. control.

3.4. UA Suppresses Cancer Stemness by Inhibiting Tumorsphere Formation in NSCLC Cells

We have found that UA can inhibit the hallmarks of cancer in NSCLC cells. However, its ability to inhibit CSCs is unknown. Tumorspheres can only form from CSCs. Thus, we conducted a tumorsphere formation assay to analyze UA's ability to inhibit cancer stemness in NSCLC cells. We cultured A549 and H460 cells in tumorsphere media in the presence of UA and 100 µM S3I-201, a STAT3 inhibitor, for 14 days, and photographed them under a microscope. After 14 days, the UA- and S3I-201-treated cells showed a significant reduction in tumorsphere compared with untreated A549 and H460 cells (Figure 4A). These data suggest that UA can inhibit CSCs through STAT3 signaling. We confirmed CSC formation by analyzing the expression levels of CSC markers, *NANOG*, *POU5F1* and *SOX2*, in the UA-treated NSCLC cells, and found these genes to be significantly inhibited (Figure 4B). The inhibition of the CSC markers by UA was confirmed by probing the levels of NANOG, OCT4 and SOX2 in the tumorspheres using western blotting (Figure 4C). These results suggest UA's ability to target CSCs.

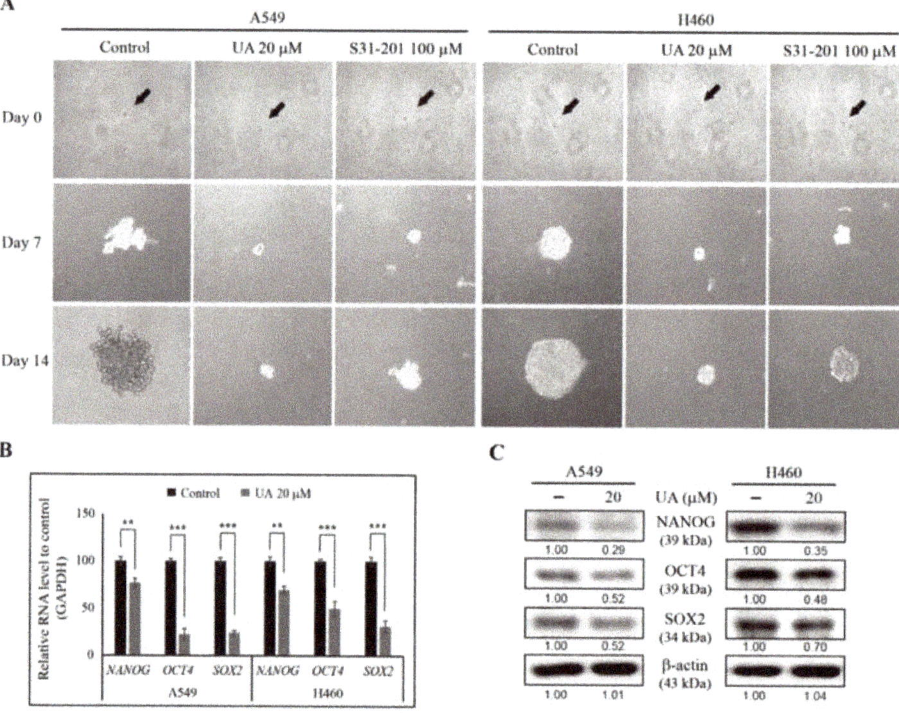

Figure 4. UA inhibited tumorsphere formation from NSCLC cells. (**A**) Tumorsphere formation was inhibited by 20 µM UA or 100 µM S3I-201 treatment after the A549 and H460 cells were cultured in DMEM/F12 media containing epidermal growth factor (EGF), basic fibroblast growth factor (bFGF) and B27 for 14 days. Photographs were taken on days 0, 7, and 14. Black arrows indicate the single cells on day 0. (**B**) RT-qPCR of the mRNA isolated from the A549 and H460 tumorspheres to demonstrate the expression of CSC marker genes after treatment with 20 µM UA for 24 h. The representative expressions of *NANOG*, *OCT4* and *SOX2* are shown. The Cp values were normalized to the level of *GAPDH* mRNA, which was set to 100. (**C**) Western blotting of the CSC marker proteins, NANOG, OCT4, and SOX2, in A549 and H460 tumorspheres after treatment with 20 µM UA for 24 h. The relative levels of proteins were determined by densitometry and normalized to that of β-actin, which was set to 100. The data were confirmed after repeating the experiment 3 times. ** $p < 0.01$ and *** $p < 0.001$ (Student's *t*-test).

3.5. UA Inhibits PD-L1 Expression Through the EGFR/JAK2/STAT3 Pathway in NSCLC Cells

We observed UA's ability to inhibit hallmarks of cancer as well as tumorsphere formation. Here, we investigated the molecular signaling involved in these mechanisms. First, we examined whether UA could bind to the EGF receptor. Molecular docking analysis suggested an interaction between UA and EGFR with a strong binding affinity of -8.6 kcal/mol, suggesting that UA acted through EGFR signaling (Figure 5A). We tested this possibility by treating A549 and H460 cells with recombinant human EGF for 1 h, then treated the cells with or without UA, and analyzed the cell lysates using western blotting. We found an increase in the level of phosphorylated EGFR in the EGF-treated cells; such an increase was significantly reduced by 20 µM UA (Figure 5B). The level of EGFR remained the same for the control, UA-treated, and EGF-treated cells, suggesting that UA was bound to the EGF receptor and blocked its activation. Next, we analyzed UA's effect on the JAK2/STAT3 pathway, one of the critical downstream targets of EGFR signaling. We found a suppression in the levels of phosphorylated EGFR, JAK2, and STAT3 by UA treatment in NSCLC cells. At the same time, the amount of total protein remained unchanged with increasing UA concentrations (Figure 5C). PD-L1 is also considered a downstream target for STAT3, which acts as a transcription factor for PD-L1. The inhibition of the expression of *POU5F1* by UA treatment also indicated that UA might act by regulating EGFR/JAK2/STAT3/PD-L1 signaling. Our previous results showed that UA could inhibit tumor angiogenesis, migration, and invasion (Figure 3). MMPs and VEGF play a vital role in tumor invasion and angiogenesis, respectively. As such, we analyzed the levels of MMP2, MMP3, MMP9, and VEGF in the NSCLC cells treated with UA and found a decreasing pattern for all these proteins (Figure 5D). We confirmed these results by analyzing the mRNA of these genes using real-time PCR, and found all these genes were similarly inhibited, much like PD-L1 *(CD274)* (Figure 5E). These results have delineated the molecular mechanism underlying the anticancer activities of UA.

3.6. UA Downregulates the Binding of STAT3 to MMP2 and PD-L1 Promoters

Previously, we showed that UA acted through the EGFR/JAK2/STAT3 signaling pathway through MMP2 or PD-L1 (Figure 5). Here, we hypothesized that the activated STAT3 was translocated to the nucleus and bound to the *MMP2* promoter to stimulate tumor cell invasion, and was bound to the PD-L1 promoter to boost tumor metastasis or regulate immune escape. We extracted chromatin DNA from the NSCLC cells with or without UA treatment, and analyzed the binding of STAT3 to the MMP2 and PD-L1 promoters. We found that UA significantly inhibited the binding of STAT3 to both promoters in the NSCLC cells (Figure 6A). These results suggest the vital role of STAT3 in the anticancer activity of UA. We tested STAT3's role in UA's activity by comparing the effect of *STAT3* silencing with that of UA treatment. We found a similar inhibition of phosphor STAT3, PD-L1, and MMP2 proteins in the cells treated with UA and the cells treated with STAT3 siRNA to that exhibited by the untreated control cells (Figure 6B). These results confirm that UA's mechanism underlying its anticancer activity depends on STAT3 signaling, and that STAT3 acts as a bridge molecule between EGFR signaling and PD-L1 expression.

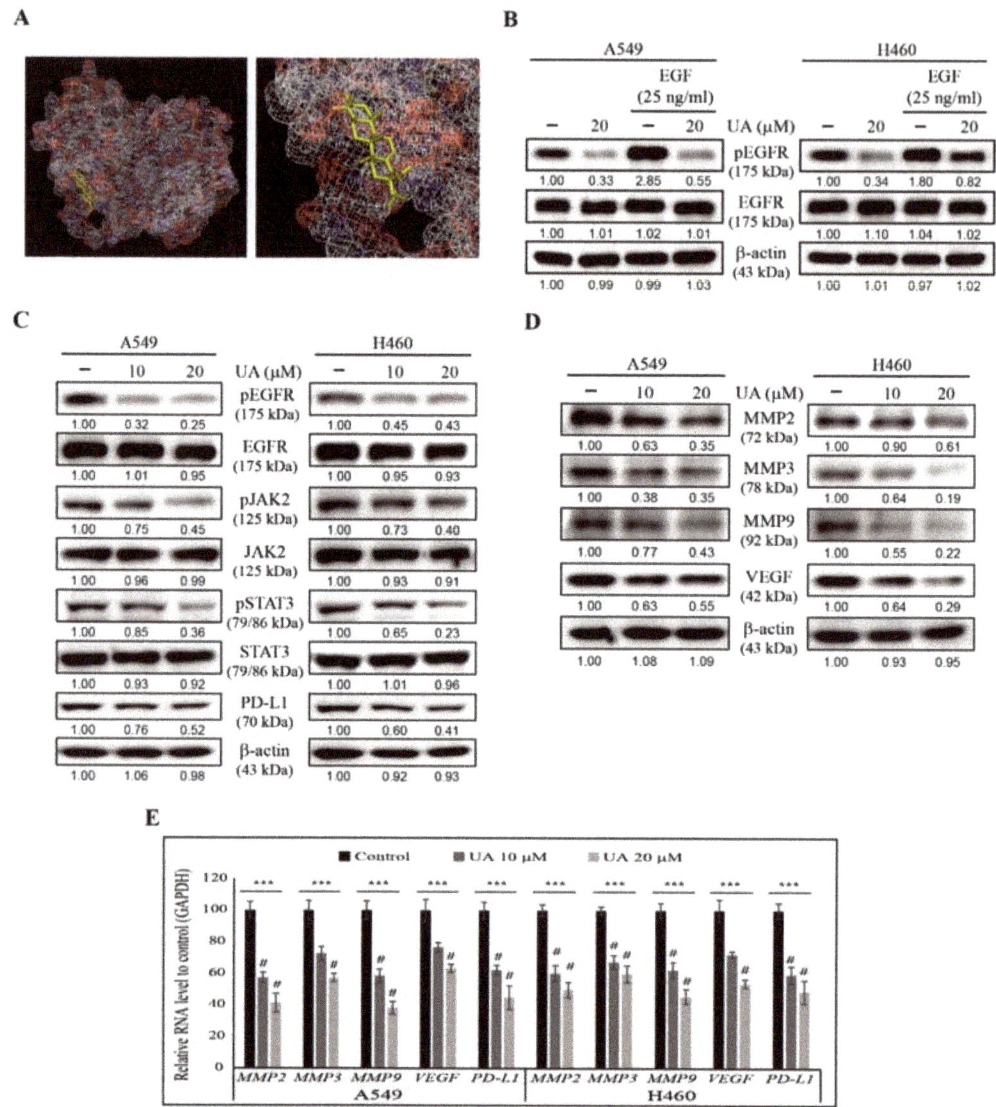

Figure 5. UA regulated EGFR/JAK2/STAT3 signaling in NSCLC cells. (**A**) Molecular docking using AutoDock Vina software showed UA binding (PubChem ID: 64945) to the ATP-binding domain of EGFR (PDB ID: 2GS2). (**B**) Western blotting of the lysate of the A549 and H460 cells pre-treated with recombinant EGF (25 ng/mL) for 1 h and then with 20 μM UA for 24 h, showing the levels of phosphorylated EGFR and total EGFR. The representative levels of the proteins were determined by densitometry and normalized to β-actin, which was set to 100. The data were confirmed after repeating the experiment 3 times. (**C**) Western blotting of the proteins involved in EGFR/JAK2/STAT3 signaling and PD-L1 in the A549 and H460 cells treated with 10 and 20 μM UA, respectively, for 24 h. β-Actin was used as the housekeeping protein. (**D**) Western blotting for MMPs and VEGF in the A549 and H460 cells treated with 10 and 20 μM UA, respectively, for 24 h. β-actin used as a housekeeping protein. (**E**) RT-qPCR analysis for the expression of *MMP2*, *MMP3*, *MMP9*, *VEGF*, and *PD-L1* in the A549 and H460 cells treated with 10 and 20 μM UA, respectively, for 24 h. The representative expressions of *MMP2*, *MMP3*, *MMP9*, *VEGF* and *PD-L1* were shown. The Cp values were normalized to that of *GAPDH*, which was set to 100. *** $p < 0.001$ (ANOVA test) and # $p < 0.001$ vs. control.

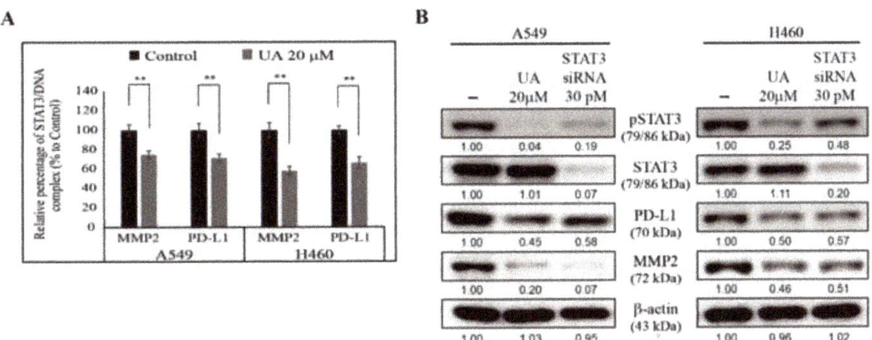

Figure 6. UA inhibited the binding of STAT3 to the MMP2 and PD-L1 promoters. (**A**) ChIP assay showed that treatment with 20 μM UA inhibited the formation of the STAT3/MMP2 and STAT3/PD-L1 complexes in A549 and H460 cells. The relative binding of STAT3 to the MMP2 and PD-L1 promoters was expressed as a percentage of the control. Statistical analysis was performed using Student's *t*-test (** $p < 0.01$). (**B**) Western blotting of A549 and H460 cells treated with 30 pM STAT3 siRNA or 20 μM UA for 24 h showing the patterns of phosphorylated STAT3, STAT3, PD-L1, and MMP2 protein levels. β-actin was used as a housekeeping protein.

Altogether, the molecular mechanism underlying UA's anticancer activity depends on EGFR signaling, which then signals to JAK2/STAT3, causing STAT3 to translocate to the nucleus to bind with the VEGF, MMP2, and PD-L1 promoters in order to block their transcription, thus inhibiting tumor angiogenesis, invasion, metastasis, and tumorsphere formation (Figure 7).

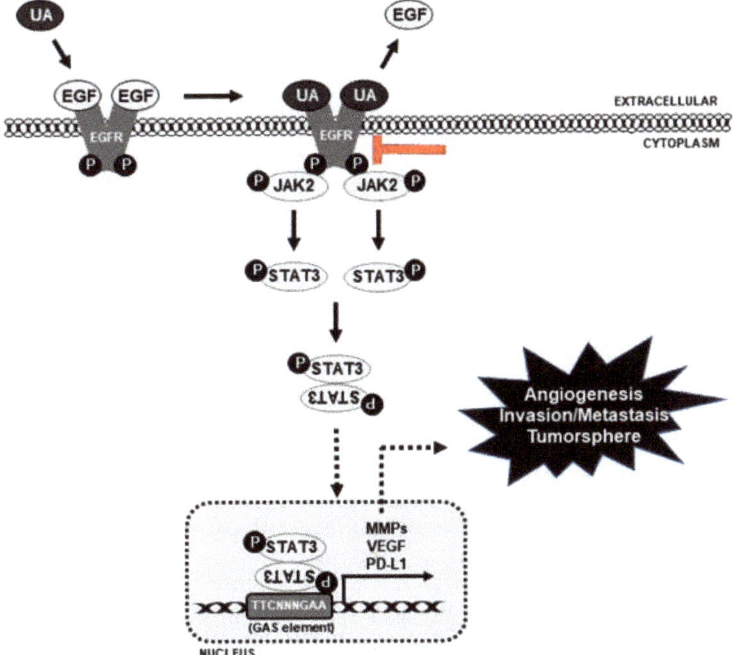

Figure 7. The molecular regulatory mechanism underlying UA's anticancer activity in NSCLC cells, and UA's inhibition of the EGFR/JAK2/STAT3 signaling pathway and PD-L1. Continuous black point arrow denotes the activity of UA, dotted point arrows denotes blockage of signals, and red block arrow denotes the inhibition of EGFR signaling by UA.

4. Discussion

Natural compounds have been used for cancer treatment for a long time. They could be used for targeted therapy and have the added advantage of causing no or fewer side effects [38,39]. UA is derived from several medicinal herbs, such as *Oldenlandia diffusa*, *Calluna vulgaris*, *Rosemarinus officinalis*, and *Eribotrya japonica* [40]. Many studies have demonstrated UA's anticancer activities, such as inhibiting tumor angiogenesis and tumorigenesis, and promoting and inducing cell cycle arrest and apoptosis against various cancer cells, such as NSCLC, hepatocellular carcinoma, breast cancer, and gastric cancer [41–43]. UA also has an in vivo role; it has been reported to inhibit tumor growth in various animal models [44,45]. Although these studies suggest UA to be a drug candidate against NSCLC, the exact molecular mechanism of UA's STAT3-dependent anticancer activity remains unclear.

The hallmarks of cancer mainly consist of the lost ability to induce apoptosis and the induction of angiogenesis, resulting in metastasis via tumor cell migration, invasion, and EMT. A drug that can inhibit these processes in a cancer cell will be suitable for anticancer studies [46]. We found that UA can induce cell death in A549 and H460 cells, which are both considered metastatic cancer cells. We observed that 20 and 30 µM UA induced almost 50% cell death in the A549 and H460 cells, respectively. From these, we chose 10 and 20 µM UA for the A549 and H460 cells, respectively, for further studies to demonstrate the concentration-dependent effects of UA. UA was also shown to alter the morphology of cells, inducing cell death. These results suggest UA as a drug candidate against NSCLC.

Uncontrolled cell cycle progression and escape from apoptosis leads to tumorigenesis. The ability to control cell cycle progression and apoptosis induction decides a cancer cell's fate [47]. A natural compound that could induce cell cycle arrest and apoptosis against metastatic lung cancer cell lines is an appropriate candidate for anticancer activities. UA's induction of G0/G1 arrest in A549 and H460 cells suggests its ability to act as an anticancer drug; cell cycle arrest leads to cell death. Inducing cell death with a natural compound is an effective therapeutic approach. Our results showed that UA also induced apoptosis in both NSCLC cells. Thus, these results provide strong evidence for UA's anticancer activity.

When a tumor starts to grow, it may lack an adequate supply of oxygen due to its increasing size, thus creating hypoxic conditions through the hypoxia-inducible factor 1α, and then activating VEGF for angiogenesis. As a result of angiogenesis, oxygen will reach the tumor microenvironment through the blood supply [48]. Once the primary tumor forms, it starts to move to another location to create a secondary tumor. The tumor cells migrate and invade into blood vessels, and traverse through the bloodstream to spread the tumor. As such, tumor cell migration and invasion are considered as the initial steps in metastasis [49]. Our results showed that UA decreased in vitro angiogenesis by inhibiting tube formation in HUVECs. UA may disrupt tube formation, block vessel formation, reduce the oxygen supply to the tumor cells, and cause hypoxia. The analysis of UA's molecular mechanism in HUVECs also showed UA's inhibition of the VEGF and phospho-STAT3 expression. STAT3 plays a significant role in tumor angiogenesis by regulating VEGF activity [50]. UA was also shown to inhibit the migration and invasion of A549 and H292 cells, suggesting its role in suppressing tumor metastasis.

Anticancer drugs can induce apoptosis and inhibit angiogenesis and metastasis, while CSCs may remain inside silently. Upon stimulation, the CSCs differentiate into the primary tumor again. This condition is dangerous because the new tumor may be resistant to that particular drug [51]. Tumorsphere assays are efficient methods for studying CSCs and performing anticancer drug screening, as tumorspheres have the same characteristics as CSCs [52]. Here, we seeded A549 and H460 cells into DMEM/F12 media containing EGF, bFGF, and B27 in low-attachment plates with or without UA for the selective growth of tumorsphere. The bulk formation of the tumorsphere was observed in the control cells, indicating the presence of CSCs in the NSCLC cells. The presence of UA or

STAT3 siRNA reduced the size of the tumorspheres, suggesting UA's ability to inhibit CSCs via a STAT3-dependent signaling mechanism. SOX2, OCT4, and NANOG are considered the stem cell markers for determining tumorsphere formation [53]. The downregulation of the mRNA and protein levels of these genes in A549- and H460-derived tumorspheres by UA treatment indicated that UA could inhibit CSC markers. These results provide evidence supporting UA's capability to target cancer cells as well as CSCs.

The knowledge of the molecular signaling mechanism underlying a natural compound's activity is as essential as understanding its pharmaceutical effects. Molecular signaling begins with receptor binding, followed by the addition of the drug. EGFR is a crucial receptor in tumorigenesis, and mutations in EGFR result in tumorigenesis. It also acts as an intracellular tyrosine kinase domain essential for the signaling pathways that regulate cancer cell proliferation [54]. EGFR overexpression was observed in most NSCLC cells; regulating EGFR signaling could help control tumorigenesis or tumor progression. We hypothesized that UA could bind to EGFR, blocking its downstream signals. Molecular docking analysis suggested a strong binding between UA and EGFR. This prediction was confirmed by the observation that the level of phospho-EGFR was increased by treatment with human recombinant EGF, but was decreased by UA treatment. Thus, UA likely replaces EGF in its binding to the EGFR, thus blocking the activation of EGFR and abolishing downstream signaling through tyrosine kinases. JAK2/STAT3, a critical signaling pathway in tumorigenesis, is mediated through EGFR signaling; STAT3 can act as a transcription factor for tumor progression [55]. Consistent with our prediction of EGFR's binding to UA, a reduction in the levels of phospho-JAK2 and STAT3 caused by UA was observed, whereas the total amount of JAK2 and STAT3 remained unchanged. These results demonstrate that the EGFR/JAK2/STAT3 signaling pathway is responsible for the anticancer activity of UA, and STAT3 plays a crucial role in the UA-dependent transcription processes.

STAT3 is an oncogene that participates in most of the tumor development phases as a transcription factor. It is involved in tumor angiogenesis by regulating VEGF [56] and in tumor invasion by mediating MMP proteins [57]. The downregulation of activated STAT3 by UA was observed in NSCLC cells, suggesting that UA might block *VEGF* mRNA expression as it inhibited angiogenesis. Furthermore, UA might block *MMP* mRNA expression, as it inhibited tumor migration and invasion. As predicted, UA significantly downregulated the expression of *VEGF*, *MMP2*, *MMP3*, and *MMP9* mRNA, confirming its anticancer ability. *CD274*, encoding the immune checkpoint ligand PD-L1, is often overexpressed in NSCLC cells. Studies have shown that PD-L1 actively participates in tumor metastasis processes other than immune escape regulation [28]. Because of STAT3's role in tumor metastasis, we hypothesized that STAT3 might also act as a transcription factor for PD-L1. Our results demonstrated the concentration-dependent inhibition of PD-L1 levels in UA-treated A549 and H460 cells, suggesting a possible relationship between STAT3 and PD-L1. As such, we analyzed the DNA-binding activity of STAT3 to the MMP2 promoter for tumor invasion, and the PD-L1 promoter for tumor metastasis.

We observed a significant reduction in STAT3/MMP2 and STAT3/PD-L1 complex formation in UA-treated NSCLC cells. Therefore, STAT3 plays an essential role in regulating PD-L1 in NSCLC cells. Moreover, UA exhibits anticancer activity by regulating EGFR/JAK2/STAT3 signaling and the expression of MMP2 and PD-L1.

5. Conclusions

In this study, we have demonstrated UA's anticancer activity against NSCLC cells A549 and H460. We have determined that UA induces G0/G1 cell cycle arrest and apoptosis, thus inhibiting tumor angiogenesis, migration, invasion, and metastasis. Furthermore, UA has been found to inhibit the expression levels of VEGF, MMP2, and PD-L1 by regulating the EGFR/JAK2/STAT3 pathway. STAT3 has been found to act as a critical regulator in the anticancer activity of UA. Altogether, UA can be a drug candidate for PD-L1-based targeted therapy for NSCLC.

Supplementary Materials: The following are available online at https://www.mdpi.com/2227-9059/9/3/297/s1, Table S1: q-PCR primer sequences and annealing temperature.

Author Contributions: K.-J.J. designed the experiments. D.Y.K., N.S., and K.-J.J. performed all the experiments. J.-M.L. served as scientific advisors and participated in the technical editing of the manuscript. D.Y.K., N.S. and K.-J.J. wrote the manuscript. D.Y.K., N.S., and K.-J.J. analyzed the data. All authors helped in revising the manuscript and approved the final version for publication. All authors have read and agreed to the published version of the manuscript.

Funding: This work was supported by the National Research Foundation of Korea (NRF) grant funded by the Ministry of Education (No.2019R1I1A1A01060399, 2019R1I1A1A01060537, and 2020R1I1A2073517).

Institutional Review Board Statement: Not applicable.

Informed Consent Statement: Not applicable.

Data Availability Statement: The data presented in this study are available on request from the corresponding author. The data are not publicly available due to privacy.

Conflicts of Interest: The authors declare no conflict of interest.

References

1. Molina, J.R.; Yang, P.; Cassivi, S.D.; Schild, S.E.; Adjei, A.A. Non-small cell lung cancer: Epidemiology, risk factors, treatment, and survivorship. *Mayo Clin. Proc.* **2008**, *83*, 584–594. [CrossRef]
2. Mendelsohn, J.; Baselga, J. Status of epidermal growth factor receptor antagonists in the biology and treatment of cancer. *J. Clin. Oncol.* **2003**, *21*, 2787–2799. [CrossRef] [PubMed]
3. Thatcher, N.; Chang, A.; Parikh, P.; Rodrigues Pereira, J.; Ciuleanu, T.; von Pawel, J.; Thongprasert, S.; Tan, E.H.; Pemberton, K.; Archer, V.; et al. Gefitinib plus best supportive care in previously treated patients with refractory advanced non-small-cell lung cancer: Results from a randomised, placebo-controlled, multicentre study (Iressa Survival Evaluation in Lung Cancer). *Lancet* **2005**, *366*, 1527–1537. [CrossRef]
4. Chatterjee, S.; Heukamp, L.C.; Siobal, M.; Schottle, J.; Wieczorek, C.; Peifer, M.; Frasca, D.; Koker, M.; Konig, K.; Meder, L.; et al. Tumor VEGF:VEGFR2 autocrine feed-forward loop triggers angiogenesis in lung cancer. *J. Clin. Investig.* **2013**, *123*, 1732–1740. [CrossRef]
5. Deng, C.; Zhang, P.; Harper, J.W.; Elledge, S.J.; Leder, P. Mice lacking p21CIP1/WAF1 undergo normal development, but are defective in G1 checkpoint control. *Cell* **1995**, *82*, 675–684. [CrossRef]
6. Abbas, T.; Dutta, A. p21 in cancer: Intricate networks and multiple activities. *Nat. Rev. Cancer* **2009**, *9*, 400–414. [CrossRef]
7. Musgrove, E.A.; Hunter, L.J.; Lee, C.S.; Swarbrick, A.; Hui, R.; Sutherland, R.L. Cyclin D1 overexpression induces progestin resistance in T-47D breast cancer cells despite p27(Kip1) association with cyclin E-Cdk2. *J. Biol. Chem.* **2001**, *276*, 47675–47683. [CrossRef]
8. Pfeffer, C.M.; Singh, A.T.K. Apoptosis: A target for anticancer therapy. *Int. J. Mol. Sci.* **2018**, *19*, 448. [CrossRef]
9. Sp, N.; Kang, D.Y.; Kim, D.H.; Yoo, J.S.; Jo, E.S.; Rugamba, A.; Jang, K.J.; Yang, Y.M. Tannic acid inhibits non-small cell lung cancer (NSCLC) stemness by inducing G0/G1 cell cycle arrest and intrinsic apoptosis. *Anticancer Res.* **2020**, *40*, 3209–3220. [CrossRef]
10. Sp, N.; Kang, D.Y.; Kim, D.H.; Park, J.H.; Lee, H.G.; Kim, H.J.; Darvin, P.; Park, Y.M.; Yang, Y.M. Nobiletin inhibits CD36-dependent tumor angiogenesis, migration, invasion, and sphere formation through the Cd36/Stat3/Nf-Kb signaling axis. *Nutrients* **2018**, *10*, 772. [CrossRef] [PubMed]
11. Sp, N.; Kang, D.Y.; Joung, Y.H.; Park, J.H.; Kim, W.S.; Lee, H.K.; Song, K.D.; Park, Y.M.; Yang, Y.M. Nobiletin inhibits angiogenesis by regulating Src/FAK/STAT3-mediated signaling through PXN in ER(+) breast cancer cells. *Int. J. Mol. Sci.* **2017**, *18*, 935. [CrossRef] [PubMed]
12. Folkman, J. Tumor angiogenesis: Therapeutic implications. *N. Engl. J. Med.* **1971**, *285*, 1182–1186. [CrossRef]
13. Parangi, S.; O'Reilly, M.; Christofori, G.; Holmgren, L.; Grosfeld, J.; Folkman, J.; Hanahan, D. Antiangiogenic therapy of transgenic mice impairs de novo tumor growth. *Proc. Natl. Acad. Sci. USA* **1996**, *93*, 2002–2007. [CrossRef] [PubMed]
14. Nishida, N.; Yano, H.; Nishida, T.; Kamura, T.; Kojiro, M. Angiogenesis in cancer. *Vasc. Health Risk Manag.* **2006**, *2*, 213–219. [CrossRef] [PubMed]
15. Nelson, A.R.; Fingleton, B.; Rothenberg, M.L.; Matrisian, L.M. Matrix metalloproteinases: Biologic activity and clinical implications. *J. Clin. Oncol.* **2000**, *18*, 1135–1149. [CrossRef]
16. P, N.S.; Darvin, P.; Yoo, Y.B.; Joung, Y.H.; Kang, D.Y.; Kim, D.N.; Hwang, T.S.; Kim, S.Y.; Kim, W.S.; Lee, H.K.; et al. The combination of methylsulfonylmethane and tamoxifen inhibits the Jak2/STAT5b pathway and synergistically inhibits tumor growth and metastasis in ER-positive breast cancer xenografts. *BMC Cancer* **2015**, *15*, 474. [CrossRef]
17. Yang, L.; Shi, P.; Zhao, G.; Xu, J.; Peng, W.; Zhang, J.; Zhang, G.; Wang, X.; Dong, Z.; Chen, F.; et al. Targeting cancer stem cell pathways for cancer therapy. *Signal Transduct. Target. Ther.* **2020**, *5*, 8. [CrossRef]

18. Shibue, T.; Weinberg, R.A. EMT, CSCs, and drug resistance: The mechanistic link and clinical implications. *Nat. Rev. Clin. Oncol.* **2017**, *14*, 611–629. [CrossRef]

19. Ma, X.L.; Sun, Y.F.; Wang, B.L.; Shen, M.N.; Zhou, Y.; Chen, J.W.; Hu, B.; Gong, Z.J.; Zhang, X.; Cao, Y.; et al. Sphere-forming culture enriches liver cancer stem cells and reveals Stearoyl-CoA desaturase 1 as a potential therapeutic target. *BMC Cancer* **2019**, *19*, 760. [CrossRef] [PubMed]

20. Larzabal, L.; El-Nikhely, N.; Redrado, M.; Seeger, W.; Savai, R.; Calvo, A. Differential effects of drugs targeting cancer stem cell (CSC) and non-CSC populations on lung primary tumors and metastasis. *PLoS ONE* **2013**, *8*, e79798. [CrossRef] [PubMed]

21. Gomez-Casal, R.; Bhattacharya, C.; Ganesh, N.; Bailey, L.; Basse, P.; Gibson, M.; Epperly, M.; Levina, V. Non-small cell lung cancer cells survived ionizing radiation treatment display cancer stem cell and epithelial-mesenchymal transition phenotypes. *Mol. Cancer* **2013**, *12*, 94. [CrossRef]

22. Yakisich, J.S.; Azad, N.; Venkatadri, R.; Kulkarni, Y.; Wright, C.; Kaushik, V.; Iyer, A.K. Formation of tumorspheres with increased stemness without external mitogens in a lung cancer model. *Stem Cells Int.* **2016**, *2016*, 5603135. [CrossRef]

23. Marotta, L.L.; Almendro, V.; Marusyk, A.; Shipitsin, M.; Schemme, J.; Walker, S.R.; Bloushtain-Qimron, N.; Kim, J.J.; Choudhury, S.A.; Maruyama, R.; et al. The JAK2/STAT3 signaling pathway is required for growth of CD44(+)CD24(-) stem cell-like breast cancer cells in human tumors. *J. Clin. Investig.* **2011**, *121*, 2723–2735. [CrossRef] [PubMed]

24. Jin, W. Role of JAK/STAT3 signaling in the regulation of metastasis, the transition of cancer stem cells, and chemoresistance of cancer by epithelial-mesenchymal transition. *Cells* **2020**, *9*, 217. [CrossRef] [PubMed]

25. Ikeda, S.; Okamoto, T.; Okano, S.; Umemoto, Y.; Tagawa, T.; Morodomi, Y.; Kohno, M.; Shimamatsu, S.; Kitahara, H.; Suzuki, Y.; et al. PD-L1 is upregulated by simultaneous amplification of the PD-L1 and JAK2 genes in non-small cell lung cancer. *J. Thorac. Oncol.* **2016**, *11*, 62–71. [CrossRef] [PubMed]

26. Zhang, N.; Zeng, Y.; Du, W.; Zhu, J.; Shen, D.; Liu, Z.; Huang, J.A. The EGFR pathway is involved in the regulation of PD-L1 expression via the IL-6/JAK/STAT3 signaling pathway in EGFR-mutated non-small cell lung cancer. *Int. J. Oncol.* **2016**, *49*, 1360–1368. [CrossRef]

27. Garcia-Diez, I.; Hernandez-Ruiz, E.; Andrades, E.; Gimeno, J.; Ferrandiz-Pulido, C.; Yebenes, M.; Garcia-Patos, V.; Pujol, R.M.; Hernandez-Munoz, I.; Toll, A. PD-L1 expression is increased in metastasizing squamous cell carcinomas and their metastases. *Am. J. Dermatopathol.* **2018**, *40*, 647–654. [CrossRef] [PubMed]

28. Wang, H.B.; Yao, H.; Li, C.S.; Liang, L.X.; Zhang, Y.; Chen, Y.X.; Fang, J.Y.; Xu, J. Rise of PD-L1 expression during metastasis of colorectal cancer: Implications for immunotherapy. *J. Dig. Dis.* **2017**, *18*, 574–581. [CrossRef]

29. Wei, X.L.; Luo, X.; Sheng, H.; Wang, Y.; Chen, D.L.; Li, J.N.; Wang, F.H.; Xu, R.H. PD-L1 expression in liver metastasis: Its clinical significance and discordance with primary tumor in colorectal cancer. *J. Transl. Med.* **2020**, *18*, 475. [CrossRef]

30. Feng, X.M.; Su, X.L. Anticancer effect of ursolic acid via mitochondria-dependent pathways. *Oncol. Lett.* **2019**, *17*, 4761–4767. [CrossRef]

31. Kashyap, D.; Tuli, H.S.; Sharma, A.K. Ursolic acid (UA): A metabolite with promising therapeutic potential. *Life Sci.* **2016**, *146*, 201–213. [CrossRef]

32. Wei, Z.Y.; Chi, K.Q.; Wang, K.S.; Wu, J.; Liu, L.P.; Piao, H.R. Design, synthesis, evaluation, and molecular docking of ursolic acid derivatives containing a nitrogen heterocycle as anti-inflammatory agents. *Bioorganic Med. Chem. Lett.* **2018**, *28*, 1797–1803. [CrossRef] [PubMed]

33. Yin, R.; Li, T.; Tian, J.X.; Xi, P.; Liu, R.H. Ursolic acid, a potential anticancer compound for breast cancer therapy. *Crit. Rev. Food Sci. Nutr.* **2018**, *58*, 568–574. [CrossRef]

34. Wang, X.; Zhang, F.; Yang, L.; Mei, Y.; Long, H.; Zhang, X.; Zhang, J.; Qimuge, S.; Su, X. Ursolic acid inhibits proliferation and induces apoptosis of cancer cells in vitro and in vivo. *J. Biomed. Biotechnol.* **2011**, *2011*, 419343. [CrossRef] [PubMed]

35. Leng, S.; Hao, Y.; Du, D.; Xie, S.; Hong, L.; Gu, H.; Zhu, X.; Zhang, J.; Fan, D.; Kung, H.F. Ursolic acid promotes cancer cell death by inducing Atg5-dependent autophagy. *Int. J. Cancer* **2013**, *133*, 2781–2790. [CrossRef]

36. Ruan, J.S.; Zhou, H.; Yang, L.; Wang, L.; Jiang, Z.S.; Sun, H.; Wang, S.M. Ursolic acid attenuates TGF-beta1-induced epithelial-mesenchymal transition in NSCLC by targeting integrin alphaVbeta5/MMPs signaling. *Oncol. Res.* **2019**, *27*, 593–600. [CrossRef]

37. Wang, M.; Yu, H.; Wu, R.; Chen, Z.Y.; Hu, Q.; Zhang, Y.F.; Gao, S.H.; Zhou, G.B. Autophagy inhibition enhances the inhibitory effects of ursolic acid on lung cancer cells. *Int. J. Mol. Med.* **2020**, *46*, 1816–1826. [CrossRef] [PubMed]

38. Sp, N.; Kang, D.Y.; Jo, E.S.; Rugamba, A.; Kim, W.S.; Park, Y.M.; Hwang, D.Y.; Yoo, J.S.; Liu, Q.; Jang, K.J.; et al. Tannic acid promotes TRAIL-induced extrinsic apoptosis by regulating mitochondrial ROS in human embryonic carcinoma cells. *Cells* **2020**, *9*, 282. [CrossRef]

39. Nippin, S.P.; Kang, D.Y.; Kim, B.J.; Joung, Y.H.; Darvin, P.; Byun, H.J.; Kim, J.G.; Park, J.U.; Yang, Y.M. Methylsulfonylmethane induces G1 arrest and mitochondrial apoptosis in YD-38 gingival cancer cells. *Anticancer Res.* **2017**, *37*, 1637–1646. [CrossRef]

40. Wu, S.; Zhang, T.; Du, J. Ursolic acid sensitizes cisplatin-resistant HepG2/DDP cells to cisplatin via inhibiting Nrf2/ARE pathway. *Drug Des. Dev. Ther.* **2016**, *10*, 3471–3481. [CrossRef] [PubMed]

41. Lai, M.Y.; Leung, H.W.; Yang, W.H.; Chen, W.H.; Lee, H.Z. Up-regulation of matrix metalloproteinase family gene involvement in ursolic acid-induced human lung non-small carcinoma cell apoptosis. *Anticancer Res.* **2007**, *27*, 145–153. [PubMed]

42. Liu, T.; Ma, H.; Shi, W.; Duan, J.; Wang, Y.; Zhang, C.; Li, C.; Lin, J.; Li, S.; Lv, J.; et al. Inhibition of STAT3 signaling pathway by ursolic acid suppresses growth of hepatocellular carcinoma. *Int. J. Oncol.* **2017**, *51*, 555–562. [CrossRef]

43. Wang, S.; Meng, X.; Dong, Y. Ursolic acid nanoparticles inhibit cervical cancer growth in vitro and in vivo via apoptosis induction. *Int. J. Oncol.* **2017**, *50*, 1330–1340. [CrossRef]
44. Zhang, J.; Wang, W.; Qian, L.; Zhang, Q.; Lai, D.; Qi, C. Ursolic acid inhibits the proliferation of human ovarian cancer stem-like cells through epithelial-mesenchymal transition. *Oncol. Rep.* **2015**, *34*, 2375–2384. [CrossRef] [PubMed]
45. Jin, H.; Pi, J.; Yang, F.; Jiang, J.; Wang, X.; Bai, H.; Shao, M.; Huang, L.; Zhu, H.; Yang, P.; et al. Folate-chitosan nanoparticles loaded with ursolic acid confer anti-breast cancer activities in vitro and in vivo. *Sci. Rep.* **2016**, *6*, 30782. [CrossRef] [PubMed]
46. Kang, D.Y.; Sp, N.; Jo, E.S.; Rugamba, A.; Hong, D.Y.; Lee, H.G.; Yoo, J.S.; Liu, Q.; Jang, K.J.; Yang, Y.M. The inhibitory mechanisms of tumor PD-L1 expression by natural bioactive gallic acid in non-small-cell lung cancer (NSCLC) cells. *Cancers* **2020**, *12*, 727. [CrossRef]
47. Kim, D.H.; Sp, N.; Kang, D.Y.; Jo, E.S.; Rugamba, A.; Jang, K.J.; Yang, Y.M. Effect of methylsulfonylmethane on proliferation and apoptosis of A549 lung cancer cells through G2/M cell-cycle arrest and intrinsic cell death pathway. *Anticancer Res.* **2020**, *40*, 1905–1913. [CrossRef]
48. Krock, B.L.; Skuli, N.; Simon, M.C. Hypoxia-induced angiogenesis: Good and evil. *Genes Cancer* **2011**, *2*, 1117–1133. [CrossRef] [PubMed]
49. Van Zijl, F.; Krupitza, G.; Mikulits, W. Initial steps of metastasis: Cell invasion and endothelial transmigration. *Mutat. Res.* **2011**, *728*, 23–34. [CrossRef]
50. Niu, G.; Wright, K.L.; Huang, M.; Song, L.; Haura, E.; Turkson, J.; Zhang, S.; Wang, T.; Sinibaldi, D.; Coppola, D.; et al. Constitutive Stat3 activity up-regulates VEGF expression and tumor angiogenesis. *Oncogene* **2002**, *21*, 2000–2008. [CrossRef] [PubMed]
51. Phi, L.T.H.; Sari, I.N.; Yang, Y.G.; Lee, S.H.; Jun, N.; Kim, K.S.; Lee, Y.K.; Kwon, H.Y. Cancer stem cells (CSCs) in drug resistance and their therapeutic implications in cancer treatment. *Stem Cells Int.* **2018**, *2018*, 5416923. [CrossRef]
52. Lee, C.H.; Yu, C.C.; Wang, B.Y.; Chang, W.W. Tumorsphere as an effective in vitro platform for screening anti-cancer stem cell drugs. *Oncotarget* **2016**, *7*, 1215–1226. [CrossRef]
53. Amini, S.; Fathi, F.; Mobalegi, J.; Sofimajidpour, H.; Ghadimi, T. The expressions of stem cell markers: Oct4, Nanog, Sox2, nucleostemin, Bmi, Zfx, Tcl1, Tbx3, Dppa4, and Esrrb in bladder, colon, and prostate cancer, and certain cancer cell lines. *Anat. Cell Biol.* **2014**, *47*, 1–11. [CrossRef] [PubMed]
54. Bethune, G.; Bethune, D.; Ridgway, N.; Xu, Z. Epidermal growth factor receptor (EGFR) in lung cancer: An overview and update. *J. Thorac. Dis.* **2010**, *2*, 48–51. [PubMed]
55. Colomiere, M.; Ward, A.C.; Riley, C.; Trenerry, M.K.; Cameron-Smith, D.; Findlay, J.; Ackland, L.; Ahmed, N. Cross talk of signals between EGFR and IL-6R through JAK2/STAT3 mediate epithelial-mesenchymal transition in ovarian carcinomas. *Br. J. Cancer* **2009**, *100*, 134–144. [CrossRef] [PubMed]
56. Chen, S.H.; Murphy, D.A.; Lassoued, W.; Thurston, G.; Feldman, M.D.; Lee, W.M. Activated STAT3 is a mediator and biomarker of VEGF endothelial activation. *Cancer Biol. Ther.* **2008**, *7*, 1994–2003. [CrossRef]
57. Byun, H.J.; Darvin, P.; Kang, D.Y.; Sp, N.; Joung, Y.H.; Park, J.H.; Kim, S.J.; Yang, Y.M. Silibinin downregulates MMP2 expression via Jak2/STAT3 pathway and inhibits the migration and invasive potential in MDA-MB-231 cells. *Oncol. Rep.* **2017**, *37*, 3270–3278. [CrossRef]

 *biomedicines*

MDPI

Review

Antitumoral Activities of Curcumin and Recent Advances to ImProve Its Oral Bioavailability

Marta Claudia Nocito [†], Arianna De Luca [†], Francesca Prestia, Paola Avena, Davide La Padula, Lucia Zavaglia, Rosa Sirianni, Ivan Casaburi, Francesco Puoci , Adele Chimento [*,‡] and Vincenzo Pezzi [*,‡]

Department of Pharmacy and Health and Nutritional Sciences, University of Calabria, Via Pietro Bucci, Arcavacata di Rende, 87036 Cosenza, Italy; nocitomarta90@tiscali.it (M.C.N.); ariannadl@hotmail.it (A.D.L.); francescaprestia908@gmail.com (F.P.); paola.avena@unical.it (P.A.); davidelapadula@live.it (D.L.P.); luciazavaglia@hotmail.it (L.Z.); rosa.sirianni@unical.it (R.S.); ivan.casaburi@unical.it (I.C.); francesco.puoci@unical.it (F.P.)
* Correspondence: adele.chimento@unical.it (A.C.); v.pezzi@unical.it (V.P.); Tel.: +39-0984-493184 (A.C.); +39-0984-493148 (V.P.)
† These authors contributed equally to this work.
‡ These authors shared senior authorship.

Abstract: Curcumin, a main bioactive component of the *Curcuma longa* L. rhizome, is a phenolic compound that exerts a wide range of beneficial effects, acting as an antimicrobial, antioxidant, anti-inflammatory and anticancer agent. This review summarizes recent data on curcumin's ability to interfere with the multiple cell signaling pathways involved in cell cycle regulation, apoptosis and the migration of several cancer cell types. However, although curcumin displays anticancer potential, its clinical application is limited by its low absorption, rapid metabolism and poor bioavailability. To overcome these limitations, several curcumin-based derivatives/analogues and different drug delivery approaches have been developed. Here, we also report the anticancer mechanisms and pharmacokinetic characteristics of some derivatives/analogues and the delivery systems used. These strategies, although encouraging, require additional in vivo studies to support curcumin clinical applications.

Keywords: curcumin; cancer cells; bioavailability; curcumin derivatives; curcumin analogues; curcumin delivery systems

Citation: Nocito, M.C.; De Luca, A.; Prestia, F.; Avena, P.; La Padula, D.; Zavaglia, L.; Sirianni, R.; Casaburi, I.; Puoci, F.; Chimento, A.; et al. Antitumoral Activities of Curcumin and Recent Advances to ImProve Its Oral Bioavailability. *Biomedicines* **2021**, *9*, 1476. https://doi.org/10.3390/biomedicines9101476

Academic Editor: Pavel B. Drašar

Received: 22 September 2021
Accepted: 7 October 2021
Published: 14 October 2021

Publisher's Note: MDPI stays neutral with regard to jurisdictional claims in published maps and institutional affiliations.

1. Introduction

Curcuma is one of the largest genera in the Zingiberaceae family which comprises approximately 133 species [1]. It is widely distributed in the tropical regions spanning India to Southern China and Northern Australia [1]. The species of greatest interest is *Curcuma longa* L., which is cultivated particularly in India. It consists of an underground root (rhizome) from which, once dried and ground, a powder with a characteristic yellow-orange color is obtained. Curcumin extract is composed of three curcuminoids at different proportions such as curcumin (1, 7-bis (hydroxyl-3-methoxyphenyl)-1,6-heptadiene-3, 5-dione) (curcumin I) (~ 77%), demethoxy curcumin (curcumin II) (17–18%) and bis-demethoxycurcumin (curcumin III) (3–5%) (Figure 1) [2]. The volatile fraction is quantitatively important, which mainly contains several terpenic compounds including α-zingiberene, curlone and α-turmerone [3]. Among the curcuminoids, the polyphenol curcumin is the most active. It possesses a wide range of pharmacological proprieties [4,5], acting as an antimicrobial [6], antiviral [7], antifungal [8], antioxidant [9], antimalarial [10], anti-inflammatory [11], antiaging [12] and antitumoral [13] agent. In recent years, extensive research indicated that this polyphenol can prevent and suppress tumor initiation, promotion and progression and can be used to treat cancer by interfering with several signaling pathways [13,14]. At a molecular level, curcumin's anticarcinogenic effects are attributed to its ability to modulate transcription factors, growth regulators, adhesion molecules, apoptotic genes, angiogenesis regulators, and more [14]. In vitro studies demonstrated that curcumin suppresses the proliferation of a wide variety of tumor cells; downregulates the expression of transcription factors (NF-kB, AP-1 and Egr-1), COX2, MMPs, TNF, chemokines, cell surface adhesion molecules and cyclins, growth factor receptors (such as EGFR and HER2); and inhibits tyrosine and serine/threonine kinases (i.e., JNK) activity [14]. By inhibiting STAT3, NF-kB and WNT/β-catenin signaling pathways, curcumin interferes with cancer development and progression [15–17]. In addition, the inhibition of Sp-1 and its regulated genes may serve as an important mechanism to prevent cancer formation, migration, and invasion [14]. Additionally, curcumin induces apoptosis through both mitochondria-dependent [18] as well as mitochondria-independent mechanisms [19], depending on the cell type.

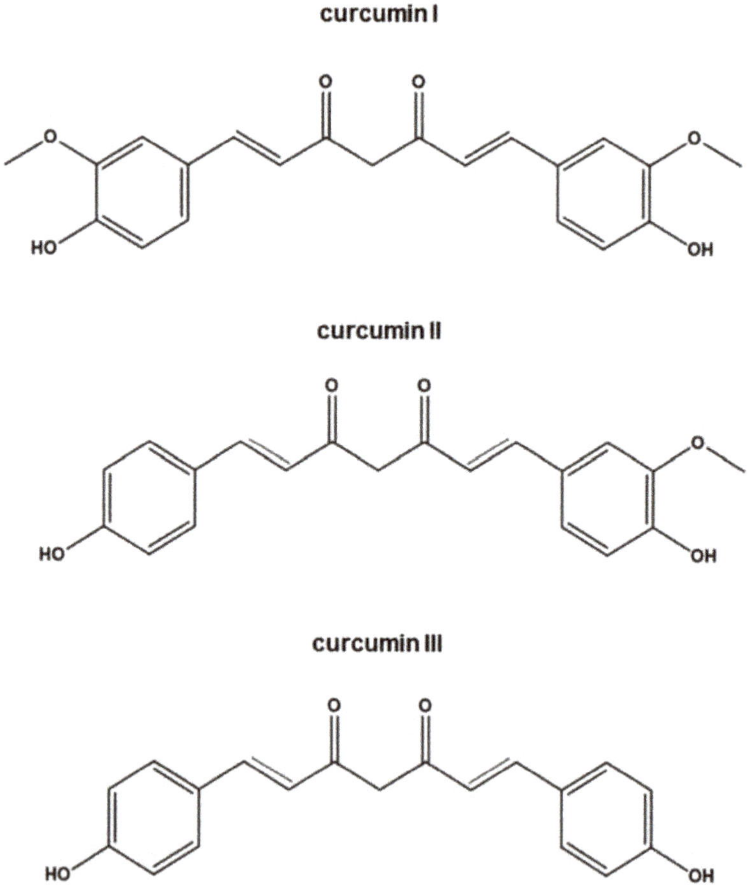

curcumin I

curcumin II

curcumin III

Figure 1. Chemical structures of curcumin I, II, III.

Despite curcumin's anticancer potential, its therapeutic application is limited by its poor bioavailability, due to the low or very low absorption after oral intake. After oral administration, curcumin is metabolized into curcumin glucuronide and curcumin sulfonate by the liver during phase II reactions. Glucuronidation and sulfation increase curcumin's water solubility, but decrease its effectiveness and accelerate its removal via urine [20]. Specifically, the different factors contributing to the low bioavailability include the low plasma level, tissue distribution, poor absorption, high rate of metabolism, inactivity of metabolic products, rapid elimination and clearance from the body [21]. In order to improve curcumin pharmacokinetic characteristics and then biological activity, synthetic derivatives/analogues [22–24] and various drug delivery strategies have been developed [25–30].

In this review, we summarize the recent advances in molecular mechanisms activated by curcumin and the elicited inhibition of cancer cell growth and progression. Additionally, the improved pharmacokinetic proprieties of curcumin derivatives/analogues, and the attempted targeted and triggered drug delivery systems, are reviewed.

2. Antitumoral Activities of Curcumin

Dietary polyphenolic phytochemicals are able to block tumor initiation or progression [31]. Several studies indicated that curcumin can effectively prevent and treat cancer at the initiation, promotion and progression level [32]. This propriety is due to its ability to target critical processes primarily involved in cancer development, such as proliferation, apoptosis and metastasis.

2.1. Antiproliferative Effects of Curcumin

Some molecular alterations associated with carcinogenesis occur in signaling pathways regulating cell proliferation [31]. Accumulating data show that curcumin displays in vitro antiproliferative effects on several types of cancer such as breast, colorectal, bladder, brain and gastrointestinal through several mechanisms of action [33] (see Table 1 for a summary of the data). Generally, the antiproliferative effects of curcumin could be attributed to its ability to regulate cell cycle progression, protein kinases activity and transcription factor expression.

In breast cancer cell lines, curcumin inhibits proliferation by reducing cyclin D1 expression at mRNA and protein levels, consequently decreasing CDK4 activity [34,35]. In MCF-7 breast cancer cells, curcumin caused G1 cell cycle arrest, cyclin E proteosomal degradation, p53 upregulation and an increase in p21 and p27 CDK inhibitor levels [36]. In MCF-7 and MDA-MB-231 breast cancer cells, a G2/M cell cycle arrest by curcumin was also reported. Specifically, following GSK3β upregulation, a loss of nuclear β-catenin produced a downregulation of its downstream target, cyclin D1 [37]. Additionally, curcumin is a potent inhibitor of NF-kB [16], a critical regulator of cell proliferation and survival [38]. Liu et al. demonstrated that in MCF-7 cells, curcumin markedly decreased cell proliferation in a concentration- and time-dependent manner by regulating the NF-kB signaling, as shown by a decrease in p65 and an increase in IκB protein expression [16]. Moreover, Zhou et al. demonstrated that, in MCF-7 cells and MCF-7 xenografts, a combination of curcumin with the chemotherapeutic agent mitomycin C produced cell cycle arrest at the G1 phase as a consequence of decreased cyclin D1, cyclin E, cyclin A, CDK2, and CDK4 expression, along with the induction of the cell cycle inhibitor p21 and p27 [35]. The potent inhibitory effect produced by curcumin and mitomycin C combination on MCF-7 growth was also observed in MCF-7- and MDA-MB-231-derived cancer stem cells [39]. In addition, curcumin was proven to be effective to sensitize MDA-MB-231 and MCF-7 cells to commonly used chemotherapeutic agents paclitaxel, cisplatin and doxorubicin [39]. In a recent work, it has been shown that curcumin inhibited the growth of MDA-MB-231 and MDA-MB-468 triple-negative breast cancer cells by reducing the expression of histone methyl transferase EZH2 and reactivating that of DLC1 [40]. EZH2 is an oncogenic factor commonly upregulated in human cancers [41], while DLC1 is a downregulated tumor suppressor in many malignant tumors [42].

In vitro studies showed that curcumin significantly inhibited colon cancer cell growth modulating multiple molecular targets and distinct signaling pathways [43,44]. A study performed by Lim et al. demonstrated that in a HCT-116 human colon cancer cell line, curcumin downregulated CDK2 expression, causing G1 phase arrest [45]. Moreover, in the same cell line, curcumin enhanced ROS generation and downregulated the transcription factor E2F4 and its target genes, including cyclin A, p21 and p27 [46]. Another study by Watson et al. investigated curcumin cytotoxicity in HCT-116 and HT29 cell lines, showing a time- and dose-dependent cell proliferation inhibition when p53 was upregulated [47]. Curcumin prevented cell proliferation by cell cycle arrest at the G2/M phase in both HCT116 p53+/+ and p53−/− cells in a concentration-dependent manner and induced senescence accompanied by autophagy [48]. Recently, Calibasi-Kocal et al. demonstrated antiproliferative, wound healing, anti-invasive and antimigratory effects of curcumin not only on HCT-116 cells, but also on the LoVo metastatic colorectal cancer cell line, confirming its anticancer activity [44].

The antitumoral effects of curcumin were also demonstrated in bladder cancer cells. Curcumin's inhibitory effects were observed in T24 [49,50], 5637 [50] and in RT4 [49] human bladder cancer cell lines. In T24 and RT4 cells, a decrease in Trop2, a cell surface receptor that transduces via calcium signals, is required for curcumin-mediated effects including inhibition of cell proliferation and migration, apoptosis and cell cycle arrest. A Trop2 decrease caused a reduction in the expression levels of its downstream target cyclin E1, and an increase in p27 levels [49].

Curcumin has been shown to decrease malignant characteristics of glioblastoma cells acting on several targets, including growth factor receptors, kinases, transcriptional factors and inflammatory cytokines [51]. In the U87 human glioma cell line, curcumin inhibited proliferation-suppressing cyclin D1 and induced p21 expression. In these cells, curcumin via ERK and JNK signaling induced expression of the transcription factor Egr-1 which, in turn, increased p21 gene expression [52]. In the same cells, curcumin induced G2/M cell cycle arrest and apoptosis by increasing FoxO1 expression [53]. In U251 and SNB19 human glioblastoma cells curcumin reduced the expression of Skp2 and upregulated p57 [54]. Skp2, a component of the ubiquitin proteasome system, is responsible for ubiquitin-mediated degradation of the cell cycle inhibitors p21, p27 and p57 [55]. Recently, Luo et al. demonstrated that curcumin decreased proliferation of U251 and LN229 human glioblastoma cells via inhibition of HDGF [56], an angiogenesis-promoting growth factor commonly upregulated in gliomas. In human glioma cell lines U251 and LN229, HDGF forms a complex with β-catenin promoting tumor growth and metastasis. Curcumin significantly reduced HDGF expression, and consequently its binding to β-catenin [56]. Moreover, several studies provided evidence that curcumin potentiated the effects of chemotherapy and radiation therapy while protecting normal tissue, selectively inducing cell death in glioblastoma cancer cells [57,58].

In gastrointestinal cancer cell lines, curcumin inhibited ODC and increased SMOX activity, two key enzymes in polyamine synthesis and catabolism, respectively. Importantly, elevated levels of polyamines have been associated with several cancers, and altered levels of the rate limiting enzymes in both biosynthesis and catabolism have been observed. The combination of curcumin with ODC inhibitor DFMO significantly potentiated ODC inhibition decreasing AGS gastric adenocarcinoma cell growth [59]. Evidence regarding curcumin's potential in gastric cancer prevention has been accumulating. Curcumin induced cell-cycle arrest at the G1 phase by downregulating cyclin D1 expression in BGC823, SGC7901, MKN1 and MGC803 gastric cancer cells by inhibiting EGF/PAK1/NF-kB pathway [60]. Moreover, curcumin exerted antiproliferative effects by inhibiting the Wnt/β-catenin signaling pathway. In SNU1, SNU-5 and AGS human gastric cancer cells, curcumin caused a reduction in Wnt3a, LRP6, phospho-LRP6, β-catenin, phospho-β-catenin, c-Myc and survivin [61]. The antiproliferative effects of curcumin were also demonstrated in SGC-7901 and BGC-823 gastric cancer cells by activating p53/p21 and inhibiting PI3K signaling pathways [62]. In another work, SGC7901 gastric cancer cell proliferation was reduced after curcumin treatment via c-Myc/long non-coding RNA (lncRNA) H19 downregulation and p53 upregulation [63]. Studies reported that H19, an oncogenic lncRNA [64], is abnormally upregulated in gastric cancer and contributes to cellular proliferation by directly inactivating p53 [65]. Furthermore, the oncogene c-Myc was shown to directly induce H19 expression by binding to the H19 promoter, and thereby promoting proliferation of gastric cancer cells [66]. A recent investigation defined curcumin effects on microRNAs expression related to gastric cancer cell proliferation [67]. In SGC-7901 cells, curcumin inhibited cell cycle progression and induced apoptosis by upregulating miR-34a, which decreased CDK4 and cyclin D1 protein expression. The same effects were obtained by transfecting the cells with specific miR-34a agomir [67]. In BGC-823 and SGC7901 gastric cancer cells, curcumin inhibited cell proliferation and increased cell death by upregulating miR-33b which, in turn, decreased expression of the apoptosis inhibitor XIAP by targeting its 3′ UTR [68].

Mounting evidence indicates that curcumin affects several molecular pathways involved in melanoma pathogenesis, making it a promising therapeutic agent to be used against this type of cancer [69]. This phytochemical compound was able to arrest cell cycle at the G2/M phase by inhibiting NF-kB and iNOS activity in human melanoma A375 and MeWo cells [70]. Another study postulated that curcumin's effect on cell cycle arrest at the G2/M phase was dependent on PDE inhibition, an enzyme that catalyzed cAMP and/or cGMP hydrolysis. Specifically, curcumin decreased cell proliferation and cell cycle progression by inhibiting PDE1A, cyclin A, UHRF1 and DNMT1 expression while increasing that of p21 and p27 in B16F10 murine melanoma cells [71]. Another report showed that curcumin caused cell cycle arrest at the G2/M phase, and induced autophagy by downregulating Akt/mTOR axis in human melanoma A375 and C8161 cell lines [72]. Curcumin antiproliferative effects were also observed in other three melanoma cell lines (C32, G-361, and WM 266-4), all of which are characterized by B-Raf mutations. In these cells, curcumin antitumor effects were associated with the suppression of NF-kB and IKK activity but were independent from the inhibition of B-Raf/MEK/ERK and Akt pathways [73].

2.2. Pro-Apoptotic Effects of Curcumin

In addition to antiproliferative properties, curcumin shows extensive therapeutic potential as an apoptotic inducer through several mechanisms demonstrated in different cancer cell models (see Table 1 for a summary of the data). In MCF-7 breast cancer cells, a sub-cytotoxic dose of curcumin induced apoptotic cell death through an increase in histone H3 acetylation and glutathionylation which, in turn, promoted the transcriptional activation of different proapoptotic genes [74]. Several reports indicated that, in breast cancer cells, curcumin induced apoptosis via p53-dependent and -independent pathways [75,76]. Patel et al. demonstrated that, in MCF-7 cells, curcumin enhanced p53 expression and activated Parp-1-mediated apoptotic pathways [77]. The p53-independent effects of curcumin were observed in cancer cells lacking a functional p53 protein such as MDA-MB-231. In these cells, curcumin induced ROS generation which altered mitochondrial membrane permeability, reduced intracellular GSH levels, increased Bax/Bcl-2 ratio and cleaved-caspase 3 expression [78]. In MDA-MB-231 cells, curcumin inhibited NF-kB p65, triggering apoptosis [79]. Moreover, in the same cell line, curcumin was able to mediate apoptotic cell death through FAS inhibition [80].

The apoptotic effects of curcumin were evident in melanoma. In A375 cells, curcumin promoted tumor cell apoptosis by inhibiting the JAK-2/STAT-3 signaling pathway and downregulating Bcl-2 [81]. Moreover, the ability of curcumin to induce apoptosis was demonstrated in four human melanoma cell lines (PMWK, Sk-mel-2, Sk-mel-28, and Mewo) with mutant p53 through the activation of FAS/caspase 8 pathway [82]. Curcumin-dependent apoptosis was also observed in HEY ovarian cancer cells where p53 knockdown or p53 inhibition did not prevent curcumin's inhibitory effect. In these cells, apoptosis occurred through the activation of p38 MAPK, the inhibition of pro-survival Akt signaling, along with a decreased expression of Bcl-2 and survivin [83]. Additionally, in the multiple myeloma RPMI 8226 cell line, curcumin upregulated p53 and Bax protein levels and downregulated MDM2, a known p53 inhibitor [84].

In colorectal cancer cells, curcumin triggered the apoptotic process via the subsequent modulation of various target molecules [43]. In HT-29 colon cancer cells, curcumin-induced apoptosis was related to a decreased COX2 expression and AKT phosphorylation along with an increased activation of AMPK signaling [85]. Moreover, this polyphenol promoted apoptosis in HCT116, HT29, and SW620 colorectal cancer cell lines by suppressing constitutive and inducible NF-kB activity and NF-kB-regulated gene products such as Bcl-2, Bcl-xL, IAP-2, COX2 and cyclin D1 [86]. Additionally, Narayan et al. showed that curcumin inhibited the Wnt/β-catenin pathway by inducing the caspase 3-mediated cleavage of beta-catenin, E-cadherin and APC, leading to loss of cell–cell adhesion and apoptosis in HCT-116 colon cancer cells [87]. It has been reported that curcumin can activate extrinsic apoptotic pathway, upregulating DR5 protein in HCT-116 and HT-29 colon cancer cells [88]. Furthermore, curcumin triggered Fas-mediated caspase 8 activation in HT-29 cells [89]. In these cells, treatment with curcumin caused a mithocondrial $[Ca^{2+}]$ increase, cytochrome c release from mitochondria to cytosol, mithocondrial membrane potential reduction, Bax increase and Bcl-2 as well as caspase 3 and 7 activation [89]. Curcumin-induced Bcl-2 downregulation and Bax upregulation in both HCT-116 [90] and COLO-205 colon cancer cells [91].

Caspase 3/7 activity was investigated by Shi et al. in bladder cancer cells. The authors demonstrated the ability of curcumin to induce apoptosis through a caspase-dependent mechanism in two human urinary bladder carcinoma cells [50]. The same apoptotic mechanism was also observed in other cancer types such as glioblastoma [92]. A similar feature also occurs in human osteosarcoma (HOS) cells, where curcumin caused cell cycle arrest determining apoptosis, as demonstrated by caspase 3 and PARP-1 cleavage [93]. Recently, it was observed that curcumin inhibited ODC activity and polyamine biosynthesis in AGS gastric adenocarcinoma cells. In this type of cell, curcumin caused an increase in ROS levels responsible for DNA damage and, thus, apoptosis [59].

The pro-apoptotic effects of curcumin are also mediated by the modulation of miRNAs in several cancer cells. Curcumin reduced Bcl-2 expression by upregulating miR-15a and miR-16 in MCF-7 cells [94] and promoted apoptosis through the miR-186 signaling pathway in human lung adenocarcinoma cells [95]. Recently, in RT4 schwannoma cells, Sohn et al. demonstrated that curcumin enhanced the expression of apoptotic protein Bax and decreased Bcl-2, as well as determined caspase 3/9 activation. All of these events were related to curcumin-mediated miRNA 344a-3p upregulation [96].

2.3. Antimetastatic Effects of Curcumin

In addition to the antiproliferative and apoptotic effects, curcumin acts as an antimetastatic agent (see Table 1 for a summary of the data) [15,44]. The metastatic cascade starts with the loss of cell-to-cell and cell-to-substrate adhesion, a feature of the EMT process, which allows the acquisition of a mobile phenotype, the dissociation of cells from primary tumor and the spread to distant tissues and organs [97].

The effects of curcumin on genes involved in EMT was evaluated in breast cancer cells. It was demonstrated that curcumin decreased the gene transcription and protein expression of Axl, Slug, CD24 and RhoA, which regulate EMT and, consequently, migration and invasion of MCF-10F and MDA-MB-231 breast cancer cells. Curcumin elicited these effects through the upregulation of miR-34a, which acts as a tumor suppressor gene in both cell lines [98]. The antimetastatic effects of curcumin occur through the modulation of several signaling pathways, including NF-kB. NF-kB/p65 transcriptionally regulates TWIST1, SLUG and SIP1 which, in turn, repress E-cadherin while activating the mesenchymal markers N-cadherin and MMP11, resulting in EMT progression [99]. In MCF-7 cells, curcumin was able to inhibit uPA production by preventing NF-kB activation [100]. Through the same mechanism, curcumin downregulated CXCL1 and 2, two inflammatory cytokines involved in MDA-MB-231 breast cancer cells migration [101]. A critical event in tumor cell invasion and metastasis is the degradation of the extracellular matrix by MMPs, enzymes that degrade a range of extracellular matrix proteins, allowing cancer cells to migrate and invade [102]. Curcumin was able to inhibit LPA activated invasion by attenuating the RhoA/ROCK/MMPs pathway in MCF-7 cells [103]. Similarly, in SO-Rb50 and Y79 human retinoblastoma cell lines, curcumin reduced migration and invasion by decreasing MMP2, RhoA, ROCK1 and vimentin expression. The authors provided evidence that, in these cells, curcumin's antitumor activity requires miR-99a upregulation and JAK/STAT3 pathway inhibition [104]. MMP's decrease after curcumin treatment was also observed in T24 and 5637 human bladder cancer cell lines [50]. Additionally, in T24 and RT4 bladder cancer cells, curcumin's antimetastatic mechanism included a Trop2 decrease [49], a gene also involved in tumor aggressiveness and metastasis formation [105]. In prostate cancer cells, curcumin treatment suppressed EGF, heregulin-stimulated PC-3 and androgen-induced LNCaP cell invasion. Particularly, curcumin significantly reduced MMP9 activity and downregulated cellular matriptase, a membrane-anchored serine protease involved in tumor formation and invasion [106]. Recently, in an HCT-116 human colorectal carcinoma cell line, it has been demonstrated that the expression of proteins related to cell migration, including MMP9 and claudin-3, was downregulated by increasing doses of curcumin [107]. These data were confirmed by an independent group using the same cell line in addition to LoVo human metastatic colon cancer cells, establishing curcumin anti-invasive and antimigratory properties [44]. Furthermore, in human melanoma A375 cells, curcumin decreased MMP2 and MMP9 expression while increasing TIMP-2, a tissue inhibitor of metalloproteinases [81]. Activation of melanoma cell migration and invasion by OPN was also counteracted by curcumin. Specifically, curcumin was able to downregulate pro-MMP2 activation by preventing OPN-mediated NF-kB nuclear translocation [108]. The ability of curcumin to interfere with NFk-B pathway was also evident in Hela cervical cancer cells, where it was also demonstrated an effect on Wnt/β-catenin signaling, two pathways involved in proliferation and invasion of cervical cancer [17]. It has been shown that STAT3 activation is associated with metastasis formation in several tumors [109]. In SCLC cells curcumin inhibited cell migration, invasion and angiogenesis by inhibiting JAK/STAT3 signaling activated in response to IL-6. As a consequence, curcumin downregulated the expression of ICAM, VEGF, MMP2 and MMP7 STAT3-regulated genes involved in tumor invasion [110]. Curcumin inhibited the JAK/STAT3 signaling pathway also in SKOV3 human ovarian cancer cells, causing a decrease in fascin, an actin-binding protein involved in cell adhesion, migration, and invasion [15]. Curcumin is able to prevent invasion by inhibiting AKT, mTOR and P70S6K phosphorylation, as demonstrated in human melanoma A375 and C8161 cells [72] and TC1889 human thymic carcinoma cells [111]. Additionally, it has been demonstrated that curcumin reduced cell invasion and migration in NSCLC A549 cells by increasing miR-206, which further suppressed PI3K/AKT/mTOR pathway activation [112]. The observation that curcumin inhibited NEDD4-mediated signaling in SNB19 and A1207 glioma cells, thus interfering with cell motility, is very interesting [113]. NEDD4 is a E3-ubiquitin ligase involved in the degradation of CNrasGEFs, guanine nucleotide exchange factors (GEFs), that serve as RAS activators, thus, promoting glioma cell migration and invasion [114].

Table 1. Antitumoral activities of curcumin.

Biological Effects	Mechanisms of Action	Cancer Type	References
Antiproliferative			
	CDK2 decrease		[35]
	CDK4 decrease		[34,35]
	Cell cycle arrest at G1 phase		[16,35,36]
	Cell cycle arrest at G2/M phase		[37]
	Cell viability decrease		[39]
	Cyclin A decrease		[35]
	Cyclin D1 decrease		[34,35,37]
	Cyclin E decrease		[35,36]
	DLC1 increase	Breast	[40]
	EZH2 decrease		[40]
	GSK3β increase		[37]
	Increased sensibility to chemotherapeutic agents		[39]
	IκB increase		[16]

Table 1. *Cont.*

Biological Effects	Mechanisms of Action	Cancer Type	References
Antiproliferative			
	NF-kB p65 decrease		[16]
	p21 and p27 increase		[35,36]
	p53 increase		[36]
	β-catenin decrease		[37]
	CDK2 decrease		[45]
	Cell cycle arrest at G1 phase		[45]
	Cell cycle arrest at G2/M phase		[47,48]
	Cell viability decrease		[44]
	Cyclin A decrease	Colon	[46]
	E2F4, decrease		[46]
	p21 and p27 increase		[46]
	p53 increase		[47]
	ROS increase		[46]
	Cell cycle arrest at G2/M phase		[49]
	Cell viability decrease		[49,50]
	Cyclin E1 decrease	Bladder	[49]
	p27 increase		[49]
	Trop2 decrease		[49]
	Cell cycle arrest at G2/M phase		[53]
	Cyclin D1 decrease		[52]
	Egr-1 increase	Glioma	[52]
	FoxO1 increase		[53]
	p21 increase		[52]
	Cell cycle arrest at G2/M phase		[54]
	HDGF / β-catenin complex inhibition		[56]
	p57 increase	Glioblastoma	[54]
	Skp2 decrease		[54]
	CDK4 decrease		[67]
	Cell cycle arrest at G1 phase		[60]
	Cell cycle arrest at G0/G1-S phase		[67]
	c-myc decrease		[61]
	c-myc/(lncRNA) H19 pathway downregulation		[63]
	Cyclin D1 decrease		[67]
	EGF/PAK1/NF-kB/cyclin D1 pathway inhibition		[60]
	LRP6 ans phospho-LRP6 decrease		[61]
	miR-33b increase		[68]
	miR-34a increase	Gastric	[67]
	ODC activity decrease		[59]
	p21 increase		[62]
	p53 increase		[62,63]
	PI3K signaling inhibition		[62]
	SMOX mRNA and activity increase		[59]
	Wnt3a decrease		[61]
	XIAP decrease		[68]
	β-catenin and phospho β-catenin decrease		[61]
	Akt/mTOR pathway downregulation		[72]
	Cell cycle arrest at G2/M phase		[70–72]
	Cyclin A decrease		[71]
	DNMT1 decrease		[71]
	IKK inhibition		[73]
	iNOS inhibition	Melanoma	[70]
	NF-kB inhibition		[70,73]
	p21 and p27 increase		[71]
	PDE decrease		[71]
	UHRF1 decrease		[71]

Table 1. *Cont.*

Biological Effects	Mechanisms of Action	Cancer Type	References
Pro-apoptotic			
	Bad increase	Breast	[77]
	Bax increase		[79]
	Bax/Bcl-2 ratio increase		[78]
	Bcl-2 decrease		[79,94]
	Cleaved caspase 3 increase		[78]
	Cleaved Parp-1 increase		[77]
	FAS inhibition		[80]
	GSH decrease		[78]
	Histone H3 acetylation and glutathionylation increase		[74]
	miR-15a and miR-16 increase		[94]
	NF-kBp65 decrease		[79]
	p53 increase		[77]
	ROS increase		[78]
	Bcl-2 decrease	Melanoma	[81]
	JAK-2/STAT-3 signaling inhibition		[81]
	p53-independent Fas/caspase 8 pathway activation		[82]
	Akt signaling inhibition	Ovarian	[83]
	Bcl-2 and survivin decrease		[83]
	p38 MAPK activation		[83]
	Bax and p53 increase	Myeloma	[84]
	MDM2 decrease		[84]
	AMPK increase	Colon	[85]
	APC decrease		[87]
	Bax increase		[89–91]
	Bcl-2 decrease		[86,89–91]
	Bcl-xL decrease		[86]
	Caspase 3 activation		[87,89]
	Caspase 7 activation		[89]
	COX2 decrease		[86]
	Cyclin D1 decrease		[86]
	DR5 upregulation		[88]
	E-cadherin decrease		[87]
	Fas-mediated caspase 8 activation		[89]
	IAP-2 decrease		[86]
	Mithocondrial [Ca^{2+}] increase		[89]
	Mithocondrial cytochrome c release		[89]
	Mithocondrial membrane potential reduction		[89]
	pAKT decrease		[85]
	β-catenin decrease		[87]
	Caspase 3/7 activation	Bladder	[50]
	Caspase 3/7 activation	Glioblastoma	[92]
	Bax increase		[92]
	Bcl-2 decrease		[92]
	Caspase 3 activation	Hosteosarcoma	[93]
	Parp-1 cleavage		[93]
	DNA damage	Gastric	[59]
	ODC activity decrease		[59]
	ROS production		[59]
	miR-186 pathway activation	Lung	[95]
	Bax and cleaved caspase 3/9 increase	Bladder	[96]
	Bcl2 decrease		[96]
	miRNA 344a-3p increase		[96]

Table 1. *Cont.*

Biological Effects	Mechanisms of Action	Cancer Type	References
Antimetastatic			
	Axl decrease		[98]
	CD24 decrease		[98]
	CXCL1 and 2 decrease		[101]
	miR-34a increase		[98]
	NF-kB inhibition	Breast	[100,101]
	Rho-A decrease		[98]
	RhoA/ROCK/MMPs/Vimentin pathway inhibition		[103]
	Slug decrease		[98]
	uPA decrease		[100]
	JAK/STAT3 pathway inhibition		[104]
	miR-99a increase		[104]
	MMP2 decrease		[104]
	RhoA decrease	Retinoblastoma	[104]
	ROCK1 decrease		[104]
	Vimentin decrease		[104]
	MMPs signaling pathways inhibition		[50]
	Trop2 decrease	Bladder	[49]
	Cellular matriptase downregulation		[106]
	MMP9 decrease	Prostate	[106]
	Angiogenesis inhibition		[44]
	Claudin-3 decrease		[107]
	Metastasis inhibition	Colon	[44]
	MMP9 decrease		[107]
	MMP2 decrease		[81,108]
	MMP9 decrease		[81]
	NF-kB signaling pathways inhibition	Melanoma	[108]
	TIMP-2 increase		[81]
	NF-kB and Wnt/βcatenin pathways inhibition	Cervical	[17]
	ICAM decrease		[110]
	VEGF decrease		[110]
	MMP2 and MMP7 decrease	SCLC	[110]
	STAT3 decrease		[110]
	IL-6-inducible JAK/STAT3 phosphorylation reduction		[110]
	Fascin decrease		[15]
	JAK/STAT3 signaling pathway inhibition	Ovarian	[15]
	pAkt, pmTOR, pP70S6K downregulation	Melanoma	[72]
	miR-27a decrease		[111]
	mTOR and Notch-1 pathways inhibition	Thymic	[111]
	miR-206 increase		[112]
	PI3K/AKT/mTOR pathway inhibition	NSCLC	[112]
	NEDD4 signaling pathways inhibition	Glioma	[113]

3. Bioavailability of Curcumin and Therapeutic Promises

Although the beneficial effects of curcumin are known, it has not yet been approved as a therapeutic agent due to its low bioavailability [21]. Among factors contributing to this limit, the following can be included: low water solubility, poor absorption, low tissue distribution, high rate of metabolism, inactivity of metabolic products and/or rapid elimination and clearance from the body [115]. Curcumin undergoes extensive phase I and II biotransformation [116]. The primary site of metabolism for curcumin is the liver, together with the intestine and gut microbiota; curcumin double bonds are reduced in enterocytes and hepatocytes by a reductase to produce dihydrocurcumin, tetrahydrocurcumin, hexahydrocurcumin and octahydrocurcumin. Phase II metabolism that occurs in the intestinal and hepatic cytosol is quite active on both curcumin and its phase I metabolites, especially through conjugation reaction with glucuronic acid and sulfate at the phenolic site catalyzed by UGTs and SULTs enzymes, respectively [116].

Over the years, in order to improve curcumin pharmacokinetic profile and cellular uptake, several strategies have been developed. These include curcumin structural derivatives, analogues preparation and novel drug delivery systems that could enhance its solubility and extend its plasma residence time.

3.1. Curcumin Structural Derivatives and Analogues

Structural modifications on the curcumin chemical backbone led to curcumin derivatives and analogues [22–24,117,118]. The derivatives category includes all those compounds that maintain the basic structure of pharmacophore and, specifically, the two phenolic rings and the α, β-unsaturated dichetonic bridge which are the portions responsible for the molecule pharmacological activities (Figure 2) [119]. Curcumin derivatives are generally synthesized by modification of the hydroxyl group of the phenol ring, which can be acylated, alkylated, glycosylated and aminoacylated (Figure 2A). Studies on the kinetic stability of synthetic curcumin derivatives showed that glycosylation of the pharmacophore aromatic ring improves the water solubility of the compound, which increases its kinetic stability and leads to a better therapeutic response. In the II phase of curcumin metabolism, conjugation reactions take place on the hydroxyl groups (4-OH) attached to the phenyl rings of curcumin [116]. Thus, curcumin stability can be increased by masking the 4-OH groups, thereby extending the active molecule retention time in the body. Benzyl rings are also crucial for inhibiting tumor growth; their modification with hydrophobic substituents such as CH3 groups have been linked with an enhancement of the curcumin derivatives antitumor activity [120]. O-methoxy substitution was found to be more effective in suppressing the NF-kB activity, although this modification simultaneously affected curcumin lipophilicity [2]. Methoxy groups could be demethylated to hydroxyl groups. The reactive chain methylene group, responsible for conformational flexibility, is important for its antitumor/anticancer activity but not for redox regulatory or apoptotic activities. This group could be acylated, alkylated or substituted with an arylidene group (Ar-CH), thereby introducing substituents on the C7 chain [121]. The hydrogenation reaction of double bonds and carbonyl groups on the C7 chain allows the simplest derivatives to be obtained (Figure 2B), such as DHC, THC, HHC and OHC [119]. A comparative study on curcumin and its derivatives demonstrated greater antioxidant activity for several hydrogenated curcumin derivatives compared to the original compound [122]. Tetrahydrocurcumin, a non-electrophilic curcumin derivative, showed greater antioxidant activity than DHC and unmodified curcumin, although it failed to suppress STAT3 signaling pathway and to induce apoptosis [123]. This is evidence that the electrophilic nature of curcumin is essential for STAT3 signaling pathway inhibition. Other curcumin derivatives also include those obtained by exploiting the reactivity of the central β-diketone with hydrazine (Figure 2C). Such heterocyclizations reactions lead to a masking and stiffening of the central 1,3-diketone 1,3-ketoenol system [124] and, after evaluation of the antioxidant activity of these compounds, it is possible to assert that several of these azoles are better antioxidants than curcumin [125]. Oxidation and cleavage are further possible modifications to the β-diketone functional group, all suitably operated to improve the characteristics of the original pharmacophore (Figure 2C). Additionally, it is possible to adopt another approach which may help to increase the curcumin bioavailability, such as the formation of metal complexes, or coordination compounds, which are adducts formed by the reaction of Lewis acids and bases [126]. Metal complexes are generally obtained by reacting curcumin, which exploits the chelating capacity of the β-diketone group with a metal salt. The metals most used for this purpose are Boron, Copper, Iron, Gallium, Manganese, Palladium, Vanadium, Zinc and Magnesium. By complexing curcumin with metal ions, such as Zn^{2+}, Cu^{2+}, Mg^{2+}, an increase in water/glycerol solubility (1:1) and fair stability to light and heat were observed [22,126].

Numerous curcumin analogues have been synthesized and tested to study their interaction with known biological targets and improve the pharmacological profile of this natural product [24,127]. Some curcumin analogues are not obtained starting from the original molecule, but they are synthesized following a condensation reaction between arylaldehydes and acetylacetone; through this biosynthetic route, many curcumin analogues have been obtained. The use of acetylacetone derivatives, bearing substituents on the central carbon, leads to analogues with alkyl substituents on the central carbon of the C7 chain [128]. Another strategy concerns the modification of the carbon atom number that makes up the central C7 chain [128]. A greater antitumor activity than curcumin has been observed with the use of a variety of newly synthesized DAP curcumin analogues. These compounds, which possess two aromatic rings (aryl groups) joined by five carbon atoms, were able to suppress cancer growth through modulation of several factors such as NF-kB, MAPK, STAT, AKT-PTEN [127]. DAPs anticancer effects in different cancer cell lines were summarized by Paulraj and co-workers [127].

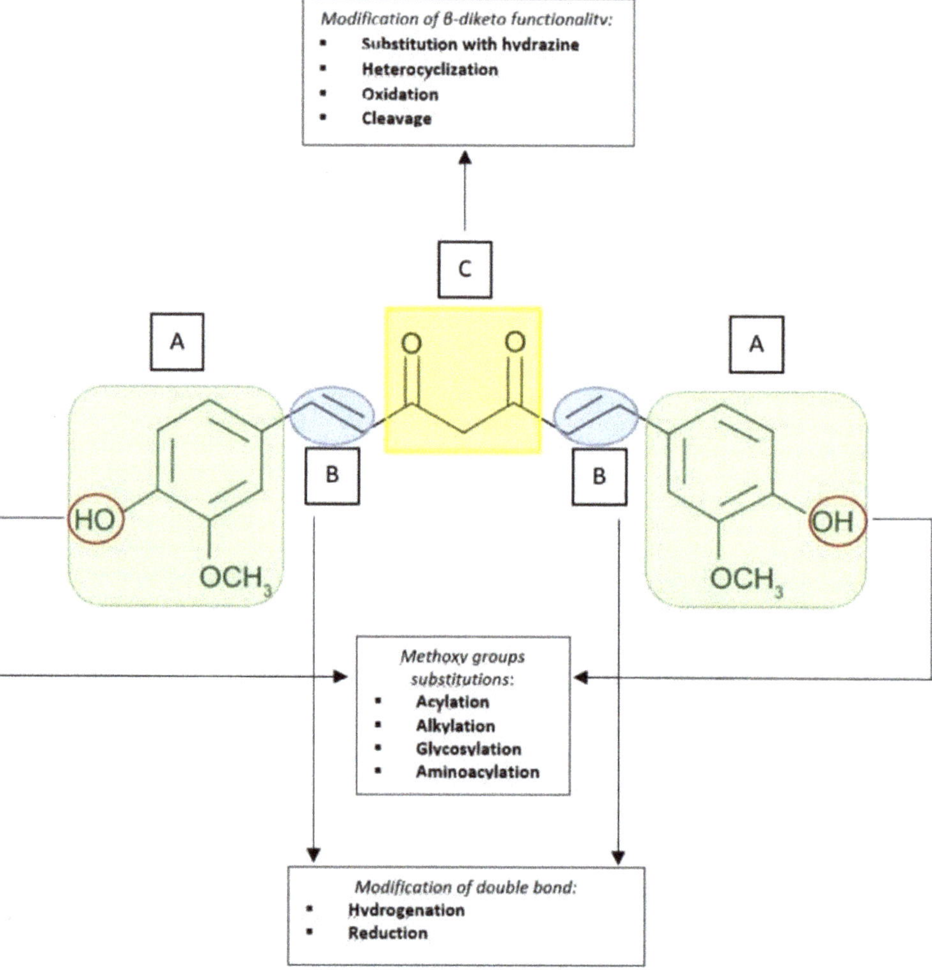

Figure 2. General chemical structure of curcumin derivatives. Curcumin chemical structure include two aromatic rings (**A**) linked to a β-dichetonic group (**C**) through a double bond (**B**).

A particular scientific interest has been shown towards the EF24 analogue, which displayed a better antitumor activity, a lower toxicity in normal cells, a marked increase in bioavailability and a lower metabolic rate compared to the natural compound [23,24]. In vivo studies have shown that, while dietary curcumin is poorly absorbed through the intestinal tract and therefore does not have a therapeutic effect at low doses, on the contrary, EF24 has greater oral bioavailability in mice [129], explaining, to some extent, its improved in vivo activity compared to curcumin. Several studies indicated that EF24 reduces cancer cell growth by inducing cell cycle arrest followed by caspase-mediated apoptosis [130,131]. These actions occur by modulating multiple pathways that determining the inhibition of NF-kB [132] and HIF-1α activity [133] and regulating reactive oxygen species (ROS) [131,134]. NF-kB signaling suppression has been found to be a fundamental aspect for its anticancer activity, since NF-kB is a transcription factor involved in the regulation of genes that monitor cell proliferation, differentiation, cell cycle control and metastasis [135]. According to a discovery by Yin et al. [136], EF24 is able to inhibit the catalytic activity of the IKK protein complex, which blocks the phosphorylation of IkB and causes its degradation while preventing the nuclear translocation of the p65 subunit. EF24 conferred radiation-induced cell death mainly by inhibiting radiation-induced NF-kB signaling in MCF-7 cells [137]. Similar effects were also observed in human neuroblastoma cells [138]. EF24 also regulates HIF-1α expression, which is closely associated with the outcome of chemotherapy in cancer treatment. EF24 overcomes sorafenib resistance through VHL tumor suppressor-dependent HIF-1α degradation and NF-kB inactivation in hepatocellular carcinoma cells [139]. Another notable EF24 antitumor mechanism is ROS production regulation. Tan et al. showed that EF24 inhibited ROS generation and activated ARE-dependent gene transcription in platinum-sensitive (IGROV1) and platinum-resistant (SK-OV-3) human ovarian cancer cells [134]. On the contrary, the ability of EF24 to increase ROS production and then induce apoptosis in cancer cells via a redox-dependent mechanism was found in breast MDA-MB-231, prostate DU-145 [140], gastric SGC-7901 and BGC-823 [141], and colon HCT-116 and SW-620 human cancer cells [131], suggesting that the EF24 role in ROS induction may be cell type dependent. EF24 ability to interfere with the targeted inhibition of antiapoptotic proteins belonging to the Bcl-2 family is being exploited in clinical settings for the treatment of age-related diseases. Although this mechanism has not been fully elucidated, the function of EF24 and other curcumin analogues as senolytic agents is well demonstrated, as they are able to selectively kill the senescent cells (cells that are no longer able to replicate) that accumulate in various organs and tissues as a result of the progress of the damage that occurs with aging. EF24 exerts its senolytic effect by inducing apoptotic death in target cells via proteasome-mediated downregulation from Bcl-2 family proteins, which represents a protection factor for senescent cells, as they are resistant to the induction of apoptosis as a result of the expression of these proteins [23].

Selvendiran et al. [142] demonstrated the ability of curcumin analog HO-3867 to reduce in vitro and in vivo ovarian cancer growth. In particular, this compound enhanced the therapeutic potential of cisplatin in A2780R drug-resistant ovarian cancer cells. The results confirmed that the co-administration of HO-3867 with cisplatin resulted in greater inhibitory effects than cisplatin alone, with cell cycle arrest and apoptotic mechanism being significantly induced by targeting the STAT3 pathway in both in vitro cells and in vivo xenograft tumors [142].

A recent study investigated the antitumor properties of MS13, another diarylpentanoid curcumin analog, on primary (SW480) and metastatic (SW620) human colon cancer cells [143]. MS13 was more cytotoxic in a dose-dependent manner and had a higher growth inhibitory effect towards SW480 and SW620 cells compared to curcumin. Several factors may explain its increased cytotoxicity activity; these include removal of β-diketone, the reduction in the electron donation capacity of OH at position 4′ [144] or the 3′,4′-dimethoxy or 3′-methoxy-4′-hydroxy substituents on the phenyl rings [145].

Recently, Shen et al. [146] tested the efficacy of the B14 analog in MCF-7 and MDA-MB-2310 breast cancer cells. The results indicated that B14 was more potent than curcumin in inhibiting cell viability, colony formation, migration and invasion. Particularly, this analog, functioning simultaneously in multiple pathways, displayed a selective antitumor activity on MCF-7 and MDA-MB-231 cells, but not on MCF-10A breast epithelial cells. Furthermore, in tumor-bearing mice, analog B14 significantly reduced tumor growth and inhibited cell proliferation and angiogenesis. Additionally, pharmacokinetic tests revealed that B14 was more stable and bioavailable than curcumin in vivo [146].

3.2. Curcumin Delivery Systems

As already stated, despite curcumin's remarkable beneficial biological effects, it showed low water solubility in acid and neutral conditions, chemical instability in neutral and alkaline environments, and rapid enzymatic metabolism, which limit its bioavailability. Curcumin bioavailability can be enhanced by:

(a). Delaying metabolism through its entrapment within the hydrophobic phases that isolate it from aqueous phase or cell membranes enzymes;

(b). Improving its bioaccessibility through an increase in the quantity that is solubilized inside the mixed micelles present in the small intestine; this can be achieved by inserting surfactants, phospholipids, fatty acids or monoglycerides into the curcumin-loaded carrier particles;

(c). Promoting its absorption by loading curcumin into particles carrier that contain substances able to increase epithelium cell membranes permeability or block efflux transporters [147]. Therefore, in order to ameliorate curcumin's pharmacokinetic characteristics, various methodological approaches have been attempted, such as polymeric approaches, magnetic approaches, solid lipid nanoparticles, liposomes, phytosomes, micelles, β-cyclodextrins and solid dispersions [21,25–30,148–150] (Figure 3) [151–154]. In addition to these approaches, curcumin conjugation with substances, such as piperine, which is able to inhibit its metabolism [27,155], has emerged as a prominent solution to increase curcumin serum concentration.

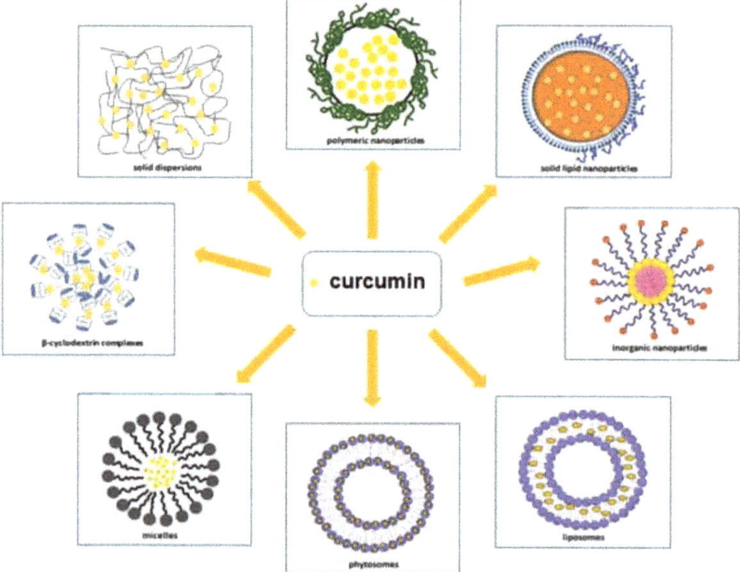

Figure 3. Principal delivery systems to enhance curcumin oral bioavailability. In the figure the polymeric nanoparticles, inorganic nanoparticles and micelles images were adapted from Praditya et al. [151], the phytosomes and liposomes images were adapted from Yang et al. [152], the solid lipid na-noparticles image was adapted from Li et al. [153] and the β cyclodextrin complexes image was adapted from Yallapu et al. [154].

3.2.1. Nanoparticles

Nanotechnology is a fast-developing field that is attracting an increasing amount of attention in the fields of drug delivery and cancer therapy, which provides an important route to develop the aqueous formulations of hydrophobic drugs [156]. Nanoparticles, which are 1000 times smaller than the human cell average, possess unique physical, chemical and biological properties that can be useful for both controlled and targeted drug delivery, and for improving the pharmacokinetics and solubility of drugs [156]. It is important to note that the particle sizes of the neosynthesized carriers may influence the therapeutic effects and drug biocompatibility, as well as the chemical-physical characteristics of the devices used [156]. Various types of nanoparticle, such as polymeric, solid lipid and inorganic nanoparticles [156], are widely used to enhance the therapeutic applications of curcumin. Using nanoparticle formulations, it is possible to increase the curcumin water solubility, ensure its intracellular delivery [157], improve the efficacy and limit the toxicity in the settings of cancer therapy, as well as inducing the chemo and radio sensitization of cancer cells [158].

Polymeric Nanoparticles

Polymeric nanoparticles (NPs), which possess the advantage of being small and biocompatible, are prepared using either natural or synthetic biodegradable polymers such as silk fibroin, chitosan, PEG, PLA), PGA, PCL and PLGA [156,157,159] (see Table 2 for a summary of the data).

Formulations of nanoparticles based on polymer PLGA have been shown to be effective for enhancing curcumin's therapeutic effects against cervical cancer [160]. Specifically, the results provided show that nano-curcumin effectively inhibited the growth of Caski and SiHa cervical cancer cells, arrested the cell cycle in the G1-S transition phase and induced apoptosis. Both cell lines revealed a significantly more evident inhibitory effect of curcumin/nano-curcumin on cell proliferation at higher concentrations (20 and 25 μM) and nano-curcumin was found to be more effective than free curcumin at reducing the clonogenic capacity of cervical cancer cells. Furthermore, treatment with nano-curcumin caused a marked decrease in miRNA-21 levels, an onco-miRNA associated with chemoresistance, in vitro and in vivo models, and improved expression of miRNA-214 tumor suppressor [160]. In another study, the anticancer potential of curcumin-encapsulated PLGA nano-formulation was investigated [161]. This formulation showed a superior cell uptake, retention and release in A2780CP highly metastatic ovarian and MDA-MB-231 breast cancer cells [161]. Additionally, it has a greater antiproliferative effect than free curcumin, as demonstrated by the IC50 values; the IC50 of nano-curcumin is 13.9 μM and 9.1 μM in A2780CP and MDA-MB-231 tumor cells, respectively, while the free curcumin showed higher IC50 values (15.2 μM in A2780CP and 16.4 μM in MDA-MB-231 cells compared to nano-curcumin in metastatic tumor cells [161]. The curcumin PLGA nanoparticles effect on cellular viability have been evaluated in LNCaP, PC3 and DU145 cancer and PWR1E non-tumorigenic prostate cell lines [162]. The results showed that the IC50 of this formulation ranged from 20 μM to 22.5 μM, while that of free curcumin ranged from 32 μM to 34 μM across all tumor cell lines, representing an almost 35% reduction in the IC50 value with curcumin-loaded nanoparticles. The evaluation of molecular mechanism of curcumin PLGA nanoparticles displayed that they were able to arrest cell cycle and induce apoptosis by interfering with the NF-kB activity [162]. To increase the anticancer efficiency of curcumin, Bisht et al. designed curcumin polymeric nanoparticles using the micellar aggregates of cross-linked and random copolymers of NIPAAM, with VP and PEG-A [163]. The authors revealed that these NPs were heavily absorbed and were able to block the clonogenicity of the MiaPaca pancreatic cancer cell lines compared to untreated cells or cells exposed to empty polymeric nanoparticles [163]. The superior anticancer effects of curcumin-loaded nanoparticles prepared with amphilic methoxy poly(ethylene glycol)-polycaprolactone (mPEG–PCL) block copolymers compared to free curcumin was confirmed in a human lung adenocarcinoma A549 transplanted mice model [164]. Moreover, curcumin NPs showed little toxicity to normal tissues including bone marrow, liver and kidney at a therapeutic dose. Curcumin silk fibroin nanoparticles (CUR-SF NPs) provided a more stable release and generated much more emphasized anticancer effects than the results obtained with the curcumin free in HCT116 colon cancer cells [165]. In this study, the methods underlying how the controlled release of CUR-SF NPs is able to improve the curcumin cellular uptake in tumor cells were elucidated. Interestingly, the anticancer effect of CUR-SF NPs that involved cell-cycle arrest in the G0/G1 and G2/M phases and apoptosis induction in cancer cells, was improved, while the side effect on normal human colon mucosal epithelial cells was reduced [165]. In another work, curcumin- and piperine-loaded zein-chitosan nanoparticles were characterized by investigating their shapes, morphologies, particle sizes and cell cytotoxicity [166]. The results showed that zein-chitosan nanoparticles loaded with curcumin and piperine have an average size of about 500 nm and high encapsulation efficiencies for curcumin (89%) and piperine (87%). Furthermore, this formulation showed good cytotoxic effects on SH-SY5Y neuroblastoma cell line [166]. A recent study confirmed the enhanced solubility and bioavailability of curcumin within chitosan nanoparticles [167]. This formulation exhibited a sustained release of drug from NPs and a four-fold higher cytotoxic activity on HeLa cervical cancer cells; in fact, DNA damage, cell cycle blockage and elevated ROS levels confirmed the anticancer activity following apoptotic pathways caused by this formulation.

Solid Lipid Nanoparticles

Solid lipid nanoparticles (SLNs) are colloidal lipid carriers of size between 50 and 1000 nm and composed by biodegradable physiological lipids [168]. Unlike liposomes, they are rigid particles used both for hydrophobic drugs loading and for their controlled and targeted delivery to the reticuloendothelial system. SLNs possess the advantages of high drug load capacity, good stability, excellent biocompatibility and increased bioavailability [169]. In SLN preparation, lipids are used with a low melting point and solids at room or body temperature and surfactants through various methods including HPH [170]. Among the solid lipids used in SLN preparation there are monostearin, glyceryl monostearate, precyrene ATO 5 (mono, di, triglycerides of C16-C18 fatty acids), compritol ATO 888, stearic acid and glyceryl trioleate which can improve the curcumin chemical stability [168]. Surfactants such as poloxamer 188, Tween 80 and DDAB, are able to reduce interfacial tension between lipid hydrophobic surface and aqueous environment by acting as surface stabilizers [168]. Several studies have evaluated not only curcumin SLNs physico-chemical properties, stability, bioenhancement, bioavailability, but also cellular uptake in vitro and antitumor efficacy in vitro and in vivo [171–174] (see Table 2 for a summary of the data). Curcumin SLNs using tristearin and PEGylated surfactants have recently been prepared [174]. The study showed that SLN-loaded curcumin with long PEGylates showed increased absorption and long-term stability after oral administration in rats. Furthermore, the bioavailability of curcumin was also 12 times higher in SLNs formulated with long PEGylates than in those formulated with shorter PEGylates [174]. In another study, it has been demonstrated that the presence of sodium caseinate (NaCas) and sodium caseinate-lactose (NaCas-Lac) conjugates as bioemulsifiers to stabilize curcumin SLNs provided a steric hindrance, allowing dispersibility and greater curcumin stability at pH acid. In addition, this formulation displayed a better antioxidant activity compared to that of free curcumin [175]. However, the SLNs use as oral delivery system is limited by drug burst release from SLNs in acid environment. In order to inhibit curcumin rapid release in acid conditions and improve curcumin bioavailability, Baek et al. prepared curcumin SLNs coated with N-carboxymethyl chitosan [171]. The results showed that this formulation showed a prolonged release in simulated intestinal fluid and greater absorption and oral bioavailability compared to free curcumin. Furthermore, curcumin exhibited a strong cytotoxicity compared to curcumin solution in MCF-7 breast cancer cells [171]. Using glyceryl monostearate and poloxamer 188 surfactant were developed curcumin SLNs and evaluated its efficiency in MDA-MB-231 breast cancer cells [176]. The results confirmed that curcumin was stably encapsulated in the lipid matrix and its solubility and release were improved compared to the free curcumin solution. Moreover, curcumin SLNs exhibited higher cellular uptake and higher cytotoxicity by apoptosis induction compared to the free drug [176]. Guorgui et al. demonstrated that plasma levels of curcumin encapsulated in SLNs and in d-α-Tocopheryl polyethylene glycol 1000 succinate-stabilized curcumin (TPGS-CUR) increased when administrated in mice [172]. Additionally, these formulations reduced growth of Hodgkin's lymphoma xenograft models by 50.5% and 43.0%, respectively, compared to free curcumin [172]. Similarly, inhibitory effects were observed in Hodgkin lymphoma L-540 cells, as proven by curcumin SLNs ability to reduce the expression of XIAP and Mcl-1, proteins involved in cell proliferation and apoptosis, and of cytokines IL-6 and TNF-α as well as to enhance the growth inhibitory effect of bleomycin, doxorubicin and vinblastine chemotherapeutic drugs [172].

Inorganic Nanoparticles

Recently, inorganic nanoparticles have received considerable attention, particularly in the oncology field for diagnostic and therapeutic applications [177,178]. They have specific physical size-dependent properties, such as the contrasting effect and magnetism as well as good microbial resistance and excellent storage capacity [179].

The in vitro and in vivo therapeutic efficacy of curcumin-loaded magnetic nanoparticles (MNPs) was evaluated in several tumors including pancreatic cancer [180] (see Table 2 for a summary of the data). Specifically, HPAF-II and Panc-1 human pancreatic cancer cells exhibited efficient internalization of this formulation in a dose-dependent manner. Moreover, they inhibited both in vitro HPAF-II and Panc-1 cell proliferation and in vivo tumor growth in an HPAF-II xenograft mouse model. The growth-inhibitory effect of MNPs-CUR formulation correlated with the suppression of PCNA, Bcl-xL, induced Mcl-1, MUC1, collagen I, and enhanced membrane β-catenin expression. Interestingly, MNPs-CUR formulation improved serum bioavailability of curcumin in mice up to 2.5-fold as compared with free curcumin [180]. Folic-acid-tagged aminated-starch-/ZnO-coated iron oxide nanoparticles were prepared for the targeted delivery of curcumin. Particularly, the authors provided evidence that ZnO-incorporated aminated-starch-coated iron oxide nanoparticles showed a significant controlled release of curcumin in vitro and reduced HepG2 liver and MCF-7 breast cancer cell viability without toxic effects on human lymphocytes [181]. In addition, folic acid taging to the nanoparticles led to both an increase in cellular uptake and in ROS generation [181]. The curcumin pluronic stabilized Fe3O4 magnetic nanoparticles (CUR-PSMNPs) showed aqueous colloidal stability, biocompatibility as well high loading affinity for curcumin and better curcumin release in acidic conditions [182]. Additionally, cell viability studies confirmed that curcumin and CUR–PSMNPs significantly reduced MCF-7 cell proliferation with IC50 values of 25.1 and 18.4 μM, respectively. The higher toxicity of CUR–PSMNPs were due to its significant greater cellular uptake respect to pure drug [182]. PEGylated curcumin was used as the surface modification of magnetic nanoparticles (MNP@PEG-CUR) in order to simultaneously ameliorate magnetic targeting characteristic of nanoparticles and PEG conjugated drug. The results indicated that MNP@PEG-CUR showed higher drug release in acidic conditions, biocompatibility and low cytotoxicity at physiological pH [183]. Gangwar et al. prepared curcumin conjugated silica coated nanoparticles and observed good stability in aqueous medium, sustained drug release and greater anticancer properties. Particularly, cytotoxicity analysis revealed that conjugate was more toxic on HeLa cervical cancer cells compared to normal fibroblasts [184]. In another work, Manju et al. reported the synthesis of water-soluble HA-CUR@AuNPs attached to curcumin and explored their targeted delivery onto cancer cells (C6 glyoma, HeLa cervical and Caco-2 colon cancer cells). The results indicated that this formulation displayed good aqueous solubility, superior cellular uptake and anticancer effects respect to free curcumin. The authors highlighted how the greater targeting efficacy via hyaluronic acid- and folate receptor-mediated endocytosis could increase the overall intracellular accumulation of HA-CUR@AuNPs [185].

3.2.2. Liposomes

Liposomes are self-assembled spherical vesicles of different sizes ranging from 20 nm to several microns in diameter and composed of one or more bilayers surrounding aqueous unit [186]. Liposomes that can carry both hydrophilic and hydrophobic molecules are excellent drug delivery systems characterized by high biocompatibility and biodegradability, high stability, low toxicity, better solubility, targeting for specific cells, controlled distribution, flexibility and ease of preparation [186]. However, in some cases, they can undergo rapid elimination from the bloodstream, physical and chemical instability, aggregation, fusion, degradation, hydrolysis and phospholipid oxidation [187]. The liposomal systems are designed with the aim of allowing curcumin to be distributed over aqueous medium and increasing its bioavailability and therapeutic effect [188] (see Table 2 for a summary of the data). The preparation method used affects the nature of the formed liposomes, as well as curcumin-carrying capacity. One study compared the characteristics of curcumin-loaded liposomes prepared using three different methods such as thin film, ethanol injection and pH-based methods [189]. The results indicated that both liposomes initial diameters and encapsulation efficiency decreased with the following trend, respectively: thin film (453 nm; 78%)> pH-driven (217 nm; 66%)> injection of ethanol (115 nm; 39%). Furthermore, it has also been shown that the source of phospholipids can also affect the physicochemical characteristics of curcumin-loaded liposomes. Liposomes produced from milk fat globule membranes (MFGMs) offered better protection than those of soy lecithin [190]. However, the instability of conventional liposomes under the physiological conditions found in the gastrointestinal tract hindered clinical applications. Indeed, coated liposomes with thiolated chitosan or other agents are made in order to improve liposome stability and then curcumin bioavailability [187,191]. It has been observed that liposome formulations improved curcumin aqueous solubility and bioavailability by 700-fold and to 8–20-fold, respectively, in tumor-bearing mice [192]. Moreover, antiproliferative activity of DMPC liposomal curcumin was evaluated in LNCaP and C4-2B prostate cancer cell lines [193]. The results showed that liposomal curcumin caused a greater inhibition of cell proliferation than free curcumin and even at doses 10 times lower [193]. Dhule et al. showed that the 2-hydroxypropyl-γ-cyclodextrin/curcumin-liposome complex was able to inhibit KHOS osteosarcoma cancer cell proliferation by inducing apoptosis in vitro as well as reduce the growth of a xenograft osteosarcoma model in vivo [194]. Furthermore, liposomes encapsulating doxorubicin and curcumin, reduced C26 colon cancer cell proliferation to a greater extent than those loaded with doxorubicin alone [195]. A liposomal curcumin formulation also resulted in the dose-dependent inhibition of the proliferation and motility and apoptosis induction in Ishikawa and HEC-1 endometrial cancer cells. These inhibitory effects occurred probably through negative regulation of the NF-kB pathway [196]. Moreover, a recent study demonstrated that small curcumin molecules encapsulated in liposome nanocarriers significantly increased the blue-light-emitting diode (BLED)-induced photodynamic therapy (BLED-PDT) effect, by increasing intracellular ROS levels and apoptosis in human lung A549 cancer cells [197].

3.2.3. Phytosomes

Curcumin is also often complexed with different types of phospholipids such as phosphatidylcholine, or other lipid carriers, in order to create phytosomal formulations. The use of phosphatidylcholine in the molecule complexation is useful in allowing the efficient transport across cell membranes and then absorption. In the phytosome, it is possible to observe an interaction between the active substance (i.e., curcumin) and the polar head of the phospholipid; this bond permits curcumin to become an integral part of the membrane [198]. Phytosomes application for the curcumin delivery has the advantage of improved its absorption and bioavailability as well as enhanced its therapeutic benefits [199–201] (see Table 2 for a summary of the data). The relative absorption of standardized curcuminoid mixture (curcumin, demethoxycurcumin, and bisdemethoxycurcumin) and its corresponding lecithin formulation (Meriva) was investigated in a randomized, double-blind, crossover human study. The results indicated that total curcuminoid absorption was about 29-fold higher for Meriva compared to unformulated curcuminoid mixture. Therefore, the improved curcuminoids absorption could explain the Meriva clinical efficacy at significantly lower doses than unformulated curcuminoids [202]. A recent study, which a CPC was prepared and solidified with Soluplus [polyvinyl caprolactam-polyvinyl acetate polyethylene glycol graft copolymer] carrier, demonstrated that solydifing process enhanced CPC flowability and dissolution rate as well as curcumin oral bioavailabilty in rats [203].

3.2.4. Micelles

Micelles consist of self-assembled aggregates of surfactants or block copolymers ranging in size from 10 to 100 nm [204]; they can be formed by dissolution, dialysis, emulsion, solvent evaporation and lyophilization [204]. Micelles, given their small molecular size that ensure an effective macromolecules transport across cell membrane, are prepared to solubilize curcumin [205]. Micelles, are systems capable to improve the gastro-intestinal absorption of curcumin, increasing plasma levels and decreasing the kinetics of elimination, with a consequent enhancement in bioavailability and a hepato-protective effect likened to free curcumin [205] (see Table 2 for a summary of the data). In addition, an increase in water solubility facilitates the development of a stable and homogenous solution for intravenous applications. Additionally, the nanoscale of micelles and presentation of the hydrophilic stabilizing interface can prolong their circulation time in vivo and enhance the cellular uptake [205]. Letchford et al. formulated curcumin micelles that contained copolymers of deblock MePEG-b-PCL that achieved an approximately 13×10^5-fold increase in water solubility [206]. Similarly, a previous study confirmed a 60-fold increase in the biological half-life of curcumin polymeric micellar compared to the free form when administered orally in rat models [207]. Schiborr et al. also observed that the use of curcumin micelles in volunteers determined a significant increase in drug plasma concentration formulated in micelles compared to that of native curcuminoids [208]. The uptake and transepithelial transport of native curcumin and micellar formulation (Sol-CUR) have been evaluated using a Caco-2 cell model [209]. The results showed that Sol-CUR caused a 185-fold increase in AUC compared to native curcumin. This curcumin oral bioavailability improvement was dependent on Sol-CUR's higher intestinal absorption and distribution [209]. Recently, pharmacokinetic studies confirmed a significant enhancement of aqueous solubility, as well as stability, dissolution, and permeability of curcumin formulated in micelles compared to free drug [210]. A study by Chen et al. focused on the synthesis, physico-chemical characterization and in vitro evaluation of curcumin loaded in the super hydrophilic zwitterionic polymers PSBMA micelles. The results demonstrated that this innovative formulation exhibited greater stability, cellular uptake and tumor cytotoxicity compared to native curcumin [211]. Interestingly, various studies using curcumin polymer micelles may provide an indication of very promising therapeutic potential in cancer treatment. Curcumin delivery from nano-sized polymer micelles formulated with MPEG-P [CL-co-PDO] were evaluated in PC-3 human prostate cancer cell line. The results indicated that mixed micelle copolymers possessed a higher encapsulation efficiency (>95%), a prolonged drug release profile, and a dose-dependent cytotoxicity effect on tumor cells in contrast to that of free curcumin [212]. Patil et al. prepared CUR-MM of Pluronic F-127 (PF127) and Gelucire® 44/14 (GL44) in order to enhance its oral bioavailability and cytotoxicity in A549 human lung cancer cell line [213]. The in vitro dissolution profile of CUR-MMs revealed a controlled curcumin release. Furthermore, this formulation showed a significant improvement in in vitro cytotoxic activity and in vivo oral bioavailability of curcumin by approximately 3 and 55 times, respectively, compared to curcumin alone. These effects could be attributed to solubilization of hydrophobic curcumin into micelle core as well as to the ability of PF127 and GL44 to inhibit P-gp mediated efflux. In another work, it has been demonstrated that CUR-MPP-TPGS-MMs showed small size, high drug-loading and sustained release profile.

Furthermore, they improved the intestinal absorption of curcumin after oral administration and then oral bioavailability in rats [214].

3.2.5. Curcumin/β-Cyclodextrin and Solid Dispersions Formulations

Among various strategies performed to improve oral curcumin solubility there are cyclodextrin (CD) inclusion complexes and solid dispersions (SDs) [215] (see Table 2 for a summary of the data). Cyclodextrins are bucket-shaped oligosaccharides, consisting of six (a-), seven (b-) or eight (g-) units of D-glucopyranose linked through an a-1,4-glycosidic bond to form macrocycles and are widely known as solubilizing and stabilizing agents [216]. β-CD, with its hydrophilic outer surface and hydrophobic inner cavity, has become a benchmark for nanotechnology research due to its modern lumen size, high drug load, low cost of production, best stability of the lipophilic drug, easy modification by molecules and good biocompatibility characteristics [217]. Mangolim et al. demonstrated that curcumin-β-CD complex exhibited a sunlight stability 18% higher and a 31-fold increased solubility compared to the pure drug [218]. Liquid-type β-cyclodextrins curcumin delivery carrier is one of the most widely used types in the food industry [219]. Particularly, the delivery carrier of hydroxypropyl β-CD curcumin prepared by saturated aqueous solution increased curcumin solubility in water by 276 times [220] and oral bioavailability by 3 times [221]. In addition, the solid granule-based β-cyclodextrins curcumin delivery vector had very good storage stability and increased its bioavailability significantly [222]. The formulations of CD combined with curcumin (CUR-CD) have been shown to have enhanced the antiproliferative, anti-inflammatory and anticancer effects of the nutraceutical agent [22]. The therapeutic effect of CUR-CD encapsulated into positively charged biodegradable chitosan (CUR-CD-CS) nanoparticles was investigated on the SCC25 skin cancer cell line. The study demonstrated that CD presence not only increased curcumin solubility and cellular absorption, but also promoted cell cytotoxicity [223]. Yallapu et al. showed that β-CD-curcumin inclusion complex (CD30) possessed an improved uptake and a greater potent therapeutic efficacy in DU145 prostate cancer cells compared to the free drug [154]. Similarly, the formulation of β-cyclodextrin-curcumin complex (CD15) enhanced curcumin delivery and improved its therapeutic efficacy compared to free curcumin in in vitro A549, NCI-H446 and NCI-H520 human lung carcinoma cell lines and in vivo mouse hepatoma H22 xenograft models. Particularly, through regulation of MAPK/NF-kB pathway, CD15 upregulated p53/p21 pathway, downregulated Cyclin E-CDK2 combination and increased Bax/caspase 3 expression to induce cellar apoptosis and G1-phase arrest [224]. A study demonstrated that in HeLa cervical cancer cells, curcumin encapsulated in crosslinked β-CD nanoparticles acted extremely rapidly on cell metabolism resulting in a significant cancer cell growth inhibition [225]. Recently, a nano-drug system namely FA-CUR-NPs consisting of FA, β-CD, εCL and curcumin was performed to improve curcumin delivery in cervical cancer tissues which overexpress FRs and to achieve controllable release in vitro and in vivo. In this system, FA binding to FRs was used as a targeting molecule, β-CD modified by ε-CL as delivery carrier and for controlling drug release and curcumin as a model drug to limit multidrug resistance after administration [226]. The study demonstrated that in vitro curcumin release rate from FA-CUR-NPs under tumor microenvironment conditions (pH 6.4), was three times faster than that under systemic circulation conditions (pH 7.4). Additionally, in vitro cytotoxicity was proportional to cellular uptake efficiency and the in vivo marked accumulation in tumor site was responsible of greater antitumor activity. These findings indicated FA-CUR-NPs could represent a promising approach for improving cancer therapy through active targeting and controllable release [226]. Moreover, in another work the use of hydroxypropyl-β-CD as a carrier-solubilizer improved solubility of the curcumin–piperine system, its permeability through biological membranes (gastrointestinal tract, blood–brain barrier), the antioxidant and antimicrobial activities and as well as the enzymatic inhibition against acetylcholinesterase and butyrylcholinesterase [227].

The SDs technology transforms crystalline materials into amorphous materials: an active substance is incorporated into a carrier, which is generally selected from suitable polymers [228]. The active ingredient, originally crystalline, is usually transformed into amorphous form in the dispersion and it is for these reasons that SD is a consolidated method applied to improve the solubility and bioavailability of drugs with poor water solubility such as curcumin [228]. The major outcomes of this technique include prolonged survival, antitumor and antimetastasis, anti-inflammatory, antibacterial activity, enhanced stability and bioavailability [229,230]. Seo et al. prepared solid dispersions of curcumin with Solutol® HS15 SD showing greater solubility and bioavailability compared to free curcumin [231]. Similarly, curcumin SDs with cellulose acetate and mannitol showed a better solubility in water and an improvement in oral bioavailability of about seven times compared to curcumin [232]. In another work, Texeira et al. demonstrated how solid dispersions of Gelucire®50/13-Aerosil curcumin possess stability over time up to 9 months and an improvement in water solubility and in dissolution rate of about 3600- and 7.3-fold, respectively. Furthermore, the major curcumin gastrointestinal absorption resulted in a systemic bioavailability increase and greater anti-inflammatory activity in rats [229]. Antioxidant and antigenotoxic effects of curcumin formulated in SDs compared to unmodified drug were evaluated in Wistar rats. The results showed that curcumin SDs, even if they did not alter the antigenotoxic effects observed with free curcumin, displayed a better water solubility, with a maximum of absorption in the gastrointestinal tract [233]. Recently, curcumin SDs were obtained using Poloxamer 407 as the encapsulant agent and their cytotoxic effects were evaluated against tumoral (breast, lung, cervical and hepatocellular carcinoma) and PLP2 non-tumoral cells. These formulations that were readily dispersible in water had cytotoxicity against all the tested tumor cell lines without toxic effects for non-tumor cells [234].

3.2.6. Curcumin Conjugates Formulations

Combinations of curcumin with other compounds increase its solubility and cellular absorption, extend the residence in plasma improving the pharmacokinetic profile and then oral bioavailability (see Table 2 for a summary of the data). Curcumin conjugation by covalent bond with piperine, an alkaloid of black pepper and known as an inhibitor of hepatic and intestinal glucuronidation [235], enhanced curcumin serum level and a significantly reduced elimination half-life and clearance, producing 154% and 2000% increase in bioavailability in rats and humans, respectively [236]. Another study also showed that piperine (20 mg/kg orally) when administered with curcumin (2 g/kg orally) increased the bioavailability of the latter up to 20 times more in epileptic rats [237]. Similarly, both intestinal absorption and curcumin bioavailability has been found to be improved following concomitantly oral intake with piperine in rats [238]. In a clinical study, the curcumin-lecithin-piperine formulation, namely BCM-95CG (Biocurcumax®), administrated to healthy subjects aged 28–50 years significantly improved curcumin bioavailability and showed a better pharmacokinetic profile than pure curcumin [239]. An in vitro study conducted by Sing et al. demonstrated that conjugates of curcumin, piperic acid and glycine, prepared by esterifying the sites of metabolic conjugation 4 and 4' phenolic hydroxyls, triggered a mitochondrion-dependent apoptosis in MCF-7 and MDA-MB-231 breast cancer cell lines; however, the authors highlighted that the conjugation did not affect the efficacy of parent molecule, while these compounds could work in vivo as prodrugs with improved pharmacokinetic profile. This suggested other in vivo studies were needed to elucidate the therapeutic potential of curcumin conjugates in breast cancer [240].

Table 2. Strategies to enhance bioavailability and/or antitumoral effects of curcumin.

Composition	Outcomes	References
Polymeric Nanoparticles (NPs)		
PLGA + CUR	Superior anticancer effects respect to free curcumin in cervical cancer cells	[160]
PLGA + CUR	Superior cellular uptake and anticancer effects respect to free curcumin in ovarian and breast cancer cells	[161]
PLGA + CUR	Superior cellular uptake and anticancer effects respect to free curcumin in prostate cancer cells	[162]
NIPAAM+ VP + PEG-A + CUR	Superior cellular absorbition and anticancer effects respect to free curcumin in pancreatic cancer cells	[163]
mPEG-PCL + CUR	Superior anticancer effects respect to free curcumin in human lung adenocarcinoma cancer cells	[164]
silk fibroin + CUR	Superior cellular uptake and anticancer effects respect to free curcumin in colon cancer cells	[165]
zein-chitosan +CUR	High encapsulation efficiencies for curcumin. Superior anticancer effects respect to free curcumin in neuroblastoma cell line	[166]
chitosan +CUR	Enhanced curcumin solubility and bioavailability. Sustained drug release from NPs and anticancer effects with respect to free curcumin in cervical cancer cells	[167]
Solid Lipid Nanoparticles (SLNs)		
SLNs + N-carboxymethyl chitosan+ CUR	Prolonged release in simulated intestinal fluid, greater absorption and oral bioavailability compared to free curcumin. Superior anticancer effects respect to free curcumin in breast cancer cells	[171]
SLNs + CUR, d-α-Tocopheryl polyethylene glycol 1000 succinate-stabilized curcumin (TPGS) + CUR	Superior curcumin plasma levels in mice. Superior anticancer effects respect to free curcumin in Hodgkin lymphoma cells and in Hodgkin's lymphoma xenograft models	[172]
SLNs + tristearin + PEGylated + CUR	Superior bioavailability, absorption and long-term stability after oral administration in the rats	[174]
SLNs + NaCas + NaCas-Lac + CUR	Superior stability at pH acid and antioxidant activity with respect to free curcumin	[175]
SLNs + glyceryl monostearate + poloxamer 188 + CUR	Superior stability, solubility, cellular uptake, release and anticancer effects respect to free curcumin in breast cancer cells	[176]
Inorganic Nanoparticles		
MNPs + CUR	Superior cellular uptake in vitro. Superior bioavailability in vivo. Superior in vitro and in vivo therapeutic efficacy respect to free curcumin in pancreatic cancer cells and in pancreatic cancer xenografts model	[180]
Folic-acid-tagged aminated-starch-/ZnO-coated iron oxide nanoparticles + CUR	Significant controlled release of curcumin and reduced hepatic and breast cancer cells viability in vitro. Cellular uptake increase in vitro	[181]
PSMNPs + CUR	Aqueous colloidal stability, biocompatibility, high loading affinity for curcumin and better curcumin release in acidic conditions. Superior cellular uptake and anticancer effects respect to free curcumin in breast cancer cells	[182]

Table 2. *Cont.*

Composition	Outcomes	References
Inorganic Nanoparticles		
MNP@PEG + CUR	Higher drug release in acidic conditions, biocompatibility and low cytotoxicity at physiological pH	[183]
Silica + CUR	Good stability in aqueous medium, sustained drug release and greater anticancer properties in cervical cancer cells compared to normal fibroblasts	[184]
HA-CUR@AuNPs	Good aqueous solubility, superior cellular uptake and anticancer effects respect to free curcumin in cervical, glioma and colon cancer cells	[185]
Liposomes		
Liposome + CUR	Improved curcumin aqueous solubility and bioavailability in tumor-bearing mice	[192]
DMPC + CUR	Superior anticancer effects respect to free curcumin in prostate cancer cells	[193]
2-hydroxypropyl-γ-cyclodextrin/ liposome + CUR	Superior in vitro and in vivo anticancer effects respect to free curcumin in osteosarcoma cancer cells	[194]
Liposome + doxorubicin + CUR	Superior anticancer effects respect to those loaded with doxorubicin alone in colon cancer cells	[195]
Liposome + CUR	Superior anticancer effects respect to free curcumin in endometrial cancer cells	[196]
Liposome + CUR + BLED-PDT therapy	Enhancement of BLED-PDT therapy effect by curcumin liposome in lung cancer cells	[197]
Phytosomes		
Curcuminoids + lecithin (Meriva ®)	Improved absorption and clinical efficacy respect to unformulated curcuminoid mixtures	[198]
Soluplus® [polyvinyl caprolactam-polyvinyl acetate polyethylene glycol graft copolymer] + CPC	Improved flowability, dissolution rate and oral bioavailability in rats	[203]
Micelles		
MePEG-b-PCL + CUR	Improved water solubility	[206]
MePEG-b-PCL + CUR	Improved biological half-life with respect to the free curcumin in rat models	[207]
Tween-80 micelles + CUR	Improved drug plasma concentration with respect to free curcumin in volunteers	[208]
Micellar formulation + CUR (Sol-CUR)	Superior uptake, transepithelial transport, distribution and bioavailability in colon cancer cell model	[209]
Micelles + CUR	Enhancement of aqueous solubility, stability, dissolution and permeability of curcumin formulated in micelles compared to free drug	[210]
PSBMA + CUR	Greater stability, cellular uptake and tumor cytotoxicity compared to free curcumin	[211]
MPEG-P [CL-co-PDO] + CUR	Superior encapsulation efficiency, prolonged drug release profile and antitumor effects respect to free curcumin in prostate cancer cells	[212]
Pluronic F-127 + Gelucire® 44/14 micelles + CUR	Controlled curcumin release, superior oral bioavailability in vivo and in vitro antitumor effects in lung cancer cells respect to free curcumin	[213]
CUR-MPP-TPGS-MMs	Small size, high drug-loading and sustained release. Improved intestinal absorption and oral bioavailabil-ity in rats	[214]

Table 2. Cont.

Composition	Outcomes	References
Curcumin/β-Cyclodextrin (β-CD) and Solid Dispersions (SDs)		
β-CD + CUR	Superior sunlight stability and solubility respect to pure colourant	[218]
Liquid-type β-CD + CUR	Improved solubility in water and bioavailability	[220,221]
Solid type β-CD + CUR	Improved storage stability and biovailability	[222]
CUR-CD-CS	Superior solubility, cellular absorption and antitumor effects compared to free curcumin in skin cancer cells	[223]
β-CD + CUR	Improved uptake and therapeutic efficacy in prostate cancer cells	[154]
β-CD + CUR	Improved delivery and therapeutic efficacy compared to free curcumin in both in vitro lung carcinoma cell lines and in vivo mouse hepatoma xenograft models	[224]
β-CD + CUR	Superior anticancer effects respect to free curcumin in cervical cancer cells	[225]
FA-CUR-NPs	Improved drug release rate, cellular uptake efficiency and in vitro and in vivo antitumor activity respect to free curcumin in cervical cancer cells	[226]
Hydroxypropyl-β-CD+ CUR+piperine	Improved solubility of the curcumin–piperine system, its permeability through biological membranes, antioxidant and antimicrobial activities and enzymatic inhibition	[227]
SDs (with Gelucire®50/13-Aerosil) + CUR	Improved stability, water solubility, dissolution rate, bioavailability and anti-inflammatory activity in rats	[229]
Solutol® HS15 SDs + CUR	Improved solubility and bioavailability compared to free curcumin	[231]
SDs (with cellulose acetate and mannitol) + CUR	Improved water solubility and oral bioavailability compared to free curcumin	[232]
SDs + CUR	Superior water solubility and gastrointestinal absorption in rats	[233]
SDs (with Poloxamer 407) + CUR	Improved water dispersibility and cytotoxic effects against breast, lung, cervical and hepatocellular cancer cells	[234]
Curcumin Conjugates Formulations		
Piperine + CUR	Curcumin enhanced serum levels, reduced elimination half-life and clearance, increased bioavailability in rats and humans	[236]
Piperine + CUR	Curcumin increased bioavailability in epileptic rats	[237]
Piperine + CUR	Curcumin enhanced intestinal absorption and bioavailability in rats	[238]
BCM-95CG (Biocurcumax®): piperine + lecithin + CUR	Curcumin improved bioavailability and pharmacokinetic profile respect to free drug in healthy subjects	[239]

4. In Situ Implant Systems

Unlike blood circulation nanosystems that are injected directly into blood vessels, nanoimplant systems have been used as large containers of drugs that must be released directly and exclusively at the application site for the treatment of topical diseases. When loaded into a membrane made with biodegradable nanofibers, curcumin may be helpful in preventing post-operative relapse of solid tumor. Different researchers have attempted to charge curcumin into nanofiber membranes by electrospinning and, by adjusting the curcumin dosage, they have found that it is possible to control the drug's loading capacity [241]. Curcumin-loaded nanofiber membrane showed a good local delivery of hydrophobic drug, sustained release, biocompatibility and biodegradability and strong cytotoxicity against rat glioma 9L cells [242]. Therefore, this formulation could represent an ideal candidate for the prevention of postoperative tumor relapse after excision [243]. Further to the application of nanocrystallized curcumin in cancer therapy, tissue engineering is another biomedical field where curcumin could also have a particular impact on bone formation. The choice of hydrogels as the base material could be appropriate due to their porosity and practical operation in clinical application. The hybrid hydrogel-micelle system was used to control malignant pleural effusion, which reduced the number of pleural tumor foci and prolonged survival time to malignant pleural effusion of carrier mice [244]. Angiogenesis is also effectively inhibited using this approach, and the micelle-hydrogel hybrid has been shown to be an exceptional transport system.

5. Clinical Trials with Curcumin

In recent decades, several clinical studies have been performed to evaluate the effect of curcumin in cancer patients. Depending on the tumor type, curcumin can act by causing a decrease in the patients symptoms or, in other cases, it can lead to an improvement in tumor markers and vital parameters [245]. Currently, 71 clinical studies reported in cancer patients treated with curcumin alone or in combination with other compounds were listed in the United States National Library of Medicine (clinicaltrials.gov). Pharmacologically, curcumin is shown to be well-tolerated and relatively safe to use in patients. The clinical trials conducted, thus far, have reported relatively no toxicity [246]. Phase I clinical trial conducted by Cheng et al. showed that oral administration of 8 g/day of curcumin for 3 months is non-toxic to patients with high-risk or pre-malignant lesions [247].

A phase II trial of curcumin conducted in twenty-five patients with advanced pancreatic cancer showed that oral curcumin was tolerated without toxicity at doses of 8 g/d for up to 18 months [248]. Particularly, it has been observed that two of them showed clinical biological activity; one patient reported having a stable disease for up to 18 months and another had a marked tumor regression along with an increase in the serum levels of cytokines (IL-6, IL-8, IL-10, and IL-1 receptor antagonists) [248]. Additionally, curcumin downregulated expression of the NF-kB, COX-2, and phosphorylated signal transducer and activator of transcription 3 in peripheral blood mononuclear cells from patients [248]. In another phase II pilot study, a combination of docetaxel, prednisone and curcumin was well-tolerated in patients with castration-resistant prostate cancer [249]. Dützmann et al. investigated the intratumoral concentrations and clinical tolerance of micellar curcuminoids composed by 57.4 mg curcumin, 11.2 mg demethoxycurcumin, and 1.4 mg bis-demethoxycurcumin administrated three times day for 4 days in glioblastoma patients [250]. The results revealed that oral administration of this formulation generated quantifiable concentrations of total curcuminoids in glioblastomas with likely effects on intratumoral energy metabolism [250]. Furthermore, in patients with orbital pseudotumors, head and neck squamous carcinoma, breast, lung and prostate cancers, curcumin application demonstrated beneficial effects with reductions in tumors volume and tumor markers [245]. Another study carried out on 11 volunteer patients with osteosarcoma aimed to quantitatively evaluate curcumin levels in the bloodstream following the intake of SLNs; this study highlighted an improvement in the bioavailability of this polyphenol, when compared with the results obtained on subjects who received unformulated curcuminoids extract [173]. However, further studies are needed to evaluate both the long-term tolerability after chronic administration and the relationship between plasma curcumin levels and disease markers. In a phase I clinical trial, an oral formulation of curcumin was evaluated in fifteen patients with advanced colorectal cancer refractory to standard chemotherapies. The results showed an absence of toxicity with curcumin, while 2 of the 15 patients showed stable disease after 2 months of curcumin treatment [251]. Carroll et al., in a phase II clinical study conducted on patients with colon cancer lesions, demonstrated that curcumin administration for 30 days determined a significant 40% reduction in aberrant crypt foci (ACF) [252]. These data suggested curcumin use as a cancer prevention agent. However, although curcumin in cancer patients has often improved life quality and reduced tumor markers, it is also true that curcumin has sometimes shown limited effect in some patients with disease advanced stage. Preclinical studies conducted so far confirmed the important role of curcumin and, in particular, the benefits that its synthetic derivatives or nanoformulations could bring to human health.

New efforts are needed to confirm the efficacy of curcumin-based products for the treatment of human diseases and to ensure that the products are non-toxic once introduced into the body. Several ongoing clinical trials should provide a deeper understanding of curcumin-based formulation efficacy and mechanism of action against human diseases.

6. Conclusions

Curcumin is one of the most promising clinical compounds of the last few decades. Its therapeutic benefits have been demonstrated in various chronic diseases and, above all, in various cancers. Precisely in the antitumor field, curcumin is able to modulate the action of growth factors, cytokines, transcription factors and genes that regulate cell proliferation, apoptosis and the metastatic process. Some of curcumin's multiple pharmacological activities have been used experimentally and clinically in both humans and animals. Notable among these are the antioxidant, anti-inflammatory and anticarcinogenic properties, which all seem related. It is encouraging that curcumin is of low toxicity. Most of the data supporting the antitumour activity of curcumin was obtained in vitro. Unfortunately, the challenges of low solubility and rapid elimination and then poor bioavailability, have delayed its adoption as a therapeutic agent. These efforts have led to the development of several curcumin formulations that have been systematically prepared to improve the absorption, water solubility, distribution and, hence, the bioavailability of curcumin. These methods have been promising with respect to preclinical and clinical efficacy and involve the use of both synthetic derivatives and analogues of curcumin as well as formulations of nanoparticles, liposomes, phytosomes, micelles and natural adjuvants such as piperine. All of these approaches generated a significant improvement in the bioavailability, absorption and retention time of curcumin, along with increased delivery to target tissues, as widely reported. However, further investigations are needed to fulfil the significant promise they currently hold and reveal new perspectives for the further enhancement of the therapeutic capacity of this interesting natural molecule. In particular, more epidemiological and clinical trials involving large numbers of subjects should be conducted to support to the obtained in vivo and in vitro results and to confirm curcumin's usefulness in cancer treatment and/or prevention.

Author Contributions: Conceptualization, original draft preparation, writing, search and identification of articles, A.C. and V.P.; screening articles, writing, M.C.N., A.D.L., F.P. (Francesca Prestia); figures preparation, M.C.N., F.P. (Francesca Prestia), F.P. (Francesco Puoci); tables preparation, M.C.N., A.D.L., F.P. (Francesca Prestia) and A.C.; writing and editing, R.S., I.C., F.P. (Francesco Puoci); editing, P.A., D.L.P., L.Z., A.D.L. and A.C.; supervision, A.C. and V.P. All authors have read and agreed to the published version of the manuscript.

Funding: This work was supported by a special award (Department of Excellence, Italian Law 232/2016) from the Italian Ministry of Research and University (MIUR) to the Department of Pharmacy, Health and Nutritional Sciences of the University of Calabria (Italy), and by MIUR ex 60% (VP, AC) and by AIRC (Associazione Italiana per la Ricerca sul Cancro), projects n. IG20122. A.D.L. was supported by a fellowship from PAC (Progetto Strategico Regionale Calabria Alta Formazione) Calabria 2014/2020—Asse Prioritario 12, Linea B, Azione 10.5.12; P.A. was supported by a post-doc fellowship 2021 from Fondazione Umberto Veronesi (FUV).

Data Availability Statement: Not applicable.

Conflicts of Interest: The authors declare no conflict of interest.

Abbreviations

AKT	protein kinase B
AMPK	adenosine monophosphate (AMP)-activated protein kinase
AP-1	activator protein1
APC	adenomatous polyposis coli

ARE	antioxidant response element
AUC	area under curve
AuNPs	gold nanoparticles
Bcl-2	B-cell lymphoma 2
Bcl-xL	B-cell lymphoma-extra large
cAMP	cyclic adenosine monophosphate
CDK2	cyclin-dependent kinase 2
CDK4	cyclin-dependent kinase 4
CD	cyclodextrin
cGMP	cyclic guanosin monophosphate
CNrasGEFs	cyclic nucleotide-ras guanine nucleotide exchange factors
COX2	cyclooxygenase-2
CPC	curcumin phospholipid complex
CUR	curcumin
CUR-CD-CS	curcumin-loaded cyclodextrin chitosan
CUR-MMs	curcumin mixed micelles
CUR-MPP-TPGS-MMs	curcumin-loaded methoxy poly(ethylene glycol)-poly(lactide) (mPEG-PLA)/D-α-tocopherol polyethylene glycol 1000 succinate (TPGS) mixed micelles
CUR-PSMNPs	curcumin pluronic stabilized Fe3O4 magnetic nanoparticles
CUR-SF NPs	curcumin silk fibroin nanoparticles
DAPs	diarylpentanoids
DDAB	dimethyl dioctadecyl ammonium bromide
DFMO	difluoro-methylornithine
DHC	dihydrocurcumin
DMPC	dimyristoyl phosphatidylcholine
DNMT1	DNA methyltransferase 1
E2F4	E2F trascription factor 4
EGFR	epidermal growth factor receptor
Egr-1	early growth response 1
EMT	epithelial mesenchymal transition
ERK	extracellulare signal-regulated kinase
EZH2	enhancer of zeste homolog 2
FA	folic acid
FA-CUR-NPs	folate receptor-targeted β-cyclodextrin nanoparticles
Fas	fas-associated protein with death domain
FAS	fatty acid synthase
FRs	folic receptors
GSH	glutathione
GSK3β	glycogen synthase kinase 3 beta
HA-CUR@AuNPs	hyaluronic acid curcumin gold nanoparticles
HDGF	hepatoma-derived growth factor
HER2	human epidermal growth factor receptor 2
HHC	hexahydrocurcumin
HIF-1α	hypoxia-inducible factor 1-alpha
HPH	high pressure homogenization
IAP-2	inhibitor of apoptosis protein-2
ICAM1	intercellular adhesion molecule 1
IC50	half-maximal inhibitory concentration
IFN-g	interferon gamma
IkB	inhibitor of KB
IKK	IκB kinase
IL-6	interleukin-6
JAK2	janus kinase 2
JNK	c-jun-N-terminal kinase
LPA	lysophosphatidic acid
LRP6	LDL (low density lipoprotein) receptor related protein 6

MAPK	mitogen-activated protein kinase
Mcl-1	myeloid cell leukemia 1
MDM2	mouse double minute 2 homolog
MePEG-b-PCL	methoxy poly (ethylene glycol) -block-polycaprolactone
miRNA	microRNA
MMPs	matrix metalloproteinases
MNPs	magnetic nanoparticles
MPEG-P [CL-co-PDO]	methoxy poly (ethylene glycol) -b-poly (ε-caprolactone-co-dioxanone)
mPEG–PCL	methoxy poly(ethylene glycol)–polycaprolactone
mTOR	mechanistic target of rapamycin
MUC1	mucin 1
NaCas	sodium caseinate
NaCas-Lac	sodium caseinate-lactose
NF-κB	nuclear factor κB
NIPAAM	N-isopropylacrylamide
NPs	nanoparticles
NSCLC	non-small cell lung cancer
NVP	N-vinyl-2-pyrrolidone
ODC	ornithine decarboxylase
OHC	octahydrocurcumin
OPN	osteopontin
PAK1	p21 (RAC1) activated kinase 1
PARP-1	poly (ADP-ribose) polymerase 1
PCL	poly-(ε-caprolactone)
PCNA	proliferating cell nuclear antigen
PDE	cyclic nucleotide phosphodiesterase
PDE1A	phosphodiesterase 1A
PEG	polyethylene glycol
PEG-A	poly(ethylene glycol) monoacrylate
PGA	polyglycolide acid
P-gp	p-glycoprotein 1
PI3K	phosphoinositide 3-kinase
PLA	polylactide
PLGA	polylactic-co-glycolic acid
PSMNPs	pluronic stabilized Fe_3O_4 magnetic nanoparticles
PSBMA	poly (sulphobetaine methacrylate)
RhoA	ras homolog family member A
ROCK1	rho associated coiled-coil containing protein kinase 1
ROS	reactive oxygen species
SCLC	small cell lung cancer
SDs	solid dispersions
SIP1	smad interacting protein 1
SLNs	solid lipid nanoparticles
SMOX	spermine oxidase
STAT3	signal transducer and activator of transcription 3
SULT	sulfotransferase
THC	tetrahydrocurcumin
TNF-α	tumor necrosis factor-a
TPGS	d-α-Tocopheryl polyethylene glycol 1000 succinate
UGT	UDP-glucurosyltransferase
UHRF1	ubiquitin like with PHD and ring finger domain 1
u-PA	urokinase-type plasminogen activator
VEGF	vascular endothelial growth factor
VHL	Von Hippel-Lindau tumor suppressor
VP	N-vinyl-2-pyrrolidone
XIAP	X-linked inhibitor of apoptosis protein
β CD	β- cyclodextrins
εCL	ε-caprolactone

References

1. Prasad, S.; Aggarwal, B.B. Turmeric, the golden spice: From traditional medicine to modern medicine. In *Herbal Medicine: Biomolecular and Clinical Aspects*; Benzie, I.F.F., Wachtel-Galor, S., Eds.; CRC Press: Boca Raton, FL, USA, 2011.
2. Sandur, S.K.; Pandey, M.K.; Sung, B.; Ahn, K.S.; Murakami, A.; Sethi, G.; Limtrakul, P.; Badmaev, V.; Aggarwal, B.B. Curcumin, demethoxycurcumin, bisdemethoxycurcumin, tetrahydrocurcumin and turmerones differentially regulate anti-inflammatory and anti-proliferative responses through a ROS-independent mechanism. *Carcinogenesis* **2007**, *28*, 1765–1773. [CrossRef]
3. Dosoky, N.S.; Satyal, P.; Setzer, W.N. Variations in the volatile compositions of Curcuma species. *Foods* **2019**, *8*, 53. [CrossRef] [PubMed]
4. Kunnumakkara, A.B.; Bordoloi, D.; Padmavathi, G.; Monisha, J.; Roy, N.K.; Prasad, S.; Aggarwal, B.B. Curcumin, the golden nutraceutical: Multitargeting for multiple chronic diseases. *Br. J. Pharm.* **2017**, *174*, 1325–1348. [CrossRef] [PubMed]
5. Patel, S.S.; Acharya, A.; Ray, R.S.; Agrawal, R.; Raghuwanshi, R.; Jain, P. Cellular and molecular mechanisms of curcumin in prevention and treatment of disease. *Crit. Rev. Food Sci. Nutr.* **2020**, *60*, 887–939. [CrossRef] [PubMed]
6. Adamczak, A.; Ozarowski, M.; Karpinski, T.M. Curcumin, a natural antimicrobial agent with strain-specific activity. *Pharmaceuticals* **2020**, *13*, 153. [CrossRef]
7. Jennings, M.R.; Parks, R.J. Curcumin as an antiviral agent. *Viruses* **2020**, *12*, 1242. [CrossRef]
8. Chen, C.; Long, L.; Zhang, F.; Chen, Q.; Chen, C.; Yu, X.; Liu, Q.; Bao, J.; Long, Z. Antifungal activity, main active components and mechanism of Curcuma longa extract against *Fusarium graminearum*. *PLoS ONE* **2018**, *13*, e0194284. [CrossRef]
9. Jagetia, G.C. Antioxidant activity of curcumin protects against the radiation-induced micronuclei formation in cultured human peripheral blood lymphocytes exposed to various doses of gamma-Radiation. *Int. J. Radiat. Biol.* **2021**, *97*, 485–493. [CrossRef]
10. Busari, Z.A.; Dauda, K.A.; Morenikeji, O.A.; Afolayan, F.; Oyeyemi, O.T.; Meena, J.; Sahu, D.; Panda, A.K. Antiplasmodial activity and toxicological assessment of curcumin PLGA-encapsulated nanoparticles. *Front. Pharm.* **2017**, *8*, 622. [CrossRef]
11. Fadus, M.C.; Lau, C.; Bikhchandani, J.; Lynch, H.T. Curcumin: An age-old anti-inflammatory and anti-neoplastic agent. *J. Tradit. Complement. Med.* **2017**, *7*, 339–346. [CrossRef]
12. Zia, A.; Farkhondeh, T.; Pourbagher-Shahri, A.M.; Samarghandian, S. The role of curcumin in aging and senescence: Molecular mechanisms. *Biomed. Pharmacother. Biomed. Pharmacother.* **2021**, *134*, 111119. [CrossRef]
13. Wang, H.; Zhang, K.; Liu, J.; Yang, J.; Tian, Y.; Yang, C.; Li, Y.; Shao, M.; Su, W.; Song, N. Curcumin regulates cancer progression: Focus on ncRNAs and molecular signaling pathways. *Front. Oncol.* **2021**, *11*, 660712. [CrossRef] [PubMed]
14. Kunnumakkara, A.B.; Bordoloi, D.; Harsha, C.; Banik, K.; Gupta, S.C.; Aggarwal, B.B. Curcumin mediates anticancer effects by modulating multiple cell signaling pathways. *Clin. Sci.* **2017**, *131*, 1781–1799. [CrossRef] [PubMed]
15. Kim, M.J.; Park, K.S.; Kim, K.T.; Gil, E.Y. The inhibitory effect of curcumin via fascin suppression through JAK/STAT3 pathway on metastasis and recurrence of ovary cancer cells. *BMC Women's Health* **2020**, *20*, 256. [CrossRef]
16. Liu, J.L.; Pan, Y.Y.; Chen, O.; Luan, Y.; Xue, X.; Zhao, J.J.; Liu, L.; Jia, H.Y. Curcumin inhibits MCF-7 cells by modulating the NF-kappaB signaling pathway. *Oncol. Lett.* **2017**, *14*, 5581–5584. [CrossRef] [PubMed]
17. Ghasemi, F.; Shafiee, M.; Banikazemi, Z.; Pourhanifeh, M.H.; Khanbabaei, H.; Shamshirian, A.; Amiri Moghadam, S.; ArefNezhad, R.; Sahebkar, A.; Avan, A.; et al. Curcumin inhibits NF-kB and Wnt/beta-catenin pathways in cervical cancer cells. *Pathol. Res. Pract.* **2019**, *215*, 152556. [CrossRef] [PubMed]
18. Kuttikrishnan, S.; Siveen, K.S.; Prabhu, K.S.; Khan, A.Q.; Ahmed, E.I.; Akhtar, S.; Ali, T.A.; Merhi, M.; Dermime, S.; Steinhoff, M.; et al. Curcumin induces apoptotic cell death via inhibition of PI3-kinase/AKT pathway in B-precursor acute lymphoblastic leukemia. *Front. Oncol.* **2019**, *9*, 484. [CrossRef]
19. Mortezaee, K.; Salehi, E.; Mirtavoos-Mahyari, H.; Motevaseli, E.; Najafi, M.; Farhood, B.; Rosengren, R.J.; Sahebkar, A. Mechanisms of apoptosis modulation by curcumin: Implications for cancer therapy. *J. Cell. Physiol.* **2019**, *234*, 12537–12550. [CrossRef]
20. Wahlstrom, B.; Blennow, G. A study on the fate of curcumin in the rat. *Acta Pharmacol. Toxicol.* **1978**, *43*, 86–92. [CrossRef]
21. Anand, P.; Kunnumakkara, A.B.; Newman, R.A.; Aggarwal, B.B. Bioavailability of curcumin: Problems and promises. *Mol. Pharm.* **2007**, *4*, 807–818. [CrossRef]
22. Tomeh, M.A.; Hadianamrei, R.; Zhao, X. A review of curcumin and its derivatives as anticancer agents. *Int. J. Mol. Sci.* **2019**, *20*, 1033. [CrossRef] [PubMed]
23. Li, W.; He, Y.; Zhang, R.; Zheng, G.; Zhou, D. The curcumin analog EF24 is a novel senolytic agent. *Aging* **2019**, *11*, 771–782. [CrossRef] [PubMed]
24. He, Y.; Li, W.; Hu, G.; Sun, H.; Kong, Q. Bioactivities of EF24, a novel curcumin analog: A review. *Front. Oncol.* **2018**, *8*, 614. [CrossRef] [PubMed]
25. Jamwal, R. Bioavailable curcumin formulations: A review of pharmacokinetic studies in healthy volunteers. *J. Integr. Med.* **2018**, *16*, 367–374. [CrossRef] [PubMed]
26. Adiwidjaja, J.; McLachlan, A.J.; Boddy, A.V. Curcumin as a clinically-promising anti-cancer agent: Pharmacokinetics and drug interactions. *Expert Opin. Drug Metab. Toxicol.* **2017**, *13*, 953–972. [CrossRef] [PubMed]
27. Olotu, F.; Agoni, C.; Soremekun, O.; Soliman, M.E.S. An update on the pharmacological usage of curcumin: Has it failed in the drug discovery pipeline? *Cell Biochem. Biophys.* **2020**, *78*, 267–289. [CrossRef]
28. Yallapu, M.M.; Nagesh, P.K.; Jaggi, M.; Chauhan, S.C. Therapeutic applications of curcumin nanoformulations. *AAPS J.* **2015**, *17*, 1341–1356. [CrossRef]

29. Karthikeyan, A.; Senthil, N.; Min, T. Nanocurcumin: A promising candidate for therapeutic applications. *Front. Pharmacol.* **2020**, *11*, 487. [CrossRef]

30. Ma, Z.; Wang, N.; He, H.; Tang, X. Pharmaceutical strategies of improving oral systemic bioavailability of curcumin for clinical application. *J. Control. Release Off. J. Control. Release Soc.* **2019**, *316*, 359–380. [CrossRef]

31. Kotecha, R.; Takami, A.; Espinoza, J.L. Dietary phytochemicals and cancer chemoprevention: A review of the clinical evidence. *Oncotarget* **2016**, *7*, 52517–52529. [CrossRef]

32. Rahmani, A.H.; Al Zohairy, M.A.; Aly, S.M.; Khan, M.A. Curcumin: A potential candidate in prevention of cancer via modulation of molecular pathways. *BioMed Res. Int.* **2014**, *2014*, 761608. [CrossRef] [PubMed]

33. Song, X.; Zhang, M.; Dai, E.; Luo, Y. Molecular targets of curcumin in breast cancer (Review). *Mol. Med. Rep.* **2019**, *19*, 23–29. [CrossRef] [PubMed]

34. Mukhopadhyay, A.; Banerjee, S.; Stafford, L.J.; Xia, C.; Liu, M.; Aggarwal, B.B. Curcumin-induced suppression of cell proliferation correlates with down-regulation of cyclin D1 expression and CDK4-mediated retinoblastoma protein phosphorylation. *Oncogene* **2002**, *21*, 8852–8861. [CrossRef] [PubMed]

35. Zhou, Q.M.; Wang, X.F.; Liu, X.J.; Zhang, H.; Lu, Y.Y.; Su, S.B. Curcumin enhanced antiproliferative effect of mitomycin C in human breast cancer MCF-7 cells in vitro and in vivo. *Acta Pharmacol. Sin.* **2011**, *32*, 1402–1410. [CrossRef]

36. Aggarwal, B.B.; Banerjee, S.; Bharadwaj, U.; Sung, B.; Shishodia, S.; Sethi, G. Curcumin induces the degradation of cyclin E expression through ubiquitin-dependent pathway and up-regulates cyclin-dependent kinase inhibitors p21 and p27 in multiple human tumor cell lines. *Biochem. Pharmacol.* **2007**, *73*, 1024–1032. [CrossRef]

37. Prasad, C.P.; Rath, G.; Mathur, S.; Bhatnagar, D.; Ralhan, R. Potent growth suppressive activity of curcumin in human breast cancer cells: Modulation of Wnt/beta-catenin signaling. *Chem. Biol. Interact.* **2009**, *181*, 263–271. [CrossRef]

38. Colomer, C.; Marruecos, L.; Vert, A.; Bigas, A.; Espinosa, L. NF-kappaB members left home: NF-kappaB-independent roles in cancer. *Biomedicines* **2017**, *5*, 26. [CrossRef]

39. Zhou, Q.; Ye, M.; Lu, Y.; Zhang, H.; Chen, Q.; Huang, S.; Su, S. Curcumin improves the tumoricidal effect of mitomycin C by suppressing ABCG2 expression in stem cell-like breast cancer cells. *PLoS ONE* **2015**, *10*, e0136694. [CrossRef]

40. Zhou, X.; Jiao, D.; Dou, M.; Zhang, W.; Lv, L.; Chen, J.; Li, L.; Wang, L.; Han, X. Curcumin inhibits the growth of triple-negative breast cancer cells by silencing EZH2 and restoring DLC1 expression. *J. Cell. Mol. Med.* **2020**, *24*, 10648–10662. [CrossRef]

41. Chase, A.; Cross, N.C. Aberrations of EZH2 in cancer. *Clin. Cancer Res.* **2011**, *17*, 2613–2618. [CrossRef] [PubMed]

42. Ren, G.; Li, G. Tumor suppressor gene DLC1: Its modifications, interactive molecules, and potential prospects for clinical cancer application. *Int. J. Biol. Macromol.* **2021**, *182*, 264–275. [CrossRef] [PubMed]

43. Pricci, M.; Girardi, B.; Giorgio, F.; Losurdo, G.; Ierardi, E.; Di Leo, A. Curcumin and colorectal cancer: From basic to clinical evidences. *Int. J. Mol. Sci.* **2020**, *21*, 2364. [CrossRef]

44. Calibasi-Kocal, G.; Pakdemirli, A.; Bayrak, S.; Ozupek, N.M.; Sever, T.; Basbinar, Y.; Ellidokuz, H.; Yigitbasi, T. Curcumin effects on cell proliferation, angiogenesis and metastasis in colorectal cancer. *J. BUON Off. J. Balk. Union Oncol.* **2019**, *24*, 1482–1487.

45. Lim, T.G.; Lee, S.Y.; Huang, Z.; Lim, D.Y.; Chen, H.; Jung, S.K.; Bode, A.M.; Lee, K.W.; Dong, Z. Curcumin suppresses proliferation of colon cancer cells by targeting CDK2. *Cancer Prev. Res.* **2014**, *7*, 466–474. [CrossRef]

46. Kim, K.C.; Lee, C. Curcumin induces downregulation of E2F4 expression and apoptotic cell death in HCT116 human colon cancer cells; Involvement of reactive oxygen species. *Korean J. Physiol. Pharmacol. Off. J. Korean Physiol. Soc. Korean Soc. Pharmacol.* **2010**, *14*, 391–397. [CrossRef]

47. Watson, J.L.; Hill, R.; Yaffe, P.B.; Greenshields, A.; Walsh, M.; Lee, P.W.; Giacomantonio, C.A.; Hoskin, D.W. Curcumin causes superoxide anion production and p53-independent apoptosis in human colon cancer cells. *Cancer Lett.* **2010**, *297*, 1–8. [CrossRef] [PubMed]

48. Mosieniak, G.; Adamowicz, M.; Alster, O.; Jaskowiak, H.; Szczepankiewicz, A.A.; Wilczynski, G.M.; Ciechomska, I.A.; Sikora, E. Curcumin induces permanent growth arrest of human colon cancer cells: Link between senescence and autophagy. *Mech. Ageing Dev.* **2012**, *133*, 444–455. [CrossRef] [PubMed]

49. Zhang, L.; Yang, G.; Zhang, R.; Dong, L.; Chen, H.; Bo, J.; Xue, W.; Huang, Y. Curcumin inhibits cell proliferation and motility via suppression of TROP2 in bladder cancer cells. *Int. J. Oncol.* **2018**, *53*, 515–526. [CrossRef]

50. Shi, J.; Zhang, X.; Shi, T.; Li, H. Antitumor effects of curcumin in human bladder cancer in vitro. *Oncol. Lett.* **2017**, *14*, 1157–1161. [CrossRef] [PubMed]

51. Walker, B.C.; Mittal, S. Antitumor activity of curcumin in glioblastoma. *Int. J. Mol. Sci.* **2020**, *21*, 9435. [CrossRef]

52. Choi, B.H.; Kim, C.G.; Bae, Y.S.; Lim, Y.; Lee, Y.H.; Shin, S.Y. P21(Waf1/Cip1) expression by curcumin in U-87MG human glioma cells: Role of early growth response-1 expression. *Cancer Res.* **2008**, *68*, 1369–1377. [CrossRef] [PubMed]

53. Cheng, C.; Jiao, J.T.; Qian, Y.; Guo, X.Y.; Huang, J.; Dai, M.C.; Zhang, L.; Ding, X.P.; Zong, D.; Shao, J.F. Curcumin induces G2/M arrest and triggers apoptosis via FoxO1 signaling in U87 human glioma cells. *Mol. Med. Rep.* **2016**, *13*, 3763–3770. [CrossRef] [PubMed]

54. Wang, L.; Ye, X.; Cai, X.; Su, J.; Ma, R.; Yin, X.; Zhou, X.; Li, H.; Wang, Z. Curcumin suppresses cell growth and invasion and induces apoptosis by down-regulation of Skp2 pathway in glioma cells. *Oncotarget* **2015**, *6*, 18027–18037. [CrossRef] [PubMed]

55. Wu, J.; Su, H.K.; Yu, Z.H.; Xi, S.Y.; Guo, C.C.; Hu, Z.Y.; Qu, Y.; Cai, H.P.; Zhao, Y.Y.; Zhao, H.F.; et al. Skp2 modulates proliferation, senescence and tumorigenesis of glioma. *Cancer Cell Int.* **2020**, *20*, 71. [CrossRef]

56. Luo, Q.; Luo, H.; Fu, H.; Huang, H.; Huang, H.; Luo, K.; Li, C.; Hu, R.; Zheng, C.; Lan, C.; et al. Curcumin suppresses invasiveness and migration of human glioma cells in vitro by inhibiting HDGF/beta-catenin complex. *Nan Fang Yi Ke Da Xue Xue Bao J. South. Med. Univ.* **2019**, *39*, 911–916. [CrossRef]

57. Fratantonio, D.; Molonia, M.S.; Bashllari, R.; Muscara, C.; Ferlazzo, G.; Costa, G.; Saija, A.; Cimino, F.; Speciale, A. Curcumin potentiates the antitumor activity of Paclitaxel in rat glioma C6 cells. *Phytomed. Int. J. Phytother. Phytopharm.* **2019**, *55*, 23–30. [CrossRef]

58. Seyithanoglu, M.H.; Abdallah, A.; Kitis, S.; Guler, E.M.; Kocyigit, A.; Dundar, T.T.; Gundag Papaker, M. Investigation of cytotoxic, genotoxic, and apoptotic effects of curcumin on glioma cells. *Cell. Mol. Biol.* **2019**, *65*, 101–108. [CrossRef]

59. Murray-Stewart, T.; Dunworth, M.; Lui, Y.; Giardiello, F.M.; Woster, P.M.; Casero, R.A., Jr. Curcumin mediates polyamine metabolism and sensitizes gastrointestinal cancer cells to antitumor polyamine-targeted therapies. *PLoS ONE* **2018**, *13*, e0202677. [CrossRef]

60. Cai, X.Z.; Wang, J.; Li, X.D.; Wang, G.L.; Liu, F.N.; Cheng, M.S.; Li, F. Curcumin suppresses proliferation and invasion in human gastric cancer cells by downregulation of PAK1 activity and cyclin D1 expression. *Cancer Biol. Ther.* **2009**, *8*, 1360–1368. [CrossRef]

61. Zheng, R.; Deng, Q.; Liu, Y.; Zhao, P. Curcumin inhibits gastric carcinoma cell growth and induces apoptosis by suppressing the Wnt/beta-catenin signaling pathway. *Med. Sci. Monit. Int. Med. J. Exp. Clin. Res.* **2017**, *23*, 163–171. [CrossRef]

62. Fu, H.; Wang, C.; Yang, D.; Wei, Z.; Xu, J.; Hu, Z.; Zhang, Y.; Wang, W.; Yan, R.; Cai, Q. Curcumin regulates proliferation, autophagy, and apoptosis in gastric cancer cells by affecting PI3K and P53 signaling. *J. Cell. Physiol.* **2018**, *233*, 4634–4642. [CrossRef]

63. Liu, G.; Xiang, T.; Wu, Q.F.; Wang, W.X. Curcumin suppresses the proliferation of gastric cancer cells by downregulating H19. *Oncol. Lett.* **2016**, *12*, 5156–5162. [CrossRef]

64. Ghafouri-Fard, S.; Esmaeili, M.; Taheri, M. H19 lncRNA: Roles in tumorigenesis. *Biomed. Pharmacother.* **2020**, *123*, 109774. [CrossRef]

65. Yang, F.; Bi, J.; Xue, X.; Zheng, L.; Zhi, K.; Hua, J.; Fang, G. Up-regulated long non-coding RNA H19 contributes to proliferation of gastric cancer cells. *FEBS J.* **2012**, *279*, 3159–3165. [CrossRef]

66. Yoshimura, H.; Matsuda, Y.; Yamamoto, M.; Kamiya, S.; Ishiwata, T. Expression and role of long non-coding RNA H19 in carcinogenesis. *Front. Biosci.* **2018**, *23*, 614–625. [CrossRef]

67. Sun, C.; Zhang, S.; Liu, C.; Liu, X. Curcumin promoted miR-34a expression and suppressed proliferation of gastric cancer cells. *Cancer Biother. Radiopharm.* **2019**, *34*, 634–641. [CrossRef]

68. Sun, Q.; Zhang, W.; Guo, Y.; Li, Z.; Chen, X.; Wang, Y.; Du, Y.; Zang, W.; Zhao, G. Curcumin inhibits cell growth and induces cell apoptosis through upregulation of miR-33b in gastric cancer. *Tumor Biol. J. Int. Soc. Oncodev. Biol. Med.* **2016**, *37*, 13177–13184. [CrossRef] [PubMed]

69. Nabavi, S.M.; Russo, G.L.; Tedesco, I.; Daglia, M.; Orhan, I.E.; Nabavi, S.F.; Bishayee, A.; Nagulapalli Venkata, K.C.; Abdollahi, M.; Hajheydari, Z. Curcumin and melanoma: From chemistry to medicine. *Nutr. Cancer* **2018**, *70*, 164–175. [CrossRef] [PubMed]

70. Zheng, M.; Ekmekcioglu, S.; Walch, E.T.; Tang, C.H.; Grimm, E.A. Inhibition of nuclear factor-kappaB and nitric oxide by curcumin induces G2/M cell cycle arrest and apoptosis in human melanoma cells. *Melanoma Res.* **2004**, *14*, 165–171. [CrossRef] [PubMed]

71. Abusnina, A.; Keravis, T.; Yougbare, I.; Bronner, C.; Lugnier, C. Anti-proliferative effect of curcumin on melanoma cells is mediated by PDE1A inhibition that regulates the epigenetic integrator UHRF1. *Mol. Nutr. Food Res.* **2011**, *55*, 1677–1689. [CrossRef]

72. Zhao, G.; Han, X.; Zheng, S.; Li, Z.; Sha, Y.; Ni, J.; Sun, Z.; Qiao, S.; Song, Z. Curcumin induces autophagy, inhibits proliferation and invasion by downregulating AKT/mTOR signaling pathway in human melanoma cells. *Oncol. Rep.* **2016**, *35*, 1065–1074. [CrossRef] [PubMed]

73. Siwak, D.R.; Shishodia, S.; Aggarwal, B.B.; Kurzrock, R. Curcumin-induced antiproliferative and proapoptotic effects in melanoma cells are associated with suppression of I kappa B kinase and nuclear factor KB activity and are independent of the B-Raf/mitogen-activated/extracellular signal-regulated protein kinase pathway and the Akt pathway. *Cancer* **2005**, *104*, 879–890. [CrossRef] [PubMed]

74. Cianfruglia, L.; Minnelli, C.; Laudadio, E.; Scire, A.; Armeni, T. Side effects of curcumin: Epigenetic and antiproliferative implications for normal dermal fibroblast and breast cancer cells. *Antioxidants* **2019**, *8*, 382. [CrossRef] [PubMed]

75. Talib, W.H.; Al-Hadid, S.A.; Ali, M.B.W.; Al-Yasari, I.H.; Ali, M.R.A. Role of curcumin in regulating p53 in breast cancer: An overview of the mechanism of action. *Breast Cancer* **2018**, *10*, 207–217. [CrossRef]

76. Bae, Y.H.; Ryu, J.H.; Park, H.J.; Kim, K.R.; Wee, H.J.; Lee, O.H.; Jang, H.O.; Bae, M.K.; Kim, K.W.; Bae, S.K. Mutant p53-Notch1 signaling axis is involved in curcumin-induced apoptosis of breast cancer cells. *Korean J. Physiol. Pharmacol. Off. J. Korean Physiol. Soc. Korean Soc. Pharmacol.* **2013**, *17*, 291–297. [CrossRef]

77. Patel, P.B.; Thakkar, V.R.; Patel, J.S. Cellular effect of curcumin and citral combination on breast cancer cells: Induction of apoptosis and cell cycle arrest. *J. Breast Cancer* **2015**, *18*, 225–234. [CrossRef]

78. Moghtaderi, H.; Sepehri, H.; Attari, F. Combination of arabinogalactan and curcumin induces apoptosis in breast cancer cells in vitro and inhibits tumor growth via overexpression of p53 level in vivo. *Biomed. Pharmacother.* **2017**, *88*, 582–594. [CrossRef]

79. Chiu, T.L.; Su, C.C. Curcumin inhibits proliferation and migration by increasing the Bax to Bcl-2 ratio and decreasing NF-kappaBp65 expression in breast cancer MDA-MB-231 cells. *Int. J. Mol. Med.* **2009**, *23*, 469–475. [CrossRef]

80. Fan, H.; Liang, Y.; Jiang, B.; Li, X.; Xun, H.; Sun, J.; He, W.; Lau, H.T.; Ma, X. Curcumin inhibits intracellular fatty acid synthase and induces apoptosis in human breast cancer MDA-MB-231 cells. *Oncol. Rep.* **2016**, *35*, 2651–2656. [CrossRef]

81. Zhang, Y.P.; Li, Y.Q.; Lv, Y.T.; Wang, J.M. Effect of curcumin on the proliferation, apoptosis, migration, and invasion of human melanoma A375 cells. *Genet. Mol. Res.* **2015**, *14*, 1056–1067. [CrossRef]

82. Bush, J.A.; Cheung, K.J., Jr.; Li, G. Curcumin induces apoptosis in human melanoma cells through a Fas receptor/caspase-8 pathway independent of p53. *Exp. Cell Res.* **2001**, *271*, 305–314. [CrossRef] [PubMed]

83. Watson, J.L.; Greenshields, A.; Hill, R.; Hilchie, A.; Lee, P.W.; Giacomantonio, C.A.; Hoskin, D.W. Curcumin-induced apoptosis in ovarian carcinoma cells is p53-independent and involves p38 mitogen-activated protein kinase activation and downregulation of Bcl-2 and survivin expression and Akt signaling. *Mol. Carcinog.* **2010**, *49*, 13–24. [CrossRef] [PubMed]

84. Li, W.; Wang, Y.; Song, Y.; Xu, L.; Zhao, J.; Fang, B. A preliminary study of the effect of curcumin on the expression of p53 protein in a human multiple myeloma cell line. *Oncol. Lett.* **2015**, *9*, 1719–1724. [CrossRef] [PubMed]

85. Lee, Y.K.; Park, S.Y.; Kim, Y.M.; Park, O.J. Regulatory effect of the AMPK-COX-2 signaling pathway in curcumin-induced apoptosis in HT-29 colon cancer cells. *Ann. N. Y. Acad. Sci.* **2009**, *1171*, 489–494. [CrossRef]

86. Sandur, S.K.; Deorukhkar, A.; Pandey, M.K.; Pabon, A.M.; Shentu, S.; Guha, S.; Aggarwal, B.B.; Krishnan, S. Curcumin modulates the radiosensitivity of colorectal cancer cells by suppressing constitutive and inducible NF-kappaB activity. *Int. J. Radiat. Oncol. Biol. Phys.* **2009**, *75*, 534–542. [CrossRef]

87. Narayan, S. Curcumin, a multi-functional chemopreventive agent, blocks growth of colon cancer cells by targeting beta-catenin-mediated transactivation and cell-cell adhesion pathways. *J. Mol. Histol.* **2004**, *35*, 301–307. [CrossRef]

88. Jung, E.M.; Lim, J.H.; Lee, T.J.; Park, J.W.; Choi, K.S.; Kwon, T.K. Curcumin sensitizes tumor necrosis factor-related apoptosis-inducing ligand (TRAIL)-induced apoptosis through reactive oxygen species-mediated upregulation of death receptor 5 (DR5). *Carcinogenesis* **2005**, *26*, 1905–1913. [CrossRef]

89. Cao, A.; Li, Q.; Yin, P.; Dong, Y.; Shi, H.; Wang, L.; Ji, G.; Xie, J.; Wu, D. Curcumin induces apoptosis in human gastric carcinoma AGS cells and colon carcinoma HT-29 cells through mitochondrial dysfunction and endoplasmic reticulum stress. *Apoptosis* **2013**, *18*, 1391–1402. [CrossRef]

90. Moragoda, L.; Jaszewski, R.; Majumdar, A.P. Curcumin induced modulation of cell cycle and apoptosis in gastric and colon cancer cells. *Anticancer Res.* **2001**, *21*, 873–878.

91. Su, C.C.; Lin, J.G.; Li, T.M.; Chung, J.G.; Yang, J.S.; Ip, S.W.; Lin, W.C.; Chen, G.W. Curcumin-induced apoptosis of human colon cancer colo 205 cells through the production of ROS, Ca^{2+} and the activation of caspase-3. *Anticancer Res.* **2006**, *26*, 4379–4389.

92. Klinger, N.V.; Mittal, S. Therapeutic potential of curcumin for the treatment of brain tumors. *Oxidative Med. Cell. Longev.* **2016**, *2016*, 9324085. [CrossRef]

93. Lee, D.S.; Lee, M.K.; Kim, J.H. Curcumin induces cell cycle arrest and apoptosis in human osteosarcoma (HOS) cells. *Anticancer Res.* **2009**, *29*, 5039–5044. [PubMed]

94. Yang, J.; Cao, Y.; Sun, J.; Zhang, Y. Curcumin reduces the expression of Bcl-2 by upregulating miR-15a and miR-16 in MCF-7 cells. *Med. Oncol.* **2010**, *27*, 1114–1118. [CrossRef]

95. Zhang, J.; Du, Y.; Wu, C.; Ren, X.; Ti, X.; Shi, J.; Zhao, F.; Yin, H. Curcumin promotes apoptosis in human lung adenocarcinoma cells through miR-186* signaling pathway. *Oncol. Rep.* **2010**, *24*, 1217–1223. [CrossRef]

96. Sohn, E.J.; Bak, K.M.; Nam, Y.K.; Park, H.T. Upregulation of microRNA 344a-3p is involved in curcumin induced apoptosis in RT4 schwannoma cells. *Cancer Cell Int.* **2018**, *18*, 199. [CrossRef]

97. Gao, D.; Mittal, V.; Ban, Y.; Lourenco, A.R.; Yomtoubian, S.; Lee, S. Metastatic tumor cells—Genotypes and phenotypes. *Front. Biol.* **2018**, *13*, 277–286. [CrossRef] [PubMed]

98. Gallardo, M.; Kemmerling, U.; Aguayo, F.; Bleak, T.C.; Munoz, J.P.; Calaf, G.M. Curcumin rescues breast cells from epithelialmesenchymal transition and invasion induced by antimiR34a. *Int. J. Oncol.* **2020**, *56*, 480–493. [CrossRef] [PubMed]

99. Pires, B.R.; Mencalha, A.L.; Ferreira, G.M.; de Souza, W.F.; Morgado-Diaz, J.A.; Maia, A.M.; Correa, S.; Abdelhay, E.S. NF-kappaB is involved in the regulation of EMT genes in breast cancer cells. *PLoS ONE* **2017**, *12*, e0169622. [CrossRef] [PubMed]

100. Zong, H.; Wang, F.; Fan, Q.X.; Wang, L.X. Curcumin inhibits metastatic progression of breast cancer cell through suppression of urokinase-type plasminogen activator by NF-kappa B signaling pathways. *Mol. Biol. Rep.* **2012**, *39*, 4803–4808. [CrossRef]

101. Bachmeier, B.E.; Mohrenz, I.V.; Mirisola, V.; Schleicher, E.; Romeo, F.; Hohneke, C.; Jochum, M.; Nerlich, A.G.; Pfeffer, U. Curcumin downregulates the inflammatory cytokines CXCL1 and -2 in breast cancer cells via NFkappaB. *Carcinogenesis* **2008**, *29*, 779–789. [CrossRef]

102. Irani, S. Emerging insights into the biology of metastasis: A review article. *Iran. J. Basic Med. Sci.* **2019**, *22*, 833–847. [CrossRef] [PubMed]

103. Sun, K.; Duan, X.; Cai, H.; Liu, X.; Yang, Y.; Li, M.; Zhang, X.; Wang, J. Curcumin inhibits LPA-induced invasion by attenuating RhoA/ROCK/MMPs pathway in MCF7 breast cancer cells. *Clin. Exp. Med.* **2016**, *16*, 37–47. [CrossRef] [PubMed]

104. Li, Y.; Sun, W.; Han, N.; Zou, Y.; Yin, D. Curcumin inhibits proliferation, migration, invasion and promotes apoptosis of retinoblastoma cell lines through modulation of miR-99a and JAK/STAT pathway. *BMC Cancer* **2018**, *18*, 1230. [CrossRef]

105. Zaman, S.; Jadid, H.; Denson, A.C.; Gray, J.E. Targeting Trop-2 in solid tumors: Future prospects. *Onco Targets Ther.* **2019**, *12*, 1781–1790. [CrossRef] [PubMed]

106. Cheng, T.S.; Chen, W.C.; Lin, Y.Y.; Tsai, C.H.; Liao, C.I.; Shyu, H.Y.; Ko, C.J.; Tzeng, S.F.; Huang, C.Y.; Yang, P.C.; et al. Curcumin-targeting pericellular serine protease matriptase role in suppression of prostate cancer cell invasion, tumor growth, and metastasis. *Cancer Prev. Res.* **2013**, *6*, 495–505. [CrossRef] [PubMed]

107. Xiang, L.; He, B.; Liu, Q.; Hu, D.; Liao, W.; Li, R.; Peng, X.; Wang, Q.; Zhao, G. Antitumor effects of curcumin on the proliferation, migration and apoptosis of human colorectal carcinoma HCT116 cells. *Oncol. Rep.* **2020**, *44*, 1997–2008. [CrossRef]

108. Philip, S.; Kundu, G.C. Osteopontin induces nuclear factor kappa B-mediated promatrix metalloproteinase-2 activation through I kappa B alpha /IKK signaling pathways, and curcumin (diferulolylmethane) down-regulates these pathways. *J. Biol. Chem.* **2003**, *278*, 14487–14497. [CrossRef]

109. Jin, W. Role of JAK/STAT3 signaling in the regulation of metastasis, the transition of cancer stem cells, and chemoresistance of cancer by epithelial-mesenchymal transition. *Cells* **2020**, *9*, 217. [CrossRef]

110. Yang, C.L.; Liu, Y.Y.; Ma, Y.G.; Xue, Y.X.; Liu, D.G.; Ren, Y.; Liu, X.B.; Li, Y.; Li, Z. Curcumin blocks small cell lung cancer cells migration, invasion, angiogenesis, cell cycle and neoplasia through Janus kinase-STAT3 signalling pathway. *PLoS ONE* **2012**, *7*, e37960. [CrossRef]

111. Han, Z.; Zhang, J.; Zhang, K.; Zhao, Y. Curcumin inhibits cell viability, migration, and invasion of thymic carcinoma cells via downregulation of microRNA-27a. *Phytother. Res. PTR* **2020**, *34*, 1629–1637. [CrossRef]

112. Wang, N.; Feng, T.; Liu, X.; Liu, Q. Curcumin inhibits migration and invasion of non-small cell lung cancer cells through up-regulation of miR-206 and suppression of PI3K/AKT/mTOR signaling pathway. *Acta Pharm.* **2020**, *70*, 399–409. [CrossRef] [PubMed]

113. Wang, X.; Deng, J.; Yuan, J.; Tang, X.; Wang, Y.; Chen, H.; Liu, Y.; Zhou, L. Curcumin exerts its tumor suppressive function via inhibition of NEDD4 oncoprotein in glioma cancer cells. *Int. J. Oncol.* **2017**, *51*, 467–477. [CrossRef]

114. Zhang, H.; Nie, W.; Zhang, X.; Zhang, G.; Li, Z.; Wu, H.; Shi, Q.; Chen, Y.; Ding, Z.; Zhou, X.; et al. NEDD4-1 regulates migration and invasion of glioma cells through CNrasGEF ubiquitination in vitro. *PLoS ONE* **2013**, *8*, e82789. [CrossRef]

115. Ravindranath, V.; Chandrasekhara, N. Absorption and tissue distribution of curcumin in rats. *Toxicology* **1980**, *16*, 259–265. [CrossRef]

116. Pan, M.H.; Huang, T.M.; Lin, J.K. Biotransformation of curcumin through reduction and glucuronidation in mice. *Drug Metab. Dispos.* **1999**, *27*, 486–494.

117. Basile, V.; Ferrari, E.; Lazzari, S.; Belluti, S.; Pignedoli, F.; Imbriano, C. Curcumin derivatives: Molecular basis of their anti-cancer activity. *Biochem. Pharmacol.* **2009**, *78*, 1305–1315. [CrossRef]

118. Sribalan, R.; Kirubavathi, M.; Banuppriya, G.; Padmini, V. Synthesis and biological evaluation of new symmetric curcumin derivatives. *Bioorg. Med. Chem. Lett.* **2015**, *25*, 4282–4286. [CrossRef]

119. Bayomi, S.M.; El-Kashef, H.A.; El-Ashmawy, M.B.; Nasr, M.N.A.; El-Sherbeny, M.A.; Badria, F.A.; Abou-Zeid, L.A.; Ghaly, M.A.; Abdel-Aziz, N.I. Synthesis and biological evaluation of new curcumin derivatives as antioxidant and antitumor agents. *Med. Chem. Res.* **2013**, *22*, 1147–1162. [CrossRef]

120. Xu, G.; Chu, Y.; Jiang, N.; Yang, J.; Li, F. The three dimensional quantitative structure activity relationships (3D-QSAR) and docking studies of curcumin derivatives as androgen receptor antagonists. *Int. J. Mol. Sci.* **2012**, *13*, 6138–6155. [CrossRef] [PubMed]

121. Zhao, C.; Liu, Z.; Liang, G. Promising curcumin-based drug design: Mono-carbonyl analogues of curcumin (MACs). *Curr. Pharm. Des.* **2013**, *19*, 2114–2135. [PubMed]

122. Somparn, P.; Phisalaphong, C.; Nakornchai, S.; Unchern, S.; Morales, N.P. Comparative antioxidant activities of curcumin and its demethoxy and hydrogenated derivatives. *Biol. Pharm. Bull.* **2007**, *30*, 74–78. [CrossRef] [PubMed]

123. Anand, P.; Thomas, S.G.; Kunnumakkara, A.B.; Sundaram, C.; Harikumar, K.B.; Sung, B.; Tharakan, S.T.; Misra, K.; Priyadarsini, I.K.; Rajasekharan, K.N.; et al. Biological activities of curcumin and its analogues (Congeners) made by man and Mother Nature. *Biochem. Pharmacol.* **2008**, *76*, 1590–1611. [CrossRef] [PubMed]

124. Reddy, A.R.; Dinesh, P.; Prabhakar, A.S.; Umasankar, K.; Shireesha, B.; Raju, M.B. A comprehensive review on SAR of curcumin. *Mini Rev. Med. Chem.* **2013**, *13*, 1769–1777. [CrossRef] [PubMed]

125. Sherin, D.R.; Rajasekharan, K.N. Mechanochemical synthesis and antioxidant activity of curcumin-templated azoles. *Arch. Pharm.* **2015**, *348*, 908–914. [CrossRef]

126. Wanninger, S.; Lorenz, V.; Subhan, A.; Edelmann, F.T. Metal complexes of curcumin–synthetic strategies, structures and medicinal applications. *Chem. Soc. Rev.* **2015**, *44*, 4986–5002. [CrossRef]

127. Paulraj, F.; Abas, F.; Lajis, N.H.; Othman, I.; Naidu, R. Molecular pathways modulated by curcumin analogue, diarylpentanoids in cancer. *Biomolecules* **2019**, *9*, 270. [CrossRef]

128. Yerdelen, K.O.; Gul, H.I.; Sakagami, H.; Umemura, N.; Sukuroglu, M. Synthesis and cytotoxic activities of a curcumin analogue and its bis-mannich derivatives. *Lett. Drug Des. Discov.* **2015**, *12*, 643–649. [CrossRef]

129. Reid, J.M.; Buhrow, S.A.; Gilbert, J.A.; Jia, L.; Shoji, M.; Snyder, J.P.; Ames, M.M. Mouse pharmacokinetics and metabolism of the curcumin analog, 4-piperidinone,3,5-bis[(2-fluorophenyl)methylene]-acetate(3E,5E) (EF-24; NSC 716993). *Cancer Chemother. Pharmacol.* **2014**, *73*, 1137–1146. [CrossRef]

130. Selvendiran, K.; Tong, L.; Vishwanath, S.; Bratasz, A.; Trigg, N.J.; Kutala, V.K.; Hideg, K.; Kuppusamy, P. EF24 induces G2/M arrest and apoptosis in cisplatin-resistant human ovarian cancer cells by increasing PTEN expression. *J. Biol. Chem.* **2007**, *282*, 28609–28618. [CrossRef]

131. He, G.; Feng, C.; Vinothkumar, R.; Chen, W.; Dai, X.; Chen, X.; Ye, Q.; Qiu, C.; Zhou, H.; Wang, Y.; et al. Curcumin analog EF24 induces apoptosis via ROS-dependent mitochondrial dysfunction in human colorectal cancer cells. *Cancer Chemother. Pharmacol.* **2016**, *78*, 1151–1161. [CrossRef]

132. Kasinski, A.L.; Du, Y.; Thomas, S.L.; Zhao, J.; Sun, S.Y.; Khuri, F.R.; Wang, C.Y.; Shoji, M.; Sun, A.; Snyder, J.P.; et al. Inhibition of IkappaB kinase-nuclear factor-kappaB signaling pathway by 3,5-bis(2-flurobenzylidene)piperidin-4-one (EF24), a novel monoketone analog of curcumin. *Mol. Pharmacol.* **2008**, *74*, 654–661. [CrossRef]

133. Thomas, S.L.; Zhong, D.; Zhou, W.; Malik, S.; Liotta, D.; Snyder, J.P.; Hamel, E.; Giannakakou, P. EF24, a novel curcumin analog, disrupts the microtubule cytoskeleton and inhibits HIF-1. *Cell Cycle* **2008**, *7*, 2409–2417. [CrossRef]

134. Tan, X.; Sidell, N.; Mancini, A.; Huang, R.P.; Shenming, W.; Horowitz, I.R.; Liotta, D.C.; Taylor, R.N.; Wieser, F. Multiple anticancer activities of EF24, a novel curcumin analog, on human ovarian carcinoma cells. *Reprod. Sci.* **2010**, *17*, 931–940. [CrossRef]

135. Yu, H.; Lin, L.; Zhang, Z.; Zhang, H.; Hu, H. Targeting NF-kappaB pathway for the therapy of diseases: Mechanism and clinical study. *Signal Transduct. Target. Ther.* **2020**, *5*, 209. [CrossRef] [PubMed]

136. Yin, D.L.; Liang, Y.J.; Zheng, T.S.; Song, R.P.; Wang, J.B.; Sun, B.S.; Pan, S.H.; Qu, L.D.; Liu, J.R.; Jiang, H.C.; et al. EF24 inhibits tumor growth and metastasis via suppressing NF-kappaB dependent pathways in human cholangiocarcinoma. *Sci. Rep.* **2016**, *6*, 32167. [CrossRef] [PubMed]

137. Aravindan, S.; Natarajan, M.; Herman, T.S.; Awasthi, V.; Aravindan, N. Molecular basis of 'hypoxic' breast cancer cell radio-sensitization: Phytochemicals converge on radiation induced Rel signaling. *Radiat. Oncol* **2013**, *8*, 46. [CrossRef] [PubMed]

138. Aravindan, S.; Natarajan, M.; Awasthi, V.; Herman, T.S.; Aravindan, N. Novel synthetic monoketone transmute radiation-triggered NFkappaB-dependent TNFalpha cross-signaling feedback maintained NFkappaB and favors neuroblastoma regression. *PLoS ONE* **2013**, *8*, e72464. [CrossRef]

139. Liang, Y.; Zheng, T.; Song, R.; Wang, J.; Yin, D.; Wang, L.; Liu, H.; Tian, L.; Fang, X.; Meng, X.; et al. Hypoxia-mediated sorafenib resistance can be overcome by EF24 through Von Hippel-Lindau tumor suppressor-dependent HIF-1alpha inhibition in hepatocellular carcinoma. *Hepatology* **2013**, *57*, 1847–1857. [CrossRef] [PubMed]

140. Adams, B.K.; Cai, J.; Armstrong, J.; Herold, M.; Lu, Y.J.; Sun, A.; Snyder, J.P.; Liotta, D.C.; Jones, D.P.; Shoji, M. EF24, a novel synthetic curcumin analog, induces apoptosis in cancer cells via a redox-dependent mechanism. *Anti-Cancer Drugs* **2005**, *16*, 263–275. [CrossRef] [PubMed]

141. Zou, P.; Xia, Y.; Chen, W.; Chen, X.; Ying, S.; Feng, Z.; Chen, T.; Ye, Q.; Wang, Z.; Qiu, C.; et al. EF24 induces ROS-mediated apoptosis via targeting thioredoxin reductase 1 in gastric cancer cells. *Oncotarget* **2016**, *7*, 18050–18064. [CrossRef] [PubMed]

142. Selvendiran, K.; Ahmed, S.; Dayton, A.; Kuppusamy, M.L.; Rivera, B.K.; Kalai, T.; Hideg, K.; Kuppusamy, P. HO-3867, a curcumin analog, sensitizes cisplatin-resistant ovarian carcinoma, leading to therapeutic synergy through STAT3 inhibition. *Cancer Biol. Ther.* **2011**, *12*, 837–845. [CrossRef]

143. Ismail, N.I.; Othman, I.; Abas, F.; Lajis, N.H.; Naidu, R. The curcumin analogue, MS13 (1,5-Bis(4-hydroxy-3- methoxyphenyl)-1,4-pentadiene-3-one), inhibits cell proliferation and induces apoptosis in primary and metastatic human colon cancer cells. *Molecules* **2020**, *25*, 3798. [CrossRef] [PubMed]

144. Liang, G.; Shao, L.; Wang, Y.; Zhao, C.; Chu, Y.; Xiao, J.; Zhao, Y.; Li, X.; Yang, S. Exploration and synthesis of curcumin analogues with improved structural stability both in vitro and in vivo as cytotoxic agents. *Bioorg. Med. Chem.* **2009**, *17*, 2623–2631. [CrossRef]

145. Lin, L.; Shi, Q.; Nyarko, A.K.; Bastow, K.F.; Wu, C.C.; Su, C.Y.; Shih, C.C.; Lee, K.H. Antitumor agents. 250. Design and synthesis of new curcumin analogues as potential anti-prostate cancer agents. *J. Med. Chem.* **2006**, *49*, 3963–3972. [CrossRef] [PubMed]

146. Shen, H.; Shen, J.; Pan, H.; Xu, L.; Sheng, H.; Liu, B.; Yao, M. Curcumin analog B14 has high bioavailability and enhances the effect of anti-breast cancer cells in vitro and in vivo. *Cancer Sci.* **2021**, *112*, 815–827. [CrossRef] [PubMed]

147. Zheng, B.; McClements, D.J. Formulation of more efficacious curcumin delivery systems using colloid science: Enhanced solubility, stability, and bioavailability. *Molecules* **2020**, *25*, 2791. [CrossRef]

148. Bansal, S.S.; Goel, M.; Aqil, F.; Vadhanam, M.V.; Gupta, R.C. Advanced drug delivery systems of curcumin for cancer chemoprevention. *Cancer Prev. Res.* **2011**, *4*, 1158–1171. [CrossRef] [PubMed]

149. Munjal, B.; Pawar, Y.B.; Patel, S.B.; Bansal, A.K. Comparative oral bioavailability advantage from curcumin formulations. *Drug Deliv. Transl. Res.* **2011**, *1*, 322–331. [CrossRef]

150. Sun, M.; Su, X.; Ding, B.; He, X.; Liu, X.; Yu, A.; Lou, H.; Zhai, G. Advances in nanotechnology-based delivery systems for curcumin. *Nanomedicine* **2012**, *7*, 1085–1100. [CrossRef]

151. Praditya, D.; Kirchhoff, L.; Bruning, J.; Rachmawati, H.; Steinmann, J.; Steinmann, E. Anti-infective Properties of the Golden Spice Curcumin. *Front. Microbiol.* **2019**, *10*, 912. [CrossRef]

152. Yang, S.; Liu, L.; Han, J.; Tang, Y. Encapsulating plant ingredients for dermocosmetic application: An updated review of delivery systems and characterization techniques. *Int. J. Cosmet. Sci.* **2020**, *42*, 16–28. [CrossRef] [PubMed]

153. Li, R.; Lim, S.J.; Choi, H.G.; Lee, M.K. Solid lipid nanoparticles as drug delivery system for water-insoluble drugs. *J. Pharm. Investig.* **2010**, *40*, 63–73. [CrossRef]

154. Yallapu, M.M.; Jaggi, M.; Chauhan, S.C. beta-Cyclodextrin-curcumin self-assembly enhances curcumin delivery in prostate cancer cells. *Colloids Surf. B Biointerfaces* **2010**, *79*, 113–125. [CrossRef] [PubMed]

155. Hewlings, S.J.; Kalman, D.S. Curcumin: A review of its effects on human health. *Foods* **2017**, *6*, 92. [CrossRef] [PubMed]

156. Mitchell, M.J.; Billingsley, M.M.; Haley, R.M.; Wechsler, M.E.; Peppas, N.A.; Langer, R. Engineering precision nanoparticles for drug delivery. *Nat. Rev. Drug Discov.* **2021**, *20*, 101–124. [CrossRef] [PubMed]

157. Umerska, A.; Gaucher, C.; Oyarzun-Ampuero, F.; Fries-Raeth, I.; Colin, F.; Villamizar-Sarmiento, M.G.; Maincent, P.; Sapin-Minet, A. Polymeric nanoparticles for increasing oral bioavailability of curcumin. *Antioxidants* **2018**, *7*, 46. [CrossRef]

158. Ferrari, R.; Sponchioni, M.; Morbidelli, M.; Moscatelli, D. Polymer nanoparticles for the intravenous delivery of anticancer drugs: The checkpoints on the road from the synthesis to clinical translation. *Nanoscale* **2018**, *10*, 22701–22719. [CrossRef] [PubMed]

159. Li, Z.; Shi, M.; Li, N.; Xu, R. Application of functional biocompatible nanomaterials to improve curcumin bioavailability. *Front. Chem.* **2020**, *8*, 589957. [CrossRef]

160. Zaman, M.S.; Chauhan, N.; Yallapu, M.M.; Gara, R.K.; Maher, D.M.; Kumari, S.; Sikander, M.; Khan, S.; Zafar, N.; Jaggi, M.; et al. Curcumin nanoformulation for cervical cancer treatment. *Sci. Rep.* **2016**, *6*, 20051. [CrossRef]

161. Yallapu, M.M.; Gupta, B.K.; Jaggi, M.; Chauhan, S.C. Fabrication of curcumin encapsulated PLGA nanoparticles for improved therapeutic effects in metastatic cancer cells. *J. Colloid Interface Sci.* **2010**, *351*, 19–29. [CrossRef]

162. Mukerjee, A.; Vishwanatha, J.K. Formulation, characterization and evaluation of curcumin-loaded PLGA nanospheres for cancer therapy. *Anticancer Res.* **2009**, *29*, 3867–3875.

163. Bisht, S.; Feldmann, G.; Soni, S.; Ravi, R.; Karikar, C.; Maitra, A.; Maitra, A. Polymeric nanoparticle-encapsulated curcumin ("nanocurcumin"): A novel strategy for human cancer therapy. *J. Nanobiotechnol.* **2007**, *5*, 3. [CrossRef] [PubMed]

164. Yin, H.T.; Zhang, D.G.; Wu, X.L.; Huang, X.E.; Chen, G. In vivo evaluation of curcumin-loaded nanoparticles in a A549 xenograft mice model. *Asian Pac. J. Cancer Prev. APJCP* **2013**, *14*, 409–412. [CrossRef] [PubMed]

165. Xie, M.; Fan, D.; Li, Y.; He, X.; Chen, X.; Chen, Y.; Zhu, J.; Xu, G.; Wu, X.; Lan, P. Supercritical carbon dioxide-developed silk fibroin nanoplatform for smart colon cancer therapy. *Int. J. Nanomed.* **2017**, *12*, 7751–7761. [CrossRef] [PubMed]

166. Baspinar, Y.; Ustundas, M.; Bayraktar, O.; Sezgin, C. Curcumin and piperine loaded zein-chitosan nanoparticles: Development and in-vitro characterisation. *Saudi Pharm. J. SPJ Off. Publ. Saudi Pharm. Soc.* **2018**, *26*, 323–334. [CrossRef] [PubMed]

167. Yadav, P.; Bandyopadhyay, A.; Chakraborty, A.; Sarkar, K. Enhancement of anticancer activity and drug delivery of chitosan-curcumin nanoparticle via molecular docking and simulation analysis. *Carbohydr. Polym.* **2018**, *182*, 188–198. [CrossRef] [PubMed]

168. Manjunath, K.; Reddy, J.S.; Venkateswarlu, V. Solid lipid nanoparticles as drug delivery systems. *Methods Find. Exp. Clin. Pharmacol.* **2005**, *27*, 127–144. [CrossRef]

169. Mukherjee, S.; Ray, S.; Thakur, R.S. Solid lipid nanoparticles: A modern formulation approach in drug delivery system. *Indian J. Pharm. Sci.* **2009**, *71*, 349–358. [CrossRef]

170. Jenning, V.; Lippacher, A.; Gohla, S.H. Medium scale production of solid lipid nanoparticles (SLN) by high pressure homogenization. *J. Microencapsul.* **2002**, *19*, 1–10. [CrossRef]

171. Baek, J.S.; Cho, C.W. Surface modification of solid lipid nanoparticles for oral delivery of curcumin: Improvement of bioavailability through enhanced cellular uptake, and lymphatic uptake. *Eur. J. Pharm. Biopharm.* **2017**, *117*, 132–140. [CrossRef]

172. Guorgui, J.; Wang, R.; Mattheolabakis, G.; Mackenzie, G.G. Curcumin formulated in solid lipid nanoparticles has enhanced efficacy in Hodgkin's lymphoma in mice. *Arch. Biochem. Biophys.* **2018**, *648*, 12–19. [CrossRef] [PubMed]

173. Gota, V.S.; Maru, G.B.; Soni, T.G.; Gandhi, T.R.; Kochar, N.; Agarwal, M.G. Safety and pharmacokinetics of a solid lipid curcumin particle formulation in osteosarcoma patients and healthy volunteers. *J. Agric. Food Chem.* **2010**, *58*, 2095–2099. [CrossRef]

174. Ban, C.; Jo, M.; Park, Y.H.; Kim, J.H.; Han, J.Y.; Lee, K.W.; Kweon, D.H.; Choi, Y.J. Enhancing the oral bioavailability of curcumin using solid lipid nanoparticles. *Food Chem.* **2020**, *302*, 125328. [CrossRef]

175. Huang, S.; He, J.; Cao, L.; Lin, H.; Zhang, W.; Zhong, Q. Improved physicochemical properties of curcumin-loaded solid lipid nanoparticles stabilized by sodium caseinate-lactose Maillard conjugate. *J. Agric. Food Chem.* **2020**, *68*, 7072–7081. [CrossRef]

176. Bhatt, H.; Rompicharla, S.V.K.; Komanduri, N.; Aashma, S.; Paradkar, S.; Ghosh, B.; Biswas, S. Development of curcumin-loaded solid lipid nanoparticles utilizing glyceryl monostearate as single lipid using QbD approach: Characterization and evaluation of anticancer activity against human breast cancer cell line. *Curr. Drug Deliv.* **2018**, *15*, 1271–1283. [CrossRef]

177. Wang, F.; Li, C.; Cheng, J.; Yuan, Z. Recent advances on inorganic nanoparticle-based cancer therapeutic agents. *Int. J. Environ. Res. Public Health* **2016**, *13*, 1182. [CrossRef]

178. Huynh, K.H.; Pham, X.H.; Kim, J.; Lee, S.H.; Chang, H.; Rho, W.Y.; Jun, B.H. Synthesis, properties, and biological applications of metallic alloy nanoparticles. *Int. J. Mol. Sci.* **2020**, *21*, 5174. [CrossRef]

179. Kouhpanji, M.R.Z.; Stadler, B.J.H. A guideline for effectively synthesizing and characterizing magnetic nanoparticles for advancing nanobiotechnology: A review. *Sensors* **2020**, *20*, 2554. [CrossRef] [PubMed]

180. Yallapu, M.M.; Ebeling, M.C.; Khan, S.; Sundram, V.; Chauhan, N.; Gupta, B.K.; Puumala, S.E.; Jaggi, M.; Chauhan, S.C. Novel curcumin-loaded magnetic nanoparticles for pancreatic cancer treatment. *Mol. Cancer Ther.* **2013**, *12*, 1471–1480. [CrossRef] [PubMed]

181. Saikia, C.; Das, M.K.; Ramteke, A.; Maji, T.K. Evaluation of folic acid tagged aminated starch/ZnO coated iron oxide nanoparticles as targeted curcumin delivery system. *Carbohydr. Polym.* **2017**, *157*, 391–399. [CrossRef] [PubMed]

182. Barick, K.C.; Ekta; Gawali, S.L.; Sarkar, A.; Kunwar, A.; Priyadarsinicd, K.I.; Hassan, P.A. Pluronic stabilized Fe3O4 magnetic nanoparticles for intracellular delivery of curcumin. *RSC Adv.* **2016**, *6*, 98674–98681. [CrossRef]

183. Ayubi, M.; Karimi, M.; Abdpour, S.; Rostamizadeh, K.; Parsa, M.; Zamani, M.; Saedi, A. Magnetic nanoparticles decorated with PEGylated curcumin as dual targeted drug delivery: Synthesis, toxicity and biocompatibility study. *Mater. Sci. Eng. C Mater. Biol. Appl.* **2019**, *104*, 109810. [CrossRef] [PubMed]

184. Gangwar, R.K.; Tomar, G.B.; Dhumale, V.A.; Zinjarde, S.; Sharma, R.B.; Datar, S. Curcumin conjugated silica nanoparticles for improving bioavailability and its anticancer applications. *J. Agric. Food Chem.* **2013**, *61*, 9632–9637. [CrossRef]

185. Manju, S.; Sreenivasan, K. Gold nanoparticles generated and stabilized by water soluble curcumin-polymer conjugate: Blood compatibility evaluation and targeted drug delivery onto cancer cells. *J. Colloid Interface Sci.* **2012**, *368*, 144–151. [CrossRef]

186. Sercombe, L.; Veerati, T.; Moheimani, F.; Wu, S.Y.; Sood, A.K.; Hua, S. Advances and challenges of liposome assisted drug delivery. *Front. Pharm.* **2015**, *6*, 286. [CrossRef]

187. Li, R.; Deng, L.; Cai, Z.; Zhang, S.; Wang, K.; Li, L.; Ding, S.; Zhou, C. Liposomes coated with thiolated chitosan as drug carriers of curcumin. *Mater. Sci. Eng. C Mater. Biol. Appl.* **2017**, *80*, 156–164. [CrossRef]

188. Huang, M.G.; Liang, C.P.; Tan, C.; Huang, S.; Ying, R.F.; Wang, Y.S.; Wang, Z.J.; Zhang, Y.F. Liposome co-encapsulation as a strategy for the delivery of curcumin and resveratrol. *Food Funct.* **2019**, *10*, 6447–6458. [CrossRef]

189. Cheng, C.; Peng, S.; Li, Z.; Zou, L.; Liu, W.; Liu, C. Improved bioavailability of curcumin in liposomes prepared using a pH-driven, organic solvent-free, easily scalable process. *RSC Adv.* **2017**, *7*, 25978–25986. [CrossRef]

190. Jin, H.H.; Lu, Q.; Jiang, J.G. Curcumin liposomes prepared with milk fat globule membrane phospholipids and soybean lecithin. *J. Dairy Sci.* **2016**, *99*, 1780–1790. [CrossRef] [PubMed]

191. Cuomo, F.; Cofelice, M.; Venditti, F.; Ceglie, A.; Miguel, M.; Lindman, B.; Lopez, F. In-vitro digestion of curcumin loaded chitosan-coated liposomes. *Colloids Surf. B Biointerfaces* **2018**, *168*, 29–34. [CrossRef] [PubMed]

192. Hamano, N.; Bottger, R.; Lee, S.E.; Yang, Y.; Kulkarni, J.A.; Ip, S.; Cullis, P.R.; Li, S.D. Robust microfluidic technology and new lipid composition for fabrication of curcumin-loaded liposomes: Effect on the anticancer activity and safety of cisplatin. *Mol. Pharm.* **2019**, *16*, 3957–3967. [CrossRef]

193. Thangapazham, R.L.; Puri, A.; Tele, S.; Blumenthal, R.; Maheshwari, R.K. Evaluation of a nanotechnology-based carrier for delivery of curcumin in prostate cancer cells. *Int. J. Oncol.* **2008**, *32*, 1119–1123. [CrossRef]

194. Dhule, S.S.; Penfornis, P.; Frazier, T.; Walker, R.; Feldman, J.; Tan, G.; He, J.; Alb, A.; John, V.; Pochampally, R. Curcumin-loaded gamma-cyclodextrin liposomal nanoparticles as delivery vehicles for osteosarcoma. *Nanomed. Nanotechnol. Biol. Med.* **2012**, *8*, 440–451. [CrossRef]

195. Tefas, L.R.; Sylvester, B.; Tomuta, I.; Sesarman, A.; Licarete, E.; Banciu, M.; Porfire, A. Development of antiproliferative long-circulating liposomes co-encapsulating doxorubicin and curcumin, through the use of a quality-by-design approach. *Drug Des. Dev. Ther.* **2017**, *11*, 1605–1621. [CrossRef]

196. Xu, H.; Gong, Z.; Zhou, S.; Yang, S.; Wang, D.; Chen, X.; Wu, J.; Liu, L.; Zhong, S.; Zhao, J.; et al. Liposomal curcumin targeting endometrial cancer through the NF-kappaB pathway. *Cell. Physiol. Biochem.* **2018**, *48*, 569–582. [CrossRef]

197. Swami Vetha, B.S.; Oh, P.S.; Kim, S.H.; Jeong, H.J. Curcuminoids encapsulated liposome nanoparticles as a blue light emitting diode induced photodynamic therapeutic system for cancer treatment. *J. Photochem. Photobiol. B: Biol.* **2020**, *205*, 111840. [CrossRef] [PubMed]

198. Lu, M.; Qiu, Q.; Luo, X.; Liu, X.; Sun, J.; Wang, C.; Lin, X.; Deng, Y.; Song, Y. Phyto-phospholipid complexes (phytosomes): A novel strategy to improve the bioavailability of active constituents. *Asian J. Pharm. Sci.* **2019**, *14*, 265–274. [CrossRef] [PubMed]

199. Marczylo, T.H.; Verschoyle, R.D.; Cooke, D.N.; Morazzoni, P.; Steward, W.P.; Gescher, A.J. Comparison of systemic availability of curcumin with that of curcumin formulated with phosphatidylcholine. *Cancer Chemother. Pharmacol.* **2007**, *60*, 171–177. [CrossRef] [PubMed]

200. Maiti, K.; Mukherjee, K.; Gantait, A.; Saha, B.P.; Mukherjee, P.K. Curcumin-phospholipid complex: Preparation, therapeutic evaluation and pharmacokinetic study in rats. *Int. J. Pharm.* **2007**, *330*, 155–163. [CrossRef]

201. Stohs, S.J.; Ji, J.; Bucci, L.R.; Preuss, H.G. A comparative pharmacokinetic assessment of a novel highly bioavailable curcumin formulation with 95% curcumin: A randomized, double-blind, crossover study. *J. Am. Coll. Nutr.* **2018**, *37*, 51–59. [CrossRef]

202. Cuomo, J.; Appendino, G.; Dern, A.S.; Schneider, E.; McKinnon, T.P.; Brown, M.J.; Togni, S.; Dixon, B.M. Comparative absorption of a standardized curcuminoid mixture and its lecithin formulation. *J. Nat. Prod.* **2011**, *74*, 664–669. [CrossRef]

203. Wang, J.; Wang, L.; Zhang, L.; He, D.; Ju, J.; Li, W. Studies on the curcumin phospholipid complex solidified with Soluplus®. *J. Pharm. Pharm.* **2018**, *70*, 242–249. [CrossRef] [PubMed]

204. Ghezzi, M.; Pescina, S.; Padula, C.; Santi, P.; Del Favero, E.; Cantu, L.; Nicoli, S. Polymeric micelles in drug delivery: An insight of the techniques for their characterization and assessment in biorelevant conditions. *J. Control Release* **2021**, *332*, 312–336. [CrossRef] [PubMed]

205. Kheiri Manjili, H.; Ghasemi, P.; Malvandi, H.; Mousavi, M.S.; Attari, E.; Danafar, H. Pharmacokinetics and in vivo delivery of curcumin by copolymeric mPEG-PCL micelles. *Eur. J. Pharm. Biopharm.* **2017**, *116*, 17–30. [CrossRef]

206. Letchford, K.; Liggins, R.; Burt, H. Solubilization of hydrophobic drugs by methoxy poly(ethylene glycol)-block-polycaprolactone diblock copolymer micelles: Theoretical and experimental data and correlations. *J. Pharm. Sci.* **2008**, *97*, 1179–1190. [CrossRef] [PubMed]

207. Ma, Z.; Shayeganpour, A.; Brocks, D.R.; Lavasanifar, A.; Samuel, J. High-performance liquid chromatography analysis of curcumin in rat plasma: Application to pharmacokinetics of polymeric micellar formulation of curcumin. *Biomed. Chromatogr. BMC* **2007**, *21*, 546–552. [CrossRef]

208. Schiborr, C.; Kocher, A.; Behnam, D.; Jandasek, J.; Toelstede, S.; Frank, J. The oral bioavailability of curcumin from micronized powder and liquid micelles is significantly increased in healthy humans and differs between sexes. *Mol. Nutr. Food Res.* **2014**, *58*, 516–527. [CrossRef]

209. Frank, J.; Schiborr, C.; Kocher, A.; Meins, J.; Behnam, D.; Schubert-Zsilavecz, M.; Abdel-Tawab, M. Transepithelial transport of curcumin in Caco-2 cells is significantly enhanced by micellar solubilisation. *Plant Foods Hum. Nutr.* **2017**, *72*, 48–53. [CrossRef]
210. Parikh, A.; Kathawala, K.; Song, Y.; Zhou, X.F.; Garg, S. Curcumin-loaded self-nanomicellizing solid dispersion system: Part I: Development, optimization, characterization, and oral bioavailability. *Drug Deliv. Transl. Res.* **2018**, *8*, 1389–1405. [CrossRef]
211. Chen, S.; Li, Q.; Li, H.; Yang, L.; Yi, J.Z.; Xie, M.; Zhang, L.M. Long-circulating zein-polysulfobetaine conjugate-based nanocarriers for enhancing the stability and pharmacokinetics of curcumin. *Mater. Sci. Eng. C Mater. Biol. Appl.* **2020**, *109*, 110636. [CrossRef]
212. Song, L.; Shen, Y.Y.; Hou, J.W.; Lei, L.; Guo, S.R.; Qian, C.Y. Polymeric micelles for parenteral delivery of curcumin: Preparation, characterization and in vitro evaluation. *Colloid Surf. A* **2011**, *390*, 25–32. [CrossRef]
213. Patil, S.; Choudhary, B.; Rathore, A.; Roy, K.; Mahadik, K. Enhanced oral bioavailability and anticancer activity of novel curcumin loaded mixed micelles in human lung cancer cells. *Phytomed. Int. J. Phytother. Phytopharm.* **2015**, *22*, 1103–1111. [CrossRef]
214. Duan, Y.; Zhang, B.; Chu, L.; Tong, H.H.; Liu, W.; Zhai, G. Evaluation in vitro and in vivo of curcumin-loaded mPEG-PLA/TPGS mixed micelles for oral administration. *Colloids Surf. B Biointerfaces* **2016**, *141*, 345–354. [CrossRef] [PubMed]
215. Hani, U.; Shivakumar, H.G. Solubility enhancement and delivery systems of curcumin a herbal medicine: A review. *Curr. Drug Deliv.* **2014**, *11*, 792–804. [CrossRef]
216. Szejtli, J. Introduction and general overview of cyclodextrin chemistry. *Chem. Rev.* **1998**, *98*, 1743–1754. [CrossRef]
217. Bakshi, P.R.; Londhe, V.Y. Widespread applications of host-guest interactive cyclodextrin functionalized polymer nanocomposites: Its meta-analysis and review. *Carbohydr. Polym.* **2020**, *242*, 116430. [CrossRef]
218. Mangolim, C.S.; Moriwaki, C.; Nogueira, A.C.; Sato, F.; Baesso, M.L.; Neto, A.M.; Matioli, G. Curcumin-beta-cyclodextrin inclusion complex: Stability, solubility, characterisation by FT-IR, FT-Raman, X-ray diffraction and photoacoustic spectroscopy, and food application. *Food Chem.* **2014**, *153*, 361–370. [CrossRef] [PubMed]
219. Guo, S. Encapsulation of curcumin into β-cyclodextrins inclusion: A review. *E3S Web Conf.* **2019**, *131*, 01100. [CrossRef]
220. Jantarat, C.; Sirathanarun, P.; Ratanapongsai, S.; Watcharakan, P.; Sunyapong, S.; Wadu, A. Curcumin-hydroxypropyl-β-cyclodextrin inclusion complex preparation methods: Effect of common solvent evaporation, freeze drying, and pH shift on solubility and stability of curcumin. *Trop. J. Pharm. Res.* **2014**, *13*, 1215–1223. [CrossRef]
221. Li, N.; Wang, N.; Wu, T.; Qiu, C.; Wang, X.; Jiang, S.; Zhang, Z.; Liu, T.; Wei, C.; Wang, T. Preparation of curcumin-hydroxypropyl-beta-cyclodextrin inclusion complex by cosolvency-lyophilization procedure to enhance oral bioavailability of the drug. *Drug Dev. Ind. Pharm.* **2018**, *44*, 1966–1974. [CrossRef]
222. Ghanghoria, R.; Kesharwani, P.; Agashe, H.B.; Jain, N.K. Transdermal delivery of cyclodextrin-solubilized curcumin. *Drug Deliv. Transl. Res.* **2013**, *3*, 272–285. [CrossRef] [PubMed]
223. Popat, A.; Karmakar, S.; Jambhrunkar, S.; Xu, C.; Yu, C. Curcumin-cyclodextrin encapsulated chitosan nanoconjugates with enhanced solubility and cell cytotoxicity. *Colloids Surf. B Biointerfaces* **2014**, *117*, 520–527. [CrossRef] [PubMed]
224. Zhang, L.; Man, S.; Qiu, H.; Liu, Z.; Zhang, M.; Ma, L.; Gao, W. Curcumin-cyclodextrin complexes enhanced the anti-cancer effects of curcumin. *Environ. Toxicol. Pharmacol.* **2016**, *48*, 31–38. [CrossRef] [PubMed]
225. Moller, K.; Macaulay, B.; Bein, T. Curcumin encapsulated in crosslinked cyclodextrin nanoparticles enables immediate inhibition of cell growth and efficient killing of cancer cells. *Nanomaterials* **2021**, *11*, 489. [CrossRef] [PubMed]
226. Hong, W.; Guo, F.; Yu, N.; Ying, S.; Lou, B.; Wu, J.; Gao, Y.; Ji, X.; Wang, H.; Li, A.; et al. A novel folic acid receptor-targeted drug delivery system based on curcumin-loaded beta-cyclodextrin nanoparticles for cancer treatment. *Drug Des. Dev. Ther.* **2021**, *15*, 2843–2855. [CrossRef]
227. Stasilowicz, A.; Tykarska, E.; Lewandowska, K.; Kozak, M.; Miklaszewski, A.; Kobus-Cisowska, J.; Szymanowska, D.; Plech, T.; Jenczyk, J.; Cielecka-Piontek, J. Hydroxypropyl-beta-cyclodextrin as an effective carrier of curcumin–piperine nutraceutical system with improved enzyme inhibition properties. *J. Enzym. Inhib. Med. Chem.* **2020**, *35*, 1811–1821. [CrossRef]
228. Tran, P.H.L.; Tran, T.T.D. Developmental strategies of curcumin solid dispersions for enhancing bioavailability. *Anti-Cancer Agents Med. Chem.* **2020**, *20*, 1874–1882. [CrossRef]
229. Teixeira, C.C.; Mendonca, L.M.; Bergamaschi, M.M.; Queiroz, R.H.; Souza, G.E.; Antunes, L.M.; Freitas, L.A. Microparticles containing curcumin solid dispersion: Stability, bioavailability and anti-inflammatory activity. *AAPS PharmSciTech* **2016**, *17*, 252–261. [CrossRef]
230. Basniwal, R.K.; Khosla, R.; Jain, N. Improving the anticancer activity of curcumin using nanocurcumin dispersion in water. *Nutr. Cancer* **2014**, *66*, 1015–1022. [CrossRef]
231. Seo, S.W.; Han, H.K.; Chun, M.K.; Choi, H.K. Preparation and pharmacokinetic evaluation of curcumin solid dispersion using Solutol® HS15 as a carrier. *Int. J. Pharm.* **2012**, *424*, 18–25. [CrossRef] [PubMed]
232. Wan, S.; Sun, Y.; Qi, X.; Tan, F. Improved bioavailability of poorly water-soluble drug curcumin in cellulose acetate solid dispersion. *AAPS PharmSciTech* **2012**, *13*, 159–166. [CrossRef]
233. Mendonca, L.M.; Machado Cda, S.; Teixeira, C.C.; Freitas, L.A.; Bianchi, M.L.; Antunes, L.M. Comparative study of curcumin and curcumin formulated in a solid dispersion: Evaluation of their antigenotoxic effects. *Genet. Mol. Biol.* **2015**, *38*, 490–498. [CrossRef]
234. Silva de Sa, I.; Peron, A.P.; Leimann, F.V.; Bressan, G.N.; Krum, B.N.; Fachinetto, R.; Pinela, J.; Calhelha, R.C.; Barreiro, M.F.; Ferreira, I.; et al. In vitro and in vivo evaluation of enzymatic and antioxidant activity, cytotoxicity and genotoxicity of curcumin-loaded solid dispersions. *Food Chem. Toxicol.* **2019**, *125*, 29–37. [CrossRef]

235. Han, H.K. The effects of black pepper on the intestinal absorption and hepatic metabolism of drugs. *Expert Opin. Drug Metab. Toxicol.* **2011**, *7*, 721–729. [CrossRef]
236. Shoba, G.; Joy, D.; Joseph, T.; Majeed, M.; Rajendran, R.; Srinivas, P.S. Influence of piperine on the pharmacokinetics of curcumin in animals and human volunteers. *Planta Med.* **1998**, *64*, 353–356. [CrossRef] [PubMed]
237. Sharma, V.; Nehru, B.; Munshi, A.; Jyothy, A. Antioxidant potential of curcumin against oxidative insult induced by pentylenetetrazol in epileptic rats. *Methods Find. Exp. Clin. Pharmacol.* **2010**, *32*, 227–232. [CrossRef]
238. Suresh, D.; Srinivasan, K. Tissue distribution & elimination of capsaicin, piperine & curcumin following oral intake in rats. *Indian J. Med. Res.* **2010**, *131*, 682–691.
239. Antony, B.; Merina, B.; Iyer, V.S.; Judy, N.; Lennertz, K.; Joyal, S. A pilot cross-over study to evaluate human oral bioavailability of BCM-95®CG (Biocurcumax™), a novel bioenhanced preparation of curcumin. *Indian J. Pharm. Sci.* **2008**, *70*, 445–449. [CrossRef] [PubMed]
240. Singh, D.V.; Agarwal, S.; Singh, P.; Godbole, M.M.; Misra, K. Curcumin conjugates induce apoptosis via a mitochondrion dependent pathway in MCF-7 and MDA-MB-231 cell lines. *Asian Pac. J. Cancer Prev. APJCP* **2013**, *14*, 5797–5804. [CrossRef] [PubMed]
241. Peng, J.R.; Qian, Z.Y. Drug delivery systems for overcoming the bioavailability of curcumin: Not only the nanoparticle matters. *Nanomed. Nanotechnol. Biol. Med.* **2014**, *9*, 747–750. [CrossRef]
242. Guo, G.; Fu, S.; Zhou, L.; Liang, H.; Fan, M.; Luo, F.; Qian, Z.; Wei, Y. Preparation of curcumin loaded poly(epsilon-caprolactone)-poly(ethylene glycol)-poly(epsilon-caprolactone) nanofibers and their in vitro antitumor activity against Glioma 9L cells. *Nanoscale* **2011**, *3*, 3825–3832. [CrossRef]
243. Wang, C.; Ma, C.; Wu, Z.K.; Liang, H.; Yan, P.; Song, J.; Ma, N.; Zhao, Q.H. Enhanced bioavailability and anticancer effect of curcumin-loaded electrospun nanofiber: In vitro and in vivo study. *Nanoscale Res. Lett.* **2015**, *10*, 439. [CrossRef] [PubMed]
244. Qin, X.; Xu, Y.; Zhou, X.; Gong, T.; Zhang, Z.R.; Fu, Y. An injectable micelle-hydrogel hybrid for localized and prolonged drug delivery in the management of renal fibrosis. *Acta Pharm. Sin. B* **2021**, *11*, 835–847. [CrossRef] [PubMed]
245. Salehi, B.; Stojanovic-Radic, Z.; Matejic, J.; Sharifi-Rad, M.; Anil Kumar, N.V.; Martins, N.; Sharifi-Rad, J. The therapeutic potential of curcumin: A review of clinical trials. *Eur. J. Med. Chem.* **2019**, *163*, 527–545. [CrossRef] [PubMed]
246. Gupta, S.C.; Patchva, S.; Aggarwal, B.B. Therapeutic roles of curcumin: Lessons learned from clinical trials. *AAPS J.* **2013**, *15*, 195–218. [CrossRef]
247. Cheng, A.L.; Hsu, C.H.; Lin, J.K.; Hsu, M.M.; Ho, Y.F.; Shen, T.S.; Ko, J.Y.; Lin, J.T.; Lin, B.R.; Ming-Shiang, W.; et al. Phase I clinical trial of curcumin, a chemopreventive agent, in patients with high-risk or pre-malignant lesions. *Anticancer Res.* **2001**, *21*, 2895–2900.
248. Dhillon, N.; Aggarwal, B.B.; Newman, R.A.; Wolff, R.A.; Kunnumakkara, A.B.; Abbruzzese, J.L.; Ng, C.S.; Badmaev, V.; Kurzrock, R. Phase II trial of curcumin in patients with advanced pancreatic cancer. *Clin. Cancer Res.* **2008**, *14*, 4491–4499. [CrossRef]
249. Mahammedi, H.; Planchat, E.; Pouget, M.; Durando, X.; Cure, H.; Guy, L.; Van-Praagh, I.; Savareux, L.; Atger, M.; Bayet-Robert, M.; et al. The new combination docetaxel, prednisone and curcumin in patients with castration-resistant prostate cancer: A pilot phase II study. *Oncology* **2016**, *90*, 69–78. [CrossRef]
250. Dutzmann, S.; Schiborr, C.; Kocher, A.; Pilatus, U.; Hattingen, E.; Weissenberger, J.; Gessler, F.; Quick-Weller, J.; Franz, K.; Seifert, V.; et al. Intratumoral concentrations and effects of orally administered micellar curcuminoids in glioblastoma patients. *Nutr. Cancer* **2016**, *68*, 943–948. [CrossRef]
251. Sharma, R.A.; Euden, S.A.; Platton, S.L.; Cooke, D.N.; Shafayat, A.; Hewitt, H.R.; Marczylo, T.H.; Morgan, B.; Hemingway, D.; Plummer, S.M.; et al. Phase I clinical trial of oral curcumin: Biomarkers of systemic activity and compliance. *Clin. Cancer Res. Off. J. Am. Assoc. Cancer Res.* **2004**, *10*, 6847–6854. [CrossRef]
252. Carroll, R.E.; Benya, R.V.; Turgeon, D.K.; Vareed, S.; Neuman, M.; Rodriguez, L.; Kakarala, M.; Carpenter, P.M.; McLaren, C.; Meyskens, F.L., Jr.; et al. Phase IIa clinical trial of curcumin for the prevention of colorectal neoplasia. *Cancer Prev. Res.* **2011**, *4*, 354–364. [CrossRef] [PubMed]

Review

Natural Products, Alone or in Combination with FDA-Approved Drugs, to Treat COVID-19 and Lung Cancer

Liyan Yang [1] and Zhonglei Wang [2,3,4,*]

[1] School of Physics and Physical Engineering, Qufu Normal University, Qufu 273165, China; yangly@iccas.ac.cn
[2] Key Laboratory of Green Natural Products and Pharmaceutical Intermediates in Colleges and Universities of Shandong Province, School of Chemistry and Chemical Engineering, Qufu Normal University, Qufu 273165, China
[3] Key Laboratory of Life-Organic Analysis of Shandong Province, School of Chemistry and Chemical Engineering, Qufu Normal University, Qufu 273165, China
[4] School of Pharmaceutical Sciences, Tsinghua University, Beijing 100084, China
* Correspondence: wangzl16@tsinghua.org.cn

Abstract: As a public health emergency of international concern, the highly contagious coronavirus disease 2019 (COVID-19) pandemic has been identified as a severe threat to the lives of billions of individuals. Lung cancer, a malignant tumor with the highest mortality rate, has brought significant challenges to both human health and economic development. Natural products may play a pivotal role in treating lung diseases. We reviewed published studies relating to natural products, used alone or in combination with US Food and Drug Administration-approved drugs, active against severe acute respiratory syndrome coronavirus 2 (SARS-CoV-2) and lung cancer from 1 January 2020 to 31 May 2021. A wide range of natural products can be considered promising anti-COVID-19 or anti-lung cancer agents have gained widespread attention, including natural products as monotherapy for the treatment of SARS-CoV-2 (ginkgolic acid, shiraiachrome A, resveratrol, and baicalein) or lung cancer (daurisoline, graveospene A, deguelin, and erianin) or in combination with FDA-approved anti-SARS-CoV-2 agents (cepharanthine plus nelfinavir, linoleic acid plus remdesivir) and anti-lung cancer agents (curcumin and cisplatin, celastrol and gefitinib). Natural products have demonstrated potential value and with the assistance of nanotechnology, combination drug therapies, and the codrug strategy, this "natural remedy" could serve as a starting point for further drug development in treating these lung diseases.

Keywords: natural product; SARS-CoV-2; lung cancer; United States Food and Drug Administration-approved drug; natural remedy

Citation: Yang, L.; Wang, Z. Natural Products, Alone or in Combination with FDA-Approved Drugs, to Treat COVID-19 and Lung Cancer. *Biomedicines* **2021**, *9*, 689. https://doi.org/10.3390/biomedicines9060689

Academic Editors: Pavel B. Drašar and Fabio Altieri

Received: 18 May 2021
Accepted: 15 June 2021
Published: 18 June 2021

Publisher's Note: MDPI stays neutral with regard to jurisdictional claims in published maps and institutional affiliations.

1. Introduction

As a traditional source for modern pharmaceutical discovery and potential drug leads, natural products have played an integral role in treating patients due to their unique structural, chemical, and biological diversity [1–3]. The current race to identify efficacious drugs, natural products with promising therapeutic effects has attracted significant attention, especially for the prevention and treatment of lung diseases, such as pulmonary fibrosis [4], asthma [5], acute lung injury [6], chronic obstructive pulmonary disease [7], defective pulmonary innate immunity [8], coronavirus disease 2019 (COVID-19) [9], and lung cancer [10]. Among the myriad of known lung maladies, COVID-19 and lung cancer are currently the most important public health concerns and burdens worldwide [11,12].

The highly contagious COVID-19 pandemic, caused by severe acute respiratory syndrome coronavirus 2 (SARS-CoV-2), has spread quickly across all continents [13,14]. Presently, this global pandemic has posed a significant threat to the lives of billions of individuals through human-to-human transmission [15,16]. In this scenario, the rapid discovery of efficacious agents against the fast-spreading COVID-19 pandemic is currently a top priority of research across the world [17]. Lung cancer, globally, is a malignant tumor with the highest mortality rate (accounting for 18% of all cancer deaths), and the five-year survival rate is very low (only 10% to 20%) [18]. Non-small cell lung cancer (NSCLC), a subtype of lung cancer with the highest incidence rate (accounting for about 85% of lung cancer [19]), has brought significant threats and challenges to human life and health as well as social and economic development. In this context, more aggressive drug trial protocols investigating anti-lung cancer agents are another top research priority [20].

Significant progress had been made in the understanding of natural products active against COVID-19 and lung cancer. However, there has been no hierarchical review (natural product, monotherapy, or in combination with a US Food and Drug Administration (FDA)-approved drug) covering the use of natural products (including natural product-based nanoparticles) as high-quality therapeutic agents for the treatment of COVID-19 or lung cancer in the literature. To underline systematically the potential importance of natural products, including their biological activity and underlying molecular mechanisms, this review will focus on the current knowledge of potential anti-COVID-19 or anti-lung cancer agents. To explore the therapeutic value of natural products better, we have focused on the current progress in representative chemical components against SARS-CoV-2 and lung cancer based on evidence from promising in vitro studies published from 1 January 2020 to 31 May 2021 by interrogating online databases (such as Google Scholar, ACS Publications, Wiley, MDPI, Web of Science, Science Direct, Springer, PubMed, and X-MOL), rather than taking an exhaustively literature-driven approach. Our purpose is to provide a promising "natural remedy" for the treatment of lung cancer and COVID-19.

2. Natural Products as Monotherapy for the Treatment of SARS-CoV-2

Natural products have demonstrated potential value, which supports this strategy as an indispensable research focus in the fight against the COVID-19 epidemic [21,22]. The chemical structures of the components described in this section are shown in Figure 1. The SARS-CoV-2 main protease (M^{pro}), also called the 3C-like protease ($3CL^{pro}$), has a vital function in viral replication and is, therefore, a preferred drug target [23]. The papain-like protease (PL^{pro}), another prime therapeutic target, plays an essential role in maturing viral RNA polyproteins and dysregulation of host inflammation [24]. Ginkgolic acid, a phenolic acid, is an essential component of the traditional herbal medicine *Ginkgo biloba* (EGb) [25]. A study has demonstrated that ginkgolic acid is characterized by half-maximal inhibitory concentration (IC_{50}) values of 1.79 μM and 16.3 μM against SARS-CoV-2 M^{pro} and SARS-CoV-2 PL^{pro}, respectively [26]. The study unambiguously showed that ginkgolic acid exerts good dual-inhibitory effects through its irreversible binding to SARS-CoV-2 cysteine proteases [26].

Angiotensin-converting enzyme 2 (ACE2), an essential ingredient of the renin–angiotensin–aldosterone system (RAAS), is a critical host cell surface receptor for viral infection [27]. The glycosylated spike protein (S protein) plays an essential role in mediating viral entry via interactions with the ACE2 cell surface receptor [28]. Hypocrellin A and shiraiachrome A, two-axial chiral perylenequinones, have been reported to exhibit potent effects on the infected monkey Vero E6 cell line by inhibiting the activity of the SARS-CoV-2 S protein at EC_{50} values of 0.22 μM and 0.21 μM, respectively, while at doses of up to 10 μM, these presented no observable cytotoxicity against these cells [29].

Transmembrane protease serine 2 (TMPRSS2), a critical factor enabling SARS-CoV-2 infection, can interact with ACE2 [30]. It has been reported that platycodin D, a triterpenoid saponin isolated from *Platycodon grandiflorum*, prevents TMPRSS2-driven infection in vitro by impairing membrane fusion [31]. Platycodin D has IC_{50} values of 0.69 μM and 0.72 μM for SARS-CoV-2 pseudovirus (pSARS-CoV-2) overexpression of ACE2 (ACE2$^+$) and ACE2/TMPRSS2$^+$, respectively, and IC_{50} values of 1.19 μM and 4.76 μM for SARS-CoV-2 in TMPRSS2-negative Vero cells and TMPRSS2-positive Calu-3 cells, respectively [31]. Resveratrol, a remarkable phytoalexin, may effectively inhibit the replication of SARS-CoV-

2 S protein in Vero E6 cells at an EC_{50} of 4.48 µM [32], and has an excellent safety tracking record, with no cytotoxicity even up to a concentration of 150 µM [33].

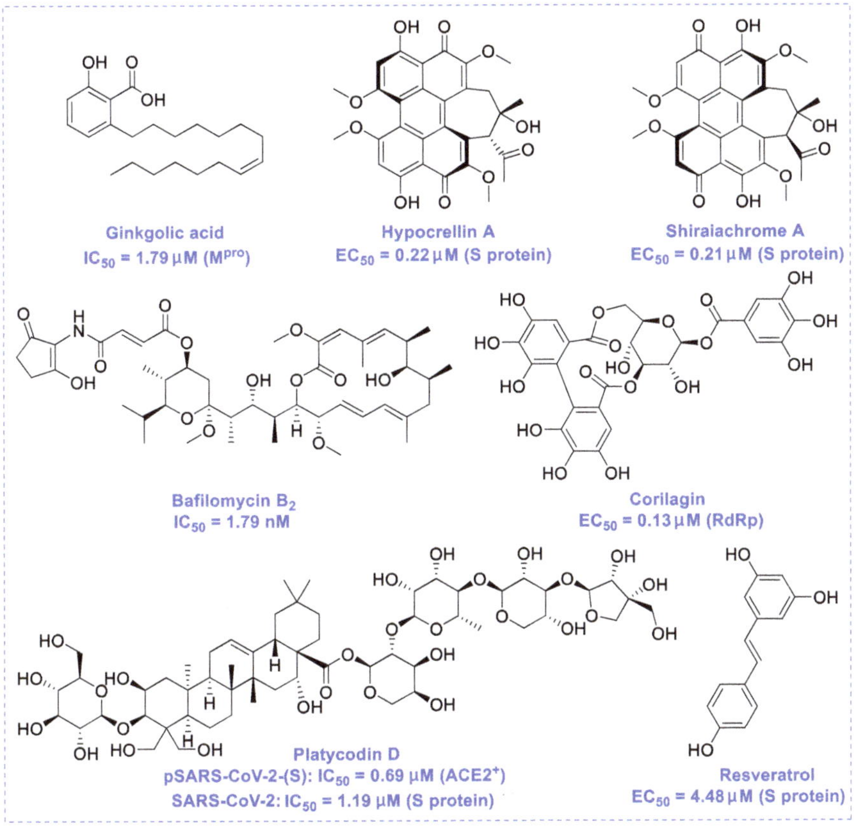

Ginkgolic acid
IC_{50} = 1.79 µM (M^{pro})

Hypocrellin A
EC_{50} = 0.22 µM (S protein)

Shiraiachrome A
EC_{50} = 0.21 µM (S protein)

Bafilomycin B_2
IC_{50} = 1.79 nM

Corilagin
EC_{50} = 0.13 µM (RdRp)

Platycodin D
pSARS-CoV-2-(S): IC_{50} = 0.69 µM (ACE2+)
SARS-CoV-2: IC_{50} = 1.19 µM (S protein)

Resveratrol
EC_{50} = 4.48 µM (S protein)

Figure 1. Promising natural products for treating SARS-CoV-2.

The RNA-dependent RNA polymerase (RdRp) of SARS-CoV-2 is another promising target that regulates the replication of the viral genome [34]. Corilagin, a non-nucleoside inhibitor, is a gallotannin isolated from the medicinal plant *Phmllanthi Fructus* [35]. Corilagin has been reported to inhibit SARS-CoV-2 infection with an EC_{50} value of 0.13 µM in a concentration-dependent manner by preventing the conformational change of RdRp and inhibits SARS-CoV-2 replication [36]. Furthermore, corilagin, as identified via molecular dynamics simulation-guided studies, could also be used as an endogenous M^{pro} candidate, with an 88% anti-SARS-CoV-2 M^{pro} activity at concentrations of 20 µM in vitro [37].

Bafilomycin B_2, which can be isolated from *Streptomyces* sp. HTL16, indicates enhanced inhibitory potency against SARS-CoV-2 at IC_{50} values of 5.11 nM (in the full-time approach) and 8.32 nM (in the pretreatment-of-virus approach) in Vero E6 cells, respectively [38]. While bafilomycin B_2 has demonstrated potential effectiveness in inhibiting the viral entry process, evidence of its utility as anti-SARS-CoV-2 agents in vivo is currently insufficient.

The above evidence supports the potential value of the above natural products as therapeutic agents for the treatment of the novel SARS-CoV-2 infection, suggesting more validation studies (both in vitro and in animal models as well as on humans) could be encouraged to perform. Besides the above-mentioned molecules, several other natural products have also been shown to exhibit potent anti-SARS-CoV-2 activities in vitro. Table 1 summarizes a range of studies investigating the in vitro effects of anti-SARS-CoV-2 agents since 2020.

Table 1. Other natural products with anti-SARS-CoV-2 activities in vitro.

No.	Name	Structure	EC$_{50}$ or IC$_{50}$ (µM)	Strain	Refs
1	Acetoside	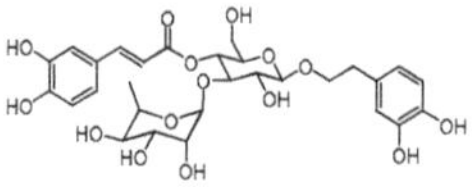	0.043	Vero E6 cells	[39]
2	Anacardic acid	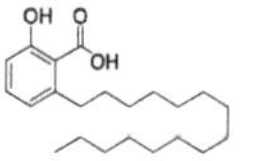	2.07	USA-WA1/2020	[26]
3	Andrographolide	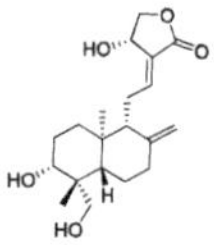	0.034	Calu-3 cells	[40,41]
4	Apigenin-7-*O*-glucoside	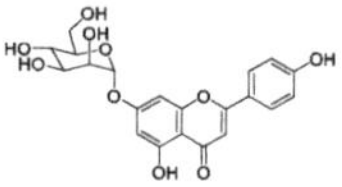	0.074	Vero E6 cells	[39]
5	Artemisinin	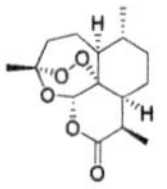	64.45	Vero E6 cells	[41,42]
6	Azithromycin	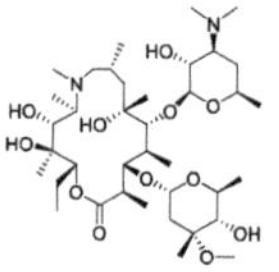	2.12	Caco-2 cells	[43]
7	Baicalin	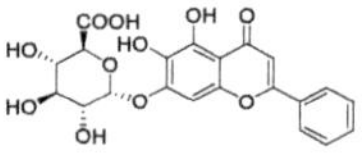	7.98	Vero E6 cells	[44]
8	Cannabidiol	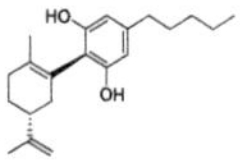	7.91	Vero E6 cells	[45,46]
9	Catechin-3-*O*-gallate	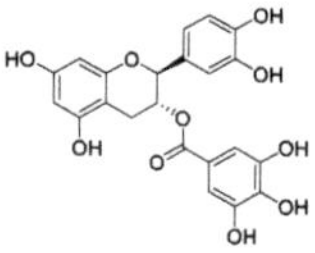	2.98	Vero E6 cells	[47]
10	Chebulagic acid	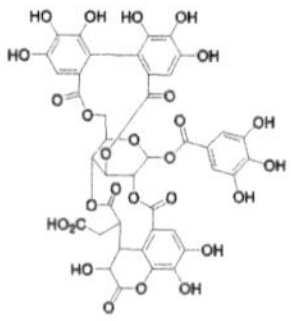	9.76	Vero E6 cells	[48]

Table 1. *Cont.*

No.	Name	Structure	EC$_{50}$ or IC$_{50}$ (µM)	Strain	Refs
11	Daurisoline	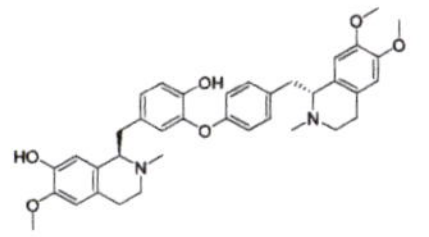	3.66	Vero E6 cells	[49]
12	EGCG	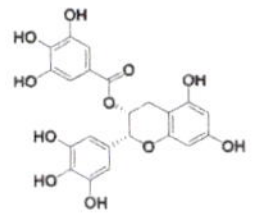	0.874	Vero E6 cells	[44,50]
13	Emetine	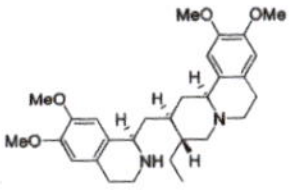	0.46	Vero E6 cells	[51,52]
14	Epicatechin-3-*O*-gallate	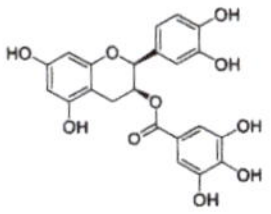	5.21	Vero E6 cells	[47]
15	Gallinamide A	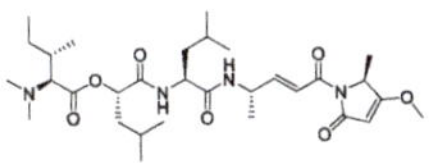	0.028	Vero E6 cells	[53]
16	Gallocatechin-3-*O*-gallate	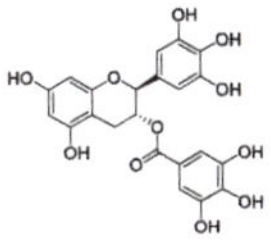	6.38	Vero E6 cells	[47]
17	Hopeaphenol	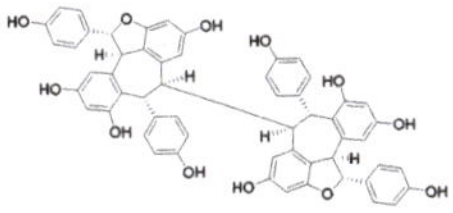	2.3	B.1.351	[54]
18	Ipomoeassin F	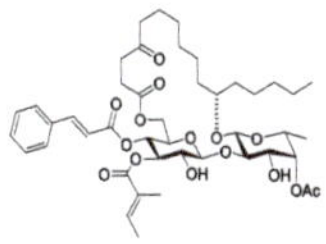		semi-permeabilized mammalian cells	[55]
19	Kobophenol A	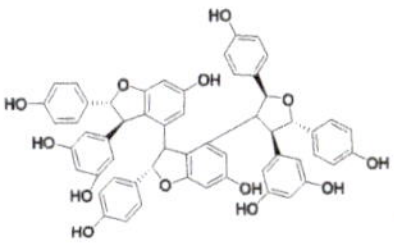	1.81	Vero E6 cells	[56]
20	Myricetin	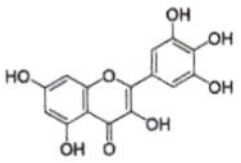	0.22	Vero E6 cells	[57,58]

Table 1. *Cont.*

No.	Name	Structure	EC$_{50}$ or IC$_{50}$ (μM)	Strain	Refs
21	Naringenin		0.092	Vero E6 cells	[39,59]
22	Osajin		3.87	Vero E6 cells	[60]
23	2,3′,4,5′,6-Pentahydroxybenzophenone		0.102	Vero E6 cells	[39]
24	Procyanidin B2		75.3	Vero E6 cells	[47]
25	Punicalagin		7.20	Vero E6 cells	[48]
26	Sennoside B		0.104	Vero E6 cells	[39]
27	Shikonin		15.75	Vero E6 cells	[61]
28	Δ9-Tetrahydrocannabinol		10.25	Vero E6 cells	[45]
29	Tetrandrine		3.00	Vero E6 cells	[60]
30	Theaflavin		8.44	HEK293T human embryonic kidney cells	[62]

Traditional Chinese medicines have attracted considerable attention due to their ability to effectively inhibit SARS-CoV-2 [63–65]. For example, the Qingfei Paidu decoction (QFPD) has shown an ability to treat COVID-19 patients at all stages with excellent clinical efficacy (cure rate >90%) [66,67]. Shuanghuanglian oral liquid or injection (SHL), another well-known traditional Chinese medicine, dose-dependently inhibits SARS-CoV-2 M^{pro} replication [68]. In addition to the above-mentioned QFPD and SHL, several other traditional Chinese medicines (such as Kegan Liyan oral liquid and Toujie Quwen granule) listed in Table 2 contain *Scutellaria baicalensis* Georgi (Chinese name: Huangqin), whose major component is baicalein, exerts a marked anti-SARS-CoV-2 effect (IC$_{50}$ of 0.94 μM, and SI > 212) [69]. Furthermore, it is crucial to investigate how herbal medicine affects SARS-CoV-2 infection by studying its active ingredients. To elucidate the underlying molecular mechanisms, a crystal structure of SARS-CoV-2 M^{pro} complexed with baicalein was constructed at a resolution of 2.2 Å (the Protein Data Bank (PDB) ID: 6M2N) [68]. Analysis of the core of the substrate-binding pocket revealed multiple interactions (such as hydrogen bonding with Leu141/Gly143 and Ser144/His163, π–π interactions with Cys145 and His4, and hydrophobic interactions with Met49 and His41), which effectively blocked SARS-CoV-2 replication via noncovalent incorporation [68]. The relevant studies [70–72] provided direct data for a better understanding of the molecular mechanisms of Chinese herbal medicine by studying its active ingredients.

Table 2. Registered clinical trials relating to traditional Chinese medicine prescriptions containing baicalein (active ingredient of Huangqin) for treatment of COVID-19 patients (Chinese Clinical Trial Registry, www.chictr.org/cn/ (accessed on 31 January 2021).

Baicalein (The Active Ingredient of Huangqin)	Molecular Mechanisms of Baicalein	Herbal Formula Containing Huangqin	Registration Number	Sample Size of the Control Group
IC$_{50}$ = 0.94 μM PDB ID: 6M2N	RdRp inhibitor via noncovalent incorporation [73], potent antagonists against TMPRSS2 [70], improving respiratory function, decreasing IL-1β and TNF-α levels, and inhibiting cell infiltration [71,72].	Qingfei Paidu decoction	ChiCTR2000029433	120
		Qingfei Paidu decoction	ChiCTR2000030883	100
		Qingfei Paidu decoction	ChiCTR2000032767	782
		Xinguan I decoction	ChiCTR2000029637	50
		Tanreqing capsules	ChiCTR2000029813	36
		Tanreqing injection	ChiCTR2000029432	72
		Kegan Liyan oral liquid	ChiCTR2000033720	240
		Kegan Liyan oral liquid	ChiCTR2000033745	240
		Kegan Liyan oral liquid	ChiCTR2000031982	240
		Shuanghuanglian oral liquid	ChiCTR2000033133	30
		Shuanghuanglian oral liquid	ChiCTR2000029605	100
		Toujie Quwen granule	ChiCTR2000031888	150

3. Natural Products as Monotherapy for the Treatment of Lung Cancer

There is no doubt that natural products have always been recognized as promising anti-lung cancer agents. Daurisoline, an autophagy blocker, is a bisbenzylisoquinoline alkaloid extracted from the herbal medicine *Nelumbo nucifera* Gaertn [74]. The chemical structures of the molecules discussed in this section are shown in Figure 2. Daurisoline increases the degradation of β-catenin by targeting heat shock protein 90 (HSP90) directly and decreases the expression of MYC proto-oncogene (c-MYC) and cyclin D1, which resulted in cell cycle arrest at the G1 phase in human lung cancer A549 cells and Hop62 cells lines to exert its anti-lung cancer activity [75]. More importantly, in animals, daurisoline has been reported to be a promising anti-lung cancer agent (by inhibiting tumor growth in lung cancer xenografts) with no observable side effects, thus highlighting a potential role for daurisoline in the treatment of lung cancer [75]. Another recent study has shown that daurisoline can effectively inhibit SARS-CoV-2 replication at IC$_{50}$ values of 3.664 μM and 0.875 μM in Vero E6 cells and in human pulmonary alveolar epithelial cells (HPAEpiC), respectively [49].

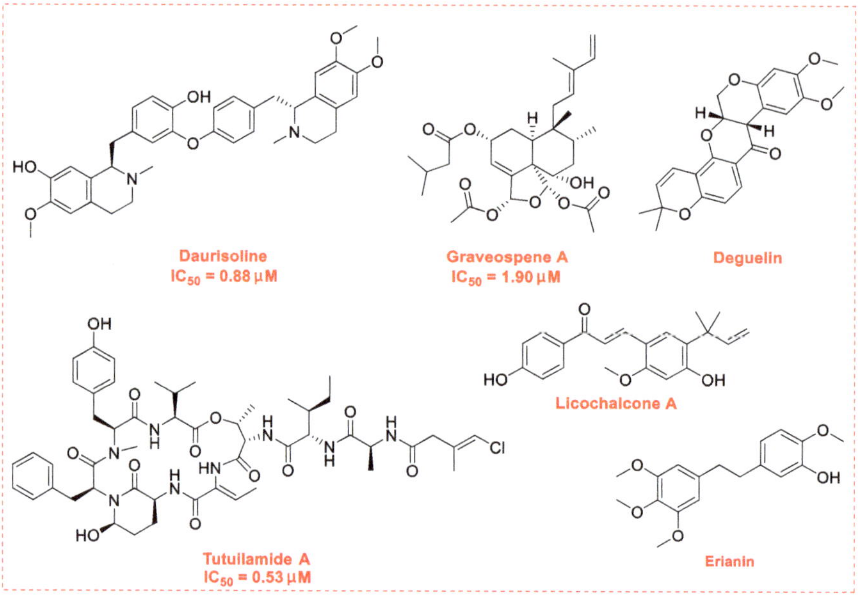

Daurisoline
IC$_{50}$ = 0.88 μM

Graveospene A
IC$_{50}$ = 1.90 μM

Deguelin

Licochalcone A

Tutuilamide A
IC$_{50}$ = 0.53 μM

Erianin

Figure 2. Promising natural products for treating lung cancer.

Graveospene A, isolated from the leaves of *Casearia graveolens*, is a new clerodane diterpenoid that has been reported to induce apoptosis in A549 cells with an IC$_{50}$ value of 1.9 μM by inducing cell cycle arrest in phase G0/G1 [76]. Deguelin, a protein kinase B (AKT) kinase inhibitor, is isolated from the African plant *Mundulea sericea* (Leguminosae) and is commonly used to inhibit the growth of several types of human cancer cell lines [77]. Deguelin promoted the phosphorylation of myeloid cell leukemia sequence-1 (Mcl-1) protein and induced the inhibition of the wildtype and mutated epidermal growth factor receptor (EGFR)-Akt signaling pathway, which resulted in activation of downstream GSK3β/FBW7 and profound anti-NSCLC activity with no obvious side effects in vivo [78].

Licochalcone A is a natural flavonoid derived from *Xinjiang licorice* and *Glycyrrhiza inflata*. Licochalcone A is known to possess a broad spectrum of activities with important pharmacological effects in various cancer cell lines [79]. Licochalcone A can significantly increase autophagic cytotoxicity (in both A549 and H460 cell lines) and downregulated the expression of c-IAP1, c-IAP2, XIAP, survivin, c-FLIPL, and RIP1, apoptosis-related proteins via inhibiting the activity of phosphorylated extracellular signal-regulated kinase (ERK) and autophagy [80]. In addition, licochalcone A has been reported to abolish the expression of programmed death ligand-1 (PD-L1) by increasing reactive oxygen species (ROS) levels in a time-dependent manner and interfering with protein translation in cancer cells [81]. Further, licochalcone A can inhibit PD-L1 translation likely through the inhibition of the phosphorylation of 4EBP1 and activation of the PERK-eIF2α signaling pathway [81]. Licochalcone A plays a vital role in reversing the ectopic expression of key microRNA (miR-144-3p, miR-20a-5p, miR-29c-3p, let-7d-3p, and miR-328-3p) to elicit lung cancer chemopreventive activities both in vivo and in vitro [82]. In addition, licochalcone A has been reported to inhibit EGFR signaling and reduced the expression of Survivin protein in a cap-dependent translation manner to exhibit profound activity in mutated NSCLC cells [83].

Erianin, a novel dibenzyl compound, can be isolated from the traditional herbal medicine *Dendrobium chrysotoxum* Lindl and has been proposed as an apoptosis-inducing agent in human lung cancer cells [84]. The main mechanisms of its anti-lung cancer activity involve the induction of ferroptosis by activating Ca^{2+}/calmodulin signaling, inhibition of cell proliferation and metastasis, and induction of cell cycle arrest in phase G2/M [85].

Tutuilamide A, isolated from marine cyanobacteria *Schizothrix* sp., is a novel cyclic peptide reported to exhibit moderate cytotoxicity activity in the H-460 human lung cancer cell line with an IC_{50} value of 0.53 μM [86]. Tutuilamide A, with the help of the vinyl chloride side chain, showed enhanced inhibitory potency with high selectivity (IC_{50} 0.73 nM) for human neutrophil elastase, which is associated mainly with the migration and metastasis of lung cancer cells [87]. Besides the above-mentioned molecules, Table 3 also exhibits other natural products (including their underlying molecular mechanisms) with notable anti-lung cancer activities reported since 2020.

Table 3. The mechanism involved in anticancer activities of other natural products (reported since 2020).

No.	Name	Structure	Mechanism of Anti-Lung Cancer	Refs
1	Acovenoside A		Inhibit the adenosine triphosphate (ATP)-dependent Na+/K+ exchange through the Na+/K+-ATPase	[88]
2	Asiatic acid		Inhibited the ionizing radiation-induced migration and invasion	[89]
3	Baicalein		Restrained ezrin tension by decreasing inducible nitric oxide synthase expression levels, suppress invasion, reduced vasculogenic mimicry formation	[90–92]
4	Baicalin		Inhibited the invasion, migration, angiogenesis, and Akt/mTOR pathway	[93,94]
5	Casticin		Induced the expressions and nuclear translocation of phosphorylation of H2AX	[95]
6	Dioscin		Down-regulated signal transducer and activator of transcription 3 and c-Jun N-terminal kinase signaling pathways	[96]
7	EGCG		Regulated CTR1 expression through the ERK1/2/NEAT1 signaling pathway	[97,98]
8	Ellagic acid		Inhibited tumor growth, increased p-AMPK, and suppressed hypoxia-inducible factor 1α levels	[99]

Table 3. *Cont.*

No.	Name	Structure	Mechanism of Anti-Lung Cancer	Refs
9	Erianthridin	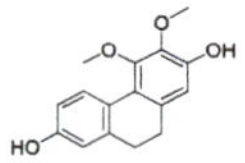	Attenuated extracellular signal-regulated kinase activity and mediated apoptosis, matrix-degrading metalloproteinases (MMPs) expression	[100,101]
10	Eugenol	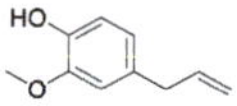	Restriction of β-catenin nuclear transportation	[102]
11	Formononetin	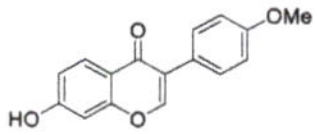	Inhibited EGFR-Akt signaling, which in turn activates GSK3β and promotes Mcl-1 phosphorylation in NSCLC cells	[103,104]
12	Gallic Acid	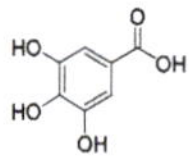	Inhibited of EGFR activation and impairment, inhibition of phosphoinositide 3-kinase (PI3K) and AKT phosphorylation	[105,106]
13	Glochidiol	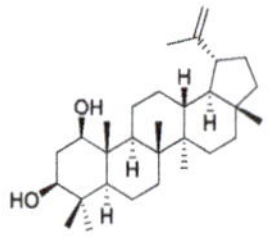	Inhibited tubulin polymerization	[107]
14	Gracillin	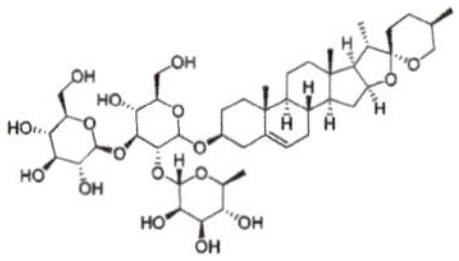	Inhibited both glycolysis and mitochondria-mediated bioenergetics, induced apoptosis through the mitochondrial pathway	[108,109]
15	Hispidulin	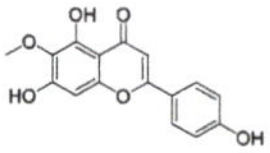	Promoted apoptosis by hispidulin via increased generation of ROS	[110]
16	Icaritin	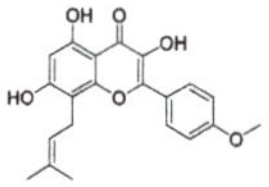	Downregulated the immunosuppressive cytokine (TNF-α, IL10, IL6) and upregulated chemotaxis (CXCL9 and CXCL10)	[111]
17	Isoharringtonine	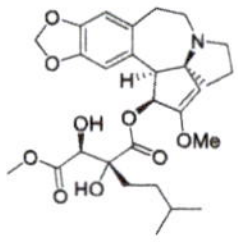	Induced death tumor spheroids by activating the intrinsic apoptosis pathway	[112]
18	Kaempferol	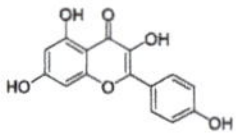	Inhibitor of nuclear factor erythroid 2-related factor 2	[113]
19	Liriopesides B	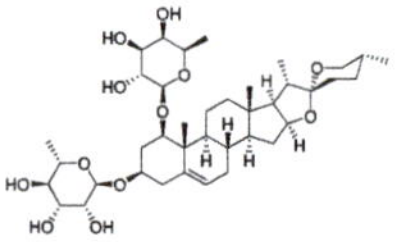	Reduced proliferation, and induced apoptosis and cell cycle arrest, inhibited the progression of the cell cycle from the G1 to the S phase	[114]

Table 3. *Cont.*

No.	Name	Structure	Mechanism of Anti-Lung Cancer	Refs
20	Nagilactone E		Activated the c-Jun N-terminal kinases, increased the phosphorylation, and promoted the localization of c-Jun in the nucleus	[115,116]
21	8-Oxo-epiberberine		Inhibited TGF-β1-induced epithelial-mesenchymal transition (EMT) possibly by interfering with Smad3	[117]
22	Parthenolide		Reduced the phosphorylation of EGFR and downstream signaling pathways mitogen-activated protein kinase (MAPK)/ERK, inhibited PI3K/Akt/FoxO3α signaling	[118–120]
23	PDB-1		Suppressed lung cancer cell migration and invasion via FAK/Src and MAPK signaling pathways	[121]
24	Polyphyllin I		Induced autophagy by activating AMPK and then inhibited mTOR signaling, promoted apoptosis, modulated the PI3K/Akt signaling	[122,123]
25	Quercetin		Inhibited proliferation and induced apoptosis	[124]
26	Silibinin		Inhibited cell proliferation, migration, invasion, and EMT expression	[125]
27	Sinomenine		Downregulated expression of MMPs and miR-21, suppressed α7 nicotinic acetylcholine receptors expression	[126–128]
28	Toxicarioside O		Decreased the expression of trophoblast cell surface antigen 2, resulting in inhibition of the PI3K/Akt pathway and EMT program	[129]
29	Vincamine		Interaction with the apoptotic protein caspase-3	[130]
30	Xanthohumol		Suppressed ERK1/2 signaling and reduced the protein levels of FOS-related antigen 1, decreased the mRNA level of cyclin D1	[131]

4. Natural Products in Combination with the FDA-Approved Drugs Inhibit SARS-CoV-2

The bisbenzylisoquinoline alkaloid cepharanthine can be isolated from the traditional herbal medicine *Stephania cephalantha* Hayata [132]. Cepharanthine exhibits a range of promising bioactivity. It has IC_{50} values of 0.026 μM, 9.5 μg/mL, and 0.83 μM against the human immunodeficiency virus type 1 (HIV-1) [133], SARS-CoV [134], and human coronavirus OC43 (HCoV-OC43) [135], respectively. This alkaloid inhibits SARS-CoV-2 entry in vitro at an IC_{50} of 0.35 μM without any evident toxicity profile (selectivity index, [SI] > 70) [136]. Furthermore, the cell death cascade induced by the cellular stress response is another key target for SARS-CoV-2 [137]. It is worth noting that this bisbenzylisoquinoline alkaloid, with a good safety profile, is an approved drug in Japan since the 1950s and is used to treat acute and chronic diseases [132], highlighting that cepharanthine can serve as a potential therapeutic candidate for the treatment of SARS-CoV-2 infection.

Nelfinavir (Viracept), the first HIV-1 protease inhibitor developed by Agouron Pharmaceuticals, was approved by the FDA in March 1997 for the treatment of HIV-AIDS [138]. Recently, nelfinavir was shown to be effective at inhibiting SARS-CoV-2 M^{pro} infection (IC_{50} = 3.3 μM) with a low level of toxicity (SI = 3.7) [139]. In addition, nelfinavir inhibited SARS-CoV-2 replication in vitro with an EC_{50} of 1.13 μM [140]. Nelfinavir was also effective at dose-dependently inhibiting SARS-CoV-2 S protein—complete inhibition at the concentration of 10 μM—with no evidence of cellular cytotoxicity [141]. Remarkably, nelfinavir can also improve lung pathology caused by SARS-CoV-2 infection [142]. Nonetheless, nelfinavir might not benefit SARS-CoV-2-infected patients by reducing viral loads in the lungs, just as it does not reduce viral load in hamsters [142].

Taken together, numerous studies have demonstrated the in vitro anti-SARS-CoV-2 activity of cepharanthine (via inhibition of SARS-CoV-2 S protein) and nelfinavir (via inhibition SARS-CoV-2 M^{pro} and partly S protein). To reveal the synergistic efficacy (Figure 3) of the above two molecules in SARS-CoV-2 infected patients, based on models of pharmacokinetics, pharmacodynamics, and viral-dynamics, Ohashi et al. constructed a mathematical prediction model of the therapeutic effects and revealed that the combination of cepharanthine (intravenous) and nelfinavir (oral) showed excellent synergistic effects in COVID-19 patients (with viral clearance occurring 1.23 days earlier than with nelfinavir alone; cepharanthine alone showed a minimal effect) [136]. Considering all these factors, including the critical value of cepharanthine and nelfinavir in anti-SARS-CoV-2 infection, both in vitro and in animal models and mathematical prediction modeling, further research is needed to explore whether these molecules exert synergistically augmented activity for the treatment of SARS-CoV-2 infection in patients. It is worth noting that further research is needed to explore whether they have anti-SARS-CoV-2 activity in vivo.

Remdesivir (GS-5734, Veklury®), an RdRp inhibitor developed by Gilead Science, was the first, and currently the only, anti-SARS-CoV-2 drug approved by the FDA (approval on 22 October 2020) for the treatment of COVID-19 [143–145]. Remdesivir exhibits broad-spectrum activity against multiple viral infections in vitro, including SARS-CoV, Middle East respiratory syndrome coronavirus (MERS-CoV), Ebola virus (EBOV), and SARS-CoV-2, with EC_{50} values of 0.069 μM, 0.090 μM, 0.012 μM, and 0.77 μM, respectively [146–149]. Furthermore, remdesivir has also been thoroughly explored in animal models. Remdesivir reduced lung viral loads in MERS-CoV-infected rhesus monkeys [150] and transgenic *Ces1c^{-}/$^{-}$ hDPP4* mice [147], protected Nipah virus-infected African green monkeys [151] and rhesus macaques from SARS-CoV-2 infection [152]. Moreover, since 2016, the efficacy and safety of remdesivir have been clinically investigated for the treatment of EBOV infection [153]. Nonetheless, the FDA-approved remdesivir does not appear highly effective in the fight against the COVID-19 pandemic [154–156]. In this scenario, the combination of remdesivir with other small molecules, including natural products and natural-product-inspired potential anti-SARS-CoV-2 agents, may exhibit a synergistic effect, compared to remdesivir alone in COVID-19 patients.

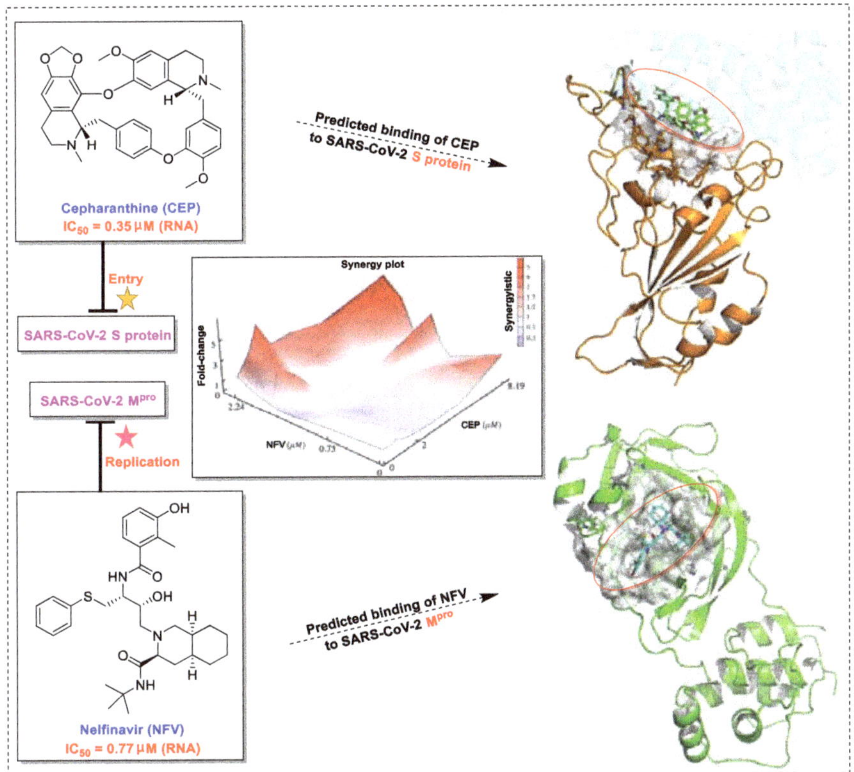

Figure 3. Cepharanthine in combination with FDA-approved nelfinavir inhibits SARS-CoV-2.

Linoleic acid, an inflammatory response modulator [157] isolated from the traditional meal *Vicia faba* [158], significantly suppresses MERS-CoV replication [159]. Toelzer et al. hypothesized that the combination of remdesivir and linoleic acid, an essential diunsaturated fatty acid, may be superior for treating COVID-19 patients over remdesivir alone [160]. Indeed, the combination of linoleic acid (50 μM) and remdesivir (20 to 200 nM) exerted a synergistic effect on SARS-CoV-2 replication in human Caco-2 ACE2+ cells in vitro [160].

The synergistic mechanisms involved in the combination of linoleic acid and remdesivir shown in Figure 4. To clarify the underlying inhibitory mechanisms of action of linoleic acid, a cryo-electron microscopy (cryo-EM) model of SARS-CoV-2 S protein complexed with linoleic acid was determined at 2.85 Å resolution (Electron Microscopy Data (EMD) ID: 11145) [160]. Further analysis of the linoleic acid binding pocket within the S protein revealed that the hydrocarbon tail of linoleic acid binds to hydrophobic amino acids. At the same time, the acidic headgroup interacts with a positively charged anchor (Arg408 and Gln409) to lock the S protein irreversibly. The hydrophobic pocket with a tube-like shape of the S protein allows a good fit for linoleic acid, and results in reduced ACE2 interactions, and thus sets the stage for an intervention strategy that targets linoleic acid binding to SARS-CoV-2 S protein [160].

As for remdesivir, it is a phosphoramidate prodrug, which requires conversion from the parent drug into the active triphosphate form (GS-443902) [161]. In cells, the triphosphate form, GS-443902, can block SARS-CoV-2 replication by evading the "proofreading" activity of viral RNA sequences [162]. In addition, Yin et al. [34] revealed the cryo-EM structure of SARS-CoV-2 RdRp in complex with remdesivir (using its triphosphate metabolite GS-443902) at 2.5 Å resolution (PDB ID: 7BV2) [34]. The cryo-EM structure unambiguously demonstrated that GS-443902 could positioned itself at the center of the catalytic site of the primer RNA, covalently binding to the primer at the 1+ position of the template strand to terminate chain elongation. Three strong H-bonds with active site residues (ribose -OH groups: Asp623, Ser682, and Asn691; sugar 2′-OH: Asn691) were identified [34]. Further research is warranted to establish whether linoleic acid and remdesivir exert synergistic anti-SARS-CoV-2 effects in vivo. At present, a more well-designed combination drug therapy that exhibits better additive or synergistic effects against COVID-19 is a promising strategy. However, for COVID-19, the nanodrug strategy (containing natural products and FDA-approved drugs) remains an open question, and undoubtedly, it has a long way to go.

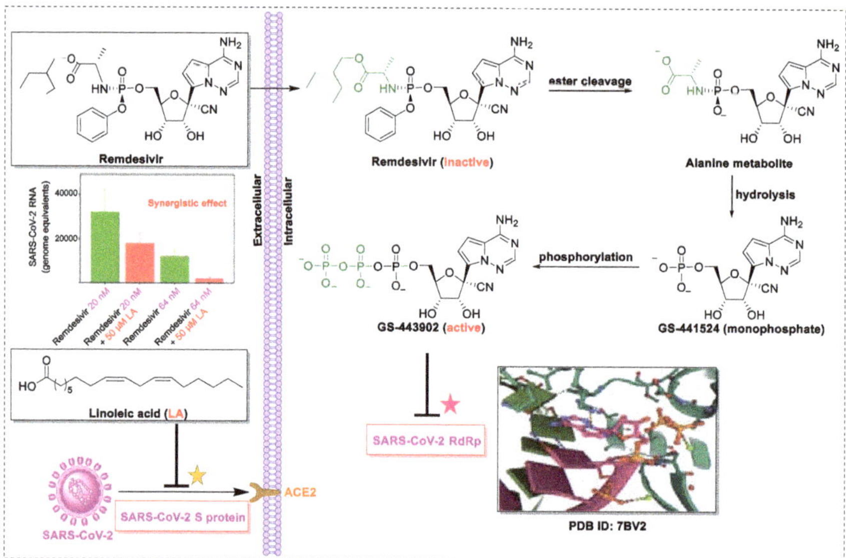

Figure 4. Linoleic acid in combination with FDA-approved remdesivir inhibit SARS-CoV-2.

5. Natural Products in Combination with the FDA-Approved Anti-Lung Cancer Drugs

As regards lung cancer, significant progress has been made in the research of natural product-based nanomedicines [163,164] and combination drug therapies [165,166], which can provide some reference for the related drug discovery and development for COVID-19. In this section, we mainly focused on the nanodrug strategy (containing natural products and FDA-approved drugs) to reveal its unique advantage in the research and development of anti-lung cancer drugs.

Curcumin is one of the main products of the *Curcuma longa* L. (turmeric) rhizome extract and has been proposed for its antimicrobial, antimutagenic, antiproliferative, and neuroprotective activities [167]. Curcumin is considered an ideal scaffold for lung cancer drug discovery due to its potent antitumor effects against NSCLC [168]. In particular, several crucial molecular pathways involved in the efficacy of curcumin as an anti-lung cancer drug involve the vascular endothelial growth factor (VEGF), EGFR, nuclear factor-κB (NF-κB), and mammalian target of rapamycin (mTOR) pathways [169]. Nonetheless, the biomedical application of curcumin is currently hindered by its poor aqueous solubility and low bioavailability [170]. In contrast, cisplatin, already marketed as the first platinum-based complex approved by the US FDA, has been used therapeutically for a broad range of cancers such as lung, lymphomas, melanoma, head, and neck cancer [171]. Unfortunately, the routine clinical practice of cisplatin is often coupled with severe toxic side effects (such as nephrotoxicity [172], severe hearing loss [173], and cardiotoxicity [174]) and intrinsic or acquired drug resistance [175].

Indeed, an efficacy study in NSCLC cells evidenced improved effects of the drug combination of curcumin and cisplatin [176]. An in vitro study showed that curcumin enhanced cisplatin-induced therapeutic efficacy in lung cancer cell lines A549, H460, and H1299 by regulating the Cu-Sp1-CTR1 regulatory loop. Furthermore, the promotion of active targeting ability with β-cyclodextrin (β-CD)-modified hyaluronic acid (HA) was identified as an effective strategy to address cellular uptake, intracellular trafficking, and therapy performance of the drug delivery systems [177]. Taking all these factors into account, Bai et al. [178] designed and constructed a β-cyclodextrin-modified hyaluronic acid-based pH- and esterase-dual-responsive supramolecular codrug combining curcumin and cisplatin (Figure 5). In detail, the designed guest moiety Cur-Pt was prepared via esterification reactions between curcumin, oxoplatin, and a molecule of succinic acid. The scheduled host moiety β-CD-modified hyaluronic acid (HA-CD) was prepared via amidation of the carboxylate salt sodium hyaluronate with free amine mono-6-deoxy-6-ethylenediamino-β-CD (β-CD-EDA). Eventually, the desired curcumin and cisplatin nanoparticles (HCPNs) were developed through a host–guest inclusion strategy and subsequent self-assembled. Interestingly, in this targeting system, curcumin acted as both the guest molecule and the chemical anticancer drug.

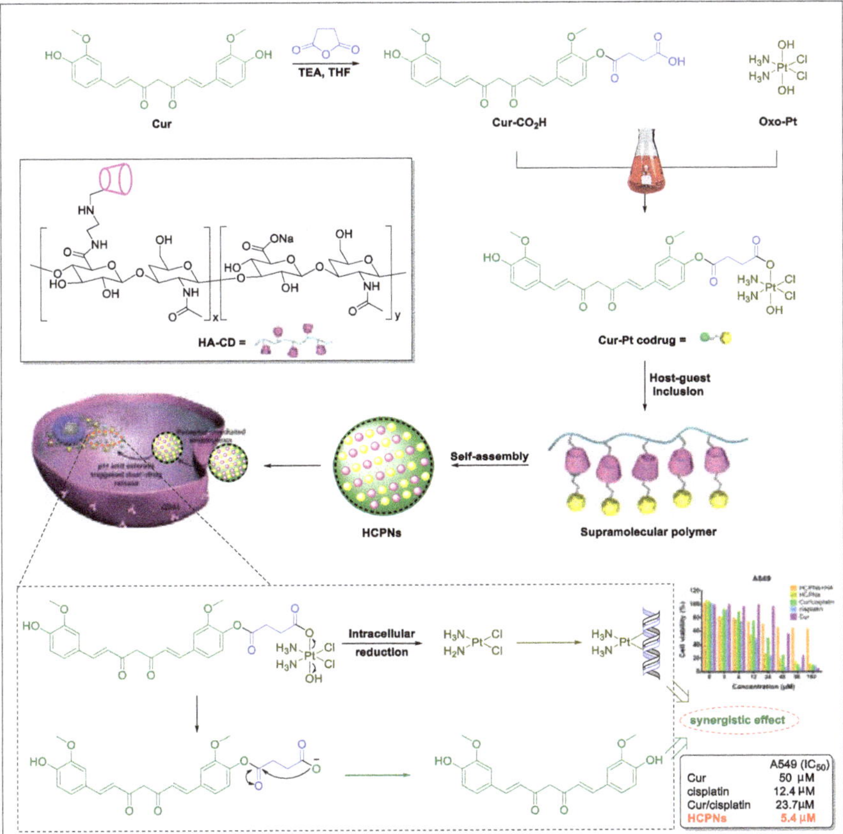

Figure 5. Synthesis, bioconversion, and synergistic effects of the nanoparticles HCPNs (image reproduced with permission from [178].

In vitro evaluation revealed that the HCPNs could be internalized by cancer cells. Once inside the cell, curcumin is released under acidic endosomal conditions (pH-responsive), and cisplatin is released via reducing of oxoplatin under higher expressed glutathione (GSH) conditions (esterase-responsive). Moreover, cell-based experiments revealed the effective cellular toxicity (high efficiency, the IC_{50} value of 5.4 μM in A549 cells) and active targeting ability (low toxicity, with low expression levels in normal LO-2 cells) of this novel drug-delivery system. Given the observed positive synergistic effect in the study, the authors concluded that HCPNs exhibited improved effects, compared with either monotherapy with curcumin or cisplatin [178]. The drug delivery and sustained release behavior of Cur from HCPNs were investigated in vitro at pH 7.4 after 48 h (11% Cur was released) and pH 5.0 after 48 h (79% Cur was released), respectively, proving the better stability than Cur alone [178]. Meanwhile, although the Tian group did not proceed further with their in vivo studies; we suggest additional in vivo studies should be performed to identify the pharmacokinetic or pharmacodynamic profile of the HCPNs and the synergistic activity against lung cancer of this codrug.

The disulfide bond, a promising redox-reactive switch in vivo, plays an essential role in many biological processes [179]. To reduce adverse effects resulting from chemotherapy regimens, the disulfide-based drug design has attracted great enthusiasm in the synthesis of prodrug or codrug, and especially for the preparation of functional nanodrugs due to their high selectivity and biocompatibility [180,181]. The nontoxic nanodrugs are activated by the excess of GSH in the tumor microenvironment, which provides an essential strategy for lung cancer-targeting treatment [182].

Celastrol, a typical pentacyclic triterpenoid, can be extracted from traditional herbal medicines of the Celastraceae family [183]. Celastrol is considered another up-and-coming natural product for lung cancer treatment due to its potent anti-NSCLC activity via its suppression of Axl protein expression [184], initiating tumor necrosis factor-related apoptosis-inducing ligand (TRAIL)-mediated apoptotic cell death [185], and suppressing cell invasion [186]. However, the clinical translation and biomedical application of celastrol are hindered due to its low bioavailability and physiological instability [187].

Gefitinib, approved by US FDA, has been used therapeutically as the first-line agent in patients with advanced lung cancer [188]. Unfortunately, the routine clinical practice of gefitinib is often coupled with severe adverse effects, such as pulmonary toxicity [189], respiratory failure, and severe comorbidities [190]. Following a reasonable design, Wu et al. developed a GSH-responsive nanodrug (identified as CEL@G-SS-NIR in Figure 6), which possesses unique therapeutic efficacy for NSCLC in mice models by inhibiting upstream and downstream EGFR signaling pathways [191]. The nanodrug CEL@G-SS-NIR was prepared in two steps: preparation of the prodrug and acquisition of the nanocomplex. As shown in Figure 6, the main molecule G-SS-NIR of the nanodrug CEL@G-SS-NIR was synthesized through a two-step reaction. First, the key intermediate G-SS was synthesized successfully in the presence of gefitinib (G), 2-hydroxyethyl disulfide (-SS-), and tiphosgene via covalent linkage. Next, the near-infrared (NIR-OH) chromophore was bound to the side chain of the G-SS to form the prodrug G-SS-NIR. The amphiphilic G-SS-NIR readily self-assembled into spherical nanomicelles in an aqueous medium (driven by the disulfide bond and the π–π interaction) and was encapsulated concomitantly the hydrophobic serine-threonine protein kinase (Akt) inhibitor celastrol (marked as CEL) to form CEL@G-SS-NIR.

This novel nanodrug CEL@G-SS-NIR possesses a suitable size (average diameter 119 ± 6 nm), outstanding overall drug loading (64.0 ± 1.4 wt.%), and excellent stability in the blood circulation, and has a rapid release rate of the free molecules (gefitinib, celastrol, and NIR-OH) at tumor region due to the breaking of the disulfide bonds in the presence of high levels of GSH [191].

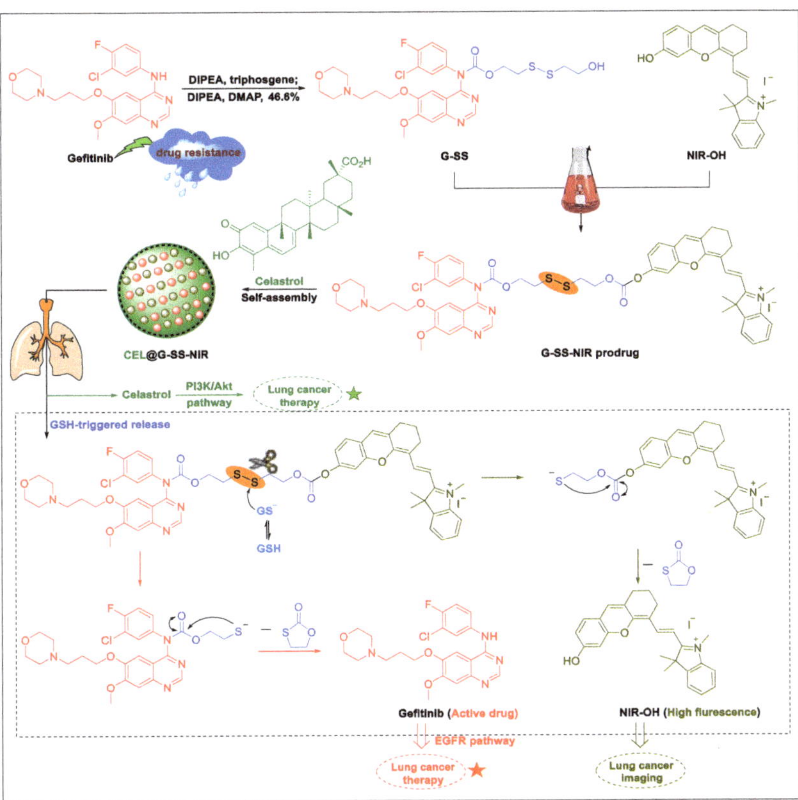

Figure 6. Synthesis, bioconversion, and synergistic effect of the nanodrug CEL@G-SS-NIR.

In vitro, the nanodrug CEL@G-SS-NIR formulation could effectively target the tumor region due to its enhanced permeability and retention effect and also allowed fluorescent imaging in vivo, at a predetermined timepoint after tail vein injection, in orthotopic lung tumors [191]. In the treatment protocol, the mice were randomly divided into five groups (five mice per treatment group), and after a single treatment cycle, the CEL@G-SS-NIR group (13.4 mg/kg, intravenously, for 20 days), compared to the control groups, exhibited stronger NSCLC tumor-suppressive effects [191]. As for the response mechanism involved, the entire process can be divided into four steps: (i) CEL@G-SS-NIR accumulates in the lung tumor region, (ii) CEL@G-SS-NIR releases the drug celastrol and the protonated intermediates (and) through the deprotonated glutathione (GS$^-$) nucleophilic attack of the disulfide bond on G-SS-NIR bonds, (iii) this further induces the synchronous releases of the parent drug gefitinib and the fluorescent dye NIR-OH via an intramolecular cyclization reaction (thiolate anion moiety reacts with the adjacent carbonyl group), and (iv) finally, the synergistic anticancer activity is activated by suppressing the phosphatidylinositol 3-kinase/serine threonine protein kinase (PI3K/Akt) signaling pathway by celastrol and downregulating EGFR signaling pathway by gefitinib. Simultaneously, a fluorescent and multispectral optoacoustic tomography imaging signal is generated by NIR-OH [192]. This study showed that disulfide-based and targeted fluorescent nano-prodrugs for treating NSCLC and tracking drug delivery systems are particularly advantageous.

6. Conclusions and Future Perspectives

COVID-19 and lung cancer, the two most critical lung diseases presenting high mortality rates, have posed a great challenge and a serious threat to human health and economic development. Since 2020, as is well-known, the scientific community has made great efforts and remarkable inroads in developing promising anti-SARS-CoV-2 and anti-lung cancer agents through various approaches. In this scenario, numerous natural products have fueled significant attention and have shown good results as potential therapeutics for the above-mentioned lung diseases. This review highlighted state-of-the-art of important natural products (including their underlying molecular mechanisms), covering studies published between 1 January 2020 and 31 May 2021, in the treatment of the above-mentioned lung diseases. We found that natural products can be applied in vitro as monotherapy for the treatment of SARS-CoV-2 (ginkgolic acid, resveratrol, and baicalein) and lung cancer (graveospene A, deguelin, and erianin), as well as in combination with the FDA-approved drug inhibit SARS-CoV-2 (cepharanthine plus nelfinavir, linoleic acid plus remdesivir) and as codrug formulations with anti-lung cancer activity in vitro (codrug of curcumin and cisplatin). The evidence revealed herein that natural products could serve as a starting point for further drug development both in COVID-19 and lung cancer. It is worth noting, however, that some natural products could be pan-assay interference compounds, which can give false readouts, and close attention should be paid to decrease futile attempts [193,194]. There is currently very little direct data associated with the clinical effect of natural products against SARS-CoV-2 infection. To understand better and explore systematically the activity of natural products, more validation studies, with high-quality evidence (both in vitro and in animal models as well as on humans), are now needed.

To improve the use of natural products, many intensive research efforts (both in vitro and in vivo) are still needed to explore the limitations of these agents, such as poor water solubility, limited oral absorption, low bioavailability, and the poor first-pass effect, which represent the first step to develop promising anti-COVID-19 or anti-lung cancer agents. It is clear that a long way is still ahead for us to realize natural product-based drug discovery and development, as only phase 1–3 clinical trials can ensure that any small molecule inhibitor can be used as a drug.

More aggressive and well-designed combination drug therapies that exhibit better additive or synergistic effects against COVID-19 and lung cancer are a promising strategy. For example, shiraiachrome A exhibits potent effects in Vero E6 cells by inhibiting the activity of the SARS-CoV-2 S protein at EC_{50} values of 0.21 μM; bafilomycin B_2 presents enhanced inhibitory potency against SARS-CoV-2 at IC_{50} values of 5.11 nM in Vero E6 cells by inhibiting the viral entry process; ginkgolic acid has IC_{50} values of 1.79 μM and 16.3 μM against SARS-CoV-2 M^{pro} and SARS-CoV-2 PL^{pro}. Combining the properties of the above-mentioned natural products with FDA-approved drugs (for example, with nelfinavir or remdesivir) could achieve optimal COVID-19 treatment through multitargeted mechanisms of action. In addition, a codrug of a natural product with an FDA-approved drug could achieve a combination booster through multitargeted activity. However, the codrug strategy remains an open question in the treatment of patients with COVID-19. Thus, we suggest researchers pay considerable attention to the development of emerging codrug therapy strategies.

In contrast, precisely fabricated nanodrugs may be a more potent weapon to enhance biocompatibility, minimize toxicity as well as side effects, achieve long-term circulation in the body, as well as sustained release, overcome undesired adverse effects, and expand the modalities of administration (intravenous injection or inhalation). However, for COVID-19, the nanodrug strategy (containing natural products and FDA-approved drugs) remains another open question. Fortunately, significant progress has been made in the research of lung cancer nanomedicines, which can provide some reference for the related drug discovery and development for COVID-19. There is no doubt that there is a long way to go and many difficulties to overcome. Nonetheless, natural products have their advantages. We sincerely hope natural products will be proven a safe and effective "natural remedy" for the treatment of the above-mentioned lung diseases with the assistance of multiple techniques and strategies.

Author Contributions: Conceptualization, Z.W.; investigation, L.Y.; resources, L.Y.; writing—original draft preparation, L.Y.; writing—review and editing, Z.W.; project administration, Z.W. All authors have read and agreed to the published version of the manuscript.

Funding: This work was supported by the project of the PhD research start-up funds of Qufu Normal University, China (Grant No. 614901, and 615201). The authors, therefore, gratefully acknowledge the Qufu Normal University for the financial supports.

Institutional Review Board Statement: Not applicable.

Informed Consent Statement: Not applicable.

Data Availability Statement: Not applicable.

Conflicts of Interest: The authors declare no conflict of interest.

Abbreviations

ACE2	angiotensin-converting enzyme 2
ACE2+	overexpression of ACE2
AKT	protein kinase B
ATP	adenosine triphosphate
β-CD	β-cyclodextrin
3CLpro	3C-Like protease
COVID-19	coronavirus disease 2019
cryo-EM	cryo-electron microscopy
EBOV	Ebola virus
EC50	half-maximal effective concentration
EGFR	epidermal growth factor receptor
EMT	epithelial-mesenchymal transition
ERK	extracellular signal-regulated kinase
FDA	US Food and Drug Administration
GSH	glutathione
HA	hyaluronic acid
HCoV-229E	human coronavirus 229E
HCPNs	curcumin and cisplatin nanoparticles
HIV-1	human immunodeficiency virus type 1
IC50	half-maximal inhibitory concentration
MAPK	mitogen-activated protein kinase
Mcl-1	myeloid cell leukemia sequence-1
MERS-CoV	Middle East respiratory syndrome coronavirus
MMPs	matrix-degrading metalloproteinases
Mpro	main protease
mTOR	mammalian target of rapamycin
NSCLC	non-small cell lung cancer
PDB	Protein Data Bank
PD-L1	programmed death ligand-1
PI3K	phosphoinositide 3-kinase
PLpro	papain-like protease
pSARS-CoV-2	SARS-CoV-2 pseudovirus
QFPD	Qingfei Paidu decoction
RdRp	RNA-dependent RNA polymerase
ROS	reactive oxygen species
S protein	spike protein
SARS-CoV	severe acute respiratory syndrome coronavirus
SARS-CoV-2	severe acute respiratory syndrome coronavirus 2
SHL	Shuanghuanglian oral liquid or injection
SI	selectivity index
TMPRSS2	transmembrane protease serine 2

References

1. Newman, D.J.; Cragg, G.M. Natural products as sources of new drugs over the nearly four decades from 01/1981 to 09/2019. *J. Nat. Prod.* **2020**, *83*, 770–803. [CrossRef] [PubMed]
2. Atanasov, A.G.; Zotchev, S.B.; Dirsch, V.M.; Supuran, C.T. Natural products in drug discovery: Advances and opportunities. *Nat. Rev. Drug Discov.* **2021**, *20*, 200–216. [CrossRef]
3. Porras, G.; Chassagne, F.; Lyles, J.T.; Marquez, L.; Dettweiler, M.; Salam, A.M.; Samarakoon, T.; Shabih, S.; Farrokhi, D.R.; Quave, C.L. Ethnobotany and the role of plant natural products in antibiotic drug discovery. *Chem. Rev.* **2021**, *121*, 3495–3560. [CrossRef]
4. Wang, L.; Li, S.; Yao, Y.; Yin, W.; Ye, T. The role of natural products in the prevention and treatment of pulmonary fibrosis: A review. *Food Funct.* **2021**, *12*, 990–1007. [CrossRef]
5. Wang, W.; Yao, Q.; Teng, F.; Cui, J.; Dong, J.; Wei, Y. Active ingredients from Chinese medicine plants as therapeutic strategies for asthma: Overview and challenges. *Biomed. Pharmacother.* **2021**, *137*, 111383. [CrossRef]
6. He, Y.Q.; Zhou, C.C.; Yu, L.Y.; Wang, L.; Deng, J.L.; Tao, Y.L.; Zhang, F.; Chen, W.S. Natural product derived phytochemicals in managing acute lung injury by multiple mechanisms. *Pharmacol. Res.* **2021**, *163*, 105224. [CrossRef]
7. Santoro, A.; Tomino, C.; Prinzi, G.; Cardaci, V.; Fini, M.; Macera, L.; Russo, P.; Maggi, F. Microbiome in chronic obstructive pulmonary disease: Role of natural products against microbial pathogens. *Curr. Med. Chem.* **2020**, *27*, 2931–2948. [CrossRef] [PubMed]
8. Belchamber, K.B.R.; Donnelly, L.E. Targeting defective pulmonary innate immunity-A new therapeutic option? *Pharmacol. Therapeut.* **2020**, *209*, 107500. [CrossRef]
9. Christy, M.P.; Uekusa, Y.; Gerwick, L.; Gerwick, W.H. Natural products with potential to treat RNA virus pathogens including SARS-CoV-2. *J. Nat. Prod.* **2021**, *84*, 161–182. [CrossRef] [PubMed]
10. Wen, T.; Song, L.; Hua, S. Perspectives and controversies regarding the use of natural products for the treatment of lung cancer. *Cancer Med.* **2021**, *10*, 2396–2422. [CrossRef]
11. Nalbandian, A.; Sehgal, K.; Gupta, A.; Madhavan, M.V.; McGroder, C.; Stevens, J.S.; Cook, J.R.; Nordvig, A.S.; Shalev, D.; Sehrawat, T.S.; et al. Post-acute COVID-19 syndrome. *Nature Med.* **2021**, *27*, 601–615. [CrossRef] [PubMed]
12. Siegel, R.L.; Miller, K.D.; Fuchs, H.E.; Jemal, A. Cancer Statistics. *CA Cancer J. Clin.* **2021**, *71*, 7–33.
13. Eurosurveillance Editorial Team. Note from the editors: World Health Organization declares novel coronavirus (2019-nCoV) sixth public health emergency of international concern. *Euro. Surveil.* **2020**, *25*, 200131e.
14. Wang, C.; Horby, P.W.; Hayden, F.G.; Gao, G.F. A novel coronavirus outbreak of global health concern. *Lancet* **2020**, *395*, 470–473. [CrossRef]
15. Chan, J.F.; Yuan, S.; Kok, K.H.; To, K.K.; Chu, H.; Yang, J.; Xing, F.F.; Liu, J.L.; Yip, C.C.; Poon, R.W.; et al. A familial cluster of pneumonia associated with the 2019 novel coronavirus indicating person-to-person transmission: A study of a family cluster. *Lancet* **2020**, *395*, 514–523. [CrossRef]
16. Phan, L.T.; Nguyen, T.V.; Luong, Q.C.; Nguyen, T.V.; Nguyen, H.T.; Le, H.Q.; Nguyen, T.T.; Cao, T.M.; Pham, Q.D. Importation and human-to-human transmission of a novel coronavirus in Vietnam. *N. Engl. J. Med.* **2020**, *382*, 872–874. [CrossRef]
17. Asselah, T.; Durantel, D.; Pasmant, E.; Lau, G.; Schinazi, R.F. COVID-19: Discovery, diagnostics and drug development. *J. Hepatol.* **2021**, *74*, 168–184. [CrossRef] [PubMed]
18. Sung, H.; Ferlay, J.; Siegel, R.L.; Laversanne, M.; Soerjomataram, I.; Jemal, A.; Bray, F. Global cancer statistics 2020: GLOBOCAN estimates of incidence and mortality worldwide for 36 cancers in 185 countries. *CA Cancer J. Clin.* **2021**, *71*, 209–249. [CrossRef]
19. Singh, S.S.; Mattheolabakis, G.; Gu, X.; Withers, S.; Dahal, A.; Jois, S. A grafted peptidomimetic for EGFR heterodimerization inhibition: Implications in NSCLC models. *Eur. J. Med. Chem.* **2021**, *216*, 113312. [CrossRef]
20. Dawkins, J.B.N.; Webster, R.M. The small-cell lung cancer drug market. *Nat. Rev. Drug Discov.* **2020**, *19*, 507–508. [CrossRef] [PubMed]
21. Verma, S.; Twilley, D.; Esmear, T.; Oosthuizen, C.B.; Reid, A.M.; Nel, M.; Lall, N. Anti-SARS-CoV natural products with the potential to inhibit SARS-CoV-2 (COVID-19). *Front. Pharmacol.* **2020**, *11*, 561334. [CrossRef] [PubMed]
22. Wang, Z.; Yang, L. Turning the tide: Natural products and natural-product-inspired chemicals as potential counters to SARS-CoV-2 infection. *Front. Pharmacol.* **2020**, *11*, 1013. [CrossRef] [PubMed]
23. Qiao, J.; Li, Y.S.; Zeng, R.; Liu, F.L.; Luo, R.H.; Huang, C.; Wang, Y.F.; Zhang, J.; Quan, B.; Shen, C.; et al. SARS-CoV-2 M^{pro} inhibitors with antiviral activity in a transgenic mouse model. *Science* **2021**, *371*, 1374–1378. [CrossRef]
24. Shin, D.; Mukherjee, R.; Grewe, D.; Bojkova, D.; Baek, K.; Bhattacharya, A.; Schulz, L.; Widera, M.; Mehdipour, A.R.; Tascher, G.; et al. Papain-like protease regulates SARS-CoV-2 viral spread and innate immunity. *Nature* **2020**, *587*, 657–662. [CrossRef] [PubMed]
25. Yang, J.; Feng, Z.; Liu, W.; Wang, Y.; Wang, G.; Yu, W.; Yang, G.; Yang, T.; Wang, Y.; Li, M. Exogenous hormone on episperm development and ginkgolic acid accumulation in *Ginkgo biloba* L. *Ind. Crop. Prod.* **2021**, *160*, 113140. [CrossRef]
26. Chen, Z.; Cui, Q.; Cooper, L.; Zhang, P.; Lee, H.; Chen, Z.; Wang, Y.; Liu, X.; Rong, L.; Du, R. Ginkgolic acid and anacardic acid are specific covalent inhibitors of SARS-CoV-2 cysteine proteases. *Cell Biosci.* **2021**, *11*, 45. [CrossRef] [PubMed]
27. Medina-Enríquez, M.M.; Lopez-León, S.; Carlos-Escalante, J.A.; Aponte-Torres, Z.; Cuapio, A.; Wegman-Ostrosky, T. ACE2: The molecular doorway to SARS-CoV-2. *Cell Biosci.* **2020**, *10*, 148. [CrossRef]
28. Benton, D.J.; Wrobel, A.G.; Xu, P.; Roustan, C.; Martin, S.R.; Rosenthal, P.B.; Skehel, J.J.; Gamblin, S.J. Receptor binding and priming of the spike protein of SARS-CoV-2 for membrane fusion. *Nature* **2020**, *588*, 327–330. [CrossRef] [PubMed]

29. Li, Y.T.; Yang, C.; Wu, Y.; Lv, J.J.; Feng, X.; Tian, X.; Zhou, Z.; Pan, X.; Liu, S.; Tian, L.W. Axial chiral binaphthoquinone and perylenequinones from the stromata of hypocrella bambusae are SARS-CoV-2 entry inhibitors. *J. Nat. Prod.* **2021**, *84*, 436–443. [CrossRef]

30. Stopsack, K.H.; Mucci, L.A.; Antonarakis, E.S.; Nelson, P.S.; Kantoff, P.W. TMPRSS2 and COVID-19: Serendipity or opportunity for intervention? *Cancer Discov.* **2020**, *10*, 779–782. [CrossRef]

31. Kim, T.Y.; Jeon, S.; Jang, Y.; Gotina, L.; Won, J.; Ju, Y.H.; Kim, S.; Jang, M.W.; Won, W.; Park, M.G.; et al. Platycodin D prevents both lysosome- and TMPRSS2-driven SARS-CoV-2 infection in vitro by hindering membrane fusion. *Exp. Mol. Med.* **2021**. [CrossRef]

32. Yang, M.; Wei, J.; Huang, T.; Lei, L.; Shen, C.; Lai, J.; Yang, M.; Liu, L.; Yang, Y.; Liu, G.; et al. Resveratrol inhibits the replication of severe acute respiratory syndrome coronavirus 2 (SARS-CoV-2) in cultured Vero cells. *Phytother. Res.* **2021**, *35*, 1127–1129. [CrossRef]

33. Ter Ellen, B.M.; Dinesh Kumar, N.; Bouma, E.M.; Troost, B.; van de Pol, D.P.I.; van der Ende-Metselaar, H.H.; Apperloo, L.; van Gosliga, D.; van den Berge, M.; Nawijn, M.C.; et al. Resveratrol and pterostilbene potently inhibit SARS-CoV-2 replication in vitro. *bioRxiv* **2020**. [CrossRef]

34. Yin, W.; Mao, C.; Luan, X.; Shen, D.D.; Shen, Q.; Su, H.; Wang, X.; Zhou, F.; Zhao, W.; Gao, M.; et al. Structural basis for inhibition of the RNA-dependent RNA polymerase from SARS-CoV-2 by remdesivir. *Science* **2020**, *368*, 1499–1504. [CrossRef] [PubMed]

35. Wu, C.; Huang, H.; Choi, H.Y.; Ma, Y.; Zhou, T.; Peng, Y.; Pang, K.; Shu, G.; Yang, X. Anti-esophageal cancer effect of corilagin extracted from Phmllanthi fructus via the mitochondrial and endoplasmic reticulum stress pathways. *J. Ethnopharmacol.* **2021**, *269*, 113700. [CrossRef] [PubMed]

36. Li, Q.; Yi, D.; Lei, X.; Zhao, J.; Zhang, Y.; Cui, X.; Xiao, X.; Jiao, T.; Dong, X.; Zhao, X.; et al. Corilagin inhibits SARS-CoV-2 replication by targeting viral RNA-dependent RNA polymerase. *Acta Pharm. Sin. B* **2021**. [CrossRef]

37. Loschwitz, J.; Jäckering, A.; Keutmann, M.; Olagunju, M.; Eberle, R.J.; Coronado, M.A.; Olubiyi, O.O.; Strodel, B. Novel inhibitors of the main protease enzyme of SARS-CoV-2 identified via molecular dynamics simulation-guided in vitro assay. *Bioorg. Chem.* **2021**, *111*, 104862. [CrossRef]

38. Xie, X.; Lu, S.; Pan, X.; Zou, M.; Li, F.; Lin, H.; Hu, J.; Fan, S.; He, J. Antiviral bafilomycins from a feces-inhabiting Streptomyces sp. *J. Nat. Prod.* **2021**, *84*, 537–543. [CrossRef]

39. Abdallah, H.; El-Halawany, A.; Sirwi, A.; El-Araby, A.; Mohamed, G.; Ibrahim, S.; Koshak, A.; Asfour, H.; Awan, Z.; Elfaky, M.A. Repurposing of Some Natural Product Isolates as SARS-COV-2 Main Protease Inhibitors via In Vitro Cell Free and Cell-Based Antiviral Assessments and Molecular Modeling Approaches. *Pharmaceuticals* **2021**, *14*, 213. [CrossRef] [PubMed]

40. Sa-ngiamsuntorn, K.; Suksatu, A.; Pewkliang, Y.; Thongsri, P.; Kanjanasirirat, P.; Manopwisedjaroen, S.; Charoensutthivarakul, S.; Wongtrakoongate, P.; Pitiporn, S.; Chaopreecha, J.; et al. Anti-SARS-CoV-2 Aactivity of Andrographis paniculata extract and its major component andrographolide in human lung epithelial cells and cytotoxicity evaluation in major organ cell representatives. *J. Nat. Prod.* **2021**, *84*, 1261–1270. [CrossRef]

41. Hu, Y.; Liu, M.; Qin, H.; Lin, H.; An, X.; Shi, Z.; Song, L.; Yang, X.; Fan, H.; Tong, Y. Artemether, artesunate, arteannuin B, echinatin, licochalcone B and andrographolide effectively inhibit SARS-CoV-2 and related viruses in vitro. *Front. Cell. Infect. Microbiol.* **2021**. [CrossRef]

42. Cao, R.; Hu, H.; Li, Y.; Wang, X.; Xu, M.; Liu, J.; Zhang, H.; Yan, Y.; Zhao, L.; Li, W.; et al. Anti-SARS-CoV-2 potential of artemisinins in vitro. *ACS Infect. Dis.* **2020**, *6*, 2524–2531. [CrossRef] [PubMed]

43. Touret, F.; Gilles, M.; Barral, K.; Nougairède, A.; Decroly, E.; de Lamballerie, X.; Coutard, B. In vitro screening of a FDA approved chemical library reveals potential inhibitors of SARS-CoV-2 replication. *Sci. Rep.* **2020**, *10*, 13093. [CrossRef] [PubMed]

44. Hong, S.; Seo, S.H.; Woo, S.J.; Kwon, Y.; Song, M.; Ha, N.C. Epigallocatechin gallate inhibits the uridylate-specific endoribonuclease Nsp15 and efficiently neutralizes the SARS-CoV-2 strain. *J. Agric. Food Chem.* **2021**. [CrossRef] [PubMed]

45. Raj, V.; Park, J.G.; Cho, K.H.; Choi, P.; Kim, T.; Ham, J.; Lee, J. Assessment of antiviral potencies of cannabinoids against SARSCoV-2 using computational and in vitro approaches. *Int. J. Biol. Macromol.* **2021**, *168*, 474–485. [CrossRef]

46. Nguyen, L.C.; Yang, D.; Nicolaescu, V.; Best, T.; Chen, S.; Friesen, J.B.; Drayman, N.; Mohamed, A.; Dann, C.; Silva, D.; et al. Cannabidiol inhibits SARS-CoV-2 replication and promotes the host innate immune response. *bioRxiv* **2021**. [CrossRef]

47. Zhu, Y.; Xie, D.Y. Docking characterization and in vitro inhibitory activity of flavan-3-ols and dimeric proanthocyanidins against the main protease activity of SARS-CoV-2. *Front. Plant Sci.* **2020**, *11*, 601316. [CrossRef] [PubMed]

48. Du, R.; Cooper, L.; Chen, Z.; Lee, H.; Rong, L.; Cui, Q. Discovery of chebulagic acid and punicalagin as novel allosteric inhibitors of SARS-CoV-2 3CLpro. *Antivir. Res.* **2021**, *190*, 105075. [CrossRef] [PubMed]

49. Wang, P.; Luo, R.; Zhang, M.; Wang, Y.; Song, T.; Tao, T.; Li, Z.; Jin, L.; Zheng, H.; Chen, W.; et al. A cross-talk between epithelium and endothelium mediates human alveolar-capillary injury during SARS-CoV-2 infection. *Cell Death Dis.* **2020**, *11*, 1042. [CrossRef]

50. Du, A.; Zheng, R.; Disoma, C.; Li, S.; Chen, Z.; Li, S.; Liu, P.; Zhou, Y.; Shen, Y.; Liu, S.; et al. Epigallocatechin-3-gallate, an active ingredient of Traditional Chinese Medicines, inhibits the 3CLpro activity of SARS-CoV-2. *Int. J. Biol. Macromol.* **2021**, *176*, 1–12. [CrossRef]

51. Choy, K.T.; Wong, A.Y.L.; Kaewpreedee, P.; Sia, S.F.; Chen, D.D.; Hui, K.P.Y.; Chu, D.K.W.; Chan, M.C.W.; Cheung, P.P.H.; Huang, X.; et al. Remdesivir, lopinavir, emetine, and homoharringtonine inhibit SARS-CoV-2 replication in vitro. *Antivir. Res.* **2020**, *178*, 104786. [CrossRef] [PubMed]

52. Kumar, R.; Afsar, M.; Khandelwal, N.; Chander, Y.; Riyesh, T.; Dedar, R.K.; Gulati, B.R.; Pal, Y.; Barua, S.; Tripathi, B.N.; et al. Emetine suppresses SARS-CoV-2 replication by inhibiting interaction of viral mRNA with eIF4E. *Antivir. Res.* **2021**, *189*, 105056. [CrossRef]

53. Ashhurst, A.; Tang, A.; Fajtova, P.; Yoon, M.; Aggarwal, A.; Stoye, A.; Larance, M.; Beretta, L.; Drelich, A.; Skinner, D.; et al. Potent in vitro anti-SARS-CoV-2 activity by gallinamide A and analogues via inhibition of cathepsin L. *bioRxiv* **2020**. [CrossRef]

54. Tietjen, I.; Cassel, J.; Register, E.T.; Zhou, X.Y.; Messick, T.E.; Keeney, F.; Lu, L.D.; Beattie, K.D.; Rali, T.; Ertl, H.C.J.; et al. The natural stilbenoid (-)-hopeaphenol inhibits cellular entry of SARS-CoV-2 USA-WA1/2020, B.1.1.7 and B.1.351 variants. *bioRxiv* **2021**. [CrossRef]

55. O'Keefe, S.; Roboti, P.; Duah, K.B.; Zong, G.; Schneider, H.; Shi, W.Q.; High, S. Ipomoeassin-F inhibits the in vitro biogenesis of the SARS-CoV-2 spike protein and its host cell membrane receptor. *J. Cell Sci.* **2021**, *134*, jcs257758. [CrossRef] [PubMed]

56. Gangadevi, S.; Badavath, V.N.; Thakur, A.; Yin, N.; Jonghe, S.D.; Acevedo, O.; Jochmans, D.; Leyssen, P.; Wang, K.; Neyts, J.; et al. Kobophenol A inhibits binding of host ACE2 receptor with Spike RBD domain of SARS-CoV-2, a lead compound for blocking COVID-19. *J. Phys. Chem. Lett.* **2021**, *12*, 1793–1802. [CrossRef]

57. Kuzikov, M.; Costanzi, E.; Reinshagen, J.; Esposito, F.; Vangeel, L.; Wolf, M.; Ellinger, B.; Claussen, C.; Geisslinger, G.; Corona, A.; et al. Identification of inhibitors of SARS-CoV-2 3CL-Pro enzymatic activity using a small molecule in-vitro repurposing screen. *ACS Pharmacol. Transl. Sci.* **2021**. [CrossRef]

58. Xiao, T.; Cui, M.; Zheng, C.; Wang, M.; Sun, R.; Gao, D.; Bao, J.; Ren, S.; Yang, B.; Lin, J.; et al. Myricetin inhibit SARS-CoV-2 viral replication by targeting Mpro and ameliorates pulmonary inflammation. *Front. Pharmacol.* **2021**. [CrossRef]

59. Clementi, N.; Scagnolari, C.; D'Amore, A.; Palombi, F.; Criscuolo, E.; Frasca, F.; Pierangeli, A.; Mancini, N.; Antonelli, G.; Clementi, M.; et al. Naringenin is a powerful inhibitor of SARS-CoV-2 infection in vitro. *Pharmacol. Res.* **2021**, *163*, 105255. [CrossRef] [PubMed]

60. Jeon, S.; Ko, M.; Lee, J.; Choi, I.; Byun, S.Y.; Park, S.; Shum, D.; Kim, S. Identification of antiviral drug candidates against SARS-CoV-2 from FDA-approved drugs. *Antimicrob. Agents Chemother.* **2020**, *64*, e00819-20. [CrossRef]

61. Jin, Z.; Du, X.; Xu, Y.; Deng, Y.; Liu, M.; Zhao, Y.; Zhang, B.; Li, X.; Zhang, L.; Peng, C.; et al. Structure of Mpro from SARS-CoV-2 and discovery of its inhibitors. *Nature* **2020**, *582*, 289–293. [CrossRef]

62. Jang, M.; Park, Y.I.; Cha, Y.E.; Park, R.; Namkoong, S.; Lee, J.I.; Park, J. Tea polyphenols EGCG and theaflavin inhibit the activity of SARS-CoV-2 3CL-protease in vitro. *Evid. Based Compl. Alt. Med.* **2020**, *2020*, 5630838. [CrossRef] [PubMed]

63. Choudhry, N.; Zhao, X.; Xu, D.; Zanin, M.; Chen, W.; Yang, Z.; Chen, J. Chinese Therapeutic Strategy for Fighting COVID-19 and Potential Small-Molecule Inhibitors against Severe Acute Respiratory Syndrome Coronavirus 2 (SARS-CoV-2). *J. Med. Chem.* **2020**, *63*, 13205–13227. [CrossRef]

64. Wang, Z.; Yang, L. Chinese herbal medicine: Fighting SARS-CoV-2 infection on all fronts. *J. Ethnopharmacol.* **2021**, *270*, 113869. [CrossRef]

65. Huang, K.; Zhang, P.; Zhang, Z.; Youn, J.Y.; Zhang, H.; Cai, H.L. Traditional Chinese Medicine (TCM) in the treatment of viral infections: Efficacies and mechanisms. *Pharm. Ther.* **2021**, *225*, 107843. [CrossRef] [PubMed]

66. National Health Commission of the People's Republic of China. Notice on the Issunance of Guidelines of Diagnosis and Treatment for 2019-nCoV Infected Pneumonia (Version 7). 2020. Available online: http://www.nhc.gov.cn/yzygj/s7653p/202003/46c929 4a7dfe4cef80dc7f5912eb1989/files/ce3e6945832a438eaae415350a8ce964.pdf (accessed on 3 March 2020).

67. Xinhua Net. Academician Xiaolin Tong: The Total Effective Rate of Qingfeipaidu Formula was 97%, none Transfer from Mild to Severe Cases. 2020. Available online: http://www.kunlunce.com/ssjj/fl1/2020-03-18/141570.html (accessed on 18 March 2020).

68. Su, H.X.; Yao, S.; Zhao, W.F.; Li, M.J.; Liu, J.; Shang, W.J.; Xie, H.; Ke, C.Q.; Hu, H.C.; Gao, M.N.; et al. Anti-SARS-CoV-2 activities in vitro of Shuanghuanglian preparations and bioactive ingredients. *Acta Pharmacol. Sin.* **2020**, *41*, 1167–1177. [CrossRef] [PubMed]

69. Liu, H.; Ye, F.; Sun, Q.; Liang, H.; Li, C.; Li, S.; Lu, R.; Huang, B.; Tan, W.; Lai, L. Scutellaria baicalensis extract and baicalein inhibit replication of SARS-CoV-2 and its 3C-like protease in vitro. *J. Enzym. Inhib. Med. Chemother.* **2021**, *36*, 497–503. [CrossRef]

70. Pooja, M.; Reddy, G.J.; Hema, K.; Dodoala, S.; Koganti, B. Unravelling high-affinity binding compounds towards transmembrane protease serine 2 enzyme in treating SARS-CoV-2 infection using molecular modelling and docking studies. *Eur. J. Pharmacol.* **2021**, *890*, 173688.

71. Song, J.; Zhang, L.; Xu, Y.; Yang, D.; Zhang, L.; Yang, S.; Zhang, W.; Wang, J.; Tian, S.; Yang, S.; et al. The comprehensive study on the therapeutic effects of baicalein for the treatment of COVID-19 in vivo and in vitro. *Biochem. Pharmacol.* **2021**, *183*, 114302. [CrossRef]

72. Ibrahim, M.A.A.; Mohamed, E.A.R.; Abdelrahman, A.H.M.; Allemailem, K.S.; Moustafa, M.F.; Shawky, A.M.; Mahzari, A.; Hakami, A.R.; Abdeljawaad, K.A.A.; Atia, M.A.M. Rutin and flavone analogs as prospective SARS-CoV-2 main protease inhibitors: In silico drug discovery study. *J. Mol. Graph. Model.* **2021**, *105*, 107904. [CrossRef] [PubMed]

73. Zandi, K.; Musall, K.; Oo, A.; Cao, D.; Liang, B.; Hassandarvish, P.; Lan, S.; Slack, R.L.; Kirby, K.A.; Bassit, L.; et al. Baicalein and baicalin inhibit SARS-CoV-2 RNA-dependent-RNA polymerase. *Microorganisms* **2021**, *9*, 893. [CrossRef] [PubMed]

74. Qian, J.Q. Cardiovascular pharmacological effects of bisbenzylisoquinoline alkaloid derivatives. *Acta Pharm. Sin.* **2002**, *23*, 1086–1092.

75. Huang, X.H.; Yan, X.; Zhang, Q.H.; Hong, P.; Zhang, W.X.; Liu, Y.P.; Xu, W.W.; Li, B.; He, Q.Y. Direct targeting of HSP90 with daurisoline destabilizes β-catenin to suppress lung cancer tumorigenesis. *Cancer Lett.* **2020**, *489*, 66–78. [CrossRef]

76. Liu, F.; Ma, J.; Shi, Z.; Zhang, Q.; Wang, H.; Li, D.; Song, Z.; Wang, C.; Jin, J.; Xu, J.; et al. Clerodane diterpenoids isolated from the leaves of Casearia graveolens. *J. Nat. Prod.* **2020**, *83*, 36–44. [CrossRef] [PubMed]

77. Tuli, H.S.; Mittal, S.; Loka, M.; Aggarwal, V.; Aggarwal, D.; Masurkar, A.; Kaur, G.; Varol, M.; Sak, K.; Kumar, M.; et al. Deguelin targets multiple oncogenic signaling pathways to combat human malignancies. *Pharmacol. Res.* **2021**, *166*, 105487. [CrossRef]

78. Gao, F.; Yu, X.; Li, M.; Zhou, L.; Liu, W.; Li, W.; Liu, H. Deguelin suppresses non-small cell lung cancer by inhibiting EGFR signaling and promoting GSK3β/FBW7-mediated Mcl-1 destabilization. *Cell Death Dis.* **2020**, *11*, 143. [CrossRef] [PubMed]

79. Wu, C.P.; Lusvarghi, S.; Hsiao, S.H.; Liu, T.C.; Li, Y.Q.; Huang, Y.H.; Hung, T.H.; Ambudka, S.V. Licochalcone A selectively resensitizes ABCG2-overexpressing multidrug-resistant cancer cells to chemotherapeutic drugs. *J. Nat. Prod.* **2020**, *83*, 1461–1472. [CrossRef]

80. Luo, W.; Sun, R.; Chen, X.; Li, J.; Jiang, J.; He, Y.; Shi, S.; Wen, H. ERK activation-mediated autophagy induction resists licochalcone A-induced anticancer activities in lung cancer cells in vitro. *OncoTargets Ther.* **2020**, *13*, 13437–13450. [CrossRef]

81. Yuan, L.W.; Jiang, X.M.; Xu, Y.L.; Huang, M.Y.; Chen, Y.C.; Yu, W.B.; Su, M.X.; Ye, Z.H.; Chen, X.; Wang, Y.; et al. Licochalcone A inhibits interferon-gamma-induced programmed death-ligand 1 in lung cancer cells. *Phytomedicine* **2021**, *80*, 153394. [CrossRef]

82. Li, B.; Zhou, D.; Li, S.; Feng, Y.; Li, X.; Chang, W.; Zhang, J.; Sun, Y.; Qing, D.; Chen, G.; et al. Licochalcone A reverses NNK-induced ectopic miRNA expression to elicit in vitro and in vivo chemopreventive effects. *Phytomedicine* **2020**, *76*, 153245. [CrossRef]

83. Gao, F.; Li, M.; Yu, X.; Liu, W.; Zhou, L.; Li, W. Licochalcone A inhibits EGFR signalling and translationally suppresses survivin expression in human cancer cells. *J. Cell Mol. Med.* **2021**, *25*, 813–826. [CrossRef]

84. Yang, A.; Li, M.Y.; Zhang, Z.H.; Wang, J.Y.; Xing, Y.; Ri, M.; Jin, C.H.; Xu, G.H.; Piao, L.X.; Jin, H.L.; et al. Erianin regulates programmed cell death ligand 1 expression and enhances cytotoxic T lymphocyte activity. *J. Ethnopharmacol.* **2021**, *273*, 113598. [CrossRef]

85. Chen, P.; Wu, Q.; Feng, J.; Yan, L.; Sun, Y.; Liu, S.; Xiang, Y.; Zhang, M.; Pan, T.; Chen, X.; et al. Erianin, a novel dibenzyl compound in Dendrobium extract, inhibits lung cancer cell growth and migration via calcium/calmodulin-dependent ferroptosis. *Signal Transduct. Tar.* **2020**, *5*, 51. [CrossRef] [PubMed]

86. Keller, L.; Canuto, K.M.; Liu, C.; Suzuki, B.M.; Almaliti, J.; Sikandar, A.; Naman, C.B.; Glukhov, E.; Luo, D.; Duggan, B.M.; et al. Tutuilamides A-C: Vinyl-chloride-containing cyclodepsipeptides from marine cyanobacteria with potent elastase inhibitory properties. *ACS Chem. Biol.* **2020**, *15*, 751–757. [CrossRef] [PubMed]

87. Chen, Q.Y.; Luo, D.; Seabra, G.M.; Luesch, H. Ahp-Cyclodepsipeptides as tunable inhibitors of human neutrophil elastase and kallikrein 7: Total synthesis of tutuilamide A, serine protease selectivity profile and comparison with lyngbyastatin 7. *Bioorg. Med. Chem.* **2020**, *28*, 115756. [CrossRef] [PubMed]

88. Hafner, S.; Lang, S.J.; Gaafary, M.E.; Schmiech, M.; Simmet, T.; Syrovets, T. The cardenolide glycoside acovenoside A interferes with epidermal growth factor receptor (EGFR) trafficking in non-small cell lung cancer cells. *Front. Pharmacol.* **2021**, *11*, 611657. [CrossRef] [PubMed]

89. Han, A.R.; Lee, S.; Han, S.; Lee, Y.J.; Kim, J.B.; Seo, E.K.; Jung, C.H. Triterpenoids from the leaves of Centella asiatica inhibit ionizing radiation-induced migration and invasion of human lung cancer cells. *Evid. Based Compl. Alt.* **2020**, *2020*, 3683460. [CrossRef]

90. Zhang, X.; Ruan, Q.; Zhai, Y.; Lu, D.; Li, C.; Fu, Y.; Zheng, Z.; Song, Y.; Guo, J. Baicalein inhibits non-small-cell lung cancer invasion and metastasis by reducing ezrin tension in inflammation microenvironment. *Cancer Sci.* **2020**, *111*, 3802–3812. [CrossRef] [PubMed]

91. Zhang, Z.; Nong, L.; Chen, M.; Gu, X.; Zhao, W.; Liu, M.; Cheng, W. Baicalein suppresses vasculogenic mimicry through inhibiting RhoA/ROCK expression in lung cancer A549 cell line. *Acta Bioch. Bioph. Sin.* **2020**, *52*, 1007–1015. [CrossRef]

92. Li, J.; Yan, L.; Luo, J.; Tong, L.; Gao, Y.; Feng, W.; Wang, F.; Cui, W.; Li, S.; Sun, Z. Baicalein suppresses growth of non-small cell lung carcinoma by targeting MAP4K3. *Biomed. Pharmacother.* **2021**, *133*, 110965. [CrossRef]

93. Sui, X.; Han, X.; Chen, P.; Wu, Q.; Feng, J.; Duan, T.; Chen, X.; Pan, T.; Yan, L.; Jin, T.; et al. Baicalin induces apoptosis and suppresses the cell cycle progression of lung cancer cells through downregulating Akt/mTOR signaling pathway. *Front. Mol. Biosci.* **2021**, *7*, 602282. [CrossRef] [PubMed]

94. Yan, Y.; Yao, L.; Sun, H.; Pang, S.; Kong, X.; Zhao, S.; Xu, S. Effects of wogonoside on invasion and migration of lung cancer A549 cells and angiogenesis in xenograft tumors of nude mice. *J. Thorac. Dis.* **2020**, *12*, 1552–1560. [CrossRef] [PubMed]

95. Cheng, Z.Y.; Hsiao, Y.T.; Huang, Y.P.; Peng, S.F.; Huang, W.W.; Liu, K.C.; Hsia, T.C.; Way, T.D.; Chung, J.G. Casticin induces DNA damage and affects DNA repair associated protein expression in human lung cancer A549 cells. *Molecules* **2020**, *25*, 341. [CrossRef] [PubMed]

96. Cui, L.; Yang, G.; Ye, J.; Yao, Y.; Lu, G.; Chen, J.; Fang, L.; Lu, S.; Zhou, J. Dioscin elicits anti-tumour immunity by inhibiting macrophage M2 polarization via JNK and STAT3 pathways in lung cancer. *J. Cell. Mol. Med.* **2020**, *24*, 9217–9230. [CrossRef]

97. Chen, A.; Jiang, P.; Zeb, F.; Wu, X.; Xu, C.; Chen, L.; Feng, Q. EGCG regulates CTR1 expression through its pro-oxidative property in non-small-cell lung cancer cells. *J. Cell. Physiol.* **2020**, *235*, 7970–7981. [CrossRef]

98. Wei, R.; Wirkus, J.; Yang, Z.; Machuca, J.; Esparza, Y.; Mackenzie, G.G. EGCG sensitizes chemotherapeutic-induced cytotoxicity by targeting the ERK pathway in multiple cancer cell lines. *Arch. Biochem. Biophys.* **2020**, *692*, 108546. [CrossRef]

99. Duan, J.; Li, Y.; Gao, H.; Yang, D.; He, X.; Fang, Y.; Zhou, G. Phenolic compound ellagic acid inhibits mitochondrial respiration and tumor growth in lung cancer. *Food Funct.* **2020**, *11*, 6332–6339. [CrossRef]

100. Boonjing, S.; Pothongsrisit, S.; Wattanathamsan, O.; Sritularak, B.; Pongrakhananon, V. Erianthridin induces non-small cell lung cancer cell apoptosis through the suppression of extracellular signal-regulated kinase activity. *Planta Med.* **2021**, *87*, 283–293.

101. Pothongsrisit, S.; Arunrungvichian, K.; Hayakawa, Y.; Sritularak, B.; Mangmool, S.; Pongrakhananon, V. Erianthridin suppresses non-small-cell lung cancer cell metastasis through inhibition of Akt/mTOR/p70 S6K signaling pathway. *Sci. Rep.* **2021**, *11*, 6618. [CrossRef]

102. Choudhury, P.; Barua, A.; Roy, A.; Pattanayak, R.; Bhattacharyya, M.; Saha, P. Eugenol emerges as an elixir by targeting β-catenin, the central cancer stem cell regulator in lung carcinogenesis: An in vivo and in vitro rationale. *Food Funct.* **2021**, *12*, 1063–1078. [CrossRef] [PubMed]

103. Yu, X.; Gao, F.; Li, W.; Zhou, L.; Liu, W.; Li, M. Formononetin inhibits tumor growth by suppression of EGFR-Akt-Mcl-1 axis in non-small cell lung cancer. *J. Exp. Clin. Canc. Res.* **2020**, *39*, 62. [CrossRef]

104. Yang, B.; Yang, N.; Chen, Y.; Maomao, Z.; Lian, Y.; Xiong, Z.; Wang, B.; Feng, L.; Jia, X. An integrated strategy for effective-component discovery of Astragali Radix in the treatment of lung cancer. *Front. Pharmacol.* **2021**, *12*, 580978. [CrossRef]

105. Kang, D.Y.; Sp, N.; Jo, E.S.; Rugamba, A.; Hong, D.Y.; Lee, H.G.; Yoo, J.S.; Liu, Q.; Jang, K.J.; Yang, Y.M. The inhibitory mechanisms of tumor PD-L1 expression by natural bioactive gallic acid in non-small-cell lung cancer (NSCLC) cells. *Cancers* **2020**, *12*, 727. [CrossRef] [PubMed]

106. Wang, D.; Bao, B. Gallic acid impedes non-small cell lung cancer progression via suppression of EGFR-dependent CARM1-PELP1 complex. *Drug Des. Dev. Ther.* **2020**, *14*, 1583–1592. [CrossRef] [PubMed]

107. Chen, H.; Miao, L.; Huang, F.; Yu, Y.; Peng, Q.; Liu, Y.; Li, X.; Liu, H. Glochidiol, a natural triterpenoid, exerts its anti-cancer effects by targeting the colchicine binding site of tubulin. *Invest. N. Drug.* **2021**, *39*, 578–586. [CrossRef]

108. Min, H.Y.; Pei, H.; Hyun, S.Y.; Boo, H.J.; Jang, H.J.; Cho, J.; Kim, J.H.; Son, J.; Lee, H.Y. Potent anticancer effect of the natural steroidal saponin gracillin is produced by inhibiting glycolysis and oxidative phosphorylation-mediated bioenergetics. *Cancers* **2020**, *12*, 913. [CrossRef]

109. Yang, J.; Cao, L.; Li, Y.; Liu, H.; Zhang, M.; Ma, H.; Wang, B.; Yuan, X.; Liu, Q. Gracillin isolated from Reineckia carnea induces apoptosis of A549 Cells via the mitochondrial pathway. *Drug Des. Dev. Ther.* **2021**, *2021*, 233–243. [CrossRef]

110. Lv, L.; Zhang, W.; Li, T.; Jiang, L.; Lu, X.; Lin, J. Hispidulin exhibits potent anticancer activity in vitro and in vivo through activating ER stress in non-small-cell lung cancer cells. *Oncol. Rep.* **2020**, *43*, 1995–2003. [CrossRef]

111. Zhao, X.; Lin, Y.; Jiang, B.; Yin, J.; Lu, C.; Wang, J.; Zeng, J. Icaritin inhibits lung cancer-induced osteoclastogenesis by suppressing the expression of IL-6 and TNF-a and through AMPK/mTOR signaling pathway. *Anti-Cancer Drug.* **2020**, *31*, 1004–1011. [CrossRef] [PubMed]

112. Lee, J.H.; Park, S.Y.; Hwang, W.; Sung, J.Y.; Shim, M.L.C.J.; Kim, Y.N.; Yoon, K. Isoharringtonine induces apoptosis of non-small cell lung cancer cells in tumorspheroids via the intrinsic pathway. *Biomolecules* **2020**, *10*, 1521. [CrossRef] [PubMed]

113. Fouzder, C.; Mukhuty, A.; Kundu, R. Kaempferol inhibits Nrf2 signalling pathway via downregulation of Nrf2 mRNA and induces apoptosis in NSCLC cells. *Arch. Biochem. Biophys.* **2021**, *697*, 108700. [CrossRef]

114. Sheng, H.; Lv, W.; Zhu, L.; Wang, L.; Wang, Z.; Han, J.; Hu, J. Liriopesides B induces apoptosis and cell cycle arrest in human non-small cell lung cancer cells. *Int. J. Mol. Med.* **2020**, *46*, 1039–1050. [CrossRef] [PubMed]

115. Chen, Y.C.; Huang, M.Y.; Zhang, L.L.; Feng, Z.L.; Jiang, X.M.; Yuan, L.W.; Huang, R.Y.; Liu, B.; Yu, H.; Wang, Y.T.; et al. Nagilactone E increases PD-L1 expression through activation of c-Jun in lung cancer cells. *Chin. J. Nat. Med.* **2020**, *18*, 517–525. [CrossRef]

116. Zhang, L.L.; Guo, J.; Jiang, X.M.; Chen, X.P.; Wang, Y.T.; Li, A.; Lin, L.G.; Li, H.; Lu, J.J. Identification of nagilactone E as a protein synthesis inhibitor with anticancer activity. *Acta Pharmacol. Sin.* **2020**, *41*, 698–705. [CrossRef] [PubMed]

117. Liu, X.; Zhang, Y.; Zhou, G.J.; Hou, Y.; Kong, Q.; Lu, J.J.; Zhang, Q.; Chen, X. Natural alkaloid 8-oxo-epiberberine inhibited TGF-β1-triggred epithelial-mesenchymal transition by interfering Smad3. *Toxicol. Appl. Pharm.* **2020**, *404*, 115179. [CrossRef] [PubMed]

118. Li, X.; Huang, R.; Li, M.; Zhu, Z.; Chen, Z.; Cui, L.; Luo, H.; Luo, L. Parthenolide inhibits the growth of non-small cell lung cancer by targeting epidermal growth factor receptor. *Cancer Cell Int.* **2020**, *20*, 561. [CrossRef] [PubMed]

119. Sun, L.; Yuan, W.; Wen, G.; Yu, B.; Xu, F.; Gan, X.; Tang, J.; Zeng, Q.; Zhu, L.; Chen, C.; et al. Parthenolide inhibits human lung cancer cell growth by modulating the IGF-1R/PI3K/Akt signaling pathway. *Oncol. Rep.* **2020**, *44*, 1184–1193. [CrossRef] [PubMed]

120. Wu, L.M.; Liao, X.Z.; Zhang, Y.; He, Z.R.; Nie, S.Q.; Ke, B.; Shi, L.; Zhao, J.F.; Chen, W.H. Parthenolide augments the chemosensitivity of non-small-cell lung cancer to cisplatin via the PI3K/AKT signaling pathway. *Front. Cell Dev. Biol.* **2021**, *8*, 610097. [CrossRef] [PubMed]

121. Meng, N.; Zhang, R.; Liu, C.; Wang, Q.; Wang, X.; Guo, X.; Wang, P.; Sun, J. PDB-1 from potentilla discolor bunge suppresses lung cancer cell migration and invasion via FAK/Src and MAPK signaling pathways. *Med. Chem. Res.* **2020**, *29*, 887–896. [CrossRef]

122. Wu, Y.; Si, Y.; Xiang, Y.; Zhou, T.; Liu, X.; Wu, M.; Li, W.; Zhang, T.; Xiang, K.; Zhang, L.; et al. Polyphyllin I activates AMPK to suppress the growth of non-small-cell lung cancer via induction of autophagy. *Arch. Biochem. Biophys.* **2020**, *687*, 108285. [CrossRef]

123. Lai, L.; Shen, Q.; Wang, Y.; Chen, L.; Lai, J.; Wu, Z.; Jiang, H. Polyphyllin I reverses the resistance of osimertinib in non-small cell lung cancer cell through regulation of PI3K/Akt signaling. *Toxicol. Appl. Pharm.* **2021**, *419*, 115518. [CrossRef] [PubMed]

124. Guo, H.; Ding, H.; Tang, X.; Liang, M.; Li, S.; Zhang, J.; Cao, J. Quercetin induces pro-apoptotic autophagy via SIRT1/AMPK signaling pathway in human lung cancer cell lines A549 and H1299 in vitro. *Thoracic Cancer* **2021**, *12*, 1415–1422. [CrossRef] [PubMed]

125. Xu, S.; Zhang, H.; Wang, A.; Ma, Y.; Gan, Y.; Li, G. Silibinin suppresses epithelial-mesenchymal transition in human non-small cell lung cancer cells by restraining RHBDD1. *Cell. Mol. Biol. Lett.* **2020**, *25*, 36. [CrossRef]

126. Shen, K.H.; Hung, J.H.; Liao, Y.C.; Tsai, S.T.; Wu, M.J.; Chen, P.S. Sinomenine inhibits migration and invasion of human lung cancer cell through downregulating expression of miR-21 and MMPs. *Int. J. Mol. Sci.* **2020**, *21*, 3080. [CrossRef] [PubMed]

127. Bai, S.; Wen, W.; Hou, X.; Wu, J.; Yi, L.; Zhi, Y.; Lv, Y.; Tan, X.; Liu, L.; Wang, P.; et al. Inhibitory effect of sinomenine on lung cancer cells via negative regulation of α7 nicotinic acetylcholine receptor. *J. Leukocyte Biol.* **2021**, *109*, 843–852. [CrossRef]

128. Liu, W.; Yu, X.; Zhou, L.; Li, J.; Li, M.; Li, W.; Gao, F. Sinomenine inhibits non-small cell lung cancer via downregulation of hexokinases II-mediated aerobic glycolysis. *OncoTargets Ther.* **2020**, *13*, 3209–3221. [CrossRef]

129. Zheng, W.; Huang, F.Y.; Dai, S.Z.; Wang, J.Y.; Lin, Y.Y.; Sun, Y.; Tan, G.H.; Huang, Y.H. Toxicarioside O inhibits cell proliferation and epithelial-mesenchymal transition by downregulation of Trop2 in lung cancer cells. *Front. Oncol.* **2020**, *10*, 609275. [CrossRef] [PubMed]

130. Al-Rashed, S.; Baker, A.; Ahmad, S.S.; Syed, A.; Bahkali, A.H.; Elgorban, A.M.; Khan, M.S. Vincamine, a safe natural alkaloid, represents a novel anticancer agent. *Bioorg. Chem.* **2021**, *107*, 104626. [CrossRef] [PubMed]

131. Gao, F.; Li, M.; Zhou, L.; Liu, W.; Zuo, H.; Li, W. Xanthohumol targets the ERK1/2-Fra1 signaling axis to reduce cyclin D1 expression and inhibit non-small cell lung cancer. *Oncol. Rep.* **2020**, *44*, 1365–1374. [CrossRef] [PubMed]

132. Bailly, C. Cepharanthine: An update of its mode of action, pharmacological properties and medical applications. *Phytomedicine* **2019**, *62*, 152956. [CrossRef]

133. Okamoto, M.; Ono, M.; Baba, M. Potent inhibition of HIV type 1 replication by an antiinflammatory alkaloid, cepharanthine, in chronically infected monocytic cells. *AIDS Res. Hum. Retrovir.* **1998**, *14*, 1239–1245. [CrossRef]

134. Zhang, C.H.; Wang, Y.F.; Liu, X.J.; Lu, J.H.; Qian, C.W.; Wan, Z.Y. Antiviral activity of cepharanthine against severe acute respiratory syndrome coronavirus in vitro. *Chin. Med. J.* **2005**, *118*, 493–496.

135. Kim, D.E.; Min, J.S.; Jang, M.S.; Lee, J.Y.; Shin, Y.S.; Song, J.H.; Kim, H.R.; Kim Se Jin, Y.H.; Kwon, S. Natural bis-benzylisoquinoline alkaloids-tetrandrine, fangchinoline, and cepharanthine, inhibit human coronavirus OC43 infection of MRC-5 human lung cells. *Biomolecules* **2019**, *9*, 696. [CrossRef]

136. Ohashi, H.; Watashi, K.; Saso, W.; Shionoya, K.; Iwanami, S.; Hirokawa, T.; Shirai, T.; Kanaya, S.; Ito, Y.; Kim, K.S.; et al. Potential anti-COVID-19 agents, cepharanthine and nelfinavir, and their usage for combination treatment. *Science* **2021**, *24*, 102367. [CrossRef]

137. Li, S.; Liu, W.; Chen, Y.; Wang, L.; An, W.; An, X.; Song, L.; Tong, Y.; Fan, H.; Lu, C. Transcriptome analysis of cepharanthine against a SARS-CoV-2-related coronavirus. *Brief. Bioinform.* **2021**, *22*, 1378–1386. [CrossRef] [PubMed]

138. Bihani, S.C.; Hosur, M.V. Molecular basis for reduced cleavage activity and drug resistance in D30N HIV-1 protease. *bioRxiv* **2021**. [CrossRef]

139. Jan, J.T.; Cheng, T.J.R.; Juang, Y.P.; Ma, H.H.; Wu, Y.T.; Yang, W.B.; Cheng, C.W.; Chen, X.; Chou, T.H.; Shie, J.J.; et al. Identification of existing pharmaceuticals and herbal medicines as inhibitors of SARS-CoV-2 infection. *Proc. Natl. Acad. Sci. USA* **2021**, *118*, e2021579118. [CrossRef] [PubMed]

140. Yamamoto, N.; Matsuyama, S.; Hoshino, T.; Yamamoto, N. Nelfinavir inhibits replication of severe acute respiratory syndrome coronavirus 2 in vitro. *bioRxiv* **2020**. [CrossRef]

141. Musarrat, F.; Chouljenko, V.; Dahal, A.; Nabi, R.; Chouljenko, T.; Jois, S.D.; Kousoulas, K.G. The anti-HIV drug nelfinavir mesylate (Viracept) is a potent inhibitor of cell fusion caused by the SARSCoV-2 spike (S) glycoprotein warranting further evaluation as an antiviral against COVID-19 infections. *J. Med. Virol.* **2020**, *92*, 2087–2095. [CrossRef]

142. Foo, C.S.; Abdelnabi, R.; Kaptein, S.J.F.; Zhang, X.; ter Horst, S.; Mols, R.; Delang, L.; Rocha-Pereira, J.; Coelmont, L.; Leyssen, P.; et al. Nelfinavir markedly improves lung pathology in SARS-CoV-2-infected Syrian hamsters despite lack of an antiviral effect. *bioRxiv* **2021**. [CrossRef]

143. Warren, T.K.; Jordan, R.; Lo, M.K.; Ray, A.S.; Mackman, R.L.; Soloveva, V.; Siegel, D.; Perron, M.; Bannister, R.; Hui, H.C.; et al. Therapeutic efficacy of the small molecule GS-5734 against Ebola virus in rhesus monkeys. *Nature* **2016**, *531*, 381–385. [CrossRef] [PubMed]

144. Li, Y.; Cao, L.; Li, G.; Cong, F.; Li, Y.; Sun, J.; Luo, Y.; Chen, G.; Li, G.; Wang, P.; et al. Remdesivir metabolite GS-441524 effectively inhibits SARS-CoV-2 infection in mouse models. *J. Med. Chem.* **2021**. [CrossRef]

145. Rubin, D.; Chan-Tack, K.; Farley, J.; Sherwat, A. FDA approval of remdesivir—A step in the right direction. *N. Engl. J. Med.* **2020**, *383*, 2598–2600. [CrossRef] [PubMed]

146. Sheahan, T.P.; Sims, A.C.; Graham, R.L.; Menachery, V.D.; Gralinski, L.E.; Case, J.a.B.; Leist, S.R.; Pyrc, K.; Feng, J.Y.; Trantcheva, I.; et al. *Broad-spectrum antiviral GS*-5734 inhibits both epidemic and zoonotic coronaviruses. *Sci. Transl. Med.* **2017**, *9*, eaal3653. [CrossRef] [PubMed]

147. Sheahan, T.P.; Sims, A.C.; Leist, S.R.; Schäfer, A.; Won, J.; Brown, A.J.; Montgomery, S.A.; Hogg, A.; Babusis, D.; Clarke, M.O.; et al. Comparative therapeutic efficacy of remdesivir and combination lopinavir, ritonavir, and interferon beta against MERS-CoV. *Nat. Commun.* **2020**, *11*, 222. [CrossRef]

148. McMullan, L.K.; Flint, M.; Chakrabarti, A.; Guerrero, L.; Lo, M.K.; Porter, D.; Nichol, S.T.; Spiropoulou, C.F.; Albariño, C. Characterisation of infectious Ebola virus from the ongoing outbreak to guide response activities in the Democratic Republic of the Congo: A phylogenetic and in vitro analysis. *Lancet Infect. Dis.* **2019**, *19*, 1023–1032. [CrossRef]

149. Wang, M.; Cao, R.; Zhang, L.; Yang, X.; Liu, J.; Xu, M.; Shi, Z.; Hu, Z.; Zhong, W.; Xiao, G. Remdesivir and chloroquine effectively inhibit the recently emerged novel coronavirus (2019-nCoV) in vitro. *Cell Res.* **2020**, *30*, 269–271. [CrossRef] [PubMed]

150. De Wit, E.; Feldmann, F.; Cronin, J.; Jordan, R.; Okumura, A.; Thomas, T.; Scott, D.; Cihlar, T.; Feldmann, H. Prophylactic and therapeutic remdesivir (GS-5734) treatment in the rhesus macaque model of MERS-CoV infection. *Proc. Natl. Acad. Sci. USA* **2020**, *117*, 6771–6776. [CrossRef] [PubMed]

151. Lo, M.K.; Feldmann, F.; Gary, J.M.; Jordan, R.; Bannister, R.; Cronin, J.; Patel, N.R.; Klena, J.D.; Nichol, S.T.; Cihlar, T.; et al. Remdesivir (GS-5734) protects African green monkeys from Nipah virus challenge. *Sci. Transl. Med.* **2019**, *11*, eaau9242. [CrossRef]

152. Williamson, B.N.; Feldmann, F.; Schwarz, B.; Meade-White, K.; Porter, D.P.; Schulz, J.; van Doremalen, N.; Leighton, I.; Yinda, C.K.; Pérez-Pérez, L.; et al. Clinical benefit of remdesivir in rhesus macaques infected with SARS-CoV-2. *Nature* **2020**, *585*, 273–276. [CrossRef]

153. Jacobs, M.; Rodger, A.; Bell, D.J.; Bhagani, S.; Cropley, I.; Filipe, A.; Gifford, R.J.; Hopkins, S.; Hughes, J.; Jabeen, F.; et al. Late Ebola virus relapse causing meningoencephalitis: A case report. *Lancet* **2016**, *388*, 498–503. [CrossRef]

154. Schwartz, I.S.; Heil, E.L.; McCreary, E.K. Remdesivir: A pendulum in a pandemic. *BMJ* **2020**, *371*, m4560. [CrossRef]

155. Wang, Y.; Zhang, D.; Du, G.; Du, R.; Zhao, J.; Jin, Y.; Fu, S.; Gao, L.; Cheng, Z.; Lu, Q.; et al. Remdesivir in adults with severe COVID-19: A randomised, double-blind, placebo-controlled, multicentre trial. *Lancet* **2020**, *395*, 1569–1578. [CrossRef]

156. McCreary, E.K.; Angus, D.C. Efficacy of remdesivir in COVID-19. *JAMA* **2020**, *324*, 1041–1042. [CrossRef] [PubMed]

157. Yan, B.; Chu, H.; Yang, D.; Sze, K.H.; Lai, P.M.; Yuan, S.; Shuai, H.; Wang, Y.; Kao, R.Y.T.; Chan, J.F.W.; et al. Characterization of lipidomic profile of human coronavirus -infected cells: Implications for lipid metabolism remodeling upon coronavirus replication. *Viruses* **2019**, *11*, 73. [CrossRef]

158. Khalil, M.I.; Salih, M.A.; Mustafa, A.A. Broad beans (Vicia faba) and the potential to protect from COVID-19 coronavirus infection. *Sudan J. Paediatr.* **2020**, *20*, 10–12. [CrossRef] [PubMed]

159. Simopoulos, A.P.; Serhan, C.N.; Bazinet, R.P. The need for precision nutrition, genetic variation and resolution in Covid-19 patients. *Mol. Aspects Med.* **2021**, *77*, 100943. [CrossRef]

160. Toelzer, C.; Gupta, K.; Yadav, S.K.N.; Borucu, U.; Davidson, A.D.; Williamson, M.K.; Shoemark, D.K.; Garzoni, F.; Staufer, O.; Milligan, R.; et al. Free fatty acid binding pocket in the locked structure of SARS-CoV-2 spike protein. *Science* **2020**, *370*, 725–730. [CrossRef] [PubMed]

161. Wang, Z.; Yang, L. GS-5734: A potentially approved drug by FDA against SARS-CoV-2. *N. J. Chem.* **2020**, *44*, 12417–12429. [CrossRef]

162. Yan, V.C.; Muller, F.L. Advantages of the parent nucleoside GS-441524 over remdesivir for Covid-19 treatment. *ACS Med. Chem. Lett.* **2020**, *11*, 1361–1366. [CrossRef] [PubMed]

163. Yuan, R.; Huang, Y.; Chan, L.; He, D.; Chen, T. Engineering EHD1-targeted natural borneol nanoemulsion potentiates therapeutic efficacy of gefitinib against non-small lung cancer. *ACS Appl. Mater. Interfaces* **2020**, *12*, 45714–45727. [CrossRef] [PubMed]

164. Wang, Y.; Yu, H.; Wang, S.; Gai, C.; Cui, X.; Xu, Z.; Li, W.; Zhang, W. Targeted delivery of quercetin by nanoparticles based on chitosan sensitizing paclitaxel-resistant lung cancer cells to paclitaxel. *Mater. Sci. Eng. C* **2021**, *119*, 111442. [CrossRef]

165. Zang, H.; Qian, G.; Arbiser, J.; Owonikoko, T.K.; Ramalingam, S.S.; Fan, S.; Sun, S.Y. Overcoming acquired resistance of EGFR-mutant NSCLC cells to the third generation EGFR inhibitor, osimertinib, with the natural product honokiol. *Mol. Oncol.* **2020**, *14*, 882–895. [CrossRef]

166. Dai, C.H.; Zhu, L.R.; Wang, Y.; Tang, X.P.; Du, Y.J.; Chen, Y.C.; Li, J. Celastrol acts synergistically with afatinib to suppress non-small cell lung cancer cell proliferation by inducing paraptosis. *J. Cell. Physiol.* **2021**, *236*, 4538–4554. [CrossRef]

167. Silva, A.C.D.; Santos, P.D.D.F.; Silva, J.T.D.P.; Leimann, F.V.; Bracht, L.; Gonçalves, O.H. Impact of curcumin nanoformulation on its antimicrobial activity. *Trends Food Sci. Tech.* **2018**, *72*, 74–82. [CrossRef]

168. Zhang, L.; Qiang, P.; Yu, J.; Miao, Y.M.; Chen, Z.Q.; Qu, J.; Zhao, Q.B.; Chen, Z.; Liu, Y.; Yao, X.; et al. Identification of compound CA-5f as a novel late-stage autophagy inhibitor with potent anti-tumor effect against non-small cell lung cancer. *Autophagy* **2019**, *15*, 391–406. [CrossRef]

169. Ashrafizadeh, M.; Najafi, M.; Makvandi, P.; Zarrabi, A.; Farkhondeh, T.; Samarghandian, S. Versatile role of curcumin and its derivatives in lung cancer therapy. *J. Cell Physiol.* **2020**, *235*, 9241–9268. [CrossRef] [PubMed]

170. Ganguly, R.; Kumar, S.; Kunwar, A.; Nath, S.; Sarma, H.D.; Tripathi, A.; Verma, G.; Chaudhari, D.P.; Aswal, V.K.; Melo, J.S. Structural and therapeutic properties of curcumin solubilized pluronic F127 micellar solutions and hydrogels. *J. Mol. Liq.* **2020**, *314*, 113591. [CrossRef]

171. Ghosh, S. Cisplatin: The first metal based anticancer drug. *Bioorg. Chem.* **2019**, *88*, 102925. [CrossRef]

172. Zha, M.; Tian, T.; Xu, W.; Liu, S.; Jia, J.; Wang, L.; Yan, Q.; Li, N.; Yu, J.; Huang, L. The circadian clock gene Bmal1 facilitates cisplatin-induced renal injury and hepatization. *Cell Death Dis.* **2020**, *11*, 446. [CrossRef]

173. Hazlitt, R.A.; Min, J.; Zuo, J. Progress in the development of preventative drugs for cisplatin-induced hearing loss. *J. Med. Chem.* **2018**, *61*, 5512–5524. [CrossRef] [PubMed]

174. Ma, W.; Wei, S.; Zhang, B.; Li, W. Molecular mechanisms of cardiomyocyte death in drug-induced cardiotoxicity. *Front. Cell Dev. Biol.* **2020**, *8*, 434. [CrossRef] [PubMed]

175. Wu, C.; Chen, L.; Tao, H.; Kong, L.; Hu, Y. Ring finger protein 38 induces the drug resistance of cisplatin in non-small cell lung cancer. *Cell Biol. Int.* **2020**. [CrossRef] [PubMed]
176. Zhang, W.; Shi, H.; Chen, C.; Ren, K.; Xu, Y.; Liu, X.; He, L. Curcumin enhances cisplatin sensitivity of human NSCLC cell lines through influencing Cu-Sp1-CTR1 regulatory loop. *Phytomedicine* **2018**, *48*, 51–61. [CrossRef] [PubMed]
177. Yang, Y.; Zhang, Y.M.; Chen, Y.; Chen, J.T.; Liu, Y. Targeted polysaccharide nanoparticle for adamplatin prodrug delivery. *J. Med. Chem.* **2013**, *56*, 9725–9736. [CrossRef]
178. Bai, Y.; Liu, C.P.; Chen, D.; Liu, C.F.; Zhuo, L.H.; Li, H.; Wang, C.; Bu, H.T.; Tian, W. β-Cyclodextrin-modified hyaluronic acid-based supramolecular self-assemblies for pH- and esterase- dual-responsive drug delivery. *Carbohydr. Polym.* **2020**, *246*, 116654. [CrossRef] [PubMed]
179. Gao, J.; Ma, X.; Zhang, L.; Yan, J.; Cui, H.; Zhang, Y.; Wang, D.; Zhang, H. Self-Assembled disulfide bond bearing paclitaxel-camptothecin prodrug nanoparticle for lung cancer therapy. *Pharmaceutics* **2020**, *12*, 1169. [CrossRef] [PubMed]
180. Lou, X.; Zhang, D.; Ling, H.; He, Z.; Sun, J.; Sun, M.; Liu, D. Pure redox-sensitive paclitaxel-maleimide prodrug nanoparticles: Endogenous albumin-induced size switching and improved antitumor efficiency. *Acta Pharm. Sin. B* **2021**. [CrossRef]
181. Wang, J.; Pei, Q.; Xia, R.; Liu, S.; Hu, X.; Xie, Z.; Jing, X. Comparison of redox responsiveness and antitumor capability of paclitaxel dimeric nanoparticles with different linkers. *Chem. Mater.* **2020**, *32*, 10719–10727. [CrossRef]
182. Lan, J.S.; Liu, L.; Zeng, R.F.; Qin, Y.H.; Hou, J.W.; Xie, S.S.; Yue, S.; Yang, J.; Ho, R.J.Y.; Ding, Y.; et al. Tumor-specific carrier-free nanodrugs with GSH depletion and enhanced ROS generation for endogenous synergistic anti-tumor by a chemotherapy-photodynamic therapy. *Chem. Eng. J.* **2021**, *407*, 127212. [CrossRef]
183. Venkatesha, S.H.; Dudics, S.; Astry, B.; Moudgil, K.D. Control of autoimmune inflammation by celastrol, a natural triterpenoid. *Pathog. Dis.* **2016**, *74*, ftw059. [CrossRef] [PubMed]
184. Lee, Y.J.; Kim, S.Y.; Lee, C. Axl is a novel target of celastrol that inhibits cell proliferation and migration, and increases the cytotoxicity of gefitinib in EGFR mutant non-small cell lung cancer cells. *Mol. Med. Rep.* **2019**, *19*, 3230–3236. [CrossRef]
185. Nazim, U.M.; Yin, H.; Park, S.Y. Autophagy flux inhibition mediated by celastrol sensitized lung cancer cells to TRAIL-induced apoptosis via regulation of mitochondrial transmembrane potential and reactive oxygen species. *Mol. Med. Rep.* **2019**, *19*, 984–993. [CrossRef] [PubMed]
186. Wang, Y.; Liu, Q.; Chen, H.; You, J.; Peng, B.; Cao, F.; Zhang, X.; Chen, Q.; Uzan, G.; Xu, L.; et al. Celastrol improves the therapeutic efficacy of EGFR-TKIs for non-small-cell lung cancer by overcoming EGFR T790M drug resistance. *Anti-Cancer Drug.* **2018**, *29*, 748–755. [CrossRef] [PubMed]
187. Shukla, S.K.; Chan, A.; Parvathaneni, V.; Kanabar, D.D.; Patel, K.; Ayehunie, S.; Muth, A.; Gupta, V. Enhanced solubility, stability, permeation and anti-cancer efficacy of Celastrol-β-cyclodextrin inclusion complex. *J. Mol. Liq.* **2020**, *318*, 113936. [CrossRef]
188. Cao, L.; Hong, W.; Cai, P.; Xu, C.; Bai, X.; Zhao, Z.; Huang, M.; Jin, J. Cryptotanshinone strengthens the effect of gefitinib against non-small cell lung cancer through inhibiting transketolase. *Eur. J. Pharmacol.* **2021**, *890*, 173647. [CrossRef] [PubMed]
189. Du, J.; Li, G.; Jiang, L.; Zhang, X.; Xu, Z.; Yan, H.; Zhou, Z.; He, Q.; Yang, X.; Luo, P. Crosstalk between alveolar macrophages and alveolar epithelial cells/fibroblasts contributes to the pulmonary toxicity of gefitinib. *Toxicol. Lett.* **2021**, *338*, 1–9. [CrossRef]
190. Xie, X.; Wang, X.; Wu, S.; Yang, H.; Liu, J.; Chen, H.; Ding, Y.; Ling, L.; Lin, H. Fatal toxic effects related to EGFR tyrosine kinase inhibitors based on 53 cohorts with 9569 participants. *J. Thorac. Dis.* **2020**, *12*, 4057–4069. [CrossRef] [PubMed]
191. Xie, X.; Zhan, C.; Wang, J.; Zeng, F.; Wu, S. An activatable nano-prodrug for treating tyrosine-kinase-inhibitor-resistant non-small cell lung cancer and for optoacoustic and fluorescent imaging. *Small* **2020**, *16*, 2003451. [CrossRef]
192. Xu, S.; Zhou, J.; Dong, X.; Zhao, W.; Zhu, Q. Fluorescent probe for sensitive discrimination of Hcy and Cys/GSH in living cells via dual-emission. *Anal. Chim. Acta* **2019**, *1074*, 123–130. [CrossRef]
193. Baell, J.; Walters, M.A. Chemistry: Chemical con artists foil drug discovery. *Nature* **2014**, *513*, 481–483. [CrossRef] [PubMed]
194. Heinrich, M.; Appendino, G.; Efferth, T.; Fürst, R.; Izzo, A.A.; Kayser, O.; Pezzuto, J.M.; Viljoen, A. Best practice in research—Overcoming common challenges in phytopharmacological research. *J. Ethnopharmacol.* **2020**, *246*, 112230. [CrossRef] [PubMed]

Review

20-Hydroxyecdysone, from Plant Extracts to Clinical Use: Therapeutic Potential for the Treatment of Neuromuscular, Cardio-Metabolic and Respiratory Diseases

Laurence Dinan [1], Waly Dioh [1], Stanislas Veillet [1] and Rene Lafont [1,2,*]

[1] Biophytis, Sorbonne Université, BC9, 4 place Jussieu, 75005 Paris, France; laurence.dinan@biophytis.com (L.D.); waly.dioh@biophytis.com (W.D.); stanislas.veillet@biophytis.com (S.V.)
[2] BIOSIPE, IBPS, Sorbonne Université, UPMC, 75005 Paris, France
* Correspondence: rene.lafont@sorbonne.universite.fr

Abstract: There is growing interest in the pharmaceutical and medical applications of 20-hydroxyecdysone (20E), a polyhydroxylated steroid which naturally occurs in low but very significant amounts in invertebrates, where it has hormonal roles, and in certain plant species, where it is believed to contribute to the deterrence of invertebrate predators. Studies in vivo and in vitro have revealed beneficial effects in mammals: anabolic, hypolipidemic, anti-diabetic, anti-inflammatory, hepatoprotective, etc. The possible mode of action in mammals has been determined recently, with the main mechanism involving the activation of the Mas1 receptor, a key component of the renin–angiotensin system, which would explain many of the pleiotropic effects observed in the different animal models. Processes have been developed to produce large amounts of pharmaceutical grade 20E, and regulatory preclinical studies have assessed its lack of toxicity. The effects of 20E have been evaluated in early stage clinical trials in healthy volunteers and in patients for the treatment of neuromuscular, cardio-metabolic or respiratory diseases. The prospects and limitations of developing 20E as a drug are discussed, including the requirement for a better evaluation of its safety and pharmacological profile and for developing a production process compliant with pharmaceutical standards.

Keywords: anabolic; diabetes; Duchenne muscular dystrophy; β-ecdysone; ecdysteroid; ecdysterone; Mas1; osteoporosis; sarcopenia; COVID-19; cardiometabolic diseases; respiratory diseases

Citation: Dinan, L.; Dioh, W.; Veillet, S.; Lafont, R. 20-Hydroxyecdysone, from Plant Extracts to Clinical Use: Therapeutic Potential for the Treatment of Neuromuscular, Cardio-Metabolic and Respiratory Diseases. *Biomedicines* **2021**, *9*, 492. https://doi.org/10.3390/biomedicines9050492

Academic Editor: Pavel B. Drašar

Received: 2 April 2021
Accepted: 26 April 2021
Published: 29 April 2021

Publisher's Note: MDPI stays neutral with regard to jurisdictional claims in published maps and institutional affiliations.

1. Purpose

The aim of this review is to describe the potential and challenges in developing 20-hydroxyecdysone (20E) as a pharmaceutical agent for the future. It is a natural plant steroid with low mammalian toxicity and accumulated data from academic and pre-clinical studies showing that it possesses many beneficial pharmacological effects in mammals, including humans, but few of these properties have yet been substantiated by clinical trials. Further, the bioavailability of oral 20E is low. We shall describe and critically assess the approaches to the sourcing, purification, quality control, and assessment of the activity of the compound in mammalian systems and the clinical trials which are currently underway.

2. Ecdysteroids

Ecdysteroids are a family of invertebrate steroid hormones which are involved in the regulation of moulting, development and reproduction [1]. They differ significantly in their structure from vertebrate steroid hormones, since they are characteristically polyhydroxylated, retain the full C_8 sterol side-chain, possess a 14α-hydroxy-7-en-6-one chromophoric group located in the B-ring and possess an A/B-*cis*-ring junction. Thus, they markedly differ from vertebrate steroid hormones in their polarity, bulk and shape, and there is no convincing evidence that ecdysteroids interact with nuclear receptors for the vertebrate steroids in mammals. It is generally accepted that 20E is the major biologically active form in insects, but other analogues act as biosynthetic intermediates (e.g., 2-deoxyecdysone), pro-hormones (ecdysone and/or 3-dehydroecdysone) or metabolites (e.g., 20,26-dihydroxyecdysone) or storage forms (e.g., ecdysteroid phosphates). Other ecdysteroids may be hormonally active in other invertebrates (e.g., ponasterone A in crustaceans). In accord with their hormonal role, the concentrations of ecdysteroids found in arthropods and other invertebrates are generally rather low (nM to μM), with the storage forms being present in the highest amounts where they occur.

In addition to ecdysteroids occurring in invertebrates (zooecdysteroids), they are also present in certain plant species as phytoecdysteroids, where they are believed to contribute to the deterrence of invertebrate predators. They are present in detectable amounts in the seeds of 5–6% of investigated plant species, and in leaves of an even greater proportion of species [2]. Concentrations vary from just detectable to high, depending on the species, the plant part and the stage of development, accounting for 1–2% of the dry weight in high accumulators. Phytoecdysteroid profiles may vary from simple (the presence of one or two major components), through intermediate (a mixture of major and minor components) to complex (a cocktail of many analogues) [3]. 20E (Figure 1) is the most frequently encountered phytoecdysteroid and very frequently is the major phytoecdysteroid present in the plant. Currently, 526 natural ecdysteroid analogues have been identified (Ecdybase [4]), most of which have been isolated only from plants, probably in large part because of the higher concentrations found in plant sources; some are found in invertebrates and plants, and a few have only so far been detected in invertebrates. To date, the only ecdysteroid which is commercially available at a reasonable cost and in large (kg) amounts is 20E (Figure 1), which can be isolated efficiently from specific high accumulating species (see later).

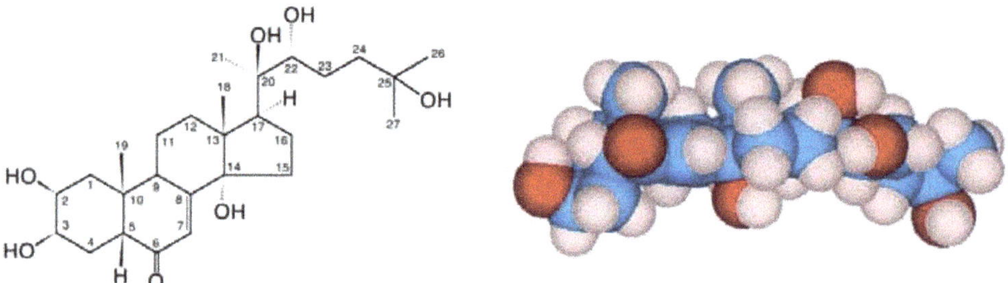

Figure 1. 20-hydroxyecdysone (20E; β-ecdysone; crustecdysone; ecdysterone; BIO101; CAS 5289-74-7; IUPAC 2β,3β,14α,20R,22R,25-hexahydroxy-5β-cholest-7-en-6-one).

3. A Traditional Medicinal Product

Research in this area has a long history and started soon after the structural identification of ecdysone (E) and 20-hydroxyecdysone (20E) from arthropods and the discovery of analogues in plants in 1965 [5], when it was thought that the compounds might be useful invertebrate control agents. As a consequence, considerable effort was put into the total chemical synthesis of E and 20E (reviewed in [6]), determining the toxicity and initial pharmacokinetic studies in mammals. Synthesis proved possible but inefficient, and the physico-chemical properties (polarity, UV stability, etc.) were not really compatible with field application. Western scientists tended to focus on the metabolism and mode of action of ecdysteroids within invertebrates, while Eastern European and Central Asian scientists continued to explore the pharmaceutical uses of ecdysteroids. The findings of these latter studies were published in Slavic and Turkic languages, so their significance was not fully appreciated for quite some time [7]. In the past two decades, there has been a considerable increase in studies focusing on the effects and potential applications around the world, which has been enabled by the concommitant commercial availability of larger amounts of pure 20E, such that it is now fair to say that this is currently the major focus of ecdysteroid research, leaving invertebrate ecdysteroid research rather in the shade.

As the phytochemical analysis of traditional medicinal plants from around the world has advanced, it has become apparent that some of them contain significant amounts of ecdysteroids (Table 1). This interesting association may indicate that ecdysteroids, alone or in conjunction with other plant secondary compounds, may have very wide therepeutic applications, but it must be borne in mind that in very few of these cases have the ecdysteroids been proven to be responsible for the pharmacological activity.

Table 1. Selected examples of ecdysteroid-rich medicinal plants and their traditional uses. The names of ecdysteroids are underlined (see www.ecdybase.org for structures, accessed on 1 April 2021). Note that it is not established that all the reported biological activities are due totally or partly to the presence of ecdysteroids.

Scientific Name	Plant Part	Major Constituents	Claimed Therapeutical Value	World Area	References
Achyranthes bidentata	root	20E, Inokosterone, polysaccharides	Anticancer, anti-inflammatory, diuretic, anti-osteoporotic	India, China Taiwan	[8,9]
Achyranthes japonica	root, leaf	Inokosterone, 20E, saponin, oleanolic acid, calcium oxalate	Antirheumatic, for amenorrhea, carbuncles, fever, dystocia, urinary ailments	Korea, Japan, China	[10]
Ajuga bracteosa A. decumbens	whole plant	20E, ajugasterone C ajugalactone, cyasterone, kiransin, β-sitosterol, cerotic acid, palmitic acid, luteolin	Antitussive, antipyretic, anti-inflammatory, antiphlogistic, antibacterial; treats bladder ailments, diarrhea, bronchitis	Taiwan	[11]
Ajuga iva	aerial parts	20E, cyasterone, ajugasterone C, apigenin, apigenin dihexoside, carvacrol	Diabetes, rheumatism, allergy, cancer, renal, metabolic disorders, digestive, cardiovascular, and respiratory disorders	Africa	[12,13]
Ajuga turkestanica	aerial parts	20E, turkesterone, cyasterone, iridoids	Weight deficiency, ulcers, burns, wound healing, heart protective, hair growth	Uzbekistan, Tadzhikistan	[14]
Boerhaavia diffusa	root, aerial parts	20E, flavonoids, rotenoids, punarnavoside	Immunomodulatory, anticancer, antidiabetic, anti-inflammatory, diuretic, hepatoprotective, antimicrobial, antifungal, anticonvulsant, antioxidant	Brazil, India, Iran, Angola, Ghana, Congo	[15]
Cyathula prostrata	leaf, root	20E, cyasterone, terpenoids, flavonols, tannins	Laxative, antitoxic, analgesic, alleviates flu, cough, rheumatism, dysentery, syphilis	Tropical Africa, China, Australia	[8,16,17]
Cyanotis arachnoidea	aerial parts, root	20E, ajugasterone C, poststerone, rubrosterone, daucosterol	Skin diseases, psoriasis, gastritis, tuberctulosis	China	[18]
Diploclisia glaucescens	stem, leaf	paristerone, 20E, capitasterone, oleanane glycosides	Rheumatism, snake venom, biliousness, venereal diseases	China	[19]
Helleborus niger	rhizomes, leaves	20E, polypodine B, bufadienolides,	Stomachic, tonic, analgesic, antirheumatic	Romania	[20]

Table 1. *Cont.*

Scientific Name	Plant Part	Major Constituents	Claimed Therapeutical Value	World Area	References
Microsorum membranifolium	fronds	20E, ecdysone, 2-deoxy-20E, 2-deoxyecdysone, various conjugates	Asthma, purgative, antiemetic, healing of fractured bones	French Polynesia	[21]
Paris polyphylla	aerial parts	20E, calonysterone, steroidal saponins, luteolin, quercetin	Antibiotic, anti-inflammatory, liver cancer	Southwest China	[22]
Pfaffia glomerata	root	20E, pterosterone, polypodine B, ginsenosides	General stimulant, analgesic, anabolic, anti-inflammatory, immunostimulant, sedative, hypocholesterolemic	Brazil	[23]
Podocarpus macrophyllus var. *nakaii*	stem bark, leaf, root, fruit	20E, ponasterone A, makisterones, pinene, camphene, cadinene, podocarpene, kaurene	Antihelminthic, blood disorders; tonic for heart, kidneys, lungs, stomach	South Africa	[8,24]
Polypodium vulgare	rhizome	20E, polypodine B, polypodaurein, polypodosaponin, flavonoids	Expectorant, cough, pertussis, diuretic	Poland	[25]
Rhaponticum carthamoides	leaf, root	20E, makisterones, triterpenes, sesquiterpene lactones, phenolic acids, flavonoids, thiophenes, lignans	Tonic, roborant, adaptogenic, antidepressive, antiparasitic	Eastern Europe	[26,27]
Serratula chinensis	roots	20E glycosides, sphingolipids, cerebrosides	Pharyngitis, measles, anti-inflammatory, hypocholesterolemic, anti-cancer,	Southern China	[8,28]
Sida rhombifolia	root, seed	20E, ecdysone, ecdysteroid glycosides, cyclopropenoid fatty acids	Enteritis, hepatitis, flu, pneumonia, improves blood circulation, resolves phlegm, alleviates pain	India	[8,29]
Tinospora cordifolia	aerial parts	20E, polypodine B, alkaloids, diterpenoid lactones, sinapic acid	Anti-osteoporotic, anti-inflammatory, anti-stress, immuno-modulator, anti-spasmodic, chemo- and radio-protective, anti-anxiety, neuroprotective	India	[30,31]
Vitex scabra	bark, leaf	20E, turkesterone, khainaoside, syringaresinol	Astringent, antihelminthic, gastrointestinal disorders, wound healing, sexual enhancer	Thailand	[32]

4. Pharmacologial Effects of 20E in Animals

Table 2 provides a summary of the effects attributed to 20E in mammals. This topic has been extensively reviewed previously [7,33–45], so here we shall just present some data from more recent studies on the anabolic and hypolipidemic effects by way of example.

Table 2. Summary of the literature concerning 20E's pharmacological effects in mammals.

Effect	20E and/or Other Ecdysteroid Preparations
Anabolic (muscle)	[46–54]
Fat-reducing/Hypolipidaemic	[55–62]
Anti-diabetic	[56,62–66]
Anti-fibrotic	[67]
Anti-inflammatory	[68,69]
Anti-oxidant	[70]
Anti-thrombotic	[71,72]
Vasorelaxant	[73]
Hematopoiesis stimulation	[74]
Angiogenic	[75,76]
Cardioprotective	[62,77–79]
Neuromuscular protective	[80,81]
Neuroprotective	[81–84]
Liver protective	[85,86]
Lung protective	[69,72,87,88]
Kidney protective	[58,67,89]
Gastric protective	[90,91]
Bone, cartilage protective	[92–99]
Skin protective/repairing	[100–102]

5. Protein Synthesis Stimulatory Effect In Vitro

The murine C2C12 cell line is an accepted model system for myotubule formation and action and consequently for the investigation of the activity of anabolic and anti-sarcopenic agents. The cells in culture can be induced to differentiate into myotubules and then be treated with the test agents. The incorporation of radioactive leucine into protein is a measure of the anabolic activity of the test compound. Figure 2A illustrates the dose–response protein synthesis stimulatory activity of 20E, which is equally active as insulin-like growth factor-1 (IGF-1) in this system. Alternatively, the anabolic effects can be evaluated by measuring myotube diameters. Similar results were observed, e.g., by [49,54] (Figure 2B).

Myotube growth in C2C12 cells is mediated negatively by myostatin, and IGF-1 markedly reduces myostatin gene expression (Figure 3). 20E mimics the effect of IGF-1 in a dose-dependent manner, with significant difference from the control occurring at 1–10 µM. Such an effect was also observed by [53].

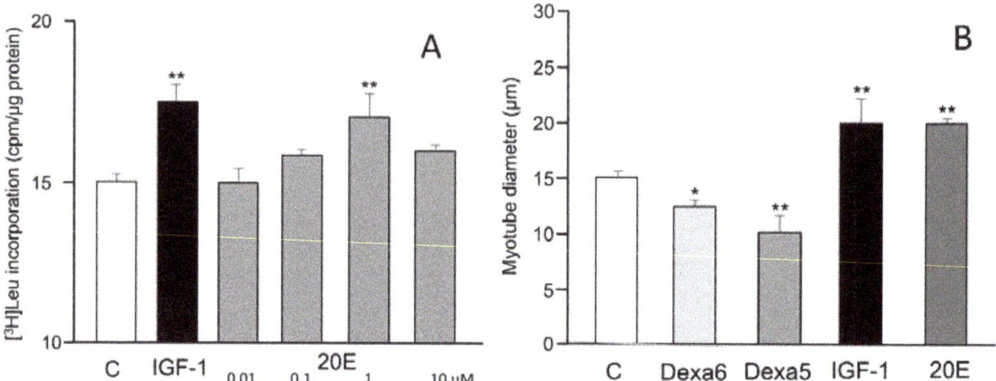

Figure 2. (**A**) The effects of 20E (0.01–10 μM) on protein synthesis in C2C12 cells, showing the anabolic effect of 20E on optimal value for 1 μM 20E. C2C12 cells were grown on 24-well plates at a density of 30,000 cells/well in 0.5 mL of growth medium (Dulbecco's Modified Eagle Medium (DMEM) 4.5 g/L glucose supplemented with 10% fetal bovine serum). Twenty-four hours after plating, the differentiation induction into multinucleated myotubes was carried out, and after 5 days, cells were pre-incubated in Krebs medium 1 h at 37 °C before being incubated in DMEM media without serum for 2.5 h in the presence of [^{3}H]-Leucine (5 μCi/mL) and DMSO (control condition) or IGF-1 (100 ng/mL) or 20E (0.01–0.1–1–10 μM). At the end of incubation, supernatants were discarded and cells were lysed in 0.1 N NaOH for 30 min. The cell soluble fraction-associated radioactivity was then counted and protein was quantified using the Lowry method (after [103]). (**B**) Effects of dexamethasone (Dexa 6 = 10^{-6} M, Dexa 5 = 10^{-5} M), IGF-1 (10 ng/mL), and 20E (10^{-6} M) on the diameter of C2C12 myotubes. Four- to six-day-old myotubes were incubated for 48 h with test chemicals, and were fixed and photographed by glutaraldehyde-induced autofluorescence. *: $p = < 0.05$; **: $p = 0.01$. (redrawn and modified from [54]).

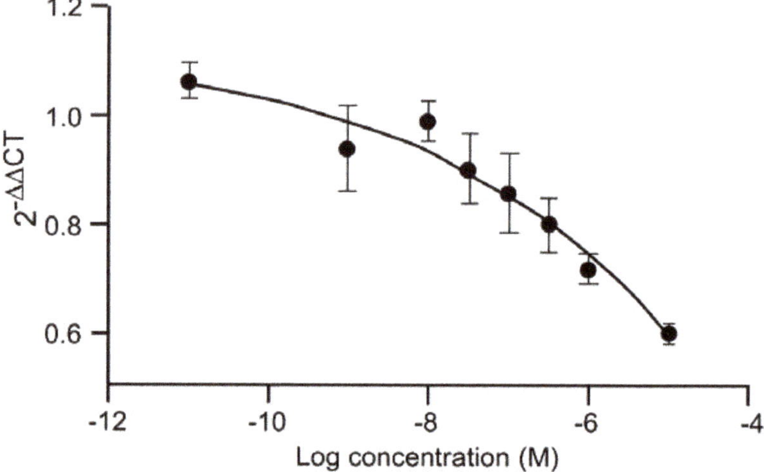

Figure 3. Dose-dependent inhibition of myostatin gene expression in C2C12 cells by 20E. C2C12 mouse myoblasts were differentiated for 6 days into myotubes. They were then treated for 6 h with concentrations of 20E ranging from 0.001 to 10 μM. Myostatin gene expression was detected by qRT-PCR. Results are shown as means ± standard error of the mean (SEM) ([103]).

6. Anti-Obesity Effect

The effects of 20E on adiposity are well documented (Table 2). Mice fed on a high-fat diet tend to adiposity and this is reflected in the mass of epididymal fat pads and the diameter of the adipocytes. Figure 4 compares these parameters in mice on low-fat, high-fat and high-fat +20E diets. 20E counteracts both the increase in the mass of the fat pads and retains the median adipocyte diameter close to that in mice fed a low-fat diet. Similar findings have been reported by other authors, e.g., [49]. 20E is also efficient in high-carbohydrate (high fructose) fed gerbils [61].

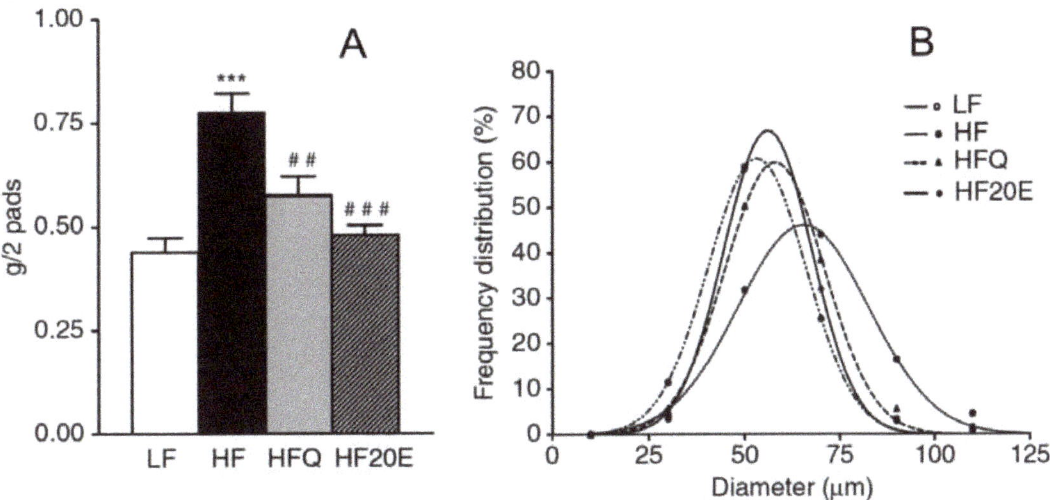

Figure 4. Effect of 20E on mice fed a high-fat diet (HF), when compared to mice fed a low-fat diet (LF). The animals received either pure 20E (50 mg/kg/day) or the same amount of 20E as a quinoa extract (Q). Panel (**A**) shows the impact on the mass of epididymal adipose tissue (*** $p < 0.01$ when compared to LF; ## $p < 0.01$ and ### $p < 0.001$ when compared to HF) and panel (**B**) shows the effect on adipocyte diameter (reproduced, with permission, from [59]).

7. Mechanism of Action of 20E in Mammals

Early studies showed that 20E displays pleiotropic effects on mammals [104]:

- Increases protein synthesis in skeletal muscles and heart [46,105]
- Increases ATP synthesis in muscles [106]
- Stimulates production of erythrocytes [74]
- Decreases hyperglycemia in diabetic animals [63]
- Reduces plasma cholesterol levels [107]
- Decreases the activity of triglyceride lipase [108]
- Protects against experimental atherosclerosis in rabbits [109]
- Activates acetylcholinesterase in the brain [110]
- Activates glutamate decarboxylase (=GABA synthesis) in the brain [111]
- Possesses immunomodulatory activity [112].

In view of these diverse effects, it could be expected that 20E might have various modes of action at the cellular level. There are no homologues to the arthropod ecdysteroid receptor (EcR) in mammals, and attempts to show high affinity binding of ecdysteroids to vertebrate nuclear receptors have not been successful [88,113], but see the possible involvement of estrogen receptor-β (ERβ) below. The mode(s) of action of 20E in mammals has not been fully elucidated for any responsive cell type. In part, this is because the action appears to be non-genomic and also because relatively high concentrations (0.1–10 μM) of 20E are required to observe effects, implying relatively low affinity interaction of 20E with its cellular target(s).

Several models for the modes of action of 20E at responding mammalian cells are being pursued. The models are not mutually exclusive, and it is possible that they could operate simultaneously or in different tissues. Given the pleiotropic effects described for 20E on mammalian cells, it seems possible that it operates through several signalling pathways, which could be more or less tissue-specific. Gorelick-Feldman et al. [50] clearly established that 20E acts through a membrane GPCR receptor, and this has been confirmed using albumin-bound 20E, which cannot cross cell membranes (Figure 5).

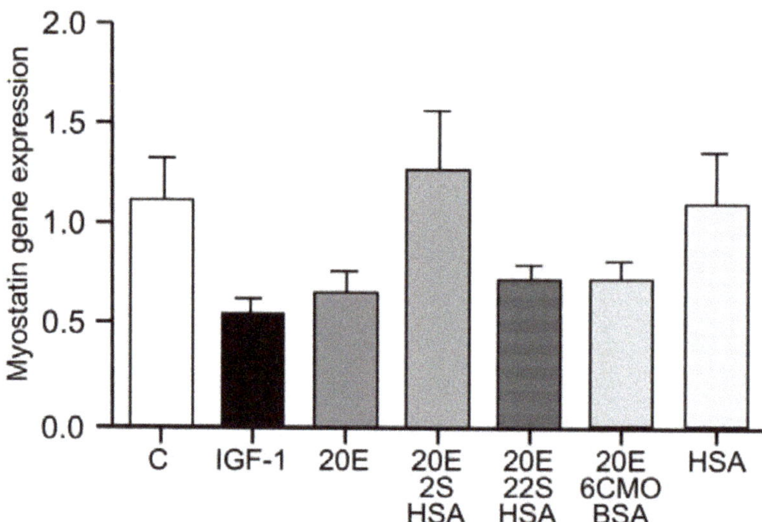

Figure 5. 20E reduction of myostatin gene expression in C2C12 cells (differentiated for 6 days into myotubules) is mediated by binding to receptor sites on the external surface of the cells. The histogram compares the activities of IGF-1 (100 nM) and 20E (10 μM) with those of conjugates of 20E covalently bound to protein (HSA or BSA) through different C-atoms (C-2 and C-22-hemisuccinates or C-6 [6-carboxymethoxime]), all at nominal 10 μM hapten concentration. BSA bovine serum albumin; HSA: human serum albumin; error bars = standard error of the mean [103,114].

The renin–angiotensin system (RAS) is strongly implicated in maintaining muscle function, and one of the peptide products of this system, angiotensin II, targets skeletal muscle cells via the AT1R receptor and has been implicated in the development of sarcopenia [115], both directly through AT1R, which results in increased resistance to insulin and IGF-1, and indirectly by increased production of myostatin, glucocorticoids, TNF-α, and IL-6. Consequently, ACE inhibitors (which inhibit the production of angiotensin II) have found some application in the treatment of sarcopenia [116]. Another component of the RAS is angiotensin 1-7, which has been identified as the natural ligand for another GPCR, Mas1, the activation of which has been hypothesized to enhance protein synthesis in muscle cells. Thus, RAS would have a "harmful arm" acting through AT1R where the activation of the receptor enhances proteolysis, but this can be counteracted by a "protective arm", acting through Mas1, where the activation of the receptor enhances protein synthesis. According to this hypothesis, protection against muscle wasting can be achieved by reducing the activation of AT1R, or the activation of Mas1, or a combination of the two. There are reasons to believe that the activation of Mas1 would be more effective at stimulating muscle anabolism and reducing adipose tissue than an ATIR antagonist, and it would have fewer side-effects than ACE inhibitors. The Mas1 receptor is expressed in many tissues, and its activation in different tissues (e.g., heart, kidney, CNS) may evoke various protective effects [117]. The activation of Mas1 by 20E could explain the anabolic effects of 20E on muscle cells [103,113,114].

Angiotensin(1-7) partially inhibits myostatin gene expression and this inhibition is abolished by the angiotensin(1-7) antagonists D-Pro^7-Ang(1-7) and D-Ala^7-Ang(1-7)

(A779). These antagonists are also effective at preventing the inhibition brought about by 20E, indicating that Ang(1-7) and 20E operate through Mas1 activation [113]. The gene interference of the Mas1 receptor using silencing RNA (siRNAMas1) provided further evidence for the involvement of Mas1 in the mode of action of 20E: siRNAMas1 was found to reverse the inhibition in myostatin gene expression brought about by Ang(1-7) or 20E [113].

Parr et al. [54] have put forward an alternative model for the mode of action of 20E in bringing about the hypertrophy of mammalian skeletal muscle cells, which involves the interaction of the steroid with nuclear estrogen receptor-β (ERβ). Using C2C12 cells, they observed that both 20E (1 μM) and E2 (1 nM) enhanced myotube diameter. Co-treatment with the anti-estrogen ZK191703 (ZK) antagonized the hypertrophy brought about not only by E2, but also by 20E, indicating that E2 and 20E share a common mode of action. The authors next showed that a reporter gene under the control of an estrogen response element could be activated dose dependently by E2 or 20E interacting with ERα or Erβ, and that this activation could be prevented if ZK was also present. Use of the estrogen receptor-specific agonists, ALPHA (for ERα) and BETA (for ERβ), indicated that ERβ mediated the hypertrophy of the C2C12 myotubes, but a note of caution is required because these agonists are only selective at low concentrations and the BETA dose–response curve was bell-shaped. Finally, the selective ERβ-antagonist (ANTIBETA) antagonized the effects of E2 and 20E in the C2C12 cells. The authors suggest that, as it had previously been shown that ERβ can modulate Akt phosphorylation, this could be the link by which 20E brings about hypertrophy in the C2C12 cells. In silico docking studies suggest that 20E can interact with the ligand-binding domains of ERα and ERβ but has the potential for stronger interaction with ERβ [118]. However, the exact nature of the receptor involved in the observed effects can be questioned, because all binding studies performed with nuclear receptors ERα and ERβ were negative [e.g., 38], and truncated membrane-bound forms of ER receptors have been described that may display different ligand specificity [119]. Finally, the recent article by Sobrino et al. [120] provides evidence that in primary human umbilical vein endothelial cells cultures, E2 acts via a membrane-bound ER to bring about NO-dependent vasodilation, and they show that this effect also requires Mas activation as it is abolished by a Mas inhibitor. Although this is a different system to the muscle cells above, it might perhaps give an indication of how both 20E and E2 can bring about muscle hypertrophy.

Further experiments with C2C12 cells [113] indicate that the two hypotheses for the mode of action of 20E on these cells might be unified. Estradiol (E2) was found to inhibit myostatin gene expression in a dose-dependent manner (0.1–10 μM). To determine if the action of E2 is dependent on interaction with a membrane receptor, the activity of E2 was compared with that of E2-CMO (estradiol 6-carboxymethyloxime) and E2-CMO-BSA (E2-CMO covalently coupled to BSA). E2-CMO was found to inhibit myostatin gene expression in C2C12 cells like E2, showing that the carboxymethyl oxime derivative retained biological activity. In contrast, the E2-CMO-BSA conjugate was inactive, so it appears that E2 must enter the cell to be active. In addition to the classical nuclear receptors (ERα and ERβ), several non-nuclear forms of ER are known, and it is presumably one of these which is involved in the regulation of myostatin gene expression in C2C12 cells. Additionally, 17-*epi*-estradiol (17αE2), a close analogue of E2 which does not bind the nuclear receptors ERα and ERβ, was found to inhibit myostatin gene expression in C2C12 cells, providing further evidence for the involvement of a non-nuclear ER [113]. By combining the available experimental data, it has been possible to put together a hypothesis for how both 20E and E2 can regulate myostatin gene expression, whereby 20E interacts externally to the cell with the integral membrane Mas1 receptor, which in turn interacts with ERx located on the inner side of the plasma membrane, while E2 enters the cell to interact with ERx, bringing about myostatin regulation in a more direct way (Figure 6).

This finding may have important consequences because the Mas receptor is expressed in many tissues, which fits with the large array of physiological effects of its endogenous ligand Ang(1-7), which are also observed for 20E [113].

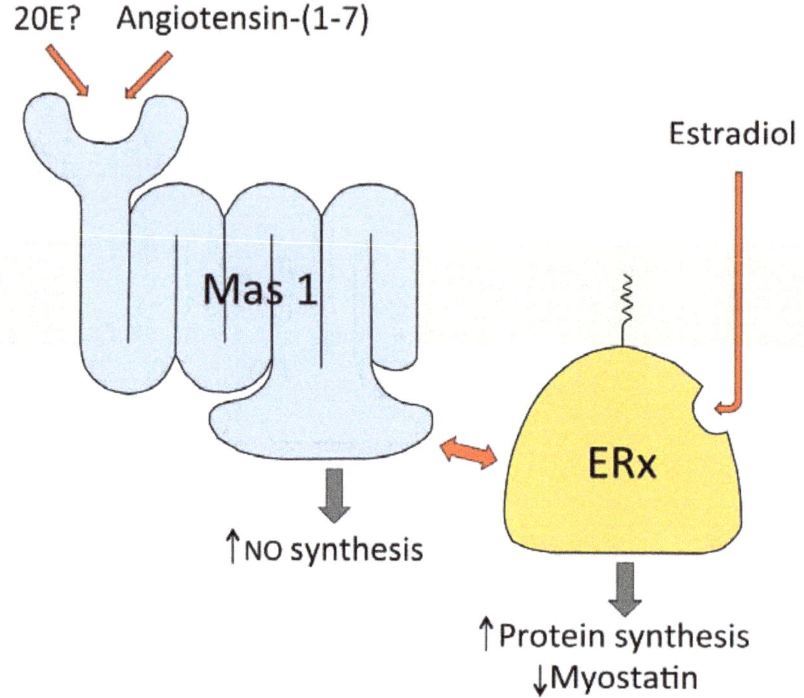

Figure 6. Diagrammatic representation of the proposed mode of action of 20E in the regulation of protein synthesis in C2C12 muscle cells in vitro ([113]).

8. Toxicity, Bioavailability, Pharmacokinetics and Metabolism in Animal Models

Toxicity

Ogawa et al. [121] determined LD_{50}s for ingested 20E or inokosterone in mice of >9 g/kg and LD_{50}s of 6.4 g/kg and 7.8 g/kg for i.p.-injected 20E and inokosterone, respectively. Thus, ecdysteroids are regarded as non-toxic to mammals. Similarly, no subacute toxicity was observed in long-term feeding experiments (0.2–2 g/kg/day) in rats. Additionally, no effects were seen after the administration of these two ecdysteroids to bullfrogs (up to 600 mg/kg in the lymph sinus) or rabbits (up to 100 mg/kg). The two ecdysteroids did not have sex hormonal activity or, interestingly in the context of more recent studies, anti-inflammatory or anabolic effects in rats. Additionally, rats and birds can feed on seeds of *Leuzea carthamoides*, which contain 2% (*w*/*w*) ecdysteroids (mainly 20E), and thrive [7].

Additionally, Seidlova-Wuttke et al. [57,94] fed ovariectomized rats with up to 500 mg/kg daily for 3 months with no stated detrimental effects, but no data on survival are presented in the articles. More recently, a purified to pharmaceutical grade 20E ($\geq$97% 20E purity) was used as a drug candidate in preclinical regulatory studies including safety pharmacology, genotoxicity and repeated toxicology in rodents (rats) and non-rodents (dog). This drug candidate depicted a good safety profile with no genotoxic effects and No Observed Adverse Effect Levels (NOAEL) established at 1000 mg/kg in repeated dose toxicity studies in rats (26-week) and dogs (39-week) (Biophytis, unpublished data).

9. Pharmacokinetics and Metabolism in Mice—Early Studies

The earliest study on the fate of radiolabelled 20E in a mammalian species was performed on mice [122] and showed after oral administration that the bioavailability was low and that elimination was essentially fecal, while after i.p. administration, a somewhat higher level was found in the urine, even though fecal excretion still predominated. Of the radioactivity crossing the intestinal tract after oral administration (<2% of that administered), most was associated with the liver and bile duct, providing the first evidence for an entero-hepatic cycle for ingested ecdysteroids.

Lafont et al. [123] studied the metabolism of ecdysone, the only commercially available radioactive molecule when the study was made, in mice after intraperitoneal injections of the radioactive molecule. Ecdysone levels increased transiently in the liver and then accumulated in the intestine, which, within one or two hours after injection, contained almost all the radioactivity. One day after injection, most of the radioactivity had been eliminated from the body. Moreover, the metabolism of 20E was faster after intraperitoneal injection than after ingestion [122]. As with ecdysone, and whatever the mode of administration, radioactivity was rapidly taken up by the liver and then excreted into the intestine via the bile [122]. Excretion was again primarily fecal in mice [123]. The distribution of radioactivity from ^{3}H-labelled 20E after injection into the caudal vein of mice has been studied by Wu et al. [124]. The kinetics of distribution showed a rapid elimination in urine and an uptake by the liver and bile, then most radioactivity accumulated into the intestine lumen before being eliminated in feces. Unfortunately, no information was provided about the associated metabolism [124].

Studies on the metabolism radiolabelled ecdysone in i.p.-injected mice [123,125] showed that the radioactivity was excreted fully within 24h unless it was diluted with unlabelled ecdysone, when the excretion of radioactivity was extended over 3 days. The injected radioactivity was rapidly captured by the liver and transferred to the intestine, where essentially all of it was located within 45–60 min. The principal metabolites formed were single or multiple products of 14-dehydroxylation, reduction of the 6-oxo group (6α-OH), reduction of the 7,8-double bond and 3-epimerisation. It is significant that ecdysone does not undergo side-chain cleavage because it does not possess a 20,22-diol and cannot be readily hydroxylated at C-20 after injection or ingestion in mice.

10. Pharmacokinetics after Oral Administration in Mice—More Recent Studies

The metabolism of radiolabelled 20E has been more extensively studied [126,127]. These studies involved 20E radiolabelled in the nucleus, so as to be able to follow side-chain cleavage. It was found that the fate of ingested 20E was initially simple, with most of the 20E remaining unaltered as it passed through the esophagus, stomach and small intestine, with only a small proportion crossing into the blood and being captured by the liver (Figure 7).

After ca. 30 min, 20E starts to reach the large intestine, where it begins to be significantly metabolized primarily to 14d20E. Further, the 20E and 14d20E start to undergo complex metabolism including side-chain cleavage. Absorbed 20E and its metabolites are taken by up the liver, enter the bile duct and are brought back to the intestine. Entero-hepatic cycling then occurs, during which 20E and the metabolites leave the gut, enter the blood, are taken up by the liver and returned through the bile duct to the intestine. This result is fully consistent with the observations of Wu et al. [124] of a sustained high radioactivity in bile for up to 12 h (Table 3). As a consequence, the duration of 20E presence in the plasma is prolonged at a low, but physiologically significant, concentration (in the μM range—depending on the dose administered), and the extent and complexity of the metabolism of 20E increase markedly, such that the excreted feces contain unaltered 20E and a wide range of metabolites, produced by permutations of several biochemical reactions.

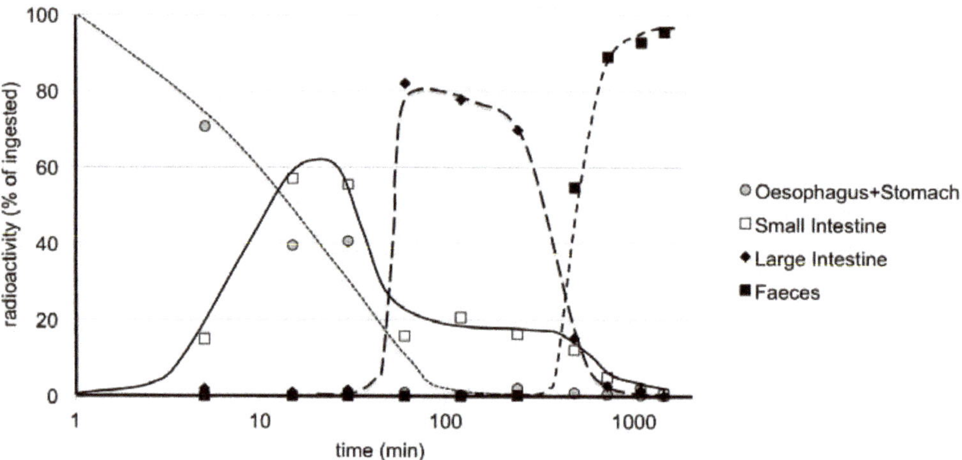

Figure 7. Time-course of the distribution of radioactivity in the stomach/intestine/feces in mice after the oral application of [1α,2α-^{3}H$_2$]20E. Note the logarithmic scale for abscissa. Each value is a mean of 2 animals. Note the plateau of small intestine content that is best explained by an entero-hepatic cycle and consistent with the prolonged concentration of radioactivity in bile as observed by Wu et al. [124] (reproduced, with permission, from [127]).

Table 3. The effect of ^{3}H-*Achyranthes bidentata* ecdysterone on the quantitative distribution of different tissues in different tracking phases of mice (mean, $n = 5$) expressed in µg g^{-1} (translated from Wu et al. [124]).

	5 min	10 min	30 min	1 h	3 h	6 h	12 h	24 h
Blood plasma	0.061	0.057	0.052	0.047	0.032	0.023	0.015	0.011
Urine	0.482	0.921	1.534	0.281	0.102	0.096	0.087	0.046
Feces	0.015	0.019	0.035	0.068	0.099	0.931	0.312	0.041
Bile	0.421	0.456	1.042	0.901	0.301	0.209	0.198	0.094
Liver	0.312	0.251	0.213	0.112	0.078	0.061	0.056	0.046
Heart	0.062	0.052	0.035	0.033	0.031	0.029	0.023	0.022
Spleen	0.041	0.026	0.024	0.018	0.023	0.027	0.032	0.027
Lungs	0.116	0.069	0.057	0.053	0.049	0.045	0.036	0.028
Kidney	0.137	0.123	0.098	0.071	0.055	0.043	0.037	0.032
Adrenals	0.184	0.139	0.098	0.081	0.073	0.065	0.052	0.043
Testis	0.048	0.036	0.029	0.026	0.023	0.021	0.019	0.017
Skeletal muscle	0.021	0.023	0.028	0.017	0.015	0.013	0.011	0.010
Spinal cord	0.019	0.028	0.067	0.041	0.033	0.024	0.020	0.018
Brain	0.013	0.014	0.015	0.013	0.012	0.01	0.008	0.007
Small intestine	0.254	0.139	0.094	0.083	0.076	0.061	0.05	0.033
Stomach	0.034	0.047	0.075	0.061	0.053	0.041	0.032	0.023
Large intestine	0.011	0.023	0.029	0.031	0.035	0.048	0.038	0.023
Bladder	0.013	0.024	0.038	0.026	0.023	0.018	0.015	0.013

The bioavailabilities of 20E (ca. 1%) and Post (ca. 19%) have been determined [127,128]. As in mice, the initial fate of ingested 20E is uncomplicated, remaining almost completely associated with the gut in unchanged form, until it reaches the large intestine, where microbial metabolism and metabolism associated with the enterohepatic cycle come into play to rapidly increase the extent and diversity of metabolites. Evidence has been provided that glucuronide conjugates of 20E are present in the bile fluid, but these are hydrolysed on reaching the large intestine. The major metabolites have been conclusively identified as 14d20E, Post, 14dPost, 3-epi-Post, 16αOHPost, and 21OHPost [128], but other permutations of these biochemical reactions almost certainly exist. Additionally, other metabolites where the 6-one-7-ene chromatophore has been modified, so that the products are no longer UV-absorbing, are likely to be present amongst the complex mixture of metabolites present in the large intestine and excreted in the feces. The elimination of ingested or injected 20E is predominantly fecal (Figure 8A,B). The microbial origin of the 14d20E metabolite is demonstrated by the absence of this metabolite in axenic rats fed 20E [129] and by its formation from 20E in incubations of intestinal contents under anaerobic conditions [127]. That 14-dehydroxylation is associated with the microbiote of the large intestine is in accord with previous findings that dehydroxylations of steroids and bile acids are only bacterial [130,131]. The locations of the other metabolic reactions are not currently known with certainty, but it is expected that they will be associated with specific enzymes of vertebrate steroid catabolism.

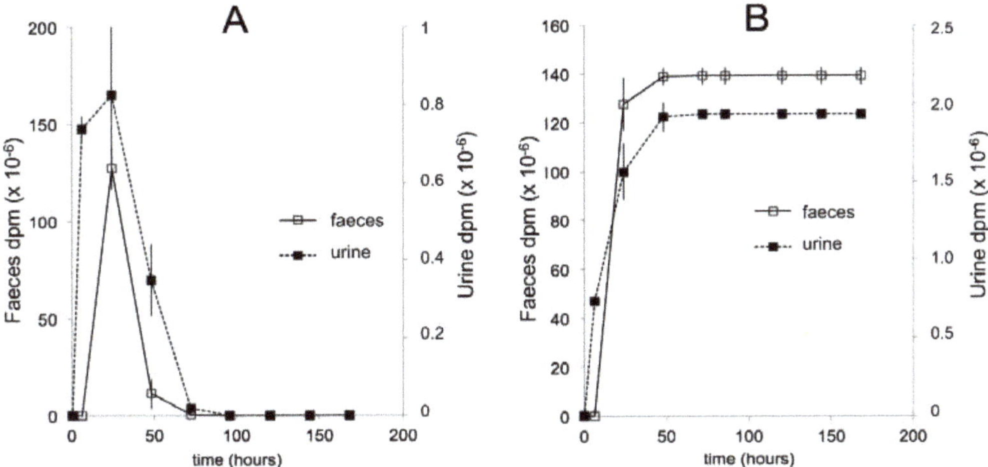

Figure 8. Comparison of the temporal (**A**) and cumulative (**B**) urinary and fecal elimination of radioactivity after the oral application of [5,7,9-^{3}H]20E to 6–7 week-old male Wistar rats. After oral application of [^{3}H]20E, elimination of radioactively labelled 20E and its metabolites is completed within 48 h. The amount of radioactivity recovered from the feces is by far the major route of excretion, since the amount in the urine corresponds to only 1.40% of the total radioactivity recovered. (Reproduced, with permission, from [127]).

11. Pharmacokinetics and Metabolism in Humans

11.1. Early Studies

Although ecdysteroids are not biosynthesized in mammals, sensitive methods can detect them in mammalian body fluids (plasma, urine). Thus, ecdysteroid-specific radioimmunoassay (RIA) was used to detect immunoreactive ecdysteroids in the serum of seven mammals (dog, rhesus monkey, sheep, cow, rabbit, mouse and rat) and humans in nM concentrations [132], which originate from the food. Ingested ecdysteroids (20E or E) reach a plasma titre maximum after ca. 30 min in humans and sheep and are rapidly cleared from the blood and urine, but the proportions of the applied dose found in these are very low, accounting only for a few percent. The effective half-time of elimination from the blood in humans was found to be 4 h for E and 9 h for 20E, and 3.1 h for E and 3.3 h for 20E for elimination from the urine [132].

Pharmacokinetic studies in humans are more restricted owing to the limitations on the use of radiolabelled molecules. Consequently, only a few 20E pharmacokinetic studies have been performed. In humans, Simon and Koolman [132] showed that ingestion of ca. 15 mg 20E induced a urinary peak level of approximately 0.5 µM. Significant immunoreactivity was detected during the first 8 h following intake, but weak urinary levels were still observed until 24 h after ingestion [133]. After a single oral intake of 20 mg 20E, a more detailed study by RIA and liquid chromatography-mass spectrometry (LC-MS) showed the existence in urine of a peak of detectable ecdysteroids after 3–4 h and a total quantity of 1–2 mg—the exact nature of the compounds was not fully determined, although one could be tentatively identified as 14d20E. In another study [134], the ingestion of 434 mg 20E resulted in the urinary excretion of a total of 5 mg of 20E, corresponding to approximately 1% of the ingested amount. This study also indicated that urinary excretion seems somewhat more significant in humans than in rodents [135], even though the fecal content was not quantified and small amounts were recovered from urine as compared to the administered dose.

11.2. Recent Studies

Biophytis has recently undertaken the first in-human, randomized, double-blind, single-centre study to determine the safety of 20E and its major metabolites in healthy young (18–55 years) and elderly (>65 years) adult subjects [136,137]. The study comprised two parts: a single ascending dose component (SAD; 16 subjects) and a multiple ascending dose component (MAD; 30 subjects). 20E showed a good safety and pharmacokinetic profile, which allowed the doses for a Phase 2 Sarcopenia clinical trial to be defined. Pharmacokinetics in subjects treated with increasing amounts of orally delivered 20E (Table 4) showed that the level found in the plasma is dose-related and also apparently saturable. The kinetics of plasma 20E was similar between young and older subjects (Figure 9).

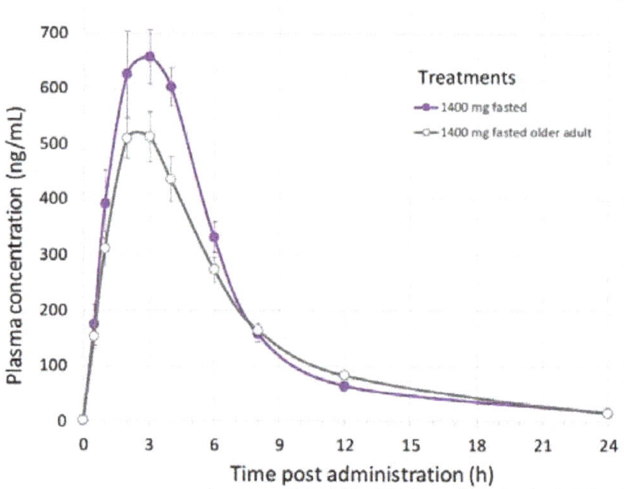

Figure 9. Pharmacokinetics of 20E in young and elderly humans given a single dose of 1400 mg. 20E was quantified by HPLC-MS [136,137].

About 2% of the administered 20E was excreted in the urine within 12 h of ingestion of a single dose, and half of this was excreted within 4 h. 20E and its metabolites were identified and quantified in the plasma of individual patients who had received oral 20E. These metabolites corresponded to the major ones previously found in rodents (14dPost, 14d20E and Post).

Table 4. Kinetic parameters of BIO101 (=97% pure 20E) after a single oral administration ($n = 6$, fasted). Values in parentheses indicate the SEM (C_{max}, AUC) or the range (t_{max}).

B1O101 PK Parameter (Unit)	100 mg	350 mg	700 mg	1400 mg
Cmax (ng/mL)	141 (16.6)	317 (37.9)	399 (24.7)	710 (20.2)
tmax (h)	2.03 (1.00–3.02)	3.00 (1. 05–4.00)	3.00 (2.00–4.02)	3.50 (2.00–4.02)
AUC0-t (ng·h/mL)	767 (31.1)	1924 (40.1)	2578 (22.9)	4148 (15.9)

12. The Major Routes of 20E Metabolism in Rodents and Humans

Figure 10 summarizes the metabolic reactions currently known to occur for 20E in mammals to produce a complex array of metabolites arising from combinations of the metabolic possibilities, although the flux through the network of combinations will depend on the exposure to and the activity of the different enzymes depending on the mammalian species. On ingestion, bioavailability is very low, as is initial metabolism, and it is only on reaching the large intestine that 14-dehydroxylation occurs, but this is a reaction performed by the gut flora. Only ecdysteroids with a 20,22-diol (such as 20E) undergo side-chain cleavage, as demonstrated by the absence of this reaction with oral or injected E. Side-chain cleavage and 14-dehydroxylation significantly enhance bioavailability (see below) and an enterohepatic cycle maintains ecdysteroid plasma levels and enables more extensive metabolism (e.g., hydroxylations, reductions, epimerizations—the two latter already observed with ecdysone [125]).

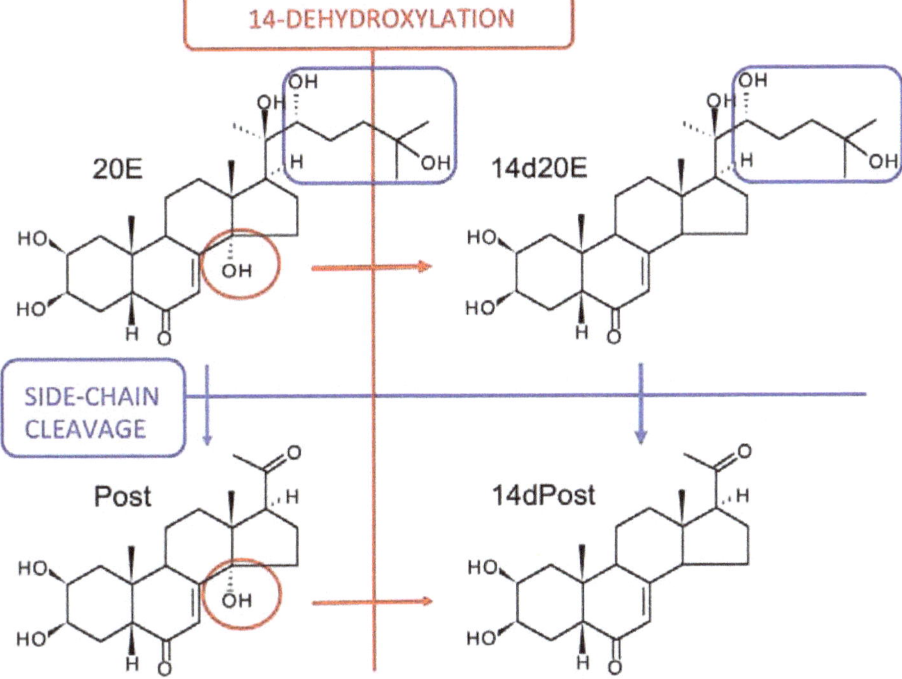

Figure 10. The principal routes of metabolism of 20E in rodents and humans (modified from Kumpun et al., 2011). The circles and arrows in red highlight the changes associated with 14-dehydroxylation, while the boxes and arrows in blue highlight the changes associated with side-chain cleavage.

13. Large-Scale Production of Pure 20E for Drug Development

General Considerations

Various theoretical possibilities for ensuring the supply of 20E in adequate amounts and at reasonable cost have to be considered:

1. Total chemical synthesis
2. Isolation from the current plant source, e.g., *Cyanotis* sp. following cultivation and collection in China (Yunnan province)
3. Cultivation of another species already identified as a good accumulator with a simple ecdysteroid profile (*Rhaponticum*, *Serratula* or *Pfaffia* sp.)
4. Use plant cell or hairy-root cultures of an ecdysteroid-producing plant
5. Generate recombinant yeasts using insect and/or plant enzymes of ecdysteroid biosynthesis.

Of these, possibilities 1 and 4 can be discounted; 1 because of the complexity ($\geq$18 steps) and poor overall yield (ca. 1%) of the process, and the rate-limiting access to a suitable starting material, and 4 probably because of the high cost of the media and low yields normally attained for plant secondary compounds, even when using elicitors [138]. Various researchers have started to explore a range of culture methods in vitro (e.g., micropropagation, callus cultures, plant cell culture, hairy-root cultures, transformed yeast fermentation) in the hope of obtaining more amenable ecdysteroid-producing systems. To date, only hairy roots of *Ajuga reptans* var. *atropurpurea* (Fujimoto et al. [139]) consistently and reliably produce ecdysteroids, but the culture is too expensive to provide a commercial source of 20E. Additionally, hairy roots of some other ecdysteroid-producing species do not contain ecdysteroids, so this cannot be viewed as a general approach.

Possibility 5 will for sure face a number of difficulties, but, once established, it would provide relatively easy scale-up. The use of modified yeasts to produce bioactive molecules is rapidly developing [140] and proved efficient even for multiple-step biosyntheses (e.g., [141]). Transforming yeast to biosynthesize ecdysteroids, as has been previously performed for certain vertebrate steroids [142,143], is an attractive prospect, but is currently confounded by our incomplete knowledge of the ecdysteroid biosynthetic pathways in either invertebrates or plants and the probable high number of genes involved in the complete pathway(s) [3].

Possibility 2 may make the supply dependent on a single country and on the difficulty to secure access to the material in case of competition for a limited supply. Possibility 3 might displace the problem to another country and requires the development of a large-scale culture.

At the moment, the molecule is extracted from suitable plant materials. It is perhaps worth reminding ourselves of the criteria that an ideal plant source should fulfil:

- The plant should accumulate a high amount of 20E
- The plant should have a simple ecdysteroid profile (ideally just 20E)
- The plant should be easy and rapid to grow in accessible areas of the world
- The plant matrix should be amenable to the ready purification of ecdysteroids
- The purification and isolation of 20E should not involve expensive chromatographic methods
- The plant should not be susceptible to pests and diseases
- The species should not be rare or protected
- Culture, harvesting and processing costs should be low; initial processing should take place close to the culture site

Clearly, no plant species will fulfil all these criteria, but the closer it comes, the more culturally and comercially viable the challenge of isolating adequate amonts of 20E will become. Major screening of plants for the presence of ecdysteroids began by using very time-consuming, semi-quantitative whole insect bioassays (*Chilo* dipping test; *Musca* assay; *Sarcophaga* assay), moved on to the combined use of ecdysteroid-specific RIAs and a microplate assay based on an ecdysteroid-responsive cell line, and most recently progressed to analysis by HPLC/DAD/MS-MS, which allows the simultaneous identification and quantification of common phytoecdysteroids and dereplication to detect the presence of previously unknown ecdysteroids [144].

Most ecdysteroid-accumulating species contain between 0.01 and 0.1% of the dry weight as ecdysteroids, whereas the few identified high-accumulators contain 1% and above. The roots of *Cyanotis* spp. can contain up to 5% of the dry weight as ecdysteroids, largely as 20E [145]. Mature stems of *Diploclisia glaucescens* were found to contain 3.2% of their dry weight as 20E [146], but the collection of the stems of this South Asian climber precludes this as a feasible source of large amounts of 20E.

14. Plants Species Presently Used to Produce Purified 20E

The relationship between the presence of phytoecdysteroids and plant taxonomy is complex, but high accumulation is associated with certain species in the genera *Achyranthes* (Amaranthacea), *Cyanotis* (Commelinaceae), *Pfaffia* (Amaranthaceae), *Rhaponticum* (syn. *Leuzea/Stemmacantha*; Asteraceae), and *Serratula* (Asteraceae), which can be considered as good sources of 20E.

Achyranthes (*A. aspera*, *A. bidentata*, *A. fauriei*, *A. japonica*): Perennial or HHA; warm temperate and sub-tropical plants; native to South-east Asia and/or Africa; all plant parts contain ecdysteroids with seeds containing the highest level (0.25%) in *A. aspera* and roots (1.74%) being the best source in *A. bidentata*;

Cyanotis (*C. arachnoidea*, *C. vaga*): Perennial; tropical and sub-tropical plants; growing in tropical Africa, the Indian subcontinent and southern China; roots accumulate very high levels of 20E (up to 5.5% of the dw; Wang et al. [145]); simple ecdysteroid profile—major source for commercial 20E cultivated on a large scale in China;

Pfaffia (*P. glomerata*, *P. iresinoides*): Perennial; tropical plants; native to South America; *P. glomerata* contains ecdysteroids throughout the plant; roots have a very simple ecdysteroid profile, consisting solely of 20E (0.9% of the dw);

Rhaponticum (*R. carthamoides*): Perennial; sub-alpine plant; native to the Altai and Sayan Mountains in Central Asia; roots, flowers and seeds accumulate high levels of 20E (1–2% of dw); complex ecdysteroid profile, but >80% as 20E—used to prepare ECDYSTEN pills containing 5 mg of 20E (OPIH, Uzbekistan) [147];

Serratula (*S. centauroides*, *S. coronata*, *S. tinctoria*, *S. wolfii*): Perennial; temperate plants; native to western and central Europe; *S. tinctoria* accumulates up to 2% of dw as ecdysteroids; the major ecdysteroids are 20E 3-Acetate, 20E and PolB; *S. coronata* is used to prepare SERPISTEN, a 8:1 mixture of 20E and 25S-inokosterone [148].

We should mention that only 2% (ca. 6000) of terrestrial plant species have been assessed for the presence of phytoecdysteroids and far fewer than this have undergone extensive phytochemical analysis to identify the ecdysteroid analogues present [144]. Thus, there are certainly many other as yet unidentified species which are high accumlators of 20E, but it has to be borne in mind that the species already tested represent the most readily available species and accessing samples of new untested, but validated, species would require the investment of a large amount of time and effort.

Owing to the wide taxonomic diversity of plant species known to accumulate phytoecdysteroids, it has been proposed that most, if not all, plants have the genetic capacity to produce ecdysteroids, but in the majority (the non-producers) the expression of the biosynthetic pathway is down-regulated [2]. Additionally, it has been shown in several ecdysteroid-accumulating species that ecdysteroid levels are additionally influenced by environmental factors (temperature, nutritional factors, invertebrate predation, etc.) [149,150]. Thus, to maximize 20E content in a chosen species, it is not only necessary to select a high producing genetic line (cultivar), but also to optimize the growth conditions and to consider treating the plants with appropriate elicitors (e.g., methyl jasmonate to mimic insect or fungal attack) at suitable times in the development.

15. Purification Process

The major requirements in the processing of plant material to obtain a drug product in adequate amounts for therapeutic use are simplicity, efficiency of extraction and purification, reproducibility and cost-effectiveness. As explained above, the choice of plant material is key, since not only should it contain a large amount of the target molecule, but the plant matrix should be readily extractable and not contain components (e.g., large amounts of polysaccharides) which make processing difficult. Given that a suitable plant source will contain typically 1–2% of the dry weight as 20E, and that a pharmaceutical/medicinal dose of 20E is likely to be in the 100 mg–1 g/day range, it is necessary to be able to process tonnes of plant material. Clearly, it would be highly advantageous if the harvested plant material can be cleaned, dried, broken up (to reduce volume and increase the surface area) and subjected to initial extraction as close to the site of harvesting as possible, as the mass of the initial extract is probably only 1–2% of the fresh weight of the plant material and, therefore, much more readily and cost-effectively transportable.

The extraction and purification have to be optimized with regard to the physico-chemical properties of the target molecule (generally already known or readily determinable) and the nature of the plant matrix (generally only vaguely understood). Owing to the large number of hydroxyl groups in 20E, it is highly soluble in alcohols, so the dried, powdered plant material will usually be extracted with methanol (preferred as it is cheaper) or ethanol (if the product is to be BIO/Organic), with or without prior extraction with a non-polar solvent, such as petroleum ether, to de-fat the plant material. Extraction may occur with heating, stirring or maceration for a defined time, all of which need to be optimized. Other methods, such as super critical fluid extraction or bi-phasic extraction, may be used, but these are generally more expensive and would only be cost-effective if the target molecule is potent (low daily dose required) and high-value, which is not the case for 20E.

Figure 11 provides a flow-diagram of a representative processing method for *Cyanotis* sp. roots, where the dried plant material is refluxed thrice with ethanol and the pooled extracts are filtered before being passed through a macroporous resin to absorb plant compounds and vacuum-dried to yield a powder containing 90% 20E. This preparation can then be recrystallized twice to bring the purity of the 20E to >97%.

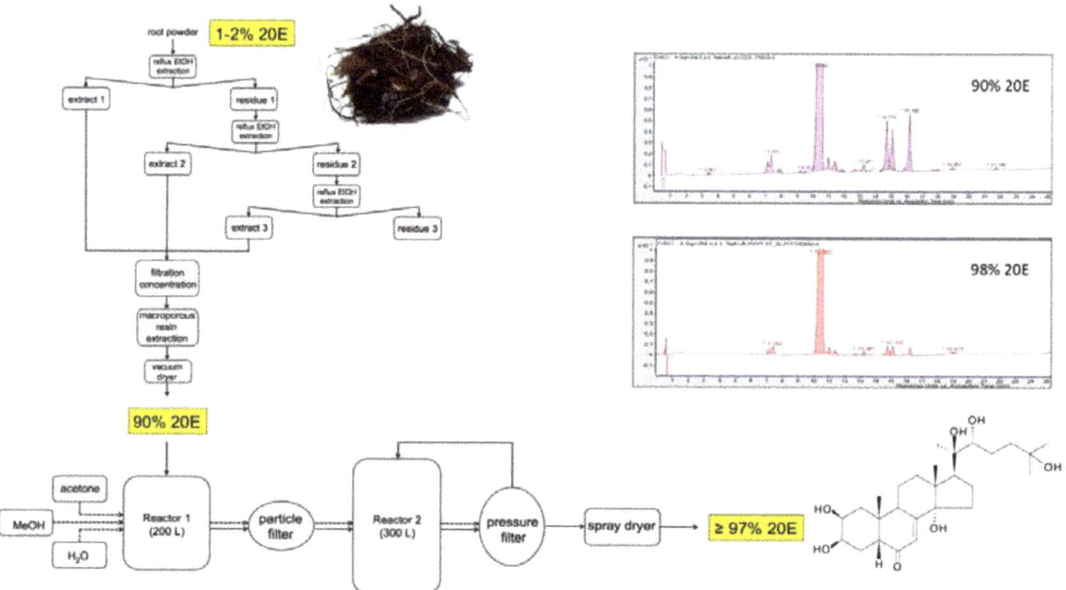

Figure 11. A representative flow-diagram for the large-scale extraction and purification of 20E from roots of *Cyanotis* sp.

A comparison of the reversed-phase (RP)-HPLC profiles of the 90% and 98% 20E preparations is shown in the inset to Figure 11. Most of the minor peaks in the chromatograms correspond to other ecdysteroids, which are very difficult to separate fully from 20E by crystallization. This underlines the need to start with plant material which contains essentially only 20E. Whenever the plant contains a high proportion of a very close molecule (e.g., inokosterone or polypodine B), it would be very difficult to achieve the required purity.

Owing to the need for optimization at each stage of the extraction and purification, the process is developed in stages going from small-scale (100 g—a few kg dry plant material) in the laboratory through increasing medium-scale stages (e.g., 10–500 kg) before being applied at the industrial scale (>1 tonne), so that difficulties can be identified and resolved early on and any problems of scale-up can be dealt with. A thorough cost analysis needs to be performed throughout the scale-up procedure to ensure that the target molecule can be brought to market at a viable cost.

16. Quality Control and Stability

Using 20E as the active ingredient of a pharmaceutical grade drug candidate requires the complete control of the raw material (roots, aerial parts or whole plant harvested from the field) following the GAP (Good Agricultural Practices; http://www.cfsan.fda.gov/~dms/prodguid.html, accessed on 1 April 2021) or the guidelines on the quality of herbal medicinal products/traditional medicinal products. This involves the description of the active substance preparation including geographical origin, cultivation, harvesting, and postharvest treatments (possible pesticides and fumigants used and/or formation of genotoxic impurities, solvents, presence of aflatoxins, traces of heavy metals and possible radioactive contamination).

The pharmaceutical grade compound (purity $\geq$ 97%) must be fully characterized, and all its impurities $\geq$ 0.1% must be identified, with none of them being above 0.5%, otherwise it would require specific toxicological studies. In the present case, given the process used and the fact that ecdysteroid-containing plants contain a complex cocktail of these molecules (e.g., [26,151]), most remaining impurities correspond to other phytoecdysteroids, the nature of which will depend on the plant used.

Compound stability studies also have to be performed, which does not raise problems in this case, as ecdysteroids are very stable molecules when stored in dry state in the dark at room temperature.

17. Regulatory Preclinical Studies

Preclinal Regulatory Requirements

Nutritional studies can be allowed with limited prerequisite regulatory elements, provided that 20E comes from an edible plant (e.g., spinach, quinoa or *Leuzea*—a plant on the Belgian list) and the amounts administered are not too high as compared with those present in a diet comprising the above plants (e.g., 50 g quinoa contains 20–30 mg 20E). Such studies were performed in relation to anabolic or anti-obesity effects on healthy subjects.

On the other hand, clinical studies aimed at treating human diseases with possibly higher doses and for extended periods require a lot of information about the molecule administered and its long-term effects on model animals, typically one rodent and one non-rodent species (Table 5).

Safety pharmacology studies must thus be conducted to investigate acute effects on the central nervous system, respiratory function, cardiovascular system including hERG potassium channel on blood pressure, heart rate, core body temperature and electro-cardiogram.

Long-term toxicity studies also have to be performed in one rodent and one non-rodent species, the duration of which must correspond to the expected duration of administration during the clinical trial.

Table 5. Non-clinical regulatory requirements and submission dossier for clinical trial approval.

Non-Clinical Activities	Studies	Clinical Trial		
		Non Medicinal Drug Trial	Phase 1	Phase 2
Regulatory requirements				
Absorption	Pharmacokinetics	NA	✓	✓
	Toxicokinetics	NA	✓	✓
	Transporters (Caco-2 cells)	NA	✓	✓
Metabolism	Microsomal metabolism	NA	NA	✓
	ADME $^{\$}$ study	NA	NA	✓
	CYP inhibition/induction	NA	✓	✓
	Metabolite identification	NA	✓	✓
Toxicology	Phototoxicity	NA	✓	✓
	Genotoxicity tests (Ames; micronuclei)	NA	✓	✓
	Repeated toxicology studies	NA	✓	✓
	Micronuclei tests	NA	✓	✓
Distribution	Red blood cell partitioning	NA	✓	✓
	Plasma protein binding	NA	✓	✓
Pharmacology	In vitro studies	✓	✓	✓
	In vivo studies	✓	✓	✓
Safety pharmacology	Cardiovascular system	NA	✓	✓
	Central nervous system and respiratory function	NA	✓	✓
Documents required for submission dossier				
Investigational Medicinal Product Dossier (IMPD)		NA	✓	✓
Investigation Brochure (IB)		NA	✓	✓
Technical Product Dossier		✓	NA	NA
Literature related to the product		✓	NA	NA
Clinical Trial Protocol		✓	✓	✓
Informed consent		✓	✓	✓
Investigative New Drug (IND) Package for the FDA		NA	✓	✓

NA: not-applicable; ✓: required; $^{\$}$ "adsorption, distribution, metabolism and excretion".

18. Regulatory Studies with 20E Metabolites

20E undergoes an extensive metabolism after administration, in part at least performed by the gut microbiote, and probably also in the liver, which is known to metabolize endogenous steroid hormones. Whenever metabolites would circulate at significant levels (area under the curve (AUC) $\geq$ 10% that of 20E), regulatory studies would be required to assess their lack of toxicity. In addition, it is relevant to determine their biological activity, i.e., whether they have a similar/lower/higher activity to that of the parent compound, possess a different (specific?) activity, or if they just represent inactivation products.

18.1. Some Drugability Calculations

As indicated above, the bioavailability of 20E is low, while that of its metabolites may be significantly higher. In a first step to relate the bioavailability of ecdysteroids to their physicochemical properties and structures, we have calculated the ADMET Traffic Light Scores for ecdysteroids according to Lobell et al. [152] based on values for MW, LogP, the polar surface area (PSA), and the number of rotatable bonds. However, rather than using solubility (which has to be experimentally determined), the number of N + O atoms (as a measure of H-acceptors) and the number of NH + OH groups (as a measure of H-donor groups) were used as in the original method of Lipinski et al. [153]. The procedure generates a score (explained in the legend of Table 6) between 0 and 10 for each compound; the lower the value, the more suitable the physicochemical properties are for an orally administered drug.

Table 6. Drugability scores for ecdysteroids.

Compound	MW	LogP	PSA (Å^2)	Rotatable Bonds	H-Acceptors	H-Donors	TL Score
20E	480.30	1.36	138.45	5	7	6	3
20,26E	496.64	0.35	158.67	6	8	7	4
14d20E	464.64	2.30	118.21	5	6	5	1
6αOH20E	482.66	1.54	141.60	5	7	7	4
6αOH14d20E	466.66	2.49	121.37	5	6	6	3
Post	362.46	1.03	94.83	1	5	3	0
14dPost	346.47	1.97	74.60	1	4	2	0
6αOHPost	364.48	1.22	97.98	1	5	4	0
6αOH14dPost	348.48	2.16	77.75	1	4	3	0

The canonical isomeric SMILES string was created for each ecdysteroid at http://endocrinedisruptome.ki.si (accessed on 15 July 2019), which allows full stereochemical specification of the structure (including C-5 and C-9), and pasted into the Molinspiration calculator (www.molinspiration.com, accessed on 15 July 2019) to obtain values for the above parameters. Scoring was as follows: MW: <400 = 0; 400–500 = 1; >500 = 2; Calculated LogP (calculated by fragment summation method): <3 = 0; 3–5 = 1; >5 = 2; Polar surface area PSA (Å^2): <120 = 0; 120–140 = 1; >140 = 2; Rotatable bonds (excluding C-O-H): <7 = 0; 8–10 = 1; >11 = 2; Number of NH + OH (H-bond donors): $\leq$5 = 0; >5 = 1; Number of N+O (H-bond acceptors): $\leq$10 = 0; >10 = 1.

Each parameter (MW, LogP, PSA, Rotatable Bonds, H-acceptors and H-donors) was scored as described in the legend to Table 6. The allocated score is represented by the background colour in the above table: 0 = green, 1 = orange and 2 = red. The scores are added to give the Traffic Light Score (0–10), where 0–2 is deemed green (good drugability), 3–6 orange (moderate drugability) and 7–10 red (poor drugability).

Table 6 indicates that, once formed, C_{21}-metabolites should be readily absorbed from the gastrointestinal tract, since they all have a TL Score of zero, whereas 20E and most of its C_{27}-metabolites should be absorbed far less readily (with scores of 4 or 5). The only exception to this is 14d20E (which is the major initial intestinal metabolite of ingested 20E), with a TL Score of 1, owing to its lower PSA and lower number of H-bond donor groups after the removal of the 14α-hydroxyl group. This is in accord with the experimental observation that this metabolite is found in the urine of rats and mice after the ingestion of 20E, implying that at least a portion of the 14d20E formed in the large intestine is absorbed into the blood to reach the kidneys [127].

18.2. SAR Studies

Only a few structure-activities studies are already available for some natural phytoecdysteroids (e.g., [35,38,47]), but almost no data are available for the above-mentioned metabolites, except for Post and for rubrosterone, a potential C_{19} metabolite not yet observed in metabolic studies. Both Post and rubrosterone seem to retain some anabolic activity [154,155]. Thus, data are still lacking for 20E metabolites.

For those major human metabolites, specific toxicological studies may be required, unless they would be formed in similar amounts in an animal model. Such studies may include off-target binding studies as previously performed with 20E [113]. In addition, it appears relevant to assess their biological activity at least in vitro in order to understand whether they participate to the overall effects observed in vivo.

19. Clinical Studies

Early experimental studies and clinical trials on humans have indicated that 20E or plant extracts containing significant amounts of ecdysteroids [104]:

- Prevent sleep disorders [156]
- Improve nerve functioning in the CNS [157]
- Improve sexual function [158]
- Improve hepatic function in hepatitis [159]
- Improve learning and memory [160]
- Improve cardio-vascular function [161]

A major aim of early human studies was to improve physical performances of healthy young people, and they showed a significant improvement in muscle mass and/or endurance after a few days/weeks. Such experiments paved the way, directly or indirectly, for use as a "doping substance" by sportsmen or bodybuilders based on the consumption of uncontrolled dietary supplements [162]. The recent confirmation of these anabolic results [163] led the authors to propose the inscription of 20E on the list of doping prohibited substances [164].

We have chosen to concentrate here on trials aimed at improving human health from a medicinal perspective using pure 20E, as concentrated plant extracts might contain other bioactive substances, e.g., flavonoids. Those selected trials are summarized in Table 7.

Table 7. Clinical studies on humans using pure 20E.

Aim	Age	N	Dose	Duration	Output	Ref
Sexual dis-adaptation	27–61	93 20F, 73M	7.5–10 mg/day	1 month	Improvement of libido and sexual activity in 75% of patients	[165]
Chonic glomeru-lonephritis	35 ± 7	35	15 mg/day	10 days	Improvement of kidney function and of microcirculation	[166]
Sexual function	N/A, M 40–60 M	60 48	5 mg, 2×/day 5 mg 3×/day	30 days	Improved sperm quality and copulative function in patients with disturbed spermatogenesis as a complication of urologic diseases Improvement of sexual function during recovery from myocardial infarction	[158]
Giardiasis	N/A	35	5 mg 3× or 4×/day	10 days	Parasite elimination in 68.7% of patients	[167]
Hepatitis	N/A	N/A	5 mg b.i.d.	30 days	In case of hepatitis B, improvement of liver state	[159]
Lambliasis	N/A	N/A	5 mg 4×/day	10 days	Therapy eliminated most parasites within 10 days	[168]
Hymenolepiasis	N/A	22	5 mg 3×/day	2 weeks	Reduction in symptoms and parasitological efficacy of 36.4%	[169]
Metabolic syndrome	Overweight	39	2 × 50 mg/day	3 months?	Reduction in body weight (-1.3%) waist circumference (-3.2%), body fat (-7.6%), C-reactive protein (-38%), total cholesterol (-17%), triglycerides (-37%), muscle increase ($+2.9\%$)	[170]
Menopause disorders	Overweight	N/A	100 or 200 mg/day	3 months	Prevention of metabolic syndrome and osteoporosis, reduction in body weight, reduction in plasma cholesterol and CRP; proposed for hormone replacement therapy	[171, 172]
Sarcopenia	≥65	231	175/350 mg b.i.d.	6–9 months	*Expected*: change from baseline of gait speed (400MW test), appendicular lean mass and handgrip strength	[173]
Metabolic syndrome	>18	64 test 28 Control	40 or 90 mg/day	3–6–9 months	Reductions in body mass, proportion of body fat, waist cicumference and hsCRP, retention of muscle mass	[174]
Prediabetes	30–60	34	300 mg/day	12 weeks	*Expected*: changes in micronuclei, reduction in fasted glycemia, glycated hemoglobin	[175]
ARDS in COVID 19	≥55	310	350 mg b.i.d.	28 days	*Expected*: Prevention of respiratory deterioration in severe COVID-19 patients	[176]
Celiac disease	Children 3–14	25	2.5 mg/kg/day	14 days	Reduction in symptoms Improvement of energy metabolism	[177]

M: male; F: female; b.i.d.: bis in die (twice a day); N/A: information not available.

20. Neuro-Muscular Diseases

Sarcopenia

Sarcopenia is an age-related skeletal muscle disorder characterized by a progressive loss of muscle mass and strength that leads to reduced mobility. Sarcopenia also has a neuronal component linked to the progressive death of motoneurones [178,179]. 20-Hydroxyecdysone has previously shown beneficial effects in several disease animal models including aging/sarcopenia. It has been purified to pharmaceutical grade ($\geq$97% purity = BIO101) for use as a drug candidate and proved safe in rat and dog models upon chronic treatment. For Single Ascending Dose (SAD) study, BIO101 was administered orally to 24 subjects from two age groups: young adults (18 $\leq$ age $\leq$ 55 years) at escalating doses (100 to 1400 mg), and older adults (65 $\leq$ age $\leq$ 85 years) at 1400 mg. For Multiple Ascending Dose (MAD) study, three doses of BIO101 (350 mg once a day; 350 mg twice a day and 450 mg twice a day) were administered to older adults over 14 days. The primary objective was to evaluate the safety and pharmacokinetics of BIO101 and to identify its main metabolites. The effects of BIO101 were also investigated by measuring selected biomarkers during MAD. BIO101 showed a good safety profile with no serious adverse events during SAD and MAD. All adverse events were of mild or moderate intensity. The bioavailability of BIO101 was rather low, and plasma levels increased less that dose-proportionally. Plasma BIO101 half-life ranged between 2.4 h and 4.9 h, and mean renal clearance ranged between 4.05 and 5.05 L/h. Several metabolites were identified and measured in plasma. Elimination used mainly the fecal route. After 14 days of administration, serum biomarkers of muscular catabolism (myoglobin, CK-MB) were reduced in subjects administered the highest dose. This Phase 1 study confirmed the safety of BIO101 and allowed us to define the most appropriate oral doses for the ongoing interventional Phase 2 clinical trial (ClinicalTrials #NCT03452488) [180].

SARA-INT is a Phase 2 interventional study performed in Europe and the USA aimed to evaluate the clinical benefits, safety and tolerability of the investigational drug BIO101 administered orally for a six-month (26 weeks) duration to older patients, community dwelling men and women aged $\geq$65 years, suffering from age-related sarcopenia (including sarcopenic obesity), and at risk of mobility disability. The double-blind, placebo controlled clinical trial will collect and analyze data on physical performance and body composition and will specifically focus on the change of one functional measurement, the gait speed measured during the 400 metre walking test plus the change of a highly standardized patient reported outcome, the physical function domain PF-10 at the SF-36 auto-evaluation questionnaire, in order to estimate the efficacy of BIO101 administered over 26 weeks in preventing mobility disability in the target population [173].

21. Cardio-Metabolic Diseases

21.1. Pre-Diabetes

Prediabetes is the term used for people whose glucose levels do not meet the criteria for diabetes but are too high to be considered normal. This is defined by the presence of blood glucose between 100 and 125 mg/dL, values per glucose tolerance curve of 140–199 mg/dL and/or HbA1c 5.7–6.4%. Prediabetes should not be considered as a clinical entity in itself, but as a risk factor for diabetes and cardiovascular disease. Prediabetes is associated with obesity (especially abdominal or visceral obesity), dyslipidemia with elevated triglycerides and/or low HDL cholesterol, and hypertension. Subjects with a diagnosis of prediabetes are included according to the criteria of the American Diabetes Association in its version 2019, between 30 and 60 years old and residents of the city of Guadalajara, Jalisco, Mexico who come to clinical nutrition consultation in the University Hospital Fray Antonio Alcalde from the city of Guadalajara, Jalisco, Mexico. The study design is a randomized clinical trial with a control group in two groups: an intervention group with 20E 300 mg every 24 h for 12 weeks and an approved placebo control group (magnesium stearate) at 300 mg every 24 h for 12 weeks [175].

21.2. Obesity

20E prevents obesity and osteoporosis following ovariectomy in rats. Whether it also protects joint cartilage was compared with the effects of estradiol (E2). Furthermore, the effects of 20E in 20 slightly overweight male and female persons were also determined. In ovariectomized rats, 20E reduced the amount of abdominal, bone marrow and joint fat ($p < 0.05$), and this resulted in better trabecular and joint cartilage architecture. In the open clinical trial with slightly overweight persons, a daily intake of 200 mg 20E reduced waist circumference by 2.7 cm, serum cholesterol by 10.2%, LDL by 13.9% and triglycerides by 31.1%, whereas HDL increased by 10.8%. It is concluded that 20E has similar osteo- and chondro-protective effects as E2 and prevents fat accumulation in the abdomen, bone marrow and joints. Hence, 20E may prevent metabolic syndrome and accompanying osteoporosis and osteoarthritis [171,172].

21.3. Menopause

Many postmenopausal women, but increasingly also young females and males, develop obesity. Today, two types of obesity are differentiated: the gynoid "pear" type with fat large gluteal and thigh fat and the android "apple" type with large visceral fat depots. Metabolic syndrome develops primarily in obese people with large visceral fat depots. Adipocytes of the visceral type secrete proinflammatory cytokines which, in the apple-type obese, set the whole body in an inflammatory condition with high oxidative stress with harmful effects in many organs including the arteries, which results in hypertension. Insulin receptors are desensitized, leading to the development of type-2 diabetes. Obese persons also suffer often from hyperlipidemia, which may lead to arteriosclerosis and heart attacks. Metabolic syndrome is not only dangerous for the cardiovascular system, but also for bones, joints, and musculature because there are accumulating fat cells in bone marrow, fat pads in joints, and adipocytes in muscles also secrete proinflammatory cytokines which inhibit the formation of bone osteoblasts, cartilage chondroblasts, and muscle myoblasts. In bone, cytokines even stimulate the formation of bone-resorbing osteoclasts. These effects of adipocytes result in epiphenomena of metabolic syndrome namely in osteoporosis, arthrosis, and sarcopenia. 20E has proven inhibitory effects on the formation of adipocytes. Therefore, 20E is able to reduce the amount of body fat and thereby simultaneously inhibit the reduction in muscle, bone and joint cartilage [172].

21.4. Metabolic Syndrome

The aim of the study was to analyze the interactions between body fat, especially visceral fat depots, on parameters such as high-sensitivity C-reactive protein (hsCRP) and the lipid profile. The effects of 20E on the body parameters fat percentage, muscle mass, body weight and abdominal circumference as well as on the serum parameters cholesterol, triglycerides, LDL, HDL and the hsCRP of obese people with metabolic syndrome were retrospectively investigated. The control group included data from a patient population with metabolic syndrome who did not take 20E. In addition, the vitamin D metabolism of obese people was analyzed. From this retrospective study, it is clear that the abdominal circumference, as an indicator of visceral obesity, correlates positively with TG and hsCRP. Patients with an abdominal circumference of $\geq$100 cm have vitamin D deficiency, whereby the deficiency correlates positively with the size of the abdominal girth. By taking 20E, a continuous and significant loss of body mass, fat percentage, and waist diameter can be achieved while preserving muscle mass in patients with metabolic syndrome compared to the control group. 20E has no significant effect on serum lipids. In contrast, the hsCRP level undergoes a marked reduction. In summary, 20E exerts a positive influence on body parameters, in particular the visceral fat depot and on the hsCRP level. Influence on the lipid profile of patients with metabolic syndrome requires further investigation. Patients with a high visceral fat depot are also exposed to a risk of vitamin D deficiency or undersupply [174].

22. Respiratory Diseases

22.1. Respiratory Failure in COVID-19 Patients

The clinical picture of COVID-19 including acute respiratory distress syndrome (ARDS), related to interstitial pulmonary fibrotic inflammation, cardiomyopathy and shock, strongly suggests that the renin–angiotensin system (RAS) is dysfunctional owing to interaction between angiotensin converting enzyme 2 (ACE2) and SARS-CoV2 spike protein. It is proposed that RAS balance could be restored in COVID-19 patients through Mas1 activation downstream of ACE2 activity, with 20E (BIO101), a non-peptidic Mas receptor (Mas1) activator developed by Biophytis (Figure 12). Indeed, Mas1 activation by 20-hydroxyecdysone harbours anti-inflammatory, anti-thrombotic, and anti-fibrotic properties. BIO101, a 97% pharmaceutical grade 20E, could then offer a new therapeutic option by improving the respiratory function and ultimately promoting survival in COVID-19 patients that develop severe forms of this devastating disease. Therefore, the objective of this COVA study is to evaluate the safety and efficacy of BIO101 in COVID-19 patients with severe pneumonia [180].

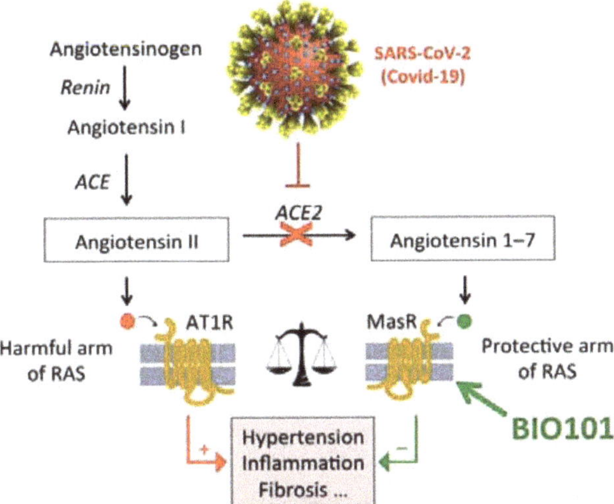

Figure 12. SARS-CoV-2 infection will result in a strong impairment of the activity of angiotensin converting enzyme 2 (ACE2), hence a lack of angiotensin-(1-7) production and a disequilibrium between the harmful and protective arms of the renin–angiotensin system (RAS). Treatment with BIO101 (20E) is expected to activate Mas receptor and the protective arm of RAS, thus preventing inflammation and lung damage.

22.2. COVID-19

The COVA clinical study is a global multicentric, double-blind, placebo-controlled, group sequential and adaptive two parts phase 2–3 study targeting patients with SARS-CoV-2 pneumonia. Part 1 is a Phase 2 exploratory Proof of Concept study to provide preliminary data on the activity, safety and tolerability of BIO101 in the target population. Part 2 is a Phase 3 pivotal randomized study to provide further evidence of the safety and efficacy of BIO101 after 28 days of double-blind dosing. BIO101 is the investigational new drug that activates the Mas1 through the protective arm of the renin–angiotensin system (RAS) [176,181].

23. Other Diseases

23.1. Hepatitis

The inclusion of Ecdysten into a complex therapy scheme for patients with chronic viral hepatitis B (5-mg tablets twice per day over a period of 30 days) substantially improved the clinical and biochemical indices of the functional state of the liver, positively influenced the humoral, cell immunity and the resistance factors, and normalized the course of autoimmune processes accompanying the liver pathology [159].

23.2. Chronic Glomerulonephritis

The administration of Ecdysten (3 tablets/day (=15 mg 20E/day) for 10 days) to patients (18 vs. 17 controls) with chronic glomerulonephritis (inflammation of the glomeruli of the kidney) improved the morphometric indices of the microcirculation of the bulbo-conjunctiva in the majority of patients examined. The frequency of vascular, intravascular and perivascular changes in the bulbar conjunctiva decreased. The frequency of twisting, irregularity, aneurism, and reticulation of the vessels and the phenomena of stagnation of the venous networks and of zones of degeneration fell substantially. In addition, a distinct tendency to a normalization of the diameter of the arterioles, capillaries, and venules appeared. The most pronounced effect was observed in patients with the nephrotic form of the disease [166].

23.3. Celiac Disease

This autoimmune disorder is due to gluten intolerance and primarily affects the small intestine, causing chronic diarrhea, impaired absorption and loss of appetite. It is expected that reduced intestinal absorption is caused by energy metabolism impairment in intestinal cells, and that it results in overall negative effects on general metabolism [177]. The treatment with Ecdysten for 2 weeks resulted in a tendency to improve energy metabolism as assessed by lowered lactic acid plasma concentrations.

23.4. Sexual Dysfunction

Ecdysten has been shown to improve sexual function in men suffering from low sperm count, poor sperm motility, erectile dysfunction or low libido when recovering from heart attack [158,165].

23.5. Parasitoses (Giardiasis, Hymenolepiasis, Lambliasis)

This area has been documented by several trials using Ecdysten [167–169]. We expect that the mechanism involved here is totally different, as it is most probably a direct effect of 20E on the parasites themselves.

24. Conclusions and Prospects

20E shows many beneficial pharmaceutical effects in mammals and is non-toxic. To date, the only ecdysteroid which has been investigated to any extent is 20E, for the simple reason that it is the only analogue which is readily commercially available at a reasonable cost. Where it has been investigated, the effects are brought about by low μM concentrations of 20E, but to achieve maintained concentrations of this level in the plasma it is necessary that 20E should be supplied orally in large amounts (100–1000 mg) at least twice a day, because the bioavailability of 20E is poor and the half-life in the plasma is short. In the future, it will be necessary to seek ways to overcome the poor bioavailability by alternative application routes (e.g., nasal) and improve formulation by identifying biologically active analogues with greater inherent bioavailability. It is fortunate that a wide range of ecdysteroid structural analogues are already known (Ecdybase [4]) and samples of many of these are available by isolation from plants or semi-synthesis, at least in the amounts required for method development and structure-activity relationship studies. Thus, it is to be expected that one of the areas of focus in the near future will be the development of more amenable and specific bioassays for each of the major effects of ecdysteroids in mammals, leading to a clearer understanding of the SARs for activity, potency and uptake from the gut.

A further driving force for such studies will be the growing need for an increased supply of 20E, since extraction from natural plant sources is a viable approach for kg amounts of pure substance (as required for thorough clinical trials), but it is difficult to envisage how this could produce multi-tonne amounts of 20E (as could be required to treat sarcopenia or metabolic syndrome on a world-wide basis). This will provide a spur to seek ways of elevating 20E accumulation in ecdysteroid-accumulating plants above the 1–2% of the dry weight typical of the sources currently used, or to develop much more efficient and greener chemical syntheses of 20E or its analogues or molecular approaches transforming yeast and developing high-yielding lines for fermentation, and to seek more effective analogues and better ways of delivery. Fundamental to both the improved yield in plants and the yeast fermentation is a thorough understanding of phytoecdysteroid biosynthesis, not only the biochemistry and molecular biology of the individual steps, but also the integration and regulation of these steps to form the complete pathway(s) and to understand the flux and dynamics through the pathway to maximize 20E formation, all of which are currently only sketchy.

In the past, the most extensively studied effect of 20E in mammals was the anabolic effect, which was exploited initially rather empirically to improve the stamina and performance of sportsmen and -women, but is now starting to gain a mechanistic understanding. The regulatory agencies have recently started to analyze those uses and those will possibly become prohibited in the near future. 20E is presently included in the 2020 Monitoring Program of the World Antidoping Agency (WADA [182]). They do not represent the most promising uses for this molecule, and the recent clinical trials aim to treat different human diseases with a rationale based on 20E mechanism of action.

Indeed, the protective arm of the RAS system has a wide number of target tissues, which may explain the multiple effects of 20E as listed in Table 2 and the activity of some medicinal plants listed in Table 1. Thus, the potential of exploiting the anabolic effects of 20E to contribute to medical conditions and syndromes where declining muscle mass and performance are serious components (sarcopenia, cachexia, Duchenne muscular dystrophy) is being realized and the first clinical trials are being performed to validate the use of 20E as a drug for these conditions. Metabolic syndrome is a further area of immense interest where the anabolic, hypolipidemic and anti-diabetic effects of 20E may find application. Moreover, it has recently appeared that given its cellular target, 20E could represent an adequate molecule to prevent the appearance of severe forms of COVID-19, and a clinical trial is under way.

20E is an intriguing molecule which appears to have significant potential to contribute to human health, both nutritionally and in the combat against chronic diseases and conditions. Aspects, such as its low oral bioavailability, need improvement, and it may be that more effective analogues can be identified in the future, but for the moment 20E is leading the way.

Author Contributions: Conceptualization, R.L., L.D.; Literature, R.L., L.D.; Graphics, R.L., W.D.; Original draft, L.D.; Review and editing, R.L., L.D., W.D., S.V. All authors have read and agreed to the published version of the manuscript.

Funding: The preparation of the manuscript was funded by Biophytis SA.

Institutional Review Board Statement: Not applicable.

Informed Consent Statement: Not applicable.

Data Availability Statement: Not applicable.

Acknowledgments: The authors wish to thank their Biophytis colleagues for scientific discussions.

Conflicts of Interest: The authors declare no conflict of interest.

Abbreviations

E	ecdysone
20E	20-hydroxyecdysone
14d20E	14-deoxy-20-hydroxyecdysone
Post	poststerone
14dPost	14-deoxypoststerone
3-epi-Post	3-*epi*-poststerone
16αOHPost	16α-hydroxypoststerone
20R/SPost	20-dihydropoststerone
21OHPost	21-hydroxypoststerone
20,26E	20,26-dihydroxyecdysone
PolB	polypodine B

References

1. Koolman, J. (Ed.) *Ecdysone: From Chemistry of Mode of Action*; Thieme Verlag: Stuttgart, Germany, 1989; 482p.
2. Dinan, L.; Savchenko, T.; Whiting, P. On the distribution of phytoecdysteroids in plants. *Cell Mol. Life Sci* **2001**, *58*, 1121–1132. [CrossRef] [PubMed]
3. Dinan, L.; Harmatha, J.; Volodin, V.; Lafont, R. Phytoecdysteroids: Diversity, biosynthesis and distribution. In *Ecdysone: Structures and Functions*; Smagghe, G., Ed.; Springer Science & Business Media B.V.: Berlin, Germany, 2009; pp. 3–45.
4. Lafont, R.; Harmatha, J.; Marion-Poll, F.; Dinan, L.; Wilson, I.D. The Ecdysone Handbook, 3rd ed. 2002. continuously updated. Available online: http://ecdybase.org/ (accessed on 1 April 2021).
5. Dinan, L.; Mamadalieva, N.; Lafont, R. Dietary Phytoecdysteroids. In *Handbook of Dietary Phytochemicals*; Xiao, J., Sarker, S.D., Asakawa, Y., Eds.; Springer Nature Singapore Pte Ltd.: Singapore, 2020; pp. 1–54.
6. Kametani, T.; Tsubuki, M. Strategies for the synthesis of ecdysteroids. In *Ecdysone. From Chemistry to Mode of Action*; Koolman, J., Ed.; Thieme Verlag: Stuttgart, Germany, 1989; pp. 74–96.
7. Sláma, K.; Lafont, R. Insect hormones—ecdysteroids: Their presence and action in vertebrates. *Eur J. Entomol* **1995**, *92*, 355–377.
8. Li, T.S.C. *Taiwanese Native Medicinal Plants*; Taylor & Francis: London, UK, 2006; 328p.
9. He, X.; Wang, X.; Fang, J.; Chang, Y.; Ning, N.; Guo, H.; Huang, L.; Huang, X. The genus *Achyranthes*: A review on traditional uses, phytochemistry, and pharmacological activities. *J. Ethnopharmacol.* **2017**, *203*, 260–278. [CrossRef] [PubMed]
10. Choi, H.S.; Lee, M.J.; Na, M.S.; Lee, M.Y.; Choi, D. Antioxidant properties of *Achyranthis* radix extract in rats. *J. Ind. Eng. Chem.* **2019**, *15*, 275–280. [CrossRef]
11. Hsieh, W.T.; Liu, Y.T.; Lin, W.C. Anti-inflammatory properties of *Ajuga bracteosa* in vivo and in vitro study and their effects on mouse models of liver fibrosis. *J. Ethnopharmacol.* **2011**, *135*, 116–125. [CrossRef]
12. Israili, Z.H.; Lyoussi, B. Ethnopharmacology of the plants of genus *Ajuga*. *Pak. J. Pharm Sci* **2009**, *22*, 425–462.
13. Bouyahya, A.; El Omari, N.; Elmenyiy, N.; Guaouguaou, F.E.; Balahbid, A.; El-Shazly, M.; Chamkhi, I. Ethnopharmacological use, phytochemistry, pharmacology, and toxicology of *Ajuga iva* (L.) Schreb. *J. Ethnopharmacol.* **2020**, *258*, 112875. [CrossRef]
14. Cheng, D.M.; Yousef, G.G.; Grace, M.H.; Rogers, R.B.; Gorelick-Feldman, J.; Raskin, I.; Lila, M.A. In vitro production of metabolism-enhancing phytoecdysteroids from *Ajuga turkestanica*. *Plant Cell Tiss Organ Cult.* **2008**, *93*, 73–83. [CrossRef]
15. Patil, K.S.; Bhaising, S.R. Ethnomedicinal uses, phytochemistry and pharmacological properties of the genus *Boerhaavia*. *J. Ethnopharmacol.* **2016**, *182*, 200–220. [CrossRef]
16. Ibrahim, B.; Sowemimo, A.; van Rooyen, A.; Van de Venter, M. Antiinflammatory, analgesic and antioxidant activities of *Cyathula prostrata* (Linn.) Blume (Amaranthaceae). *J. Ethnopharmacol.* **2012**, *141*, 282–289. [CrossRef]
17. Ajuogu, P.K.; Ere, R.; Nodu, M.B.; Nwachikwu, C.U.; Lgbere, O.O. The influence of graded levels of *Cyathula prostrata* (Linn.) Blume on semen quality characteristics of adult New Zealand white bucks. *Transl. Anim. Sci.* **2020**, *4*, 1134–1139. [CrossRef]
18. Lee, S.; Xiao, C.; Pei, S. Ethnobotanical survey of medicinal plants at periodic market of Honhe Prefecture in Yunnan Province, SW China. *J. Ethnopharmacol.* **2008**, *117*, 362–377. [CrossRef]
19. Fang, L.; Li, J.; Zhou, J.; Wang, X.; Guo, L. Isolation and purification of three ecdysteroids from the stems of *Diploclisia glaucescens* by high-speed countercurrent chromatography and their anti-inflammatory activities *in vitro*. *Molecules* **2017**, *22*, 1310. [CrossRef]
20. Schink, M.; Garcia-Käufer, M.; Bertrams, J.; Duckstein, S.M.; Müller, M.B.; Huber, R.; Stintzing, F.C.; Gründemann, C. Differential cytotoxic properties of *Helleborus niger* L. on tumour and immunocompetent cells. *J. Ethnopharmacol.* **2015**, *159*, 129–136. [CrossRef]
21. Ho, H.; Teai, T.; Bianchini, J.P.; Lafont, R.; Raharivelomanana, P. Ferns: From traditional uses to pharmacological development, chemical identification of active principles. In *Working with Ferns: Issues and Applications*; Fernández, H., Ed.; Springer Science+Business Media B.V.: Berlin, Germany, 2011; pp. 321–346.
22. Zhu, L.; Tan, J.; Wang, B.; Guan, L.; Liu, Y.; Zheng, C. In vitro antitumor activity and antifungal activity of Pennogenin steroidal saponins from *Paris polyphylla* var. yunnanensis. *Iranian J. Pharm. Res.* **2011**, *10*, 279–286.
23. Franco, R.R.; de Almeida Takata, L.; Chagas, K.; Justino, A.B.; Saraiva, A.L.; Goulart, L.R.; de Melo Rodrigues Ávila, V.; Otoni, W.C.; Espindola, F.A.; da Silva, C.R. A 20-hydroxyecdysone-enriched fraction from *Pfaffia glomerata* (Spreng.) pedersen roots alleviates stress, anxiety, and depression in mice. *J. Ethnopharmacol.* **2021**, *267*, 113599.

24. Abdillahi, H.S.; Finnie, J.F.; Van Staden, J. Anti-inflammatory, antioxidant, anti-tyrosinase and phenolic contents of four *Podocarpus* species used in traditional medicine in South Africa. *J. Ethnopharmacol.* **2011**, *136*, 496–503. [CrossRef]

25. Gleńsk, M.; Dudek, M.K.; Ciach, M.; Wlodarczyk, M. Isolation and structural determination of flavan-3-ol derivatives from the *Polypodium vulgare* L. rhizomes water extract. *Nat. Prod. Res.* **2019**, in press. [CrossRef]

26. Kokoska, L.; Janikova, D. Chemistry and pharmacology of *Rhaponticum carthamoides*: A review. *Phytochemistry* **2009**, *70*, 842–855. [CrossRef]

27. Shikov, A.N.; Narkevich, I.A.; Flisyuk, E.V.; Luzhanin, V.G.; Pozharitskaya, O.N. Medicinal plants from the 14th edition of the Russia, Pharmacopoeia, recent updates. *J. Ethnopharmacol.* **2021**, *268*, 113685. [CrossRef]

28. Zhang, Z.-Y.; Yang, W.-Q.; Fan, C.-L.; Zhao, H.-N.; Huang, X.-J.; Wang, Y.; Ye, W.-C. New ecdysteroid and ecdysteroid glycosides from the roots of Serratula chinensis. *J. Asian Nat. Prod. Res.* **2017**, *19*, 208–214. [CrossRef] [PubMed]

29. Dinda, B.; Das, N.; Dinda, S.; Dinda, M.; SilSarma, I. The genus *Sida*, L. A traditional medicine: Its ethnopharmacological phytochemical and pharmacological data for commercial exploitation in herbal drug industry. *J. Ethnopharmacol.* **2015**, *176*, 135–176. [CrossRef] [PubMed]

30. Bala, M.; Pratap, K.; Verma, P.K.; Singh, B.; Padwad, Y. Validation of ethnopharmacological potential of *Tinospora cordifolia* for anticancer and immunomodulatory activities and quantification of bioactive molecules by HPTLC. *J. Ethnopharmacol.* **2015**, *175*, 131–137. [CrossRef] [PubMed]

31. Sharma, B.; Dutt, V.; Kaur, N.; Mittal, A.; Dabur, R. *Tinospora cordifolia* protects from skeletal muscle atrophyby alleviating oxidative stress and inflammation induced by sciatic denervation. *J. Ethnopharmacol.* **2020**, *254*, 112720. [CrossRef]

32. Suksamrarn, A.; Kumpun, S.; Yingyongnarongkul, B. Ecdysteroids from *Vitex scabra* bark. *J. Nat. Prod.* **2002**, *65*, 1690–1692. [CrossRef]

33. Báthori, M. Phytoecdysteroids effects on mammalians, isolation and analysis. *Mini Rev. Med. Chem.* **2002**, *2*, 285–293. [CrossRef]

34. Lafont, R.; Dinan, L. Practical uses for ecdysteroids in mammals including humans: An update. *J. Insect. Sci.* **2003**, *3/7*, 30.

35. Báthori, M.; Pongrácz, Z. Phytoecdysteroids—From isolation to their effects on Humans. *Curr. Med. Chem.* **2005**, *12*, 153–172. [CrossRef]

36. Dinan, L.; Lafont, R. Effects and application of arthropod steroid hormones (ecdysteroids) in mammals. *J. Endocrinol.* **2006**, *191*, 1–8. [CrossRef]

37. Zou, D.; Cao, L.; Chen, Q. Advances in pharmacological research on ecdysteroids. *Zhongguo Xinyao Zazhi* **2008**, *17*, 371–374.

38. Báthori, M.; Tóth, N.; Hunyadi, A.; Márki, A.; Zador, E. Phytoecdysteroids and anabolic-androgenic steroids—Structure and effects on Humans. *Curr. Med. Chem.* **2008**, *15*, 75–91. [CrossRef]

39. Lafont, R.; Dinan, L. Innovative and future applications of ecdysteroids. In *Ecdysone: Structures and Functions*; Smagghe, G., Ed.; Springer Science & Business Media B.V.: Berlin, Germany, 2009; pp. 551–578.

40. Cahliková, L.; Macáková, K.; Chlebek, J.; Hošt'álková, A.; Kulkhánková, A.; Opletal, L. Ecdysterone and its activity on some degenerative diseases. *Nat. Prod. Commun.* **2011**, *6*, 707–718. [CrossRef]

41. Lafont, R. Recent progress in ecdysteroid pharmacology. *Theor Appl. Ecol.* **2012**, 6–12.

42. Laekeman, G.; Vlietinck, A. Phytoecdysteroids: Phytochemistry and pharmacological activity. In *Natural Products*; Ramawat, K.G., Mérillon, J.M., Eds.; Springer: Berlin, Germany, 2013; pp. 3827–3849.

43. Bajguz, A.; Bakala, I.; Talarek, M. Ecdysteroids in plants and their pharmacological effects in vertebrates and Humans. In *Studies in Natural Product Chemistry*; Atta-ur-Rahman, F.R.S., Ed.; Elsevier, B.V.: Amsterdam, The Netherlands, 2015; Volume 45, Chapter 5; pp. 121–145.

44. Dinan, L.; Lafont, R. Phytoecdysteroids occurrence, distribution, biosynthesis, metabolism, mode of action and applications: Developments from 2005 to 2015. *Pharm Bull. J. (Kazakhstan)* **2015**, 9–37.

45. Das, N.; Mishra, S.K.; Bishayee, A.; Ali, E.S.; Bishayee, A. The phytochemical, biological, and medicinal attributes of phytoecdysteroids: An updated review. *Acta Pharm. Sinica B* **2020**, in press. [CrossRef]

46. Chermnykh, N.S.; Shimanovsky, N.L.; Shutko, G.V.; Syrov, V.N. Effects of methandrostenolone and ecdysterone on physical endurance of animals and protein metabolism in the skeletal muscles. *Farmakol. Toksikol.* **1988**, *6*, 57–62.

47. Tóth, N.; Szabó, A.; Kacsala, P.; Héger, J.; Zádor, E. 20-Hydroxyecdysone increases fiber size in a muscle-specific fashion in rat. *Phytomedicine* **2008**, *15*, 691–698. [CrossRef]

48. Syrov, V.N. Comparative experimental investigations of the anabolic activity of ecdysteroids and steranabols. *Pharm. Chem. J.* **2000**, *34*, 193–197. [CrossRef]

49. Gorelick-Feldman, J.; MacLean, D.; Ilic, N.; Poulev, A.; Lila, M.A.; Cheng, D.; Raskin, I. Phytoecdysteroids increase protein synthesis in skeletal muscle cells. *J. Agric. Food Chem.* **2008**, *56*, 3532–3537. [CrossRef]

50. Gorelick-Feldman, J.; Cohick, W.; Raskin, I. Ecdysteroids elicit a rapid Ca^{2+} flux leading to Akt activation and increased protein synthesis in skeletal muscle cells. *Steroids* **2010**, *70*, 632–637. [CrossRef]

51. Goreclick-Feldman, J.I. Phytoecdysteroids: Understanding their Anabolic Activity. Ph.D. Thesis, Rutgers University, Newark, NJ, USA, 2009. Available online: https://rucore.libraries.rutgers.edu/rutgers-lib/25806/ (accessed on 1 April 2021).

52. Lawrence, M.M. *Ajuga turkestanica* as a Countermeasure against Sarcopenia and Dynapenia. Master's Thesis, Appalachian State University, Boone, NC, USA, 2012.

53. Zubeldia, J.M.; Hernández-Santana, A.; Jiménez-del-Rio, M.; Pérez-López, V.; Pérez-Machín, R.; Garcia-Castellano, J.M. In vitro characterization of the efficacy and safety profile of a proprietary *Ajuga turkestanica* extract. *Chinese Med.* **2012**, *3*, 215–222. [CrossRef]

54. Parr, M.K.; Zhao, P.; Haupt, O.; Tchoukouegno Ngueu, S.; Hengevoss, J.; Fritzemeier, K.H.; Piechotta, M.; Schlörer, N.; Muhn, P.; Zheng, W.Y.; et al. Estrogen receptor beta is involved in skeletal muscle hypertrophy induced by the phytoecdysteroid ecdysterone. *Mol. Nutr. Food Res.* **2014**, *58*, 1861–1872. [CrossRef] [PubMed]

55. Syrov, V.N.; Khushbaktova, Z.A.; Abzalova, M.K.; Sultanov, M.B. On the hypolipidemic and antiatherosclerotic action of phytoecdysteroids. *Doklady Akademii Nauk Uzbekistan SSR* **1983**, 44–45.

56. Kizelsztein, P.; Govorko, D.; Komarnytsky, S.; Evans, A.; Wang, Z.; Cefalu, W.T.; Raskin, I. 20-Hydroxyecdysone decreases weight and hyperglycemia in a diet-induced obesity mice model. *Am. J. Physiol. Endocrinol. Metab.* **2009**, *296*, E433–E439. [CrossRef]

57. Seidlova-Wuttke, D.; Ehrhardt, C.; Wuttke, W. Metabolic effects of 20-OH-ecdysone in ovariectomized rats. *J. Steroid Biochem Mol. Biol* **2010**, *119*, 121–126. [CrossRef]

58. Syrov, V.N.; Khushbaktova, Z.A.; Nabiev, A.N. An experimental study of the hepatoprotective properties of phytoecdysteroids and nerobol in carbon tetrachloride—Induced liver injury. *Eksp. Klin. Farmakol.* **1992**, *55*, 61–65.

59. Foucault, A.-S.; Mathé, V.; Lafont, R.; Even, P.; Dioh, W.; Veillet, S.; Tomé, D.; Huneau, D.; Hermier, D.; Quignard-Boulangé, A. Quinoa extract enriched in 20-hydroxyecdysone protects mice from diet-induced obesity and modulates adipokines expression. *Obesity* **2012**, *20*, 270–277. [CrossRef]

60. Foucault, A.-S.; Even, P.; Lafont, R.; DIoh, W.; Veillet, S.; Tomé, D.; Huneau, J.F.; Hermier, D.; Quignard-Boulangé, A. Quinoa extract enriched in 20-hydroxyecdysone affects energy homeostasis and intestinal fat absorption in mice fed a high-fat diet. *Physiol. Behav.* **2014**, *128*, 226–231. [CrossRef]

61. Agoun, H.; Semiane, N.; Mallek, A.; Bellahreche, Z.; Hammadi, S.; Madjerab, M.; Abdlalli, M.; Khalkhal, A.; Dahmani, Y. High-carbohydrate diet-induced metabolic disorders in *Gerbillus tarabuli* (a new model of non-alcoholic fatty liver disease). Protective effects of 20-hydroxyecdysone. *Arch. Physiol. Biochem.* **2019**, *127*, 127–135. [CrossRef]

62. Buniam, J.; Chukijrungroat, N.; Rattanavichit, Y.; Surapongchai, J.; Weerachayaphom, J.; Bupha-Intr, T.; Saengsirisuwan, V. 20-Hydroxyecdysone ameliorates metabolic and cardiovascular dysfunction in high-fat-high-fructose-fed ovariectomized rats. *BMC Complement. Med. Ther.* **2020**, *20*, 140. [CrossRef]

63. Yoshida, T.; Otaka, T.; Uchiyama, M.; Ogawa, S. Effect of ecdysterone on hyperglycemia in experimental animals. *Biochem. Pharmacol.* **1971**, *20*, 3263–3268. [CrossRef]

64. Sundaram, R.; Naresh, R.; Shanthi, P.; Sachdanandam, P. Efficacy of 20-OH-ecdysone on hepatic key enzymes of carbohydrate metabolism in streptozotocin induced diabetic rats. *Phytomedicine* **2012**, *19*, 725–729. [CrossRef]

65. Sundaram, R.; Naresh, R.; Shanthi, P.; Sachdanandam, P. Ameliorative effect of 20-OH ecdysone on streptozotocin induced oxidative stress and β-cell damage in experimental hyperglycemic rats. *Process. Biochem.* **2012**, *47*, 2072–2080. [CrossRef]

66. Chen, L.; Zheng, S.; Huang, M.; Ma, X.; Yang, J.; Deng, S.; Huang, Y.; Wen, Y.; Yang, X. β-edysterone from *Cyanotis arachnoidea* exerts hypoglycemic effects through activation of IRS-1/Akt/GLUT4 and IRS-1/Akt/GLUT2 signal pathways in KK-Ay mice. *J. Funct. Foods* **2017**, *39*, 123–132. [CrossRef]

67. Hung, T.J.; Chen, W.M.; Liu, S.F.; Liao, T.N.; Lee, T.C.; Chuang, L.Y.; Guh, J.Y.; Hung, C.Y.; Hung, H.J.; Chen, P.Y.; et al. 20-Hydroxyecdysone attenuates TGF-β1-induced renal cellular fibrosis in proximal tubule cells. *J. Diabetes Complicat.* **2012**, *26*, 463–469. [CrossRef]

68. Kurmukov, A.G.; Syrov, V.N. Anti-inflammatory properties of ecdysterone. *Meditsinskii Zhurnal Uzb.* **1988**, 68–70.

69. Song, G.; Xia, X.C.; Zhang, K.; Yu, R.; Li, B.; Li, M.; Yu, X.; Zhang, J.; Xue, S. Protective effect of 20-hydroxy-ecdysterone against lipopolysaccharides-induced acute lung injury in mice. *J. Pharm. Drug Res.* **2019**, *2*, 109–114.

70. Cai, Y.J.; Dai, J.Q.; Fang, J.G.; Ma, L.P.; Hou, L.F.; Yang, L.; Liu, Z.L. Antioxidative and free radical scavenging effects of ecdysteroids from *Serratula strangulata*. *Can. J. Physiol. Pharmacol.* **2002**, *80*, 1187–1194. [CrossRef]

71. Azizov, A.P. Effects of eleutherococcus, elton, leuzea, and leveton on the blood coagulation system during training in athletes. *Eksp Klin Farmakol* **1997**, *60*, 58–60.

72. Xia, X.C.; Xue, S.P.; Wang, X.Y.; Liu, R. Effects of 20-hydroxyecdysone on expression of inflammatory cytokines in acute lung injury mice. *Mod. Prev. Med.* **2016**, 5.

73. Dongmo, A.B.; Nkeng-Efouet, P.A.; Devkota, K.P.; Wegener, J.W.; Sewald, N.; Wagner, H.; Vierling, W. Tetra-acetylajugasterone a new constituent of *Vitex cienkowskii* with vasorelaxant activity. *Phytomed* **2014**, *21*, 787–792. [CrossRef]

74. Syrov, V.N.; Nasyrova, S.S.; Khushbaktova, Z.A. The results of experimental study of phytoecdysteroids as erythropoiesis stimulators in laboratory animals. *Eksperimental'naia i Klin. Farmakol.* **1997**, *60*, 41–44.

75. Chen, Z.; Zhu, G.; Zhang, J.H.; Liu, Z.; Tang, W.; Feng, H. Ecdysterone-sensitive smooth muscle cell proliferation stimulated by conditioned medium of endothelial cells cultured with bloody cerebrospinal fluid. *Acta Neurochir Suppl.* **2008**, *104*, 183–187.

76. Luo, H.; Yi, B.; Fan, W.; Chen, K.; Gui, L.; Chen, Z.; Li, L.; Feng, H.; Chi, L. Enhanced angiogenesis and astrocyte activation by ecdysterone treatment in a focal cerebral ischemia rat model. *Acta Neurochirurgica Suppl.* **2011**, *110*, 151–155.

77. Kurmukov, A.G.; Ermishina, O.A. The effect of ecdysterone on experimental arrhythmias and changes in the hemodynamics and myocardial contractility induced by coronary artery occlusion. *Farmakol. Toksikol.* **1991**, *54*, 27–29. [PubMed]

78. Korkach, Y.P.; Kotsyuruba, A.V.; Psryslazhnam, O.D.; Mohyl'nyts'ka, L.D.; Sahach, V.F. NO-dependent mechanisms of ecdysterone protective action on the heart and vessels in streptozotocin-induced diabetes mellitus in rats. *Fiziol. Zhurnal* **2007**, *53*, 3–8.

79. Xia, X.; Zhang, Q.; Liang, G.; Lu, S.; Yang, Y.; Tian, Y. Role of 20-hydroxyecdysone in protecting rats against diabetic cardiomyopathy. *Chln J. Geriatr Heart Brain Vessel Dis.* **2013**, *15*, 412–415.

80. Dilda, P.; Latil, M.; Didry-Barca, B.; On, S.; Serova, M.; Veillet, S.; Lafont, R. BIO101 demonstrates combined beneficial effects on skeletal muscle and respiratory functions in a mouse model of Duchenne muscular dystrophy. World Muscle Society WMS 2019, Copenhagen, Denmark (1-5/10/2019). *Neuromuscul. Disord.* **2019**, *29*, S158. [CrossRef]

81. Latil, M.; Bézier, C.; Cottin, S.; Lafont, R.; Veillet, S.; Dilda, P.; Charbonnier, F.; Biondi, O. BIO101 demonstrates combined beneficial effects on muscle and motor neurons in a mouse model of severe spinal muscular atrophy. World Muscle Society WMS 2019, Copenhagen, (1-5/10/2019). *Neuromuscul. Disord.* **2019**, *29*, S189. [CrossRef]

82. Luo, H.; Luo, C.; Zhang, Y.; Chi, L.; Li, L.; Chen, K. Effect of ecdysterone on injury of lipid peroxidation following focal cerebral ischemia in rats. *Zhongguo Yaoye* **2009**, *18*, 12–14.

83. Liu, Z.; Chen, Y.; Chen, Z.; Tang, W.; Zhu, G.; Wang, X.; Feng, H. Effect of ecdysterone on the nervous lesions of rabbits acquired after subarachnoid hemorrhage. *Med. J. Chin. People's Lib. Army* **2011**, *36*, 1351–1353.

84. Hu, J.; Luo, C.X.; Chu, W.H.; Shen, Y.A.; Qian, Z.-M.; Zhu, G.; Yu, Y.B.; Feng, H. 20-Hydroxyecdysone protects against oxidative stress-induced neuronal injury by scavenging free radicals and modulating NF-kB and JNK pathways. *PLoS ONE* **2012**, *7*, e50764. [CrossRef]

85. Shakhmurova, G.A.; Khushbaktova, Z.A.; Syrov, V.N. Estimation of hepatoprotective and immunocorrecting effects of the sum of phytoecdysteroids from *Silene viridiflora* in experimental animals treated with tetrachlormethan. *O'zbekiston Biol. J.* **2010**, 16–20.

86. Xia, X.; Zhang, Q.; Wang, Z.; Gui, G.; Liang, G.; Liu, R. Protective effect of 20-hydroxyecdysone on diabetic hepathopathy of rats. *Xiandai Yufang Yixue* **2013**, *40*, 4031–4034.

87. Wu, X.; Jiang, Y.; Fan, S.; Wang, R.; Xiang, M.; Niu, H.; Li, T. Effects of ecdysterone on rat lung reperfusion injury. *Chin. Pharm. Bull.* **1998**, *14*, 256–258.

88. Li, J.; Wu, X.; Zhang, J.; Wu, X.; Gao, D.; Shen, T.; Gu, C. Effect of ecdysterone on the expression of toll-like receptor 4 and surfactant protein A in lung tissue of rats with acute lung injury. *Infect. Inflamm. Repair* **2013**, 22–26.

89. Zou, D.; Xu, Z.; Cao, L.; Chen, Q. Effects of ecdysterone on early stage diabetic nephropathy in streptozotocin-induced diabetic rats. *Chin. J. New Drugs Clin. Remedies* **2010**, *29*, 842–846.

90. Shakhmurova, G.A.; Syrov, V.N.; Khushbaktova, Z.A. Immunomodulating and antistress activity of ecdysterone and turkesterone under immobilization-induced stress conditions in mice. *Pharm. Chem. J.* **2010**, *44*, 7–9. [CrossRef]

91. Zhou, Y.; Wu, X.; Liao, J.; Wu, C.; Zhang, Y.; Zhang, Z. Effect of ecdysterone on the healing of gastric ulcer in model rats. *China Pharm.* **2010**, *21*, 2332–2335.

92. Gao, L.; Cai, G.; Shi, X. β-Ecdysterone induces osteogenic differentiation in mouse mesenchymal stem cells and relieves osteoporosis. *Biol. Pharm. Bull.* **2009**, *31*, 2245–2249. [CrossRef]

93. Kapur, P.; Wuttke, W.; Jarry, H.; Seidlova-Wuttke, D. Beneficial effects of β-ecdyson on the joint, epiphyseal cartilage tissue and trabecular bone in ovariectomized rats. *Phytomed* **2010**, *17*, 350–355. [CrossRef] [PubMed]

94. Seidlova-Wuttke, D.; Christel, D.; Kapur, P.; Nguyen, B.T.; Jarry, H.; Wuttke, W. β-Ecdyson has bone protective but no estrogenic effects in ovariectomized rats. *Phytomedicine* **2010**, *17*, 884–899. [CrossRef]

95. Dai, W.; Zhang, H.; Zhong, Z.A.; Jiang, L.; Chen, H.; Lay, Y.A.; Kot, A.; Ritchie, R.O.; Lane, N.E.; Yao, W. β-Ecdysone augments peak bone mass in mice of both sexes. *Clin. Orthop Relat Res.* **2015**, *473*, 2495–2504. [CrossRef] [PubMed]

96. Dai, W.; Jiang, L.; Evan Lay, Y.-A.; Chen, H.; Jin, G.; Zhang, H.; Kot, A.; Ritchie, R.O.; Lane, N.E.; Yao, W. Prevention of glucocorticoid induced bone changes with beta-ecdysone. *Bone* **2015**, *74*, 48–57. [CrossRef]

97. Wang, G.; Zhang, X.; Zhang, W.; Xia, L. Protective effect of ecdysterone on rabbits chondrocytes that is injured by lipopolysaccharide. *Tianjin Med J.* **2015**, 587–590.

98. Wen, F.; Yu, J.; He, C.J.; Zhang, Z.W.; Yang, A.F. β-ecdysterone protects against apoptosis by promoting autophagy in nucleus pulposus cells and ameliorates disc degeneration. *Mol. Med. Rep.* **2019**, *19*, 2440–2448. [CrossRef]

99. Zhao, Q.; Wang, Z.; He, J. Effect of β-ecdysterone on the proliferation, differentiation and apoptosis of rat osteoblasts induced by high glucose. *Chin. J. Clin. Pharmacol.* **2021**, 443–446.

100. Detmar, M.; Dumas, M.; Bonté, F.; Meybeck, A.; Orfanos, C.E. Effects of ecdysterone on the differentiation of normal keratinocytes *in vitro*. *Eur. J. Dermatol.* **1994**, *4*, 558–562.

101. Zhegn, G.Y.; Wu, X.; Li, Y.L.; Zfang, J.H.; Wang, W.J. Preparation and dose-effect analysis of ecdysterone cream for promoting wound healing. *J. Southern Med. Univ.* **2008**, *28*, 828–831.

102. Ehrhardt, C.; Wessels, J.T.; Wuttke, W.; Seidlova-Wuttke, D. The effects of 20-hydroxyecdysone and 17β-estradiol on the skin of ovariectomized rats. *Menopause* **2011**, *18*, 323–327. [CrossRef]

103. Dilda, P.; Foucault, A.S.; Serova, M.; On, S.; Raynal, S.; Veillet, S.; Dioh, W.; Lafont, R. BIO101, a drug candidate targeting Mas receptor for the treatment of age-reated muscle degeneration. From molecular target identification to clinical development. *J. Cachexia Sarcopenia Muscle* **2016**, *7*, 655.

104. Antoshechkin, A.G. Selective plant extracts and their combination as nutritional therapeutic remedies. *J. Nutr Ther.* **2016**, *5*, 1–11. [CrossRef]

105. Syrov, V.N.; Kurmukov, A.G. Anabolic activity of phytoecdysone—ecdysterone—isolated from *Rhaponticum carthamoides* (Will.) Illjin. *Farmakol. Toksikol.* **1967**, *39*, 690–693.

106. Kholodova, I.D.; Tugai, V.A.; Zimina, V.P. Effect of vitamin D_3 and 20-hydroxyecdysone on the content of ATP, creatine phosphate, carnosine and Ca^{2+} in skeletal muscles. *Ukr. Biokhimicheskii Zhurnal* **1997**, *69*, 3–9.

107. Lupien, P.J.; Hinse, C.; Chaudhary, K.D. Ecdysterone as a hypoholesterolemic agent. *Arch. Int. Physiol. Biochim.* **1969**, *77*, 206–212.

108. Catalan, R.E.; Martinez, A.M.; Aragones, M.D.; Miguel, B.G.; Robles, A.; Godoy, J.E. Alterations in rat lipid metabolism following ecdysterone treatment. *Comp. Biochem. Physiol.* **1985**, *81b*, 771–775. [CrossRef]

109. Matsuda, H.; Kawaba, T.; Yamamoto, Y.; Ogawa, S. Effect of ecdysterone on experimental atherosclerosis in rabbits. *Nippon Yakurigaku Zasshi* **1974**, *70*, 325–339. [CrossRef]

110. Catalan, R.E.; Aragones, M.D.; Godoy, J.E.; Martinez, A.M. Ecdysterone induces acetylcholinesterase in mammalian brain. *Comp. Biochem. Physiol.* **1984**, *78c*, 193–195. [CrossRef]

111. Chaudhary, K.D.; Lupien, P.J.; Hinse, C. Effect of ecdysone on glutamic decarboxylase in rat brain. *Experientia* **1969**, *25*, 250–251. [CrossRef]

112. Chiang, H.C.; Wang, J.J.; Wu, R.T. Immunomodulating effects of the hydrolysis products of formosamin C and β-ecdysone from *Paris formosana* Hayata. *Anticancer Res.* **1992**, *12*, 1475–1478.

113. Lafont, R.; Raynal, S.; Serova, M.; Didry-Barca, B.; Guibout, L.; Dinan, L.; Latil, M.; Dilda, P.J.; Dioh, W.; Veillet, S. 20-Hydroxyecdysone activates the protective arm of the renin angiotensin system via Mas receptor. *bioRxiv* **2020**. [CrossRef]

114. Raynal, S.; Foucault, A.-S.; Ben Massoud, S.; Dioh, W.; Lafont, R.; Veillet, S. BIO101, a drug candidate targeting sarcopenic obesity through MAS receptor activation. *J. Cachexia Sarcopenia Muscle* **2015**, *6*, 429.

115. Yoshida, T.; Galvez, S.; Tiwari, S.; Rezk, B.M.; Semprun-Prieto, L.; Higashi, Y.; Sukhanov, S.; Yablonka-Reuveni, Z.; Delafontaine, P. Angiotensin II inhibits satellite cell proliferation and prevents skeletal muscle regeneration. *J. Biol. Chem.* **2013**, *288*, 23823–23832. [CrossRef]

116. Band, M.M.; Sumukadas, D.; Struthers, A.D.; Avenell, A.; Donnan, P.T.; Kemp, P.R.; Smith, K.T.; Hume, C.L.; Hapca, A.; Witham, M.D. Leucine and ACE inhibitors as therapies for sarcopenia (LACE trial): Study protocol for a randomised controlled trial. *Trials* **2018**, *19*, 6. [CrossRef]

117. Höcht, C.; Mayer, M.; Taira, C.A. Therapeutic perspectives of angiotensin-(1-7) in the treatment of cardiovascular diseases. *Open Pharmacol. J.* **2009**, *3*, 21–31. [CrossRef]

118. Parr, M.K.; Botré, F.; Naß, A.; Hengevoss, J.; Diel, P.; Wolber, G. Ecdysteroids: A novel class of anabolic agents? *Biol. Sport* **2015**, *32*, 169–173. [CrossRef]

119. Schreihofer, D.A.; Duong, P.; Cunningham, R.L. N-terminal truncations in sex steroid receptors and rapid steroid actions. *Steroids* **2018**, *133*, 15–20. [CrossRef]

120. Sobrino, A.; Vallejo, S.; Novella, S.; Lázaro-Franco, M.; Mompeón, A.; Bueno-Betí, C.; Walther, T.; Sánchez-Ferrer, C.; Peiró, C. Mas receptor is involved in the estrogen-receptor induced nitric oxide-dependent vasorelaxation. *Biochem. Pharmacol.* **2017**, *129*, 67–72. [CrossRef]

121. Ogawa, S.; Nishimoto, N.; Matsuda, H. Pharmacology of ecdysones un vertebrates. In *Invertebrate Endocrinology and Hormonal Heterophylly*; Burdette, W.J., Ed.; Springer: Berlin, Germany, 1974; pp. 341–344.

122. Hikino, H.; Oizumi, Y.; Takemoto, T. Absorption, distribution, metabolism, and excretion of insect-metamorphosing hormone ecdysterone in mice. *Chem. Pharm. Bull. (Tokyo)* **1972**, *20*, 2454–2458. [CrossRef]

123. Lafont, R.; Girault, J.-P.; Kerb, U. Excretion and metabolism of injected ecdysone in the white mouse. *Biochem. Pharmacol.* **1988**, *37*, 1174–1177. [CrossRef]

124. Wu, M.; Zhao, S.; Ren, L.; Wang, R.; Bai, X.; Han, H.; Li, B.; Chen, H. Research on the relationship between tissue quantitative distribution of ^{3}H-*Achyranthes bidentata* ecdysterone and channel-tropism of herbal drugs in mice. *China J. Chin. Mater. Med.* **2011**, *36*, 3018–3022.

125. Girault, J.-P.; Lafont, R.; Kerb, U. Ecdysone catabolism in the white mouse. *Drug Metab. Dispos.* **1988**, *16*, 716–720. [PubMed]

126. Kumpun, S.; Girault, J.-P.; Dinan, L.; Blais, C.; Maria, A.; Dauphin-Villemant, C.; Yingyongnarongkul, B.; Suksamrarn, A.; Lafont, R. The metabolism of 20-hydroxyecdysone in mice: Relevance to pharmacological effects and gene switch applications of ecdysteroids. *J. Steroid Biochem. Mol. Biol.* **2011**, *126*, 1–9. [CrossRef] [PubMed]

127. Dinan, L.; Balducci, C.; Guibout, L.; Foucault, A.S.; Bakrim, A.; Kumpun, S.; Girault, J.-P.; Tourette, C.; Dioh, W.; Dilda, P.J.; et al. Ecdysteroid metabolism in mammals: The fate of ingested 20-hydroxyecdysone in mice and rats. *J. Steroid Biochem. Mol. Biol.* **2021**, in press. [CrossRef]

128. Balducci, C.; Dinan, L.; Guibout, L.; Foucault, A.S.; Carbonne, C.; Durand, J.-D.; Caradeux, C.; Bertho, G.; Girault, J.-P.; Lafont, R. The complex metabolism of poststerone in male rats. *J. Steroid Biochem. Mol. Biol.* **2021**, in press. [CrossRef]

129. Gharib, B.; Nugon-Baudon, L.; Lafont, R.; De Reggi, M. Ecdysteroids, a new pathological marker in man. In *Biologie Prospective: Compte-rendus du 8è colloque de Pont-à-Mousson*; John Libbey Eurotext: Arcueil, France, 1993; pp. 203–206.

130. Schiffer, L.; Barnard, L.; Baranowski, E.S.; Gilligan, L.C.; Taylor, A.E.; Arit, W.; Shackleton, C.H.L.; Storbeck, K.-H. Human steroid biosynthesis, metabolism and excretion are differentially reflected by serum and urine steroid metabolomes: A comprehensive review. *J. Steroid Biochem. Mol. Biol.* **2019**, *194*, 105439. [CrossRef]

131. Wells, J.E.; Hylemon, P.B. dentification and characterization of a bile acid 7alpha-dehydroxylation operon in *Clostridium* sp. strain TO-931, a highly active 7alpha-dehydroxylating strain isolated from human feces. *Appl. Environ. Microbiol.* **2000**, *66*, 1107–1113. [CrossRef]

132. Simon, P.; Koolman, J. Ecdysteroids in vertebrates: Pharmalogical aspects. In *Ecdysone—from Chemistry to Mode of Action*; Koolman, J., Ed.; Georg Thieme Verlag: Stuttgart, Germany, 1989; pp. 254–259.
133. Simon, P. Ecdysteroids in the mammalian organism and their detection as a means of diagnosis antihelmintic infections (Ecdysteroide im Säugerorganismus und ihr Nachweis als Möglichkeit der Diagnose helmintischer Infektionen). Ph.D. Thesis, University of Marburg, Marburg, Germany, 1988.
134. Bolduc, T.M. Human Urinary Excretion Profiles after Exposure to ecdysterone. Master's Thesis, University of Utah, Salt Lake City, UT, USA, 2008.
135. Brandt, W. Pharmakokinetik und Metabolismus des 20-Hydroxyecdysons im Menschen. Ph.D. Thesis, Marburg, Germany, 2003.
136. Dioh, W.; Del Signore, S.; Dupont, P.; Dilda, P.; Lafont, R.; Veillet, S. *SARA-PK: A Single and Multiple Ascending Oral Doses Study to Assess the Safety and Evaluate the Pharmacokinetics of BIO101 in Healthy Young and Older Volunteers*; ICFSR: Barcelona, Spain, 2017.
137. Dioh, W.; Tourette, C.; Del Signore, S.; Daudigny, L.; Balducci, C.; Dupont, P.; Dilda, P.; Agus, S.; Lafont, R.; Veillet, S. A phase I, combined study of the safety and pharmacokinetics of BIO101 (20-hydroxyecdysone) in healthy young and elderly adult volunteers after single ascending and multiple ascending oral doses for 14 days. **2021**. Manuscript in preparation.
138. Thiem, B.; Kikowska, M.; Malinski, M.P.; Kruszka, D.; Napierala, M.; Florek, E. Ecdysteroids: Production in plant in vitro cultures. *Phytochem. Rev.* **2017**, *16*, 603–622. [CrossRef]
139. Fujimoto, Y.; Ohyama, K.; Nomura, K.; Hyodo, R.; Takahashi, K.; Yamada, J.; Morisaki, M. Biosynthesis of sterols and ecdysteroids in *Ajuga* hairy roots. *Lipids* **2000**, *35*, 279–288. [CrossRef]
140. Chen, R.; Yang, S.; Zhang, L.; Zhou, Y.J. Advanced strategies for the production of natural products in yeast. *iScience* **2000**, *23*, 100879. [CrossRef]
141. Yan, X.; Yan, Y.; Wei, W.; Wang, P.; Liu, Q.; Wei, Y.; Zhang, L.; Zhao, G.; Yue, J.; Zhou, Z. Production of bioactive ginsenoside compound K in metabolically engineered yeast. *Cell Res.* **2014**, *24*, 770–773. [CrossRef]
142. Duport, C.; Spagnoli, R.; Degryse, E.; Pompon, D. Self-sufficient biosynthesis of pregnenolone and progesterone in engineered yeast. *Nat. Biotechnol.* **1998**, *16*, 186–189. [CrossRef]
143. Szczebara, F.M.; Chandelier, C.; Villeret, C.; Masurel, A.; Bourot, S.; Duport, C.; Blanchard, S.; Groisillier, A.; Testet, E.; Costaglioli, P.; et al. Total biosynthesis of hydrocortisone from a simple carbon source in yeast. *Nat. Biotechnol.* **2003**, *21*, 143–149. [CrossRef]
144. Dinan, L.; Balducci, C.; Guibout, L.; Lafont, R. Small-scale analysis of phytoecdysteroids in seeds by HPLC-DAD-MS for the identification and quantification of specific analogues, dereplication and chemotaxonomy. *Phytochem. Anal.* **2020**, *31*, 643–661. [CrossRef]
145. Wang, J.-L.; Ruan, D.-C.; Cheng, Z.-Y.; Yang, C.-R. The dynamic variations of 20-hydroxyecdysone in *Cyanotis arachnoidea*. *Acta Bot. Yunnanica* **1996**, *18*, 459–464.
146. Bandara, B.M.R.; Jayasinghe, L.; Karunaratne, V.; Wanningama, G.P.; Bokel, M.; Kraus, W. Ecdysterone from stem of *Diploclisia glaucescens*. *Phytochemistry* **1989**, *28*, 1073–1075. [CrossRef]
147. Ramazonov, N.S.H.; Bobaev, I.D.; Syrov, V.N.; Sagdullaev, S.H.; Mamatkhanov, A.U. Chemistry, biology and production technology of phytoecdysteroids. *Sci. Technol. Tashkent* **2016**, 258. (In Russian)
148. Volodin, V.V.; Pchelenko, L.D.; Volodina, S.O.; Kudyasheva, A.G.; Shevchenko, O.G.; Zagorskaya, N.P. Pharmacological estimate of new ecdysteroid-containing substance "Serpisten". *Rastit Resur.* **2006**, *42*, 113–130.
149. Martnussen, I.; Volodin, V.; Volodina, S.; Uleberg, E. Effect of climate on plant growth and level od adaptogenic compounds in Maral root (*Leuzea carthamoides* Willd., DC), crowned saw-wort (*Serratula coronata* L.) and roseroot (*Rhodiola rosea* L.). *Eur. J. Plant. Sci. Biotechnol.* **2011**, *5*, 72–77.
150. Dinan, L. Phytoecdysteroids: Biological aspects. *Phytochem* **2001**, *57*, 325–339. [CrossRef]
151. Báthori, M.; Girault, J.P.; Kalasz, H.; Mathé, I.; Dinan, L.N.; Lafont, R. Complex phytoecdysteroid cocktail of *Silene otites* (Caryophyllaceae). *Arch. Insect Biochem. Physiol.* **1999**, *41*, 1–8. [CrossRef]
152. Lobell, M.; Hendrix, M.; Hinzen, B.; Keldenich, J.; Meier, H.; Schmeck, C.; Schohe-Loop, R.; Wunberg, T.; Hillisch, A. In silico ADMET traffic lights as a tool for the prioritization of HTS hits. *ChemMedChem* **2006**, *1*, 1229–1236. [CrossRef]
153. Lipinski, C.A.; Lombardo, F.; Dominy, B.W.; Feeney, P.J. Experimental and computational approaches to estimate solubility and permeability in drug discovery and developments settings. *Adv. Drug Deliv. Rev.* **2001**, *46*, 3–26. [CrossRef]
154. Otaka, T.; Uchiyama, M.; Okui, S.; Takemoto, T.; Hikino, H.; Ogawa, S.; Nishimoto, N. Stimulatory effect of insect metamorphosing steroids from *Achyranthes* and *Cyathula* on protein synthesis in mouse liver. *Chem. Pharm. Bull.* **1968**, *16*, 2426–2429. [CrossRef]
155. Issaadi, H.M.; Csábi, J.; Hsieh, T.J.; Gáti, T.; Tóth, G.; Hunyadi, A. Side-chain cleaved phytoecdysteroid metabolites as activators of protein kinase B. *Bioorg. Chem.* **2019**, *82*, 405–413. [CrossRef]
156. Novikov, V.S.; Shamarin, I.A.; Bortnovskii, V.N. A trial of the pharmacological correction of sleep disorders in sailors during a cruise. *Voen Med. Zhurnal* **1992**, 47–49.
157. Marina, T.F. Influence of CNS stimulators of plant origin on reflex activity of spinal cord. In: Stimulators of the Central Nervous System. *Tomsk* **1966**, 31–36.
158. Mirzaev, Y.R.; Syrov, V.N.; Krushev, S.A.; Iskanderova, S.D. Study of the effects of ecdysten on the sexual function under experimental and clinical conditions. *Eksp. Klin. Farm.* **2000**, *63*, 35–37.
159. Syrov, V.N.; Khushbaktova, Z.A.; Komarin, A.S.; Abidov, A.B.; Pechenitsina, T.V.; Aripkhodzhaeva, F.A. Experimental and clinical evaluation of the efficacy of ecdysten in the treatment of hepatitis. *Eksp. Klin. Farmakol.* **2004**, *67*, 56–59. [PubMed]

160. Mosharrof, A.H. Effects of extract from Rhaponticum carthamoides (Willd) Iljin (*Leuzea*) on learning and memory in rats. *Acta Physiol. Pharmacol. Bulg.* **1987**, *13*, 37–42.
161. Opletal, L.; Sovova, M.; Dittrich, M.; Solich, P.; Dvorák, J.; Krátký, F.; Cerovský, J.; Hofbauer, J. Phytotherapeutic aspects of diseases of the circulatory system. Leuzea carthamoides (WILLD.) DC: The status of research and possible use of the taxon. *Ceska Slov. Farm.* **1997**, *46*, 247–255.
162. Ambrosio, G.; Wirth, D.; Joseph, J.F.; Mazzarino, M.; de la Torre, X.; Botrè, F.; Parr, M.K. How reliable is dietary supplement labelling?—Experiences from the analysis of ecdysterone supplements. *J. Pharm. Biomed. Anal.* **2020**, *177*, 112877. [CrossRef]
163. Isenmann, E.; Ambrosio, G.; Joseph, J.F.; Mazzarino, M.; de la Torre, X.; Zimmer, P.; Kazlauskas, R.; Goebel, C.; Botrè, F.; Diel, P.; et al. Ecdysteroids as non-conventional anabolic agent: Performance enhancement by ecdysterone supplementation in humans. *Arch. Toxicol.* **2019**, *93*, 1807–1816. [CrossRef]
164. Parr, M.K.; Ambrosio, G.; Wuest, B.; Mazzarino, M.; de la Torre, X.; Sibilia, F.; Joseph, J.F.; Diel, P.; Botré, F. Targeting the administration of ecdysterone in doping control samples. *Forensic Toxicol.* **2020**, *38*, 172–184. [CrossRef]
165. Kibrik, N.D.; Reshetnyak, J.A. Therapeutical approaches to sexual disadaption. *Eur. Neuropsychopharmacol.* **1996**, *6*, 167. [CrossRef]
166. Saatov, Z.; Agzamkhodjaeva, D.A.; Syrov, V.N. Distribution of phytoecdysteroids in plants of Uzbekistan and the possibility of using drugs based on them in nephrological practice. *Chem. Nat. Comp.* **1999**, *35*, 186–191. [CrossRef]
167. Osipova, S.O.; Islamova, Z.I.; Syrov, V.N.; Badalova, N.S.; Khushbaktova, Z.A. Ecdysten in the treatment of giardiasis. *Med. Parazitol. (Mosk)* **2002**, 29–33.
168. Islamova, Z.I.; Syrov, V.N.; Khushbaktova, Z.A.; Osipova, S.O. The efficacy of ecdystene versus metronidazole in the treatment of lambliasis. *Med. Parazitol. (Mosc.)* **2010**, 14–17.
169. Makhmudova, L.B. Experience of using ecdisten in the treatment of hymenolepiasis. *Med. Parazitol. (Mosc.)* **2012**, 45–47.
170. Seidlova-Wuttke, D.; Wuttke, W. In a placebo-controlled study β-ecdysone (ECD) prevented the development of the metabolic syndrome. *Planta Med.* **2012**, *78*, CL37. [CrossRef]
171. Wuttke, M.; Seidlova-Wuttke, D. Beta-ecdysone (Ecd) prevents visceral, bone marrow and joint fat accumulation and has positive effects on serum lipids, bone and joint cartilage. *Planta Med.* **2012**, *78*, PD68. [CrossRef]
172. Wuttke, W.; Seidlova-Wuttke, D. Eine neue Alternative für die Prävention und Therapie postmenopausaler Erkrankungen, insbesondere des metabolischen Syndroms. *J. Gynäkol. Endokrinol.* **2015**, *25*, 6–12.
173. Agus, S.; Dioh, W. A Double-blind, Placebo Controlled, Randomized INTerventional Clinical Trial (SARA-INT). *Clinicaltrials.gov* NCT03452488 **2018**.
174. Thole, S.W. The Metabolic Syndrome: The Effects of β-ecdysone on Selected Body Parameters and Serum Lipids on the Metabolic Syndrome. Ph.D. Thesis, University of Göttingen, Göttingen, Germany, 2018.
175. Rayas, A.L.F. Effect of phytoecdysterone administration in subjects ith prediabetes. *Clinicaltrials.gov identifier NCT03906201* **2019**.
176. Agus, S.; Dioh, W. Testing the Efficacy and Safety of BIO101 for the Prevention of Respiratory Deterioration in COVID-19 Patients (COVA). *Clinicaltrials.gov NCT04472728* **2020**, ongoing.
177. Dustmukhamedova, D.K.H.; Kamilovz, A.T. The characteristic of energy metabolism disorders and its correction in children with celiac disease. *Am. J. Med. Medic. Sci.* **2020**, *10*, 780–783.
178. Kwan, P. Sarcopenia: Neurological point of view. *J. Bone Metab.* **2013**, *24*, 83–89. [CrossRef] [PubMed]
179. Drey, M.; Krieger, B.; Sieber, C.C.; Bauer, J.M.; Hettwer, S.; Bertsch, T.; DISARCO Study Group. Motoneuron loss is associated with sarcopenia. *J. Am. Med. Dir. Assoc.* **2014**, *15*, 435–439. [CrossRef] [PubMed]
180. Dioh, W.; Chabane, M.; Tourette, C.; Azbekyan, A.; Morelot-Panzini, C.; Hajjar, L.A.; Lins, M.; Nair, G.B.; Whitehouse, T.; Mariani, J.; et al. Testing the efficacy and safety of BIO101, for the prevention of respiratory deterioration, in patients with COVID-19 pneumonia (COVA study): A structured summary of a study protocol for a randomised controlled trial. *Trials* **2021**, *22*, 42. [CrossRef] [PubMed]
181. Latil, M.; Camelo, S.; Veillet, S.; Lafont, R.; Dilda, P.J. Developing new drugs that activate the protective arm of the renin-angiotensin system as a potential treatment for repiratory failure in COVID-19 patients. *Drug Discov. Today* **2021**, in press. [CrossRef]
182. WADA. Summary of Major Modifications and Explanatory Notes. 2020 prohibited list. Available online: https://www.wada-ama.org/sites/default/files/wada_2020_english_summary_of_modifications_.pdf (accessed on 1 April 2021).

 *biomedicines*

MDPI

Review

Natural Products That Changed Society

Søren Brøgger Christensen

The Museum of Natural Medicine & The Pharmacognostic Collection, University of Copenhagen, DK-2100 Copenhagen, Denmark; soren.christensen@sund.ku.dk; Tel.: +45-3533-6253

Abstract: Until the end of the 19th century all drugs were natural products or minerals. During the 19th century chemists succeeded in isolating pure natural products such as quinine, morphine, codeine and other compounds with beneficial effects. Pure compounds enabled accurate dosing to achieve serum levels within the pharmacological window and reproducible clinical effects. During the 20th and the 21st century synthetic compounds became the major source of drugs. In spite of the impressive results achieved within the art of synthetic chemistry, natural products or modified natural products still constitute almost half of drugs used for treatment of cancer and diseases like malaria, onchocerciasis and lymphatic filariasis caused by parasites. A turning point in the fight against the devastating burden of malaria was obtained in the 17th century by the discovery that bark from trees belonging to the genus *Cinchona* could be used for treatment with varying success. However isolation and use of the active principle, quinine, in 1820, afforded a breakthrough in the treatment. In the 20th century the synthetic drug chloroquine severely reduced the burden of malaria. However, resistance made this drug obsolete. Subsequently artemisinin isolated from traditional Chinese medicine turned out to be an efficient antimalarial drug overcoming the problem of chloroquine resistance for a while. The use of synthetic analogues such as chloroquine or semisynthetic drugs such as artemether or artesunate further improved the possibilities for healing malaria. Onchocerciasis (river blindness) made life in large parts of Africa and South America miserable. The discovery of the healing effects of the macrocyclic lactone ivermectin enabled control and partly elimination of the disease by annual mass distribution of the drug. Also in the case of ivermectin improved semisynthetic derivatives have found their way into the clinic. Ivermectin also is an efficient drug for treatment of lymphatic filariasis. The serendipitous discovery of the ability of the spindle toxins to control the growth of fast proliferating cancer cells armed physicians with a new efficient tool for treatment of some cancer diseases. These possibilities have been elaborated through preparation of semisynthetic analogues. Today vincristine and vinblastine and semisynthetic analogues are powerful weapons against cancer diseases.

Keywords: malaria; quinine; chloroquine; artemisinin; onchocerciasis; ivermectin; moxidectin; cancer; vincristine; vinblastine

Citation: Christensen, S.B. Natural Products That Changed Society. *Biomedicines* **2021**, *9*, 472. https://doi.org/10.3390/biomedicines9050472

Academic Editor: Shaker A. Mousa

Received: 15 March 2021
Accepted: 24 April 2021
Published: 26 April 2021

1. Introduction

Until the end of the 19th century only natural products and minerals were available as drugs. Nature today still continuously surprises by offering molecules outside the scope of human creativity to give drugs enabling healing of previous incurable diseases [1–3]. Natural products still are the basis for half of all new drugs, either as the parent natural products or optimized derivatives [3,4]. Marine organism and cultures of fungi including endophytes have added new possibilities for finding miraculous natural products [3,5,6]. Until the 20th century malaria was an almost worldwide burden causing death and disability. Quinine and artemisimin made treatment of malaria successful [7–9]. Onchocerciasis (river blindness) prevented cultivation of riverbanks particularly in Sub-Sahara Africa but also in Latin-America. Lymphatic filariasis made life miserable for millions of people in Africa and Latin America. The natural product ivermectin enabled treatment of onchocerciasis, lymphatic filariasis and other parasitic diseases [10]. Fast proliferating cancer diseases had a severe death rate until the middle of the 20th century, when the vinca alkaloids were discovered. Vincristine, vinblastine and semisynthetic analogues, became efficient drugs for treatment of some of these cancer diseases [8,11]. These drugs are examples of drugs that changed the society and they will be mentioned in the present review. Many more natural products that also had an impact on the society e.g., antibiotics such as penicillin and chemotherapeutics like taxols have been omitted because of limited space and time.

Some natural products are only found in trace amounts in few organisms. Today we are able in some cases to solve the problem of sustainable supply by total synthesis of the compounds. Semisynthesis enables design and preparation of more efficient drugs by modifying the structure of the parent natural product or by making prodrugs [4]. This only has been possible for the last 150 years. Actually vitalism denied the possibility of synthesis of organic compounds since a theorem in this philosophy claims that only living organism can form C-C bonds [12]. The first evidence of the limitations of this paradigm came when Wöhler in 1828 published the synthesis of urea by dissolving silver cyanate or lead cyanate in aqueous ammonia [13]. Since this reaction did not involve formation of C-C bonds this might appear a poor argument. A convincing counterevidence was obtained when tetrachloromethane (**2**) was converted in tetrachloroethene (**3**), which then was converted into acetic acid (**5**) via trichloroacetic acid (**4**). Since tetrachloroemethane was obtained from carbon disulfide (**1**) prepared from charcoal and sulfur this proved that C-C bonds could be formed in vitro using only inorganic reagents (Scheme 1) [14]:

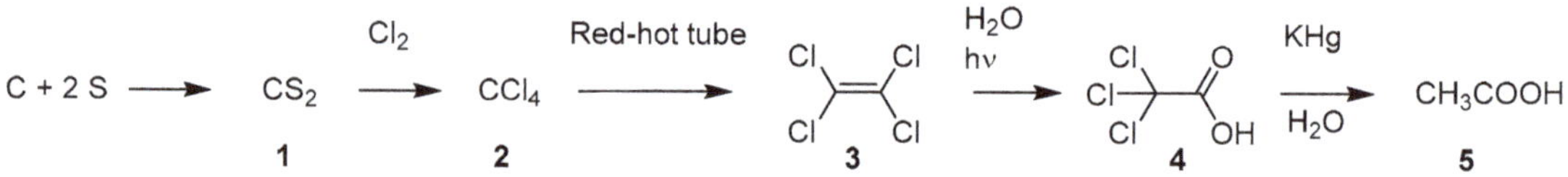

Scheme 1. Formation of acetic acid from carbon, chlorine and sulfur.

The art of organic synthesis was developed during the 19th, 20th and 21st century and today very complex organic molecules have been prepared by total synthesis. Total synthesis is a procedure for preparing highly complex natural products starting with simple commercially available materials [15–18]. Acetylsalicylic acid (**6**), an analogue of salicin the antipyretic principle isolated from willow barks, was synthesized in 1859 but not marketed under the name of aspirin before 1899 [19,20]. Phenazone (**7**) was synthesized in 1887 and marketed as antipyrin in 1889–1890 during an influenza epidemic (Figure 1) [21].

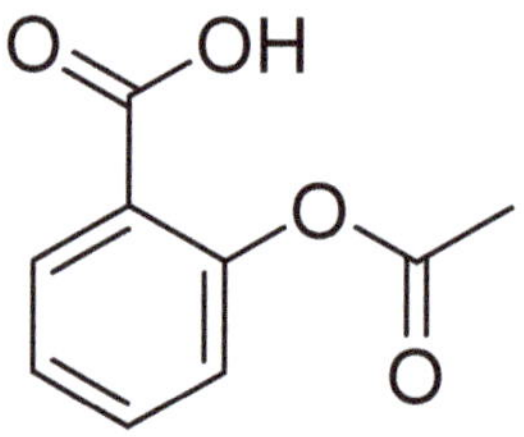

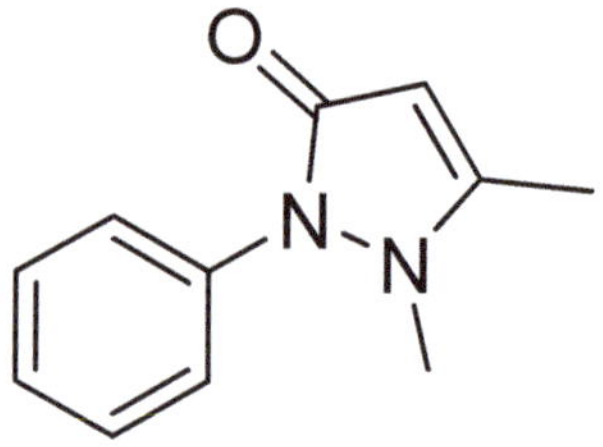

Figure 1. Acetylsalicylic acid (**6**) and phenazone (**7**).

Even though synthetic chemistry has made impressive progress, many natural products either are so complex or so easily available that the compound isolated from biological material or isolated from cell cultures is still used in pharmaceutical formulations. This is the case for e.g., taxol, morphine, codeine, vincristine, vinblastine and quinine [22]. One of the most complicated natural products prepared by synthesis is eribulin [23]. Even though a procedure for total synthesis of a natural product has been developed this frequent is so complicated and with such a poor yield that it is not economically feasible for production of the compound on kg scale. A procedure for synthesis of quinine was described in 1944. This achievement though highly praised during the war never made synthetic quinine available, because the many steps in the synthesis. It still is debated whether the synthesis from 1944 leads to quinine. An undisputed total synthesis of quinine was not developed until 2001, and even this synthesis is not feasible for production of quinine in kg scale [24].

The most important of all natural products, defined as compounds produced by a living organism, is dioxygen, normally just mentioned as oxygen. Dioxygen is formed by photosynthetic cleavage of water (Scheme 2) [25]. Dioxygen was only present in trace amount in the atmosphere of the earth before living organism developed photosynthesis and, thus, must be considered a natural product [25,26].

$$6\,H_2O + 6\,CO_2 \xrightarrow{\;h\nu\;} C_6H_{12}O_6 + 6\,O_2$$

Scheme 2. Simplified presentation of the photosynthesis [25].

2. Malaria

Malaria is a disease caused by five different protozoan parasites of the genus *Plamodium* (*P. malaria* causing quartan malaria, *P. ovale* causing ovale tertian malaria, *P. vivax* causing benign tertian malaria, *P. knowlesi*, which only recently has been shown to infect humans, and *P. falciparum* responsible for the majority of fatalities caused by malaria) [7]. The life cycle of the parasite is depicted in Figure 2.

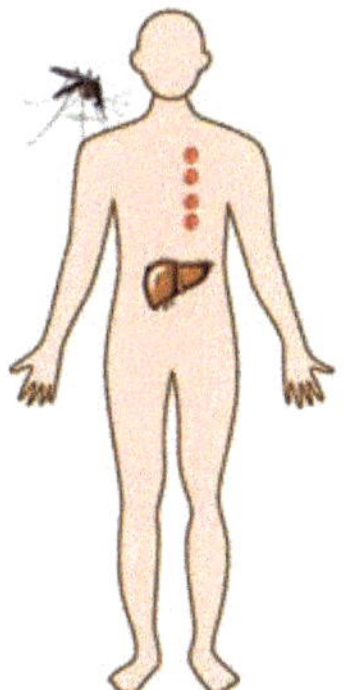

Figure 2. Life cycle of *Plasmodium* parasites: When an infected female *Anopheles* mosquito takes a blood meal approximately eight to 10 sporozoites are introduced into the blood. Either the sporozoites have entered hepatocytes after 45 min. or they are cleared. In the liver the sporozoites proliferate asexually for between 5 days (*P. falciparum*) and 15 days (*P. malariae*). A proportion of hepatic schizonts may rest in the liver as hyonozoites in the case of *P. ovale* and *P. vivax* infections. The hypnozoites may awake weeks or months after the infection. The merozoites delivered to the blood invade erythrocytes, feed on the proteins mainly hemoglobin and proliferate asexually. After 36 hours (*P. malariae*, *P. ovales*, *P. vivax* and *P. knowlesi*) or 54 hours (*P. malariae*) the cells rupture and release between 6 and 36 merozoites, which invade healthy erythrocytes. The characteristic fever attacks appear when the erythrocytes rupture. After a series of asexual proliferation gametocytes appear. Gametocytes are ingested during a blood meal of an *Anopheles* mosquito, in which they undergo a sexual proliferation [7].

Protozoan parasites are parasites living inside the cells of the host. From the *Plasmodium* genus the most deadly parasite is *P. falciparum* causing malignant tertian malaria, also known as cerebral malaria. This species is only transmitted in the tropics. Two studies formed the basis for our understanding of the disease, the first was the discovery of parasites in the erythrocytes of malaria patients by Alphonse Laveran in 1880 and the second the role of *Anopheles* mosquitoes as vectors described by Ronald Ross in 1897 [7,27,28]. Both researchers were awarded the Nobel Prize in 1907 and 1902, respectively. In contrast to *P. falciparum* the other parasites are also infectious in subtropical and temperate areas. About year 1900 malaria had its maximal distribution reaching latitudinal extremes of 64° norths and 32° south encompassing all continents Europe, the Americas, Asia and Australia. In year 2004 the area in which humans were at risk of being infected with malaria was decreased from 53% to 27% of the Earth's land surface [29]. In the period 2010 to 2018 the estimated cases of malaria has decreased from 251 million to 228 million and the number of deaths from 585,000 to 405,000, which still are scaring high numbers [30]. Previous the presence of *P. falciparum* meant that Europeans were prevented from entering tropic Africa, because of the mortality from cerebral malaria. An about seven times higher death rate for French and British soldiers were observed in the tropic African colonies in the first half of the 19th century than seen elsewhere [31]. In 1805 a party of 44 Europeans sailed up the river Niger but only five survived the journey [32] illustrating the mortality of malaria. No doubt the high mortality rate prevented colonization of Africa (Figure 3).

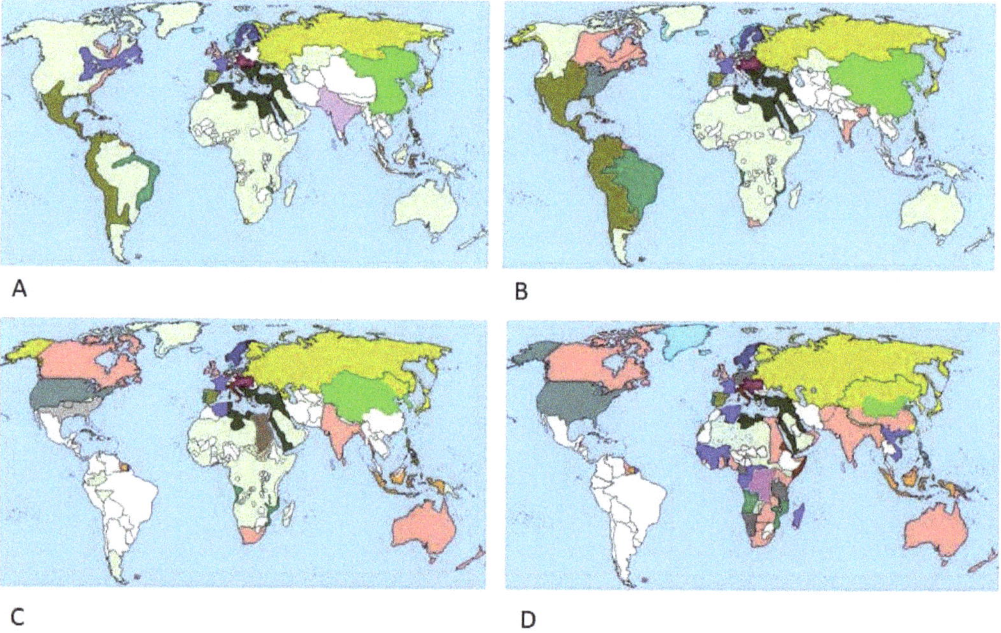

Figure 3. (**A**): World map 1660. A large part of South America was colonized by Spain (light green) and Portugal.(Green) Russia is dark blue (**B**): World map 1754. Almost all South America is colonized British colored red, whereas only tiny parts of Africa. (**C**): World map 1822. Even though Australia was known since 1606 it was not colonized until James Cook described the continent in 1770. (**D**): World map 1898. The major part of Africa is colonized and by 1914 all of Africa was colonized (Russia black, German and Belgium colonies blakc, Turkish colonies Light green , British colonies red, France colonies dark blue, Dutch colonies light red (comons.wikimedia.org/wiki/File:World_1898_empires_colonies_territory.png, Accessed 1 April 2021)

Even though mortality of malaria in Europe was not comparable with that in Africa, the disease still was a severe burden. Malaria (known as ague or intermittent fever, in Danish: *koldfeber*, cold fever) was prevalent in Southern England in Kent, Essex and parts of London in the 17th century. Rome was also reputed for malaria [32].

Malaria and a number of other diseases well known in Europe, Asia and Africa such as smallpox, influenza, measles and yellow fever were no burdens for the pre-Columbian American population [33]. Consequently the endemic population had no immunologic defense. The epidemic infections brought to America by infected Spaniards led to demographic falls. The Indian population on the island Espanola decreased from 3,770,000 in 1496 to 125 in 1570. Similar catastrophic falls were observed in Mexico and Peru [33]. Mating between Spaniards and Indians created Mestizos, who were less sensitive to the diseases and soon outnumbered the original population [33].

2.1. Drugs to Treat and Prevent Malaria

2.1.1. Cinchona Bark

After the conquest of the Inca Empire in South America the Jesuit monks followed the conquistadores as missionaries. Jesuits, who were the apothecaries at that time, noticed that the Incas used the bark of cinchona trees when working in cold streams to prevent shivering [34]. They hypothesized that the bark also might prevent shivering provoked by intermittent fever (malaria). A source mentions that the first European patient treated with cinchona bark was a Spaniard treated in the territory of Loja, in southern Ecuador in 1631 [33,35]. A successful outcome of this case study encouraged the Jesuits monks to export the bark to Europe [32,36]. In 1632 Jesuits send the first batch to Europe [37]. The bark was imported to Europe under different names, Jesuit bark acknowledging the Jesuits, who discovered the healing effects, Cardinal's bark, honoring Cardinal Lugo, or Peruvian bark. The genus name *Cinchona* was invented by Linnaeus in his Systema Naturae (1742) [37]. The genus was grouped in the plant family Rubiaceae (Figure 4)

Figure 4. *Cinchona officinalis* L. and *C. pubescens* Vahl. [38].

Already in 1649 a recipe for treatment of malaria, Schedula Romana, summarized trials carried out using cinchona bark [34,35,37]. Protestant Europe only reluctantly used the Catholic drug. Another obstacle was the doctrine of Hippocrates that well-being depends on the balance between the four humours or liquids: yellow bile, black bile, phlegm and blood [12]. The bark did not provoke bleeding, vomiting or in other ways affect the balance between the four humours and consequently, it could not have a healing effect [37]. Robert Talbor, a former apothecary and therefore not so concerned about this doctrine, became famous by successfully curing King Charles II of England and the Dauphin and Dauphine of France for malaria. In 1677 cinchona bark was included in the London pharmacopoeia as Cortex Peruvianus [37] (Figure 5).

Figure 5. Examples of *Cinchona* barks stored at The Museum of Natural Medicine & The Pharmacognostic Collection (University of Copenhagen).

From a modern scientific standard it is amazing to realize that the bark was included in a pharmacopoeia even though the botanical origin was not known until 1738 [34,37]. The missing knowledge of the mother plant facilitated trading with ineffective and even poisonous barks from other trees [34].

Cinchona bark was used so extensively by the Europeans that the source was in danger of becoming extinct by the middle of the 19th century [36]. In spite of an embargo a Dutchman claiming to be a tourist in 1852 smuggled seeds and seedlings from the South Americans cinchona territories to Java, Indonesia (Dutch East Indies), and established a plantation [39]. The plantations on Java became very successful so in 1924 Java had almost monopoly in production of *Cinchona* bark [40]. British botanists organized expeditions in South America to get the plant material to India [41]. These are early examples of what today is now known as bio-piracy [42,43]. In favor of these illegal exportations it might be argued that otherwise the *Cinchona* sources probably would have been overexploited leading to extinction in South America.

In the Danish Pharmacopoeia from 1805 are listed four different cinchona barks: Cortex Carabaeus harvested from *Cinchona caribaea*, which is a synonym of *Exostema caribaeum* (Jacq) Schult (Rubiaceae). The bark is known to be extremely bitter because of the presence of 4-phenylcoumarins, but cinchona alkaloids have not been found [44]. Cortex Peruvianus, harvested from *Cinchonae officinalis* L., Cortex Peruvianus Flavus (yellow or royal bark) and Cortex Peruvianus Ruber (red bark). No species is mentioned for the latter two but later they have been assigned to *C. calisaya* Wedd (yellow bark) and *C. pubescens* Vahl (red bark) [45]. An investigation of historical samples of cinchona barks have revealed a variation in quinine content from only trace amounts to 5% and no species stands out by possessing a high amount of quinine [46]. Similar analyses of species of *C. calisaya* collected in the field at different locations in Bolivia reveals variations in quinine content in the bark from trace amounts to 25% [45]. These findings confirmed the results of analysis from 1945 revealing variation in the quinine content in *C. officinalis* and *C. pubescens* species collected in Ecuador, Bolivia and Peru from trace amounts to 7% [47]. Such inhomogeneity must have made correct dosing of the bark a problem considering the poor analytical techniques present in the 18th and 19th centuries and considering the limited therapeutic window for quinine [7]. In the Danish Pharmacopoeia from 1907 only Cortex Chinae is mentioned, which is defined as bark originating from different preferential cultivated *Cinchona* species. The bark must contain at least 4% of alkaloids and at least 1% of quinine. Quinine chloride was included in the Danish Pharmacopoeia from 1893.

2.1.2. Quinine

In 1820 Pelletier and Caventou succeeded in isolating an active principle from *Cinchona* bark [34,48]. The compound was named quinine (Figure 6, **8**). To isolate quinine a suspension of ground bark slurried in water mixed with slaked lime was extracted with an organic solvent like hot toluene. The organic layer was extracted with diluted sulfuric acid and upon concentration of the aqueous phase quinine sulfate crystallized [37,39]. This procedure enabled large scale production. Already early comparisons of the use of *Cinchona* bark or quinine sulfate revealed advantages of using the isolated active principle [34]. Quinine sulfate was given in a pill in a dose of 1 g each day. In contrast several grams of the crude bark had to be swallowed and even though wine was offered to make swallowing easier it must have been unpleasant. Decoctions and tinctures of the bark were also used, but probably not standardized.

8 Quinine   R = -OCH$_3$
9 Cinchonidine R = -H

10 Quindiine   R = -OCH$_3$
11 Cinchonine R = -H

Figure 6. Major *Cinchona* alkaloids. Quinine and quinidine are included on WHO list of essential medicines (https://www.who.int/medicines/publications/essentialmedicines/en/ine, accessed on 1 April 2021).

In addition to being easier to swallow, the most important consequence of the use of the isolated compound was that it enabled administration of well-defined doses. Considering the variation of quinine content in the different barks and taking into account the analytical techniques available in the early 19th century the use of purified quinine facilitated correct dosing, even though a skilled apothecary probably would have a good feeling of the quality of the bark [34,35]. In Denmark a price for quinine sulfate was fixed in 1840 by the government (Interims Taxt 1840).

Inspection of the relocation costs, defined as the increased mortality among soldiers by relocation to tropical areas ("white man's grave"), reveals a dramatic fall around 1840. This increased survival has been related to the use of purified quinine [31,49] instead of cinchona bark. During the 19th century the mortality of sailors on the naval operations stopping slave trade was severely reduced after introduction of quinine for prevention and cure [31,49]. Earlier attempts using cinchona bark never gave convincing results [49]. It is interesting to notice that the Berlin Conference regulating the European colonization and trade in Africa took place 1884–1885 after the prophylactic use of quinine had been generally accepted. Quinine also played an important role during the American Civil War.

The war might have ended with a victory to the Union troops already in the first year had malaria and typhoid fever not made the soldiers unable to fight [36].

In 1928 the total consumption of quinine in India was estimated to be 160,000 lbs (72 tons) [50]. In 1924 the world consumption of quinine sulfate as 500 ton, of which 95% was produced from 10,000 ton of bark grown on Java giving the colony an income of 16–22 million £. India was the second most important supplier for 10 million US$ corresponding to 25 ton of quinine sulfate [36,39]. From the mid-19th century until 1940 quinine was the standard treatment of malaria [36]. Even though quinine had been used extensively since Pelletier and Cavantou isolated the compound in 1820 the first suggestion of the constitution was published by Paul Rabe in 1908 [51] and the relative configuration in 1950 (Figure 6, **8**) [52]. In spite of the missing knowledge about the structure, salts of quinine had been used as drugs since 1820.

2.1.3. Quinine Resistance

Resistance to quinine appears sporadically, but failure of treatment is only seen in Southeast Asia and New Guinea [53].

2.1.4. Synthetic Malaria Drugs

An early attempt to produce synthetic quinine was made by William Perkin in 1856 based on a hypothesis which today appears naïve. The attempt did not lead to quinine but instead the dye mauve. After Queen Victoria appeared public dressed in mauve dyed robes the dye became an enormous commercial success [54].

Paul Ehrlich in 1891 treated a malaria patient with methylene blue (Figure 7, **12**). However, this dye never became useful in the clinic as an antimalarial drug. It was, however, used as a template for developing chloroquine (**13**) [55,56]. The first synthetic antimalarial used in the clinic was pamaquine (plasmoquine, **16**) from 1924, which was followed by mepacrine (atabrine, quinacrine, **17**) in 1932 and resochin (chloroquine) in 1934 (**13**) [7]. Atabrine was first marketed in Germany in 1932 [57]. Chloroquine was evaluated to be too toxic for clinical use and consequently efforts to make a less toxic analogue were undertaken to give the 2-methylated analogue sontochin (**14**) (1936) [57].

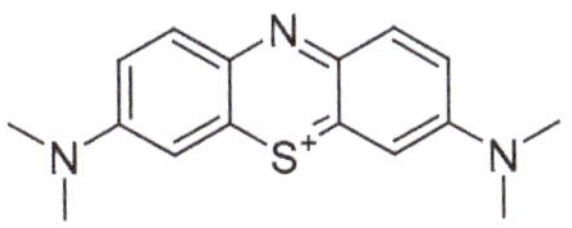

12 Methylene blue

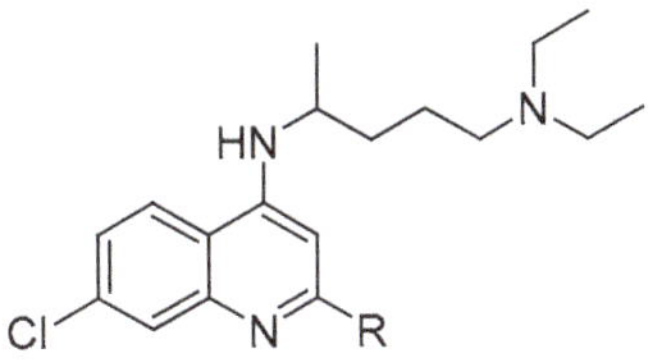

13 R = -**H**   Chloroquine
14 R = -CH$_3$ Sontochin

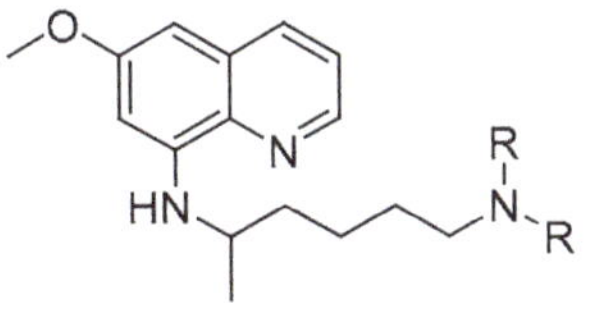

15 R = -H   Primaquine
16 R = -C$_2$H$_5$ Pamaquine

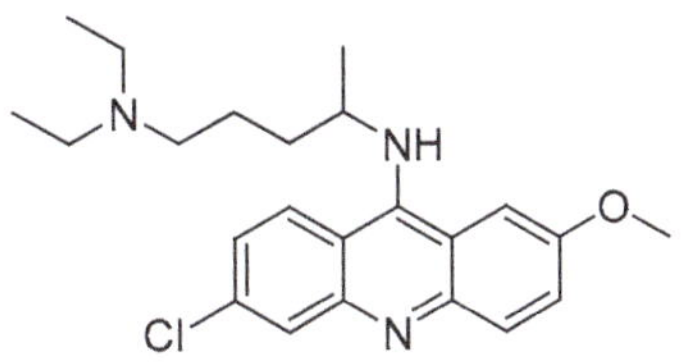

17 Mepacrine

Figure 7. Early synthetic antimalarial drugs. Chloroquine and primaquine are included on WHO list of essential medicines (https://www.who.int/medicines/publications/essentialmedicines/en/ine, accessed on 1 April 2021).

2.1.5. Malaria during the Pacific War

It is generally known that after the Japanese bombing of Pearl Harbor in 1941 the USA entered World War 2. Less known is that the mortality due to malaria on the Pacific Islands was far beyond that caused by Japanese troops [36]. The Japanese occupation of Dutch East Indies, in particular Java, meant an interruption of the supply of sufficient amounts of quinine to the Allies. Consequently intensive research programs were initiated in order to find alternative antimalarial drugs [57]. As a first attempt totaquine, a mixture of alkaloids extracted from *Cinchona* bark was investigated [57]. However, the amounts of *Cinchona* bark available could not meet the demand. Like the Allies the Japanese used the synthetic primaquine (**15**) for prophylaxis but the Japanese in insufficient doses [36]. In spite of these precautions malaria caused unacceptable losses. A group at the National institute of Health was organized in 1939 to develop alternative antimalarial drugs. The group organized a Committee on Medical Research, which collaborated with universities, industry and private individuals, the US Army and Navy and appropriate scientists in Britain and Australia. The collaboration resulted in synthesis of about 16,000 compounds. Some 80 compounds were selected. The most promising compound turned out to be chloroquine (**13**) already patented, but discarded by Bayer in Germany [57]. The 4-methyl derivative sontochin (**14**) was for a period preferred to chloroquine.

Plantations were started in Tanzania, Kenya, Cameroon and Rwanda. Also a joint venture between USA and Guatemala resulted in 400 ha with 1.75 million trees by mid-1948 [39]. These new plantations, however, did not solve the problem with quinine supply during the war since the amount of bark which can be harvested from a tree is very dependent on the age of the tree. From a four year old *C. ledgeriana* 0.25 kg, from an eight year old tree 4 kg, from a 15 year old 10 kg and from a 25 year old 20 kg of dried bark can be harvested [40]. After harvest of the bark the trees are covered up enabling continued growth.

Woodward in 1944 published a total synthesis of quinine but even though it created hope for overcoming the Japanese disruption of quinine supply during the Second World War, the many steps in this procedure, however, prevented its commercial use and even today quinine is still produced from Cinchona bark [54].

2.1.6. Malaria Control

Intensive studies performed by the Allies established in 1944 that chloroquine was a safe antimalarial. By combination of chloroquine and insecticides like DDT programs to eliminate malaria were initiated [58–60]. In Africa chloroquine became a pillar for malaria eradication [61,62]. In the late 1950s chloroquine-resistant parasites developed in six independent regions in India, Thailand, Indonesia, New Guinea, Venezuela and Guyana and spread over most of the areas infected with malaria [63]. As a consequence malaria re-emerged India [58], Kenya [64], and Africa [61,62].

A contribution to the decreased chloroquine sensitivity is mutations in the parasite genome to form a *pfcrt* gene. This gene encodes the *Plasmodium falciparum* chloroquine-resistant transporter (*Pf*CRT) a 424 amino acid protein with 10 transmembrane helices that facilitates transport of chloroquine away from ferriprotoporphyrin IX [59,63,65,66]. Today resistance has made chloroquine obsolete to control *P. falciparum* infections; but it still is used against *P. vivax*. Even though the use of chloroquine today is limited, the previous importance of chloroquine to reduce the burden of malaria should not be underestimated [60]. In the interim from appearance of chloroquine resistant-parasites to the appearance of the artemisnins synthetic a number of drugs were used to control malaria. Among these drugs are mefloquine (Figure 8, **17**), halofantrine (**22**), lumefantrine (**23**), primaquine (**15**), atovaquone (**20**), sulfadoxine (**21**), tetracycline (**25**) and proguanil (**18**), a prodrug which in the liver is transformed into the active drug cycloguanil (**19**). *Plasmodium* parasites host an endoparasite, the apicoplast. The apicoplast has a bacterial ancestry. Consequently the naturally occurring antibiotic tetracycline, which binds to the ribosome of bacteria, can be used for treatment of malaria. Terpene and type II fatty acid biosynthesis also proceed in the apicoplasts. Blockage of this system will not kill the mother parasites but daughter parasites with malfunctioning apicoplasts are affected [67]. Tafenoquine (**24**) was approved for treatment of malaria in 2018 [68,69]. Attempts to use the natural product fosfidomycin, which targets the deoxyxylose pathway in the apicoplast, have not lead to clinical useful drugs [70–72].

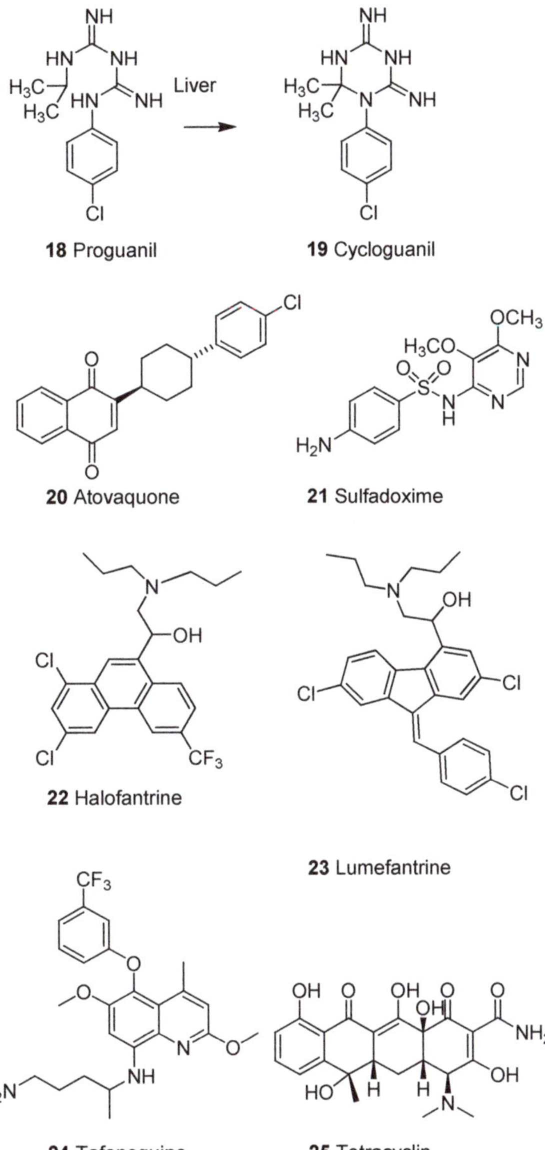

Figure 8. Antimalarial drugs. Except for tetracycline all the compounds are synthetic drugs. Proguanil, sulfadoxime, lumefanthrine and tetracycline are included in WHO list of essential medicines (https://www.who.int/medicines/publications/essentialmedicines/en/, accessed on 1 April 2021).

Ferroquine (Figure 9, **26**) possesses the same pharmacophore as chloroquine but contains a ferrocene moiety, which prevents the compound from being a substrate for the *Pf*CRT transporter [73,74]. The compound was brought to clinical trials II [75] but has not been approved as a drug by U.S. Food and Drug Administration (FDA) USA.

26 Ferroquine

Figure 9. Ferroquine (**26**).

2.1.7. Mechanism of Action of Quinolines

Both quinine and chloroquine are amines and consequently they will be protonated in acidic media. The target molecule of both compounds is heme (ferriprotoporphyrin IX, Figure 10, **29**).

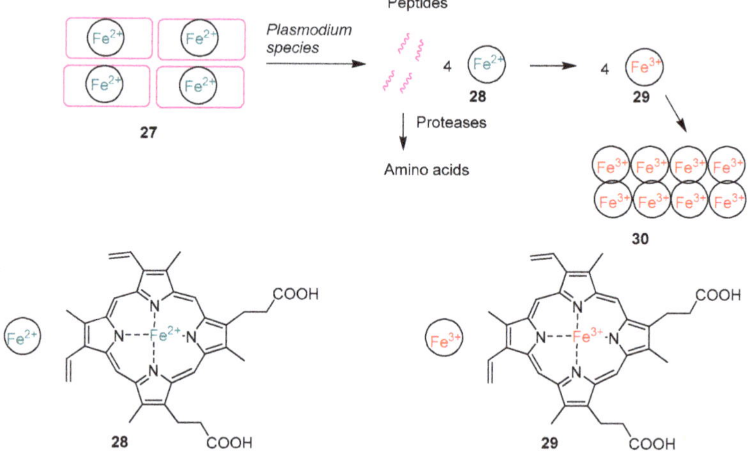

Figure 10. Digestion of hemoglobin (**27**) to give ferroprotoporphyrin (**28**), which spontaneously oxidizes to ferriprotoporphyrin IX (**29**). Precipitation of hemozoin (**30**) prevents the toxic effects of **29** [76,77].

Heme is formed in the food vacuole of the parasite when hemoglobin is digested to supply the parasite with amino acids. The pH of the food vacuole varies from 5.0 to 5.2, 1 to 2 units lower than that of the cytosol in the parasite. Consequently the neutral base will diffuse into the food vacuole where it accumulates after protonation [76]. After cleavage of hemoglobin the ferroprotoporphyrin **28** remains. In the ferrprotophorphytin the Fe^{2+} is quickly oxidized to Fe^{3+} affording ferriprotoporphyrin IX (**29**), which will provoke formation of reactive oxygen species unless inactivated [60]. Under physiologic conditions the ferriprotophorphyrin IX will polymerize to hemozoin (**30**), the characteristic malaria pigment seen in infected erythrocytes as dark spots. In the presence of chloroquine or quinine this polymerization is prevented, leading to the death of the parasite [76,77].

Even though many isoquinoline antimalarial drugs are chiral, resolution is not performed. The biological target molecule for these compounds ferriprotoporphyrin IX, (**29**) is achiral meaning that the two enantiomers have the same affinity. In contrast the volume of distribution, clearance and adverse effects do depend on the stereochemistry [78].

2.1.8. Artemisinins

According to a legend a meeting between Mao Zedong and Ho Chi Minh led to a secret research program to find a new drug for controlling malaria among the soldiers fighting for North Vietnam in the Vietnam civil war [79]. The program was initiated in 1967 as Project 523. In contrast to the program set up by the Allies in 1939 this project resulted in a novel class of antimalarial drugs. The program was managed by the Chinese chemist You-You Tu, who in 2015 was awarded the Nobel Prize for the outcome. A starting point was taken in traditional Chinese medicine. Screening 2000 herb preparations afforded 640 hits. An extract from *Artemisia annua* L. (Asteraceae) showed promising but non-reproducible antiplasmodial effect (Figure 11).

Figure 11. *Artemisia annua.* (http://www.plantsoftheworldonline.org/taxon/urn:lsid:ipni.org:names:304416-2, accessed on 1 April 2021).

Studies of the ancient Ge Hong's A Handbook of Prescriptions for Emergencies (340 AD) revealed that extraction was performed by wringing out the juice of the plant in cold water. Inspired by this information the extracts were made at low temperature and an active extract was obtained. In 1972 a colorless crystalline substance was obtained and named quinghaosu, meaning the active principle in quinghao, the Chinese name for *A. annua*. Today the name artemisinin (Scheme 3, **31**) is preferred, at least outside China. Reduction of the lactone to an acetal afforded a mixture of the two dihydroartemisnins **32a** and **32b**, which turned out to be a better drug than the mother compound [80,81]. Artemether (**33**) and arteether (**34**) are lipid soluble. Artesunate (**35**) is soluble in water.

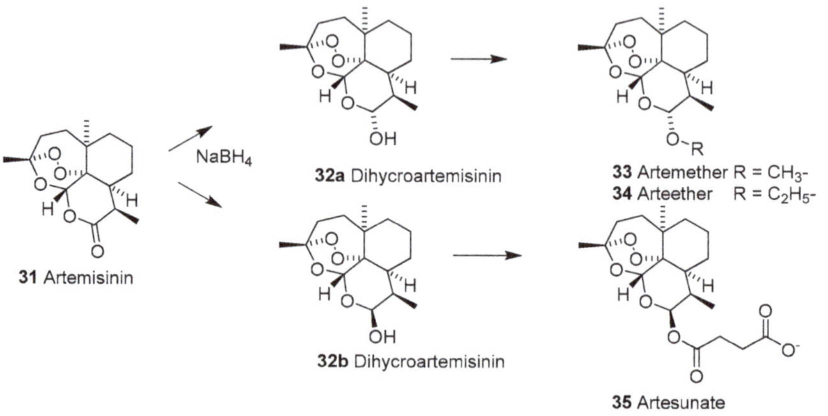

31 Artemisinin

NaBH$_4$

32a Dihycroartemisinin

32b Dihycroartemisinin

33 Artemether R = CH$_3$-
34 Arteether R = C$_2$H$_5$-

35 Artesunate

Scheme 3. Conversion of artemisinin to clinical used artemisinins: artemether (**33**), arteether (**34**) and artesunate (**35**) [80,81]. Artesunate and artemether are included in WHO list of essential medicines (https://www.who.int/medicines/publications/essentialmedicines/en/, accessed on 1 April 2021).

The first report in English of this amazing study appeared in 1979, some years after the end of the Vietnam War [82]. Later it has been realized that the structure of artemisinin had been previously published by Stefanivic et al. in 1972 at a Symposium on the Chemistry of Natural Products in New Delhi. The compound was named arteannuin. Unfortunately this group never realized the importance of the discovery [83]. This story emphasize that you do not become an outstanding scientist just because you are lucky. You have to realize that you have been lucky.

2.1.9. Mechanism of Action of Artemisinins

The pharmacophore of artemisinin is the 1,2,4-trioxane ring containing an endocyclic peroxide. Removal of this peroxide means loss of all antiplasmodial activity [77]. Whereas the artemisinins kills plasmodium parasites in low nanomolar concentration they are only toxic towards mammalian cells including fast proliferating cancer cell lines in micromolar concentration. Erythrocytes infected with plasmodium parasites have a rapid selective uptake of artemisinins in contrast to non-infected erythrocytes. After uptake the drug can be detected in the cytosol and the food vacuole. The Fe^{2+} in ferroprotoporphyrin (Scheme 4, **27**) might interact with the peroxide bridge in the artemisinins to give a reactive C-radical, which may alkylate the porphyrin skeleton preventing it from precipitation as heme (**28**) and consequently enabling the formation of ROS fatal for the parasite [77,84].

In addition to ferroprotoporphyrin (**27**) alkylation, the formed C-radicals also alkylate proteins and lipids, causing extensive cellular damage. The buildup of damaged proteins induces ER stress and attenuation of translation, which affords a lethal level of ubiquitinated proteins. Eventually merozoite death occurs. This mechanism of action differentiates artemisinins from other malaria drugs, which typically have a well-defined target [85]. The selectivity of the artemisinins originates in the Fe^{2+} needed for generation of the C-radicals.

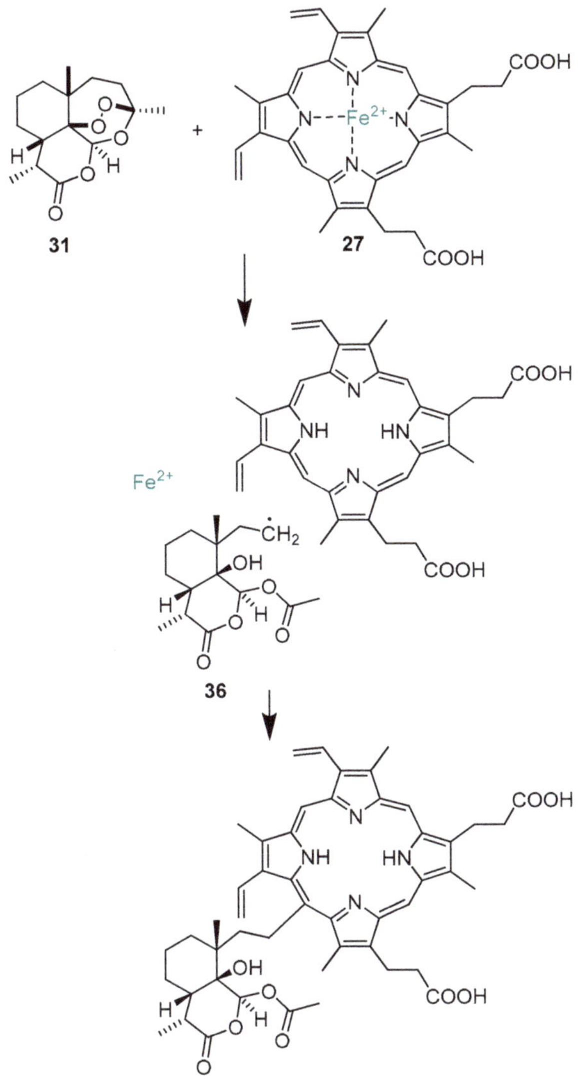

Scheme 4. The iron(II) ion of ferroprotoporphyrin (**27**) activates artemisinin (**31**) to give a reactive C-radical (**36**), which might alkylate ferroporotporphyrin (**27**) preventing precipitation [84]. The exact mechanism of alkylation, C-radial formation and the possibility of degradation of other biomolecules are still debated [77].

2.1.10. Clinical Use of Artemisinins

Poor absorbance of artemisinin (**31**) makes artesunate (**35**) the preferred drug among the arteminsinins. Artemether (**33**) and artesunate (**35**) are both prodrugs of dihydroartemisnin (**32a** and **32b**). Artesuante is soluble in water and consequently preferred. However, artemether can have advantages by rectal administration. Because of very rapid killing of parasites the artemisinins are very efficient for treatment of falciparum malaria [7,81].

In the presence of dihydroartemisin in the blood the number of parasites will be reduced 10,000 fold for each asexual cell cycle (approximately 36 h), whereas antimalarial antibiotics such as tetracycline only reveal a 10-fold reduction of parasitaemia. This rapid reduction is in particular advantageous for treatment of falciparum malaria since death might occur within hours after the first symptoms appear [7]. A drawback of the artemisinins is a very short half-live of about 1 hour. Rapid elimination from the blood also increases the risk of recrudescence [7,85]. In contrast a compound like chloroquine has a half-life of 30–60 days, proguanil 16 hours, primaquine 6 hours and lumefantrine 86 hours [7]. The short half-live of the arteminisins creates a risk for a long period with low blood level and consequently a high risk for development of resistance. To avoid this risk artemisinin combination therapy (ACT) is recommended meaning that artemisinins are given with drugs possessing a long half-live such as sulfadoxine (**21**)-pyrimethamine, mefloquine (Figure 12, **38**), amodiaquine (**39**) or lumefantrine (**23**) [7]. ACT has now become first choice for treatment of uncomplicated falciparum malaria [85]. The use of artemisinins has thus become an important tool for malaria control.

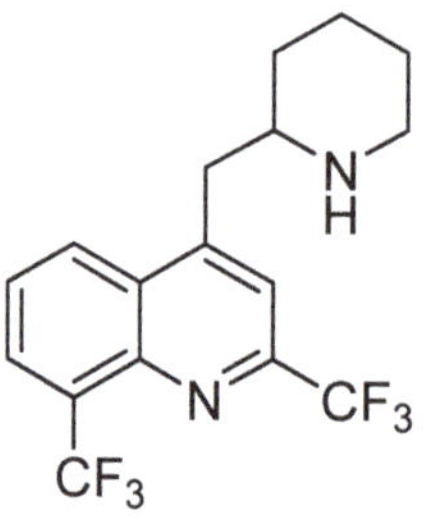

38 Mefloquine

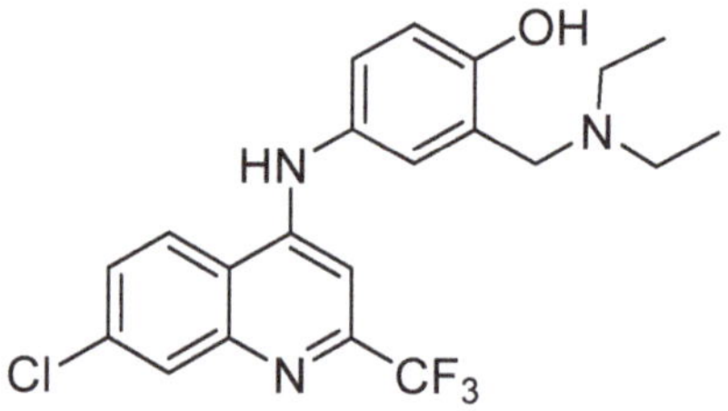

39 Amodiaquine

Figure 12. Synthetic antimalarial drugs. Mefloquine and amodiaquine are included in WHO list of essential medicines. (https://www.who.int/medicines/publications/essentialmedicines/en/, accessed on 1 April 2021).

2.1.11. Resistance toward Artemsinins

In spite of the precautions artemisinin resistance defined a decreased sensitivity for artemisinins have occurred in South East Asia from China to India, in Equitorial Guinea and Uganda in Africa [85]. Considering the burden of malaria in Africa the development on this continent is particularly alarming. Parasites resistant to artemisinins are characterized by the presence of *PfK13* mutations. The Kelch 13 proteins are substrate for an E3-ligase, which bind to phosphatidylinositol-3-kinase and marks it for ubiquitination and eventually cleavage. Mutations in Kelch13 decrease the ubiquitination and consequent an increased level of Kelch13, thereby an increase level of phosphatidylinositol-3-kinase and finally of the product phosphatidylinositol-3-phosphate. An increased level of this lipid makes the parasite less sensitive to oxidative damages [85]. In spite of the decreased response ACT still is the first choice for treatment of uncomplicated falciparum malaria.

2.1.12. Sustainable Supply of Artemisnin

WHO has recommended ACT as the drug of first choice for treatment of malaria. The recognition has created an increased demand for the agents. The current source is cultivated *A. annua*. Attempts to increase the artemisinin yield have afforded two cultivars with about 2% contents. Unfortunately some of the established varieties are not stable over generations. Attempts to produce artemisinin by bioengineering using cell cultures have not been successful at the present. Heterologous production of precursors which by semi-synthesis can be converted into the target compounds has only been partially commercially successful [86].

2.1.13. New Drugs for Treatment and Prophylaxis of Malaria

The Medicine for Malaria Venture, a non-profit, private-public partnership was established in 1999. The current portfolio contains a number of promising drug candidates like arterolane (OZ277, Figure 13, **40**), artefenomel (OZ439, **41**) and ferroquine (**30**) [73,74]. Artefenomel (**40**) and arterolane (**39**), like the artemisinins, contain a 1,2,4-trioxane ring and has the same mechanism [87]. The advantages of these drugs are longer half-lives in the body, 4 hours for arterolane [87] and 23 hours for artefenomel [73,87] in contrast to the 1 hour half-life of artemisinins. Long time exposure to the drugs is expected to make Kelch13 mutated parasites more sensitive [73]. Arterolane (**40**) is presently used clinically in combination with primaquine in India [73,87]. Other 1,2,4-trioxanes are also tested.

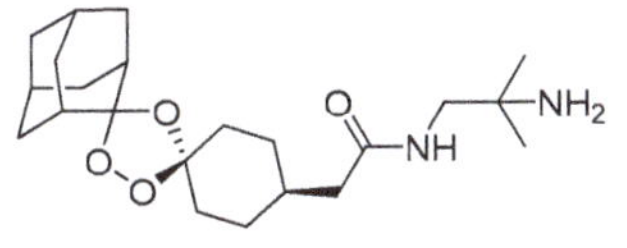

40 Arterolane

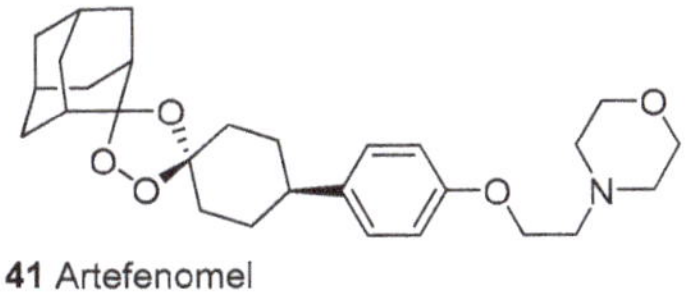

41 Artefenomel

Figure 13. Recent antimalarial drugs expected to overcome resistance among mutated parasites.

Both ferroquine (**26**) and artefenomel (**41**) are synthetic drugs, which can be prepared on a large scale.

Other drugs under development and the status of the development by Medicine for Malaria Venture can be seen on their homepage https://www.mmv.org/research-development/mmv-supported-projects, accessed on 1 April 2021. The ultimate goal of the project is to find at drug that can cure malaria in a single dose.

3. Onchocerciasis

Onchocerca volvulus a nematode living in in nodules under the skin of the patient is the causative agent for onchocerciasis. The nodules are large granulomas formed by tissue reaction around adult worms. Each nodule might host several worms of both sexes. The male worms have a length of 2 to 5 cm and the females 35 to 70 cm. The pathogenic organisms are microfilariae produced in numbers of 500 to 1500 per day from each female. The female, however, only can produce microfilariae if it lives symbiotic with bacteria belonging to the genus *Wolbachia*. The females are productive in about 10 years and the microfilariae live about 1 to 2 years. The microfilariae have a length of 300 μm. Severely infected patients may carry 2000 microfilariae in 1 mg upper dermis, whereas the microfilariae are seldom found in the blood or in other body fluids. Patients infected with *Onchocerca* nematodes suffer from dermatitis conditions such as leopard skin and depigmentation. These alterations of the skin are caused by reactions to dead or dying microfilariae [88]. Fibroblast proliferation leads to fibrosis and elastic fibers are replaced with hyalinized scar tissue. Some scars might also be caused by scratching. In advanced stages the skin resembles skin of very old individuals. Microfilariae might be visible in the cornea of patients. Dead microfilariae provoke inflammation leading to punctuate (snowflake) keratitis. Chronic inflammation causes opacification of the cornea. If not treated onchocerciasis might provoke optic nerve atrophy and chorioretinitis leading to loss of sight, hence the name river blindness [88,89]. The disease was first described as craw-craw in 1875 [88,89].

The life cycle (Figure 14) of the parasite involves as a vector blackfly belonging to the genus *Simulium*. The fly becomes infected with microfilariae when taking a blood meal. A new blood meal from another person might transfer the parasites to this person. The female flies taking blood meals restrict their flight within a few km from rivers. Thus, the transmission is limited to areas near rivers explaining the name river blindness. The original hotspots of the disease were river banks in Africa and Yemen. The slave trade probably also brought the disease to Central and South America [88]. A particular nasty consequence is that the disease makes the fertile soils often found on river banks inhabitable. The importance of *Simulium* blackflies as vectors was described in 1927 [89].

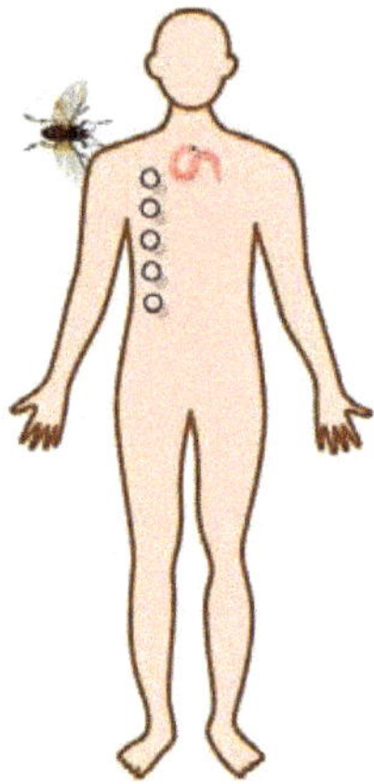

Figure 14. Life cycle of *Onchocerca volvus*. The patient is infected by a blood meal from a blackfly belonging to the genus *Similium*. The nematode (worm) is situated in nodules in the skin. The male nematode is up to 5 cm and the female nematode up to 70 cm long. The female worms produce a number of microfilariae, which cause the symptoms [88,89]. Like in the case of malaria transmission is performed by an insect.

More than 99% of persons infected with onchocerciasis are found in Sub-saharan Africa. Isolated foci are found in Latina America and in Yemen. The transmission has been stopped in Ecuador, Columbia, Mexico and parts of Guatemala [88].

3.1. Ivermectin

For decades scientists at the Kitasato Institute, Japan, have grown unusual microorganisms and isolated new compounds from the culture media leading to the discovery of the antimicrobial pyrindicin, the protein kinase C inhibitor staurosporin and the proteasome inhibitor lactacystin. In 1973s. Omura became head of the Kitasato Institute and initiated a collaboration with MSD Research Laboratory, At the Kitasato Institute extraordinary microorganisms were cultivated and bioactivity of isolated compounds screened for in vivo bioactivity. The most promising compounds were forwarded to MSD research laboratory for in vivo tests. A product isolated from the broth of the soil bacteria *Streptomyces avermectinus* was tested in mice infected with the nematode *Nematospiroides dubius*. The product showed excellent anthelmintic activity with no toxic effects on mice. The active principles turned out to be eight avermectins (Figure 15, **42–49**). The most active compounds appeared to be the compounds of the B-series containing a 5-hydroxyl group [90].

Reduction of the C-22-C-23 double bond improved the activity. A mixture of 80% hydrogenated avermectin B_{1a} and B_{1b} was named ivermectin (**50**). The mixture possesses a broad activity against a broad range of nematodes. The agent killed external and internal parasites in horses, cattle, pigs, sheep and heartworms in dogs. The uncompelled antiparasitic effects led to creation of a new term endectocide to describe compounds, which are capable of killing a wide variety of parasitic and health-threatening organisms both inside and outside the body [90–92]. Ivermectin interacts by preventing closure of glutamate and to a minor extent GABA-gated ion channels leading to hyperpolarization of the neural membrane and eventually killing of the parasite. The adult nematodes in general are not killed by the drugs. However, after administration of the drug female worms are prevented from releasing pathogenic amounts of microfilariae for 12 months even though the half live of the drug in humans is only 12 to 36 hours. GABA and glutamate receptors only occur in the central nervous system in mammalian but are commonplace in insects, nematodes and ticks. Ivermectin is unable to penetrate the blood-brain barrier in mammals, explaining the selectivity of the drug [90]. Ivermectin causes microfilariae to disappear rapidly from the skin in patients suffering from onchocerciasis. Apparently ivermectin prevents disarming the immune system enabling the host to kill microfilariae. The drug has to be taken for 18 years since even though the production of microfilariae is suppressed the adult worms are not killed. Ivermectin removes the symptoms of onchocerciasis since microfilariae are removed from the skin and after some time from the eye preventing further damage [90].

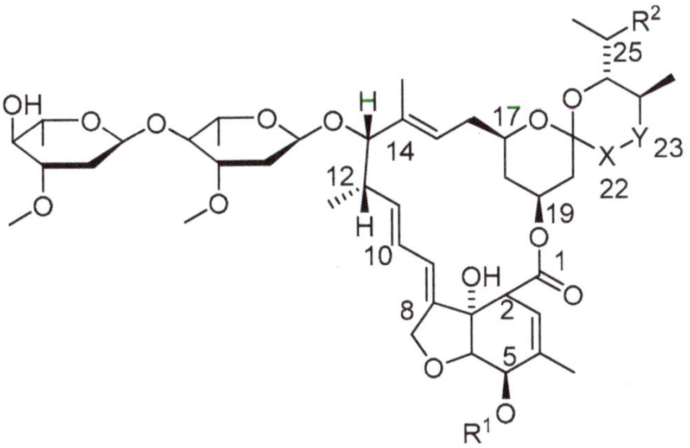

42 Avermectin A_{1a} $R^1 = CH_3$, $R^2 = C_2H_5$ X-Y = CH=CH
43 Avermectin A_{1b} $R^1 = CH_3$, $R^2 = CH_3$ X-Y = CH=CH
44 Avermectin A_{2a} $R^1 = CH_3$, $R^2 = C_2H_5$ X-Y = $CH_2CH(\alpha OH)$
45 Avermectin A_{2b} $R^1 = CH_3$, $R^2 = CH_3$ X-Y = $CH_2CH(\alpha OH)$
46 Avermectin B_{1a} $R^1 = H$, $R^2 = C_2H_5$ X-Y = CH=CH
47 Avermectin B_{1b} $R^1 = H$, $R^2 = CH_3$ X-Y = CH=CH
48 Avermectin B_{2a} $R^1 = H$, $R^2 = C_2H_5$ X-Y = $CH_2CH(\alpha OH)$
49 Avermectin B_{2b} $R^1 = H$, $R^2 = CH_3$ X-Y = $CH_2CH(\alpha OH)$
50 Ivermectin $R^1 = H$, $R^2 = C_2H_5$ X-Y = CH_2CH_2 (80 %)
$R^1 = H$, $R^2 = CH_3$ X-Y = CH_2CH_2 (20 %)

Figure 15. Avermectins and ivermectin [90]. The glycoside side chain is attached to C-13.

Additional anthelmintic macrolactones, the milbemycins, were found in the growth medium of *S.hygroscopicus*. The milbemycins possess the same macrocyclic lactone skeleton as avermectins but are not oxygenated at C-13 and consequently they are not glycosides [93].

3.1.1. Macrocyclic Latones Veterinary Use

Since 1981 Merck & Co. has marketed a number of macrolactone formulations for veterinary use. Some analogues like eprinomectin (Figure 16, **51**) [94] and selamectin (**52**) [95] (Revolt, Zoaetis) are used for oral as well as topical application. Selamectin is a semisynthetic derivative prepared by deglycosylation of doramectin, hydrogenation of the C-22-C23 double bond, transforming C5 to a ketone and reacting it with hydroxylamine [96]. In contrast to ivermectin selamectin has no toxic effects on dogs and cats. Eprinomectin is not found in the milk from cows. Macrocyclic lactones of the ivermectin type have become the most used drugs for treatment of parasites in cattle, sheep, and pets in USA and UK [91].

In depth description of the different macrocyclic lactones, use and properties have been presented by Vercruy and Rew [97].

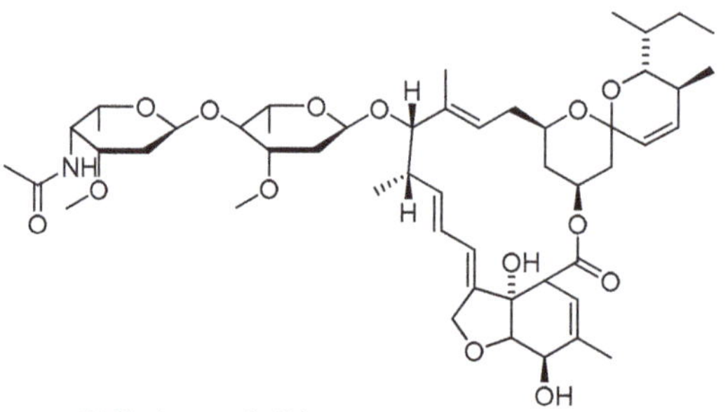

51 Eprinomectin B1a 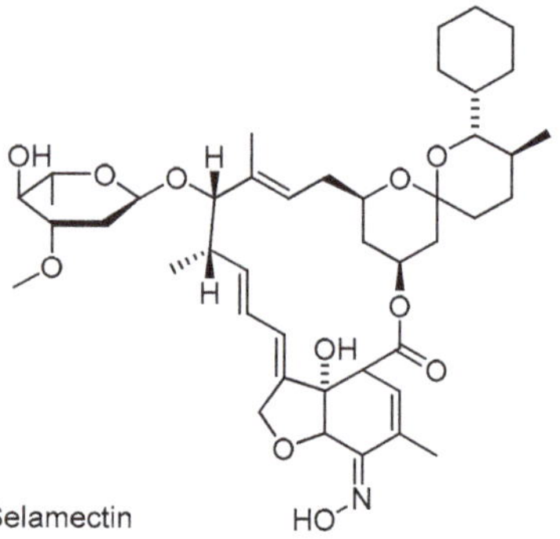   **52** Selamectin

Figure 16. Macrolactone endectocides [97].

3.1.2. Ivermectin for Human Use

After approval of the drug by the French Medical Agency in 1987 the drug was donated free of charge under the name of Mectizan by Merck & Co. Inc. (Fort Kenneworth, NJ, USA) to NGO organizations for treatment of onchocerciasis in as long as needed in amounts as much as needed [90]. The drawback of ivermectin is that treatment for keeping patients free from symptoms has to last for up to 18 years. Estimates of the number of infected persons in 1995 varied from 17 to 35 million cases globally. Not less than 0.27 million patients were blinded with 99% in in Sub-Saharan Africa [98,99]. Massive clinical trials in Africa have revealed that only sporadic side effects are observed and that the drug only needs to be administered orally once a year to keep the patient free from symptoms and to interrupt transmission.

The African Program for Onchocerciasis Control was established in 1995 to administer mass drug administration to 19 African countries. The program was closed in 2015. The number of patients was in 2017 estimated to 1.34 million [98]. The final evaluation stated that onchocerciasis had been bee reduced in all African countries including conflict areas. Success of the program has been obtained in Sudan, Mali, Senegal, Burundi Malawi, Nigeria and Ethiopia where onchocerciasis is almost eliminated [99]. In western Uganda onchocerciasis was observed among 88% of persons older than 19 years in 1991 in Kabarole and Kyenjojo districts. An investigation in 2012 found no positive cases [99,100]. The activities have been taken over by WHO African Region Expanded Special Program for Elimination of Neglected Tropical Diseases. The program is estimated to prevent 17.4 million from suffering from the symptoms of onchocerciasis. Combined use of mectizan and vector control afforded better results than only distribution of drugs [99]. The Onchocerciasis Elimination Program for the Americas has succeeded in eliminating the disease in Amerika except in a cross-border region deep in the Amazon forest between Venezuela and Brasilia. About 30,000 people are at risk. The mass drug administration of ivermectin thus has severely reduced the burden of onchocerciasis [89,99]. Appearance of resistance have called for new combination therapies including ivermectin [101].

3.1.3. Moxidectin

In 2018 FDA approved moxidectin (Figure 17, **53**) for treatment of onchocerciasis [69].

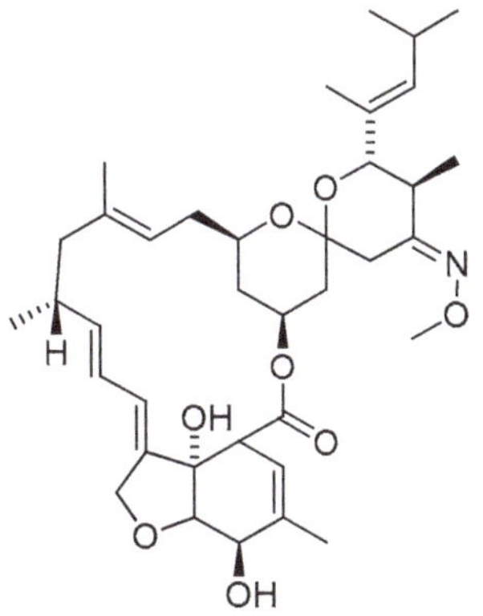

53 Moxidectin

Figure 17. Moxidectin (**53**) [102].

Moxidectin is a semisynthetic milbemycin. Milbemycins are not oxygenated at C-13 and consequently the compound does not possess the disaccharide side chain of avermectins. The absence of the carbohydrate moiety makes the compound more lipophilic. [93,98,103]. Moxidectin prepared from nemadectin isolated from the broth of *Streptomyce cyaneogriseus* subsp. *Noncyanogenous*. The 5-hydroxyl group of nemadectin is protected, the 23-hydroxy group oxidized, the methoxime introduced and the 5-hydroxy protecting group removed [103]. Moxidectin has been approved for treatment of onchocerciasis in humans and possesses a stronger and longer lasting suppression of microfilarial expression than ivermectin. Consequently it might be a better tool for elimination of onchocerciasis [98].

4. Lymphatic Filariasis

The pathogenic organisms in lymphatic filariasis caused are the nematodes *Wuchereria bancrofti, Brugia malayi* or *B. timori*. Lymphatic filariasis is characterized by lymphedema. After some years the edema may become non-pitting with thickening of the skin and loss of skin elasticity. Further progression leads to elephantitis [88] characterized by large disabling edemas. Over a billion people in more than 80 countries are threatened by this disease. The causative nematode *W. bancrofti* is found in tropical regions lf Africa, Asia, the Americas and the Pacific. The disease is particular frequent in hot and humid climates. *B. malayi* dominates in South-east Asia and South-west India. *B. tumori* is limited to islands in eastern Indonesia [88]. The life cycle of the parasite is similar to that of *Onchocerca* nematodes. The worms are found in lymphatic tissue of the human host. The female *W. bancrofti measures* 100 mm long and 0.25 mm in diameter. The microfilariae are released into the lymphatic vessels and from here they enter the veins. Female mosquitoes ingest the microfilariae and transmit them to a new human host by taking a new blood meal. The principal vector for *W. bacrofti* is *Culex quinquefasciatus*. In Africa *Anofeles* sp. also can transmit *W. bancrofti*. *B. malayi* uses bot *Mansonia* and *Anopheles* sp. as vectors. The time of development into the mosquito is about 12 days. In the human host the nematode might live and produce microfilariae for 20 years but the average lifespan is much shorter. The microfilariae have a lifespan of about 1 year. In the blood the density of microfilariae might reach 10,000/ml [88].

In addition to having reduced the burden of onchocerciasis ivermectin also is efficient in treating the parasitic diseases lymphatic filariasis. Ivermectin was approved for treatment of lymphatic filarisasis in 1998 by the French Medical Agency. Combination of albenzol and ivermectin was found 99% effective in elimination microfilaria from the blood. After GSK donated albenzol to the society and Merck extended the ivermectin donation to include lymphatic filariasis WHO initiated the Global Program to Eliminate Lymphatic Filariasis. From 2000 to 2020 7.7 billions doses against lymphatic filariasis have been distributed to 910 million persons in 68 countries reducing the number of people needing treatment from 1420 million to 597 million [104]. In the Americas Costa, Suriname, Trinidad and Tobago were removed from the WHO list of countries endemic for lymphatic filariasis in 2011 and it is hoped that Brazil, The Dominican Republic, Haiti and Guyana soon can be removed. WHO today recommends treatment with combination therapy consisting of ivermectin 200 µg/kg, diethylcarbamzine citrate (6 mg/kg) and albendazole 400 mg. In 2019 WHO has launched the ambition that onchocerciasis is eliminated in 2030 [98]. Sathoshi Omura and William C. Campbell together with You-you Tu shared the 2015 Nobel Prize in Medicine, the first two being mentioned for their work on ivermectin.

5. Cancer Diseases

A number of hallmarks are characteristic for cancer cells: (1) sustained proliferative signaling, (2) evading growth suppressors, (3) replicative immortality, (4) activated invasion and metastasis, (5) induction of angiogenesis, (6) resistance against cell death, (7) deregulation of cellular energetics and (8) not sensitive to immune destruction [105]. High rate of proliferation is characteristic for a number of cancer diseases and some of these are controlled by drugs targeting cells in the proliferative state. Spindle toxins offered a new efficient way of treating some cancer diseases like lymphomas and leukemia [106].The first spindle toxins vincristine and vinblastine were only discovered about 1960 [11].

5.1. Vinca Alkaloids

The leaves of Madagascan periwinkle [*Chataranthus roseus* (L.) G.Don Apocynaceae, named *Vinca rosea* by Linnaeus, Figure 18)] were used to control diabetes mellitus in traditional medicine on Jamaica.

Figure 18. *Catharanthus roseus* (The Museum of Natural Medicine & The Pharmacognostic Collection, University of Copenhagen).

A product named vinculin had been marketed in England for treatment of diabetes. Injection of extracts from the leaves showed no antidiabetic activity, but, unexpectedly, killed rabbits [107]. Autopsy revealed multiple abscesses, particular in the liver and kidneys [107]. The rabbits died from infections becoming fatal because of a suppressed immune system. The white blood count was severely reduced and neutrophil leukocytes had virtually disappeared. In addition the bone marrow was destroyed [108]. It was realized that the active principle was partitioned into an acidic aqueous solution from a benzene extract. Chromatographic separation on almost neutral alumina afforded a brown mixture, from which a crystalline hydrosulfate was obtained. Injection of a solution of the crystals into rats caused a severe drop in the number of granulocytes after 36 hours, which lasted for three to six days [107]. Structural elucidation revealed that he compounds belonged to a previously unknown type of dimeric indole alkaloids [109]. The relative configuration of vinblastine Figure 19, **54**) and vincristine (**55**) was published in 1964 [110] and confirmed by an X-ray analysis [111], which also revealed the absolute configuration. Among the 120 alkaloids present in the leaves only two are used in approved drugs [112]. Three semisynthetic analouges also have been approved vindesine (**56**), vinorelbine (**57**) and vinflunine (**59**) [11].

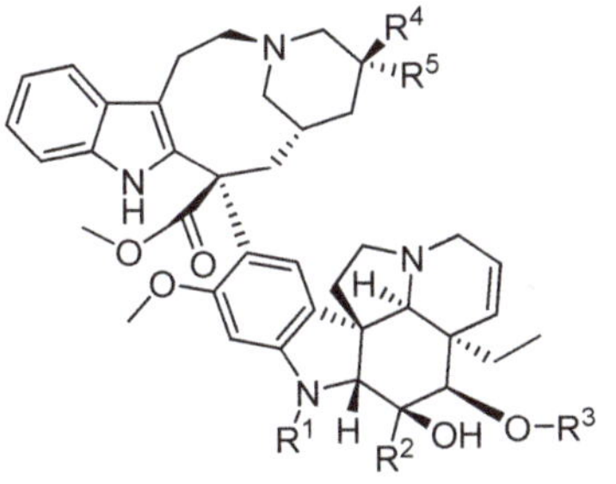

	R^1	R^2	R^3	R^4	R^5
54 Vinblastine	$-CH_3$	$-CO_2CH_3$	$-OCCH_3$	$-OH$	$-CH_2CH_3$
55 Vincristine	$-CHO$	$-CO_2CH_3$	$-OCCH_3$	$-OH$	$-CH_2CH_3$
56 Vindesine	$-CH_3$	$-CONH_2$	$-H$	$-OH$	$-CH_2CH_3$
57 Vinorelbine	$-CH_3$	$-CO_2CH_3$	$-OCCH_3$	$-H$	$-CH_2CH_3$
58 Vinflunine	$-CH_3$	$-CO_2CH_3$	$-OCCH_3$	$-H$	$-CF_2CH_3$

Figure 19. Vinblastine (**54**) and vincristine (**55**) are native natural products, whereas vindesine (**56**), vinorelbine (**57**) and vinflunine (**58**) are semisynthetic analogues [11].

Inspection of the structures of all the clinical used alkaloids reveals that they all possess the same carbon skeleton. Vincristine (**55**) is a formamide of dihydroindole, and vinblastine (**54**) is the *N*-methyldihydroindole. A formation of a bond from the α-methylester carbon of catharanthine (Scheme 5, **60**) and the carbon *ortho* to the methoxy group of vindoline (**59**) created the dimeric carbon skeleton. The biosynthetic reaction is catalyzed by α-3′.4′-anhydrovinbalse synthase (Scheme 5) [22]. The dimerization is a complex reaction, which is described in details elsewhere.

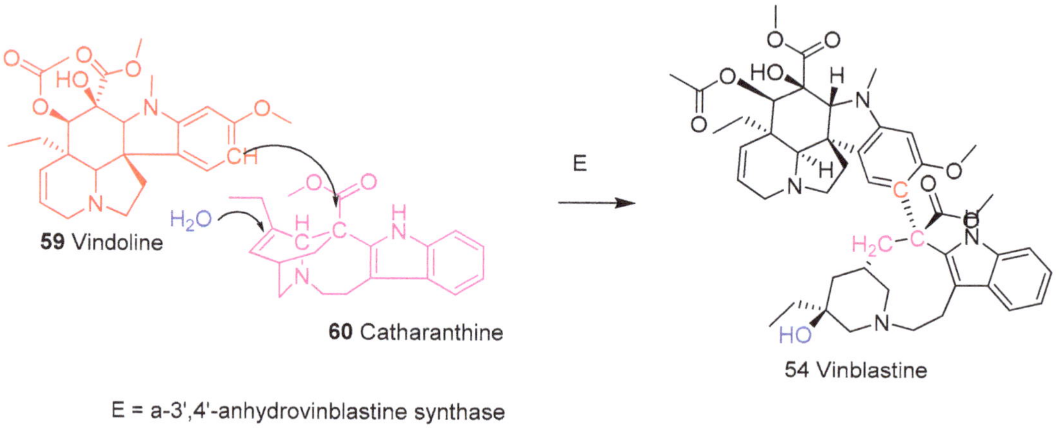

Scheme 5. Dimerization of indolalkaloids to give vincaalkaloids. For details see [22].

5.1.1. Clinical Use

The ability to reduce the white blood count inspired researchers to investigate the chemotherapeutic potential of the alkaloids. Preclinical and clinical trials afforded overwhelming results and soon vincristine was introduced as combination therapy for treatment of childhood acute lymphoblastic leukemia (ALL) and non-Hodgkin's lymphoma (NHL). The alkaloid is particularly effective in the treatment of hematological and lymphatic neoplasms, as well as of solid tumors (i.e., breast cancer, non-small cell lung cancer) [11]. Vincristine (**55**) is used for treatment of Philadelphia chromosome-negative acute lymphoblastic leukemia, β-cell lymphomas, metastatic melanoma, estrogen-receptor negative breast cancer, glioma, colorectal cancer, non-Hodgkin's lymphoma, Hodgkin's lymphoma, neuroblastoma, rhabdomyosarcsoma, multiple myeloma and Wilm's tumor [11]. A combination of vincristine (**55**) and actinomycin C for treatment of Wilm's tumor increased 2-year relapse free survival from 55% to 81% of patients with advanced disease [113]. Vincristine (**55**) is not as efficient for treatment of cancer diseases in adults as among children. However, it has some effect towards non-Hodgkin's lymphoma and Hodgkin's disease. In general the effect towards hematological malignancies are better than those against in solid tumors [113]. In particular neurotoxicity forms an upper limit for doses of vincristine (**55**) [113]. The convincing effects of vincristine for treatment of hematological malignancies particular in the childhood have led to characterization of the drug as a wonder drug [114].

The main side effects of vincristine (**55**) are neurotoxicity causing polyneuropathy [113]. Many signs of vincristine intoxication disappear weeks or months after termination of treatment. However, long term effects and even permanent damages are described [113]. The presently obtained experiences with the semisynthetic vinflunine (**58**) suggest that the side effects of this drugs are easier to manage [11].

Vinblastine (**54**) is used for treatment of leukemia, non-Hodgkin's and Hodgkin's lymphoma, breast cancer, nephroblastome, Ewing carcoma, small-cell lung cancer, testicular carcinoma and germ cell tumors [11].

5.1.2. Mechanism of Action

Microtubules are key components of the cytoskeleton. They are crucial in the development and maintenance of cell shape, in the transport of vesicles, mitochondria, cell signaling and in mitosis [114]. The microtubules are polymers of α- and β-tubulin arranged in filamentous tubes. Both tubulins have a molecular weight of 55 kDa. The tubules might be many μm long [114]. Microtubule dynamics means that tubules are assembled and disassembled constantly [113]. During mitosis, first the chromosomes attach via their kinetochores to the spindle during prometaphase. Second complex movements of the chromosomes bring them in properly aligned positions for separations and finally during the anaphase microtubules drag each of the duplicated chromosomes towards the spindle poles. In the telophase they are incorporated in the two daughter cells (Figure 20) [114]. The spindle toxins or microtubule targeting agents inhibit microtubules function and thereby prevent the cell going from metaphase into anaphase. A prolonged senescence G1-state causes death of rapidly dividing cells by apoptosis [11,113,114]. The different spindle toxins bind to different binding sites in the tubulins. Vinblastine binds to the β-tubulin at the vinca-binding domain [11]. In low but clinical relevant concentrations the mitosis is prevented but at higher concentrations de-polymerization is observed [114].

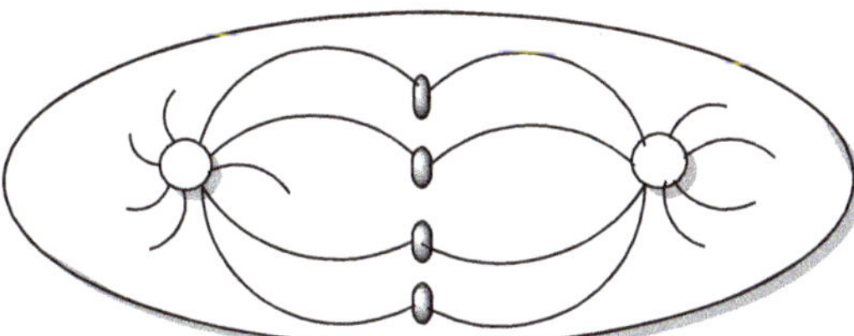

Figure 20. Cell division. In the prometaphase the nuclear envelope breaks down and the chromosomes condense and microtubules contact chromosomes at the kinetochores. In the early metaphase most chromosome have congressed to the equator. In the anaphase, depicted in Figure 20, the duplicated chromosomes have separated and move toward the spindle pools and the cell divides to form two daughter cells. In the telophase the separated chromosomes have reached the spindle poles and the cells divides into daughter cells [114]. The vinca alkaloids destroy the mitotic spindle and consequently block the mitosis [113,114].

5.1.3. Analogues of Vincalkaloids

At the present five vinca alkaloids are in clinical use: the two natural products vinblastine (**54**) and vincristine (**55**) [115], and the semisynthetic compounds vindesine (**56**), vinorelbine (**57**), and vinflunine (**56**) [11]. Vindesine (**56**) is only approved in some countries, where it is used for pediatric solid tumors, malignant melanoma, blast crisis of chronic myeloid leukemia, acute lymphocytic leukemia, metastatic colorectal-, breast-, renal- and esophageal-carcinomas. Vinorelbine (**57**) has a wide antitumor spectrum of activity such as advanced breast cancer, advanced/metastatic non-small-cell lung cancer and rhabdomyosarcoma. The newest approved analogue vinflunine (**58**) is used for metastatic and advanced urothelial cancer after failure of platin-containing therapy [11]. In order to avoid resistance the vinca alkaloids are preferentially used in combination therapy with drugs possessing a different mechanism of action. Such drugs could be cychlophosphamide, an alkylating agent, doxorubicin and bleomycin both interacting with DNA, and cisplatin causing DNA damage [11]. In order to improve the effect new formulations such as incorporation into microspheres, nanoparticles and liposomes are being developed. In addition, new prodrugs are developed to target the agents toward the tumors [4,11,116].

5.1.4. Sustainable Production of Vinca Alkaloids

It takes 2 tons of leaves to produce 1 g of vincristine and 500 kg to produce 1 kg of vinblastine (**54**). It is estimated that the production of 1 kg of vinblastine costs 1 million dollars and 1 kg of vincristine costs 3.5 million [112]. The extreme costs of production have provoked several attempts to produce the alkaloids by cell cultures. However, extensive metabolic cross talks and shuttling of about 35 intermediates synthesized via 30 enzymatic reactions in four different types of tissue and five different sub-cellular compartments have prevented the development of commercially interesting biotechnological production methods [117]. Hopefully a breakthrough can afford a feasible method for kg scale production of vinca alkaloids [115,118].

6. Conclusions

Natural products and natural products chemistry have had an impressive impact on drug development and consequently on our society [3]. An outstanding example of this is the fight against malaria. Until the discovery of quinine mankind had no efficient drug to control malaria, which was a devastating disease putting a severe burden not only on the population in tropical Africa but also on Europe and Asia [12,29]. It might be considered as a irony of history that the breakthrough to solve the problem of malaria was found in South America, where malaria was not a problem [33,35]. Open minded studies revealed that the bark of trees used to prevent muscle shivering turned out to be a warhead to combat malaria. Botanical studies let to the discovery of the mother trees, and a new genus, *Cinchona*, was created to enabling scientific description of the species with the healing bark [34,35]. A recipe for the use of the bark as a drug against malaria, Schedula Romana, was developed [34]. However, besides the unpleasant swallowing of the crude bark, an additional severe drawback of the bark was the variation of the level of the active principles. Isolation of the active alkaloid, quinine, in a pure state enabled for the first time development of an efficient drug with reproducible effects [31,49]. As with many other natural products development of a sustainable production of quinine became an absolute requirement for successful use of the drug. Until the Japanese occupation during the Second World War the sustainable production was maintained by plantations on Java and to a smaller extent in India. When the supply from Java was discontinued scientists developed synthetic antimalarial drugs, in particular chloroquine, which actually outcompeted quinine [57]. The development of chloroquine, however, was not the final stroke against malaria. After a few decades the parasites developed resistance and new drugs were desperately needed. In spite of all the disasters caused by the Vietnam War military need for a new drug led to the discovery of the artemisinins [80]. Artesunate and artemether are used in combination with other antimalarial drugs to postpone development of resistance. However, reports of lower sensitivities of parasites for artemisinin combination therapy are reported [85].

Onchocerciasis made fertile riverbanks in Africa inhospitable because of the threat of river blindness. The search for a drug to combat parasites in livestock led to the discovery of the macrocyclic lactone ivermectin [92]. Ivermectin is a very efficient drug for elimination of parasites in animals. The pathogenic agents in onchocerciasis in animals as well as in humans are *Onchocerca* nematodes. Fortunately ivermectin was found to be efficient for removal of the symptoms not only in animals but also in humans. Under the name of mectizan ivermectin is now offered free of charge for mass distribution in Africa and other parts of the world to combat onchocerciasis [90,99]. Even though impressive progresses have been made for relieving the world for the burden of onchocerciasis is still not eliminated but hopefully new drugs such as moxidectin and continued mass distribution will further reduce the burden [98].

Beside parasitic diseases cancer causes mortality and reduced life quality for a large part of mankind [119,120]. Until the 1960s hematological malignancies particular in the childhood had a high death rate [114]. Attempts to understand the antidiabetic effects of leaves from *C. rosesus* revealed that active agent had a poor effect on diabetes but severely suppressed the number of white blood cells. Intensive studies revealed that the active principles were spindle toxins and this opened a new path for treatment of diseases caused by fast proliferating cells [11,108]. Further studies revealed the structures of the compounds and enabled large scale isolation. However, as with many other natural products, sustainable supply in an economic feasible way is still an important restriction for the use of the compounds. Maybe studies may enable heterologous methods for production of the compounds in cell cultures [112,115,117]. Today many cancer diseases for which no cure existed before discovery of the spindle toxins are cured efficiently.

The importance of natural products chemistry has been revealed by discovery of drugs with new mechanisms of actions like the spindle toxins for treatment of cancer diseases, the trioxolanes for treatment of malaria and the macrocyclic lactones for treatment of parasitic diseases. Many new drugs developed by inspiration from traditional uses turns out not have a curative effect on diseases for which their mother plants are used [33,35,108]. Open minded researchers realized that the active principles enabled treatment of incurable diseases, as in the case of quinine and vinca alkaloids. A drawback of natural products, however, is their limited supply. Sustainable production on a ton scale may enforce the use of alternative sources. Bioengineering of other organisms offers alternative possibilities [121] but are only possible if the biosynthesis is known in details [86,122].

In some cases preparation of derivatives of the natural product affords optimized drugs. This has been the case for the artemisinins like artemether (**33**) and artesunate (**35**) [7,81], macrocyclic lactones like moxidectin (**53**) [93] and vinca alkaloids like vindesine (**56**), vinorelbine (**57**), and vinflunine (**56**) [11]

Modern techniques have now enabled cultivation of fungi, marine organisms, endophytes and bacteria, which might open new paths for discovery of new families of natural products [5,6,123,124].

Above all an absolute requirement for new wonder drugs is devoted interdisciplinary research. As illustrated in the cases of vinca alkaloids and artemisinin, the isolation of an active principle is only the first step on the long and complicated path to registration of a new drug.

Funding: This research received no external funding.

Institutional Review Board Statement: Not applicable.

Informed Consent Statement: Not applicable.

Conflicts of Interest: The author declares no conflict of interest.

References

1. Cragg, G.M.; Grothaus, P.G.; Newman, D.J. Impact of natural products on developing new anti-cancer agents. *Chem. Rev.* **2009**, *109*, 3012–3043. [CrossRef]
2. Newman, D.J.; Cragg, G.M. Natural products as drugs and leads to drugs: The historical perspective. In *Natural Product Chemistry for Drug Discovery*; Royal Society of Chemistry: London, UK, 2010; pp. 3–27.
3. Newman, D.J.; Cragg, G.M. Natural Products as Sources of New Drugs over the Nearly Four Decades from 01/1981 to 09/2019. *J. Nat. Prod.* **2020**, *83*, 770–803. [CrossRef] [PubMed]
4. Franzyk, H.; Christensen, S.B. Targeting Toxins towards Tumors. *Molecules* **2021**, *26*, 1292. [CrossRef] [PubMed]
5. Hyde, K.D.; Xu, J.; Rapior, S.; Jeewon, R.; Lumyong, S.; Niego, A.G.A.T.; Abeywickrama, P.D.; Aluthmuhandiram, J.V.S.; Brahamanage, R.S.; Brooks, S.; et al. The amazing potential of fungi: 50 ways we can exploit fungi industrially. *Fungal Divers.* **2019**, *97*, 1–136.
6. Blunt, J.W.; Carroll, A.R.; Copp, B.R.; Davis, R.A.; Keyzers, R.A.; Prinsep, M.R. Marine natural products. *Nat. Prod. Rep.* **2018**, *35*, 8–53. [CrossRef] [PubMed]
7. White, N.J. Malaria. In *Mansons Tropical Diseases*, 23rd ed.; Farrar, J., Hotez, P., Junghanss, T., Kang, G., Lalloo, D., White, N.J., Eds.; Elsevier: Amsterdam, The Netherlands, 2014; Volume 9, pp. 523–601.
8. Seca, A.M.L.; Pinto, D.C.G.A. Plant secondary metabolites as anticancer agents: Successes in clinical trials and therapeutic application. *Int. J. Mol. Sci.* **2018**, *19*, 263. [CrossRef]
9. Trendafilova, A.; Moujir, L.M.; Sousa, P.M.C.; Seca, A.M.L. Research Advances on Health Effects of Edible Artemisia Species and Some Sesquiterpene Lactones Constituents. *Foods* **2020**, *10*, 65. [CrossRef]
10. Omura, S. Ivermectin: 25 years and still going strong. *Int. J. Antimicrob. Agents* **2008**, *31*, 91–98. [CrossRef] [PubMed]
11. Martino, E.; Casamassima, G.; Castiglione, S.; Cellupica, E.; Pantalone, S.; Papagni, F.; Rui, M.; Siciliano, A.M.; Collina, S. Vinca alkaloids and analogues as anti-cancer agents: Looking back, peering ahead. *Bioorg. Med. Chem. Lett.* **2018**, *28*, 2816–2826. [CrossRef]
12. Singer, C.; Underwood, E.A. *A Short History of Medicine*, 2nd ed.; Clarendon Press: Oxford, UK, 1962.
13. Wöhler, F. Ueber künstliche Bildung des Harnstoffs. *Ann. Phys. Chem.* **1828**, *12*, 253–257. [CrossRef]
14. McKie, D. Wöhlers "synthetic" urea and the rejection of vitalism: A chemical legend. *Nature* **1944**, *153*, 608–610. [CrossRef]
15. Nicolaou, K.C.; Sorenson, E.J. *Classics in Total Synthesis: Targets, Strategies, Methods*; Wiley: Weinheim, Germany, 1996.
16. Nicolaou, K.C.; Snyder, S.A. *Classics in Total Synthesis II: More Targets, Stratgeis and Methods*; Wiley: Weinheim, Germany, 2003.
17. Nicolaou, K.C.; Montagnon, T. *Molcules that Changed the World*; Wiley: Weinheim, Germany, 2008; p. 366.

18. Nicolaou, K.C.; Chen, J.S. *Classics in Total Synthesis III: Further Targets, Strategies, Methods*; Wiley: Weinheim, Germany, 2011.

19. Mahdi, J.G.; Mahdi, A.J.; Bowen, I.D. The historical analysis of aspirin discovery, its relation to the willow tree and antiproliferative and anticancer potential. *Cell Prolif.* **2006**, *39*, 147–155. [CrossRef]

20. Marko, V. *Fraom Aspitin to Viagara*; Springer Nature: Cham, Schwitzerland, 2020.

21. Appendino, G.; Pollastro, F. Plants: Revamping the oldest source of medicines with modern science. In *Natural Product Chemistry for Drug Discovery*; Royal Society of Chemistry: London, UK, 2010; pp. 140–173.

22. Dewick, P.M. *Medicinal Natural Products*, 3rd ed.; John Wiley and Sons Ltd.: Chicester, UK, 2009.

23. Hayashi, Y. Time Economy in Total Synthesis. *J. Org. Chem.* **2021**, *86*, 1–23. [CrossRef] [PubMed]

24. Seeman, J.I. The Woodward-Doering/Rabe-Kindler total synthesis of quinine: Setting the record straight. *Angew. Chem. Int. Ed.* **2007**, *46*, 1378–1413. [CrossRef] [PubMed]

25. Barber, J. Photosystem II: The water splitting enzyme of photosynthesis and the origin of oxygen in our atmosphere. *Q. Rev. Biophys.* **2016**, *49*, e14. [CrossRef]

26. Oro, J. Chemical evolution and the origin of life. *Adv. Space Res.* **1983**, *3*, 77–94. [CrossRef]

27. Nye, E.R. Alphonse Laveran (1845-1922): Discoverer of the malarial parasite and Nobel laureate, 1907. *J. Med. Biogr.* **2002**, *10*, 81–87. [CrossRef]

28. Yoeli, M. Sir Ronald Ross and the evolution of malaria research. *Bull. N. Y. Acad. Med.* **1973**, *49*, 722–735.

29. Hay, S.I.; Guerra, C.A.; Tatem, A.J.; Noor, A.M.; Snow, R.W. The global distribution and population at risk of malaria: Past, present, and future. *Lancet Infect. Dis.* **2004**, *4*, 327–336. [CrossRef]

30. WHO. *World Malaria Report*; WHO: Geneva, Switzerland, 2019.

31. Curtin, P.D. The End of the "White Man's Grave"? Nineteenth-Century Mortality in West Africa. *J. Interdiscip. Hist.* **1990**, *21*, 63–88. [CrossRef]

32. Rocco, F. *The Miraculous Fever-Tree. Malaria, Medicine and the Cure that Changed the World*; HarperCollins: London, UK, 2003.

33. Naranjo, P. Epidemic hecatomb in the New World. *Allergy Proc.* **1992**, *13*, 237–241. [CrossRef]

34. Gachelin, G.; Garner, P.; Ferroni, E.; Trohler, U.; Chalmers, I. Evaluating Cinchona bark and quinine for treating and preventing malaria. *J. R. Soc. Med.* **2017**, *110*, 31–40. [CrossRef]

35. Gachelin, G.; Garner, P.; Ferroni, E.; Trohler, U.; Chalmers, I. Evaluating Cinchona bark and quinine for treating and preventing malaria. *J. R. Soc. Med.* **2017**, *110*, 73–82. [CrossRef]

36. Meshnick, S.R.; Dobson, M.J. The History of Antimalarial Drugs. In *Antimalarial Chemotherapy*; Rosenthal, P.J., Ed.; Humana Press: Towota, NJ, USA, 2001; pp. 15–25.

37. McHale, D. The cinchona tree. *Biologist* **1986**, *33*, 45–53.

38. Pabst, G. *Köhler's Medizinal-Pflanzen*; Verlag von Fr. Eugen Köhler: Gera-Untermhaus, Germany, 1887; Volume 1.

39. Nair, K.P.P. Cinchona (*Cinchona* sp.). In *The Agronomy and Economy of Important Tree Crops of the Developing World*; Nair, K.P.P., Ed.; Elsevier: Amsterdam, The Netherlands, 2010.

40. Cross, S.H. *Quinine, Production and Marketing*; Bureau of Foreign and Domestic Commerce: Washington, DC, USA, 1924.

41. Williams, D. Clements Robert Markham and the Introduction of the Cinchona Tree into British India, 1861. *Georg. J.* **1962**, *128*, 431–442. [CrossRef]

42. Efferth, T.; Banerjee, M.; Paul, N.W.; Abdelfatah, S.; Arend, J.; Hamdoun, S.; Hamm, R.; Hong, C.; Kadioglu, O.; Nass, J.; et al. Biopiracy of natural products and good bioprospecting practice. *Phytomedicine* **2016**, *23*, 166–173. [CrossRef] [PubMed]

43. Cragg, G.M.; Katz, F.; Newman, D.J.; Rosenthal, J. The impact of the United Nations Convention on Biological Diversity on natural products research. *Nat. Prod. Rep.* **2012**, *29*, 1407–1423. [CrossRef]

44. Rivero-Cruz, I.; Cristians, S.; Ovalle-Magallanes, B.; Mata, R. Mexican copalchis of the Rubiaceae family: More than a century of pharmacological and chemical investigations. *Phytochem. Rev.* **2019**, *18*, 1435–1455. [CrossRef]

45. Maldonado, C.; Barnes, C.J.; Cornett, C.; Holmfred, E.; Hansen, S.H.; Persson, C.; Antonelli, A.; Rønsted, N. Phylogeny Predicts the Quantity of Antimalarial Alkaloids within the Iconic Yellow Cinchona Bark (Rubiaceae: *Cinchona calisaya*). *Front. Plant Sci.* **2017**, *8*, 391. [CrossRef]

46. Canales, N.A.; Gress Hansen, T.N.; Cornett, C.; Walker, K.; Driver, F.; Antonelli, A.; Maldonado, C.; Nesbitt, M.; Barnes, C.J.; Roensted, N. Historical chemical annotations of *Cinchona* bark collections are comparable to results from current day high-pressure liquid chromatography technologies. *J. Ethnopharmacol.* **2020**, *249*, 112375. [CrossRef] [PubMed]

47. Martin, W.E.; Gandara, J.A. Alkaloid content of Ecuadoran and other American cinchona barks. *Bot. Gaz.* **1945**, *107*, 184–199. [CrossRef]

48. Karamanou, M.; Tsoucalas, G.; Pantos, K.; Androutsos, G. Isolating colchicine in 19th century: An old drug revisited. *Curr. Pharm. Des.* **2018**, *24*, 654–658. [CrossRef]

49. Mitcham, J.C. Patrolling the White Man's Grave: The impact of disease on Anglo-American naval operations against the slave trade, 1841–1862. *North. Mar.* **2010**, *20*, 37–56.

50. Holland, J.H. Ledger Bark and Red Bark. *Bull. Misc. Inf.* **1932**, *1932*, 1–17. [CrossRef]

51. Streller, S.; Roth, K. Eine Rinde erobert die Welt. *Chem. Unserer Zeit* **2012**, *46*, 228–247. [CrossRef]

52. Prelog, V. Häfliger, Über China-alkaloide. *Helv. Chim Acta* **1950**, *23*, 257–2578.

53. Baird, J.K. Effectiveness of antimalarial drugs. *N. Engl. J. Med.* **2005**, *352*, 1565–1577. [CrossRef]

54. Kaufman, T.S.; Ruveda, E.A. The quest for quinine: Those who won the battles and those who won the war. *Angew. Chem. Int. Ed.* **2005**, *44*, 854–885. [CrossRef] [PubMed]
55. Gensini, G.F.; Conti, A.A.; Lippi, D. The contributions of Paul Ehrlich to infectious disease. *J. Infect.* **2007**, *54*, 221–224. [CrossRef]
56. Winau, F.; Westphal, O.; Winau, R. Paul Ehrlich—In search of the magic bullet. *Microbes Infect.* **2004**, *6*, 786–789. [CrossRef] [PubMed]
57. Coatney, G.R. Pitfalls in a discovery: The chronicle of chloroquine. *Am. J. Trop. Med. Hyg.* **1963**, *12*, 121–128. [CrossRef]
58. Sharma, V.P. Re-emergence of malaria in India. *Indian J. Med. Res.* **1996**, *103*, 26–45. [PubMed]
59. Wellems, T.E.; Plowe, C.V. Chloroquine-resistant malaria. *J. Infect. Dis.* **2001**, *184*, 770–776. [CrossRef] [PubMed]
60. Ridley, R.G. Medical need, scientific opportunity and the drive for antimalarial drugs. *Nature* **2002**, *415*, 686–693. [CrossRef] [PubMed]
61. Frosch, A.E.P.; Venkatesan, M.; Laufer, M.K. Patterns of chloroquine use and resistance in sub-Saharan Africa: A systematic review of household survey and molecular data. *Malar. J.* **2011**, *10*, 116. [CrossRef] [PubMed]
62. Trape, J.F. The public health impact of chloroquine resistance in Africa. *Am. J. Trop. Med. Hyg.* **2001**, *64*, 12–17. [CrossRef]
63. Ecker, A.; Lehane, A.M.; Clain, J.; Fidock, D.A. PfCRT and its role in antimalarial drug resistance. *Trends Parasitol.* **2012**, *28*, 504–514. [CrossRef]
64. Shanks, G.D.; Biomndo, K.; Hay, S.I.; Snow, R.W. Changing patterns of clinical malaria since 1965 among a tea estate population located in the Kenyan highlands. *Trans. R. Soc. Trop. Med. Hyg.* **2000**, *94*, 253–255. [CrossRef]
65. Sidhu, A.b.S.; Verdier-Pinard, D.; Fidock, D.A. Chloroquine resistance in *Plasmodium falciparum* malaria parasites conferred by pfcrt mutations. *Science* **2002**, *298*, 210–213. [CrossRef] [PubMed]
66. Dorsey, G.; Fidock, D.A.; Wellems, T.E.; Rosenthal, P.J. Mechanisms of Quinoline Resistance. In *Antimalarial Chemotherapy*; Rosenthal, P.J., Ed.; Humana Press: Totowa, NJ, USA, 2001; pp. 153–172.
67. Dahl, E.L.; Shock, J.L.; Shenai, B.R.; Gut, J.; DeRisi, J.L.; Rosenthal, P.J. Tetracyclines specifically target the apicoplast of the malaria parasite *Plasmodium falciparum*. *Antimicrob. Agents Chemother.* **2006**, *50*, 3124–3131. [CrossRef]
68. Berman, J.; Brown, T.; Dow, G.; Toovey, S. Tafenoquine and primaquine do not exhibit clinical neurologic signs associated with central nervous system lesions in the same manner as earlier 8-aminoquinolines. *Malar. J.* **2018**, *17*, 407. [CrossRef] [PubMed]
69. Araujo, E.; Braun, M.-G.; Dragovich, P.S.; Converso, A.; Nantermet, P.G.; Roecker, A.J.; Dhar, T.G.M.; Haile, P.; Hurtley, A.; Merritt, J.R.; et al. To Market, To Market—2018: Small Molecules. In *2019 Medicinal Chemistry Reviews—2019*; Bonson, J.J., Ed.; American Chemical Society: Washington, DC, USA, 2019; Volume 54, pp. 469–596.
70. Christensen, S.B. Negelected Diseases. In *Textbook of Drug Desing and Discovery*, 4th ed.; Krogsgaard-Larsen, P., Strømgård, K., Madsen, U., Eds.; CRC Press: Boca Raton, FL, USA, 2010; pp. 314–358.
71. Uddin, T.; McFadden, G.I.; Goodman, C.D. Validation of putative apicoplast-targeting drugs using a chemical supplementation assay in cultured human malaria parasites. *Antimicrob. Agents Chemother.* **2018**, *62*, 1–17. [CrossRef]
72. Schlitzer, M. Malaria Chemotherapeutics Part I: History of Antimalarial Drug Development, Currently Used Therapeutics, and Drugs in Clinical Development. *ChemMedChem* **2007**, *2*, 944–986. [CrossRef] [PubMed]
73. Mathews, E.S.; John, A.R.O. Tackling resistance: Emerging antimalarials and new parasite targets in the era of elimination. *F1000Research* **2018**, *7*, 1170. [CrossRef] [PubMed]
74. Blank, B.R.; Gonciarz, R.L.; Talukder, P.; Gut, J.; Legac, J.; Rosenthal, P.J.; Renslo, A.R. Antimalarial Trioxolanes with Superior Drug-Like Properties and In Vivo Efficacy. *ACS Infect. Dis.* **2020**, *6*, 1827–1835. [CrossRef] [PubMed]
75. Biot, C.; Nosten, F.; Fraisse, L.; Ter-Minassian, D.; Khalife, J.; Dive, D. The antimalarial ferroquine: From bench to clinic. *Parasite* **2011**, *18*, 207–214. [CrossRef] [PubMed]
76. Tilley, L.; Loria, P.; Foley, M. Chloroquine and Other Quinoline Antimalarials. In *Antimalarial Chemotherapy*; Rosenthal, P.J., Ed.; Humana Press: Totowa, NJ, USA, 2001; pp. 87–121.
77. Haynes, R.K.; Cheu, K.-W.; N'Da, D.; Coghi, P.; Monti, D. Considerations on the Mechanism of Action of Artemisinin Antimalarials: Part 1—The 'Carbon Radical' and 'Heme' Hypotheses. *Infect. Dis. Drug Targets* **2013**, *13*, 217–277. [CrossRef] [PubMed]
78. Brocks, D.R.; Mehvar, R. Stereoselectivity in the pharmacodynamics and pharmacokinetics of the chiral antimalarial drugs. *Clin. Pharmacokinet.* **2003**, *42*, 1359–1382. [CrossRef]
79. Cooper, R.; Deakin, J. *Botanical Miracles. Chemistry of Plants that Changed the World*; CRC Press: Boca Raton, FL, USA, 2016; p. 271.
80. Tu, Y. The discovery of artemisinin (qinghaosu) and gifts from Chinese medicine. *Nat. Med.* **2011**, *17*, 1217–1220. [CrossRef]
81. Klayman, D.L. Qinhaosu (Artemisinin): An Antimalarial Drug from China. *Science* **1985**, *228*, 1049–1055. [CrossRef]
82. Group, Q.A.C.R. Antimalaria studies on qinhaosu. *Chin. Med. J.* **1979**, *92*, 811–816.
83. Vajs, V.; Jokic, A.; Milosavljevic, S. Artemisinin Story from the Balkans. *Nat. Prod. Commun.* **2017**, *12*. [CrossRef]
84. Christensen, S.B.; Bygbjerg, I.C. Neglected Diseases. In *Textbook of Drug Design and Discovery*, 5th ed.; Strømgård, K., Krogsgaard-Larsen, P., Madsen, U., Eds.; CRC Press: Boca Raton, FL, USA, 2017; pp. 347–368.
85. Lu, F.; He, X.-L.; Richard, C.; Cao, J. A brief history of artemisinin: Modes of action and mechanisms of resistance. *Chin. J. Nat. Med.* **2019**, *17*, 331–336. [CrossRef]
86. Ikram, N.K.B.K.; Simonsen, H.T. A Review of Biotechnological Artemisinin Production in Plants. *Front. Plant Sci.* **2017**, *8*, 1966. [CrossRef] [PubMed]

87. Giannangelo, C.; Siddiqui, G.; De Paoli, A.; Anderson, B.M.; Edgington-Mitchell, L.E.; Charman, S.A.; Creek, D.J. System-wide biochemical analysis reveals ozonide antimalarials initially act by disrupting Plasmodium falciparum haemoglobin digestion. *PLoS Pathog.* **2020**, *16*, e1008485. [CrossRef] [PubMed]

88. Simonsen, P.E.; Fischer, P.U.; Hoerauf, A.; Weil, G.J. The Filariases. In *Manson's Tropical Diseases*; Farrar, J., Hotez, P., Junghanss, T., Kang, G., Lalloo, D., White, N., Eds.; Elservier: Amsterdam, The Netherlands, 2014; pp. 737–765.e5.

89. Hopkins, A.D. Onchocerciasis: Status and progress towards elimination. *CAB Rev.* **2020**, *15*, 021. [CrossRef]

90. Omura, S. A Splendid Gift from the Earth: The Origins and Impact of the Avermectins (Nobel Lecture). *Angew. Chem. Int. Ed.* **2016**, *55*, 10190–10209. [CrossRef] [PubMed]

91. Laing, R.; Gillan, V.; Devaney, E. Ivermectin—Old Drug, New Tricks? *Trends Parasitol.* **2017**, *33*, 463–472. [CrossRef]

92. Omura, S.; Crump, A. The live an times of ivermectin—A success story. *Nat. Rev. Microbiol.* **2004**, *2*, 984–989. [CrossRef]

93. Prichard, R.; Menez, C.; Lespine, A. Moxidectin and the avermectins: Consanguinity but not identity. *Int. J. Parasitol. Drugs Drug Resist.* **2012**, *2*, 134–153. [CrossRef]

94. Shoop, W.L.; Egerton, J.R.; Eary, C.H.; Haines, H.W.; Michael, B.F.; Mrozik, H.; Eskola, P.; Fisher, M.H.; Slayton, L.; Ostlind, D.A.; et al. Eprinomectin: A novel avermectin for use as a topical endectocide for cattle. *Int. J. Parasitol.* **1996**, *26*, 1237–1242. [CrossRef]

95. Bishop, B.F.; Bruce, C.I.; Evans, N.A.; Goudie, A.C.; Gration, K.A.F.; Gibson, S.P.; Pacey, M.S.; Perry, D.A.; Walshe, N.D.A.; Witty, M.J. Selamectin: A novel broad-spectrum endectocide for dogs and cats. *Vet. Parasitol.* **2000**, *91*, 163–176. [CrossRef]

96. Lee, S.R.; Kim, H.U.; Kang, H.M.; Choi, I.S. Preparation of Selamectin intermediate and process for Selamectin. KR2019056863A, 2019.

97. Vercruysse, J.; Rew, R.S. *Macrocyclic Lactones in Antiparasitic Therapy*; CABI Publishing: Wallingford, UK, 2002.

98. Milton, P.; Hamley, J.I.D.; Walker, M.; Basanez, M.-G. Moxidectin: An oral treatment for human onchocerciasis. *Expert Rev. Anti Infect. Ther.* **2020**, *18*, 1067–1081. [CrossRef]

99. Brattig, N.W.; Cheke, R.A.; Garms, R. Onchocerciasis (River Blindness)—More than a Century of Research and Control. *Acta Trop.* **2020**, *218*, 105677. [CrossRef]

100. Lakwo, T.L.; Garms, R.; Rubaale, T.; Katabarwa, M.; Walsh, F.; Habomugisha, P.; Oguttu, D.; Unnasch, T.; Namanya, H.; Tukesiga, E.; et al. The disappearance of onchocerciasis from the Itwara focus, western Uganda after elimination of the vector Simulium neavei and 19 years of annual ivermectin treatments. *Acta Trop.* **2013**, *126*, 218–221. [CrossRef] [PubMed]

101. Sainas, S.; Dosio, F.; Boschi, D.; Lolli, M.L. Targeting Human Onchocerciasis: Recent Advances Beyond Ivermectin. *Ann. Rep. Med. Chem.* **2018**, *51*, 1–38.

102. Verdu, J.R.; Cortez, V.; Martinez-Pinna, J.; Ortiz, A.J.; Lumaret, J.-P.; Lobo, J.M.; Sanchez-Pinero, F.; Numa, C. First assessment of the comparative toxicity of ivermectin and moxidectin in adult dung beetles: Sub-lethal symptoms and pre-lethal consequences. *Sci. Rep.* **2018**, *8*, 14885. [CrossRef] [PubMed]

103. Awasthi, A.; Razzak, M.; Al-Kassas, R.; Harvey, J.; Garg, S. Analytical profile of moxidectin. *Profiles Drug Subst. Excipients Relat. Methodol.* **2013**, *38*, 315–366.

104. WHO. Lymphatic Filariasis. 2020. Available online: https://www.who.int/news-room/fact-sheets/detail/lymphatic-filariasis (accessed on 1 April 2021).

105. Björkling, F.; Moreira, J.; Stenvang, J. Anticancer Agents. In *Textbook of Drug Design and Discovery*, 5th ed.; Strømgaard, K., Krogsgaard-Larsen, P., Madsen, U., Eds.; CRC Press: Boca Raton, FL, USA, 2017; pp. 369–386.

106. Xie, S.; Zhou, J. Harnessing Plant Biodiversity for the Discovery of Novel Anticancer Drugs Targeting Microtubules. *Front. Plant. Sci.* **2017**, *8*, 720. [CrossRef]

107. Noble, R.L.; Beer, C.T.; Cutts, J.H. Role of chance observations in chemotherapy: *Vinca rosea*. *Ann. N. Y. Acad. Sci.* **1958**, *76*, 882–894. [CrossRef]

108. Noble, R.L. The Discovery of the Vinca Alkaloids—Chemotherapeutic-Agents Against Cancer. *Biochem. Cell Biol. Biochim. Biol. Cell.* **1990**, *68*, 1344–1351. [CrossRef]

109. Neuss, N.; Gorman, M.; Svoboda, G.H.; Maciak, G.; Beer, C.T. Vinca alkaloids. III. Characterization of leurosine and vincaleukoblastine, new alkaloids from *Vinca rosea*. *J. Am. Chem. Soc.* **1959**, *81*, 4754–4755. [CrossRef]

110. Neuss, N.; Gorman, M.; Hargrove, W.; Cone, N.J.; Biemann, K.; Buechi, G.; Manning, R.E. Vinca alkaloids. XXI. Structures of the oncolytic alkaloids vinblastine and vincristine. *J. Am. Chem. Soc.* **1964**, *86*, 1440–1442. [CrossRef]

111. Moncrief, W.J.; Lipscomb, W.N. Structure of leurocristine methiodide dihydrate by anomalous scattering methods; relation to leurocristine (vincristine) and vincaleukoblastine (vinblastine). *Acta Crystallogr.* **1966**, *21*, 322–331. [CrossRef]

112. Barrales-Cureño, H.J.; Reyes, C.R.; García, I.V.; Valdez, L.G.L.; De Jesús, A.G.; Ruíz, J.A.C.; Herrera, L.M.S.; Caballero, M.C.C.; Magallón, J.A.S.; Perez, J.E.; et al. Alkaloids of Pharmacological Importance in *Catharanthus roseus*. In *Alkaloids. Their Importance in Natur and Human Life*; Kurek, J., Ed.; IntechOpen: London, UK, 2019.

113. Gidding, C.E.; Kellie, S.J.; Kamps, W.A.; de Graaf, S.S. Vincristine revisited. *Crit. Rev. Oncol. Hematol.* **1999**, *29*, 267–287. [CrossRef]

114. Jordan, M.A.; Wilson, L. Microtubules as a target for anticancer drugs. *Nat. Rev. Cancer* **2004**, *4*, 253–265. [CrossRef] [PubMed]

115. Almagro, L.; Fernandez-Perez, F.; Pedreno, M.A. Indole alkaloids from *Catharanthus roseus*: Bioproduction and their effect of human health. *Molecules* **2015**, *20*, 2973–3000. [CrossRef] [PubMed]

116. Brady, S.F.; Pawluczyk, J.M.; Lumma, P.K.; Feng, D.M.; Wai, J.M.; Jones, R.; Feo-Jones, D.; Wong, B.K.; Miller-Stein, C.; Lin, J.H.; et al. Design and synthesis of a pro-drug of vinblastine targeted at treatment of prostate cancer with enhanced efficacy and reduced systemic toxicity. *J. Med. Chem.* **2002**, *45*, 4706–4715. [CrossRef] [PubMed]

117. Verma, P.; Khan, S.A.; Parasharami, V.; Mathur, A.K. Biotechnological Interventions to Modulate Terpenoid Indole Alkaloid Pathway in *Catharanthus roseus* Using In Vitro Tools and Approaches. In *Cathararanthus roseus. Current Research and Future Prospects*; Naeem, M., Aftab, T., Masroor, M., Khan, A., Eds.; Springer International Publishing: Berlin/Heidelberg, Germany, 2017; pp. 247–275.
118. Li, C.; Galani, S.; Hassan, F.-U.; Rashid, Z.; Naveed, M.; Fang, D.; Ashraf, A.; Qi, W.; Arif, A.; Saeed, M.; et al. Biotechnological approaches to the production of plant-derived promising anticancer agents: An update and overview. *Biomed. Pharmacother.* **2020**, *132*, 110918.
119. Torre, L.A.; Bray, F.; Siegel, R.L.; Ferlay, J.; Lortet-Tieulent, J.; Jemal, A. Global cancer statistics, 2012. *CA Cancer J. Clin.* **2015**, *65*, 87–108. [CrossRef]
120. Bray, F.; Ferlay, J.; Soerjomataram, I.; Siegel, R.L.; Torre, L.A.; Jemal, A. Global cancer statistics 2018: GLOBOCAN estimates of incidence and mortality worldwide for 36 cancers in 185 countries. *CA Cancer J. Clin.* **2018**, *68*, 394–424. [CrossRef] [PubMed]
121. Arya, S.S.; Rookes, J.E.; Cahill, D.M.; Lenka, S.K. Next-generation metabolic engineering approaches towards development of plant cell suspension cultures as specialized metabolite producing biofactories. *Biotechnol. Adv.* **2020**, *45*, 107635. [CrossRef]
122. Atanasov, A.G.; Waltenberger, B.; Pferschy-Wenzig, E.-M.; Linder, T.; Wawrosch, C.; Uhrin, P.; Temml, V.; Wang, L.; Schwaiger, S.; Heiss, E.H.; et al. Discovery and resupply of pharmacologically active plant-derived natural products: A review. *Biotechnol. Adv.* **2015**, *33*, 1582–1614. [CrossRef] [PubMed]
123. Mayer, A.M.S.; Glaser, K.B.; Cuevas, C.; Jacobs, R.S.; Kem, W.; Little, R.D.; McIntosh, J.M.; Newman, D.J.; Potts, B.C.; Shuster, D.E. The odyssey of marine pharmaceuticals: A current pipeline perspective. *Trends Pharmacol. Sci.* **2010**, *31*, 255–265. [CrossRef] [PubMed]
124. Li, S.-J.; Zhang, X.; Wang, X.-H.; Zhao, C.-Q. Novel natural compounds from endophytic fungi with anticancer activity. *Eur. J. Med. Chem.* **2018**, *156*, 316–343. [CrossRef]

Review

Cannabis sativa: Interdisciplinary Strategies and Avenues for Medical and Commercial Progression Outside of CBD and THC

Jackson M. J. Oultram [1], Joseph L. Pegler [1], Timothy A. Bowser [2], Luke J. Ney [3], Andrew L. Eamens [1] and Christopher P. L. Grof [1,2,*]

[1] Centre for Plant Science, University of Newcastle, University Drive, Callaghan, NSW 2308, Australia; Jackson.Oultram@uon.edu.au (J.M.J.O.); Joseph.Pegler@uon.edu.au (J.L.P.); Andy.Eamens@newcastle.edu.au (A.L.E.)

[2] CannaPacific Pty Ltd., 109 Ocean Street, Dudley, NSW 2290, Australia; tim@cannapacific.com.au

[3] School of Psychological Sciences, University of Tasmania, Hobart, TAS 7005, Australia; luke.ney@utas.edu.au

* Correspondence: Chris.Grof@newcastle.edu.au; Tel.: +612-4921-5858

Abstract: *Cannabis sativa* (*Cannabis*) is one of the world's most well-known, yet maligned plant species. However, significant recent research is starting to unveil the potential of *Cannabis* to produce secondary compounds that may offer a suite of medical benefits, elevating this unique plant species from its illicit narcotic status into a genuine biopharmaceutical. This review summarises the lengthy history of *Cannabis* and details the molecular pathways that underpin the production of key secondary metabolites that may confer medical efficacy. We also provide an up-to-date summary of the molecular targets and potential of the relatively unknown minor compounds offered by the *Cannabis* plant. Furthermore, we detail the recent advances in plant science, as well as synthetic biology, and the pharmacology surrounding *Cannabis*. Given the relative infancy of *Cannabis* research, we go on to highlight the parallels to previous research conducted in another medically relevant and versatile plant, *Papaver somniferum* (opium poppy), as an indicator of the possible future direction of *Cannabis* plant biology. Overall, this review highlights the future directions of cannabis research outside of the medical biology aspects of its well-characterised constituents and explores additional avenues for the potential improvement of the medical potential of the *Cannabis* plant.

Keywords: *Cannabis sativa* (*Cannabis*); cannabinoids; tetrahydrocannabinol (THC); cannabidiol (CBD); cannabinoid receptors (CB_1 and CB_2); *Papaver somniferum* (opium poppy); secondary metabolites

Citation: Oultram, J.M.J.; Pegler, J.L.; Bowser, T.A.; Ney, L.J.; Eamens, A.L.; Grof, C.P.L. *Cannabis sativa*: Interdisciplinary Strategies and Avenues for Medical and Commercial Progression Outside of CBD and THC. *Biomedicines* **2021**, *9*, 234. https://doi.org/10.3390/biomedicines9030234

Academic Editors: Pavel B. Drašar and Raffaele Capasso

Received: 1 February 2021
Accepted: 23 February 2021
Published: 26 February 2021

Publisher's Note: MDPI stays neutral with regard to jurisdictional claims in published maps and institutional affiliations.

1. Introduction

Cannabis sativa (*Cannabis*) is arguably one of the world's most versatile crops. While the genetic origin and evolution of *Cannabis* is a long-standing and heavily debated topic [1–4], in broad terms, today, *Cannabis* can be separated into two distinct categories, specifically 'hemp' and 'marijuana'. Much like other agricultural crop commodities, *Cannabis* has been domesticated and bred for thousands of years to produce phenotypic and/or chemotypic traits of value to humans [2–5]. The chemotypic distinction between hemp and marijuana predominantly stems from the abundance of the principal psychoactive cannabinoid, Δ^9-tetrahydrocannabinol (THC), present in the plant as the acidic form, Δ^9-tetrahydrocannabinolic acid (THCA) [6]. To be considered hemp, *Cannabis* must possess a low percentage of THC relative to the total dry weight of flowers, with this low THC percentage varying from country to country. In order to be legally cultivated as hemp, the cultivated plants must possess less than 0.3% THC (*w/w*) in Canada [4,7] and China [8], whereas since 2001, the European Union determined that the THC content (*w/w*) of hemp must be below 0.2% [6].

Hemp has traditionally been bred as a source for textile products due to the strong, elongated bast fibres present in the phloem of the stem. More recently, the elevated cellulosic content of hemp cell walls has garnered interest in the plant as a source for the development of sustainable biofuel production [6]. Hempseed, and hempseed oil, have historically been utilised as a food source, with more contemporary research revealing their unique dietary value. In particular, the essential polyunsaturated fatty acids (PUFAs), linoleic acid (LA) and linolenic acid (LNA), comprise 50–70% and 15–25% of the total fatty acid content of hempseed, respectively; a 3:1 ratio promoted as nutritionally optimal [9–13]. PUFAs found in hempseed oil are incorporated into phospholipid bilayers and are integral to membrane fluidity and the maintenance of its permeability [14]. Moreover, the two proteins, edestin and albumin found in hempseed, contain rich amino acid profiles comparable to that of high-quality soybean and egg white [15]. Given the functions and importance of both fatty and amino acids, hempseed and hempseed oil may have some potential, albeit minor, for reducing the incidence of certain diseases, while in parallel conferring a range of health benefits [15–17]. Alternatively, marijuana has traditionally been bred for its recreational intoxication properties derived from the THCA-containing resin produced on the protruding secretory hair-like structures known as trichomes which are predominantly located on female reproductive parts of the *Cannabis* plant [18,19]. The sticky resin produced from these specialised epidermal glands is a rich mix of cannabinoid and non-cannabinoid constituents, numbering at least 104 and 441, respectively [20,21]. Most recently, two novel cannabinoids, namely Δ^9-tetrahydrocannabiphorol (Δ^9-THCP) and cannabidiphorol (CBDP), near identical in structure to THC and cannabidiol (CBD), respectively, were identified [22]. Notably, Δ^9-THCP was demonstrated to possess higher cannabimimetic activity than THC, and its recent discovery is therefore postulated as a potential candidate cannabinoid responsible for variation in pharmacological properties observed in uncharacterised *Cannabis* varieties. This also identifies the likelihood of secondary metabolites present in *Cannabis* resin that remain to be discovered.

In addition to possessing a range of phenotypic and chemotypic traits of interest to the textile, medicinal, food and energy industries as an agricultural crop, *Cannabis* is extremely versatile and hardy, hence the application of the colloquial term for this species, 'weed'. The phenotypic flexibility of *Cannabis* provides it with the capacity to adapt and survive a range of abiotic and biotic insults, such as drought [23], heavy metal stress [24], high temperature [25], poor soil nutrient content [3], high plant density [26], and stem damage from the larva of *Ostrinia nubilalis*, the European corn borer [27]. Tolerance to a range of abiotic stress conditions is exemplified by the tap root of *Cannabis* which is able to adapt to highly variable edaphic conditions, either penetrating deep (greater than 2 metres) into dry soil, or developing an extensive lateral root network in response to its growth in soil that has a high moisture content [26]. Further, the widespread legalisation of medicinal application and recreational use of *Cannabis* is driving the growth of diverse research programs encompassing the broad scope, from plant breeding to clinical trials. In the United States of America (USA), for example, to date, 33 states have approved the medicinal use of *Cannabis*, while 14 states and territories have legalised the recreational use of marijuana by adults. At the federal level in the USA, however, *Cannabis* remains a *'Schedule I Substance'*. In direct contrast to the heavy legislation of *Cannabis* in the USA, its direct neighbour, Canada, legalised the use of *Cannabis* across the country in 2018 under the *'Cannabis Act'* [28]. As the legislative approval of *Cannabis* use increases worldwide, there will be an increasing need for interdisciplinary research to characterise secondary metabolites of interest and to increase the production of *Cannabis* to meet the demand for medicinal and recreational products.

Currently, there exists an extant literature on the medical potential for the best characterised cannabinoids, THC and CBD [29–34]. Significantly less attention in medical research has been paid to the potential for the minor phytocannabinoids to treat illnesses, and there is still the need for methods to produce these cannabinoids cost-effectively for commercial production. In particular, the medical *Cannabis* industry faces significant challenges in multiple aspects of product development. For instance, THC is associated with multiple side effects, and furthermore, pharmaceutical-standard THC and CBD are expensive to produce. Due to these hurdles, many companies around the world which have attempted to capitalise on the increasing legality of *Cannabis* have been unsuccessful [35]. Therefore, here we review the current literature describing emerging research concerning the medical potential of the minor cannabinoids, as well as to outline the agricultural and production considerations that will be necessary to meet the needs of the growing medical market. Readers interested primarily in the effects of CBD and THC should consult any of the substantial reviews on these topics that are published elsewhere and referred to here in Section 2.2. It should also be noted that there are some recent review articles on the molecular targets of the minor cannabinoids [36,37], but to the best of our knowledge, no published review of the current literature has combined this research with the potential for improving *Cannabis* yield and extraction efficacy to make these possibilities economically and logistically pragmatic. This review therefore presents a novel, interdisciplinary perspective on the practical possibilities for improving the *Cannabis* species for its utilisation in the cannabinoid industry in the near future.

2. The Endocannabinoid System and Its Associated Molecular Targets

2.1. An Overview of the Endocannabinoid System

The discovery of the endogenous cannabinoid system followed the initial isolation [38] and synthesis [39] of the primary psychoactive compound in *Cannabis*, THC. Following on from this in the late 1980s, and into the early 1990s, two cannabinoid receptors, CB_1 and CB_2, were identified [40,41]. Surprisingly, it was discovered that CB_1 was highly abundant in the central nervous system (CNS), and in the CNS, CB_1 is one of the most profuse G protein-coupled receptors [42]. The identification of these two CB receptors subsequently led to the discovery of an endogenous receptor ligand termed arachidonylethanolamide (anandamide), a receptor ligand accurately predicted to exist based on the presence of the CB receptors themselves [43]. A second receptor ligand, 2-arachidonoylglycerol (2-AG) was later identified [44,45]. Anandamide and 2-AG are both synthesised from arachidonic acid. Synthesis of anandamide is complex, and therefore remains to be elucidated, though it is thought to occur largely via the cleavage of arachidonic acid by a phospholipase D from its membrane precursor, N-arachidonoyl phosphatidylethanolamine [46]. The synthesis of 2-AG occurs following the conversion of diacylglycerol by the metabolic enzyme, diacylglycerol lipase (DAGL). Hydrolysis of anandamide occurs via the enzyme activity of fatty acid amide hydrolase (FAAH), whereas 2-AG is hydrolysed by both FAAH and monoacylglycerol lipase (MAGL) [47]. Inhibition of these enzymes increases anandamide and 2-AG concentrations and has therapeutic potential [48–50]. Similarly, it is possible that modulation of precursory compounds of anandamide and 2-AG may have therapeutic potential [51].

Previous investigations into CB receptor distribution within the fetal, neonatal and adult human brain revealed that the CB receptors were primarily localised to areas responsible for; (1) higher cognitive function; (2) movement, and; (3) control of sensory and motor functions of the autonomic nervous system [52]. Protein crystallisation has revealed the structure of CB_1 [53] and CB_2 [54] to assist in the characterisation of the molecular binding of ligands, such as THC, and potentially other key cannabinoids, both naturally or synthetically produced. Using radiolabelled synthetic cannabinoids, it was shown that the highest density of cannabinoid binding, and thus CB receptor localisation, appeared in the basal ganglia, hippocampus and cerebellum [42]. Cannabinoids were shown to function on hippocampal presynaptic receptors, via regulating the release of γ-aminobutyric acid (GABA) to modulate higher cognitive functions, while also increasing the activity of p38 mitogen-activated protein kinases [55,56]. Similarly, GABA modulation in the basal ganglia, specifically the presynaptic striatal projection neuron axons and their termini, was found to be stimulated to differing degrees by either endocannabinoids or synthetic cannabinoids [57,58]. The binding of the CB_1 receptor by both endogenous and exogenous cannabinoids also modulates excitatory synaptic transmission in Purkinje cells located in the cerebellum [59–62]. Crucially, endocannabinoid signalling was recognised as the mediatory secondary messenger responsible for long-term potentiation, and depression [49,63], which are both fundamental to the control of synaptic transmission. CB_1 receptors and endocannabinoid signalling also interacts with other systems in the brain, such as the dopaminergic [64], and glucocorticoid [65] pathways, to modulate stress response and associative learning processes.

While early understanding of receptor distribution suggested exclusive 'central' aggregation in specific regions of the brain, it is now understood that there is a more extensive presence of CB_1 type receptors in peripheral tissues. Two CB_1 receptor isoforms have since been identified, both of which display distinct expression patterns in pancreatic β-cells and liver hepatocytes [66]. Antagonism of peripheral CB receptors located in skeletal muscles was shown to trigger glucose uptake, while simultaneously initiating lipid mobilisation in white adipose tissue [67]. Though the protein expression pattern of CB_1 does show some overlap with CB_2 in peripheral tissues, and conversely some CB_2 receptors are cerebrally positioned [68–72], peripheral receptors are predominantly CB_2 type receptors. Analysis of *CB_2* transcript levels has previously revealed its expression in the tonsils, spleen, and peripheral blood mononuclear cells, where further cell isolation showed detectable *CB_2* transcript levels in polymorphonuclear neutrophils (PMN), T4 cells, T8 cells, natural killer (NK) cells, macrophages, and B cells. However, at the protein level, the CB_2 receptor appears to be restricted to B cells [73]. Similarly, CB_2 receptor binding has been observed in other immune system regions, namely the lymph node cortex, as well as in the Peyer's patches, which are areas of B lymphocyte aggregation [74]. The expression and/or localisation of functional CB_2 protein has also been reported for mast cells, modulating their initial activation, or downregulating their activity post their initial activation, an activity change which can in turn provoke an anti-inflammatory response [75]. Anandamide and 2-AG, as well as their metabolic enzymes, are detectable in blood [76,77], hair [78–80], saliva [81–83], breast milk [84,85], and reproductive fluids [84,86]. Compounded with the peripheral anti-inflammatory response, CB_2 receptor agonists can mediate peripheral antinociception without the psychotropic CNS effects associated with phytocannabinoid CB_1 receptor binding [87,88]. This characteristic of exerting medically beneficial effects, while simultaneously avoiding any psychotropic responses, is likely to form a key focus of future cannabinoid research.

2.2. The Expanded Cannabinoid System and Its Less Characterised Receptors

It has been clearly demonstrated that the collective effects of cannabinoid administration cannot be explained solely by the presence of CB receptors. Conversely, it has been increasingly recognised that cannabinoids have the potential to affect other molecular targets and receptor types, particularly given their role as presynaptic secondary messengers on various neuron species [89,90] (Table 1). One such receptor is the G protein-coupled receptor (GPCR), GPR55, with the *GPR55* transcript identified in the adrenals, jejunum, and ileum in mammalian systems [91]. Studies on canine, rat and mouse gastrointestinal systems collectively suggest that GPR55 may be involved in smooth muscle contractions and colonic motility, especially when activated by CBD, pointing to a potential target for treatment of some gastrointestinal disorders [92–95]. Human embryonic kidney 293 (HEK293) cells expressing the GPR55 protein have been assessed for their response when treated with the lysolipid, L-α-lysophosphatidylinositol (LPI), as well as following their treatment with endogenous, synthetic or phytocannabinoids. LPI was found to induce phosphorylation of the protein, extracellular signal-related kinase (ERK) in GPR55-expressing cells, while also initiating a transient Ca^{2+} signal involved in downstream messaging and intracellular processing [96]. The degree of elevation in the concentration of Ca^{2+} increases in HEK293 cells when mediated by GPR55-phospholipase C coupling varied depending on whether THC, anandamide, methanandamide or the CB_2 agonist, JWH015 was administered [97]. However, there was no Ca^{2+} response initiated by CBD, the CBD regioisomer abnormal CBD, the endogenous cannabinoids, 2-arachidonoylglycerol and *O*-arachidonoyl ethanolamine, or the synthetic cannabinoids, WIN55,212-2 and CP55,940 [97]. Beyond Ca^{2+} transients, cannabinoid ligand interaction with the GPR55 receptor promotes ERK phosphorylation, as well as the varied activation of cyclic adenosine monophosphate (cAMP) response element binding protein (CREB), nuclear factor-$_\kappa$B (NF-$_\kappa$B) and nuclear factor of activated T-cell (NFAT) transcription factors, the latter two of which are involved in inflammation of endothelial cells and irritable bowel syndrome (IBS) [98–101]. The *GPR55* transcript can also be found in the basal ganglia, hippocampus, forebrain, cerebellum, cortex and large dorsal root ganglion (DRG) [97,102–104]. The expression of *GPR55* in these tissues significantly broadens the potential for its therapeutic application. For instance, activation of the GPR55 receptor by THC enhances neuronal excitability and reduces the M-type potassium current, which when combined with the expression pattern of *GPR55* in the large DRG, indicates a nociceptive role [97]. Inflammatory pain was modulated by abnormal CBD through GPR55 antagonism in acute arthritis models in rats [105]. Evidence of pro-nociception was observed in rats when the abundance of GPR55-dependent Ca^{2+} increased in periaqueductal grey neurons and which preceded a pain threshold reduction [106]. However, another study [107] reported that GPR55 knockout mice show no difference to wild-type mice in neuropathic pain models.

Another seven-transmembrane G protein-coupled receptor, termed GPR18, was first identified in canine gastric mucosa and a human colonic cancer cell line, with a high abundance of the *GPR18* transcript detected in human testis and spleen tissue [108]. The candidate ligand was later suggested to be *N*-arachidonoyl glycine (NAGly), an anandamide metabolite, which was first detected when GPR18-expressing cell lines, including the L929, K562 and Chinese hamster ovary (CHO) cell lines produced, high levels of intracellular Ca^{2+} and inhibited the production of cAMP following NAGly exposure [109]. In addition, quantitative real-time PCR analysis revealed high levels of *GPR18* expression in peripheral lymphocytes, further supporting the suggestion of a role in immune system function [109].

The transient receptor potential vanilloid (TRPV) channels are a subfamily of transmembrane ligand-gated ion channels that mediate signal transduction processes initiated by a broad range of noxious stimuli in animals, with the TRPVs, TRPV1 through to TRPV4, activated to varying degrees via cannabinoid application. TRPV expression in several human tissues and the documented role of TRPVs in human disease is a current avenue of interest. The capsaicin and temperature (~42 °C) responsive TRPV1, displays an ambiguous expression profile. However, the weight of evidence suggests that its expression domain is rather broad in animal systems. Specifically, the TRPV1 protein was observed to be localised to the dorsal root and trigeminal ganglions [110], thermoregulatory tissue smooth muscle cells [111], urothelial cells [112], corneal fibroblasts [113], and a broad distribution profile in the brain, including the hippocampus, cortex and olfactory bulb [114]. Sharing 50% sequence identity to TRPV1, TRPV2 has been demonstrated to respond to high-intensity thermal stimuli (~52 °C). However, unlike TRPV1, TRPV2 is insensitive to capsaicin [115]. Given its sensory involvement, TRPV2 localisation in the ganglia is unsurprising. However, TRPV2 is also localised to the brain, lung, spleen, intestine, mast cells and lymphocytes [115–118], which, when considered together, infers additional TRPV2 function beyond heat sensing, and by extension, activation by non-thermal receptor modulators. The initiation of signal cascades via TRPV2 are potentially involved in diseases and physiological responses including cancer [119], the innate and adaptive immune responses [116,117,120,121], cardiomyopathy [122,123], muscular dystrophy [124,125], and insulin secretion response [126–128].

The cannabinoid-responsive TRPVs, TRPV3 and TRPV4, are also temperature sensitive proteins. The responsive temperature range (27–40 °C) for these two receptors is below that of TRPV1 and TRPV2, but they do closely overlap with one another [129–132]. Their thermosensory involvement localises these two TRPVs to keratinocytes, where they sense warmth on the skin and transmit a signal to nearby neurons [133–138]. In the tongue and nasal epithelium, TRPV3 is activated by the 'pungent' carvacrol as well as by thymol and camphor [133,139], whereas the mevalonate (MVA) pathway product and cannabinoid/terpenoid precursor, isopentenyl diphosphate (IPP), has been shown to inhibit TRPV3 activity [140]. TRPV4, in association with aquaporin 5 (AQP5), is additionally involved in osmosensing and regulatory volume decrease in cells following swelling in hypotonic environments [141–144]. Located in the brain [145,146], kidneys [147], CNS [148], and endocardium [149], TRPV4 activity is also modulated by phorbol esters and arachidonic acid expanding its activation beyond physical stimuli [150,151].

In addition to the vanilloid subtype of the transient receptor potential channels are the melastatin and ankyrin subtypes. Of the melastatin type, transient receptor potential melastatin 8 (TRPM8) is a cold/menthol-responsive channel located in the DRG and trigeminal ganglia [152,153]. Of the ankyrin subtype, transient receptor potential ankyrin 1 (TRPA1) acts similarly to TRPM8 in response to cold stimuli covering a similar temperature range (~8–28 °C). However, it is suggested that TRPA1 contributes to sensation of lower temperatures, and is also similarly localised in sensory neurons [154–157]. TRPA1 is additionally activated by formalin and allyl isothiocyanates such as mustard oil [158,159], and has further been implicated in eliciting inflammatory pain [160–163].

Multiple other targets show notable interactions with the endocannabinoid system; however, a comprehensive description of all interactions is beyond the scope of this review. Briefly, other notable molecular interactions include glycine receptors with anandamide, and in addition, CBD and THC have also been shown to activate glycine receptors [164,165]. Further, THC appears to exhibit dose-dependent effects on glycine receptor activation [166]. The activation of peroxisome proliferator-activated receptors (PPAR), in particular the α and γ subtypes, is responsible for many of the metabolic, analgesic, neuroprotective, and other health-related benefits of cannabinoids [167]. Cannabinoids have also been shown to interact with serotonergic sites, particularly with the 5-HT$_{1A}$ [168] and 5-HT$_{2A}$ [169,170] receptors, and these interactions are strongly associated with disorders such as anxiety and post-traumatic stress [171,172]. Consequently, the spectrum of potential therapeutic applications is very broad for cannabinoids and would require a specifically dedicated and lengthy review in its own right. Currently lacking are robust, double-blind in vivo and clinical studies of the constituents of the broader cannabinoid profile that target specific diseases, and/or can be used to treat the symptoms of these diseases, possibly via targeting the interactions between cannabinoids and these other putative or lesser-known receptors.

Table 1. Receptor modulation by cannabinoids and studies outlining their potential involvement in disease treatment.

Receptor	Cannabinoid	Disease/Interaction	Study Type	Reference
CB$_1$	Anandamide	Appetite	Murine models	[173,174]
	Met-F-AEA	Thyroid cancer	in vitro human	[175]
	THCB $_{(PA)}$	Pain	Murine models	[176]
	THC $_{(PA)}$	Epilepsy	Murine models	[177]
		Sleep	Various studies	[178]
	THCP $_{(Ag)}$	Pain, anxiety, hypothermia, catalepsy	Murine models	[22]
	THCV $_{(\cdot)}$	Pain, anxiety, hypothermia, catalepsy	Murine models	[179,180]
		Parkinson's disease	Murine models	[181]
		Obesity	Murine models	[182]
		Epilepsy	in vitro murine	[183]
	THC, WIN55,212-2, CP55, 940	Emesis	Animal models	[184–188]
	WIN55,212-2	Parkinson's disease	Murine model	[189]
		Prostate cancer	in vitro human	[190]
	WIN55,212-2, JWH-133	Breast, lung cancer	in vitro human	[191,192]

Table 1. *Cont.*

Receptor	Cannabinoid	Disease/Interaction	Study Type	Reference
CB$_2$	CBC $_{(Ag)}$	Inflammation	in vitro models	[193]
	CBG $_{(PA)}$	Inflammatory bowel disease	Murine models	[194]
	HU-308, AM630	Parkinson's disease	Murine models	[195,196]
	THCP $_{(Ag)}$	Pain, anxiety, hypothermia, catalepsy	Murine models	[22]
	THCV $_{(\hat{\,})}$	Inflammation	Murine models	[180]
CB$_2$	THCV $_{(\hat{\,})}$	Parkinson's disease	Murine models	[181]
		Pain, anxiety, hypothermia, catalepsy	Murine models	[179]
	WIN55,212-2	Prostate cancer	in vitro human	[190]
	WIN55,212-2, JWH-133	Breast, lung cancer	in vitro human	[191,192]
GPR55	Abnormal CBD	Parkinson's disease	Murine models	[103]
GPR55	Abnormal CBD	Pain/arthritis	Murine models	[105]
	CBD $_{(An)}$	Gastrointestinal disorders	Canine, murine models	[93–96]
	CBDV $_{(An)}$	Rett syndrome	Murine models	[197]
		LPI inhibitor	in vitro	[198]
	THC, anandamide, JWH015	Pain	in vitro HEK239	[97]
TRPV1	CBDV $_{(Ag)}$	Anti-seizure	in vitro HEK239	[199]
	CBG $_{(Ag)}$, CBGV, CBD $_{(Ag)}$, CBDV $_{(Ag)}$, THCV $_{(Ag)}$	Receptor desensitisation	in vitro HEK239	[200]
TRPV2	CBD $_{(Ag)}$, CBGV, CBG $_{(Ag)}$, THCV $_{(Ag)}$, CBDV $_{(Ag)}$, CBN $_{(Ag)}$	Receptor desensitisation	in vitro HEK239	[200]
TRPV3	CBGV, CBGA $_{(Ag)}$	Receptor desensitisation	in vitro HEK239	[201]
TRPV4	CBGV, CBGA, CBN, CBG	Receptor desensitisation	in vitro HEK239	[201]
TRPM8	CBG $_{(An)}$, CBC $_{(An)}$, CBD $_{(An)}$, CBDV $_{(An)}$, THC $_{(An)}$, THCA $_{(An)}$	Colorectal cancer	in vitro model	[200,202,203]
TRPA1	CBC $_{(Ag)}$, CBN $_{(Ag)}$, THC $_{(Ag)}$, THCV $_{(Ag)}$, THCA $_{(Ag)}$, CBDA, CBG $_{(Ag)}$	Receptor desensitisation	in vitro HEK239	[200,202]
	CBDV $_{(Ag)}$	Ulcerative colitis	in vitro human	[204]
		Muscular dystrophy	in vitro studies	[205]

PA = Partial Agonist, Ag = Agonist, ˆ = Dose Dependent, An = Antagonist.

2.3. Examples of the Potential Medicinal Use of Cannabinoids

While research into the cannabinoids and their role in human disease is still in its infancy, the field abounds in promising preliminary studies. Cannabinoids, both of the endo- and phytocannabinoid categories, have been demonstrated to provide protection against further neurodegeneration in lesioned neurons post-treatment with toxic doses of 6-hydroxydopamine, as well as the neuron degeneration linked to Parkinson's disease [189,206]. Moreover, symptoms of dyskinesia associated with Parkinson's disease and other movement disorders, originating from deficiencies in the cannabinoid receptor-rich basal ganglia in marmosets, and reserpine-treated rats, have been reduced by CB$_1$ receptor stimulation-mediated suppression of involuntary motor behaviour [189,207–210]. Central nervous system activation of the CB_2 receptor has exhibited promising results in combating the inflammation and oxidative stress of Parkinson's disease which is associated with dopaminergic neuron loss in the substantia nigra pars compacta in nonhuman models [195,196].

Studies into the treatment of a variety of cancers through cannabinoid use have also proved valuable. For example, CB_1 and CB_2 activation by either endogenous or synthetic receptor ligands has inhibited prostate [190] and pancreatic [211] adenocarcinoma growth, as well as breast [191] and thyroid [175] tumour growth. Modulation of non-CB receptors by the minor cannabinoids is also under investigation for their role in the initiation of oncogenic signalling cascades that may induce the arrest of the cell cycle, or inhibit the growth of tumours [212]. Endocannabinoid-mediated breast cancer cell proliferation has been inhibited by a reduction in prolactin action at the receptor level [213], and CB_1 and CB_2 receptor activation has induced apoptosis of cancerous cells in the breast [191] and colon [214]. In non-small-cell lung cancer cell lines, treatment with agonists targeting CB_1 and CB_2, or specifically CB_2, were demonstrated to induce apoptosis, and to attenuate chemotaxis, metastatic growth and development, metastatic proliferation, and angiogenesis [192]. Similarly, cannabinoid activity against vascularization was also observed in human grade glioma cells in mice, with CB_2 activation reducing tumour angiogenesis by inhibiting vascular endothelial cell migration and the suppression of pro-angiogenic factors in tumour cells [215].

First alluded to over 40 years ago, the use of *Cannabis* as a treatment for epilepsy has garnered traction in recent years and several comprehensive reviews have recently described the efficacy of cannabinoids in the treatment and/or management of epilepsy [216–218]. Further evidence of the involvement of the endocannabinoid systems in seizure mitigation is suggested with inactivation of the endocannabinoid degrading, FAAH, with FAAH shown to reduce both kainic acid associated seizure activity, and synaptic decline and damage to cytoskeletal elements in the hippocampus of rat models [219,220]. A double-blind, placebo-controlled study of 218 patients in which CBD was administered at a dose of 10 and 20 mg per kg reduced the frequency of drop seizures in both children and adults with Lennox-Gastaut syndrome, when compared to conventional epilepsy treatment [221]. A similar double-blind, placebo-controlled study of 120 children with the epilepsy disorder, Dravet syndrome, saw a significant reduction in the frequency of convulsive seizures when treated with CBD, as compared with those administered the placebo [222]. In a retrospective, open-labelled study, Press et al. [223] reported improvements in seizure control and frequency reduction in paediatric patients using oral *Cannabis* extracts, as well as additional improvements in some off-target metrics, including alertness and motor skill usage also observed. Use of a THC extract has attenuated seizure duration and termination via the activation of CB_1. However, inhibition of CB_1 receptor activity has also been demonstrated to increase the frequency and duration of seizures in non-human models, findings which firmly identify a role for CB_1 in seizure responses [177]. Indeed, transgenic CB_1 overexpressing mice were reported to have reduced kainic acid-induced seizure severity and mortality with reduced hippocampal neuron damage [224]. While these examples suggest promise in the efficacy of cannabinoids, or the modulation of cannabinoid receptor activity against epilepsy, there currently remains deficiencies in access to data emerging from large, controlled clinical studies.

The treatment of Parkinson's disease, cancer and epilepsy are persistently pursued and remain 'high-value' targets for researchers. However, the importance of treating other less deleterious ailments, or the treatment of the negative side effects that originate from the aggressive treatment strategies of major diseases such as cancer, chemotherapy for example, is not without utility. A suite of clinical trials have supported the ability of *Cannabis*-derived metabolite constituents to (1) act as effective antiemetics [184–188], (2) ease the spasticity symptoms associated with Motor Neuron Disease and Multiple Sclerosis [225], (3) stimulate appetite [173,174,226–229], (4) help regulate sleep patterns [178,230–232], (5) initiate analgesia [233–236], (6) act as an anxiolytic to alleviate the psychotic symptoms of schizophrenia [237–241], (7) treat anxiety and post-traumatic stress disorders [31,171,242],

(8) be utilised as palliative care agents [243,244], (9) aid in the acute inflammatory response and its protracted recovery [245], and (10) mitigate the effects of opioid addiction [246,247].

A full review of the current understanding of cannabis in the medical sphere is beyond the scope of this review and has been published elsewhere [90,248]. Despite much of the current research remaining in the preliminary stages, requiring a greater amount of more stringent, double-blind studies, the medicinal promise of *Cannabis* is readily evident. Meta-analyses relating to the legitimacy of medical *Cannabis*, specifically the use of CBD and THC in control randomised trials, have been conducted. Studies surrounding the use of CBD indicate that the drug is well tolerated with minimal serious adverse side effects and drug–drug interactions [249]. CBD is described as effective in the treatment of refractory seizures, but scientifically stringent data are lacking to claim effectiveness for other indications, with concerns remaining about the quality control in drug preparation and long-term safety [250]. It has been noted that inconsistencies across current studies relating to dosage and administration methods limit the conclusions that can be drawn to direct medical intervention using CBD [251]. Currently, cannabinoid therapies for sleep quality and mental health-related disorders also suggest that while preliminary evidence may indicate positive outcomes, the collation of eligible studies provides insufficient evidence to suggest efficacy or promote usage until additional, and more stringent studies have been conducted [252,253]. Although more stringent studies on the effectiveness of cannabinoids to control pain and spasticity exist, additional comprehensive studies demonstrating improvements in the treatment of chemotherapy associated nausea, sleep disorders, weight gain, and Tourette's syndrome, and which also note the risk of short-term adverse events of cannabinoid treatment, are still required [32].

3. The Cannabinoid and Terpene Pathways of *Cannabis*

It is clear that modulation of the endocannabinoid system can be achieved outside of THC, CBD, and their CB receptors. Despite this, the majority of research conducted to date has sought to understand how these two cannabinoids interact with the various constituents of the expanded endocannabinoid system. However, significant knowledge exists concerning what further compounds can be extracted from *Cannabis* as well as an emerging understanding of how such compounds can be efficiently extracted from the *Cannabis* plant. To date, the most studied phytochemicals in *Cannabis* are the cannabinoids and terpenes. Together, these two classes of phytochemical comprise approximately 41% of the total number of known secondary metabolites identified in *Cannabis* [21,22]. Cannabinoid and terpenoid biosynthesis occurs in hair-like capitate stalked glandular trichomes [254,255], which cover the female floral organs, and exhibit a particularly high density on the bracts (a specialised leaf of the floral organs; Figure 1).

In trichome development, a protodermal cell is enlarged vertically out from the epidermis and subsequently undergoes anticlinal division, prior to a series of periclinal division events to create a secretory and auxiliary tier of cells atop the epidermal basal cells [256–259]. Additional division events develop the secretory tier of disc cells that form a cavity on the external surface of the trichome from a portion of the outer wall. This cavity then enlarges as the secretory vesicles that harbour a diverse payload of secondary metabolites are extruded into the expanding waxy cavity. Post their cellular release, the secreted vesicles disintegrate upon contact with the thickened outer cuticle wall to release their contents [256–259].

The complete biosynthetic pathway of how the prenylated polyketides, particularly minor cannabinoids, are derived from precursor molecules still requires further elucidation, particularly in view of the recent discovery of the two novel cannabinoids, THCP and CBDP [22]. Cannabigerolic acid (CBGA), the key intermediate substrate required for the synthesis of the three primary cannabinoids—cannabichromenic acid (CBCA), THCA and CBDA—arises from molecular products of the polyketide and methylerythritol 4-phosphate (MEP) pathways. A schematic representation of the MEP pathway is provided in Figure 2A. More specifically, the MEP pathway begins in the plastid via the condensation of the substrates, pyruvate and triose phosphate, a reaction that is catalysed by 1-deoxy-D-xylulose-5-synthase

(DXS), and which produces 1-deoxy-D-xylulose-5-phosphate (DXP) [260–262]. Via the action of 1-deoxy-D-xylulose-5-reductase (DXR) in the presence of the co-factor NADPH, DXP is next reduced to MEP [263] and subsequently, MEP is converted to CDP-ME by the action of the enzyme, 4-diphosphocytidyl-2-C-methyl-D-erythritol (CDP-ME) synthase. The kinase, DCP-ME kinase then phosphorylates CDP-ME to produce 4-diphospho-cytidyl-2-C-methyl-D-erythritol-2-phosphate (CDP-ME2P) [264,265]. CDP-ME2P is subsequently converted to 2-C-methyl-D-erythritol 2,4-cyclodiphosphate (ME-2,4cPP) via the activity of the enzyme, ME-2,4cPP synthase, prior to another synthase, 4-hydroxy-3-methylbut-2-enyl diphosphate synthase (HDS), converting ME-2,4cPP to 1-hydroxy-2-methyl-2-(E)-butenyl 4-diphosphate (HDMPP). In the final step of the MEP pathway, HDMPP is used as a substrate by 4-hydroxy-3-methylbut-2-enyl diphosphate reductase (HDR) to produce IPP and dimethylallyl diphosphate (DMAPP) [264–266].

The HDR enzyme is essential for the in planta production of IPP and DMAPP, with over 98% of these two molecules produced by the MEP pathway. IPP and DMAPP both form essential precursor substrates for the biosynthesis of cannabinoids and terpenoids [261]. In the cytosol, IPP is also produced by the MVA pathway (Figure 2B). At the start of the MVA pathway, acetyl-CoA is converted to 3-hydroxy-3-methylglutaryl-CoA (HMG-CoA) by the enzyme, HMG-CoA synthase. Next, HMG-CoA is converted to MVA in the highly rate-limiting step of the MVA pathway, a step that is regulated via the activity of 3-hydroxy-3-methylglutaryl-CoA reductase (HMGR) [267–269]. MVA is then converted to MVA phosphate by MVA kinase (MVK), and subsequently, MVA phosphate is converted to its diphosphate form via the activity of phospho-MVA kinase (PMK). MVA diphosphate is subsequently converted to IPP via its decarboxylation by mevalonate 5-diphosphate decarboxylase (MVD) [270–272]. Via the use of yellow fluorescent protein (YFP) fusion constructs, the activity of PMK and MVD has been observed in the peroxisome in *Catharanthus roseus* (Madagascar periwinkle) and *Arabidopsis thaliana* (*Arabidopsis*) to strongly indicate peroxisomal localisation of these two enzymes in planta, and not in the cytosol [270,271,273]. IPP isomerase catalyses the conversion between IPP and DMAPP, a conversion reaction that provides the building blocks for terpene biosynthesis [274–276]. Geranyl diphosphate synthase (GPPS) catalyses the production of the ten-carbon (C_{10}) molecule, geranyl diphosphate (GPP), via the condensation of one molecule each of DMAPP and IPP [277,278]. Similarly, formation of the C_{15} molecule, farnesyl diphosphate (FPP), and the C_{20} molecule, geranylgeranyl-diphosphate (GGPP), is catalysed by their specific synthases, farnesyl diphosphate synthase (FPPS) and geranylgeranyl diphosphate synthase (GGPPS), respectively, which condense either 2 or 3 molecules of IPP together with a single molecule of DMAPP [279–281]. Together, GPP, FPP and GGPP form the precursors necessary for monoterpene or CBGA biosynthesis (GPP precursor), or the numerous sesqui-, di-, tri-, or tetra-terpene products (FPP or GGPP precursors) found in *Cannabis* [282,283].

Figure 1. A close up of the female floral architecture of mature *Cannabis sativa* plants. The cannabinoid-containing glandular trichomes are visible in the magnified image, and are characterised by a globular head which is connected to the plant via a stalk. Colouration of the heads ranges from translucent, to a creamy white, to brown.

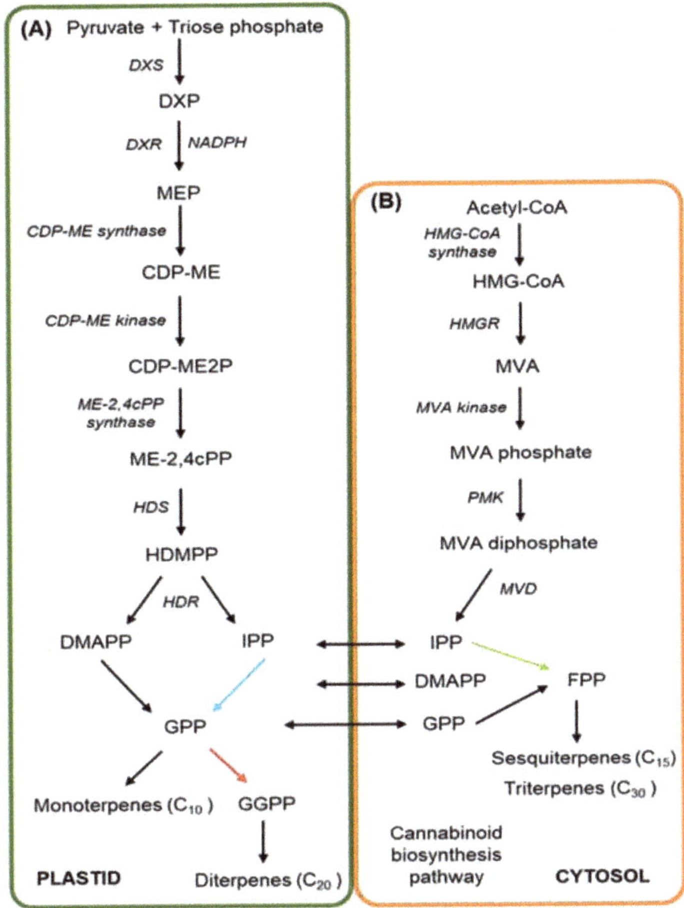

Figure 2. An overview of the mevalonate and methylerythritol 4-phosphate pathways in *Cannabis sativa*. The MEP (**A**) and MVA (**B**) pathways both produce terpenoid precursors, as well as the substrate for cannabinoid production, GPP. (**A**) The MEP pathway begins in the plastid with the condensation of pyruvate and glyceraldehyde 3-phosphate by DXS to produce DXP, prior to a series of enzymatic reactions to produce HDMPP. HDR then converts HDMPP to IPP and DMAPP, serving as the precursor to GPP, GGPP, and subsequently monoterpene and diterpene production. (**B**) The cytosolic MVA pathway is initiated by the conversion of acetyl-CoA to HMG-CoA and then to MVA, catalysed by the regulated, and rate-limiting enzyme, HMGR. MVA undergoes phosphorylation and then is decarboxylated to produce IPP, which is then converted to FPP as the basis for sesquiterpene and triterpene synthesis, or for GPP production for use in the cannabinoid biosynthesis pathway.

The polyketide pathway is initiated when acetyl-CoA is carboxylated to malonyl-CoA, which in turn serves as the precursor for the fatty acid chains used to produce hexanoate (Figure 3) [254,255,261]. The acyl-activating enzyme (AAE), which in *Cannabis* is encoded by two putative genes, termed *CsAAE1* and *CsAAE3*, with the encoded proteins localised to the cytoplasm and peroxisome, respectively, where they function to catalyse the synthesis of hexanoyl-CoA from hexanoate [255]. Condensation of hexanoyl-CoA, together with three malonyl-CoA molecules, is subsequently catalysed by the polyketide synthases, tetraketide synthase (TKS), or olivetol synthase [284,285]. The product of these two synthases, and post a final round of aldol cyclisation by the olivetolic acid cyclase (OAC) enzyme, is olivetolic acid (OA) [284]. Via the utilisation of GPP from the MVA pathway, OA is then prenylated by geranylpyrophosphate:olivetolate geranyltransferase (GOT), to produce CBGA [286–288]. The *cis* isomer of GPP, neryl diphosphate (NPP), can be used as a substrate by GOT in place of GPP, to produce cannabinolic acid (CBNA) [289]. CBGA then serves as the primary cannabinoid precursor for the synthesis of cannabichromenic acid (CBCA), THCA and CBDA, with the production of each of these three acids catalysed by a specific oxidocyclisation enzyme, namely the CBCA, THCA and CBDA synthases [289–293]. The use of divarinic acid as a substitute for OA by GOT, putatively produces the propyl cannabinoid homolog, cannabigerovarinic acid (CBGVA) [286,294]. The aforementioned cannabinoid-specific synthases that yield CBCA, CBDA, and THCA can all recruit CBGVA to produce cannabidivarinic acid (CBDVA), cannabichromevarinic acid (CBCVA) and Δ^9-tetrahydrocannabivarinic acid (THCVA), respectively [294–296]. The resulting cannabinoids are maintained in their acidic forms until they are thermally decarboxylated to convert them into their neutral forms [297–300].

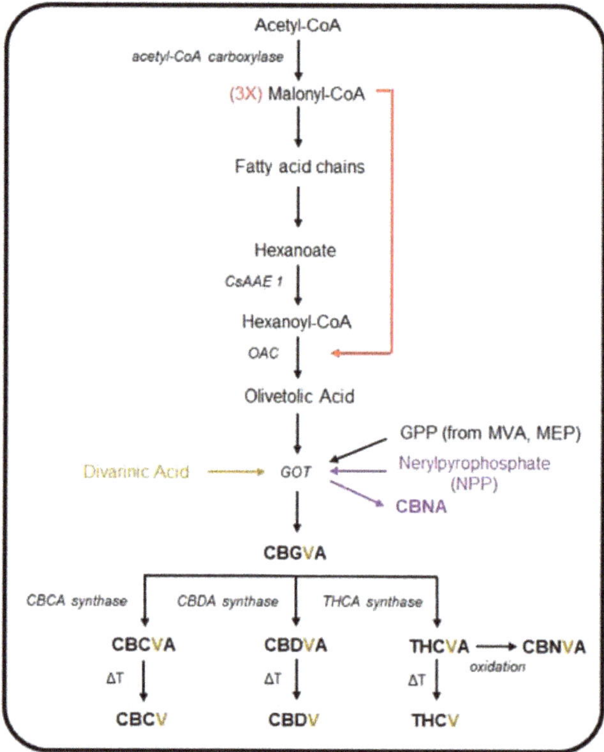

Figure 3. An overview of the cannabinoid biosynthesis pathway in *Cannabis sativa*. Malonyl-CoA, formed from acetyl-CoA, is used downstream with hexanoyl-CoA to produce olivetolic acid (OA). Next, OA is used as substrate along with other biomolecules by the GOT enzyme to produce the major cannabinoid precursor, CBGA. When GOT uses substrates additional to OA, such as divarinic acid or nerylpyrophosphate, a range of other minor cannabinoids are produced.

Research to date has primarily focused on the biosynthetic pathways and putative medical benefits of the two major cannabinoids, THC and CBD. Therefore, the medical and biological potential of the minor cannabinoids that also contribute to the total cannabinoid profile of the *Cannabis* plant have been largely overlooked. The small proportion that these minor cannabinoids contribute to the total cannabinoid profile of the *Cannabis* plant presents a significant obstacle for in-depth analysis of their effects when consumed. A comprehensive, and ever-increasing list of naturally occurring minor phytocannabinoids has been compiled based upon their derivation from THC, CBD, CBG (cannabigerol) and CBC, which represent the diversity that stems from variations to the three fundamental components of cannabinoids, including the (1) resorcinyl core; (2) isoprenyl residue, and; (3) resorcinyl side chain [20,301]. Eighty-two individual cannabinoids from 10 cannabinoid types, specifically the (1) CBG; (2) CBC; (3) CBD; (4) Δ^9-THC; (5) Δ^8-THC; (6) cannabicyclol (CBL); (7) cannabielsoin (CBE); (8) cannabinol (CBN); (9) cannabinodol (CBND), and; (10) cannabitriol (CBT) types, in addition to the miscellaneous types, and their transformation products, as well as terpenoids, hydrocarbons, sugars and fatty acids are among the constituents that comprise the chemical cornucopia of glandular trichomes. Further, several minor oxygenated cannabinoids, cannabinoid metabolites, and cannabinoid esters present in *Cannabis* have yet to be isolated and/or experimentally validated but have been identified using a variety of spectroscopic techniques [302–304]. In addition, a number of interesting structural formations have been observed in some of the minor cannabinoids. For example, cannabioxepane (CBX) has a tetracyclic skeleton with a seven-membered ring, a structure not previously reported for a characterised cannabinoid, while cannabisol is a Δ^9-THC dimer with a methylene bridge. However, it must be noted that the binding affinity for specific CB receptors for these minor cannabinoids remains unknown, with some potentially not recognised, and therefore not bound by any known CB receptor [305,306]. The CBD derivative, cannabimovone, and the farnesyl prenylogue of CBG, sesquicannabigerol, were also spectroscopically characterised, with CB receptor binding assays predicting receptor–cannabinoid affinity, highlighting the structural and potential psychoactive diversity among the minor phytocannabinoids [307,308]. In addition to the identification of their parent cannabinoid precursors, plausible biochemistry behind the synthesis of these compounds is offered. However, the actual enzymatic production of many of these minor cannabinoids remains to be determined. Furthermore, the non-enzymatic formation of some of the minor cannabinoids is certainly likely, but it remains of interest to understand whether there is a greater portion of enzyme-catalysed reactions in the production of the minor cannabinoids, or indeed whether there are alternative pathways, or even additional pathway entry points in the biosynthesis of cannabinoids, both minor and primary.

4. Minor Cannabinoids and Their Biological Interactions

There is mounting evidence that the minor cannabinoids described above share combinations of many of the same molecular targets as THC and CBD, and therefore may potentially have unique medical applications that cannot be achieved by THC or CBD alone. The THC propyl homologue, THCV, is a CB_1 and CB_2 competitive antagonist against CP55,940 and WIN55,21–2, acting with similar potency to that of THC [309,310]. THCV also antagonised anandamide and methanandamide in mice vas deferens, attenuating stimulated contractile responses [309]. More recently, THCV was shown to similarly displace CP55,940 from CB_1 and CB_2 in CHO cells, and contrary to previous assumptions, was shown to be a weak partial CB_1 agonist at high doses [179]. Moreover, Zagzoog et al. [310] showed THCV to produce anxiolytic, hypothermic, anti-nociceptive, hypolocomotive, and cataleptic effects in vivo in mice. CB_2 agonism by THCV was demonstrated to reduce inflammation and attenuate hyperalgesia in mice following injection of carrageenan and formalin, respectively [180]. Neuroprotective properties were observed in 6-hydroxydopamine lesioned rats, where THCV administration preceded maintenance of tyrosine hydroxylase-positive neurons in this Parkinsonian model [311]. Similarly, THCV delayed onset of abnormal involuntary movements associated with Parkinson's disease in mice, and reduced their severity after administration following symptom onset [181]. The in vitro demonstrated inhibition of GABA release by WIN55,21-2 at Purkinje cell synapses was reversed by THCV, which also prevented the action of WIN55,212–2 when used in pre-incubation [312,313]. In vitro studies of insulin-resistant human hepatocytes showed THCV restoration of insulin signaling mediated by CB_1, while also improving glucose tolerance and increased sensitivity to insulin in mice obesity models [182]. Antiepileptic properties were also established in vitro, specifically when THCV reduced both the frequency and amplitude of epileptiform activity in rat piriform cortex slices [183]. The majority of published studies have focused on the CB_1 and CB_2 receptors, but the in vitro activity of THCV has been observed for the TRPV1 to TRPV4 group of receptors, as well as for the TRPA1 receptor [200–202]. THCV can enhance $5\text{-}HT_{1A}$ receptor activation to produce antipsychotic-like effects in rats [314], but does not affect other endocannabinoid system constituents such as PPARγ [315], FAAH [200], or MAGL [200]. One clinical trial in humans where THCV was administered once daily for five days followed by intravenous administration of THC suggested that THCV inhibited an increase in heart rate, protected against verbal recall impairment, and reduced the subjective psychoactive intensity induced by THC [316]. Further, THCV affects brain regions associated with reward and aversive stimuli, as well as areas associated with cognitive control [317,318]

Recently, a four-carbon side chain variant, Δ^9-tetrahydrocannabutol (THCB), was isolated and which showed CB_1 and CB_2 binding affinities similar to those of THC, with in vivo mice studies suggesting potential analgesic and anti-inflammatory properties [176]. Similarly, the recently identified seven-carbon side chain variant, THCP, was shown to be able to bind to both CB_1 and CB_2 with 33 and 5 times greater affinity than THC, respectively, as well as to initiate catalepsy, hypothermia, analgesia, and reduce locomotion; all indications of potent full CB_1 agonism [22]. THCA has been shown in rodent culture supernatants to reduce the abundance of inflammatory and oxidant markers [319,320], though no other research to our knowledge of this nature has been published. In addition, Δ^8-THC has been shown to possess higher antiemetic effects than THC [188], and has been successfully trialed for repressing emesis in children [184]. Furthermore, in humans, Δ^8-THC appears required to be administered at higher doses than THC to display a similar degree of psychoactive properties [321]. THCA is a $5\text{-}HT_{1A}$ agonist [322], a PPARγ agonist [323], and displays the same properties against TRP channels as does THC [200]. However, little pharmacological, pharmacokinetic, or recent safety data are available for any of these compounds.

Improvements in seizure frequency has been reported in an epileptic patient coinciding with increased CBDV serum levels, after which in vitro studies confirmed that CBDV, at least, possesses the ability to influence GABA receptors; a finding that indicates a potential avenue for anticonvulsant properties [324]. Further, in vitro analyses revealed CBDV to have anticonvulsant effects in four seizure models, namely the (1) maximal electroshock-, (2) audiogenic-, (3) penytylenetetrazole-(PTZ), and (4) pilocarpine-induced seizure models [325,326]. Using rat brain tissue samples, PTZ-induced seizures coincided with an increase in *Early growth response 1 (Egr1)*, *Activity-regulated cytoskeleton-associated protein (Arc)*, *Chemokine (C-C motif) ligand 4 (Ccl4)*, *Brain-derived neurotrophic factor (Bdnf)*, and *FBJ osteosarcoma oncogene (Fos)* gene expression [327]. Interestingly, the administration of CBDV was shown to reduce the expression of all of these genes [327]. Additional seizure studies identified TRPV1 as the potential receptor modulating anti-seizure effects via the use of *trpv1* knockout mice which showed a reduced response to CBDV [328]. Desensitisation of TRPV1, in addition to TRPV2, by both CBDV and CBD has been observed [199], while Ca^{2+} transients were induced in TRVP2-expressing HEK293 cells more potently by CBD than by CBDV. However, THC was a more potent inducer of Ca^{2+} transients than either CBDV or CBD [200]. Another study did alternately suggest that CBD was the more potent agonist of TRPV2 than THC, but this study did not include the assessment of CBDV [329]. Cannabinoid administration improved symptoms in mice models of Rett syndrome, including motor control and sociability [197], and through TRPA1, CBDV mediates anti-inflammatory effects in intestinal tissue of humans with ulcerative colitis [204]. Similar to CBD, CBDV inhibits FAAH and anandamide reuptake [200]. However, unlike CBD, CBDV does not show affinity for the CB_1 or CB_2 receptors [180]. CBDV may confer some benefit in patients with Autism Spectrum Disorder [330] and Duchenne muscular dystrophy [205]. CBDV did, however, fail to alleviate the neuropathic pain associated with human immunodeficiency virus (HIV) [331], and in another study, the administration of CBDV induced DNA damage in human cell lines at concentrations similar to those observed in *Cannabis* consumers [332], indicating carcinogenicity potential for CBDV. However, CBDV has been safely trialed in humans at a single 600 mg oral dose [330], and it remains to be determined whether CBDV will be efficacious for other illnesses in clinical trial.

CBG has shown partial agonism of CB_1 and CB_2, α_2-adrenceptor agonism and 5-HT_{1A} antagonism, while exerting some minor anti-nociceptive and anxiolytic properties in vivo [179,333]. Mice models of inflammatory bowel disease (IBD) showed positive outcomes with CBG treatment including reductions in the level of reactive oxygen species in intestinal cells, as well as reduced nitric oxide concentration in macrophages through CB_2 modulation [194]. Further in vivo animal studies provided evidence for neuroprotectivity against symptoms of Huntington's disease in 3-nitropropionate treated mice, with improvement in motor function, reduction in proinflammatory marker upregulation and increased antioxidant defenses, with R6/2 mice showing a reduction in the expression profiles of several genes linked to the disease following CBG treatment [334]. Similarly, in vitro analysis of NSC-34 neuronal cells showed that CBG pre-treatment reduced both inflammation and the expression of pro-inflammatory cytokines, and inhibited cell death resulting from the cell culture medium of lipopolysaccharide (LPS) stimulated RAW 264.7 macrophages [335]. CBG shows a similar profile at TRP channels compared to CBD, with agonist properties at TRPV1 through to TRPV4, and at TRPA1, but antagonism at TRPM8 [200]. It is also an anandamide reuptake inhibitor [336], and an LPI inhibitor at GRP55 [97]. As for the propyl analogue of CBG, CBGV, very little information surrounding its clinical application exists, except to show that CBGV has activity at GPR55, TRPV3 and TRPV4 [198,201].

CBC use in a clinical setting, or in human trials, appears to be untested currently, and additionally, cannabichromevarin (CBCV) currently has even fewer studies dedicated to it. However, CBC has seen some use in animal models and in vitro studies. CBC has been shown to inhibit FAAH, MAGL, and anandamide reuptake [200,337], but has been demonstrated to have no effect at TRPV1 or TRPV2. Further, CBC is a very weak CB_1 agonist [338–340], and only exhibits modest agonist properties at CB_2. An early study suggested that CBC, CBCV, and a CBC variant which lacks a carbon side chain, possessed anti-inflammatory properties in rat edema models and varying anti-bacterial and anti-fungal properties [341]. More recently, CBC was seen to produce anti-inflammatory effects in LPS paw edema models in mice in CB_1- and CB_2-independent pathways and also produce hypothermia, catalepsy, and locomotor suppression [342]. The authors went on to suggest that the effects of CBC were altered in the presence of THC, with an additive effect against inflammation [342] and similarly, tail-flick tests revealed that subtle analgesic properties of CBC were potentiated by its combination with THC [343]. Selective CB_2, but not CB_1 agonism, was exhibited by CBC on mouse pituitary tumour cells, and the persistent administration of CBC caused desensitisation of CB_2 receptors [344]. Intestinal studies suggest that CBC confers some benefit against inflammation. However, this was potentially independent of CB_1, CB_2, or TRPA1, the expression of which were all downregulated in the presence of CBC in one study, but shown to be unchanged in another study [193,345]. Colorectal cancer cell viability was attenuated through TRPM8 antagonism by CBG, as well as by the administration of CBD, CBDV, and CBC, albeit to lesser degrees [203]. Other studies have indicated that CBC is not a potent antagonist of TRPM8, and instead suggest that CBD, CBG, THC, and THCA are more effective antagonists of TRPM8 [200,202]. Additionally, CBC, CBN, THC, THCV, THCA, CBDA, and CBG all induced intracellular Ca^{2+} increases in HEK293 and rat DRG neurons through TRPA1 [200,202]. CBC has also shown promise in increasing neural stem cell viability in animal models (in vitro), mediated through ERK phosphorylation [346]. However, it is concerning that large amounts of CBC are required to produce pharmacological effects [90], which implies that CBC may be difficult to implement in a human health context.

The binding affinity of CBN, and of its primary derivatives, to the two main cannabinoid receptors was established in 2000, and showed rather unsurprisingly that alterations at carbon atom positions 1, 3, and 9, resulted in significantly different affinities at both receptors [347]. An earlier study indicated CBN to have cataleptic, hypothermic, and locomotive effects, as did 11-hydroxy-CBN; a hepatic microsome CYP2C- and CYP3A4-catalysed metabolite [348,349]. Additionally, CBN directly inhibited the activity of the human cytochrome P450 family 1 (CYP1) enzymes, CYP1A2 and CYP1B1 [350]. Assays of cultured neuronal cells expressing an inducible disease conferring huntingtin (Htt) protein, suggest that CBN has protective effects against cell death in vivo, with low toxicity even at the high concentrations required for protectivity [351]. Interestingly, cannabinoid receptor loss has been indicated as a pathophysiology of Huntington's disease [352,353], which may suggest that the purported protective action of cannabinoids is independent of cannabinoid receptor binding. Subcutaneously delivered CBN delayed the onset of amyotrophic lateral sclerosis (ALS) symptoms in murine models but failed to affect survival, so was postulated to mask the early spasticity associations without affecting disease progression [354]. A synergistic effect of CBN with CBD at reducing mechanical sensitisation in rat masseter muscles was observed in one study, however high concentrations of CBD ameliorated the efficacy of CBN [355]. CBN has been reported to have no effect at FAAH, MAGL, or TRPV1, but acts as an agonist at TRPA1 and TRPV2 [200].

Via the use of in silico analyses, the even lesser-known cannabinoids, cannabiripsol (CBR) and CBT, are predicted to have cytochrome P450 inhibitor activity [356]. In another in silico study, CBL, CBT, and CBE were assessed, and ranked in this order, to have acetylcholinesterase-inhibiting function. However, their inhibitory effects were less than those of THC, CBN, and CBDV [357]. Exactly how well in silico studies translate to clinical relevance, or even to in vitro and/or in vivo studies, restricts what conclusions can be accurately drawn. Minor phytocannabinoids do represent an understudied portion of the *Cannabis* plant. Very few studies exist that have utilised an in vivo approach to ascertain the viability of minor cannabinoids to potentially produce any significant medical benefits, and fewer still cover any human clinical trials. There has been indication that some cannabinoids exhibit synergistic action, and as a result there may be value in investigating the interactions among cannabinoids or constituents of the *Cannabis* plant.

5. Directions in *Cannabis* Development for Secondary Metabolite Production

The establishment of superior varieties of *Cannabis* has been the target for plant breeders since the domestication of this species. To produce new medically relevant *Cannabis* varieties with elevated concentrations of specific minor cannabinoids, or to develop techniques to manipulate the cannabinoid biosynthetic pathway in other organisms, a deeper understanding of the genetics of the *Cannabis* plant is first required. Here we outline the progress in relation to (1) the sequencing of the *Cannabis* genome, and (2) the potential to molecularly manipulate the *Cannabis* plant itself for the altered production of specific cannabinoids. In this regard, we highlight the established success in *Papaver somniferum* (opium poppy), as a parallel example for maximising yield and the concentration of key secondary metabolites of medical and commercial relevance.

5.1. Next-Generation Sequencing of the Cannabis Plant and Its Potential for Genetic Manipulation

Over the last 25 years, various experimental approaches have been employed to unveil the wealth of information contained in the *Cannabis* genome. Using early DNA sequencing and karyotyping techniques, the X and Y sex chromosome characteristics of *Cannabis* were uncovered, as were the diploid (2n = 20) genome sizes for male and female plants [358,359]. The female *Cannabis* plant was revealed to have a genome size of 818 megabase (Mb), while the male *Cannabis* plant was determined to have a larger genome size of 843 Mb; specifically due to the larger size of the Y chromosome, compared to the X chromosome of female plants [358]. Microsatellite markers have been employed as a tool for DNA typing *Cannabis*, and these polymorphic short tandem repeat (STR) markers have been utilised as a measurement of genetic relationships among cultivars [360–362]. More recently, the rapid change in technologies surrounding Next-Generation Sequencing (NGS) platforms has meant that studies can unravel whole genomes in a fraction of the time required via the use of older methods. As a result, the first draft *Cannabis* reference genome, and transcriptome, were constructed in 2011 using the high THCA, low CBDA cultivar, 'Purple Kush', and the high CBDA, low THCA hemp strains, 'Finola' and 'USO-31' [363]. Using a PacBio long-read sequencing platform, the Purple Kush and Finola genomes were again sequenced in 2019 to generate a physical and genetic map for *Cannabis*, and further distinguish the genes, and importantly the gene products (specifically, the encoded enzymes), underpinning the secondary metabolite profiles responsible for the divergent chemotype between hemp and marijuana cultivars [364,365].

Earlier work surrounding the chemotypic variance of cannabinoids observed in *Cannabis* unveiled the relationship between THCA and CBDA synthase expression, describing a single locus (B), with two codominant alleles, B_D and B_T [295]. A 1:1:2 segregation ratio results in the production of three chemotypes of the B locus, including the (1) pure CBD (B_D/B_D homozygote), (2) pure THC (B_T/B_T homozygote), and (3) mixed CBD/THC (B_D/B_T heterozygote) chemotypes [295]. However, later studies based around NGS platforms indicated an alternate genetic model of synthase gene duplication and rearrangement at multiple linked loci, and that CBDA synthase is more ancient, has a greater affinity for the CBGA substrate, and that the CBDA synthase locus is solely responsible for the cannabinoid chemotypes observed in *Cannabis* [363,365–369]. In an attempt to classify variability in chemotypes, and to associate genotype to chemotype in a diverse germplasm collection, DNA sequence characterised amplified region (SCAR) markers associated with THCA/CBDA synthases were assessed in 22 *Cannabis* varieties representing 2 fibre and 1 drug type plants from East (n = 8), Central (n = 1), and South (n = 2) Asia, as well as from Europe (n = 7) and of mixed (n = 4) domestication status [370]. This approach revealed a variability in cannabinoid profiles (CBD:THC) across 'chemotype II', or B_D/B_T equivalent plants, more than three-fold greater than previously observed, supporting the allelic variant and multiple loci prediction, when assuming that a heterozygote plant in a single locus model would have a 1:1 CBD:THC ratio [370].

Other large-scale genetic diversity studies using NGS, and which compared the evolutionary relationships between 340 *Cannabis* varieties from existing datasets, and from other novel multiplexed libraries, highlighted the murky ancestry of the *Cannabis* plant resulting from generations of repeated rounds of selective breeding, and also provides an extensive data platform for future genotyping efforts [371]. Moreover, Lynch et al. [371] classed their assessed *Cannabis* varieties into three genetic groups, including (1) hemp, (2) narrow leaflet, and (3) broad leaflet drug types, in order to determine the genomic and genetic variation of their population for the potential use of varieties from each group in either agricultural or medicinal applications. The authors indicated unique cannabinoid and terpenoid profiles for each group, structured loosely around geographic origin of each species, and noted the requirement for the inclusion of the putative *Cannabis* species, *C. ruderalis*, in future studies to fully elucidate their genetic distinction and ancestral lineage [371]. The development of expressed sequence tag simple sequence repeat (EST-SSR) markers to assess genetic diversity of 115 *Cannabis* genotypes also revealed geographical-based clustering into 4 groupings, including the Northern China, Southern China, Central China and Europe groupings [372]. Interestingly, a genetic similarity coefficient derived from 45 of 117 randomly selected EST-SSRs markers revealed that despite physical proximity to the other Chinese varieties, Northern Chinese varieties had a greater similarity coefficient to the European grouping, predicted to be related to latitude and day length [372]. The analysis of inter simple sequence repeats (ISSR) of 27 native Chinese hemp varieties identified a similar geographic distribution to genetic distance relationship, while also revealing the hemp varieties were genetically diverse, yet primitive, a finding which adds further weight to the suggestion that the *Cannabis* plant originated in southern China and then spread north [373].

The recent assembly and annotation of the mitochondrial genome of *Cannabis* using NGS methods will also allow for similar studies to be performed to determine the extent of the genetic diversity among *Cannabis* varieties [374]. In addition, the assembly of two chloroplast genomes from different *Cannabis* varieties will aid in validating the phylogenetic relationship of *Cannabis* among the Rosales order of the Plantae kingdom [375]. However, as with all sequencing, repeated efforts across diverse genotypic populations compared against reference genomes will increase the accuracy and reliability of publicly available repositories. RNA sequencing as a tool for differentiating strains has been used with some success, where the transcriptome isolated from cannabinoid-containing glandular trichomes from different varieties allows for comparative analysis based on the cannabinoid and terpenoid chemical profiles [376,377]. As the regulatory landscape surrounding the use of *Cannabis* evolves, and the value of the unique chemical profile of specific *Cannabis* varieties is realised, breeders are likely to use these sequencing techniques to rapidly characterise and protect their 'strains'. The development of such highly targeted databases provides the platform for precise manipulation of phenotypic or chemotypic traits in *Cannabis* to deliver improved medical efficacy or novel therapeutics.

A forward and/or reverse genetics approach with the application of chemical mutagenesis agents, such as ethyl methanesulfonate (EMS), a mutagen that introduces point mutations into the plant genome, is an effective approach for functional genomic assessments and effective plant breeding regimes, and has been successfully demonstrated in a variety of plant species, including hemp [9,378–383]. The application of alkylating agents such as EMS in a time-dependent manner causes a larger number of point mutations across the genome, compared to an irradiating method such as X-ray, or fast neutron bombardment, both of which produce much larger genome deletions and/or chromosome rearrangements [384–386]. Deletions ranging from 0.8 to 12 kilobases (kb) were produced in *Arabidopsis* using fast neutron bombardment, a widely used model plant species with an average gene density of one gene per 4.8 kb. The size of the genome alterations produced by this approach can, however, potentially cause the loss of function, or significantly altered expression of more than one gene. Therefore, a considerable drawback of using such an approach is the time and effort required post-mutagenesis to identify a 'causative mutation'. While the *Arabidopsis* genome is comparatively smaller than that of *Cannabis*, a similar post-mutagenesis investigative strategy would likely be required in other plant species with nuclear genomes either of a similar or significantly larger-size [385,387]. Regardless, these types of methods require rather large numbers of plants to be effective as deletions and point mutations are not site directed, which is a considerable limitation as even rapid standard screening techniques demand intensive laboratory work [388–390].

Since the advent of the CRISPR/Cas9 gene-editing system in late 2012 [391], the ability to manipulate plant genomes has become more cost efficient and less experimentally tedious when compared to the traditional genetic engineering approaches used by plant breeders in other crop species [392]. The CRISPR/Cas9 system effectively directs site-specific genome editing using RNA-guided, microbial-derived nucleases that initiate double-stranded DNA breaks in eukaryotic and bacterial systems [391,393]. The specificity of this system greatly reduces the amount of off-target genome alterations compared to more traditional transformation techniques. However, off-targeting has also been observed with CRISPR/Cas9 use, an inherent challenge when manipulating any biological system [394–397]. Earlier work was directed towards human applications, but increasingly this system has been utilised in plant systems, with examples in *Arabidopsis*, tobacco (*Nicotiana tabacum*), rice (*Oryza sativa*), lettuce (*Lactuca sativa*), maize (*Zea mays*), soybean (*Glycine max*) and wheat (*Triticum aestivum*) now documented [398–410]. By no means an exhaustive list of CRISPR/Cas9-facilitated manipulation in plants, the above does, however, highlight the potential applicability of this targeted mutagenesis approach to modulate specific biosynthetic pathways in *Cannabis* to produce superior varieties that display phenotypic and chemotypic traits of interest, and as a tool to discover key genes involved the production of minor cannabinoids. Transformation technologies has thus far been conducted in hemp varieties only, and therefore require further development and considerable refinement for application in other *Cannabis* varieties. The first report of successful hemp transformation emerged in 2001 [411], and two years later, a protocol for successful *Agrobacterium tumefaciens*-mediated transformation of tissue cultured hemp callus was implemented [412]. More recently, Wahby et al. [413,414] successfully transformed hemp using both *A. tumefaciens* and *A. rhizogenes*, establishing the initial protocol for hairy root culture in *Cannabis*, a system used for the production of key phytochemicals. Despite these successes, *Cannabis* has proven to be a difficult plant species to transform with such variables as variety, plant age and the explant used for callus production, all demonstrated to be crucial factors underpinning transformant regeneration efficiency [415]. As with any novel plant transformation system, in order to overcome poor transformation efficiency, optimised protocols with respect to culture media, experimental approach, and selected explant material, will be required for routine and robust transformation of *Cannabis*.

5.2. Synthetic Production of Cannabinoids

Recently, the synthetic biology approach utilising microorganisms to produce high-quality cannabinoid products has removed the requirement for plant material [287,416]. Luo and colleagues [287] were successful in producing CBG, CBD, THC and Δ^9-THCV from galactose, via manipulation of the native MVA pathway of the yeast *Saccharomyces cerevisiae* post the introduction of *Cannabis* genes encoding cannabinoid synthases, olivetolic acid synthase and geranylpyrophosphate: olivetolate geranyltransferase. Production of THCA from CBGA through functional THCA synthase expression in the two yeast species, *S. cerevisiae* and *Pichia pastoris*, has been demonstrated. However, attempts to introduce the same functionality in *Escherichia coli*, a bacterium, have proved unsuccessful [293,417]. Over-expression of genes encoding enzymes in the MVA and prenyl diphosphate pathways, also in *S. cerevisiae*, produced prenyl alcohol precursors required for terpenoid and cannabinoid synthesis [418], while expression of a functional aromatic prenyltransferase from *Streptomyces* resulted in THCA production from OA and DPP in the yeast, *Komagataella phaffi* [419]. These approaches present an attractive alternative with the ability to conceivably produce large quantities of minor cannabinoids that are only found in trace amounts in planta, while also reducing and/or removing the costs, carbon emissions (associated with indoor growth; [420]) and environmental variables associated with the agricultural crop production. However, it should be noted that due to the criminalisation of *Cannabis* since the early 1930s, there are very few studies analysing water and energy use associated with the cultivation of *Cannabis*, although undoubtedly, as research in this area becomes more prevalent, efficient horticultural practices will reduce the consumption of water and energy for the large-scale cultivation of *Cannabis*.

5.3. Phenotypic Parameters Affecting Cannabis Yield and Potency

In *Cannabis* plants exhibiting an illicit drug chemotype (high THC), a primary concern, in conjunction with desired cannabinoid content, is overall biomass yield of female floral tissue. Consistent with other agriculturally significant species, *Cannabis* is sensitive to environmental variations which alter physiological characteristics affecting plant growth and yield potential. Early work on *Cannabis* flowering, uncovered the response to photoperiodism [421,422], which has subsequently been exploited, particularly by illicit indoor growers, who can cultivate *Cannabis* year-round by manipulating the response to reduced photoperiod length [423]. Photoperiodism is a well-known biological response critical for development of branching and floral architecture in *Cannabis*, and as a result, has implications for yield potential [423,424]. A reduction in day length from 18 to 12 h induces flowering, and maintenance of this regime for 8 weeks produces an acceptable floral yield [423]. Elevated light intensity from 400 watts per square metre (W m^{-2}), to 600 W m^{-2}, produced a higher yield of floral tissue per plant in several chemotypes when grown indoors [424]. In addition, an increase in plant density from 16 to 20 plants m^{-2} reduced biomass yield of floral tissue in all 600 W m^{-2} treated plants [425]; a finding that indicates that light interception is compromised at the lower canopy level in crowded growth conditions. The use of different artificial lighting systems in controlled environment greenhouse applications also affects yield, but there are 'trade offs' when using light emitting diode (LED), versus high-intensity discharge (HID) light sources. HID lighting is generally of lower cost and generates greater photon flux density between 400 and 700 nm, while LED lighting has greater configurability for specified needs and emits substantially less heat than HID lighting; with both lighting options having similar electricity to photosynthetic photon conversion efficiencies, expressed as, $\mu mol\ J^{-1}$ [426–428]. The importance of light quality has been demonstrated in cucumber (*Cucumis sativus*) where a significant increase in dry weight was measured in plants grown under an 'artificial solar spectrum', produced by sulfur plasma and quartz-halogen lamps irradiating a light spectrum that emulated standard sunlight, when compared with those plants provided with either fluorescent or HID lighting [429]. Photosynthetic photon flux density significantly affects harvestable floral biomass yield, while elevated UV-B radiation and electrical lighting power density ($W\ m^{-2}$) increased the 'potency' of *Cannabis* through an elevation in THC concentration; all of which highlight the importance of light quantity and quality capture by the photosynthetic apparatus of this species to improve the harvestable output of cultivated *Cannabis* [423,430–433].

Manipulating temperature conditions in indoor growth facilities has revealed a relationship with factors affecting plant growth and development. Rate of photosynthesis, water use efficiency, rate of transpiration, and leaf stomatal conductance, all increased in *Cannabis* plants with a temperature increase from 20 to 30 °C, suggesting an optimal temperature range for cultivation [431]. Temperature and photosynthetic rate are tightly linked with the photosynthetic apparatus sensitive to fluctuations in temperature, responding particularly with reduced Ribulose 1,5-bisphosphate (RuBP) regeneration and lowered stomatal aperture, which together decreased CO_2 uptake; both rate reducing outcomes [434–436]. It is worthwhile to note that *Cannabis* varieties are similarly sensitive to temperature where photosynthetic rate, water use efficiency, leaf number, and stem elongation, are modulated in response to temperature change [431,437,438]. Mineral supplementation via fertilizer application has produced mixed results in terms of biomass and secondary metabolite concentration and/or profile composition in *Cannabis*. *Cannabis* was shown to be sensitive to nitrogen (N), phosphorus (P) and potassium (K) (NPK) supplementation, as well as the plant biostimulant, humic acid. The application of NPK reduced THC, CBN and CBD content, but increased CBG content in the *Cannabis* inflorescence, while the application of humic acid was found to significantly lower the THC, CBD, CBG, THCV, CBC, CBL and CBT content of the *Cannabis* inflorescence [439]. However, N supplementation alone increased hemp seed yield, plant height, chlorophyll content, while decreasing fibre yield [440]. The application of exogenous hormones during distinct developmental phases of *Cannabis* growth has also produced mixed results in relation to secondary metabolite content and biomass. Gibberellic acid (GA) application to whole flowering plants with developed, resinous trichomes reduced chlorophyll levels, DXS activity, mono- and sesquiterpene levels, and THC content, while increasing HMGR activity, to suggest a degree of interference (either directly or indirectly) by GA to both the MVA and MEP pathways [441,442]. Abscisic acid (ABA) application at the vegetative stage of *Cannabis* development, increased chlorophyll *a* content, but reduced HMGR, THC and CBD content. In contrast, ABA application at the flowering stage of development decreased total chlorophyll and HMGR content, and increased DXS activity and the content of THC in the flowers of female *Cannabis* plants, findings which again indicated either direct or indirect phytohormone-mediated interference of both the MVA and MEP pathways [441,442].

Alterations of the architecture of the *Cannabis* flower via the application of molecular-assisted breeding, or genetic engineering, are potential strategies to increase the floral yield of *Cannabis*. Alternatively, directed manipulation of the biosynthetic pathways by application of similar approaches leading to increased cannabinoid or terpenoid content would provide greater value via the targeted elevation of the exact concentration of specific secondary metabolites. Currently, research describing the implementation of such strategies in *Cannabis* are scarce. However, investigations of trichome development in *Arabidopsis* and other plant species are not. The extremely well-annotated genome of *Arabidopsis*, combined with the ease that *Arabidopsis* can be genetically manipulated, identifies *Arabidopsis* for use in baseline studies that are potentially applicable to more valuable agricultural species. Indeed, *Arabidopsis*-based studies of trichome development have revealed a cohort of genes of interest. As with the development of any specialised cell type, it is underpinned by a complex gene network, and in *Arabidopsis*, the protein products encoded by the *GLABROUS1* (*GL1*), *GL2*, *GL3* and *TRANSPARENT TESTA GLABROUS* loci are responsible for various aspects of trichome morphogenesis, maturation, branching and spatial variation [443–446]. Additional gene products have been identified as essential for correct branching patterns and trichome responses to hormones, with EMS-induced mutation to the MYB encoding gene, *TRIPTYCHON*, resulting in the 'nesting', or grouping of trichomes with higher local densities [447,448]. A gene encoding a zinc-finger transcription factor from *Arabidopsis*, *GLABROUS INFLORESCENCE STEMS*, increased glandular trichome density on the leaves, sepals, inflorescence and its branches, while also increasing the content of nicotine secretion into the glandular heads when over-expressed in tobacco plants [449]. Similarly, overexpression of a serine proteinase inhibitor, *SaPIN2a*, from American nightshade (*Solanum americanum*) in transformed tobacco, significantly increased the branching and density of glandular trichomes [450]. Regulation of the expression of the gene encoding the DXS synthase 2 (*DXS2*) enzyme, which is active in the MEP pathway in *Cannabis*, and also in tomato (*Solanum lycopersicum*) via a RNA silencing approach, resulted in an increase in trichome density on tomato leaves and reduced the accumulation of the monoterpene, β-phellandrene [451]. In cotton (*Gossypium* spp.), a mutation in the *PIGMENT GLAND FORMATION* locus, resulted in the expression of the glandless phenotype: a strategy adopted to remove toxic gossypol from cotton seeds for human consumption [452]. While the opposite phenotypic outcome of increased trichome density would be the desired result in *Cannabis* experimentation, when taken together, these findings highlight the importance of targeting specific genetic networks for molecular manipulation to initiate the expression of desired and/or designer plant phenotypes.

Increasing the biomass of agriculturally valuable species is not a novel undertaking, and anthropogenic selection has perhaps inadvertently, been conducted by humans since the dawn of agriculture. Plant height is identified as a target for manipulation in relation to overall biomass yield in maize and sorghum (*Sorghum bicolor*) [453], and in *Cannabis* grown for fibre, stem length is an important parameter for fibre yield which is affected by plant density and soil N content [454,455]. The inverse is true for *Cannabis* varieties grown for their cannabinoid content, where reduced stem lengths produce a shorter overall plant stature and correlates with a greater photoassimilate input into reproductive tissues leading to the development of floral architecture with increased accumulation of cannabinoids and terpenoids [456]. Small [3] suggests that the value of drug chemotype varieties is linked to the development of 'semi-dwarf *Cannabis* germplasm', characterised by compact, congested flowers on short branches. Such plants ultimately produce more cannabinoids due to greater resource partitioning into floral and trichome development and are of short enough stature that they can be grown at high indoor densities where the artificial environment is readily manipulated to produce greater amounts of secondary metabolites. The combination of key phenotypic traits associated with increased secondary metabolite accumulation, including dense compact floral arrangements, and semi-dwarf stature, and with novel chemotypic traits that confer targeted medical efficacy epitomises the new varieties (chemovars) to be pursued as part of a highly focused research strategy. Similar strategies that use marker assisted breeding and EMS to provide the molecular basis to generate plants that produce elevated levels of desired compounds have been undertaken in other medically significant plant species. Quantitative trait loci mapping of *Artemisia annua* L. (sweet wormwood), a plant species which produces the anti-malarial compound, artemisinin, provided the platform for marker assisted breeding programs to increase artemisinin yield [457], and by extension, revealed both the pathway for similar research that would later be undertaken in opium poppy and the avenues for the future development of similar strategies in *Cannabis*.

5.4. Papaver somniferum: Potential Parallels for Future Cannabis Research

With significant change surrounding the societal views and scientific inquiry into *Cannabis* on the horizon, it is important to look at past endeavours to envisage future directions. While *Cannabis* is a unique plant for its utility, *Papaver somniferum* (*Papaver*; opium poppy) rivals the versatility seen across *Cannabis* varieties, and given its long history of human use, it is an excellent comparison to investigate. *Papaver*, otherwise known as opium poppy, is responsible for the production of the most medically significant alkaloids, including morphine, codeine, thebaine, oripavine and noscapine. These opioids accumulate in the phloem, particularly the mesocarp capsule of *Papaver* aerial tissues in specialised cells called lactifers, which join to form a latex-containing network of anastomosing vessels [458–460]. The therapeutic efficacy of *Papaver*-derived opioids is better understood than the secondary metabolites of *Cannabis*, and the scope of their effects is far reaching. Morphine has been utilised for decades as one of the most widely used analgesics, effective in the post-operative clinical setting [461–463]. Codeine has been shown to be a less effective analgesic than morphine [464,465] but has historically been accepted as the prevailing antitussive [466]. More recent evidence suggests however, that there are more effective treatments, especially for chronic coughing disorders [467–469]. Additionally, noscapine, another *Papaver* alkaloid, displays antitussive properties, and is also suggested to potentially mitigate stroke mortality and induce apoptosis in a broad set of cancers [470–473]. Thebaine and oripavine are not themselves used therapeutically. However, they are precursors for a wide range of semi-synthetic opioids including, but not limited to, hydrocodone, oxycodone and hydromorphone, as well as naloxone, which is interestingly employed to treat the acute effects of opioid overdose [474–479].

Given the multitude of efficacious compounds produced by *Papaver*, and the commercial value emanating from such, the desire to generate plant varieties that produce specific chemical profiles is one that is mirrored in *Cannabis*. While the latter is currently reliant on years of predominantly illicit breeding programs to produce plants with increased psychoactive properties, the development of novel *Papaver* varieties has already been established. EMS treatment of poppy seeds preceded the identification of a variety termed *top1* (*thebaine oripavine poppy 1*) which harboured a mutation leading to premature arrest of the morphine and codeine biosynthesis pathway. The resulting *top1* plants displayed a pigmented latex, and the enhanced accumulation of thebaine and oripavine, but failed to produce either codeine or morphine [480]. Similarly, a reduction in codeine 3-O-demethylase (CODM) activity, via either a viral-induced gene silencing (VIGS) strategy, or a fast neutron bombardment mutagenesis approach, yielded *Papaver* plants with enhanced codeine accumulation, but which were unable to synthesise morphine from a codeine substrate [481–483]. These high codeine *Papaver* varieties that harbour CODM polymorphisms, provided a basis for a marker-assisted breeding platform, and to produce *Papaver* chemotypes accumulating novel alkaloid profiles [483]. Similar actions utilising *Cannabis* may also mediate alterations to the cannabinoid biosynthesis pathways to produce varieties with elevated minor cannabinoid content. Sequencing of a high noscapine variety of *Papaver*, termed HN1, led to the discovery of a 10 gene cluster responsible for noscapine biosynthesis that was absent in either a high morphine (HM1)- or high thebaine (HT1)-producing variety of *Papaver* [484]. Generation of an F_2 mapping population from HN1 and HM1 parents showed tight linkage of this gene cluster, revealing high noscapine-producing progeny that were homozygous for the HN1 gene cluster, while heterozygosity, or absence of the HN1 gene cluster, was associated with plant lines that produced low or undetectable levels of noscapine, respectively [484]. The identification of the *STORR* (*[S]- to [R]-reticuline*) locus led to the development of high noscapine *Papaver* varieties with a non-functioning cytochrome P450-oxidoreductase fusion protein, inhibiting the [S]-reticuline conversion to [R]-reticuline necessary for completion of the morphinan pathway [485–488]. A VIGS approach has been successfully utilised to individually regulate the expression of six genes encoding enzymes involved in the final six conversion steps of [R]-reticuline to morphine, each of which were shown to alter the major alkaloid profile [489]. An RNA silencing approach which employed a chimeric hairpin RNA to target all members of the multigene codeinone reductase family produced a non-narcotic, [S]-reticuline-accumulating variety of *Papaver* [490]. In the exploitation of the versatility of *Papaver* beyond narcotics, varieties with high food value have been established through EMS and gamma ray mutagenesis breeding programs to produce increased seed yield (5.66 g/capsule versus the 3.39 g/capsule of control plants) with elevated levels of unsaturated seed oil and no narcotic production [491]. While this is not an exhaustive list of selectively bred, or engineered *Papaver* varieties, the long-standing and successful development of *Papaver* varieties with superior phenotypic and/or chemotypic traits of interest certainly provides a reference for guiding *Cannabis* research strategies, which at present are comparatively in their infancy. Development of varieties producing high levels of alkaloid biosynthetic pathway intermediates is a promising indicator for the potential production of *Cannabis* varieties that reliably produce high levels of minor cannabinoids or intermediates in the cannabinoid biosynthesis pathway.

6. Conclusions

In summary, we have reviewed the current literature of several important aspects of cannabinoid research outside of THC and CBD, which dominate discussion in the *Cannabis* research field. Emerging research has begun to reveal the pharmacology and molecular targets of the minor cannabinoids. Due to the wide spectrum of molecular effects involved with cannabinoid consumption, it is clear that there are a range of medical ailments that could be addressed through endocannabinoid augmentation using secondary metabolites of *Cannabis*. Here, we have illustrated that via the utilisation of specific minor cannabinoids, which share some, but not all targets of THC and CBD, the medical reach of cannabinoid-containing pharmaceuticals could potentially be broadened. However, there are many challenges that currently impede this possibility, even outside of the international legal environment. Firstly, there is further room for significant characterisation of minor cannabinoid pharmacology, and currently, disease-orientated preclinical and clinical trials are lacking. Critically, techniques for producing cannabinoid isolates—even CBD and THC—are still in their infancy, and this remains a clear barrier to large-scale commercialisation of pharmaceutical cannabinoids. Here, we have reviewed the currently available literature which covers the processes involved in the biosynthesis of cannabinoids, as well as the techniques involved in the production of novel *Cannabis* chemotypes, including methods of improving yield that might be adopted from historically similar cases, such as the opioid industry. Based on this historical example, and the existing literature, it is likely that a molecular genetic modification approach will be applied to *Cannabis* to generate new opportunities for the improved yield of specific minor and major cannabinoids in the near future. In conclusion, there are multiple enticing and potentially profitable opportunities for commercial and academic growth in the *Cannabis* market outside of THC and CBD, and here, we highlight some of the most important current perspectives of this growing industry.

Author Contributions: Conceptualization and design, J.M.J.O., J.L.P., T.A.B., A.L.E. and C.P.L.G.; writing, J.M.J.O., J.L.P., T.A.B., A.L.E., L.J.N. and C.P.L.G.; review and editing, J.M.J.O., J.L.P., T.A.B., A.L.E., L.J.N. and C.P.L.G. All authors have read and agreed to the published version of the manuscript.

Funding: J.M.J.O.'s research is supported by CannaPacific Pty Ltd., Australia.

Conflicts of Interest: The authors declare no conflict of interest.

References

1. Clarke, R.C.; Merlin, M.D. Letter to the Editor: Small, Ernest. 2015. Evolution and Classification of *Cannabis sativa* (Marijuana, Hemp) in Relation to Human Utilization. *Bot. Rev.* **2015**, *81*, 295–305. [CrossRef]
2. Schultes, R.E.; Klein, W.M.; Plowman, T.; Lockwood, T.E. Cannabis: An Example of Taxonomic Neglect. *Bot. Museum Leafl. Harvard Univ.* **1974**, *23*, 337–367.
3. Small, E. Evolution and Classification of *Cannabis sativa* (Marijuana, Hemp) in Relation to Human Utilization. *Bot. Rev.* **2015**, *81*, 189–294. [CrossRef]
4. Small, E.; Cronquist, A. A Practical and Natural Taxonomy for Cannabis. *Taxon* **1976**, *25*, 405–435. [CrossRef]
5. Small, E.; Naraine, S.G.U. Size Matters: Evolution of Large Drug-Secreting Resin Glands in Elite Pharmaceutical Strains of *Cannabis sativa* (Marijuana). *Genet. Resour. Crop Evol.* **2016**, *254*, 349–359. [CrossRef]
6. Salentijn, E.M.J.; Zhang, Q.; Amaducci, S.; Yang, M.; Trindade, L.M. New Developments in Fiber Hemp (*Cannabis sativa* L.) Breeding. *Ind. Crops Prod.* **2015**, *68*, 32–41. [CrossRef]
7. Mead, A. The Legal Status of Cannabis (Marijuana) and Cannabidiol (CBD) under U.S. Law. *Epilepsy Behav.* **2017**, *70*, 288–291. [CrossRef]
8. Amaducci, S.; Scordia, D.; Liu, F.H.; Zhang, Q.; Guo, H.; Testa, G.; Cosentino, S.L. Key Cultivation Techniques for Hemp in Europe and China. *Ind. Crop. Prod.* **2015**, *68*, 2–16. [CrossRef]
9. Bielecka, M.; Kaminski, F.; Adams, I.; Poulson, H.; Sloan, R.; Li, Y.; Larson, T.R.; Winzer, T.; Graham, I.A. Targeted Mutation of Δ12 and Δ15 Desaturase Genes in Hemp Produce Major Alterations in Seed Fatty Acid Composition Including a High Oleic Hemp Oil. *Plant Biotechnol. J.* **2014**, *12*, 613–623. [CrossRef] [PubMed]
10. Callaway, J.C.; Tennilä, T.; Pate, D.W. Occurence of "Omega 3" Stearidonic Acid (Cis-6,9,12,15-Ocadecatetraenoic Acid) in Hemp (*Cannabis sativa* L.) Seed. *J. Int. Hemp Assoc.* **1996**, *3*, 61–63.
11. Dimić, E.; Romanić, R.; Vujasinović, V. Essential Fatty Acids, Nutritive Value and Oxidative Stability of Cold Pressed Hempseed (*Cannabis sativa* L.) Oil from Different Varieties. *Acta Aliment.* **2009**, *38*, 229–236. [CrossRef]
12. Deferne, J.; Pate, D.W. Hemp Seed Oil: A Source of Valuable Essential Fatty Acids. *J. Int. Hemp Assoc.* **1996**, *3*, 4–7.
13. Erasmus, U. *Fats That Heal, Fats That Kill: The Complete Guide to Fats, Oils, Cholesterol, and Human Health*, 3rd ed.; Alive Books: Burnaby, BC, Canada, 1993.
14. Stubbs, C.D.; Smith, A.D. The Modification of Mammalian Membrane Polyunsaturated Fatty Acid Composition in Relation to Membrane Fluidity and Function. *Biochim. Biophys. Acta* **1984**, *779*, 89–137. [CrossRef]
15. Callaway, J.C. Hempseed as a Nutritional Resource: An Overview. *Euphytica* **2004**, *140*, 65–72. [CrossRef]
16. Harbige, L.S.; Layward, L.; Morris-Downes, M.M.; Dumonde, D.C.; Amor, S. The Protective Effects of Omega-6 Fatty Acids in Experimental Autoimmune Encephalomyelitis (EAE) in Relation to Transforming Growth Factor-Beta 1 (TGF-B1) up-Regulation and Increased Prostaglandin E2 (PGE2) Production. *Clin. Exp. Immunol.* **2000**, *122*, 445–452. [CrossRef] [PubMed]
17. Prociuk, M.A.; Edel, A.L.; Richard, M.N.; Gavel, N.T.; Ander, B.P.; Dupasquier, C.M.C.; Pierce, G.N. Cholesterol-Induced Stimulation of Platelet Aggregation Is Prevented by a Hempseed-Enriched Diet. *Can. J. Physiol. Pharmacol.* **2008**, *86*, 153–159. [CrossRef]
18. Clarke, R.C.; Merlin, M.D. Cannabis Domestication, Breeding History, Present-Day Genetic Diversity, and Future Prospects. *CRC Crit. Rev. Plant Sci.* **2016**, *35*, 293–327. [CrossRef]
19. Dayanandan, P.; Kaufman, P.B. Trichomes of *Cannabis sativa* L. (Cannabaceae). *Am. J. Bot.* **1976**, *63*, 578–591. [CrossRef]
20. ElSohly, M.A.; Slade, D. Chemical Constituents of Marijuana: The Complex Mixture of Natural Cannabinoids. *Life Sci.* **2005**, *78*, 539–548. [CrossRef]
21. Pertwee, R.G. (Ed.) *Handbook of Cannabis*; Oxford University Press: Oxford, UK, 2015. [CrossRef]
22. Citti, C.; Linciano, P.; Russo, F.; Luongo, L.; Iannotta, M.; Maione, S.; Laganà, A.; Capriotti, A.L.; Forni, F.; Vandelli, M.A.; et al. A Novel Phytocannabinoid Isolated from *Cannabis sativa* L. with an in Vivo Cannabimimetic Activity Higher than Δ9-Tetrahydrocannabinol: Δ9-Tetrahydrocannabiphorol. *Sci. Rep.* **2019**, *9*, 20335. [CrossRef] [PubMed]
23. Babaei, M.; Ajdanian, L. Screening of Different Iranian Ecotypes of Cannabis under Water Deficit Stress. *Sci. Hortic.* **2020**, *260*. [CrossRef]
24. Linger, P.; Ostwald, A.; Haensler, J. *Cannabis sativa* L. Growing on Heavy Metal Contaminated Soil: Growth, Cadmium Uptake and Photosynthesis. *Biol. Plant.* **2005**, *49*, 567–576. [CrossRef]
25. Bouquet, R.J. Cannabis. *Bull. Narc.* **1950**, *2*, 14–30.
26. Amaducci, S.; Zatta, A.; Raffanini, M.; Venturi, G. Characterisation of Hemp (*Cannabis sativa* L.) Roots under Different Growing Conditions. *Plant Soil* **2008**, *313*. [CrossRef]
27. Small, E.; Marcus, D.; Butler, G.; McElroy, A.R. Apparent Increase in Biomass and Seed Productivity in Hemp (*Cannabis sativa*) Resulting from Branch Proliferation Caused by the European Corn Borer (*Ostrinia nubilalis*). *J. Ind. Hemp* **2007**, *12*, 15–26. [CrossRef]
28. Government of Canada. Department of Justice. Available online: https://www.justice.gc.ca/eng/cj-jp/cannabis/ (accessed on 15 January 2020).
29. Ney, L.J.; Matthews, A.; Bruno, R.; Felmingham, K.L. Cannabinoid Interventions for PTSD: Where to Next? *Prog. Neuro-Psychopharmacol. Biol. Psychiatry* **2019**, *93*, 124–140. [CrossRef]

30. Lattanzi, S.; Brigo, F.; Trinka, E.; Zaccara, G.; Cagnetti, C.; Del Giovane, C.; Silvestrini, M. Efficacy and Safety of Cannabidiol in Epilepsy: A Systematic Review and Meta-Analysis. *Drugs* **2018**, *78*, 1791–1804. [CrossRef] [PubMed]

31. Patel, S.; Hill, M.N.; Cheer, J.F.; Wotjak, C.T.; Holmes, A. The Endocannabinoid System as a Target for Novel Anxiolytic Drugs. *Neurosci. Biobehav. Rev.* **2017**, *76*, 56–66. [CrossRef] [PubMed]

32. Whiting, P.F.; Wolff, R.F.; Deshpande, S.; Di Nisio, M.; Duffy, S.; Hernandez, A.V.; Keurentjes, J.C.; Lang, S.; Misso, K.; Ryder, S.; et al. Cannabinoids for Medical Use: A Systematic Review and Meta-Analysis. *JAMA J. Am. Med. Assoc.* **2015**, *313*, 2456–2473. [CrossRef] [PubMed]

33. Wade, D.T.; Collin, C.; Stott, C.; Duncombe, P. Meta-Analysis of the Efficacy and Safety of Sativex (Nabiximols), on Spasticity in People with Multiple Sclerosis. *Mult. Scler.* **2010**, *16*, 707–714. [CrossRef] [PubMed]

34. Stevens, A.J.; Higgins, M.D. A Systematic Review of the Analgesic Efficacy of Cannabinoid Medications in the Management of Acute Pain. *Acta Anaesthesiol. Scand.* **2017**, *61*, 268–280. [CrossRef]

35. Morales, P.; Jagerovic, N. Novel Approaches and Current Challenges with Targeting the Endocannabinoid System. *Expert Opin. Drug Discov.* **2020**, *15*, 917–930. [CrossRef]

36. Morales, P.; Hurst, D.P.; Reggio, P.H. Molecular Targets of the Phytocannabinoids: A Complex Picture. *Prog. Chem. Org. Nat. Prod.* **2017**, *103*, 103–131. [CrossRef]

37. Sampson, P.B. Phytocannabinoid Pharmacology: Medicinal Properties of *Cannabis sativa* Constituents Aside from the "Big Two". *J. Nat. Prod.* **2020**, *84*, 142–160. [CrossRef]

38. Gaoni, Y.; Mechoulam, R. Isolation, Structure, and Partial Synthesis of an Active Constituent of Hashish. *J. Am. Chem. Soc.* **1964**, *86*, 1646–1647. [CrossRef]

39. Mechoulam, R.; Gaoni, Y. A Total Synthesis of Dl-Δ1-Tetrahydrocannabinol, the Active Constituent of Hashish. *J. Am. Chem. Soc.* **1965**, *87*, 3273–3275. [CrossRef] [PubMed]

40. Devane, W.A.; Dysarz, F.A.; Johnson, M.R.; Melvin, L.S.; Howlett, A.C. Determination and Characterization of a Cannabinoid Receptor in Rat Brain. *Mol. Pharmacol.* **1988**, *34*, 605–613.

41. Munro, S.; Thomas, K.L.; Abu-Shaar, M. Molecular Characterization of a Peripheral Receptor for Cannabinoids. *Nature* **1993**, *365*, 61–65. [CrossRef]

42. Herkenham, M.; Lynn, A.B.; Little, M.D.; Johnson, M.R.; Melvin, L.S.; De Costa, B.R.; Rice, K.C. Cannabinoid Receptor Localization in Brain. *Proc. Natl. Acad. Sci. USA* **1990**, *87*, 1932–1936. [CrossRef] [PubMed]

43. Devane, W.A.; Hanuš, L.; Breuer, A.; Pertwee, R.G.; Stevenson, L.A.; Griffin, G.; Gibson, D.; Mandelbaum, A.; Etinger, A.; Mechoulam, R. Isolation and Structure of a Brain Constituent That Binds to the Cannabinoid Receptor. *Science* **1992**, *258*, 1946–1949. [CrossRef]

44. Mechoulam, R.; Ben-Shabat, S.; Hanus, L.; Ligumsky, M.; Kaminski, N.E.; Schatz, A.R.; Gopher, A.; Almog, S.; Martin, B.R.; Compton, D.R.; et al. Identification of an Endogenous 2-Monoglyceride, Present in Canine Gut, That Binds to Cannabinoid Receptors. *Biochem. Pharmacol.* **1995**, *50*, 83–90. [CrossRef]

45. Sugiura, T.; Kondo, S.; Sukagawa, A.; Nakane, S.; Shinoda, A.; Itoh, K.; Yamashita, A.; Waku, K. 2-Arachidonoylglycerol: A Possible Endogenous Cannabinoid Receptor Ligand in Brain. *Biochem. Biophys. Res. Commun.* **1995**, *215*, 89–97. [CrossRef] [PubMed]

46. Liu, J.; Wang, L.; Harvey-White, J.; Osei-Hyiaman, D.; Razdan, R.; Gong, Q.; Chan, A.C.; Zhou, Z.; Huang, B.X.; Kim, H.Y.; et al. A Biosynthetic Pathway for Anandamide. *Proc. Natl. Acad. Sci. USA* **2006**, *103*, 13345–13350. [CrossRef]

47. Mechoulam, R.; Parker, L.A. The Endocannabinoid System and the Brain. *Annu. Rev. Psychol.* **2013**, *64*, 21–47. [CrossRef]

48. Di Marzo, V. The Endocannabinoid System: Its General Strategy of Action, Tools for Its Pharmacological Manipulation and Potential Therapeutic Exploitation. *Pharmacol. Res.* **2009**, *60*, 77–84. [CrossRef] [PubMed]

49. Kano, M.; Ohno-Shosaku, T.; Hashimotodani, Y.; Uchigashima, M.; Watanabe, M. Endocannabinoid-Mediated Control of Synaptic Transmission. *Physiol. Rev.* **2009**, *89*, 309–380. [CrossRef] [PubMed]

50. Tripathi, R.K.P. A Perspective Review on Fatty Acid Amide Hydrolase (FAAH) Inhibitors as Potential Therapeutic Agents. *Eur. J. Med. Chem.* **2020**, *188*, 111953. [CrossRef] [PubMed]

51. Ligresti, A.; Cascio, M.G.; Pryce, G.; Kulasegram, S.; Beletskaya, I.; De Petrocellis, L.; Saha, B.; Mahadevan, A.; Visintin, C.; Wiley, J.L.; et al. New Potent and Selective Inhibitors of Anandamide Reuptake with Antispastic Activity in a Mouse Model of Multiple Sclerosis. *Br. J. Pharmacol.* **2006**, *147*, 83–91. [CrossRef] [PubMed]

52. Glass, M.; Dragunow, M.; Faull, R.L.M. Cannabinoid Receptors in the Human Brain: A Detailed Anatomical and Quantitative Autoradiographic Study in the Fetal, Neonatal and Adult Human Brain. *Neuroscience* **1997**, *77*, 299–318. [CrossRef]

53. Hua, T.; Vemuri, K.; Pu, M.; Qu, L.; Han, G.W.; Wu, Y.; Zhao, S.; Shui, W.; Li, S.; Korde, A.; et al. Crystal Structure of the Human Cannabinoid Receptor CB1. *Cell* **2016**, *167*, 750–762. [CrossRef]

54. Li, X.; Hua, T.; Vemuri, K.; Ho, J.H.; Wu, Y.; Wu, L.; Popov, P.; Benchama, O.; Zvonok, N.; Locke, K.; et al. Crystal Structure of the Human Cannabinoid Receptor CB2. *Cell* **2019**, *176*, 459–467. [CrossRef]

55. Katona, I.; Sperlágh, B.; Sík, A.; Käfalvi, A.; Vizi, E.S.; Mackie, K.; Freund, T.F. Presynaptically Located CB1 Cannabinoid Receptors Regulate GABA Release from Axon Terminals of Specific Hippocampal Interneurons. *J. Neurosci.* **1999**, *19*, 4544–4558. [CrossRef]

56. Derkinderen, P.; Ledent, C.; Parmentier, M.; Girault, J.A. Cannabinoids Activate P38 Mitogen-Activated Protein Kinases through CB1 Receptors in Hippocampus. *J. Neurochem.* **2001**, *77*, 957–960. [CrossRef]

57. Herkenham, M.; Lynn, A.B.; de Costa, B.R.; Richfield, E.K. Neuronal Localization of Cannabinoid Receptors in the Basal Ganglia of the Rat. *Brain Res.* **1991**, *547*, 267–274. [CrossRef]

58. Mátyás, F.; Yanovsky, Y.; Mackie, K.; Kelsch, W.; Misgeld, U.; Freund, T.F. Subcellular Localization of Type 1 Cannabinoid Receptors in the Rat Basal Ganglia. *Neuroscience* **2006**, *137*, 337–361. [CrossRef]

59. Bidaut-Russell, M.; Devane, W.A.; Howlett, A.C. Cannabinoid Receptors and Modulation of Cyclic AMP Accumulation in the Rat Brain. *J. Neurochem.* **1990**, *55*, 21–26. [CrossRef]

60. Kawamura, Y.; Fukaya, M.; Maejima, T.; Yoshida, T.; Miura, E.; Watanabe, M.; Ohno-Shosaku, T.; Kano, M. The CB1 Cannabinoid Receptor Is the Major Cannabinoid Receptor at Excitatory Presynaptic Sites in the Hippocampus and Cerebellum. *J. Neurosci.* **2006**, *26*, 2991–3001. [CrossRef]

61. Kreitzer, A.C.; Regehr, W.G. Cerebellar Depolarization-Induced Suppression of Inhibition Is Mediated by Endogenous Cannabinoids. *J. Neurosci.* **2001**, *21*, RC174. [CrossRef] [PubMed]

62. Lévénès, C.; Daniel, H.; Soubrié, P.; Crépel, F. Cannabinoids Decrease Excitatory Synaptic Transmission and Impair Long-Term Depression in Rat Cerebellar Purkinje Cells. *J. Physiol.* **1998**, *510*, 867–879. [CrossRef]

63. Ohno-Shosaku, T.; Maejima, T.; Kano, M. Endogenous Cannabinoids Mediate Retrograde Signals from Depolarized Postsynaptic Neurons to Presynaptic Terminals. *Neuron* **2001**, *29*, 729–738. [CrossRef]

64. Ney, L.J.; Akhurst, J.; Bruno, R.; Laing, P.A.F.; Matthews, A.; Felmingham, K.L. Dopamine, Endocannabinoids and Their Interaction in Fear Extinction and Negative Affect in PTSD. *Prog. Neuro-Psychopharmacol. Biol. Psychiatry* **2021**, *105*, 110118. [CrossRef] [PubMed]

65. Balsevich, G.; Petrie, G.N.; Hill, M.N. Endocannabinoids: Effectors of Glucocorticoid Signaling. *Front. Neuroendocrinol.* **2017**, *47*, 86–108. [CrossRef] [PubMed]

66. González-Mariscal, I.; Krzysik-Walker, S.M.; Doyle, M.E.; Liu, Q.R.; Cimbro, R.; Santa-Cruz Calvo, S.; Ghosh, S.; Cieala, A.; Moaddel, R.; Carlson, O.D.; et al. Human CB1 Receptor Isoforms, Present in Hepatocytes and β-Cells, Are Involved in Regulating Metabolism. *Sci. Rep.* **2016**, *6*, 33302. [CrossRef]

67. Nogueiras, R.; Veyrat-Durebex, C.; Suchanek, P.M.; Klein, M.; Tschöp, J.; Caldwell, C.; Woods, S.C.; Wittmann, G.; Watanabe, M.; Liposits, Z.; et al. Peripheral, but Not Central, CB1 Antagonism Provides Food Intake-Independent Metabolic Benefits in Diet-Induced Obese Rats. *Diabetes* **2008**, *57*, 2977–2991. [CrossRef]

68. Ashton, J.C.; Friberg, D.; Darlington, C.L.; Smith, P.F. Expression of the Cannabinoid CB2 Receptor in the Rat Cerebellum: An Immunohistochemical Study. *Neurosci. Lett.* **2006**, *396*, 113–116. [CrossRef] [PubMed]

69. Núñez, E.; Benito, C.; Pazos, M.R.; Barbachano, A.; Fajardo, O.; González, S.; Tolón, R.M.; Romero, J. Cannabinoid CB2 Receptors Are Expressed by Perivascular Microglial Cells in the Human Brain: An Immunohistochemical Study. *Synapse* **2004**, *53*, 208–213. [CrossRef] [PubMed]

70. Ross, R.A.; Coutts, A.A.; McFarlane, S.M.; Anavi-Goffer, S.; Irving, A.J.; Pertwee, R.G.; MacEwan, D.J.; Scott, R.H. Actions of Cannabinoid Receptor Ligands on Rat Cultured Sensory Neurones: Implications for Antinociception. *Neuropharmacology* **2001**, *40*, 221–232. [CrossRef]

71. Van Sickle, M.D.; Duncan, M.; Kingsley, P.J.; Mouihate, A.; Urbani, P.; Mackie, K.; Stella, N.; Makriyannis, A.; Piomelli, D.; Davison, J.S.; et al. Identification and Functional Characterization of Brainstem Cannabinoid CB2 Receptors. *Science* **2005**, *310*, 329–332. [CrossRef]

72. Wotherspoon, G.; Fox, A.; McIntyre, P.; Colley, S.; Bevan, S.; Winter, J. Peripheral Nerve Injury Induces Cannabinoid Receptor 2 Protein Expression in Rat Sensory Neurons. *Neuroscience* **2005**, *135*, 235–245. [CrossRef]

73. Galiègue, S.; Mary, S.; Marchand, J.; Dussossoy, D.; Carrière, D.; Carayon, P.; Bouaboula, M.; Shire, D.; LE Fur, G.; Casellas, P. Expression of Central and Peripheral Cannabinoid Receptors in Human Immune Tissues and Leukocyte Subpopulations. *Eur. J. Biochem.* **1995**, *232*, 54–61. [CrossRef] [PubMed]

74. Lynn, A.B.; Herkenham, M. Localization of Cannabinoid Receptors and Nonsaturable High-Density Cannabinoid Binding Sites in Peripheral Tissues of the Rat: Implications for Receptor-Mediated Immune Modulation by Cannabinoids. *J. Pharmacol. Exp. Ther.* **1994**, *268*, 1612–1623.

75. Facci, L.; Dal Toso, R.; Romanello, S.; Buriani, A.; Skaper, S.D.; Leon, A. Mast Cells Express a Peripheral Cannabinoid Receptor with Differential Sensitivity to Anandamide and Palmitoylethanolamide. *Proc. Natl. Acad. Sci. USA* **1995**, *92*, 3376–3380. [CrossRef] [PubMed]

76. Lam, P.M.W.; Marczylo, T.H.; El-Talatini, M.; Finney, M.; Nallendran, V.; Taylor, A.H.; Konje, J.C. Ultra Performance Liquid Chromatography Tandem Mass Spectrometry Method for the Measurement of Anandamide in Human Plasma. *Anal. Biochem.* **2008**, *380*, 195–201. [CrossRef] [PubMed]

77. Fanelli, F.; Di Lallo, V.D.; Belluomo, I.; De Iasio, R.; Baccini, M.; Casadio, E.; Gasparini, D.I.; Colavita, M.; Gambineri, A.; Grossi, G.; et al. Estimation of Reference Intervals of Five Endocannabinoids and Endocannabinoid Related Compounds in Human Plasma by Two Dimensional-LC/MS/MS. *J. Lipid Res.* **2012**, *53*, 481–493. [CrossRef] [PubMed]

78. Krumbholz, A.; Anielski, P.; Reisch, N.; Schelling, G.; Thieme, D. Diagnostic Value of Concentration Profiles of Glucocorticosteroids and Endocannabinoids in Hair. *Ther. Drug Monit.* **2013**, *35*, 600–607. [CrossRef]

79. Mwanza, C.; Chen, Z.; Zhang, Q.; Chen, S.; Wang, W.; Deng, H. Simultaneous HPLC-APCI-MS/MS Quantification of Endogenous Cannabinoids and Glucocorticoids in Hair. *J. Chromatogr. B Anal. Technol. Biomed. Life Sci.* **2016**, *1028*, 1–10. [CrossRef]

80. Voegel, C.D.; Baumgartner, M.R.; Kraemer, T.; Wüst, S.; Binz, T.M. Simultaneous Quantification of Steroid Hormones and Endocannabinoids (ECs) in Human Hair Using an Automated Supported Liquid Extraction (SLE) and LC-MS/MS—Insights into EC Baseline Values and Correlation to Steroid Concentrations. *Talanta* **2021**, *222*, 121499. [CrossRef]

81. Ney, L.J.; Felmingham, K.L.; Bruno, R.; Matthews, A.; Nichols, D.S. Simultaneous Quantification of Endocannabinoids, Oleoylethanolamide and Steroid Hormones in Human Plasma and Saliva. *J. Chromatogr. B Anal. Technol. Biomed. Life Sci.* **2020**, *1152*, 122252. [CrossRef]

82. Ney, L.J.; Stone, C.; Nichols, D.; Felmingham, K.L.; Bruno, R.; Matthews, A. Endocannabinoid Reactivity to Acute Stress: Investigation of the Relationship between Salivary and Plasma Levels. *Biol. Psychol.* **2021**, *159*, 108022. [CrossRef]

83. Matias, I.; Gatta-Cherifi, B.; Tabarin, A.; Clark, S.; Leste-Lasserre, T.; Marsicano, G.; Piazza, P.V.; Cota, D. Endocannabinoids Measurement in Human Saliva as Potential Biomarker of Obesity. *PLoS ONE* **2012**, *7*, e42399. [CrossRef] [PubMed]

84. Schuel, H.; Burkman, L.J.; Lippes, J.; Crickard, K.; Forester, E.; Piomelli, D.; Giuffrida, A. N-Acylethanolamines in Human Reproductive Fluids. *Chem. Phys. Lipids* **2002**, *121*, 211–227. [CrossRef]

85. Lam, M.P.Y.; Siu, S.O.; Lau, E.; Mao, X.; Sun, H.Z.; Chiu, P.C.N.; Yeung, W.S.B.; Cox, D.M.; Chu, I.K. Online Coupling of Reverse-Phase and Hydrophilic Interaction Liquid Chromatography for Protein and Glycoprotein Characterization. *Anal. Bioanal. Chem.* **2010**, *398*, 791–804. [CrossRef] [PubMed]

86. Lewis, S.E.M.; Rapino, C.; Di Tommaso, M.; Pucci, M.; Battista, N.; Paro, R.; Simon, L.; Lutton, D.; Maccarrone, M. Differences in the Endocannabinoid System of Sperm from Fertile and Infertile Men. *PLoS ONE* **2012**, *7*, e47704. [CrossRef]

87. Malan, T.P.; Ibrahim, M.M.; Deng, H.; Liu, Q.; Mata, H.P.; Vanderah, T.; Porreca, F.; Makriyannis, A. CB2 Cannabinoid Receptor-Mediated Peripheral Antinociception. *Pain* **2001**, *93*, 239–245. [CrossRef]

88. Jaggar, S.I.; Hasnie, F.S.; Sellaturay, S.; Rice, A.S.C. The Anti-Hyperalgesic Actions of the Cannabinoid Anandamide and the Putative CB2 Receptor Agonist Palmitoylethanolamide in Visceral and Somatic Inflammatory Pain. *Pain* **1998**, *76*, 189–199. [CrossRef]

89. Katona, I.; Freund, T.F. Multiple Functions of Endocannabinoid Signaling in the Brain. *Annu. Rev. Neurosci.* **2012**, *35*, 529–558. [CrossRef] [PubMed]

90. Ligresti, A.; De Petrocellis, L.; Di Marzo, V. From Phytocannabinoids to Cannabinoid Receptors and Endocannabinoids: Pleiotropic Physiological and Pathological Roles through Complex Pharmacology. *Physiol. Rev.* **2016**, *96*, 1593–1659. [CrossRef]

91. Ryberg, E.; Larsson, N.; Sjögren, S.; Hjorth, S.; Hermansson, N.O.; Leonova, J.; Elebring, T.; Nilsson, K.; Drmota, T.; Greasley, P.J. The Orphan Receptor GPR55 Is a Novel Cannabinoid Receptor. *Br. J. Pharmacol.* **2007**, *152*, 1092–1101. [CrossRef] [PubMed]

92. Li, K.; Fichna, J.; Schicho, R.; Saur, D.; Bashashati, M.; MacKie, K.; Li, Y.; Zimmer, A.; Göke, B.; Sharkey, K.A.; et al. A Role for O-1602 and G Protein-Coupled Receptor GPR55 in the Control of Colonic Motility in Mice. *Neuropharmacology* **2013**, *71*, 255–263. [CrossRef]

93. Ross, G.R.; Lichtman, A.; Dewey, W.L.; Akbarali, H.I. Evidence for the Putative Cannabinoid Receptor (GPR55)-Mediated Inhibitory Effects on Intestinal Contractility in Mice. *Pharmacology* **2012**, *90*, 55–65. [CrossRef]

94. Lin, X.H.; Yuece, B.; Li, Y.Y.; Feng, Y.J.; Feng, J.Y.; Yu, L.Y.; Li, K.; Li, Y.N.; Storr, M. A Novel CB Receptor GPR55 and Its Ligands Are Involved in Regulation of Gut Movement in Rodents. *Neurogastroenterol. Motil.* **2011**, *23*, 862-e342. [CrossRef] [PubMed]

95. Galiazzo, G.; Giancola, F.; Stanzani, A.; Fracassi, F.; Bernardini, C.; Forni, M.; Pietra, M.; Chiocchetti, R. Localization of Cannabinoid Receptors CB1, CB2, GPR55, and PPARα in the Canine Gastrointestinal Tract. *Histochem. Cell Biol.* **2018**, *150*, 187–205. [CrossRef] [PubMed]

96. Oka, S.; Nakajima, K.; Yamashita, A.; Kishimoto, S.; Sugiura, T. Identification of GPR55 as a Lysophosphatidylinositol Receptor. *Biochem. Biophys. Res. Commun.* **2007**, *362*, 928–934. [CrossRef] [PubMed]

97. Lauckner, J.E.; Jensen, J.B.; Chen, H.Y.; Lu, H.C.; Hille, B.; Mackie, K. GPR55 Is a Cannabinoid Receptor That Increases Intracellular Calcium and Inhibits M Current. *Proc. Natl. Acad. Sci. USA* **2008**, *105*, 2699–2704. [CrossRef] [PubMed]

98. Waldeck-Welermair, M.; Zoratti, C.; Osibow, K.; Balenga, N.; Goessnitzer, E.; Waldhoer, M.; Malli, R.; Graier, W.F. Integrin Clustering Enables Anandamide-Induced Ca2+ Signaling in Endothelial Cells via GPR55 by Protection against CB1-Receptor-Triggered Repression. *J. Cell Sci.* **2008**, *121*, 1704–1717. [CrossRef]

99. Liu, Z.; Lee, J.; Krummey, S.; Lu, W.; Cai, H.; Lenardo, M.J. The Kinase LRRK2 Is a Regulator of the Transcription Factor NFAT That Modulates the Severity of Inflammatory Bowel Disease. *Nat. Immunol.* **2011**, *12*, 1063–1070. [CrossRef]

100. Henstridge, C.M.; Balenga, N.A.; Schröder, R.; Kargl, J.K.; Platzer, W.; Martini, L.; Arthur, S.; Penman, J.; Whistler, J.L.; Kostenis, E.; et al. GPR55 Ligands Promote Receptor Coupling to Multiple Signalling Pathways. *Br. J. Pharmacol.* **2010**, *160*, 604–614. [CrossRef]

101. Badrichani, A.Z.; Stroka, D.M.; Bilbao, G.; Curiel, D.T.; Bach, F.H.; Ferran, C. Bcl-2 and Bcl-X(L) Serve an Anti-Inflammatory Function in Endothelial Cells through Inhibition of NF-KB. *J. Clin. Investig.* **1999**, *103*, 543–553. [CrossRef]

102. Wu, C.S.; Chen, H.; Sun, H.; Zhu, J.; Jew, C.P.; Wager-Miller, J.; Straiker, A.; Spencer, C.; Bradshaw, H.; Mackie, K.; et al. GPR55, a G-Protein Coupled Receptor for Lysophosphatidylinositol, Plays a Role in Motor Coordination. *PLoS ONE* **2013**, *8*, e60314. [CrossRef]

103. Celorrio, M.; Rojo-Bustamante, E.; Fernández-Suárez, D.; Sáez, E.; Estella-Hermoso de Mendoza, A.; Müller, C.E.; Ramírez, M.J.; Oyarzábal, J.; Franco, R.; Aymerich, M.S. GPR55: A Therapeutic Target for Parkinson's Disease? *Neuropharmacology* **2017**, *125*, 319–332. [CrossRef]

104. Sawzdargo, M.; Nguyen, T.; Lee, D.K.; Lynch, K.R.; Cheng, R.; Heng, H.H.Q.; George, S.R.; O'Dowd, B.F. Identification and Cloning of Three Novel Human G Protein-Coupled Receptor Genes GPR52, ΨGPR53 and GPR55: GPR55 Is Extensively Expressed in Human Brain. *Mol. Brain Res.* **1999**, *64*, 193–198. [CrossRef]

105. Schuelert, N.; McDougall, J.J. The Abnormal Cannabidiol Analogue O-1602 Reduces Nociception in a Rat Model of Acute Arthritis via the Putative Cannabinoid Receptor GPR55. *Neurosci. Lett.* **2011**, *500*, 72–76. [CrossRef]

106. Deliu, E.; Sperow, M.; Console-Bram, L.; Carter, R.L.; Tilley, D.G.; Kalamarides, D.J.; Kirby, L.G.; Brailoiu, G.C.; Brailoiu, E.; Benamar, K.; et al. The Lysophosphatidylinositol Receptor GPR55 Modulates Pain Perception in the Periaqueductal Gray. *Mol. Pharmacol.* **2015**, *88*, 265–272. [CrossRef]

107. Carey, L.M.; Gutierrez, T.; Deng, L.; Lee, W.H.; Mackie, K.; Hohmann, A.G. Inflammatory and Neuropathic Nociception Is Preserved in GPR55 Knockout Mice. *Sci. Rep.* **2017**, *7*, 944. [CrossRef]

108. Gantz, I.; Muraoka, A.; Yang, Y.K.; Samuelson, L.C.; Zimmerman, E.M.; Cook, H.; Yamada, T. Cloning and Chromosomal Localization of a Gene (GPR18) Encoding a Novel Seven Transmembrane Receptor Highly Expressed in Spleen and Testis. *Genomics* **1997**, *42*, 462–466. [CrossRef] [PubMed]

109. Kohno, M.; Hasegawa, H.; Inoue, A.; Muraoka, M.; Miyazaki, T.; Oka, K.; Yasukawa, M. Identification of N-Arachidonylglycine as the Endogenous Ligand for Orphan G-Protein-Coupled Receptor GPR18. *Biochem. Biophys. Res. Commun.* **2006**, *347*, 827–832. [CrossRef] [PubMed]

110. Kobayashi, K.; Fukuoka, T.; Obata, K.; Yamanaka, H.; Dai, Y.; Tokunaga, A.; Noguchi, K. Distinct Expression of TRPM8, TRPA1, and TRPV1 MRNAs in Rat Primary Afferent Neurons with Aδ/C-Fibers and Colocalization with Trk Receptors. *J. Comp. Neurol.* **2005**, *493*, 596–606. [CrossRef]

111. Cavanaugh, D.J.; Chesler, A.T.; Jackson, A.C.; Sigal, Y.M.; Yamanaka, H.; Grant, R.; O'Donnell, D.; Nicoll, R.A.; Shah, N.M.; Julius, D.; et al. Trpv1 Reporter Mice Reveal Highly Restricted Brain Distribution and Functional Expression in Arteriolar Smooth Muscle Cells. *J. Neurosci.* **2011**, *31*, 5067–5077. [CrossRef] [PubMed]

112. Avelino, A.; Cruz, F. TRPV1 (Vanilloid Receptor) in the Urinary Tract: Expression, Function and Clinical Applications. *Naunyn. Schmiedebergs. Arch. Pharmacol.* **2006**, *373*, 287–299. [CrossRef]

113. Yang, Y.; Yang, H.; Wang, Z.; Mergler, S.; Wolosin, J.M.; Reinach, P.S. Functional TRPV1 Expression in Human Corneal Fibroblasts. *Exp. Eye Res.* **2013**, *107*, 121–129. [CrossRef] [PubMed]

114. Tóth, A.; Boczán, J.; Kedei, N.; Lizanecz, E.; Bagi, Z.; Papp, Z.; Édes, I.; Csiba, L.; Blumberg, P.M. Expression and Distribution of Vanilloid Receptor 1 (TRPV1) in the Adult Rat Brain. *Mol. Brain Res.* **2005**, *135*, 162–168. [CrossRef]

115. Caterina, M.J.; Rosen, T.A.; Tominaga, M.; Brake, A.J.; Julius, D. A Capsaicin-Receptor Homologue with a High Threshold for Noxious Heat. *Nature* **1999**, *398*, 436–441. [CrossRef]

116. Frederick, J.; Buck, M.E.; Matson, D.J.; Cortright, D.N. Increased TRPA1, TRPM8, and TRPV2 Expression in Dorsal Root Ganglia by Nerve Injury. *Biochem. Biophys. Res. Commun.* **2007**, *358*, 1058–1064. [CrossRef] [PubMed]

117. Shimosato, G.; Amaya, F.; Ueda, M.; Tanaka, Y.; Decosterd, I.; Tanaka, M. Peripheral Inflammation Induces Up-Regulation of TRPV2 Expression in Rat DRG. *Pain* **2005**, *119*, 225–232. [CrossRef] [PubMed]

118. Saunders, C.I.; Kunde, D.A.; Crawford, A.; Geraghty, D.P. Expression of Transient Receptor Potential Vanilloid 1 (TRPV1) and 2 (TRPV2) in Human Peripheral Blood. *Mol. Immunol.* **2007**, *44*, 1429–1435. [CrossRef]

119. Santoni, G.; Amantini, C.; Maggi, F.; Marinelli, O.; Santoni, M.; Nabissi, M.; Morelli, M.B. The TRPV2 Cation Channels: From Urothelial Cancer Invasiveness to Glioblastoma Multiforme Interactome Signature. *Lab. Investig.* **2020**, *100*, 186–198. [CrossRef] [PubMed]

120. Link, T.M.; Park, U.; Vonakis, B.M.; Raben, D.M.; Soloski, M.J.; Caterina, M.J. TRPV2 Has a Pivotal Role in Macrophage Particle Binding and Phagocytosis. *Nat. Immunol.* **2010**, *11*, 232–241. [CrossRef]

121. Zhang, D.; Spielmann, A.; Wang, L.; Ding, G.; Huang, F.; Gu, Q.; Schwarz, W. Mast-Cell Degranulation Induced by Physical Stimuli Involves the Activation of Transient-Receptor-Potential Channel TRPV2. *Physiol. Res.* **2012**, *61*, 113–124. [CrossRef]

122. Iwata, Y.; Ohtake, H.; Suzuki, O.; Matsuda, J.; Komamura, K.; Wakabayashi, S. Blockade of Sarcolemmal TRPV2 Accumulation Inhibits Progression of Dilated Cardiomyopathy. *Cardiovasc. Res.* **2013**, *99*, 760–768. [CrossRef] [PubMed]

123. Lorin, C.; Vögeli, I.; Niggli, E. Dystrophic Cardiomyopathy: Role of TRPV2 Channels in Stretch-Induced Cell Damage. *Cardiovasc. Res.* **2015**, *106*, 153–162. [CrossRef]

124. Iwata, Y.; Katanosaka, Y.; Arai, Y.; Shigekawa, M.; Wakabayashi, S. Dominant-Negative Inhibition of Ca2+ Influx via TRPV2 Ameliorates Muscular Dystrophy in Animal Models. *Hum. Mol. Genet.* **2009**, *18*, 84–834. [CrossRef]

125. Iwata, Y.; Wakabayashi, S.; Ito, S.; Kitakaze, M. Production of TRPV2-Targeting Functional Antibody Ameliorating Dilated Cardiomyopathy and Muscular Dystrophy in Animal Models. *Lab. Investig.* **2020**, *100*, 324–337. [CrossRef]

126. Hisanaga, E.; Nagasawa, M.; Ueki, K.; Kulkarni, R.N.; Mori, M.; Kojima, I. Regulation of Calcium-Permeable TRPV2 Channel by Insulin in Pancreatic β-Cells. *Diabetes* **2009**, *58*, 174–184. [CrossRef] [PubMed]

127. Kanzaki, M.; Zhang, Y.Q.; Mashima, H.; Li, L.; Shibata, H.; Kojima, I. Translocation of a Calcium-Permeable Cation Channel Induced by Insulin-like Growth Factor-I. *Nat. Cell Biol.* **1999**, *1*, 165–170. [CrossRef] [PubMed]

128. Aoyagi, K.; Ohara-Imaizumi, M.; Nishiwaki, C.; Nakamichi, Y.; Nagamatsu, S. Insulin/Phosphoinositide 3-Kinase Pathway Accelerates the Glucose-Induced First-Phase Insulin Secretion through TrpV2 Recruitment in Pancreatic β-Cells. *Biochem. J.* **2010**, *432*, 375–386. [CrossRef] [PubMed]

129. Caterina, M.J.; Schumacher, M.A.; Tominaga, M.; Rosen, T.A.; Levine, J.D.; Julius, D. The Capsaicin Receptor: A Heat-Activated Ion Channel in the Pain Pathway. *Nature* **1997**, *389*, 816–824. [CrossRef] [PubMed]

130. Smith, G.D.; Gunthorpe, M.J.; Kelsell, R.E.; Hayes, P.D.; Reilly, P.; Facer, P.; Wright, J.E.; Jerman, J.C.; Walhin, J.P.; Ooi, L.; et al. TRPV3 Is a Temperature-Sensitive Vanilloid Receptor-like Protein. *Nature* **2002**, *418*, 186–190. [CrossRef]

131. Güler, A.D.; Lee, H.; Iida, T.; Shimizu, I.; Tominaga, M.; Caterina, M. Heat-Evoked Activation of the Ion Channel, TRPV4. *J. Neurosci.* **2002**, *22*, 6408–6414. [CrossRef]

132. Xu, H.; Ramsey, I.S.; Kotecha, S.A.; Moran, M.M.; Chong, J.A.; Lawson, D.; Ge, P.; Lilly, J.; Silos-Santiago, I.; Xie, Y.; et al. TRPV3 Is a Calcium-Permeable Temperature-Sensitive Cation Channel. *Nature* **2002**, *418*, 181–186. [CrossRef]

133. Moqrich, A.; Hwang, S.W.; Earley, T.J.; Petrus, M.J.; Murray, A.N.; Spencer, K.S.R.; Andahazy, M.; Story, G.M.; Patapoutian, A. Impaired Thermosensation in Mice Lacking TRPV3, a Heat and Camphor Sensor in the Skin. *Science* **2005**, *307*, 1468–1472. [CrossRef] [PubMed]

134. Mandadi, S.; Sokabe, T.; Shibasaki, K.; Katanosaka, K.; Mizuno, A.; Moqrich, A.; Patapoutian, A.; Fukumi-Tominaga, T.; Mizumura, K.; Tominaga, M. TRPV3 in Keratinocytes Transmits Temperature Information to Sensory Neurons via ATP. *Pflugers Arch. Eur. J. Physiol.* **2009**, *458*, 1093–1102. [CrossRef] [PubMed]

135. Chung, M.K.; Lee, H.; Mizuno, A.; Suzuki, M.; Caterina, M.J. TRPV3 and TRPV4 Mediate Warmth-Evoked Currents in Primary Mouse Keratinocytes. *J. Biol. Chem.* **2004**, *279*, 21569–21575. [CrossRef]

136. Todaka, H.; Taniguchi, J.; Satoh, J.I.; Mizuno, A.; Suzuki, M. Warm Temperature-Sensitive Transient Receptor Potential Vanilloid 4 (TRPV4) Plays an Essential Role in Thermal Hyperalgesia. *J. Biol. Chem.* **2004**, *279*, 35133–35138. [CrossRef]

137. Watanabe, H.; Vriens, J.; Suh, S.H.; Benham, C.D.; Droogmans, G.; Nilius, B. Heat-Evoked Activation of TRPV4 Channels in a HEK293 Cell Expression System and in Native Mouse Aorta Endothelial Cells. *J. Biol. Chem.* **2002**, *277*, 47044–47051. [CrossRef] [PubMed]

138. Peier, A.M.; Reeve, A.J.; Andersson, D.A.; Moqrich, A.; Earley, T.J.; Hergarden, A.C.; Story, G.M.; Colley, S.; Hogenesch, J.B.; McIntyre, P.; et al. A Heat-Sensitive TRP Channel Expressed in Keratinocytes. *Science* **2002**, *296*, 2046–2049. [CrossRef] [PubMed]

139. Xu, H.; Delling, M.; Jun, J.C.; Clapham, D.E. Oregano, Thyme and Clove-Derived Flavors and Skin Sensitizers Activate Specific TRP Channels. *Nat. Neurosci.* **2006**, *9*, 628–635. [CrossRef]

140. Bang, S.; Yoo, S.; Yang, T.J.; Cho, H.; Hwang, S.W. Isopentenyl Pyrophosphate Is a Novel Antinociceptive Substance That Inhibits TRPV3 and TRPA1 Ion Channels. *Pain* **2011**, *152*, 1156–1164. [CrossRef]

141. Liu, X.; Bandyopadhyay, B.; Nakamoto, T.; Singh, B.; Liedtke, W.; Melvin, J.E.; Ambudkar, I. A Role for AQP5 in Activation of TRPV4 by Hypotonicity: Concerted Involvement of AQP5 and TRPV4 in Regulation of Cell Volume Recovery. *J. Biol. Chem.* **2006**, *281*, 15485–15495. [CrossRef]

142. Becker, D.; Blase, C.; Bereiter-Hahn, J.; Jendrach, M. TRPV4 Exhibits a Functional Role in Cell-Volume Regulation. *J. Cell Sci.* **2005**. [CrossRef]

143. Vriens, J.; Watanabe, H.; Janssens, A.; Droogmans, G.; Voets, T.; Nilius, B. Cell Swelling, Heat, and Chemical Agonists Use Distinct Pathways for the Activation of the Cation Channel TRPV4. *Proc. Natl. Acad. Sci. USA* **2004**, *101*, 396–401. [CrossRef] [PubMed]

144. Liedtke, W.; Choe, Y.; Martí-Renom, M.A.; Bell, A.M.; Denis, C.S.; Šali, A.; Hudspeth, A.J.; Friedman, J.M.; Heller, S. Vanilloid Receptor-Related Osmotically Activated Channel (VR-OAC), a Candidate Vertebrate Osmoreceptor. *Cell* **2000**, *103*, 525–535. [CrossRef]

145. Shibasaki, K.; Tominaga, M.; Ishizaki, Y. Hippocampal Neuronal Maturation Triggers Post-Synaptic Clustering of Brain Temperature-Sensor TRPV4. *Biochem. Biophys. Res. Commun.* **2015**, *458*, 168–173. [CrossRef] [PubMed]

146. Zhang, L.; Papadopoulos, P.; Hamel, E. Endothelial TRPV4 Channels Mediate Dilation of Cerebral Arteries: Impairment and Recovery in Cerebrovascular Pathologies Related to Alzheimer's Disease. *Br. J. Pharmacol.* **2013**, *170*, 661–670. [CrossRef]

147. Wissenbach, U.; Bödding, M.; Freichel, M.; Flockerzi, V. Trp12, a Novel Trp Related Protein from Kidney. *FEBS Lett.* **2000**, *485*, 127–134. [CrossRef]

148. Liedtke, W.; Friedman, J.M. Abnormal Osmotic Regulation in Trpv4-/- Mice. *Proc. Natl. Acad. Sci. USA* **2003**, *100*, 13698–13703. [CrossRef] [PubMed]

149. Heckel, E.; Boselli, F.; Roth, S.; Krudewig, A.; Belting, H.G.; Charvin, G.; Vermot, J. Oscillatory Flow Modulates Mechanosensitive Klf2a Expression through Trpv4 and Trpp2 during Heart Valve Development. *Curr. Biol.* **2015**, *25*, 1354–1361. [CrossRef] [PubMed]

150. Watanabe, H.; Davis, J.B.; Smart, D.; Jerman, J.C.; Smith, G.D.; Hayes, P.; Vriens, J.; Cairns, W.; Wissenbach, U.; Prenen, J.; et al. Activation of TRPV4 Channels (HVRL-2/MTRP12) by Phorbol Derivatives. *J. Biol. Chem.* **2002**, *277*, 13569–13577. [CrossRef]

151. Watanabe, H.; Vriens, J.; Prenen, J.; Droogmans, G.; Voets, T.; Nillus, B. Anandamide and Arachidonic Acid Use Epoxyeicosatrienoic Acids to Activate TRPV4 Channels. *Nature* **2003**, *424*, 434–438. [CrossRef] [PubMed]

152. McKemy, D.D.; Neuhausser, W.M.; Julius, D. Identification of a Cold Receptor Reveals a General Role for TRP Channels in Thermosensation. *Nature* **2002**, *416*, 52–58. [CrossRef]

153. Peier, A.M.; Moqrich, A.; Hergarden, A.C.; Reeve, A.J.; Andersson, D.A.; Story, G.M.; Earley, T.J.; Dragoni, I.; McIntyre, P.; Bevan, S.; et al. A TRP Channel That Senses Cold Stimuli and Menthol. *Cell* **2002**, *108*, 705–715. [CrossRef]

154. Del Camino, D.; Murphy, S.; Heiry, M.; Barrett, L.B.; Earley, T.J.; Cook, C.A.; Petrus, M.J.; Zhao, M.; D'Amours, M.; Deering, N.; et al. TRPA1 Contributes to Cold Hypersensitivity. *J. Neurosci.* **2010**, *30*, 15165–15174. [CrossRef] [PubMed]

155. Story, G.M.; Peier, A.M.; Reeve, A.J.; Eid, S.R.; Mosbacher, J.; Hricik, T.R.; Earley, T.J.; Hergarden, A.C.; Andersson, D.A.; Hwang, S.W.; et al. ANKTM1, a TRP-like Channel Expressed in Nociceptive Neurons, Is Activated by Cold Temperatures. *Cell* **2003**, *112*, 819–829. [CrossRef]

156. Bandell, M.; Story, G.M.; Hwang, S.W.; Viswanath, V.; Eid, S.R.; Petrus, M.J.; Earley, T.J.; Patapoutian, A. Noxious Cold Ion Channel TRPA1 Is Activated by Pungent Compounds and Bradykinin. *Neuron* **2004**, *41*, 849–857. [CrossRef]

157. Karashima, Y.; Talavera, K.; Everaerts, W.; Janssens, A.; Kwan, K.Y.; Vennekens, R.; Nilius, B.; Voets, T. TRPA1 Acts as a Cold Sensor in Vitro and in Vivo. *Proc. Natl. Acad. Sci. USA* **2009**, *106*, 1273–1278. [CrossRef]

158. McNamara, C.R.; Mandel-Brehm, J.; Bautista, D.M.; Siemens, J.; Deranian, K.L.; Zhao, M.; Hayward, N.J.; Chong, J.A.; Julius, D.; Moran, M.M.; et al. TRPA1 Mediates Formalin-Induced Pain. *Proc. Natl. Acad. Sci. USA* **2007**, *104*, 13525–13530. [CrossRef] [PubMed]

159. Jordt, S.E.; Bautista, D.M.; Chuang, H.H.; McKemy, D.D.; Zygmunt, P.M.; Högestätt, E.D.; Meng, I.D.; Julius, D. Mustard Oils and Cannabinoids Excite Sensory Nerve Fibres through the TRP Channel ANKTM1. *Nature* **2004**, *427*, 260–265. [CrossRef]

160. Dai, Y.; Wang, S.; Tominaga, M.; Yamamoto, S.; Fukuoka, T.; Higashi, T.; Kobayashi, K.; Obata, K.; Yamanaka, H.; Noguchi, K. Sensitization of TRPA1 by PAR2 Contributes to the Sensation of Inflammatory Pain. *J. Clin. Investig.* **2007**, *117*, 1979–1987. [CrossRef] [PubMed]

161. Nagata, K.; Duggan, A.; Kumar, G.; García-Añoveros, J. Nociceptor and Hair Cell Transducer Properties of TRPA1, a Channel for Pain and Hearing. *J. Neurosci.* **2005**, *25*, 4052–4061. [CrossRef]

162. Bautista, D.M.; Jordt, S.E.; Nikai, T.; Tsuruda, P.R.; Read, A.J.; Poblete, J.; Yamoah, E.N.; Basbaum, A.I.; Julius, D. TRPA1 Mediates the Inflammatory Actions of Environmental Irritants and Proalgesic Agents. *Cell* **2006**, *124*, 1269–1282. [CrossRef]

163. Kremeyer, B.; Lopera, F.; Cox, J.J.; Momin, A.; Rugiero, F.; Marsh, S.; Woods, C.G.; Jones, N.G.; Paterson, K.J.; Fricker, F.R.; et al. A Gain-of-Function Mutation in TRPA1 Causes Familial Episodic Pain Syndrome. *Neuron* **2010**, *66*, 671–680. [CrossRef]

164. Xiong, W.; Cui, T.; Cheng, K.; Yang, F.; Chen, S.R.; Willenbring, D.; Guan, Y.; Pan, H.L.; Ren, K.; Xu, Y.; et al. Cannabinoids Suppress Inflammatory and Neuropathic Pain by Targeting A3 Glycine Receptors. *J. Exp. Med.* **2012**, *209*, 1121–1134. [CrossRef]

165. Hejazi, N.; Zhou, C.; Oz, M.; Sun, H.; Jiang, H.Y.; Zhang, L. Δ9-Tetrahydrocannabinol and Endogenous Cannabinoid Anandamide Directly Potentiate the Function of Glycine Receptors. *Mol. Pharmacol.* **2006**, *69*, 991–997. [CrossRef] [PubMed]

166. Xiong, W.; Cheng, K.; Cui, T.; Godlewski, G.; Rice, K.C.; Xu, Y.; Zhang, L. Cannabinoid Potentiation of Glycine Receptors Contributes to Cannabis-Induced Analgesia. *Nat. Chem. Biol.* **2011**, *7*, 296–303. [CrossRef]

167. O'Sullivan, S.E. An Update on PPAR Activation by Cannabinoids. *Br. J. Pharmacol.* **2016**, *173*, 1899–1910. [CrossRef]

168. Russo, E.B.; Burnett, A.; Hall, B.; Parker, K.K. Agonistic Properties of Cannabidiol at 5-HT1a Receptors. *Neurochem. Res.* **2005**, *30*, 1037–1043. [CrossRef] [PubMed]

169. Franklin, J.M.; Carrasco, G.A. Cannabinoid-Induced Enhanced Interaction and Protein Levels of Serotonin 5-HT2A and Dopamine D2 Receptors in Rat Prefrontal Cortex. *J. Psychopharmacol.* **2012**, *26*, 1333–1347. [CrossRef]

170. Franklin, J.M.; Carrasco, G.A. Cannabinoid Receptor Agonists Upregulate and Enhance Serotonin 2A (5-HT2A) Receptor Activity via ERK1/2 Signaling. *Synapse* **2013**, *67*, 145–159. [CrossRef] [PubMed]

171. Hill, M.N.; Campolongo, P.; Yehuda, R.; Patel, S. Integrating Endocannabinoid Signaling and Cannabinoids into the Biology and Treatment of Posttraumatic Stress Disorder. *Neuropsychopharmacology* **2018**, *43*, 80–102. [CrossRef]

172. Blessing, E.M.; Steenkamp, M.M.; Manzanares, J.; Marmar, C.R. Cannabidiol as a Potential Treatment for Anxiety Disorders. *Neurotherapeutics* **2015**, *12*, 825–836. [CrossRef] [PubMed]

173. Harrold, J.A.; Elliott, J.C.; King, P.J.; Widdowson, P.S.; Williams, G. Down-Regulation of Cannabinoid-1 (CB-1) Receptors in Specific Extrahypothalamic Regions of Rats with Dietary Obesity: A Role for Endogenous Cannabinoids in Driving Appetite for Palatable Food? *Brain Res.* **2002**, *952*, 232–238. [CrossRef]

174. Jamshidi, N.; Taylor, D.A. Anandamide Administration into the Ventromedial Hypothalamus Stimulates Appetite in Rats. *Br. J. Pharmacol.* **2001**, *134*, 1151–1154. [CrossRef]

175. Portella, G.; Laezza, C.; Laccetti, P.; De Petrocellis, L.; Di Marzo, V.; Bifulco, M. Inhibitory Effects of Cannabinoid CB1 Receptor Stimulation on Tumor Growth and Metastatic Spreading: Actions on Signals Involved in Angiogenesis and Metastasis. *FASEB J.* **2003**, *17*, 1771–1773. [CrossRef]

176. Linciano, P.; Citti, C.; Luongo, L.; Belardo, C.; Maione, S.; Vandelli, M.A.; Forni, F.; Gigli, G.; Laganà, A.; Montone, C.M.; et al. Isolation of a High-Affinity Cannabinoid for the Human CB1 Receptor from a Medicinal *Cannabis sativa* Variety: Δ9-Tetrahydrocannabutol, the Butyl Homologue of Δ9-Tetrahydrocannabinol. *J. Nat. Prod.* **2020**, *83*, 88–98. [CrossRef]

177. Wallace, M.J.; Blair, R.E.; Falenski, K.W.; Martin, B.R.; DeLorenzo, R.J. The Endogenous Cannabinoid System Regulates Seizure Frequency and Duration in a Model of Temporal Lobe Epilepsy. *J. Pharmacol. Exp. Ther.* **2003**, *307*, 129–137. [CrossRef]

178. Murillo-Rodríguez, E. The Role of the CB1 Receptor in the Regulation of Sleep. *Prog. Neuro-Psychopharmacol. Biol. Psychiatry* **2008**, *32*, 1420–1427. [CrossRef]

179. Zagzoog, A.; Mohamed, K.A.; Kim, H.J.J.; Kim, E.D.; Frank, C.S.; Black, T.; Jadhav, P.D.; Holbrook, L.A.; Laprairie, R.B. In Vitro and in Vivo Pharmacological Activity of Minor Cannabinoids Isolated from *Cannabis sativa*. *Sci. Rep.* **2020**, *10*, 20405. [CrossRef]

180. Bolognini, D.; Costa, B.; Maione, S.; Comelli, F.; Marini, P.; Di Marzo, V.; Parolaro, D.; Ross, R.A.; Gauson, L.A.; Cascio, M.G.; et al. The Plant Cannabinoid Δ 9-Tetrahydrocannabivarin Can Decrease Signs of Inflammation and Inflammatory Pain in Mice. *Br. J. Pharmacol.* **2010**, *160*, 677–687. [CrossRef] [PubMed]

181. Espadas, I.; Keifman, E.; Palomo-Garo, C.; Burgaz, S.; García, C.; Fernández-Ruiz, J.; Moratalla, R. Beneficial Effects of the Phytocannabinoid Δ9-THCV in L-DOPA-Induced Dyskinesia in Parkinson's Disease. *Neurobiol. Dis.* **2020**, *141*, 104892. [CrossRef] [PubMed]

182. Wargent, E.T.; Zaibi, M.S.; Silvestri, C.; Hislop, D.C.; Stocker, C.J.; Stott, C.G.; Guy, G.W.; Duncan, M.; Di Marzo, V.; Cawthorne, M.A. The Cannabinoid Δ9-Tetrahydrocannabivarin (THCV) Ameliorates Insulin Sensitivity in Two Mouse Models of Obesity. *Nutr. Diabetes* **2013**, *3*, e68. [CrossRef] [PubMed]

183. Hill, A.J.; Weston, S.E.; Jones, N.A.; Smith, I.; Bevan, S.A.; Williamson, E.M.; Stephens, G.J.; Williams, C.M.; Whalley, B.J. 9-Tetrahydrocannabivarin Suppresses in Vitro Epileptiform and in Vivo Seizure Activity in Adult Rats. *Epilepsia* **2010**, *51*, 1522–1532. [CrossRef]

184. Abrahamov, A.; Abrahamov, A.; Mechoulam, R. An Efficient New Cannabinoid Antiemetic in Pediatric Oncology. *Life Sci.* **1995**, *56*, 2097–2102. [CrossRef]

185. Darmani, N.A. Δ9-Tetrahydrocannabinol and Synthetic Cannabinoids Prevent Emesis Produced by the Cannabinoid CB1 Receptor Antagonist/Inverse Agonist SR 141716A. *Neuropsychopharmacology* **2001**, *24*, 198–203. [CrossRef]

186. Darmani, N.A. The Cannabinoid CB1 Receptor Antagonist SR 141716A Reverses the Antiemetic and Motor Depressant Actions of WIN 55, 212–2. *Eur. J. Pharmacol.* **2001**, *430*, 49–58. [CrossRef]

187. Darmani, N.A.; Sim-Selley, L.J.; Martin, B.R.; Janoyan, J.J.; Crim, J.L.; Parekh, B.; Breivogel, C.S. Antiemetic and Motor-Depressive Actions of CP55,940: Cannabinoid CB1 Receptor Characterization, Distribution, and G-Protein Activation. *Eur. J. Pharmacol.* **2003**, *459*, 83–95. [CrossRef]

188. Darmani, N.A.; Janoyan, J.J.; Crim, J.; Ramirez, J. Receptor Mechanism and Antiemetic Activity of Structurally-Diverse Cannabinoids against Radiation-Induced Emesis in the Least Shrew. *Eur. J. Pharmacol.* **2007**, *563*, 187–196. [CrossRef]

189. Lastres-Becker, I.; Cebeira, M.; De Ceballos, M.L.; Zeng, B.Y.; Jenner, P.; Ramos, J.A.; Fernández-Ruiz, J.J. Increased Cannabinoid CB1 Receptor Binding and Activation of GTP-Binding Proteins in the Basal Ganglia of Patients with Parkinson's Syndrome and of MPTP-Treated Marmosets. *Eur. J. Neurosci.* **2001**, *14*, 1827–1832. [CrossRef] [PubMed]

190. Sarfaraz, S.; Afaq, F.; Adhami, V.M.; Mukhtar, H. Cannabinoid Receptor as a Novel Target for the Treatment of Prostate Cancer. *Cancer Res.* **2005**, *65*, 1635–1641. [CrossRef]

191. Qamri, Z.; Preet, A.; Nasser, M.W.; Bass, C.E.; Leone, G.; Barsky, S.H.; Ganju, R.K. Synthetic Cannabinoid Receptor Agonists Inhibit Tumor Growth and Metastasis of Breast Cancer. *Mol. Cancer Ther.* **2009**, *8*, 3117–3129. [CrossRef]

192. Preet, A.; Qamri, Z.; Nasser, M.W.; Prasad, A.; Shilo, K.; Zou, X.; Groopman, J.E.; Ganju, R.K. Cannabinoid Receptors, CB1 and CB2, as Novel Targets for Inhibition of Non-Small Cell Lung Cancer Growth and Metastasis. *Cancer Prev. Res.* **2011**, *4*, 65–75. [CrossRef]

193. Izzo, A.A.; Capasso, R.; Aviello, G.; Borrelli, F.; Romano, B.; Piscitelli, F.; Gallo, L.; Capasso, F.; Orlando, P.; Di Marzo, V. Inhibitory Effect of Cannabichromene, a Major Non-Psychotropic Cannabinoid Extracted from *Cannabis sativa*, on Inflammation-Induced Hypermotility in Mice. *Br. J. Pharmacol.* **2012**, *166*, 1444–1460. [CrossRef]

194. Borrelli, F.; Fasolino, I.; Romano, B.; Capasso, R.; Maiello, F.; Coppola, D.; Orlando, P.; Battista, G.; Pagano, E.; Di Marzo, V.; et al. Beneficial Effect of the Non-Psychotropic Plant Cannabinoid Cannabigerol on Experimental Inflammatory Bowel Disease. *Biochem. Pharmacol.* **2013**, *85*, 1306–1316. [CrossRef]

195. Gómez-Gálvez, Y.; Palomo-Garo, C.; Fernández-Ruiz, J.; García, C. Potential of the Cannabinoid CB2 Receptor as a Pharmacological Target against Inflammation in Parkinson's Disease. *Prog. Neuropsychopharmacol. Biol. Psychiatry* **2016**, *64*, 200–208. [CrossRef] [PubMed]

196. Javed, H.; Azimullah, S.; Haque, M.E.; Ojha, S.K. Cannabinoid Type 2 (CB2) Receptors Activation Protects against Oxidative Stress and Neuroinflammation Associated Dopaminergic Neurodegeneration in Rotenone Model of Parkinson's Disease. *Front. Neurosci.* **2016**, *10*, 321. [CrossRef]

197. Vigli, D.; Cosentino, L.; Raggi, C.; Laviola, G.; Woolley-Roberts, M.; De Filippis, B. Chronic Treatment with the Phytocannabinoid Cannabidivarin (CBDV) Rescues Behavioural Alterations and Brain Atrophy in a Mouse Model of Rett Syndrome. *Neuropharmacology* **2018**, *140*, 121–129. [CrossRef] [PubMed]

198. Anavi-Goffer, S.; Baillie, G.; Irving, A.J.; Gertsch, J.; Greig, I.R.; Pertwee, R.G.; Ross, R.A. Modulation of L-α-Lysophosphatidylinositol/GPR55 Mitogen-Activated Protein Kinase (MAPK) Signaling by Cannabinoids. *J. Biol. Chem.* **2012**, *287*, 91–104. [CrossRef]

199. Iannotti, F.A.; Hill, C.L.; Leo, A.; Alhusaini, A.; Soubrane, C.; Mazzarella, E.; Russo, E.; Whalley, B.J.; Di Marzo, V.; Stephens, G.J. Nonpsychotropic Plant Cannabinoids, Cannabidivarin (CBDV) and Cannabidiol (CBD), Activate and Desensitize Transient Receptor Potential Vanilloid 1 (TRPV1) Channels in Vitro: Potential for the Treatment of Neuronal Hyperexcitability. *ACS Chem. Neurosci.* **2014**, *5*, 1131–1141. [CrossRef]

200. De Petrocellis, L.; Ligresti, A.; Moriello, A.S.; Allarà, M.; Bisogno, T.; Petrosino, S.; Stott, C.G.; Di Marzo, V. Effects of Cannabinoids and Cannabinoid-Enriched Cannabis Extracts on TRP Channels and Endocannabinoid Metabolic Enzymes. *Br. J. Pharmacol.* **2011**, *163*, 1479–1494. [CrossRef]

201. De Petrocellis, L.; Orlando, P.; Moriello, A.S.; Aviello, G.; Stott, C.; Izzo, A.A.; di Marzo, V. Cannabinoid Actions at TRPV Channels: Effects on TRPV3 and TRPV4 and Their Potential Relevance to Gastrointestinal Inflammation. *Acta Physiol.* **2012**, *204*, 255–266. [CrossRef]

202. De Petrocellis, L.; Vellani, V.; Schiano-Moriello, A.; Marini, P.; Magherini, P.C.; Orlando, P.; Di Marzo, V. Plant-Derived Cannabinoids Modulate the Activity of Transient Receptor Potential Channels of Ankyrin Type-1 and Melastatin Type-8. *J. Pharmacol. Exp. Ther.* **2008**, *325*, 1007–1015. [CrossRef]

203. Borrelli, F.; Pagano, E.; Romano, B.; Panzera, S.; Maiello, F.; Coppola, D.; De Petrocellis, L.; Buono, L.; Orlando, P.; Izzo, A.A. Colon Carcinogenesis Is Inhibited by the TRPM8 Antagonist Cannabigerol, a Cannabis-Derived Non-Psychotropic Cannabinoid. *Carcinogenesis* **2014**, *35*, 2787–2797. [CrossRef] [PubMed]

204. Pagano, E.; Romano, B.; Iannotti, F.A.; Parisi, O.A.; D'Armiento, M.; Pignatiello, S.; Coretti, L.; Lucafò, M.; Venneri, T.; Stocco, G.; et al. The Non-Euphoric Phytocannabinoid Cannabidivarin Counteracts Intestinal Inflammation in Mice and Cytokine Expression in Biopsies from UC Pediatric Patients. *Pharmacol. Res.* **2019**, *149*, 104464. [CrossRef]

205. Iannotti, F.A.; Pagano, E.; Moriello, A.S.; Alvino, F.G.; Sorrentino, N.C.; D'Orsi, L.; Gazzerro, E.; Capasso, R.; De Leonibus, E.; De Petrocellis, L.; et al. Effects of Non-Euphoric Plant Cannabinoids on Muscle Quality and Performance of Dystrophic Mdx Mice. *Br. J. Pharmacol.* **2019**, *176*, 1568–1584. [CrossRef]

206. García-Arencibia, M.; González, S.; de Lago, E.; Ramos, J.A.; Mechoulam, R.; Fernández-Ruiz, J. Evaluation of the Neuroprotective Effect of Cannabinoids in a Rat Model of Parkinson's Disease: Importance of Antioxidant and Cannabinoid Receptor-Independent Properties. *Brain Res.* **2007**, *1134*, 162–170. [CrossRef]

207. Di Marzo, V. Enhanced Levels of Endogenous Cannabinoids in the Globus Pallidus Are Associated with a Reduction in Movement in an Animal Model of Parkinson's Disease. *FASEB J.* **2000**, *14*, 1432–1438. [CrossRef] [PubMed]

208. Fox, S.H.; Henry, B.; Hill, M.; Crossman, A.; Brotchie, J. Stimulation of Cannabinoid Receptors Reduces Levodopa-Induced Dyskinesia in the MPTP-Lesioned Nonhuman Primate Model of Parkinson's Disease. *Mov. Disord.* **2002**, *17*, 1180–1187. [CrossRef]

209. Morgese, M.G.; Cassano, T.; Cuomo, V.; Giuffrida, A. Anti-Dyskinetic Effects of Cannabinoids in a Rat Model of Parkinson's Disease: Role of CB1 and TRPV1 Receptors. *Exp. Neurol.* **2007**, *208*, 110–119. [CrossRef]

210. Sañudo-Peña, M.C.; Patrick, S.L.; Khen, S.; Patrick, R.L.; Tsou, K.; Walker, J.M. Cannabinoid Effects in Basal Ganglia in a Rat Model of Parkinson's Disease. *Neurosci. Lett.* **1998**, *248*, 171–174. [CrossRef]

211. Donadelli, M.; Dando, I.; Zaniboni, T.; Costanzo, C.; Dalla Pozza, E.; Scupoli, M.T.; Scarpa, A.; Zappavigna, S.; Marra, M.; Abbruzzese, A.; et al. Gemcitabine/Cannabinoid Combination Triggers Autophagy in Pancreatic Cancer Cells through a ROS-Mediated Mechanism. *Cell Death Dis.* **2011**, *2*, E152. [CrossRef]

212. Afrin, F.; Chi, M.; Eamens, A.L.; Duchatel, R.J.; Douglas, A.M.; Schneider, J.; Gedye, C.; Woldu, A.S.; Dun, M.D. Can Hemp Help? Low-THC Cannabis and Non-THC Cannabinoids for the Treatment of Cancer. *Cancers* **2020**, *12*. [CrossRef] [PubMed]

213. De Petrocellis, L.; Melck, D.; Palmisano, A.; Bisogno, T.; Laezza, C.; Bifulco, M.; Di Marzo, V. The Endogenous Cannabinoid Anandamide Inhibits Human Breast Cancer Cell Proliferation. *Proc. Natl. Acad. Sci. USA* **1998**, *95*, 8375–8380. [CrossRef]

214. Cianchi, F.; Papucci, L.; Schiavone, N.; Lulli, M.; Magnelli, L.; Vinci, M.C.; Messerini, L.; Manera, C.; Ronconi, E.; Romagnani, P.; et al. Cannabinoid Receptor Activation Induces Apoptosis through Tumor Necrosis Factor α-Mediated Ceramide de Novo Synthesis in Colon Cancer Cells. *Clin. Cancer Res.* **2008**, *14*, 7691–7700. [CrossRef]

215. Blázquez, C.; Casanova, M.L.; Planas, A.; Del Pulgar, T.G.; Villanueva, C.; Fernández-Aceñero, M.J.; Aragonés, J.; Huffman, J.W.; Jorcano, J.L.; Guzmán, M. Inhibition of Tumor Angiogenesis by Cannabinoids. *FASEB J.* **2003**, *17*, 529–531. [CrossRef] [PubMed]

216. Cunha, J.M.; Carlini, E.A.; Pereira, A.E.; Ramos, O.L.; Pimentel, C.; Gagliardi, R.; Sanvito, W.L.; Lander, N.; Mechoulam, R. Chronic Administration of Cannabidiol to Healthy Volunteers and Epileptic Patients. *Pharmacology* **1980**, *21*, 175–185. [CrossRef] [PubMed]

217. Rosenberg, E.C.; Tsien, R.W.; Whalley, B.J.; Devinsky, O. Cannabinoids and Epilepsy. *Neurotherapeutics* **2015**, *12*, 747–768. [CrossRef] [PubMed]

218. Stockings, E.; Zagic, D.; Campbell, G.; Weier, M.; Hall, W.D.; Nielsen, S.; Herkes, G.K.; Farrell, M.; Degenhardt, L. Evidence for Cannabis and Cannabinoids for Epilepsy: A Systematic Review of Controlled and Observational Evidence. *J. Neurol. Neurosurg. Psychiatry* **2018**, *89*, 741–753. [CrossRef]

219. Karanian, D.A.; Karim, S.L.; Wood, J.A.T.; Williams, J.S.; Lin, S.; Makriyannis, A.; Bahr, B.A. Endocannabinoid Enhancement Protects against Kainic Acid-Induced Seizures and Associated Brain Damage. *J. Pharmacol. Exp. Ther.* **2007**, *322*, 1059–1066. [CrossRef]

220. Naidoo, V.; Karanian, D.A.; Vadivel, S.K.; Locklear, J.R.; Wood, J.A.T.; Nasr, M.; Quizon, P.M.P.; Graves, E.E.; Shukla, V.; Makriyannis, A.; et al. Equipotent Inhibition of Fatty Acid Amide Hydrolase and Monoacylglycerol Lipase—Dual Targets of the Endocannabinoid System to Protect against Seizure Pathology. *Neurotherapeutics* **2012**, *9*, 801–813. [CrossRef]

221. Devinsky, O.; Patel, A.D.; Cross, J.H.; Villanueva, V.; Wirrell, E.C.; Privitera, M.; Greenwood, S.M.; Roberts, C.; Checketts, D.; VanLandingham, K.E.; et al. Effect of Cannabidiol on Drop Seizures in the Lennox–Gastaut Syndrome. *N. Engl. J. Med.* **2018**, *378*, 1888–1897. [CrossRef]

222. Devinsky, O.; Cross, J.H.; Laux, L.; Marsh, E.; Miller, I.; Nabbout, R.; Scheffer, I.E.; Thiele, E.A.; Wright, S. Trial of Cannabidiol for Drug-Resistant Seizures in the Dravet Syndrome. *N. Engl. J. Med.* **2017**, *376*, 2011–2020. [CrossRef] [PubMed]

223. Press, C.A.; Knupp, K.G.; Chapman, K.E. Parental Reporting of Response to Oral Cannabis Extracts for Treatment of Refractory Epilepsy. *Epilepsy Behav.* **2015**, *45*, 49–52. [CrossRef] [PubMed]

224. Guggenhuber, S.; Monory, K.; Lutz, B.; Klugmann, M. AAV Vector-Mediated Overexpression of CB1 Cannabinoid Receptor in Pyramidal Neurons of the Hippocampus Protects against Seizure-Induced Excitoxicity. *PLoS ONE* **2010**, *5*, e15707. [CrossRef] [PubMed]

225. Riva, N.; Mora, G.; Sorarù, G.; Lunetta, C.; Ferraro, O.E.; Falzone, Y.; Leocani, L.; Fazio, R.; Comola, M.; Comi, G.; et al. Safety and Efficacy of Nabiximols on Spasticity Symptoms in Patients with Motor Neuron Disease (CANALS): A Multicentre, Double-Blind, Randomised, Placebo-Controlled, Phase 2 Trial. *Lancet Neurol.* **2019**, *18*, 155–164. [CrossRef]
226. Beal, J.E.; Olson, R.; Laubenstein, L.; Morales, J.O.; Bellman, P.; Yangco, B.; Lefkowitz, L.; Plasse, T.F.; Shepard, K.V. Dronabinol as a Treatment for Anorexia Associated with Weight Loss in Patients with AIDS. *J. Pain Symptom Manag.* **1995**, *10*, 89–97. [CrossRef]
227. Foltin, R.W.; Fischman, M.W.; Byrne, M.F. Effects of Smoked Marijuana on Food Intake and Body Weight of Humans Living in a Residential Laboratory. *Appetite* **1988**, *11*, 1–14. [CrossRef]
228. Mattes, R.D.; Engelman, K.; Shaw, L.M.; Elsohly, M.A. Cannabinoids and Appetite Stimulation. *Pharmacol. Biochem. Behav.* **1994**, *49*, 187–195. [CrossRef]
229. Williams, C.M.; Rogers, P.J.; Kirkham, T.C. Hyperphagia in Pre-Fed Rats Following Oral Δ9-THC. *Physiol. Behav.* **1998**, *65*, 343–346. [CrossRef]
230. Feinberg, I.; Jones, R.; Walker, J.M.; Cavness, C.; March, J. Effects of High Dosage Delta-9-Tetrahydrocannabinol on Sleep Patterns in Man. *Clin. Pharmacol. Ther.* **1975**, *14*, 458–466. [CrossRef]
231. Freemon, F.R. The Effect of Chronically Administered Delta-9-Tetrahydrocannabinol upon the Polygraphically Monitored Sleep of Normal Volunteers. *Drug Alcohol Depend.* **1982**, *10*, 345–353. [CrossRef]
232. Pivik, R.T.; Zarcone, V.; Dement, W.C.; Hollister, L.E. Delta-9-Tetrahydrocannabinol and Synhexl: Effects on Human Sleep Patterns. *Clin. Pharmacol. Ther.* **1972**, *13*, 426–435. [CrossRef]
233. Agarwal, N.; Pacher, P.; Tegeder, I.; Amaya, F.; Constantin, C.E.; Brenner, G.J.; Rubino, T.; Michalski, C.W.; Marsicano, G.; Monory, K.; et al. Cannabinoids Mediate Analgesia Largely via Peripheral Type 1 Cannabinoid Receptors in Nociceptors. *Nat. Neurosci.* **2007**, *10*, 870–879. [CrossRef] [PubMed]
234. Meng, I.D.; Manning, B.H.; Martin, W.J.; Fields, H.L. An Analgesia Circuit Activated by Cannabinoids. *Nature* **1998**, *395*, 381–383. [CrossRef] [PubMed]
235. Ottani, A.; Leone, S.; Sandrini, M.; Ferrari, A.; Bertolini, A. The Analgesic Activity of Paracetamol Is Prevented by the Blockade of Cannabinoid CB1 Receptors. *Eur. J. Pharmacol.* **2006**, *531*, 280–281. [CrossRef]
236. Walker, J.M.; Hohmann, A.G.; Martin, W.J.; Strangman, N.M.; Huang, S.M.; Tsou, K. The Neurobiology of Cannabinoid Analgesia. *Life Sci.* **1999**, *65*, 665–673. [CrossRef]
237. Leweke, F.M.; Piomelli, D.; Pahlisch, F.; Muhl, D.; Gerth, C.W.; Hoyer, C.; Klosterkötter, J.; Hellmich, M.; Koethe, D. Cannabidiol Enhances Anandamide Signaling and Alleviates Psychotic Symptoms of Schizophrenia. *Transl. Psychiatry* **2012**, *2*, e94. [CrossRef]
238. Lutz, B.; Marsicano, G.; Maldonado, R.; Hillard, C.J. The Endocannabinoid System in Guarding against Fear, Anxiety and Stress. *Nat. Rev. Neurosci.* **2015**, *16*, 705–718. [CrossRef] [PubMed]
239. Navarro, M.; Hernández, E.; Muñoz, R.M.; Del Arco, I.; Villanúa, M.A.; Carrera, M.R.A.; Rodríguez De Fonseca, F. Acute Administration of the CB1 Cannabinoid Receptor Antagonist SR 141716A Induces Anxiety-like Responses in the Rat. *Neuroreport* **1997**, *8*, 491–496. [CrossRef]
240. Moreira, F.A.; Aguiar, D.C.; Guimarães, F.S. Anxiolytic-like Effect of Cannabinoids Injected into the Rat Dorsolateral Periaqueductal Gray. *Neuropharmacology* **2007**, *52*, 958–965. [CrossRef]
241. Rey, A.A.; Purrio, M.; Viveros, M.P.; Lutz, B. Biphasic Effects of Cannabinoids in Anxiety Responses: CB1 and GABA B Receptors in the Balance of Gabaergic and Glutamatergic Neurotransmission. *Neuropsychopharmacology* **2012**, *37*, 2624–2634. [CrossRef]
242. Ney, L.J.; Matthews, A.; Bruno, R.; Felmingham, K.L. Modulation of the Endocannabinoid System by Sex Hormones: Implications for Posttraumatic Stress Disorder. *Neurosci. Biobehav. Rev.* **2018**, *94*, 302–320. [CrossRef]
243. Carter, G.T.; Flanagan, A.M.; Earleywine, M.; Abrams, D.I.; Aggarwal, S.K.; Grinspoon, L. Cannabis in Palliative Medicine: Improving Care and Reducing Opioid-Related Morbidity. *Am. J. Hosp. Palliat. Med.* **2011**, *28*, 297–303. [CrossRef]
244. Bar-Sela, G.; Vorobeichik, M.; Drawsheh, S.; Omer, A.; Goldberg, V.; Muller, E. The Medical Necessity for Medicinal Cannabis: Prospective, Observational Study Evaluating the Treatment in Cancer Patients on Supportive or Palliative Care. *Evid.-Based Complement. Altern. Med.* **2013**, *2013*, 510392. [CrossRef] [PubMed]
245. Motwani, M.P.; Bennett, F.; Norris, P.C.; Maini, A.A.; George, M.J.; Newson, J.; Henderson, A.; Hobbs, A.J.; Tepper, M.; White, B.; et al. Potent Anti-Inflammatory and Pro-Resolving Effects of Anabasum in a Human Model of Self-Resolving Acute Inflammation. *Clin. Pharmacol. Ther.* **2018**, *104*, 675–686. [CrossRef]
246. Lucas, P. Rationale for Cannabis-Based Interventions in the Opioid Overdose Crisis. *Harm Reduct. J.* **2017**, *14*, 58. [CrossRef]
247. Boehnke, K.F.; Litinas, E.; Clauw, D.J. Medical Cannabis Use Is Associated with Decreased Opiate Medication Use in a Retrospective Cross-Sectional Survey of Patients with Chronic Pain. *J. Pain* **2016**, *17*, 739–744. [CrossRef]
248. Groce, E. *The Health Effects of Cannabis and Cannabinoids: The Current State of Evidence and Recommendations for Research*; National Academies Press: Washington, DC, USA, 2017. [CrossRef]
249. Dos Santos, R.G.; Guimarães, F.S.; Crippa, J.A.S.; Hallak, J.E.C.; Rossi, G.N.; Rocha, J.M.; Zuardi, A.W. Serious Adverse Effects of Cannabidiol (CBD): A Review of Randomized Controlled Trials. *Expert Opin. Drug Metab. Toxicol.* **2020**, *16*, 517–526. [CrossRef] [PubMed]
250. White, C.M. A Review of Human Studies Assessing Cannabidiol's (CBD) Therapeutic Actions and Potential. *J. Clin. Pharmacol.* **2019**, *59*, 923–934. [CrossRef] [PubMed]
251. Pauli, C.S.; Conroy, M.; Vanden Heuvel, B.D.; Park, S.H. Cannabidiol Drugs Clinical Trial Outcomes and Adverse Effects. *Front. Pharmacol.* **2020**, *11*, 63. [CrossRef]

252. Kuhathasan, N.; Dufort, A.; MacKillop, J.; Gottschalk, R.; Minuzzi, L.; Frey, B.N. The Use of Cannabinoids for Sleep: A Critical Review on Clinical Trials. *Exp. Clin. Psychopharmacol.* **2019**, *27*, 383–401. [CrossRef] [PubMed]

253. Black, N.; Stockings, E.; Campbell, G.; Tran, L.T.; Zagic, D.; Hall, W.D.; Farrell, M.; Degenhardt, L. Cannabinoids for the Treatment of Mental Disorders and Symptoms of Mental Disorders: A Systematic Review and Meta-Analysis. *Lancet Psychiatry* **2019**, *6*, 995–1010. [CrossRef]

254. Marks, M.D.; Tian, L.; Wenger, J.P.; Omburo, S.N.; Soto-Fuentes, W.; He, J.; Gang, D.R.; Weiblen, G.D.; Dixon, R.A. Identification of Candidate Genes Affecting Δ9-Tetrahydrocannabinol Biosynthesis in *Cannabis sativa*. *J. Exp. Bot.* **2009**, *60*, 3715–3726. [CrossRef]

255. Stout, J.M.; Boubakir, Z.; Ambrose, S.J.; Purves, R.W.; Page, J.E. The Hexanoyl-CoA Precursor for Cannabinoid Biosynthesis Is Formed by an Acyl-Activating Enzyme in *Cannabis sativa* Trichomes. *Plant J.* **2012**, *71*, 353–365. [CrossRef]

256. Kim, E.-S.; Mahlberg, P.G. Secretory Cavity Development in Glandular Trichomes of *Cannabis sativa* L. (Cannabaceae). *Am. J. Bot.* **1991**, *78*, 220–229. [CrossRef]

257. Kim, E.S.; Mahlberg, P.G. Secretory Vesicle Formation in the Secretory Cavity of Glandular Trichomes of *Cannabis sativa* L. (Cannabaceae). *Mol. Cells* **2003**, *15*, 387–395. [CrossRef]

258. Mahlberg, P.G.; Eun, S.K. Accumulation of Cannabinoids in Glandular Trichomes of Cannabis (Cannabaceae). *J. Ind. Hemp* **2004**, *9*, 15–36. [CrossRef]

259. Mahlberg, P.G.; Kim, E.-S. Cuticle Development on Glandular Trichomes of *Cannabis sativa* (Cannabaceae). *Am. J. Bot.* **1991**, *78*, 1113–1122. [CrossRef]

260. Arigoni, D.; Sagner, S.; Latzel, C.; Eisenreich, W.; Bacher, A.; Zenk, M.H. Terpenoid Biosynthesis from 1-Deoxy-D-Xylulose in Higher Plants by Intramolecular Skeletal Rearrangement. *Proc. Natl. Acad. Sci. USA* **1997**, *94*, 10600–10605. [CrossRef] [PubMed]

261. Fellermeier, M.; Eisenreich, W.; Bacher, A.; Zenk, M.H. Biosynthesis of Cannabinoids: Incorporation Experiments with 13C-Labeled Glucoses. *Eur. J. Biochem.* **2001**, *268*, 1596–1604. [CrossRef]

262. Schwender, J.; Zeidler, J.; Gröner, R.; Müller, C.; Focke, M.; Braun, S.; Lichtenthaler, F.W.; Lichtenthaler, H.K. Incorporation of 1-Deoxy-D-Xylulose into Isoprene and Phytol by Higher Plants and Algae. *FEBS Lett.* **1997**. [CrossRef]

263. Botella-Pavía, P.; Besumbes, Ó.; Phillips, M.A.; Carretero-Paulet, L.; Boronat, A.; Rodríguez-Concepción, M. Regulation of Carotenoid Biosynthesis in Plants: Evidence for a Key Role of Hydroxymethylbutenyl Diphosphate Reductase in Controlling the Supply of Plastidial Isoprenoid Precursors. *Plant J.* **2004**, *40*, 188–199. [CrossRef]

264. Phillips, M.A.; León, P.; Boronat, A.; Rodríguez-Concepción, M. The Plastidial MEP Pathway: Unified Nomenclature and Resources. *Trends Plant Sci.* **2008**, *13*, 619–623. [CrossRef] [PubMed]

265. Hsieh, M.H.; Chang, C.Y.; Hsu, S.J.; Chen, J.J. Chloroplast Localization of Methylerythritol 4-Phosphate Pathway Enzymes and Regulation of Mitochondrial Genes in IspD and IspE Albino Mutants in Arabidopsis. *Plant Mol. Biol.* **2008**, *66*, 663–673. [CrossRef]

266. Bick, J.A.; Lange, B.M. Metabolic Cross Talk between Cytosolic and Plastidial Pathways of Isoprenoid Biosynthesis: Unidirectional Transport of Intermediates across the Chloroplast Envelope Membrane. *Arch. Biochem. Biophys.* **2003**, *415*, 146–154. [CrossRef]

267. Buhaescu, I.; Izzedine, H. Mevalonate Pathway: A Review of Clinical and Therapeutical Implications. *Clin. Biochem.* **2007**, *40*, 575–584. [CrossRef]

268. Goldstein, J.L.; Brown, M.S. Regulation of the Mevalonate Pathway. *Nature* **1990**, *343*, 425–430. [CrossRef]

269. Miziorko, H.M. Enzymes of the Mevalonate Pathway of Isoprenoid Biosynthesis. *Arch. Biochem. Biophys.* **2011**, *505*, 131–143. [CrossRef] [PubMed]

270. Guirimand, G.; Simkin, A.J.; Papon, N.; Besseau, S.; Burlat, V.; St-Pierre, B.; Giglioli-Guivarc'h, N.; Clastre, M.; Courdavault, V. Cycloheximide as a Tool to Investigate Protein Import in Peroxisomes: A Case Study of the Subcellular Localization of Isoprenoid Biosynthetic Enzymes. *J. Plant Physiol.* **2012**, *169*, 825–829. [CrossRef]

271. Simkin, A.J.; Guirimand, G.; Papon, N.; Courdavault, V.; Thabet, I.; Ginis, O.; Bouzid, S.; Giglioli-Guivarc'h, N.; Clastre, M. Peroxisomal Localisation of the Final Steps of the Mevalonic Acid Pathway in Planta. *Planta* **2011**, *234*, 903–914. [CrossRef] [PubMed]

272. Vranová, E.; Coman, D.; Gruissem, W. Network Analysis of the MVA and MEP Pathways for Isoprenoid Synthesis. *Annu. Rev. Plant Biol.* **2013**, *64*, 665–700. [CrossRef]

273. Thabet, I.; Guirimand, G.; Courdavault, V.; Papon, N.; Godet, S.; Dutilleul, C.; Bouzid, S.; Giglioli-Guivarc'h, N.; Clastre, M.; Simkin, A.J. The Subcellular Localization of Periwinkle Farnesyl Diphosphate Synthase Provides Insight into the Role of Peroxisome in Isoprenoid Biosynthesis. *J. Plant Physiol.* **2011**, *168*, 2110–2116. [CrossRef] [PubMed]

274. Campbell, M.; Hahn, F.M.; Poulter, C.D.; Leustek, T. Analysis of the Isopentenyl Disphosphate Isomerase Gene from *Arabidopsis thaliana*. *Plant Mol. Biol.* **1998**, *36*, 323–328. [CrossRef] [PubMed]

275. Okada, K.; Kasahara, H.; Yamaguchi, S.; Kawaide, H.; Kamiya, Y.; Nojiri, H.; Yamane, H. Genetic Evidence for the Role of Isopentenyl Diphosphate Isomerases in the Mevalonate Pathway and Plant Development in Arabidopsis. *Plant Cell Physiol.* **2008**, *49*, 604–616. [CrossRef]

276. Sapir-Mir, M.; Mett, A.; Belausov, E.; Tal-Meshulam, S.; Frydman, A.; Gidoni, D.; Eya, Y. Peroxisomal Localization of Arabidopsis Isopentenyl Diphosphate Isomerases Suggests That Part of the Plant Isoprenoid Mevalonic Acid Pathway Is Compartmentalized to Peroxisomes. *Plant Physiol.* **2008**, *148*, 1219–1228. [CrossRef]

277. Burke, C.C.; Wildung, M.R.; Croteau, R. Geranyl Diphosphate Synthase: Cloning, Expression, and Characterization of This Prenyltransferase as a Heterodimer. *Proc. Natl. Acad. Sci. USA* **1999**, *96*, 13062–13067. [CrossRef]

278. Bouvier, F.; Suire, C.; D'Harlingue, A.; Backhaus, R.A.; Camara, B. Molecular Cloning of Geranyl Diphosphate Synthase and Compartmentation of Monoterpene Synthesis in Plant Cells. *Plant J.* **2000**, *24*, 241–252. [CrossRef] [PubMed]
279. Ogura, K.; Koyama, T. Enzymatic Aspects of Isoprenoid Chain Elongation. *Chem. Rev.* **1998**, *98*, 1263–1276. [CrossRef] [PubMed]
280. Ohnuma, S.I.; Hirooka, K.; Tsuruoka, N.; Yano, M.; Ohto, C.; Nakane, H.; Nishino, T. A Pathway Where Polyprenyl Diphosphate Elongates in Prenyltransferase: Insight into a Common Mechanism of Chain Length Determination of Prenyltransferases. *J. Biol. Chem.* **1998**, *273*, 26705–26713. [CrossRef] [PubMed]
281. Wang, K.; Ohnuma, S.I. Chain-Length Determination Mechanism of Isoprenyl Diphosphate Synthases and Implications for Molecular Evolution. *Trends Biochem. Sci.* **1999**, *24*, 445–451. [CrossRef]
282. Booth, J.K.; Bohlmann, J. Terpenes in *Cannabis sativa*—From Plant Genome to Humans. *Plant Sci.* **2019**, *284*, 67–72. [CrossRef]
283. Oldfield, E.; Lin, F.Y. Terpene Biosynthesis: Modularity Rules. *Angew. Chem. Int. Ed. Engl.* **2012**, *51*, 1124–1137. [CrossRef]
284. Gagne, S.J.; Stout, J.M.; Liu, E.; Boubakir, Z.; Clark, S.M.; Page, J.E. Identification of Olivetolic Acid Cyclase from *Cannabis sativa* Reveals a Unique Catalytic Route to Plant Polyketides. *Proc. Natl. Acad. Sci. USA* **2012**, *109*, 12811–12816. [CrossRef]
285. Taura, F.; Tanaka, S.; Taguchi, C.; Fukamizu, T.; Tanaka, H.; Shoyama, Y.; Morimoto, S. Characterization of Olivetol Synthase, a Polyketide Synthase Putatively Involved in Cannabinoid Biosynthetic Pathway. *FEBS Lett.* **2009**, *583*, 2061–2066. [CrossRef]
286. Fellermeier, M.; Zenk, M.H. Prenylation of Olivetolate by a Hemp Transferase Yields Cannabigerolic Acid, the Precursor of Tetrahydrocannabinol. *FEBS Lett.* **1998**, *427*, 283–285. [CrossRef]
287. Luo, X.; Reiter, M.A.; D'Espaux, L.; Wong, J.; Denby, C.M.; Lechner, A.; Zhang, Y.; Grzybowski, A.T.; Harth, S.; Lin, W.; et al. Complete Biosynthesis of Cannabinoids and Their Unnatural Analogues in Yeast. *Nature* **2019**, *567*, 123–126. [CrossRef]
288. Valliere, M.A.; Korman, T.P.; Woodall, N.B.; Khitrov, G.A.; Taylor, R.E.; Baker, D.; Bowie, J.U. A Cell-Free Platform for the Prenylation of Natural Products and Application to Cannabinoid Production. *Nat. Commun.* **2019**, *10*, 565. [CrossRef]
289. Taura, F.; Morimoto, S.; Shoyama, Y. Purification and Characterization of Cannabidiolic-Acid Synthase from *Cannabis sativa* L. Biochemical Analysis of a Novel Enzyme That Catalyzes the Oxidocyclization of Cannabigerolic Acid to Cannabidiolic Acid. *J. Biol. Chem.* **1996**, *271*, 17411–17416. [CrossRef]
290. Morimoto, S.; Komatsu, K.; Taura, F.; Shoyama, Y. Purification and Characterization of Cannabichromenic Acid Synthase from *Cannabis sativa*. *Phytochemistry* **1998**, *49*, 1525–1529. [CrossRef]
291. Shoyama, Y.; Tamada, T.; Kurihara, K.; Takeuchi, A.; Taura, F.; Arai, S.; Blaber, M.; Shoyama, Y.; Morimoto, S.; Kuroki, R. Structure and Function of Δ1-Tetrahydrocannabinolic Acid (THCA) Synthase, the Enzyme Controlling the Psychoactivity of *Cannabis sativa*. *J. Mol. Biol.* **2012**, *423*, 96–105. [CrossRef]
292. Taura, F.; Morimoto, S.; Shoyama, Y.; Mechoulam, R. First Direct Evidence for the Mechanism of Δ1-Tetrahydrocannabinolie Acid Biosynthesis. *J. Am. Chem. Soc.* **1995**, *117*, 9766–9767. [CrossRef]
293. Taura, F.; Dono, E.; Sirikantaramas, S.; Yoshimura, K.; Shoyama, Y.; Morimoto, S. Production of Δ1-Tetrahydrocannabinolic Acid by the Biosynthetic Enzyme Secreted from Transgenic Pichia Pastoris. *Biochem. Biophys. Res. Commun.* **2007**, *361*, 675–680. [CrossRef] [PubMed]
294. De Zeeuw, R.A.; Wijsbeek, J.; Brejmer, D.D.; Vree, T.B.; Van Ginneken, C.A.M.; Van Rossum, J.M. Cannabinoids with a Propyl Side Chain in Cannabis: Occurrence and Chromatographic Behavior. *Science* **1972**, *175*, 778–779. [CrossRef] [PubMed]
295. De Meijer, E.P.M.; Bagatta, M.; Carboni, A.; Crucitti, P.; Moliterni, V.M.C.; Ranalli, P.; Mandolino, G. The Inheritance of Chemical Phenotype in *Cannabis sativa* L. *Genetics* **2003**, *163*, 335–346.
296. Shoyama, Y.; Hirano, H.; Nishioka, I. Biosynthesis of Propyl Cannabinoid Acid and Its Biosynthetic Relationship with Pentyl and Methyl Cannabinoid Acids. *Phytochemistry* **1984**, *23*, 1909–1984. [CrossRef]
297. Kanter, S.L.; Musumeci, M.R.; Hollister, L.E. Quantitative Determination of Δ9-Tetrahydrocannabinol and Δ9-Tetrahydrocannabinolic Acid in Marihuana by High-Pressure Liquid Chromatography. *J. Chromatogr. A* **1979**, *171*, 504–508. [CrossRef]
298. Perrotin-Brunel, H.; Buijs, W.; Van Spronsen, J.; Roosmalen, M.J.E.V.; Peters, C.J.; Verpoorte, R.; Witkamp, G.J. Decarboxylation of Δ9-Tetrahydrocannabinol: Kinetics and Molecular Modeling. *J. Mol. Struct.* **2011**, *987*, 67–73. [CrossRef]
299. Shoyama, Y.; Yagi, M.; Nishioka, I.; Yamauchi, T. Biosynthesis of Cannabinoid Acids. *Phytochemistry* **1975**, *14*, 2189–2192. [CrossRef]
300. Veress, T.; Szanto, J.I.; Leisztner, L. Determination of Cannabinoid Acids by High-Performance Liquid Chromatography of Their Neutral Derivatives Formed by Thermal Decarboxylation. I. Study of the Decarboxylation Process in Open Reactors. *J. Chromatogr. A* **1990**. [CrossRef]
301. Hanuš, L.O.; Meyer, S.M.; Muñoz, E.; Taglialatela-Scafati, O.; Appendino, G. Phytocannabinoids: A Unified Critical Inventory. *Nat. Prod. Rep.* **2016**, *33*, 1357–1392. [CrossRef] [PubMed]
302. Ahmed, S.A.; Ross, S.A.; Slade, D.; Radwan, M.M.; Khan, I.A.; ElSohly, M.A. Structure Determination and Absolute Configuration of Cannabichromanone Derivatives from High Potency *Cannabis sativa*. *Tetrahedron Lett.* **2008**, *49*, 6050–6053. [CrossRef] [PubMed]
303. Ahmed, S.A.; Ross, S.A.; Slade, D.; Radwan, M.M.; Zulfiqar, F.; ElSohly, M.A. Cannabinoid Ester Constituents from High-Potency *Cannabis sativa*. *J. Nat. Prod.* **2008**, *71*, 536–542. [CrossRef]
304. Radwan, M.M.; Ross, S.A.; Slade, D.; Ahmed, S.A.; Zulfiqar, F.; Elsohly, M.A. Isolation and Characterization of New Cannabis Constituents from a High Potency Variety. *Planta Med.* **2008**, *74*, 267–272. [CrossRef]
305. Pagani, A.; Scala, F.; Chianese, G.; Grassi, G.; Appendino, G.; Taglialatela-Scafati, O. Cannabioxepane, a Novel Tetracyclic Cannabinoid from Hemp, *Cannabis sativa* L. *Tetrahedron* **2011**, *67*, 3369–3373. [CrossRef]

306. Zulfiqar, F.; Ross, S.A.; Slade, D.; Ahmed, S.A.; Radwan, M.M.; Ali, Z.; Khan, I.A.; Elsohly, M.A. Cannabisol, a Novel Δ 9-THC Dimer Possessing a Unique Methylene Bridge, Isolated from *Cannabis sativa*. *Tetrahedron Lett.* **2012**, *53*, 3560–3562. [CrossRef] [PubMed]

307. Pollastro, F.; Taglialatela-Scafati, O.; Allarà, M.; Muñoz, E.; Di Marzo, V.; De Petrocellis, L.; Appendino, G. Bioactive Prenylogous Cannabinoid from Fiber Hemp (*Cannabis sativa*). *J. Nat. Prod.* **2011**, *74*, 2019–2022. [CrossRef]

308. Taglialatela-Scafati, O.; Pagani, A.; Scala, F.; De Petrocellis, L.; Di Marzo, V.; Grassi, G.; Appendino, G. Cannabimovone, a Cannabinoid with a Rearranged Terpenoid Skeleton from Hemp (Eur. J. Org. Chem. 11/2010). *Eur. J. Org. Chem.* **2010**, *2010*, 2023. [CrossRef]

309. Thomas, A.; Stevenson, L.A.; Wease, K.N.; Price, M.R.; Baillie, G.; Ross, R.A.; Pertwee, R.G. Evidence That the Plant Cannabinoid Δ 9-Tetrahydrocannabivarin Is a Cannabinoid CB 1 and CB 2 Receptor Antagonist. *Br. J. Pharmacol.* **2005**, *146*, 917–926. [CrossRef] [PubMed]

310. Pertwee, R.G.; Thomas, A.; Stevenson, L.A.; Ross, R.A.; Varvel, S.A.; Lichtman, A.H.; Martin, B.R.; Razdan, R.K. The Psychoactive Plant Cannabinoid, Δ 9-Tetrahydrocannabinol, Is Antagonized by Δ 8- and Δ 9-Tetrahydrocannabivarin in Mice in Vivo. *Br. J. Pharmacol.* **2007**, *150*, 586–594. [CrossRef]

311. García, C.; Palomo-Garo, C.; García-Arencibia, M.; Ramos, J.A.; Pertwee, R.G.; Fernández-Ruiz, J. Symptom-Relieving and Neuroprotective Effects of the Phytocannabinoid Δ 9-THCV in Animal Models of Parkinson's Disease. *Br. J. Pharmacol.* **2011**, *163*, 1495–1506. [CrossRef]

312. Ma, Y.L.; Weston, S.E.; Whalley, B.J.; Stephens, G.J. The Phytocannabinoid Δ 9-Tetrahydrocannabivarin Modulates Inhibitory Neurotransmission in the Cerebellum. *Br. J. Pharmacol.* **2008**, *154*, 204–215. [CrossRef]

313. Dennis, I.; Whalley, B.J.; Stephens, G.J. Effects of Δ 9-Tetrahydrocannabivarin on [35S]GTPγS Binding in Mouse Brain Cerebellum and Piriform Cortex Membranes. *Br. J. Pharmacol.* **2008**. [CrossRef] [PubMed]

314. Cascio, M.G.; Zamberletti, E.; Marini, P.; Parolaro, D.; Pertwee, R.G. The Phytocannabinoid, Δ9-Tetrahydrocannabivarin, Can Act through 5-HT1A Receptors to Produce Antipsychotic Effects. *Br. J. Pharmacol.* **2015**, *172*, 1305–1318. [CrossRef]

315. O'Sullivan, S.E.; Bennett, A.J.; Kendall, D.A.; Randall, M.D. Cannabinoids and Peroxisome Proliferator-Activated Receptor γ (PPARg). In Proceedings of the 16th Annual Symposium on the Cannabinoids, Tihany, Hungary, 24–28 June 2006; Volume 59.

316. Englund, A.; Atakan, Z.; Kralj, A.; Tunstall, N.; Murray, R.; Morrison, P. The Effect of Five Day Dosing with THCV on THC-Induced Cognitive, Psychological and Physiological Effects in Healthy Male Human Volunteers: A Placebo-Controlled, Double-Blind, Crossover Pilot Trial. *J. Psychopharmacol.* **2016**, *30*, 140–151. [CrossRef] [PubMed]

317. Tudge, L.; Williams, C.; Cowen, P.J.; McCabe, C. Neural Effects of Cannabinoid CB1 Neutral Antagonist Tetrahydrocannabivarin on Food Reward and Aversion in Healthy Volunteers. *Int. J. Neuropsychopharmacol.* **2015**, *18*, Pyu094. [CrossRef]

318. Rzepa, E.; Tudge, L.; McCabe, C. The CB1 Neutral Antagonist Tetrahydrocannabivarin Reduces Default Mode Network and Increases Executive Control Network Resting State Functional Connectivity in Healthy Volunteers. *Int. J. Neuropsychopharmacol.* **2016**, *19*, Pyv092. [CrossRef] [PubMed]

319. Moldzio, R.; Pacher, T.; Krewenka, C.; Kranner, B.; Novak, J.; Duvigneau, J.C.; Rausch, W.D. Effects of Cannabinoids Δ(9)-Tetrahydrocannabinol, Δ(9)-Tetrahydrocannabinolic Acid and Cannabidiol in MPP+ Affected Murine Mesencephalic Cultures. *Phytomedicine* **2012**, *19*, 819–824. [CrossRef]

320. Verhoeckx, K.C.M.; Korthout, H.A.A.J.; Van Meeteren-Kreikamp, A.P.; Ehlert, K.A.; Wang, M.; Van Der Greef, J.; Rodenburg, R.J.T.; Witkamp, R.F. Unheated *Cannabis sativa* Extracts and Its Major Compound THC-Acid Have Potential Immuno-Modulating Properties Not Mediated by CB1 and CB2 Receptor Coupled Pathways. *Int. Immunopharmacol.* **2006**, *6*, 656–665. [CrossRef]

321. Hollister, L.E.; Gillespie, H.K. Delta-8- and Delta-9-Tetrahydrocannabinol Comparison in Man by Oral and Intravenous Administration. *Clin. Pharmacol. Ther.* **1973**, *14*, 353–357. [CrossRef] [PubMed]

322. Rock, E.M.; Limebeer, C.L.; Navaratnam, R.; Sticht, M.A.; Bonner, N.; Engeland, K.; Downey, R.; Morris, H.; Jackson, M.; Parker, L.A. A Comparison of Cannabidiolic Acid with Other Treatments for Anticipatory Nausea Using a Rat Model of Contextually Elicited Conditioned Gaping. *Psychopharmacology* **2014**, *231*, 3207–3215. [CrossRef] [PubMed]

323. Nadal, X.; del Río, C.; Casano, S.; Palomares, B.; Ferreiro-Vera, C.; Navarrete, C.; Sánchez-Carnerero, C.; Cantarero, I.; Bellido, M.L.; Meyer, S.; et al. Tetrahydrocannabinolic Acid Is a Potent PPARγ Agonist with Neuroprotective Activity. *Br. J. Pharmacol.* **2017**, *174*, 4263–4276. [CrossRef] [PubMed]

324. Morano, A.; Cifelli, P.; Nencini, P.; Antonilli, L.; Fattouch, J.; Ruffolo, G.; Roseti, C.; Aronica, E.; Limatola, C.; Di Bonaventura, C.; et al. Cannabis in Epilepsy: From Clinical Practice to Basic Research Focusing on the Possible Role of Cannabidivarin. *Epilepsia Open* **2016**, *1*, 145–151. [CrossRef]

325. Hill, T.D.M.; Cascio, M.G.; Romano, B.; Duncan, M.; Pertwee, R.G.; Williams, C.M.; Whalley, B.J.; Hill, A.J. Cannabidivarin-Rich Cannabis Extracts Are Anticonvulsant in Mouse and Rat via a CB1 Receptor-Independent Mechanism. *Br. J. Pharmacol.* **2013**, *170*, 679–692. [CrossRef] [PubMed]

326. Hill, A.J.; Mercier, M.S.; Hill, T.D.M.; Glyn, S.E.; Jones, N.A.; Yamasaki, Y.; Futamura, T.; Duncan, M.; Stott, C.G.; Stephens, G.J.; et al. Cannabidivarin Is Anticonvulsant in Mouse and Rat. *Br. J. Pharmacol.* **2012**, *167*, 1629–1642. [CrossRef] [PubMed]

327. Amada, N.; Yamasaki, Y.; Williams, C.M.; Whalley, B.J. Cannabidivarin (CBDV) Suppresses Pentylenetetrazole (PTZ)-Induced Increases in Epilepsy-Related Gene Expression. *PeerJ* **2013**, *1*, E214. [CrossRef] [PubMed]

328. Huizenga, M.N.; Sepulveda-Rodriguez, A.; Forcelli, P.A. Preclinical Safety and Efficacy of Cannabidivarin for Early Life Seizures. *Neuropharmacology* **2019**, *148*, 189–198. [CrossRef]

329. Qin, N.; Neeper, M.P.; Liu, Y.; Hutchinson, T.L.; Lubin, M.L.; Flores, C.M. TRPV2 Is Activated by Cannabidiol and Mediates CGRP Release in Cultured Rat Dorsal Root Ganglion Neurons. *J. Neurosci.* **2008**, *28*, 6231–6238. [CrossRef] [PubMed]

330. Pretzsch, C.M.; Voinescu, B.; Lythgoe, D.; Horder, J.; Mendez, M.A.; Wichers, R.; Ajram, L.; Ivin, G.; Heasman, M.; Edden, R.A.E.; et al. Effects of Cannabidivarin (CBDV) on Brain Excitation and Inhibition Systems in Adults with and without Autism Spectrum Disorder (ASD): A Single Dose Trial during Magnetic Resonance Spectroscopy. *Transl. Psychiatry* **2019**, *9*, 313. [CrossRef]

331. Eibach, L.; Scheffel, S.; Cardebring, M.; Lettau, M.; Özgür Celik, M.; Morguet, A.; Roehle, R.; Stein, C. Cannabidivarin for HIV-Associated Neuropathic Pain: A Randomized, Blinded, Controlled Clinical Trial. *Clin. Pharmacol. Ther.* **2020**, 1–8. [CrossRef]

332. Russo, C.; Ferk, F.; Mišík, M.; Ropek, N.; Nersesyan, A.; Mejri, D.; Holzmann, K.; Lavorgna, M.; Isidori, M.; Knasmüller, S. Low Doses of Widely Consumed Cannabinoids (Cannabidiol and Cannabidivarin) Cause DNA Damage and Chromosomal Aberrations in Human-Derived Cells. *Arch. Toxicol.* **2019**, *93*, 179–188. [CrossRef]

333. Cascio, M.G.; Gauson, L.A.; Stevenson, L.A.; Ross, R.A.; Pertwee, R.G. Evidence That the Plant Cannabinoid Cannabigerol Is a Highly Potent α 2-Adrenoceptor Agonist and Moderately Potent 5HT 1A Receptor Antagonist. *Br. J. Pharmacol.* **2010**, *159*, 129–141. [CrossRef]

334. Valdeolivas, S.; Navarrete, C.; Cantarero, I.; Bellido, M.L.; Muñoz, E.; Sagredo, O. Neuroprotective Properties of Cannabigerol in Huntington's Disease: Studies in R6/2 Mice and 3-Nitropropionate-Lesioned Mice. *Neurotherapeutics* **2015**, *12*, 185–199. [CrossRef]

335. Gugliandolo, A.; Pollastro, F.; Grassi, G.; Bramanti, P.; Mazzon, E. In Vitro Model of Neuroinflammation: Efficacy of Cannabigerol, a Non-Psychoactive Cannabinoid. *Int. J. Mol. Sci.* **2018**, *19*, 1992. [CrossRef] [PubMed]

336. Kathmann, M.; Flau, K.; Redmer, A.; Tränkle, C.; Schlicker, E. Cannabidiol Is an Allosteric Modulator at Mu- and Delta-Opioid Receptors. *Naunyn. Schmiedebergs. Arch. Pharmacol.* **2006**, *372*, 354–361. [CrossRef] [PubMed]

337. Ligresti, A.; Moriello, A.S.; Starowicz, K.; Matias, I.; Pisanti, S.; De Petrocellis, L.; Laezza, C.; Portella, G.; Bifulco, M.; Di Marzo, V. Antitumor Activity of Plant Cannabinoids with Emphasis on the Effect of Cannabidiol on Human Breast Carcinoma. *J. Pharmacol. Exp. Ther.* **2006**, *318*, 1375–1387. [CrossRef]

338. Booker, L.; Naidu, P.S.; Razdan, R.K.; Mahadevan, A.; Lichtman, A.H. Evaluation of Prevalent Phytocannabinoids in the Acetic Acid Model of Visceral Nociception. *Drug Alcohol Depend.* **2009**, *105*, 42–47. [CrossRef] [PubMed]

339. Esposito, G.; Scuderi, C.; Valenza, M.; Togna, G.I.; Latina, V.; de Filippis, D.; Cipriano, M.; Carratù, M.R.; Iuvone, T.; Steardo, L. Cannabidiol Reduces Aβ-Induced Neuroinflammation and Promotes Hippocampal Neurogenesis through PPARγ Involvement. *PLoS ONE* **2011**, *6*, e28668. [CrossRef]

340. Rosenthaler, S.; Pöhn, B.; Kolmanz, C.; Nguyen Huu, C.; Krewenka, C.; Huber, A.; Kranner, B.; Rausch, W.D.; Moldzio, R. Differences in Receptor Binding Affinity of Several Phytocannabinoids Do Not Explain Their Effects on Neural Cell Cultures. *Neurotoxicol. Teratol.* **2014**, *46*, 49–56. [CrossRef]

341. Turner, C.E.; Elsohly, M.A. Biological Activity of Cannabichromene, Its Homologs and Isomers. *J. Clin. Pharmacol.* **1981**, *21*, 283S–291S. [CrossRef] [PubMed]

342. DeLong, G.T.; Wolf, C.E.; Poklis, A.; Lichtman, A.H. Pharmacological Evaluation of the Natural Constituent of *Cannabis sativa*, Cannabichromene and Its Modulation by Δ9-Tetrahydrocannabinol. *Drug Alcohol Depend.* **2010**, *112*, 126–133. [CrossRef]

343. Davis, W.M.; Hatoum, N.S. Neurobehavioral Actions of Cannabichromene and Interactions with Δ9-Tetrahydrocannabinol. *Gen. Pharmacol.* **1983**, *14*, 247–252. [CrossRef]

344. Udoh, M.; Santiago, M.; Devenish, S.; McGregor, I.S.; Connor, M. Cannabichromene Is a Cannabinoid CB2 Receptor Agonist. *Br. J. Pharmacol.* **2019**, *176*, 4537–4547. [CrossRef]

345. Romano, B.; Borrelli, F.; Fasolino, I.; Capasso, R.; Piscitelli, F.; Cascio, M.G.; Pertwee, R.G.; Coppola, D.; Vassallo, L.; Orlando, P.; et al. The Cannabinoid TRPA1 Agonist Cannabichromene Inhibits Nitric Oxide Production in Macrophages and Ameliorates Murine Colitis. *Br. J. Pharmacol.* **2013**, *169*, 213–229. [CrossRef] [PubMed]

346. Shinjyo, N.; Di Marzo, V. The Effect of Cannabichromene on Adult Neural Stem/Progenitor Cells. *Neurochem. Int.* **2013**, *63*, 432–437. [CrossRef] [PubMed]

347. Mahadevan, A.; Siegel, C.; Martin, B.R.; Abood, M.E.; Beletskaya, I.; Razdan, R.K. Novel Cannabinol Probes for CB1 and CB2 Cannabinoid Receptors. *J. Med. Chem.* **2000**, *43*, 3778–3785. [CrossRef] [PubMed]

348. Yamamoto, I.; Watanabe, K.; Kuzuoka, K.; Narimatsu, S.; Yoshimura, H. The Pharmacological Activity of Cannabinol and Its Major Metabolite, 11-Hydroxycannabinol. *Chem. Pharm. Bull.* **1987**, *35*, 2144–2147. [CrossRef]

349. Watanabe, K.; Yamaori, S.; Funahashi, T.; Kimura, T.; Yamamoto, I. Cytochrome P450 Enzymes Involved in the Metabolism of Tetrahydrocannabinols and Cannabinol by Human Hepatic Microsomes. *Life Sci.* **2007**, *80*, 1415–1419. [CrossRef]

350. Yamaori, S.; Kushihara, M.; Yamamoto, I.; Watanabe, K. Characterization of Major Phytocannabinoids, Cannabidiol and Cannabinol, as Isoform-Selective and Potent Inhibitors of Human CYP1 Enzymes. *Biochem. Pharmacol.* **2010**, *79*, 1691–1698. [CrossRef]

351. Aiken, C.T.; Tobin, A.J.; Schweitzer, E.S. A Cell-Based Screen for Drugs to Treat Huntington's Disease. *Neurobiol. Dis.* **2004**, *16*, 546–555. [CrossRef]

352. Glass, M.; Faull, R.L.M.; Dragunow, M. Loss of Cannabinoid Receptors in the Substantia Nigra in Huntington's Disease. *Neuroscience* **1993**, *56*, 523–527. [CrossRef]

353. Blázquez, C.; Chiarlone, A.; Sagredo, O.; Aguado, T.; Pazos, M.R.; Resel, E.; Palazuelos, J.; Julien, B.; Salazar, M.; Börner, C.; et al. Loss of Striatal Type 1 Cannabinoid Receptors Is a Key Pathogenic Factor in Huntington's Disease. *Brain* **2011**, *134*, 119–136. [CrossRef] [PubMed]

354. Weydt, P.; Hong, S.; Witting, A.; Möller, T.; Stella, N.; Kliot, M. Cannabinol Delays Symptom Onset in SOD1 (G93A) Transgenic Mice without Affecting Survival. *Amyotroph. Lateral Scler. Other Mot. Neuron Disord.* **2005**, *6*, 182–184. [CrossRef] [PubMed]

355. Wong, H.; Cairns, B.E. Cannabidiol, Cannabinol and Their Combinations Act as Peripheral Analgesics in a Rat Model of Myofascial Pain. *Arch. Oral Biol.* **2019**, *104*, 33–39. [CrossRef]

356. Baroi, S.; Saha, A.; Bachar, R.; Bachar, S.C. Cannabinoid as Potential Aromatase Inhibitor through Molecular Modeling and Screening for Anti-Cancer Activity. *Dhaka Univ. J. Pharm. Sci.* **2020**, *19*, 47–58. [CrossRef]

357. Furqan, T.; Batool, S.; Habib, R.; Shah, M.; Kalasz, H.; Darvas, F.; Kuca, K.; Nepovimova, E.; Batool, S.; Nurulain, S.M. Cannabis Constituents and Acetylcholinesterase Interaction: Molecular Docking, in Vitro Studies and Association with CNR1 RS806368 and ACHE RS17228602. *Biomolecules* **2020**, *10*, 758. [CrossRef] [PubMed]

358. Sakamoto, K.; Akiyama, Y.; Fukui, K.; Kamada, H.; Satoh, S. Characterization; Genome Sizes and Morphology of Sex Chromosomes in Hemp (*Cannabis sativa* L.). *Cytologia* **1998**, *635*, 459–464. [CrossRef]

359. Sakamoto, K.; Shimomura, K.; Komeda, Y.; Kamada, H.; Satoh, S. A Male-Associated DNA Sequence in a Dioecious Plant, *Cannabis sativa* L. *Plant Cell Physiol.* **1995**, *36*, 1549–1554. [CrossRef] [PubMed]

360. Alghanim, H.J.; Almirall, J.R. Development of Microsatellite Markers in *Cannabis sativa* for DNA Typing and Genetic Relatedness Analyses. *Anal. Bioanal. Chem.* **2003**, *376*, 1225–1233. [CrossRef]

361. Gilmore, S.; Peakall, R.; Robertson, J. Short Tandem Repeat (STR) DNA Markers Are Hypervariable and Informative in *Cannabis sativa*: Implications for Forensic Investigations. *Forensic Sci. Int.* **2003**, *131*, 65–74. [CrossRef]

362. Hsieh, H.M.; Hou, R.J.; Tsai, L.C.; Wei, C.S.; Liu, S.W.; Huang, L.H.; Kuo, Y.C.; Linacre, A.; Lee, J.C.I. A Highly Polymorphic STR Locus in *Cannabis sativa*. *Forensic Sci. Int.* **2003**, *131*, 53–58. [CrossRef]

363. Van Bakel, H.; Stout, J.M.; Cote, A.G.; Tallon, C.M.; Sharpe, A.G.; Hughes, T.R.; Page, J.E. The Draft Genome and Transcriptome of *Cannabis sativa*. *Genome Biol.* **2011**, *12*, R102. [CrossRef] [PubMed]

364. Gao, S.; Wang, B.; Xie, S.; Xu, X.; Zhang, J.; Pei, L.; Yu, Y.; Yang, W.; Zhang, Y. A High-Quality Reference Genome of Wild *Cannabis sativa*. *Hortic. Res.* **2020**, *7*, 73. [CrossRef]

365. Laverty, K.U.; Stout, J.M.; Sullivan, M.J.; Shah, H.; Gill, N.; Holbrook, L.; Deikus, G.; Sebra, R.; Hughes, T.R.; Page, J.E.; et al. A Physical and Genetic Map of *Cannabis sativa* Identifies Extensive Rearrangements at the THC/CBD Acid Synthase Loci. *Genome Res.* **2019**, *29*, 146–156. [CrossRef]

366. Kojoma, M.; Seki, H.; Yoshida, S.; Muranaka, T. DNA Polymorphisms in the Tetrahydrocannabinolic Acid (THCA) Synthase Gene in "Drug-Type" and "Fiber-Type" *Cannabis sativa* L. *Forensic Sci. Int.* **2006**, *159*, 132–140. [CrossRef] [PubMed]

367. McKernan, K.; Helbert, Y.; Tadigotla, V.; McLaughlin, S.; Spangler, J.; Zhang, L.; Smith, D. Single Molecule Sequencing of THCA Synthase Reveals Copy Number Variation in Modern Drug-Type *Cannabis sativa* L. *bioRxiv* **2015**, 28654. [CrossRef]

368. Onofri, C.; De Meijer, E.P.M.; Mandolino, G. Sequence Heterogeneity of Cannabidiolic- and Tetrahydrocannabinolic Acid-Synthase in *Cannabis sativa* L. and Its Relationship with Chemical Phenotype. *Phytochemistry* **2015**, *116*, 57–68. [CrossRef] [PubMed]

369. Weiblen, G.D.; Wenger, J.P.; Craft, K.J.; ElSohly, M.A.; Mehmedic, Z.; Treiber, E.L.; Marks, M.D. Gene Duplication and Divergence Affecting Drug Content in *Cannabis sativa*. *New Phytol.* **2015**, *208*, 1241–1250. [CrossRef] [PubMed]

370. Welling, M.T.; Liu, L.; Shapter, T.; Raymond, C.A.; King, G.J. Characterisation of Cannabinoid Composition in a Diverse *Cannabis sativa* L. Germplasm Collection. *Euphytica* **2016**, *208*, 463–475. [CrossRef]

371. Lynch, R.C.; Vergara, D.; Tittes, S.; White, K.; Schwartz, C.J.; Gibbs, M.J.; Ruthenburg, T.C.; DeCesare, K.; Land, D.P.; Kane, N.C. Genomic and Chemical Diversity in Cannabis. *CRC. Crit. Rev. Plant Sci.* **2016**, *35*, 349–363. [CrossRef]

372. Gao, C.; Xin, P.; Cheng, C.; Tang, Q.; Chen, P.; Wang, C.; Zang, G.; Zhao, L. Diversity Analysis in *Cannabis sativa* based on Large-Scale Development of Expressed Sequence Tag-Derived Simple Sequence Repeat Markers. *PLoS ONE* **2014**, *9*, e110638. [CrossRef]

373. Zhang, L.G.; Chang, Y.; Zhang, X.F.; Guan, F.Z.; Yuan, H.M.; Yu, Y.; Zhao, L.J. Analysis of the Genetic Diversity of Chinese Native *Cannabis sativa* Cultivars by Using ISSR and Chromosome Markers. *Genet. Mol. Res.* **2014**, *13*, 10490–10500. [CrossRef]

374. White, K.H.; Vergara, D.; Keepers, K.G.; Kane, N.C. The Complete Mitochondrial Genome for *Cannabis sativa*. *Mitochondrial DNA Part B Resour.* **2016**, *1*, 715–716. [CrossRef]

375. Oh, H.; Seo, B.; Lee, S.; Ahn, D.H.; Jo, E.; Park, J.K.; Min, G.S. Two Complete Chloroplast Genome Sequences of *Cannabis sativa* Varieties. *Mitochondrial DNA* **2015**, *27*, 2835–2837. [CrossRef]

376. Booth, J.K.; Page, J.E.; Bohlmann, J. Terpene Synthases from *Cannabis sativa*. *PLoS ONE* **2017**, *12*, e0173911. [CrossRef]

377. Zager, J.J.; Lange, I.; Srividya, N.; Smith, A.; Markus Lange, B. Gene Networks Underlying Cannabinoid and Terpenoid Accumulation in Cannabis. *Plant Physiol.* **2019**, *180*, 1877–1897. [CrossRef] [PubMed]

378. Amano, E.; Smith, H.H. Mutations Induced by Ethyl Methanesulfonate in Maize. *Mutat. Res. Fundam. Mol. Mech. Mutagen.* **1965**, *2*, 344–351. [CrossRef]

379. Froese-Gertzen, E.E.; Konzak, C.F.; Nilan, R.A.; Heiner, R.E. The Effect of Ethyl Methanesulfonate on the Growth Response, Chromosome Structure and Mutation Rate in Barley. *Radiat. Bot.* **1964**, *4*, 61–69. [CrossRef]

380. Jander, G.; Baerson, S.R.; Hudak, J.A.; Gonzalez, K.A.; Gruys, K.J.; Last, R.L. Ethylmethanesulfonate Saturation Mutagenesis in Arabidopsis to Determine Frequency of Herbicide Resistance. *Plant Physiol.* **2003**, *131*, 139–146. [CrossRef] [PubMed]

381. Luan, Y.S.; Zhang, J.; Gao, X.R.; An, L.J. Mutation Induced by Ethylmethanesulphonate (EMS), in Vitro Screening for Salt Tolerance and Plant Regeneration of Sweet Potato (*Ipomoea batatas* L.). *Plant Cell. Tissue Organ Cult.* **2007**, *88*, 77–81. [CrossRef]

382. Stavreva, D.A.; Ptáček, O.; Plewa, M.J.; Gichner, T. Single Cell Gel Electrophoresis Analysis of Genomic Damage Induced by Ethyl Methanesulfonate in Cultured Tobacco Cells. *Mutat Res.* **1998**, *422*, 323–330. [CrossRef]

383. Watanabe, S.; Mizoguchi, T.; Aoki, K.; Kubo, Y.; Mori, H.; Imanishi, S.; Yamazaki, Y.; Shibata, D.; Ezura, H. Ethylmethanesulfonate (EMS) Mutagenesis of Solanum Lycopersicum Cv. Micro-Tom for Large-Scale Mutant Screens. *Plant Biotechnol.* **2007**, *24*, 33–38. [CrossRef]

384. Koornneeff, M.; Dellaert, L.W.M.; van der Veen, J.H. EMS- and Relation-Induced Mutation Frequencies at Individual Loci in *Arabidopsis thaliana* (L.) Heynh. *Mutat. Res. Fundam. Mol. Mech. Mutagen.* **1982**, *93*, 109–123. [CrossRef]

385. Li, X.; Song, Y.; Century, K.; Straight, S.; Ronald, P.; Dong, X.; Lassner, M.; Zhang, Y. A Fast Neutron Deletion Mutagenesis-Based Reverse Genetics System for Plants. *Plant J.* **2001**, *27*, 235–242. [CrossRef]

386. Shirley, B.W.; Hanley, S.; Goodman, H.M. Effects of Ionizing Radiation on a Plant Genome: Analysis of Two Arabidopsis Transparent Testa Mutations. *Plant Cell* **1992**, *4*, 333–347. [CrossRef] [PubMed]

387. Bevan, M.; Bancroft, I.; Bent, E.; Love, K.; Goodman, H.; Dean, C.; Bergkamp, R.; Dirkse, W.; Van Staveren, M.; Stiekema, W.; et al. Analysis of 1.9 Mb of Contiguous Sequence from Chromosome 4 of *Arabidopsis thaliana*. *Nature* **1998**, *391*, 485–488. [CrossRef]

388. Bolon, Y.T.; Haun, W.J.; Xu, W.W.; Grant, D.; Stacey, M.G.; Nelson, R.T.; Gerhardt, D.J.; Jeddeloh, J.A.; Stacey, G.; Muehlbauer, G.J.; et al. Phenotypic and Genomic Analyses of a Fast Neutron Mutant Population Resource in Soybean. *Plant Physiol.* **2011**, *156*, 240–253. [CrossRef] [PubMed]

389. Colbert, T.; Till, B.J.; Tompa, R.; Reynolds, S.; Steine, M.N.; Yeung, A.T.; McCallum, C.M.; Comai, L.; Henikoff, S. High-Throughput Screening for Induced Point Mutations. *Plant Physiol.* **2001**, *126*, 480–484. [CrossRef] [PubMed]

390. Jander, G.; Norris, S.R.; Joshi, V.; Fraga, M.; Rugg, A.; Yu, S.; Li, L.; Last, R.L. Application of a High-Throughput HPLC-MS/MS Assay to Arabidopsis Mutant Screening; Evidence That Threonine Aldolase Plays a Role in Seed Nutritional Quality. *Plant J.* **2004**, *39*, 465–475. [CrossRef] [PubMed]

391. Jinek, M.; Chylinski, K.; Fonfara, I.; Hauer, M.; Doudna, J.A.; Charpentier, E. A Programmable Dual-RNA-Guided DNA Endonuclease in Adaptive Bacterial Immunity. *Science* **2012**, *337*, 816–821. [CrossRef] [PubMed]

392. Jaganathan, D.; Ramasamy, K.; Sellamuthu, G.; Jayabalan, S.; Venkataraman, G. CRISPR for Crop Improvement: An Update Review. *Front. Plant Sci.* **2018**, *9*, 985. [CrossRef]

393. Mali, P.; Yang, L.; Esvelt, K.M.; Aach, J.; Guell, M.; DiCarlo, J.E.; Norville, J.E.; Church, G.M. RNA-Guided Human Genome Engineering via Cas9. *Science* **2013**, *339*, 823–826. [CrossRef]

394. Fu, Y.; Foden, J.A.; Khayter, C.; Maeder, M.L.; Reyon, D.; Joung, J.K.; Sander, J.D. High-Frequency off-Target Mutagenesis Induced by CRISPR-Cas Nucleases in Human Cells. *Nat. Biotechnol.* **2013**, *31*, 822–826. [CrossRef]

395. Hsu, P.D.; Scott, D.A.; Weinstein, J.A.; Ran, F.A.; Konermann, S.; Agarwala, V.; Li, Y.; Fine, E.J.; Wu, X.; Shalem, O.; et al. DNA Targeting Specificity of RNA-Guided Cas9 Nucleases. *Nat. Biotechnol.* **2013**, *31*, 827–832. [CrossRef] [PubMed]

396. Ran, F.A.; Hsu, P.D.; Wright, J.; Agarwala, V.; Scott, D.A.; Zhang, F. Genome Engineering Using the CRISPR-Cas9 System. *Nat. Protoc.* **2013**, *8*, 2281–2308. [CrossRef]

397. Tsai, S.Q.; Nguyen, N.T.; Malagon-Lopez, J.; Topkar, V.V.; Aryee, M.J.; Joung, J.K. CIRCLE-Seq: A Highly Sensitive in Vitro Screen for Genome-Wide CRISPR-Cas9 Nuclease off-Targets. *Nat. Methods* **2017**, *14*, 607–614. [CrossRef] [PubMed]

398. Woo, J.W.; Kim, J.; Kwon, S.I.; Corvalán, C.; Cho, S.W.; Kim, H.; Kim, S.G.; Kim, S.T.; Choe, S.; Kim, J.S. DNA-Free Genome Editing in Plants with Preassembled CRISPR-Cas9 Ribonucleoproteins. *Nat. Biotechnol.* **2015**, *33*, 1162–1164. [CrossRef] [PubMed]

399. Lowder, L.G.; Zhang, D.; Baltes, N.J.; Paul, J.W.; Tang, X.; Zheng, X.; Voytas, D.F.; Hsieh, T.F.; Zhang, Y.; Qi, Y. A CRISPR/Cas9 Toolbox for Multiplexed Plant Genome Editing and Transcriptional Regulation. *Plant Physiol.* **2015**, *169*, 971–985. [CrossRef] [PubMed]

400. Jiang, W.; Zhou, H.; Bi, H.; Fromm, M.; Yang, B.; Weeks, D.P. Demonstration of CRISPR/Cas9/SgRNA-Mediated Targeted Gene Modification in Arabidopsis, Tobacco, Sorghum and Rice. *Nucleic Acids Res.* **2013**, *41*, e188. [CrossRef]

401. Feng, Z.; Zhang, B.; Ding, W.; Liu, X.; Yang, D.L.; Wei, P.; Cao, F.; Zhu, S.; Zhang, F.; Mao, Y.; et al. Efficient Genome Editing in Plants Using a CRISPR/Cas System. *Cell Res.* **2013**, *23*, 1229–1232. [CrossRef]

402. Bortesi, L.; Fischer, R. The CRISPR/Cas9 System for Plant Genome Editing and Beyond. *Biotechnol. Adv.* **2015**, *33*, 41–52. [CrossRef]

403. Podevin, N.; Davies, H.V.; Hartung, F.; Nogué, F.; Casacuberta, J.M. Site-Directed Nucleases: A Paradigm Shift in Predictable, Knowledge-Based Plant Breeding. *Trends Biotechnol.* **2013**, *31*, 375–383. [CrossRef]

404. Ma, X.; Zhang, Q.; Zhu, Q.; Liu, W.; Chen, Y.; Qiu, R.; Wang, B.; Yang, Z.; Li, H.; Lin, Y.; et al. A Robust CRISPR/Cas9 System for Convenient, High-Efficiency Multiplex Genome Editing in Monocot and Dicot Plants. *Mol. Plant* **2015**, *8*, 1274–1284. [CrossRef]

405. Ali, Z.; Abulfaraj, A.; Idris, A.; Ali, S.; Tashkandi, M.; Mahfouz, M.M. CRISPR/Cas9-Mediated Viral Interference in Plants. *Genome Biol.* **2015**, *16*. [CrossRef]

406. Xing, H.L.; Dong, L.; Wang, Z.P.; Zhang, H.Y.; Han, C.Y.; Liu, B.; Wang, X.C.; Chen, Q.J. A CRISPR/Cas9 Toolkit for Multiplex Genome Editing in Plants. *BMC Plant Biol.* **2014**, *14*, 327. [CrossRef] [PubMed]

407. Wang, Y.; Cheng, X.; Shan, Q.; Zhang, Y.; Liu, J.; Gao, C.; Qiu, J.L. Simultaneous Editing of Three Homoeoalleles in Hexaploid Bread Wheat Confers Heritable Resistance to Powdery Mildew. *Nat. Biotechnol.* **2014**, *32*, 947–951. [CrossRef] [PubMed]

408. Li, J.F.; Norville, J.E.; Aach, J.; McCormack, M.; Zhang, D.; Bush, J.; Church, G.M.; Sheen, J. Multiplex and Homologous Recombination-Mediated Genome Editing in Arabidopsis and Nicotiana Benthamiana Using Guide RNA and Cas9. *Nat. Biotechnol.* **2013**, *31*, 688–691. [CrossRef] [PubMed]

409. Belhaj, K.; Chaparro-Garcia, A.; Kamoun, S.; Patron, N.J.; Nekrasov, V. Editing Plant Genomes with CRISPR/Cas9. *Curr. Opin. Biotechnol.* **2015**, *32*, 76–84. [CrossRef]

410. Li, Z.; Liu, Z.B.; Xing, A.; Moon, B.P.; Koellhoffer, J.P.; Huang, L.; Ward, R.T.; Clifton, E.; Falco, S.C.; Cigan, A.M. Cas9-Guide RNA Directed Genome Editing in Soybean. *Plant Physiol.* **2015**, *169*, 960–970. [CrossRef]

411. Mackinnon, L.; McDougall, G.; Aziz, N.; Millam, S. *Progress Towards Transformation of Fibre Hemp*; Scottish Crop Research Institute Annual Report 2000/2001; Scottish Crop Research Institute: Dundee, UK, 2000; pp. 84–86.

412. Feeney, M.; Punja, Z.K. Tissue Culture and Agrobacterium-Mediated Transformation of Hemp (*Cannabis sativa* L.). *Vitr. Cell. Dev. Biol. Plant* **2003**, *39*, 578–585. [CrossRef]

413. Wahby, I.; Caba, J.M.; Ligero, F. Agrobacterium Infection of Hemp (*Cannabis sativa* L.): Establishment of Hairy Root Cultures. *J. Plant Interact.* **2013**, *8*, 312–320. [CrossRef]

414. Srivastava, S.; Srivastava, A.K. Hairy Root Culture for Mass-Production of High-Value Secondary Metabolites. *Crit. Rev. Biotechnol.* **2007**, *27*, 29–43. [CrossRef]

415. Ślusarkiewicz-Jarzina, A.; Ponitka, A.; Kaczmarek, Z. Influence of Cultivar, Explant Source and Plant Growth Regulator on Callus Induction and Plant Regeneration of *Cannabis sativa* L. *Acta Biol. Cracoviensia Ser. Bot.* **2005**, *47*, 145–151.

416. Carvalho, Â.; Hansen, E.H.; Kayser, O.; Carlsen, S.; Stehle, F. Designing Microorganisms for Heterologous Biosynthesis of Cannabinoids. *FEMS Yeast Res.* **2017**, *17*, Fox037. [CrossRef] [PubMed]

417. Zirpel, B.; Stehle, F.; Kayser, O. Production of Δ9-Tetrahydrocannabinolic Acid from Cannabigerolic Acid by Whole Cells of Pichia (Komagataella) Pastoris Expressing Δ9-Tetrahydrocannabinolic Acid Synthase from *Cannabis sativa* L. *Biotechnol. Lett.* **2015**, *37*, 1869–1875. [CrossRef] [PubMed]

418. Ohto, C.; Muramatsu, M.; Obata, S.; Sakuradani, E.; Shimizu, S. Overexpression of the Gene Encoding HMG-CoA Reductase in Saccharomyces Cerevisiae for Production of Prenyl Alcohols. *Appl. Microbiol. Biotechnol.* **2009**, *82*, 837–845. [CrossRef] [PubMed]

419. Zirpel, B.; Degenhardt, F.; Martin, C.; Kayser, O.; Stehle, F. Engineering Yeasts as Platform Organisms for Cannabinoid Biosynthesis. *J. Biotechnol.* **2017**, *259*, 204–212. [CrossRef]

420. Mills, E. The Carbon Footprint of Indoor Cannabis Production. *Energy Policy* **2012**, *46*, 58–67. [CrossRef]

421. Borthwick, H.A.; Scully, N.J. Photoperiodic Responses of Hemp. *Bot. Gaz.* **1954**, *116*, 14–29. [CrossRef]

422. Schaffner, J.H. The Influence of Relative Length of Daylight on the Reversal of Sex in Hemp. *Ecology* **1923**, *4*, 323–334. [CrossRef]

423. Potter, D.J.; Duncombe, P. The Effect of Electrical Lighting Power and Irradiance on Indoor-Grown Cannabis Potency and Yield. *J. Forensic Sci.* **2012**, *57*, 618–622. [CrossRef]

424. Spitzer-Rimon, B.; Duchin, S.; Bernstein, N.; Kamenetsky, R. Architecture and Florogenesis in Female *Cannabis sativa* Plants. *Front. Plant Sci.* **2019**, *10*, 350. [CrossRef]

425. Vanhove, W.; Van Damme, P.; Meert, N. Factors Determining Yield and Quality of Illicit Indoor Cannabis (Cannabis Spp.) Production. *Forensic Sci. Int.* **2011**, *212*, 1. [CrossRef] [PubMed]

426. Viršile, A.; Olle, M.; Duchovskis, P. LED Lighting in Horticulture. In *Light Emitting Diodes for Agriculture: Smart Lighting*; Springer: Singapore, 2017; pp. 113–147. [CrossRef]

427. Nelson, J.A.; Bugbee, B. Economic Analysis of Greenhouse Lighting: Light Emitting Diodes vs. High Intensity Discharge Fixtures. *PLoS ONE* **2014**, *9*, e99010. [CrossRef]

428. Tamulaitis, G.; Duchovskis, P.; Bliznikas, Z.; Breive, K.; Ulinskaite, R.; Brazaityte, A.; Novičkovas, A.; Žukauskas, A. High-Power Light-Emitting Diode Based Facility for Plant Cultivation. *J. Phys. D. Appl. Phys.* **2005**, *38*, 3182–3187. [CrossRef]

429. Hogewoning, S.W.; Douwstra, P.; Trouwborst, G.; Van Ieperen, W.; Harbinson, J. An Artificial Solar Spectrum Substantially Alters Plant Development Compared with Usual Climate Room Irradiance Spectra. *J. Exp. Bot.* **2010**, *61*, 1267–1276. [CrossRef]

430. Backer, R.; Schwinghamer, T.; Rosenbaum, P.; McCarty, V.; Eichhorn Bilodeau, S.; Lyu, D.; Ahmed, M.B.; Robinson, G.; Lefsrud, M.; Wilkins, O.; et al. Closing the Yield Gap for Cannabis: A Meta-Analysis of Factors Determining Cannabis Yield. *Front. Plant Sci.* **2019**, *10*. [CrossRef]

431. Chandra, S.; Lata, H.; Khan, I.A.; Elsohly, M.A. Photosynthetic Response of *Cannabis sativa* L. to Variations in Photosynthetic Photon Flux Densities, Temperature and CO2 Conditions. *Physiol. Mol. Biol. Plants* **2008**, *14*, 299–306. [CrossRef] [PubMed]

432. Chandra, S.; Lata, H.; Mehmedic, Z.; Khan, I.A.; ElSohly, M.A. Light Dependence of Photosynthesis and Water Vapor Exchange Characteristics in Different High Δ9-THC Yielding Varieties of *Cannabis sativa* L. *J. Appl. Res. Med. Aromat. Plants* **2015**, *2*, 39–47. [CrossRef]

433. Lydon, J.; Teramura, A.H.; Coffman, C.B. UV-B Radiation Effects on Photosynthesis, Growth and Cannabinoid Production of Two *Cannabis sativa* Chemotypes. *Photochem. Photobiol.* **1987**, *46*, 201–206. [CrossRef]

434. Berry, J.; Bjorkman, O. Photosynthetic Response and Adaptation to Temperature in Higher Plants. *Annu. Rev. Plant Physiol.* **1980**, *31*, 491–543. [CrossRef]

435. Larcher, W. Photosynthesis as a Tool for Indicating Temperature Stress Events. In *Ecophysiology of Photosynthesis*; Schulze, E., Caldwell, M.M., Eds.; Springer: Berlin/Heidelberg, Germany, 1995; pp. 261–277. [CrossRef]

436. Hikosaka, K.; Ishikawa, K.; Borjigidai, A.; Muller, O.; Onoda, Y. Temperature Acclimation of Photosynthesis: Mechanisms Involved in the Changes in Temperature Dependence of Photosynthetic Rate. *J. Exp. Bot.* **2006**, *57*, 291–302. [CrossRef]

437. Chandra, S.; Lata, H.; Khan, I.A.; ElSohly, M.A. Temperature Response of Photosynthesis in Different Drug and Fiber Varieties of *Cannabis sativa* L. *Physiol. Mol. Biol. Plants* **2011**, *17*, 297–303. [CrossRef]

438. Van der Werf, H.M.G.; Brouwer, K.; Wijlhuizen, M.; Withagen, J.C.M. The Effect of Temperature on Leaf Appearance and Canopy Establishment in Fibre Hemp (*Cannabis sativa* L.). *Ann. Appl. Biol.* **1995**, *126*, 551–561. [CrossRef]

439. Bernstein, N.; Gorelick, J.; Zerahia, R.; Koch, S. Impact of N, P, K, and Humic Acid Supplementation on the Chemical Profile of Medical Cannabis (*Cannabis sativa* L). *Front. Plant Sci.* **2019**, *10*. [CrossRef] [PubMed]

440. Malceva, M.; Vikmane, M.; Stramkale, V. Changes of Photosynthesis-Related Parameters and Productivity of *Cannabis sativa* under Different Nitrogen Supply. *Environ. Exp. Biol.* **2011**, *9*, 61–69.

441. Mansouri, H.; Asrar, Z. Effects of Abscisic Acid on Content and Biosynthesis of Terpenoids in *Cannabis sativa* at Vegetative Stage. *Biol. Plant.* **2012**, *56*, 153–156. [CrossRef]

442. Mansouri, H.; Asrar, Z.; Mehrabani, M. Effects of Gibberellic Acid on Primary Terpenoids and Δ9-Tetrahydrocannabinol in *Cannabis sativa* at Flowering Stage. *J. Integr. Plant Biol.* **2009**, *51*, 553–561. [CrossRef]

443. Larkin, J.C.; Oppenheimer, D.G.; Lloyd, A.M.; Paparozzi, E.T.; Marks, M.D. Roles of the GLABROUS1 and TRANSPARENT TESTA GLABRA Genes in Arabidopsis Trichome Development. *Plant Cell* **1994**, *6*, 1065–1076. [CrossRef] [PubMed]

444. Payne, C.T.; Zhang, F.; Lloyd, A.M. GL3 Encodes a BHLH Protein That Regulates Trichome Development in Arabidopsis through Interaction with GL1 and TTG1. *Genetics* **2000**, *156*, 1349–1362.

445. Rerie, W.G.; Feldmann, K.A.; Marks, M.D. The GLABRA2 Gene Encodes a Homeo Domain Protein Required for Normal Trichome Development in Arabidopsis. *Genes Dev.* **1994**, *8*, 1388–1399. [CrossRef]

446. Szymanski, D.B.; Jilk, R.A.; Pollock, S.M.; Marks, M.D. Control of GL2 Expression in Arabidopsis Leaves and Trichomes. *Development* **1998**, *125*, 1161–1171.

447. Hülskamp, M.; Miséra, S.; Jürgens, G. Genetic Dissection of Trichome Cell Development in Arabidopsis. *Cell* **1994**, *76*, 555–566. [CrossRef]

448. Perazza, D.; Herzog, M.; Hülskamp, M.; Brown, S.; Dorne, A.M.; Bonneville, J.M. Trichome Cell Growth in *Arabidopsis thaliana* Can Be Derepressed by Mutations in at Least Five Genes. *Genetics* **1999**, *152*, 461–476.

449. Liu, Y.; Liu, D.; Hu, R.; Hua, C.; Ali, I.; Zhang, A.; Liu, B.; Wu, M.; Huang, L.; Gan, Y. AtGIS, a C2H2 Zinc-Finger Transcription Factor from Arabidopsis Regulates Glandular Trichome Development through GA Signaling in Tobacco. *Biochem. Biophys. Res. Commun.* **2017**, *483*, 209–215. [CrossRef]

450. Luo, M.; Wang, Z.; Li, H.; Xia, K.F.; Cai, Y.; Xu, Z.F. Overexpression of a Weed (Solanum Americanum) Proteinase Inhibitor in Transgenic Tobacco Results in Increased Glandular Trichome Density and Enhanced Resistance to Helicoverpa Armigera and Spodoptera Litura. *Int. J. Mol. Sci.* **2009**, *10*, 1896–1910. [CrossRef] [PubMed]

451. Paetzold, H.; Garms, S.; Bartram, S.; Wieczorek, J.; Urós-Gracia, E.M.; Rodríguez-Concepción, M.; Boland, W.; Strack, D.; Hause, B.; Walter, M.H. The Isogene 1-Deoxy-D-Xylulose 5-Phosphate Synthase 2 Controls Isoprenoid Profiles, Precursor Pathway Allocation, and Density of Tomato Trichomes. *Mol. Plant* **2010**, *3*, 904–916. [CrossRef]

452. Ma, D.; Hu, Y.; Yang, C.; Liu, B.; Fang, L.; Wan, Q.; Liang, W.; Mei, G.; Wang, L.; Wang, H.; et al. Genetic Basis for Glandular Trichome Formation in Cotton. *Nat. Commun.* **2016**, *7*, 10456. [CrossRef] [PubMed]

453. Salas Fernandez, M.G.; Becraft, P.W.; Yin, Y.; Lübberstedt, T. From Dwarves to Giants? Plant Height Manipulation for Biomass Yield. *Trends Plant Sci.* **2009**, *14*, 454–461. [CrossRef]

454. Campiglia, E.; Radicetti, E.; Mancinelli, R. Plant Density and Nitrogen Fertilization Affect Agronomic Performance of Industrial Hemp (*Cannabis sativa* L.) in Mediterranean Environment. *Ind. Crop. Prod* **2017**, *100*, 246–254. [CrossRef]

455. Van der Werf, H.M.G.; Wijlhuizen, M.; de Schutter, J.A.A. Plant Density and Self-Thinning Affect Yield and Quality of Fibre Hemp (*Cannabis sativa* L.). *Field Crop. Res.* **1995**, *40*, 153–164. [CrossRef]

456. Small, E. Dwarf Germplasm: The Key to Giant Cannabis Hempseed and Cannabinoid Crops. *Genet. Resour. Crop Evol.* **2018**, *65*, 1071–1107. [CrossRef]

457. Graham, L.A.; Besser, K.; Blumer, S.; Branigan, C.A.; Czechowski, T.; Elias, L.; Guterman, I.; Harvey, D.; Isaac, P.G.; Khan, A.M.; et al. The Genetic Map of Artemisia Annua L Identifies Loci Affecting Yield of the Antimalarial Drug Artemisinin. *Science* **2010**, *327*, 328–331. [CrossRef] [PubMed]

458. Fairbairn, J.; Kapoor, L. The Lactiferous Vessels of Papaver Somniferum L. *Planta Med.* **1960**, *8*, 49–61. [CrossRef]

459. Nessler, C.L.; Mahlberg, P.G. Laticifers in Stamens of Papaver Somniferum L. *Planta* **1976**, *129*, 83–85. [CrossRef]

460. Weid, M.; Ziegler, J.; Kutchan, T.M. The Roles of Latex and the Vascular Bundle in Morphine Biosynthesis in the Opium Poppy, Papaver Somniferum. *Proc. Natl. Acad. Sci. USA* **2004**, *101*, 13957–13962. [CrossRef] [PubMed]

461. Fuller, J.G.; McMorland, G.H.; Douglas, M.J.; Palmer, L. Epidural Morphine for Analgesia after Caesarean Section: A Report of 4880 Patients. *Can. J. Anaesth.* **1990**, *37*, 636–640. [CrossRef] [PubMed]

462. Stenseth, K.; Sellevold, O.; Breivik, H. Epidural Morphine for Postoperative Pain: Experience with 1085 Patients. *Acta Anaesthesiol. Scand.* **1985**, *29*, 148–156. [CrossRef] [PubMed]

463. Walder, B.; Schafer, M.; Henzi, I.; Tramèr, M.R. Efficacy and Safety of Patient-Controlled Opioid Analgesia for Acute Postoperative Pain. *Acta Anaesthesiol. Scand.* **2001**, *45*, 795–804. [CrossRef]

464. Goldsack, C.; Scuplak, S.M.; Smith, M. A Double-Blind Comparison of Codeine and Morphine for Postoperative Analgesia Following Intracranial Surgery. *Anaesthesia* **1996**, *51*, 1029–1032. [CrossRef]

465. Walker, D.J.; Zacny, J.P. Subjective, Psychomotor, and Analgesic Effects of Oral Codeine and Morphine in Healthy Volunteers. *Psychopharmacology* **1998**, *140*, 191–201. [CrossRef]

466. Sevelius, H.; McCoy, J.F.; Colmore, J.P. Dose Response to Codeine in Patients with Chronic Cough. *Clin. Pharmacol. Ther.* **1971**, *12*, 449–455. [CrossRef]
467. Bolser, D.C.; Davenport, P.W. Codeine and Cough: An Ineffective Gold Standard. *Curr. Opin. Allergy Clin. Immunol.* **2007**, *7*, 32–36. [CrossRef]
468. Freestone, C.; Eccles, R. Assessment of the Antitussive Efficacy of Codeine in Cough Associated with Common Cold. *J. Pharm. Pharmacol.* **1997**, *49*, 1045–1049. [CrossRef]
469. Takahama, K.; Shirasaki, T. Central and Peripheral Mechanisms of Narcotic Antitussives: Codeine-Sensitive and -Resistant Coughs. *Cough* **2007**, *152*, 349–355. [CrossRef]
470. Jackson, T.; Chougule, M.B.; Ichite, N.; Patlolla, R.R.; Singh, M. Antitumor Activity of Noscapine in Human Non-Small Cell Lung Cancer Xenograft Model. *Cancer Chemother. Pharmacol.* **2008**, *63*, 117–126. [CrossRef] [PubMed]
471. Joshi, H.C.; Zhou, J. Noscapine and Analogues as Potential Chemotherapeutic Agents. *Drug News Perspect.* **2000**, *13*, 543–546. [CrossRef] [PubMed]
472. Rida, P.C.G.; Livecche, D.; Ogden, A.; Zhou, J.; Aneja, R. The Noscapine Chronicle: A Pharmaco-Historic Biography of the Opiate Alkaloid Family and Its Clinical Applications. *Med. Res. Rev.* **2015**, *35*, 1072–1096. [CrossRef]
473. Ye, K.; Ke, Y.; Keshava, N.; Shanks, J.; Kapp, J.A.; Tekmal, R.R.; Petros, J.; Joshi, H.C. Opium Alkaloid Noscapine Is an Antitumor Agent That Arrests Metaphase and Induces Apoptosis in Dividing Cells. *Proc. Natl. Acad. Sci. USA* **1998**, *95*, 1601–1606. [CrossRef] [PubMed]
474. Carroll, R.J.; Leisch, H.; Rochon, L.; Hudlicky, T.; Cox, D.P. One-Pot Conversion of Thebaine to Hydrocodone and Synthesis of Neopinone Ketal. *J. Org. Chem.* **2009**, *74*, 747–752. [CrossRef]
475. Endoma-Arias, M.A.A.; Cox, D.P.; Hudlicky, T. General Method of Synthesis for Naloxone, Naltrexone, Nalbuphone, and Nalbuphine by the Reaction of Grignard Reagents with an Oxazolidine Derived from Oxymorphone. *Adv. Synth. Catal.* **2013**, *355*, 1869–1873. [CrossRef]
476. MacHara, A.; Werner, L.; Endoma-Arias, M.A.; Cox, D.P.; Hudlicky, T. Improved Synthesis of Buprenorphine from Thebaine and/or Oripavine via Palladium-Catalyzed N-Demethylation/Acylation and/or Concomitant O-Demethylation. *Adv. Synth. Catal.* **2012**, *354*, 613–626. [CrossRef]
477. Murphy, B.; Šnajdr, I.; Machara, A.; Endoma-Arias, M.A.A.; Stamatatos, T.C.; Cox, D.P.; Hudlický, T. Conversion of Thebaine to Oripavine and Other Useful Intermediates for the Semisynthesis of Opiate-Derived Agents: Synthesis of Hydromorphone. *Adv. Synth. Catal.* **2014**, *356*, 2679–2687. [CrossRef]
478. Orman, J.S.; Keating, G.M. Buprenorphine/Naloxone: A Review of Its Use in the Treatment of Opioid Dependence. *Drugs* **2009**, *69*, 577–607. [CrossRef] [PubMed]
479. Werner, L.; Wernerova, M.; MacHara, A.; Endoma-Arias, M.A.; Duchek, J.; Adams, D.R.; Cox, D.P.; Hudlicky, T. Unexpected N-Demethylation of Oxymorphone and Oxycodone N-Oxides Mediated by the Burgess Reagent: Direct Synthesis of Naltrexone, Naloxone, and Other Antagonists from Oxymorphone. *Adv. Synth. Catal.* **2012**, *354*, 2706–2712. [CrossRef]
480. Millgate, A.G.; Pogson, B.J.; Wilson, I.W.; Kutchan, T.M.; Zenk, M.H.; Gerlach, W.L.; Fist, A.J.; Larkin, P.J. Morphine-Pathway Block in Top1 Poppies. *Nature* **2004**, *431*, 413–414. [CrossRef]
481. Hagel, J.M.; Facchini, P.J. Dioxygenases Catalyze the O-Demethylation Steps of Morphine Biosynthesis in Opium Poppy. *Nat. Chem. Biol.* **2010**, *6*, 273–275. [CrossRef] [PubMed]
482. Fist, A.J.; Miller, J.A.C.; Gregory, D. Papaver Somniferum with High Concentration of Codeine. WO2009143574, 3 December 2009.
483. Winzer, T.; Walker, T.C.; Meade, F.; Larson, T.R.; Graham, I.A. Modified Plant. WO2017122011, 20 July 2017.
484. Winzer, T.; Gazda, V.; He, Z.; Kaminski, F.; Kern, M.; Larson, T.R.; Li, Y.; Meade, F.; Teodor, R.; Vaistij, F.E.; et al. A Papaver Somniferum 10-Gene Cluster for Synthesis of the Anticancer Alkaloid Noscapine. *Science* **2012**, *336*, 1704–1708. [CrossRef]
485. Guo, L.; Winzer, T.; Yang, X.; Li, Y.; Ning, Z.; He, Z.; Teodor, R.; Lu, Y.; Bowser, T.A.; Graham, I.A.; et al. The Opium Poppy Genome and Morphinan Production. *Science* **2018**, *362*, 343–347. [CrossRef] [PubMed]
486. Winzer, T.; Graham, I.A.; Walker, T.C. Genes Involved in Noscapine Production. WO2013136057, 19 September 2013.
487. Winzer, T.; Kern, M.; King, A.J.; Larson, T.R.; Teodor, R.I.; Donninger, S.L.; Li, Y.; Dowle, A.A.; Cartwright, J.; Bates, R.; et al. Morphinan Biosynthesis in Opium Poppy Requires a P450-Oxidoreductase Fusion Protein. *Science* **2015**, *349*, 309–3012. [CrossRef] [PubMed]
488. Winzer, T.; Graham, I.A.; Walker, T.C. Production of Noscapine. WO2016207643, 29 December 2016.
489. Wijekoon, C.P.; Facchini, P.J. Systematic Knockdown of Morphine Pathway Enzymes in Opium Poppy Using Virus-Induced Gene Silencing. *Plant J.* **2012**, *69*, 1052–1063. [CrossRef] [PubMed]
490. Allen, R.S.; Millgate, A.G.; Chitty, J.A.; Thisleton, J.; Miller, J.A.C.; Fist, A.J.; Gerlach, W.L.; Larkin, P.J. RNAi-Mediated Replacement of Morphine with the Nonnarcotic Alkaloid Reticuline in Opium Poppy. *Nat. Biotechnol.* **2004**, *22*, 1559–1566. [CrossRef] [PubMed]
491. Sharma, J.R.; Lal, R.K.; Gupta, A.P.; Misra, H.O.; Pant, V.; Singh, N.K.; Pandey, V. Development of Non-Narcotic (Opiumless and Alkaloid-Free) Opium Poppy, Papaver Somniferum. *Plant Breed.* **1999**, *118*, 449–452. [CrossRef]